THE EFFICIENT USE OF ENERGY

General Editor: I.G.C. DRYDEN

DSc PhD CEng FRIC FIChemE SFInstF

IPC Science and Technology Press
in collaboration with the Institute of Fuel
acting on behalf of the
UK Department of Energy

ISBN 0 902852 33 7

Typeset by Mid-County Press, 18 Ridgway, Wimbledon, SW19
Printed by Kingprint Ltd, Richmond, Surrey
Bound by Dorset Bookbinding Co Ltd, Ferndown, Dorset.

Foreword
by the Secretary of State for Energy

It gives me great pleasure to contribute this foreword to "The Efficient Use of Energy". It has been planned and written as a successor to the HMSO publication "The Efficient Use of Fuel", which in its two editions of 1944 and 1958 has stood for so long as an authoritative compendium of the principles and basic technology of good fuel utilization practice.

The very rapid and profound changes which have taken place world-wide during the last decade in the sources of energy supply and in the corresponding technologies of energy production, utilization, process control and materials have created the technical incentive for this new handbook, with a coverage appropriate to modern needs. The concern of the Government has been to ensure that it is a worthy successor to the HMSO publication, and the Institute of Fuel willingly accepted an invitation to act on our behalf in collaboration with the publishers to that end.

Its appearance at this time is most appropriate. The era of cheap energy is over and it would be foolhardy in the extreme for any of us to conduct our affairs in the expectation that it will ever return. We can no longer afford, individually or nationally, the wasteful practices into which we understandably fell when supplies were cheap and abundant.

There are clear indications that the new relationship between the cost of energy and the cost of resources to achieve better energy utilization is leading to the adoption of improved practices in industry and elsewhere. But it may well be that, as a legacy from the recent past, many engineers and managers are not as conversant as they might be with some of the well-established principles and disciplines of good housekeeping in the energy field, and that the increasing significance of these in the overall costing of industrial processes and services is still undervalued. The present handbook is intended to bring existing shortcomings and unrealized opportunities to the attention of those who are in a position to rectify them.

The creation of this new work has been a happy collaboration between the Department, the publishers and the Institute of Fuel. The effort which the members of the Institute and associates have deployed with the publishers in its preparation is immense and I am most grateful to all who have participated. I wish it the outstanding success which its high quality and authoritative character deserve.

Tony Benn

The Rt Hon Tony Benn, MP
Secretary of State for Energy

Preface

I would like to record my indebtedness especially to the original Panel of the Institute of Fuel, who worked out the basis of the present structure and selected many of those invited to contribute; to the present Editorial Advisory Committee; and to the many contributors. These groups are all listed elsewhere.

Special thanks are also due to the following:

Mr L. H. Leighton of the Department of Energy who first suggested my participation and gave help and encouragement throughout.

Dr N.. M. Potter whose advocacy of and enthusiasm for the project during his term of office as President of the Institute of Fuel 1970/71 were largely instrumental in forming the Editorial Panel.

Dr R. Jackson, Secretary of the Institute, who coordinated the work of this Panel, and was readily available throughout the preparation of the book with helpful suggestions and advice.

Dr J. Gibson, President of the Institute 1975/76, for the valuable introductory survey, ready cooperation in respect of other contributions, and his collaboration with the Department of Energy and IPC in launching the Handbook.

Mr Byrom Lees, Chairman of the Institute's Publications Committee, who refereed many of the contributions.

Dr D. G. Skinner, who came to the rescue in a critical period and was always ready with help and wise comment on other contributions.

Mr T. P. Flanagan of the SIRA Institute, who advised on certain scripts and suggested a suitable contributor.

Mr E. J. Wightman, who prepared Chapter 18.1,2 at extremely short notice and has been unfailingly helpful throughout.

Messrs Newnes-Butterworths, who made available artwork required to expedite Chapter 18.1,2.

Mr H. B. Weston of NIFES Ltd who arranged contributions at short notice, frequently gave valuable advice, and during a further emergency arranged permission to adapt a section of material from the Fuel Economy Handbook.

Mr A. Graham of Graham/Trotman/Dudley Publishers Ltd who gave permission to précis the above material from their series of publications 'Fuel Economy Handbook'.

Mr A. Shearer, who assisted in finding additional contributors for the book as a whole.

Messrs GEC Diesels for permission to use data incorporated in Chapter 12.1 and 12.3.

Margery Griffith, who has assisted with the project since 1973 and whose unstinted efforts, phenomenal memory for detail, and meticulous subediting and proof reading have played a tremendous part in realizing the Handbook.

Christine Vickers, who most intelligently relettered or redrew, rearranged and greatly improved many more of the diagrams than was originally intended.

Ann Preston, who prepared the index at very short notice.

My wife Dorothy, who has cheerfully borne a disproportionate share of family duties for more than a year.

It seems suitable also to mention that the editorial skill of the late Dr G. E. Foxwell in shaping the original edition of EUF under the difficult conditions of 1943 has proved an excellent foundation on which to build.

Finally, I shall be grateful if readers will point out errors and omissions so that any future edition may be improved.

Ian G. C. Dryden
June 1975

Contributors

The many contributors to *The Efficient Use of Energy* are listed below, in alphabetical order. Inevitably, some of those listed contributed more to the book than others. But in general it would be invidious to make further distinctions among the contributors beyond what is attempted here, because each has helped in making up the whole, and the usefulness of the book would have been lessened by the absence of even the smallest item. All contributions have been modified in greater or lesser measure to fit the requirements of the book as a whole. Chapter 15.2 was compiled jointly by D. G. Skinner and the Editor from material provided by the Department of Energy and others.

*H. Barber (Ch.7.1)	Loughborough University of Technology
L. H. Bell (App.D2)	Esso Research Centre, Abingdon
+J.Bettelheim (Ch.19.3)	Central Electricity Research Laboratories
J. Bolton (Ch.10.7)	Department of Health and Social Security
W. C. Carter (Ch.10.3)	Clarke Chapman Ltd, now Consultant
J. H. Chesters, FRS (Ch.17)	Consultant
*P. N. J. Chipperfield (Ch.19.5, 19.10)	ICI Ltd
+P. N. Cooper (Ch.14)	University of Aston in Birmingham
+A. T. S. Cunningham (Ch.18.3)	Central Electricity Generating Board, Marchwood Engineering Laboratories
+J. Dalmon (Ch.19.3)	Central Electricity Research Laboratories
A. DeRose (App.D1)	British Gas Corporation
R. M. E. Diamant (Ch.15.3)	University of Salford
I. G. C. Dryden (Ch.1.1, part Ch.8.1 (*It* diagram), part Ch.15.2, App.B3, part App.C, App.F)	Consultant, General Editor
P. D. Dunn (Ch.13.3, 4, 5)	University of Reading
K. C. Gandhi (Ch.8.1)	Stein Atkinson Stordy Ltd
+P. M. Gaywood (Ch.10.2)	British Gas Corporation
J. Gibson (Ch.1.2)	National Coal Board, Pres.Inst. Fuel 1975—76
D. W. Gill (Ch.6.2, App.C1—3, 5, 8, 11)	National Coal Board
E. M. Goodger (App.A7)	Cranfield Institute of Technology
J. P. Graham (App.B6, 7)	British Carbonization Research Association
W. C. Hankins (Ch.7.5)	The Electricity Council
*W. L. Harrison (Ch.8.2, part Ch.7)	Industrial Consultant, Electroheat Technology
+J. E. Harry (Ch.7.1)	Loughborough University of Technology
+A. B. Hart (Ch.20.1—3)	Central Electricity Research Laboratories
J. Highley (Ch.5)	National Coal Board
B. H. Holland (Ch.3.2)	British Gas Corporation
*A. G. Horsler (Ch.10,2)	British Gas Corporation
*S. E. Hunt (Ch.14)	University of Aston in Birmingham
*P. J. Jackson (Ch.18.3)	Central Electricity Generating Board, Marchwood Engineering Laboratories
+K. W. James (Ch.19.6)	Central Electricity Research Laboratories
L. J. Jolley (Ch.19.1)	Consultant
H. Kay (Ch.16.2)	Stein Atkinson Stordy Ltd
R. Kelley (Ch.10.6)	Clarke Chapman Ltd
+T. E. Langford (Ch.19.6)	Central Electricity Research Laboratories
W. R. Laws (Ch.16.3)	British Steel Corporation

J. Lawton (Ch.7.2, 7.4)	Electricity Council Research Centre
+J. W. Laxton (Ch.20.1—3)	Central Electricity Research Laboratories
+D. H. Lucas (Ch.19.3)	Central Electricity Research Laboratories
S. L. Lyons (Ch.7.7)	The Electricity Council
D. McNeil, OBE (App.B11, D4)	Technical Consultant
P. L. Martin (Ch.9.2)	Oscar Faber and Partners
K. J. Mitchell (Ch.10.4)	Babcock and Wilcox (Operations) Ltd
D. H. Napier (Ch.19.2)	Imperial College of Science and Technology
*National Industrial Fuel Efficiency Service Ltd (senior members of staff): (Ch.2, Ch.6.4, Ch.9.1, Ch.10.5, Ch.13.1, 2, Ch.19.4, 7, 9, App.C9, 10, and part 11)	
J. A. O'Shea (Ch.15.1)	Integrated Energy Systems Ltd
A. Parker, CBE (App.E)	Chemical Engineering Consultant; formerly Director of Fuel (including Air Pollution) Research, DSIR
G. A. Payne (Ch.4, App.C12)	Industrial Energy Adviser
A. J. Perkins (Ch.7.3)	The Electricity Council
N. M. Potter (App.B8, 12)	Consultant, Pres. Inst. Fuel at inception of book (1970—71)
J. M. Riley (Ch.10.1)	The Beeston Boiler Company Ltd
A. G. Roberts (Ch.6.3)	BCURA Ltd
G. Robertshaw (Ch.3.1)	British Gas Corporation
*D. F. Rosborough (App.A1, 2, 4, 5, 8, 9; B1, 2, 4, 5, 9, 10)	Esso Research Centre, Abingdon
P. L. Rumsby (Ch.19.8, App.A3, 6, 10)	BP Coal Ltd
A. B. Shearer (Ch.12.1, 3)	GEC Diesels Ltd
C. R. Simmons (Ch.12.2)	Gas Turbine News
D. G. Skinner (Part Ch.15.2)	Consultant
C. O. Smith (Ch.11)	NALFLOC Ltd
+P. Tate (Part Ch.10.2)	National Coal Board
G. G. Thurlow (Ch.6.1, App.C4)	National Coal Board
L. A. N. Tozer (Ch.7.6)	The Electricity Council
+J. Wagstaff (Ch.19.5, 19.10)	ICI Ltd
*J. M. Ward (Ch.19.3, 6, Ch.20.1—3)	Central Electricity Research Laboratories
E. J. Wightman (Ch.18.1, 2)	Consultant
H. C. Wilkinson (App.D3)	British Carbonization Research Association
A. Williams (Ch.8.3)	University of Leeds
+A. F. Williams (App.A1, 2, 4, 5, 8, 9, B1, 2, 4, 5, 9, 10)	Esso Research Centre, Abingdon
A. C. Yates (Ch.16.1)	Humphreys and Glasgow Ltd

*   Direct contact and coordinator
\+   Co-contributor

The chart on p 42, and some of the material in Appendix D2, are reproduced from British Standard BS 2869:1970 by permission of BSI, 2 Park Street, London W1A 2BS, from whom copies can be obtained.

Contents

Chapter 1

Plan and perspective

1.1 Structure of the book

Strategy and plan
General policy
Progress
Energy and entropy, and sensible heat
Assessment

1.2 Perspective

World energy production since World War II
World energy consumption
World energy resources
Other energy resources for the future
Factors in more efficient energy use and conservation

In this book, as with *THE EFFICIENT USE OF FUEL* first published in 1944, it was intended at least to mention most topics of relevance in energy handling and to describe in some degree of depth the more important features of current plant and equipment. Furthermore, unlike its predecessor it was decided to minimize the treatment of topics that were limited in interest to the UK (where it *happened* to be written) in favour of those with virtually worldwide appeal. Thus it was hoped to preserve much of the encyclopedic and text-book character of the former volume within the compass of a single book readily handled.

But in the intervening 30 years, so many aspects of fuel availability and the use of fuels have changed almost out of recognition, the scale and geographical distribution have so altered, the rate of change has so accelerated, and the environmental limitations of both fuel winning and fuel usage have become so much more widely recognized, that these good intentions began to seem almost unrealizable. In attempting the task nonetheless, I as general editor have been encouraged particularly by four things: the apparently feasible guidelines worked out in lengthy discussions by the original Panel of the Institute of Fuel and its President at the time, Dr N. M. Potter, the support and confidence of Mr L. H. Leighton of the Department (now) of Energy, the understanding of the publishers who finally agreed to undertake the entire financial risk and permitted me not only to supervise the assembly and co-ordination of material but also to handle much of the detail of its further progress as far as page proofs and index, and the ready co-operation of the majority of the contributors approached in compressing so much into so moderate a space. On the whole, it has proved possible to write the main text in a form that can be understood by a professional engineer or fuel technologist lacking specialized knowledge of the particular subject under discussion, or by managers without special technical training but faced with decisions on plant or policy based on the advice of qualified staff. The one restriction is the ability to read adequately the English language.

In order to compress a great deal of information into a relatively small space, fuel characteristics, inter-fuel process details, scientific and technological details — all readily available elsewhere and covered by references in this book — appear as Appendixes (A, B, C) written in checklist style or even further abbreviated. This method was devised in haste and incorporated after telephone approval in principle from a very few members of the Panel or the present Advisory Committee. It was a risk supported mainly by a personal 'hunch'; much editorial interpolation and rearrangement was needed, but most contributors understood the situation and were willing to check for accuracy proofs that bore only a superficial resemblance to the original script submitted. I hope that the attempt has been successful; in so far as it has not, and in respect of the final content of these three Appendixes, I must take much of the responsibility.

1.1 Structure of the book

Strategy and Plan

The main Chapters follow a line of thought as follows:
Raw fuel — heat production or use of heat including furnaces and boilers — mechanical energy — electrical energy — 'total energy' use — heat saving — materials and control — environmental matters (i.e. increasing complexity of concept). Fuel burners are described before complete furnaces or boilers (again complexity increasing). Individual fuels are dealt with in the order gaseous, liquid, solid. Purely technically, this again follows from simplicity to the more complicated.

It did however prove necessary to follow on from the main present-day methods of steam and mechanical-energy production to that of electrical energy. And even then the use of electricity (because of its inclusion with fuel-fired furnaces in Chapter 8) had to appear before an account of its generation. This left combined cycles and total energy (which were important concepts long before being given these names) to follow on.

No fully satisfactory place for nuclear power generation could be found. And although the efficiency of this process is less susceptible to the detailed influence of the operator than that of any other, its growing importance in several technically advanced countries was adjudged to merit fairly complete treatment if only to enable the wider technical community to appreciate its principles and problems.

General Policy

It was early decided to use solely the SI system of units — dual systems would have required considerably more space and made reading more difficult. It was expected that the UK and much of the world would change to these units (and more important, would launch technological graduates trained largely or exclusively in terms of them) between 1975 and 1980. It was further expected that the book would remain broadly up-to-date without major revision until at least 1985. Two Tables (see *end papers*) should suffice to relate the various national and international units. The advantages of the SI system are particularly marked in the measurement of pressure (see also p 235), where local units based on lbf or kgf depend on the local value of the gravity constant, and are not absolute units that can be transferred *accurately* from place to place (the conversion values given in the Tables are based on world average value of g).

It was agreed that the main purpose of the book was to discuss the efficiency with which energy was used in industry, not the domestic environment, though manufacture of domestic *fuels* would be included.   Use of steam would be restricted to a Section that would be adequate with references to *The Efficient Use of Steam*[1], which still remains largely relevant, and recent editions of the Steam Tables. Little would be said on reciprocating steam engines, electric motors, or mechanical engineering in general. Omissions and borderline cases are inevitable; for example, refrigeration of foodstuffs receives only passing mention, and crematorium furnaces none. (These form probably less than 0.1% and less than 0.01% respectively of energy consumption in the UK.)

It had been intended to include copious references to a new edition of *Technical Data on Fuel* if its publication date approximately coincided. Unfortunately this has been delayed, and references are here made to the sixth edition[2], 1961.

Progress

As mentioned earlier, the outline of this book was planned originally by a small Panel (see list after Preface) of the Institute of Fuel (UK) between July 1971 and the end of 1972. IPC was first approached with a modified version in July 1973, and at the end of that year accepted responsibility for publication.   Some members of the original Panel, and some newcomers, accepted appointment to an Editorial Advisory Committee (see list after Preface), some of whose members have provided further advice on contributors and have refereed some of the contributions, though it has held no meetings as a body.

It has proved difficult to edit the material while much still remained unwritten. Sources of a very few of the contributions were changed during the course of writing. In particular, matters of detailed overlap have often needed reconsideration; cross references have been inserted largely during the final weeks of preparing the script.

It had been decided to minimize overlap except where clearly beneficial to the immediate context. At first sight the introductory essay (Ch.1.2) covers some of the same ground as Appendix E (and portions of Ch.14). But the methods, approach and sources are all different, and it is desirable in such general discussions to illustrate the extent to which different interpretations are possible.

Owing to the urgency and complications *en route*, at the time when the book went to press only the general editor had read the whole of the content in final form. I must therefore accept responsibility for any errors that have been overlooked.

Energy and Entropy, and Sensible Heat

The 'law' of conservation of matter is a limited aspect of that of energy; though the two are now known to be interconvertible via the Einstein relation $E = Mc^2$, this process is so limited in normal life on the earth's surface that it is possible to treat them as independent principles. A given immutable quantity of energy is thus in effect associated with a given immutable quantity of matter and susceptible to quantitative treatment on this basis.   Temperature and entropy (heat energy/absolute temperature) are to an extent measures of quality rather than quantity. Sensible heat is the kinetic energy of random molecular motion in a substance. Temperature is a measure of the kinetic energy per unit mass. Specific heat is dimensionally similar to entropy. In the simplification appropriate to a perfect gas, entropy = 1.5 R units/mol, $c_v = R \times$ molecular degrees of freedom (n), $c_p = R(n + 2)$.

Fuels and Combustion

A carbonaceous fuel is a source of heat energy because, by reacting with oxygen mainly to carbon dioxide and water, it can release sufficient heat energy to raise considerably the initial temperature of the gaseous products. The source of this heat release lies in the difference between the 'potential' energies inherent in the structure of the fuel molecules and those of the product molecules. The maximum flame temperature depends on the fuel constitution as reflected in that of the combustion products. If air is the source of oxygen, the maximum temperature is reduced by dilution with nitrogen.

Spontaneous processes of this kind always increase the total *entropy of the system*, which increases further during subsequent processes such as cooling and temperature equalization.   The *potential energy of the fuel* should on the other hand be totally accountable in the output of heat energy.

Definitions of Efficiency

Combustion Efficiency.   Normal: Total fraction of the energy in the fuel used that is channelled into a form adjudged useful in the particular process.

Examples are:

1. A free-standing oil heater, 100%.
2. Electrical air or liquid immersion heater, 100%.
3. Best fuel-fired boiler, 90% (in steam produced). Even in these cases, cooling, eventually leaving the same amount of heat at an unusable low temperature, begins immediately.
4. Furnaces, much lower figures ($\approx 5\%$) depending on experience.   Here the judgement of what is useful becomes uncertain.   In a metal-smelting furnace the small proportion actually used in chemical reduction (e.g. $\approx 30 - 40 + \%$ in a blast furnace making iron) is definite, but other conditions render it unrealistic to use this as a criterion for possible improvement to anything approaching 100%.

Modified Combustion Efficiency. Some proportion of the inevitable losses is subtracted from the assessment of what is useful.

Fuel Consumption per Unit of Product (Comparison with the Current Norm). More realistic for (e.g.) furnaces.

Calorific Values. These are determined and used on four kinds of basis. The consequent efficiencies calculated may differ by moderate or negligible amounts. Reference should be made[3,4] for details, corrections and interconversion formulae.

The basic definition is: quantity of heat released by complete combustion in a calorimeter of unit quantity of fuel under given conditions.

The four bases are:
1. Gross CV at constant pressure.
2. Net CV at constant pressure.
3. Gross CV at constant volume.
4. Net CV at constant volume.

Gross signifies that water formed or liberated during combustion is in the liquid phase; net, in the vapour phase. As a convenient approximation, the gross CV at constant volume of a solid or liquid fuel may be compared without correction with the gross CV at constant pressure of a gaseous fuel. Both gross and net values are commonly quoted[2], but the constant-pressure or constant-volume basis is not always specified; for gases the net value at constant pressure is normal practice, for liquids and solids the constant-volume basis.

Efficiency of Conversion to Mechanical and Electrical Energy. If heat energy (e.g. a steam or gas turbine) is used as the intermediary, a loss additional to those observed with a combustion process is incurred owing to a thermodynamic limit imposed by entropy considerations (Ch.10.7): the maximum possible value of this subsequent efficiency is equal to $(T_1 - T_2/T_1)$ where 1 and 2 refer to the working limits of absolute temperature. For electricity, the practical maximum at present allowing for turbine efficiency is $\approx 37\%$ less a small generator loss. Once in these non-thermal forms, very little storage of energy is possible, but if used as generated very high efficiencies of *energy use* are possible.

Nonetheless, the claim sometimes made that only electricity represents pure energy savours of special pleading. There are special advantages in using electrical energy, such as time and therefore energy saving, convenience, and economy of space, that can be added to the economic benefits; but these cannot be expressed numerically as fuel or energy efficiencies.

The need to conserve the heat rejected in these conversion devices has led to the modern emphasis on total energy schemes including district heating. Alternatively, methods such as fuel cells to convert chemical to electrical energy directly are continuing to claim attention.

Cost-Benefit Analysis. This enables all advantages to be considered on a common (money) basis. It is much used in connection with total energy schemes (see Ch.15). The relation to efficiency of energy use is local, indirect, and ephemeral (e.g. see Ch.15.1).

Other Definitions. It is becoming commoner in comparisons to recognize the interdependence of manufacturing processes by calculating the energy content (i.e. used in manufacture) of materials (cf. *Table 11*) employed and thereby deduce the total use including that newly added in the process under consideration. This can be done for foodstuffs as well as industrial products, and provides in effect a means for energy costing as an alternative to financial costing. It is an interesting and potentially enlightening exercise.

The Real Significance of improving Energy Efficiency

The many alternative choices mentioned, and the vagueness of meaning in some cases, suggest either that some new standard definition should be widely adopted, or that each should continue to be employed in a limited field (there is usually a defensible reason). Perhaps the best policy is to state the precise basis employed where there is any ambiguity.

Common to all the bases employing actual energy measurements is their ephemeral nature. They are all instantaneous efficiencies. And next day what remains? A consignment of material suitably transformed; a further day's memory of warm and comfortably fed life; some additional products in shops or for export; people or goods transported elsewhere. The actual energy used, however efficiently, has been degraded and diffused; there remains a slight and undetectable, perhaps in the long term undesirable, warming of the atmosphere. For the real benefits, qualitative effects would need to be compared with quantitative energy expenditure — a common and insoluble dilemma in human affairs.

Clearly the real benefits of higher efficiency must be sought at the input end. Less energy and money need be expended to bring about the same results. The rate of depletion of fuel resources, and pollution, can be reduced. These are worthy aims at any time, and now more than ever because of the growing scarcity of and competition for raw materials, and population congestion. Looking further ahead, there may be even more lasting benefits stemming from reduced demands and improved conservation prospects. We may not yet have realized all the implications.

Assessment

Much is being published today on energy efficiency and policy, in both the long and the short term. I am conscious that this book for the most part sets (as intended) mainly an accurate background

References p 16

of current practice and fact against these discussions; apart from exceptions such as the excellent broad breakdown into forms of use provided by Dr Gibson in Ch.1.2 and somewhat more general views expressed by Dr Parker and myself in connection with Appendixes E and F, little overall guidance on desirable ways of improving efficiency is offered.

Full assessment of the detailed evidence presented in this book, in respect of even one national economy, would be a lengthy task, preferably the subject of a separate small volume. Even in the shortest term for the UK only it would be rash to give much more than general guidance, such as advice to adjust when opportunity offers an entire process to improve its energy utilization; to use steam directly where convenient if it can help to conserve electricity usage; to reduce steam and air leaks; and to improve insulation whether of buildings or plant*. Dr Gibson himself and his quotation from NIFES[20] (*Table 10*) have illustrated so far as possible such principles and the savings to which they could lead, in Chapter 1.2.

It is now more than time to turn to this account of recent trends in energy use and efficiency and possible improvements.

1.2 Perspective

World Energy Production since World War II

When *The Efficient Use of Fuel* was first published during the second World War over 60% of world energy production was coal based. To a large extent in the immediate post-war years the main energy consumers managed to be self-sufficient.

---

* For a pitfall in the lagging of small pipes, see App.C11 p 547

Table 1   World energy production (Mtce)[5]

Years	Coal and lignite	Crude oil	Natu- ral gas	Primary elec- tricity (nuclear, hydro and geo- thermal power)	Total
1950	1605	700	261	41	2607
1955	1807	1029	400	59	3295
1960	2191	1396	622	86	4295
1965	2268	2001	931	117	5318
1970	2394	3002	1436	157	6989
1972	2430	3340	1616	179	7566
1974 (estd)	2430	3740	1750	220	8140

1 Mtce = 10^6 tonne coal equivalent = 2.88×10^4 TJ

Table 2   World consumption of energy by regions (Mtce)[5]

	1950	1960	1965	1970	1972
Western Europe	600	875	1135	1489	1600
North America	1221	1627	1969	2521	2680
Japan	49	123	207	380	440
Total OECD	1870	2625	3311	4390	4720
USSR, Eastern Europe and China	560	1366	1563	1938	2100
Rest of the world	239	382	514	702	780
World total	2669	4373	5388	7030	7600

1 Mtce = 2.88×10^4 TJ

Table 3   Dependence on imported energy[5]

	1955 (%)	1960 (%)	1965 (%)	1970 (%)	1973 (%)
EEC (the Nine)	19	30	46	59	61
Japan	23	41	65	84	87
USSR	Nil	Nil	Nil	Nil	Nil
USA	1	6	7	8	16

Western Europe continued to depend almost entirely on coal and lignite — the USA and the Soviet Union had indigenous oil and gas resources in addition to vast developing coal industries. International trade in coal and oil represented a relatively small share of total world demand.

In the fifties economic growth accelerated, and to meet the world's industrial needs energy production rose three-fold. *Table 1*[5] summarizes the post-war growth of energy production in the main categories. Crude oil (44%) has overtaken and outstripped coal (32%) and natural gas (21%) has become a substantial contributor.

World Energy Consumption

World consumption data in *Table 2*[5] show similar trends for the main regions and that the OECD countries have continued to consume over 60% of the world's energy.

The Soviet Union has been able to provide from its own resources and until recent years this has been substantially true for the USA. In the EEC countries and Japan there has been a striking increase in imports (mainly oil from the Middle East and Africa), and Japan is now dependent to the extent of 87% on imported energy (*Table 3*)[5].

World Energy Resources*

Estimates of world energy resources are dependent on geological survey data, on the inherent quality

---

* This may be compared with very recently available statistics discussed in Appendix E

Table 4 World energy reserves: variation in estimates[6]

	Present known			Potential future[2]		
	Reserves	Life at 1971[1] consumption rates	Life at future consumption rates	Reserves	Life at 1971[1] consumption rates	Life at future consumption rates
Oil						
Lowest	80 × 10⁹ tons	32	16	250 × 10⁹ tons	100	30
Highest	90 × 10⁹ tons	36	18	360 × 10⁹ tons	140	40
Coal						
Lowest	130 × 10⁹ tons	60	30	1100 × 10⁹ tons[3]	500	150
Highest	2200 × 10⁹ tons	1000	190	4800 × 10⁹ tons[3]	2200	250
Natural gas						
Lowest	34 000 km³	33	15	90 000 km³	90	25
Highest	48 000 km³	45	19	340 000 km³	330	40
Uranium						
Lowest	0.9 × 10⁶ tons[4]		16/(50—100)[5]	1.3 × 10⁶ tons[6]		20/(500—100)[5]
Highest				3.2 × 10⁶ tons[6]		37/(50—100)[5]
Shale/tar sand						
Lowest	97 × 10⁹ tons	39	Extend oil by 9	280 × 10⁹ tons	110	Extend oil by 10
Highest	120 × 10⁹ tons	48	Extend oil by 11	500 × 10⁹ tons	200	Extend oil by 17

Units: Oil, and oil shale and tar sands, are expressed in 10⁹ tons of oil. Coal is expressed in 10⁹ tons of oil equivalent). Natural gas is expressed in 10¹² m³ (multiply by 0.86 to give 10⁹ tons of oil equivalent). Uranium is expressed in million (10⁶) tons of uranium. Life is in years.

(1) 1971 consumption rates assumed are: *Oil*: 2500 Mtoe; *Gas*: 900 Mtoe; *Coal*: 1500 Mtoe.
(2) Potential future reserves are usually estimates of recoverable reserves, allowing for improvements in technology of extraction and price rises, unless otherwise stated.
(3) Several sources quote 7.6 × 10¹² tons of coal as potential total reserves, which is much higher than recoverable reserves.
(4) Known reserves recoverable at less then $20/kg.
(5) The life of uranium resources is considerably increased when fast breeder reactors are considered, rather than existing light water or similar generation reactors (hence two figures for life in the table). Owing to uncertainties in the future development of nuclear power the lifetimes quoted are particularly speculative.
(6) Reserves of 3.2 Mt of uranium assume a recovery cost of not more than $30/kg. However, it has been estimated that 60 Mt are available at costs up to $200/kg. In addition there are in sea water some 4 Gt of uranium which are inaccessible with present technology and economic conditions.

Based on: *Energy for the Future*, Institute of Fuel, London, 1973

of source material, on its abundance, on the technical means and the economic costs of winning it. Technical and cost advances can materially affect the resource potential. In *Energy for the Future*[6] the difficulties inherent in estimating resources are discussed and the variation that exists between attempted estimates is presented (*Table 4*). The spread of figures is a measure of their unreliability, but even so, taking the highest estimates of source-life, the portents are ominous. 'The present growth of oil consumption could not continue for the length of a man's lifetime. Coal is available for much longer; uranium is only cheaply available for an extended period if used in a breeder reactor. Fusion fuels are unlimited — deuterium could replace the entire energy flux to the earth for many times the life of the sun, but unfortunately the technology required is not available or even at the stage where its introduction can be foreseen.'

Solid Fuels

More than 95% of the world's coal resources are found in the northern hemisphere — in areas north of latitude $30°$N. *Table 5*[7] shows that about 90% of the total resources are found in three countries — USSR, USA and China. The large amounts in the resources column are geological estimates of essentially all the coal in the ground. While Nelson *et al* draw attention to the hypothetical nature of the estimates, they also emphasize they are not as speculative as comparable estimates would be for oil or gas since the geology of coal is relatively simple, the deposits are uniform over wide areas, and they lie near the earth's surface.

World production in 1972 was 2.43 Gt and at that rate the recoverable reserves would last about 230 years. But as the total resources are 20 times and those economically exploitable are

Table 5   World resources of solid fossil fuels[7]

	Total resources	Recoverable reserves
	(Gt)	
USSR	5713	136
USA	2926	186
China, P.R. of	1011	38
Canada	109	6
Europe*	608	127
Other	388	58
	10755	551

* Excluding USSR

Note: The criteria for estimating recoverable reserves differ greatly, making comparisons between and among countries inadvisable

Table 6   Net imports of oil (Mt)[5]

	1961	1970	1973
EEC (the Nine)	175	500	590
Other Western Europe	34	112	146
Total Western Europe	209	612	736
Japan	40	210	284
USA	88	157	301
Total of these countries	337	979	1332

estimated to be twice the recoverable resources these figures should provide a challenge to increase the proportion of recoverable wealth rather than a complacent acceptance of a life for coal of up to 200 years depending on how steeply production is increased.

Coal is an extractive industry and most of it is mined underground. It is labour intensive and not without risk. The fullest realization of the resources will depend on reducing the dependence on labour and minimizing the risk while using the product in the most efficient way.

Oil

Reference has already been made to the rapid growth of oil as an energy source and the self-sufficiency of USSR, and USA until recently, in this respect. Massive additional oil reserves were discovered and exploited in the Middle East and Africa from the mid 1950s leading to cheap oil supplies on world markets. W. Europe and Japan possessed negligible resources of their own and became the principal outlets for these new supplies (*Table 6*)[5]. The imbalance in oil reserves and consumption is vividly shown (*Table 7*)[5]. The major industrialized nations consume almost 70% of the world's oil and possess only 9% of the proved reserves, while the Middle East and Africa consume only 4% of the world total, but have a dominant 70% of the present proved oil reserves within their boundaries. This imbalance between the geographical location of the world's oil reserves and its usage is the significant factor affecting the world energy market today and will influence the pattern of development for many years to come.

Bearing in mind that resource estimation is more highly speculative for oil than for coal, there is nevertheless almost unanimity in the forecast that the production of oil will peak towards the end of this century. The most recent oil field to be discovered and developed in the North Sea has hardly begun production, yet forecasts of oil production[10] from the UK Continental Shelf show peak production ($\approx$120 Mt) around 1981 with a gradual decline to below 40 Mt by 1990. There

Table 7   Imbalance in oil reserves and consumption[5]

	Proved reserves 1972		Oil consumption 1972	
	(Mt)	(% of world total)	(Mt)	(% of world total)
Middle East and Africa	56 600	66	111	4
North America, Western Europe and Japan	8 800	10	1913	69
Rest of world	20 600	24	742	27
World total	86 000	100	2766	100

Table 8   Proved recoverable reserves of natural gas and natural-gas liquids[7]

	Natural gas (km^3)	Natural-gas liquids (Mt)
Middle East	9884	n.a.
Other Asia*	22742	n.a.
Europe*	21757	67
Africa	5709	24
Oceania	586	92
Total: E. Hemisphere	40208	n.a.
N. America	10649	1085
S. America	1589	53†
	12238	n.a.
Total world	52446	n.a.

* European data include the USSR except for natural-gas liquids
† Venezuela only

is considerable optimism that other significant finds will be made, but they cannot do more than shift the peak into the 1990s. The forecast world pattern is similar — taking the most optimistic future estimate the life of oil at future consumption rates is likely to be 30—40 years (*Table 4*).

Natural Gas

Data for natural gas reserves and the condensates associated with them are listed in *Table 8*[7]. At first, difficulties of transport and storage made the development of natural gas resources slower and less efficient than for oil. Where possible, pipelines have been laid connecting production points with markets, notably in the USA, Canada, Western Europe and USSR. In many oil-producing areas where pipelines may not be feasible natural

gas is still being wasted in large quantities, although this is being partially overcome by liquefying the gas (LNG) and shipping it in specially constructed vessels. Natural gas is now a major fuel in many countries. Development can be rapid and recovery relatively high, even under unfavourable conditions as evidenced by North Sea gas. In the USA which first developed large-scale pipeline distribution, supplies are dwindling and there are ambitious plans to convert coal into substitute natural gas (SNG) supplemented by substantial imports of natural gas from Canada and LNG from abroad.

Oil Shale and Tar Sands

The estimated reserves of hydrocarbons locked in oil shale and tar sands are twice the postulated figure for ultimate world recoverable reserves of conventional crude oil. They are nearer the earth's surface than petroleum, enabling them to be located and defined without much difficulty. Unlike oil, however, they lack fluidity and consequently, in common with solid fuels, they require much effort to produce, are hard to handle and difficult to process. The tar-like material in the sands[8] is highly viscous, chemically like crude oil, and the synthetic crude material (Syncrude) extracted from them can be processed in oil refineries. The oil 'content' of the shales[9] is different chemically and more difficult to process.

The majority of these hydrocarbon reserves lie within N. America. There are fairly substantial shale reserves in USSR, China and Brazil, whilst Venezuela recently revealed considerable reserves of tar sand.

The effort required to tap these resources in the USA can be illustrated by the requirements to mine oil shale in sufficient quantities to produce 1 million barrels of oil a day by treatment in retorts — barely one-thirteenth of the present USA daily consumption and equivalent to 50 Mt a year. No less than 550 Mt of shale would have to be handled annually, almost equivalent to the amount of coal mined in the USA in a year. For processing, 184 Mt of water would be drawn annually from the nearby Upper Colorado Basin and not returned to it. Water resources are clearly a limiting factor in processing conventionally mined deposits. This has led to the investigation of *in situ* processes using explosives to loosen the shales which are then distilled from their underground caverns and collected at the surface. The water requirements are less, but the difficulties are many and far from solved. It is clear, therefore, that these sources are not going to make a significant contribution to world energy before the end of the century.

Nuclear Fuels

Nuclear fuels are regarded as the key sources of energy in the long-term. They will play an increasing role in supplying electrical energy and are expected to become the principal means of generating electricity by the end of the century. The UK took the first step in 1956 to inaugurate her nuclear generating programme and now gets 10% of her electricity from nuclear power stations using natural uranium, which is equivalent to 9 Mt of

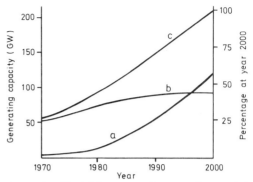

Figure 2   Electricity generating capacity trends in the UK[11].
a *nuclear;* b *fossil;* c *total*
Courtesy W. Short and Nature, Lond.

coal burnt or about 6 Mt oil a year — about 3% of the total energy consumption (*Figure 1*).

In the next generation of nuclear reactors in Britain and elsewhere in the world, uranium will be the dominant fuel. Later this century breeder reactors which use uranium but extend its usefulness through generating their own fuel by breeding plutonium from the isotope ^{238}U are expected to be in use. Until then, the demand for uranium is expected to increase rapidly. Thorium, another fissile element capable of another breeder cycle (^{232}Th → ^{233}U) will have an expanding but subordinate role in nuclear power generation. In the footnotes to *Table 4* the increased life of uranium resources when fast-breeder reactors are considered is noted, and the benefits are represented by the extended life figures.

Uranium is widely distributed throughout the world. The larger part of proven economic reserves (80%) is in the USA, Canada, S. Africa and Australia, excluding communist countries for which adequate data are not available. Nuclear fuels are the only energy resources that are commonly classified according to the relative costs of recovery. Thus the known resources of uranium recoverable for less than $20/kg are estimated at 0.9 Mt, but this becomes 3.2 Mt at a recovery cost of $30/kg (at 1973 prices) and so on.

Although there are massive nuclear construction programmes being mounted, especially in the USA, the forward estimates of the nuclear contribution will still represent a small fraction of total energy consumption by 1985. In Britain, for example, to fulfil the aims of providing an increasing share of the electricity demand (up to 80%) thus permitting a more rational use of fossil fuels in the face of severe world shortage, it has been advocated that 30 000—40 000 MW$_e$ of nuclear capacity should be ordered in the next decade and built before 1990, and a further 70 000 MW$_e$ built before the end of the century (*Figure 2*)[11]. This mammoth programme which requires a present-day investment of £8000 million for the first stage alone

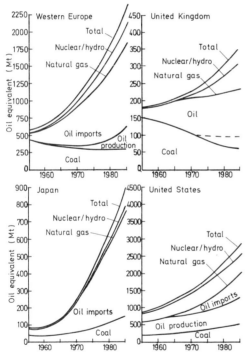

Figure 1   Patterns of energy consumption[6]. The above graphs illustrate the relative share of each fuel in the total annual energy consumption of various geographical areas. Statistics of past and predicted consumption were obtained from a variety of sources. Smooth lines were drawn through the historical figures and forecast figures to give the curves shown.
The broken line on the UK graph represents a divergent view taken by the National Coal Board.
Courtesy Institute of Fuel

poses a near-impossible task and at the end of the century will contribute around some 30% of Britain's total energy consumption at best.

Technical problems have arisen with all the different thermal reactors. The technology of breeder reactors is more difficult still, and while considerable progress has been made during the nine years successful operation of an experimental breeder reactor in the UK at Dounreay, many problems remain to be solved and such reactors could hardly be expected to make a serious contribution to the energy load before the 1990s.

Farther into the future is the hope that nuclear fusion power can be realized. This is obtained by simulating the energy-producing reactions which occur on the sun's surface. The energy resources for fusion are deuterium (heavy hydrogen) for the deuterium—deuterium reaction and lithium which is potentially the source of tritium for the deuterium—tritium fusion process. Deuterium in water (35 g/t) is expensive to concentrate, but if practical fusion devices can be ultimately developed to commercial scale the supply would suffice for millions of years. Lithium, less plentiful, is almost an unlimited source too in this respect, especially if it can be recovered from sea-water (concentration 0.15 ppm w/w). The outcome of recent work and the realization of fusion power is still far from certain.

Having said all this, the rate of nuclear-power development may not be controlled by improved reactor technology so much as by environmental considerations. Nuclear reactors can discharge up to 60% more cooling water than fossil-fuel stations; there are serious radioactive waste-disposal problems to cause disquiet and the fear of a serious nuclear incident to provoke opposition.

These problems will have to be solved if the energy for man's continued survival is to be provided in this manner. Until the answers are apparent, the practical solution is a reversion to coal and careful husbanding of oil and natural gas, together with the most effective measures to achieve the highest degree of efficiency in the use of all these energy sources.

Hydraulic Resources

These resources do not involve any chemical reactions in their usage, depend on converting potential energy into kinetic energy for mechanical work or the production of electricity and, unlike the resources of fossil or nuclear fuels, they are continuously renewable, mainly through the evaporative power of the sun.

By this means, about 330 000 km^3 of water are evaporated from the oceans, etc. each year. Two-thirds returns as precipitation to the seas and the remainder falls as rain or snow on the land and is re-evaporated except for about 37 000 km^3 which flows back to the seas. From the average elevation of the land it is possible to estimate that the total energy dissipated in this seaward flow is equivalent to about 290 × 10^6 TJ or 80 000 TW h — about equal to the present annual use of all commercial energy throughout the world[7].

For many reasons, geographical, seasonal or mechanical, only a very small fraction of this total is utilizable. Within the industrial countries most of the hydro sites have been developed — the others are too remote, too costly to harness or invalidated for environmental reasons. In geographically favourable locations, hydraulic power can provide as much as 50% of total electrical energy needs, as in Italy, but generally the contribution amounts to only 3—4% in industrialized countries. In the less highly developed regions of Africa, Asia and Latin America there is still scope and a useful contribution can be expected, especially where hydro power and irrigation schemes can be linked.

Other Energy Resources for the Future

Solar Energy

Solar energy has been the life-force behind man's evolution. It is the source of energy for agriculture and provided power indirectly before the industrial revolution for most of man's activities. About 1.5 × 10^{18} kW h (4.8 × 10^{12} TJ) of energy would be incident upon the earth's surface each year were it all constantly illuminated; but the projected surface is one-quarter of the spherical surface, and allowing for atmospheric absorption the average availability is about 10^{12} TJ — some 5000 times current world energy requirements (see also App.E1, footnote). Present limits remain to be explored, but there would appear to be a very high ultimate limit to the amount of electricity and heat that can be generated in this way since devices for these purposes are available.

Solar energy, unfortunately, has a very low flux density even before atmospheric absorption and is intermittent while solar cells are inefficient (current conversion 5%) and expensive. The use of solar heat to generate steam for use in a turbo-alternator system is currently being investigated. Such a system could yield 28 MW/km^2 of collector surface a day in a favourable solar climate[12]. Unless the efficiencies of these devices can be much improved and/or their cost substantially reduced they are unlikely to make any significant contribution in industrial areas this century. They are more likely to benefit the under-developed countries in favourable climates where encouragement to develop them would be a sensible and effective way of providing technical and financial assistance.

Geothermal, Tidal, Wave, and Wind Power

The generation of electricity by these equilibrium sources is possible and except for waves there are examples of their use to supplement conventional methods to be found where geographical conditions are favourable. The first geothermal power plant was built in Italy in 1907 and other plants now exist in New Zealand, USA, Mexico, Japan, Iceland and the USSR. Tidal energy is used at La Rance in Brittany to generate 240 MW_e, but costs are two to three times conventional costs. Geothermal and tidal resources are small, however, and are unlikely to provide more than a very small fraction of world demands — probably some 200 000 MW_e.

Windmills for local power production have been used for centuries for homes and small communities. They will continue to make small contributions to the energy situation and should be exploited wherever local conditions allow, but in common with the above they are low-intensity resources requiring large investments and relatively heavy use of construction materials.

Attention has recently been drawn to the possibility of harnessing wave energy[13], not yet attempted on any large scale. It is suggested that the useable coastline in the UK amounts to some 900 miles which could possibly sustain a generating capacity of 30 000 MW_e or save somewhere in the region of 65 Mtce. It suffers the same drawbacks as the other equilibrium sources in terms of investment and demands on construction materials. The technology is not expected to be complicated, however, and wave energy provides a seasonal peak in winter when demand for electricity is greatest.

Hydrogen and Methanol

These are substitute fuels derived from other resources or basic raw materials but requiring more energy to produce than they can in their turn provide. Hydrogen can be made by electrolytic thermal and/or chemical splitting of water; methanol can be made from fossil fuels or, less probably, other carbon-containing materials, by reaction with water or hydrogen. Their manufacture requires, if the drain on fossil fuels is to be avoided, lavish use of electricity. Until nuclear electricity is plentiful and cheap they will have premium uses — when oil is scarce they will assume greater prominence as fuels for transport.

They have several advantages including cleanliness, flexibility and (apart from liquid-hydrogen and gas cylinders) low distribution costs. As an alternative to liquefaction, schemes for the long-distance transport of natural gas have been proposed in which methanol is used as an intermediate storage medium. The methanol would be synthesized from natural gas at the point of production, transported by pipeline or conventional tankers, stored until required and then reconverted to methane. Hydrogen can be stored either as a compressed gas or in liquid form (at 20 K). For transport purposes various hydrogen compounds including ammonia, hydrazine and metal hydrides have been suggested. Hydrogen can be released or absorbed from metal hydrides by varying the pressure, but the carrier metals tend to be heavy,

resulting in a low stored-energy to weight ratio, and expensive. At the moment, hydrogen and methanol are more valuable in the chemical industry.

Factors in More Efficient Energy Use and Conservation

Patterns of Use

The use of energy in some prominent industrial regions was illustrated in *Figure 1*. The overriding point is, however, that 80% of the world's energy is consumed by 20% of the population. It is inevitable that the pattern of use by major users must change as the under-developed countries progress and require a much greater share of world energy. Some countries will attempt to become self-sufficient in energy — the original aim of the USA by 1980, now lowered to 'a high degree of self-sufficiency by mid-1980s'. Others will have to compete for resources in greater world demand, and the only sensible way of avoiding bankruptcy will be by improving efficiency, making economies, and developing newer forms of energy, e.g. nuclear.

The present pattern of use in the developing countries which encompass 80% of the world's population is not relevant to their future pattern because they will undergo major change. Incidentally, they also represent the countries where the rate of population increase will be greatest. Development of their indigenous resources, e.g. solar, hydro and possibly geothermal power, will be able to make a vital contribution to energy supplies in areas that do not have large reserves of fossil fuels. Cooking and heating loads, in particular, seem appropriate candidates to make use of these sources. This will enable energy imports to be concentrated on areas where there is no alternative — such as agriculture. The development of irrigation schemes, associated wherever possible with hydroelectric power generation, will be desirable. The more intensive use of machinery and fertilizers in the agricultural underdeveloped countries, in which the present level of use is minimal, would, however, make a sizeable increase in the world's energy demand. As far as the growth of industry in the underdeveloped nations is concerned, this should be concentrated in areas where the requirement for imported raw materials is least, and in low-energy industries in particular. The part which nuclear power will be able to play will be small owing to the intense competition for the limited capital resources available. The development of long-term substitutes for fossil fuels, other than nuclear power, will clearly be required.

The energy consumption in an industrial nation, e.g. the UK which uses 4.5% of the world's energy, is illustrated in *Figure 3* and is fairly typical. It is obvious that while there should be determined efforts to save energy wherever possible the sectors which offer the best opportunities are those which consume most — domestic, road transport and iron and steel. Savings in agriculture, on the other hand, will have no great impact since a mere 1.5% of total UK energy is consumed here.

Flow of Energy

For a fuller understanding of the situation it is necessary to look at the flow of energy from source to end use as in *Figure 4* for the UK in 1973 [14]. The energy-flow diagram for the USA is very similar but for the very much higher contribution from natural gas and a negligible amount of nuclear energy. *Figure 4* shows the loss of effective energy at every stage and the summation of heat used and lost appears towards the right-hand side of the chart. Thus we not only see where the losses occur, but how large they are. Added to the right-hand side of the chart is a histogram illustrating the efficiency of the fuel or application. The efficiency is greatest ($\approx$65%) when the fuel is used directly, falls to about 30% for electricity and drops further to 20% or less when oil is used in internal-combustion engines.

This should help decide where the fuel should go and where savings are possible. Nevertheless, it would be misleading to think that the decisions are simple or the issues clear-cut. Electricity is indispensable in very many applications, but where it is not alternatives should be considered. Transport is an example of such low efficiency in fuel usage in a major section of consumption that improvement should be mandatory — lower speeds, more efficient and less powerful engines will some day be more acceptable than no car at all. Looking at the alternatives, the efficiency for transport could be almost doubled by electric traction; improvements in battery technology could make electric cars competitive for low-speed small 'city' transport in a system rationalized for the community.

Energy Transport and Transmission

As can be seen from *Table 9*, although the range of the estimates is large, there is general agreement that electricity is the most expensive form in which to transmit energy, followed by coal, gas and petroleum[15—17].

In addition to the costs of providing the transmission system detailed above, energy is required for the transmission itself and losses will inevitably occur. The costs associated with transport and losses will depend on the form of energy involved. Williams[16] includes estimates of transport energy and losses, and these have been used as a further basis for *Table 9*.

When energy is transported to take part in further conversion processes, the efficiency of the conversion process will also influence the form in which the energy is transported. For example, the choice of site for a coal-fired power station requires consideration of the cost of transporting three times the amount of coal equivalent to the electrical energy produced. In particular, it will be preferable to site future coal processes in which coal is used to manufacture SNG or Syncrude in mining areas, so that coal transport costs are minimized.

At present, in the UK there are 19 200 km of main transmission lines which have a replacement value of £1375 million at 1972 price levels. This

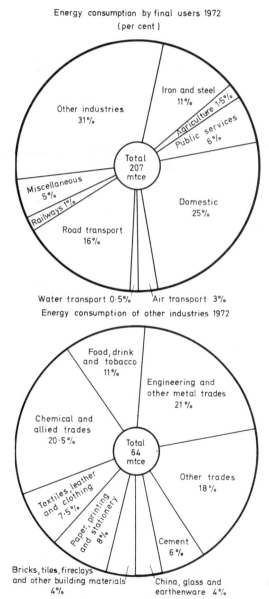

Energy consumption by final users 1972
(per cent)

Energy consumption of other industries 1972

Figure 3   Energy consumption in the UK (1972). Adapted from 'Energy Conservation in the United Kingdom', HMSO, London, 1974
Mtce = 10^6 tonne coal equivalent = 2.88×10^4 TJ
Courtesy HMSO

References p 16

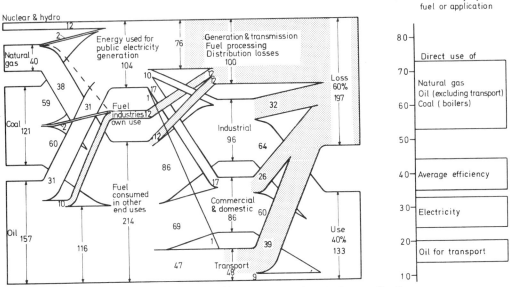

Figure 4   UK flow of energy (Mtce) *in 1973[14]; efficiency (%) by fuel or application*
Courtesy B. J. Bowden and Heating & Ventilating Engr

Table 9   Costs and efficiency of energy transmission

| Form of energy | Cost (p/therm per 100 miles*) | | | | Efficiency[16] | |
	Hottel and Howard[15]	Williams[16]	Hill[17]	Range	Energy loss (%)	Transport energy (% per 100 miles*)
Petroleum (pipeline)	0.04	0.01		0.01–0.04	Insignificant	0.06
Petroleum (rail)		0.1		0.1	Under 1%	0.15
Natural gas	0.07	0.03–0.05	0.04–0.08	0.03–0.08	1–2%	0.2
Coal (pipeline)	0.01	0.1	0.08–0.12	0.08–0.12	Under 1%	0.2
Coal (rail)	0.1–0.25	0.1–0.17	0.12	0.1–0.25	Under 1%	0.2
Electricity (overhead)	0.3–0.75	0.2–0.7	0.3–0.45	0.2–0.75	*ca.* 2%	Included in loss
Electricity (underground)			3–11†		—	—

*   1 mile = 1.61 km. Prices at 1972/73
†   Based on comparisons in CEGB statistical yearbook

represents a capital investment of approximately £25/kW. The main distribution system for natural gas consists of some 3600 km of pipeline which has an estimated replacement value of *ca.* £800 million. This is equivalent to a capital investment of approximately £11/kW.

Electricity and SNG Production

The potential of two methods of using coal to generate electricity more efficiently which are

under active development has been examined[18]. The results suggest that more regard be paid to the higher overall efficiency of energy transmission achieved by gasification compared with electricity generation.

The process of fluidized combustion offers many advantages, further enhanced in pressurized systems. The process lends itself readily to sulphur absorption techniques.

There is growing interest, also, in the potential increased efficiency of electricity generation based

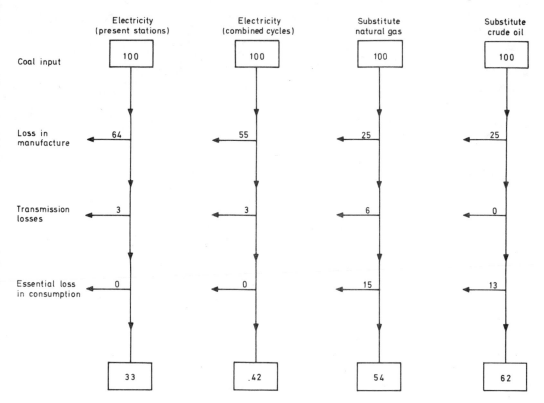

Figure 5 Relative efficiencies of electricity, SNG and SCO production[19]. Courtesy L. Grainger and Chemistry & Industry

on combined-cycle systems involving both gas turbines and steam turbines. Various arrangements of boilers and turbines are possible, but the objective in each case is to reduce the overall capital cost of the power station and to increase the proportion of fuel which is converted to electrical energy. The possible extent of these improvements depends on the highest temperatures which industrial gas turbines will accept; at present this is about $850°C$, but it is expected to rise progressively following the higher temperatures at which aircraft turbines operate.

Studies of the economics of producing SNG from coal indicate that it is a much more efficient means of supplying energy from coal than is the generation of electricity. The proportion of the original energy actually made available to the consumer could be closer to 54%, even having made due allowance for the lower efficiency of the gas-fired appliance (*Figure 5*)[19]. Conversion of coal to substitute crude oil (App.B3) for direct use could raise this further towards about 60%.

The main loss in conventional electricity generation is in the vast quantities of cooling water used which, in their turn, pose environmental problems (Ch.19.6). The utilization of this waste heat (Ch.15.2) should be encouraged where appropriate.

Smaller units would facilitate this, while contrary to popular belief they would permit a net gain in efficiency.

Magnetohydrodynamic power generation (Ch. 13.3), largely abandoned in W. Europe in the 1960s to concentrate on nuclear power generation, has attracted renewed interest in the USA, Europe and Japan. A Russian open-cycle MHD power station using natural gas as fuel is now operating at an overall efficiency of over 50%. Despite this interest there are still severe technical problems which suggest that the results of the research under way should be assessed before more effort is applied.

Industrial and Domestic Energy Consumption

In 1973 the largest industrial consumer, steel, used 11% of the world's energy. In the UK steady improvements in blast-furnace technology have reduced coke rate by 50% over the past 25 years and there are firm plans to achieve further gains in efficiency in ways which could well serve as examples to others. The potential savings in other industries can be judged by the findings of the National Industrial Fuel Efficiency Service Ltd (NIFES), once a British Government body, but

Table 10 Potential fuel savings in the UK (1972/73)[20]

Industrial group	Mean saving (%)
Ceramics, brick, glass	15
Chemicals	18
Iron and steel	20
Engineering and metals	18
Textiles and leather	15
Food, drink and tobacco	15
Other manufacturing	21

Table 11 Typical energy contents of materials and manufactured products[21]. Courtesy Scientific Affairs Division, NATO

	Energy* (MJ/kg)	(Cost of energy)*/ (value of product)
Metals		
Steel (various forms)	25—30	0.3
Aluminium (various forms)	60—270	0.4
Copper	25—30	0.05
Magnesium	80—100	0.1
Other products		
Glass (bottles)	30—50	0.3
Plastic	10	0.04
Paper	25	0.3
Inorganic chemicals (average value)	12	0.2
Cement	9	0.5
Lumber	4	0.1

* These are typical values. The actual value depends on the purity, form, manufacturing process and other variables

now a private company[20]. Heat and power surveys carried out over a sample of industrial groups indicate the potential savings shown in *Table 10.*

It is estimated that improving standards of building design, insulation and temperature control could save 6 Mtce in Britain and, at the same time, be economically viable. Similar high savings are possible in domestic premises and this leads to the conclusion that, coupled with improvements in heating systems, an energy saving of 20% of domestic energy — the largest sector of consumption — is possible.

Materials and Recycling Waste

Table 11 shows that different materials require energy in varying extents to produce them[21]. Furthermore, this energy content is discarded when they are rejected after use. In many cases their disposal uses energy and often creates environmental problems. Wood, which is derived from solar energy, is an especially versatile material which has a fuel value when its usefulness as a material is done.

The choice of materials, methods of construction and the possibilities of recycling with the object of reducing energy consumption will increase in importance as construction costs rise.

References

1   Lyle, Sir Oliver, *The Efficient Use of Steam*, HMSO, London, 1947
2   Ref.(A)
3   BS 526:1961
4   Ref.(A), pp 136—141
5   Association of the Coal Producers of the European Community, *Energy in Europe*, National Coal Board, London, 1974
6   Fells, I. *et al*, *Energy for the Future*, Institute of Fuel, London, 1973
7   Nelson, E. L. *et al*, 'Survey of World Energy Resources — 1974', *Paper to 9th World Energy Conf.*, Detroit, 1974
8   Berkowitz, N. and Speight, J. G. *Fuel, Lond.* 1975, Vol.54, 138
9   Cook, E. W. *Fuel, Lond.* 1974, Vol.53, 146
10 *Production and Reserves of Oil and Gas in the United Kingdom*, HMSO, London, 1974
11 Bainbridge, G. R. *Nature, Lond.* 1974, Vol.249, 734
12 Ref.(B), p 337
13 Ref.(C)
14 Bowden, B. J. *Heat. Vent. Engr* 1974, Vol.47, 539
15 Ref.(B), Ch.2
16 Williams, E. C., 'Energy Policy and Planning — A General View', *Paper to Inst. Fuel Conf. on Energy Transfer in the UK and its Relation to EEC Policy*, Lyndhurst, May 1973
17 Hill, G. R. *Chemtech.* 1972 (May), p 292
18 Grainger, L. *Phil. Trans. R. Soc.* 1974, Vol.A276, 527
19 Grainger, L. *Chemy Ind.* 1974, p 737
20 Short, W. *Nature, Lond.* 1974, Vol.249, 715
21 Ref.(D)

General Reading

(A) Spiers, H. M. (Ed.), *Technical Data on Fuel*, 6th edn, World Power Conf., London, 1961
(B) Hottel, H. C. and Howard, J. B., *New Energy Technology*, MIT Press, Cambridge, Mass., and London, 1971
(C) The Central Policy Review Staff, *Energy Conservation*, HMSO, London, 1974
(D) Kovach, E. G. (Ed.), *The Technology of Efficient Energy Utilization*, Scientific Affairs Division, NATO, Brussels, 1973

Chapter 2

Economic and cost factors in the choice of fuel and equipment

This brief Chapter checklists the type of appraisal that should be made when considering the installation or modification of plant if an attempt at the maximum efficiency and/or the lowest total cost is to be made. It relates to the situation of the UK in 1974, but the overall requirements are universal; the relative weights of the items will vary and should be interpreted in the context of local conditions, prices and objectives.

For the less developed countries in particular, consideration of the immediate factors listed below should sometimes be preceded by a wider survey of the availability of water, oil, gas, coal or other fuels, capital, and facilities for plant supply and fabrication; geographical situation, national and local restrictions on choice, and national and local legislation on environmental factors. This last item may be assisted by a perusal of Chapter 19.

Fuel Storage

Many plants need the insurance of reserve fuel storage to give some assurance of continuity of production or space heating if deliveries of fuel are interrupted by industrial or political action, extremely bad weather interfering with transport, or in some remoter areas the need to store between large bulk deliveries such as those supplied by sea. The costs of such storage should be considered when deciding between alternative possible fuels and also when considering what levels of stocks should be held. Some of the factors may have to be evaluated with other specialists such as the accountancy and production departments.

Factors to be considered are:
1. The cost of a storage area — for what other purposes could it have been used, and what protection by fencing etc. is needed to keep out intruders.
2. The cost of preparation work — any building work, concrete flooring, storage tanks and bund walls against spillage — any fire-fighting equipment and lighting.
3. The extra cost of laying down stocks, such as extra transport distances and supervision compared with direct delivery to boilers or furnaces — extra machinery such as bulldozers to spread and consolidate solid-fuel stocks.
4. The cost of maintaining stocks — solid fuels may lose slightly in calorific value during storage, usually mainly in the first year or so — degradation may also take place due to extra handling and action of weather which may reduce efficiency of usage on some types of stoker. — Cost of prevention measures against spontaneous combustion.
5. The cost of reclaiming stocks — diggers, loaders or lorries for solid fuels — heating up and possible transfer pumping for liquid fuels. Losses of solid fuels due to crushing into unprotected land.
6. Possible advantages of seasonal pricing against costs of storage — establishing real losses due to stoppages of production — nuisance by wind-blown dust from solid-fuel storage — time value of money locked up as stored fuel.

Although individual firms cannot at present consider storage of natural gas, they may wish to protect themselves by storing liquefied petroleum gases (LPG), which can be vaporized by suitable heaters to give a gas of somewhat similar burning properties to natural gas, or at least capable of being burned by the same burner even if with some adjustments. However, such storage requires a large pressure vessel and rigorous fire precautions, permanent water sprays for emergency use, plus the vaporizers already mentioned. The heat of vaporization must be accounted for as a storage cost. Such considerations have caused firms to move to complete dual-fuel burners and use fuel oils as the standby fuel. In the UK and other countries, sales of natural gas are often on an interruptible basis, and the user accepts that the supply may be cut off at fairly short notice for an agreed maximum number of hours or days per year. There has been a tendency for such interruption periods to increase in the 1970—74 period as sales of space-heating gas usage have increased, and also since a bottleneck has occurred in supply owing to the development problems of the early 1970s in the North Sea production areas.

As the price of natural gas has become attractive, owing to the rapid increases of 1973—74 in crude oil prices, this also has accentuated the difficulties of supply and demand. Any user on an interruptible natural gas agreement would be wise to accept that the interruption might be in one continuous midwinter period, and consider his standby fuel storage quantity accordingly. In effect, this storage is 'frozen' money which may or may not be fully justified, dependent on any winter's security and on the gas supplier's production problems. Storage tanks and pipework for light fuel oil are considerably cheaper than for heavy oils, as no outflow heating, sludge-coil provision or line tracing of importance is needed, and the insulation of tanks and feed lines does not need to be considered. These savings may well justify the higher costs of light fuel oil. However there is a much greater certainty that such stocks will be used in part or whole each year, and so their cost can be taken into every year's budget, whereas the factors previously listed for single-fuel plants and related storage allow rather more flexibility and speculation on the likelihood of interruption of supplies relative to the value of production or other activities being carried out in the factory or building.

Cost of Shut-downs

The cost of plant shut-down should be considered for any establishment. It relates not only to the question of storage of fuel or the economics of dual (or even treble) fuel boilers, but also to the amount of money worth investing in duplicate or standby plant to minimize the chance of complete breakdown. This may range from spoilage of valuable materials in a production process to the inconvenience of sending children home from school; from a situation that can be very accurately costed to the imponderable future loss due to interruptions to office work or education. Nevertheless no decision on quantities of fuel to be kept in reserve storage, or on standby plant, should be made without a serious attempt to value the losses due to shut-down.

Factors to be considered are:
1. What is the effect of a sudden, short-term stoppage—breakdown only. Usually fuels for boilers and furnaces can be seen to be running out, as even the smallest storage tank or boilerhouse bunker holds several hours' supply. (This also refers to electricity supplies and may justify standby generators.)

Short-term is considered as only up to a few hours. Some continuous processes are very badly affected by very short stoppages, e.g. continuous chemical processes, and some plastics and artificial-fibre plants, where even half-an-hour shut-down may mean many hours of dismantling, cleaning, rejection of half-processed material, etc. Other processes can make good such short-term stoppages by overtime working, e.g. food processing, textile spinning and weaving; the cost of such a stoppage is then only that of the overtime payments. In offices, schools, etc. a short-term stoppage is of nuisance value only, and often in daylight hours work can continue with some personal inconvenience.

2. What is the effect of a medium-term stoppage, say from five hours to three days — this can be sudden, due to breakdown requiring some time to repair, or can be anticipated owing to fuel stocks running out.

If it is sudden, there can be the same initial loss in certain processes as in (1) above, but other processes that can survive without spoilage under (1) may have spoilage in this case; e.g. certain food industries may buy in fresh fruit, vegetables etc. that are no longer usable after a day or so in storage; food being processed may not be edible if processing cannot be completed within a few hours. Even with these stoppages total production might be recoverable by overtime working.

3. What is the effect of a long-term stoppage — say total failure of a boiler or furnace requiring lengthy repairs or even replacement — or prolonged interruption of fuel supplies — this is usually the major question on reserve storage. Production is probably completely lost; labour costs without production might be reduced by lay-offs — but the employment market might cause permanent loss of labour if laid off. It may be possible to cancel deliveries of raw materials to conserve cash flow.

So far as boilers are concerned, and these are the most common fuel-using appliances in commerce and industry, a guide to evaluation of the value of investment in standby plant, based on experience at thousands of boiler plants, can be summarised as:
1. Breakdowns of one to five hours; up to three a year.
2. Breakdowns of one shift (or one day) up to three days; one every two to three years.
3. Breakdowns lasting over three days; one every six years.

It is assumed that stoppages of a few minutes due to blockages of fuel feed, alarm signals, electrical trip-outs etc. are dealt with by normal staff arrangements without significant effect on production. The chances of breakdown also vary with size — a small boiler with part-time attention, say in a school or office, is more likely to have breakdowns than a larger boiler in a factory where attention is more frequent and back-up maintenance staff are on hand.

Evaluation of Standby or Duplicated Plant

The cheapest boiler (or furnace) plant usually consists of a single unit capable of carrying the maximum load of the process or heating system. However this is only true of initial capital cost and not necessarily true if total costs over several years are considered, e.g. a typical packaged shell-type steam boiler, even on single-shift working, usually burns more than its initial capital cost in fuel every year, so that any increases in capital cost which significantly reduce fuel consumption may be a very good investment indeed.

Factors to be considered are:
1. What is the load pattern expected over a typical year? A heating boiler may be installed with a 'reserve capacity' in case a building has to be heated up from a completely cold state in midwinter; the design load to keep the building warm may only be needed for the two or three coldest days of the year and for much of the heating season the weather is mild so that the load may be below 20% of installed capacity for much of the season. A process load may however be fairly constant over the whole year, and there may be other plants where the boiler has a process load plus a variable space-heating load.
2. Are there one or more sudden peak loads of fairly short duration each day? If the boiler is large enough to meet these peaks, it will be working at relatively low load for the remainder of the day. Can these peaks be reduced by altering the process, or by staggering working hours, or by storing heat — the cheapest storage is hot water below 100°C and the most expensive is steam via an accumulator, but the cost of storage vessels and controls may partly be offset by initial cost savings on the boiler plant and by running at a higher load factor through the year.

3. Is the load likely to increase in future years? It is sometimes tempting to add a margin for such anticipated loads when installing a single boiler. However this ties up extra capital to no useful purpose and also reduces the running load factor on the boiler, until eventually the load does appear — also it may never take place if future business does not justify the expansion, or if a new process is introduced.
Therefore one should:
4. Bring forward the estimates of losses that might occur if a single boiler (or furnace) breaks down for short, medium or long periods.
5. Decide what is the really essential load that must be met in order to keep production at an acceptable level, or to keep occupants satisfied. For example, people will probably manage without hot water for washing in office toilets for a few days; space heating might be cut back for a few days with the co-operation of the staff and indeed, the coincidence of complete breakdown and the coldest day of the year (which would require full heating load) would be remote. Some side-process work might be shut down for a few days by storing partly finished material.
6. Obtain manufacturer's efficiency curves for their plant at different running loads, from say 25 to 100% of full rating — or better, obtain test results made by an organization independent of the manufacturer.

From all this information it would not normally be the conclusion that the plant cheapest in initial cost should be purchased. If a boiler is to spend most of its life working at part load, then a manufacturer should be chosen whose design has its highest efficiency at this part load, or at least a boiler with a very flat efficiency curve down to say 25 or 30% of full rating. A boiler with a slightly higher efficiency at full load, but with a pronounced fall-off as load drops, might be a little cheaper but could in fact be more expensive in overall costs over several years. For example, if

the average efficiency of one boiler at say 50% load is 4% greater than another, although the full-load efficiencies are identical and the cost of the one with lesser fall-off is greater, then for space-heating applications it would be the better choice. If the season's fuel bill were roughly equal to the cost of the boiler, then it would still be a good investment to pay 10% more for this boiler than for the cheaper one.

The installation of two boilers rather than one large one may well increase capital cost by 30 to 40%. However, if this enables essential production to carry on if one boiler suffers breakdown, or if the essential services of cooking and washing utensils could be carried on in say a restaurant or hotel, this extra cost may again be an excellent investment. Two boilers are also useful in space-heating loads, as one boiler could deal with demand for around two-thirds of the heating season, operating on higher load factor and almost certainly higher efficiency. The second boiler would be a standby for this period in case of breakdown. Even in very cold weather when both boilers are needed, appreciable comfort could be provided by one boiler at full load if the other stopped, whereas with a single boiler there would be no heating at all. A twin-boiler installation might well show a 10% economy in fuel compared to a single boiler plant, since at the two ends of the season one large boiler would be below its burner's modulating range and would be running very inefficiently on a stop/start basis.

Evaluation of Overall Costs

Boilers

The major factor in the cost of useful heat is the fuel cost. The initial capital cost is usually less, as an annual charge for capital and interest, than the labour, maintenance and sundries such as chemicals and spares needed to run the plant. This is borne out in *Figure 1*, which shows the typical proportions of total running cost borne by these three constituents. These are related to the Boiler Usage Factor, defined by (boiler running-hours times average load) divided by (hours-in-year times boiler rating).

As fuel cost represents such a high proportion of the total, even further research is needed to improve efficiency. For too long an efficiency of 80% has been considered good practice with modern boilers such as packaged designs. Burners for oil and gas are still being used which are set to operate with appreciable excess air in products of combustion (say 10 or 11% carbon dioxide in flue gas for oil fuel) and often the amount of excess air used increases as the burner modulates towards its minimum burning rate. With the accumulated knowledge of research and development it should be possible to provide burners that will operate at or near stoichiometric conditions, so

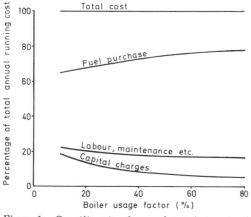

Figure 1   Constituents of annual running costs of a boiler plant (UK, 1974)

reducing (with oil fuels) the sulphur trioxide formation that is feared for corrosion reasons in lower-temperature areas. Elimination of nearly all the excess-air usage would reduce heat losses in the products of combustion leaving the boiler.

A fresh attack is needed on exit-gas temperatures from these boilers. For too long this has been kept fairly high, say 230°C, to prevent low-temperature corrosion problems. This is due to the sad experiences with sulphur-containing fuel oils and the old town gas made from coal. Natural gas however has a negligible sulphur content, and even with fuel oils modern materials such as heat-resisting glass, coated metals, acid-resistant metals, fibreglass materials etc. should enable further heat recovery to preheat combustion-air feed or process water so that gas temperatures could be further reduced.

The other loss is the heat emitted by the surface of the boiler. Although this may vary from 1½ to 2% at full load, most plants do not operate at full load, and at part load this loss, fairly constant in hourly output, may become 4 to 5% of the heat in the input fuel. Work is needed to recover this 'radiation loss' and possibly the best method would be to guide the combustion air for the boiler around the outside of it by means of some baffle or shield rapidly removable if maintenance has to be carried out. This would return much of this heat into the boiler.

With all these methods it should be possible to obtain efficiencies approaching 90%, and although capital cost may well be increased considerably, a fuel saving of say 12% (by improving efficiencies from 80% to nearly 90%) would justify this. However, the purchaser, often non-technical, must be persuaded by the fuel technologist that such an increase in capital cost may turn out to be an excellent investment. So long as boilers and similar fuel-using equipment are purchased on the basis of minimum capital cost, energy conservation will be difficult.

Furnaces and Other Energy-handling Plant

The same general principles of assessment as for boilers may be applied to furnaces, although the factors bearing on efficiency, cost of stoppages etc. will often be more complex and clearly will differ from those for a steam boiler. Some of these are discussed in Chapter 8; there is also the further option of electric furnaces, based on power generated from primary fuels at perhaps 28% efficiency but sometimes used at far higher efficiency than in a direct-fired furnace and sometimes having other advantages that cannot be expressed solely in terms of fuel efficiency. Production of mechanical energy in internal combustion engines and gas has yet a third set of efficiency factors and cost criteria, discussed in Chapter 12; combinations of these with boilers and steam turbines form the subject of Chapter 15. Miscellaneous but vitally important methods of saving or recovering heat are dealt with in Chapter 16. Finally, the importance to efficiency and general economy of suitable refractories, insulation and instrumentation is brought out in Chapters 17 and 18.

Production of heat

Chapter 3

Gas firing

3.1 Natural gas

Burner principles
Burner applications
Boilers
Melting furnaces
General heating furnaces: direct-fired
General heating furnaces: indirect-fired
Drying, curing and calcining: direct-fired
Drying, curing and calcining: indirect-fired
Tank and vat heating
Space heating
Working flame burners

3.2 Liquefied petroleum gases

3.1 Natural gas

The great majority of gaseous fuels in use are easy to burn and control using relatively simple techniques.

Fuel gases are grouped in families[1] according to their Wobbe numbers as an aid to the classification and testing of gas-burning equipment. There are three families, covering the ranges of:

Family	Wobbe Number (MJ/standard m³)
1	24.4 to 28.8
2	48.2 to 53.2
3	72.6 to 87.8

Gases manufactured from coal or oil, termed town gas (App.B1, B9), usually fall into the 1st family, natural gas (App.A1) and substitute natural gases (App.B1, B2) fall into the 2nd family, and liquefied petroleum gases, referred to as LPG (end of this Chapter and App.A9, B5), fall into the 3rd family.

The wide range of uses of gas for domestic, commercial and industrial purposes has led to the development of innumerable varieties of gas burners. The smallest burners, as used to provide pilot flames, have ratings as low as 30 W whilst the larger burners, as for example used in some water-tube boilers, have ratings up to 60 MW. This is therefore a capacity range of two million to one.

In the UK almost all of the piped gas supplies will be distributing natural gas by the mid-1970s. For this reason most of the information in this Section relates to equipment suitable for natural gas, but it may be equally applicable to liquid petroleum gases. Some burners designed for town gas are also satisfactory for use with natural gas, but these are not included where they are suitable solely for use with town gas. In isolated circumstances gases other than those in the families described require to be burned; for example process gases such as blast-furnace gas, and producer gas.

As an introduction to the subject of burner selection, a general description of the different ways of burning gas is presented to explain burner principles and the terminology used in differentiating between systems. The detailed information on burners selected for different applications forms the main part of this Section and it will be seen that this is categorized into the different processes and purposes for which gas is used. This method of presentation has been chosen as it is the logical starting point in selecting burners or understanding why particular types of burners are employed.

Burner Principles

Natural-draught Neat-gas Burners

The simplest way to burn gas is by releasing it from an orifice so that combustion occurs at the boundary with the surrounding open air, by diffusion. When the velocity of the issuing gas is very low any of the gas families will burn in this manner, but the resulting flame is lambent, is easily disturbed by draughts, and the combustion intensity is very low. Such flames therefore have few practical applications. When the issuing velocity is increased, the rate of mixing with air and consequently the combustion intensity are increased. With natural gas and LPG, however, the velocity which is necessary to provide useful flames is greater than the burning velocity of the gases and in consequence the flame lifts from the gas port and either becomes unstable or is extinguished. With town gases, which usually contain considerable hydrogen, the flames can remain stable over a pressure range which may be satisfactory for some burner requirements. As previously stated however the availability of town gases is limited.

In order to use the simple neat-gas burner with natural gas many ways of obtaining flame stability have been devised. *Figure 1* shows a selection of these. They have taken the form of shields or deflectors or impinging jets to promote a low-velocity mixing zone where combustion can be stabilized at the burner port. The need for burners of this type is predominantly for domestic and commercial appliances but they are also used for some industrial purposes where the individual burner ratings are very limited. It is unusual to use this type of burner with an individual rating greater than 50 kW, although multiples of them are used in manifold arrangement.

Low-pressure Aerated or Premix Burners

This category of burner uses the pressure energy of the gas issuing from the gas port to entrain part of the air required for combustion into a mixing tube or venturi, and the resulting air–gas mixture is burned at one or more burner ports supplied from the injection system. The most familiar example is the bunsen burner, and the most used in terms of numbers are the burners in domestic cookers and other domestic appliances. For larger applications these burners are provided with multiple burner ports in drilled pipes, in

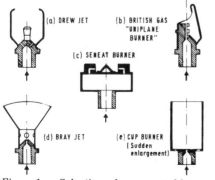

Figure 1   Selection of non-aerated burners

pressed steel box and pipe sections, and in castings.

In the majority of these burners the flames, or the combustion products following the flame zone, are used to transfer heat to the surfaces to be heated or to provide a source of hot gases. Some types use a surface of ceramic or metal gauze, or a fibre pad incorporating a catalyst, in which the air—gas mixture burns on or within the surface layers or ports so that the surface is heated and the resulting radiant heat provides a considerable proportion of the total heat output. Surface temperatures for these units are from 500°C to 800°C when supplied with air—gas mixture from low-pressure gas supplies of less than 7 kPa gauge (70 mbar). The total or most of the total combustion air is injected by the gas for this application method. Ratings of individual burners are from 500 W to 150 kW or thereabouts.

It is important with this category of burner that the design is carefully matched to the characteristics of the gases concerned. For example, for natural gas and LPG the burner ports require a means of flame retention, either piloting flames or optimized spacing between burner ports, to achieve good flame stability, or they can alternatively be used at a low rating per unit port area. These measures are needed to take account of the lower burning velocities of these gases compared with town gas.

High-pressure Aerated or Premix Burners

The majority of low-pressure aerated burners operate with up to half of the total air for combustion entrained as primary air, the remainder of the air being supplied as secondary air around the flames.

Where burners of the type described in the preceding Section are supplied with gas at pressures above approximately 20 kPa gauge (200 mbar) for natural or town gas, or above 70—120 kPa gauge (700—1200 mbar) for LPG, the increased energy available can be used to induce all of the combustion air and provide a mixture pressure suitable for combustion to take place in closed combustion chambers operating at pressures slightly above atmospheric. For example they can be used for some furnace applications and for burners which are enclosed, and where there is some restriction on the combustion-product exit. Because higher mixture pressures are available higher velocities can be obtained, and consequently the size of the burner equipment is smaller for a given output compared with that for low-pressure equipment.

This burner category is not in widespread use as pressure gas supplies are not generally available, although gas compressors can be installed. It can however be useful for small installations where there are a number of burners and there is an advantage in having simple burner controls.

Nozzle-mixing Pressure Air Burners[2]

This burner category, known also as nozzle-mixing burners, tunnel-mixing burners, or package burners,

uses an air supply at pressure which is discharged through a nozzle or nozzles or through an annulus or annuli at the burner head or face. Gas ports, which again may be of many forms, admit gas into the air stream and combustion takes place either totally or partially in a burner tunnel or at the burner face.

In some designs part of the combustion air is admitted into the gas stream prior to the main mixing zone so that the resulting flame is of this air—gas mixture, the remaining air completing combustion. Many designs of nozzle-mixing burners are suitable for all types of gases but some are suitable only for individual gas families. These burners are the most commonly used types for industrial processes and boiler firing up to and including power-station boilers.

Premix Pressure Air Burners

Two categories of premix burners have already been described, i.e. low- and high-pressure types in which the gas pressure is used to inject combustion air. This final type of burner uses air under pressure which is mixed with gas in an injector; the resulting air—gas mixture is burned either at burner nozzles or in burner tunnels. With attention to choice of the particular type, these burners can be used with all gases. These burner systems are commonly known as air-blast systems and it is usual to govern the gas supply to the injectors to zero gauge pressure, so that the air then entrains the correct volume of gas automatically. As such systems are self-proportioning, burner control is simply effected by a single valve in the air supply. However, if the pressure in the combustion chamber varies, a pressure back-loading control may be required.

Premix burners provide versatile systems which are used for many industrial processes, but the system is not normally used for boiler firing.

Burner Applications

Types of burners used for heating and process plant are described in this Section. Brief notes are included on the control equipment which is normally used together with air and gas pressure requirements. The following general notes on control safety and conservation are relevant to burner selection and should be carefully related to the individual items of plant.

Safety and Automatic Control

In the selection of burners the most important requirement is clearly that the heating method suits the particular production need or type of boiler plant. There are usually several ways in which a plant can be fired, and one most important aspect is the number of burners that are required. As an example of this point, there are some furnaces in operation which use only one burner whereas other furnaces having a similar duty are fitted with more than twenty burners. For the best application of safety and automatic control equipment it must be more economic to use the smallest number of burners. Progress in the design of plant and specialized burner equipment is now enabling this principle to be followed to a considerable extent, and it is likely that much of the plant in the future will utilize a smaller number of burners than was customary in the past.

Considerable advantage can be obtained by the inclusion of a higher standard of automatic control, which results in lower labour requirements and also improves the overall safety of operation.

Conservation and Thermal Efficiency

The need to conserve fuel resources and the trend towards higher fuel costs demands that all heating operations and systems are carried out at the highest practicable thermal efficiency and that attention is given to all other aspects of fuel economy. The following considerations are recommended for attention:

1. That all plant is used at the optimum loading capacity with respect to the minimum fuel consumption per unit of production. This loading may not be the maximum possible capacity.

2. That for intermittently operating plant the heating periods are chosen to minimize losses in heating and cooling: i.e. to run a number of cycles consecutively rather than separate cycles which allow the plant to lose all stored heat.

3. That the burners and associated controls are selected and set to provide economical operation: i.e. the heat-input rating should not be greater than that necessary to achieve an acceptable heating-up rate. Excessively fast heating-up rates can waste fuel and damage plant. The control equipment, in association with plant design, should enable the correct combustion conditions to be obtained over the full range of fuel input. Combustion and other air should not exceed that necessary for the plant or process. Lack of attention to minimizing excess air, either in initial design or in subsequent adjustment, is probably the main factor that leads to unnecessary fuel usage.

4. That the facilities for starting and stopping the plant be designed to achieve this in a straight-forward and safe manner, so that operators will be encouraged to stop plant when it is not in full use rather than keep it running because of the time and effort needed to stop and restart it.

5. That each application be examined with a view to recovery of heat from waste gases, either by the use of counterflow heating methods and/or the provision of heat-recovery methods such as recuperators (Ch.16.2) to preheat combustion air or to provide heat for other uses.

Boilers (Ch.10)

Watertube Boilers (Ch.10.3)

This Section relates to burners for watertube boilers of a few thousand to several million kilograms per hour evaporation, or the equivalent output of hot water.

Most watertube boilers are fitted with register-type burners. These comprise air-supply casings, normally known as windboxes, which are integrated into the firing walls of the boiler, and the fuel nozzles. Normally such windboxes are supplied with air from a single fan. Alternatively packaged burners, comprising combined fan and burner units, are fired through ports in the firing walls. Both of these types of burners are of the nozzle-mixing category.

Register burners may be dual-fuel, for gas or oil firing, and some designs can accept multiple fuels, for example, pulverized coal, oil or various gases. Package burners for this application are usually

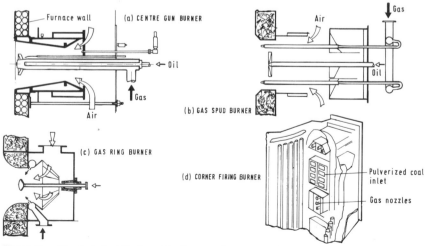

Figure 2   Typical dual-fuel register burners for watertube boilers

dual-fuel for gas or oil firing, so that either fuel can be used to suit supply arrangements.

In register burners there are four basic alternative methods of introducing the gas; these are shown in *Figure 2a—d*. In *Figure 2a* the gas is introduced through a concentric tube on the axis of the oil gun. Ports at the end of the tube direct the gas either radially or at an intermediate angle to the axis of the burner downstream of an axial swirl stabilizer in the air stream. In the larger outputs the size of the gas supply tube required can interfere with the flow pattern at the oil atomizer and this therefore limits the size range of this type.

An alternative method is shown in *Figure 2b*. The gas is injected through a number of separate tubes known as pokers or spuds, set in a circle in an air-supply annulus containing a stabilizer. Flame stabilization at each gas outlet is achieved by directing some of the gas into the reverse-flow pattern generated by the air stabilizer and aided by other stabilizer discs fitted at the end of each of the gas spuds.

A third method is shown in *Figure 2c* in which an annular gas duct is situated at the throat of the quarl of an oil burner of conventional design. Ports in the annulus direct the gas into the air stream and flame stability is achieved by the influence of the swirling air from the air register swirl vanes and from the reverse-flow zone created by the disc stabilizer.

The fourth method shown in *Figure 2d* is used for corner-fired boilers in which the direction of the flame is at a tangent to a circle around the centre of the boiler. The rotation set up by the flow from burners at each of the four corners provides a zone of swirling turbulent combustion.

There are also variations of these four types of burner which are generally applied to packaged watertube boilers.

The ratings of register burners for boilers in general industrial duty are 3—20 MW and for power station boilers up to 60 MW. The ratings of package burners are normally between 1 and 15 MW.

The gas pressures used in the lower-rating burners are normally between 30 kPa gauge (300 mbar) and up to 200 kPa gauge (2 bar) for the high-rating burners. Combustion-air pressures range from 0.5 to 2 kPa gauge (5 to 20 mbar) above the combustion-chamber pressure.

A typical controls layout for the burners described is shown in *Figure 3*. There are variations in detail to take account of specific load patterns, but the basic elements are as shown.

Installations may be either fully automatic or include some manual actions, for example the setting of dampers and the operation of valves. All systems include interlocks on all critical operations and supervision as appropriate. All systems include timed starting operations including prepurging, timed ignition periods and timed flame establishment. After a shut-down for any reason there is a timed post purge. All systems include comprehensive flame-safeguard equipment which operate two shut-off valves in series on the main gas and on the pilot gas supplies. · Both air and gas supplies are monitored for pressure and/or flow[3].

Shell Boilers

This Section relates to burners for shell boilers of a few hundred to about 25 000 kg/h evaporation or the equivalent output of hot water. It also applies to Cornish and Lancashire boilers.

Most of these boilers have one or two firetubes which are individually fired with integral fan-type package burners. These burners are usually dual-fuel with either light or heavy oil as the alternative fuel. Although this conventional arrangement is in operation on most boilers, some alternative systems are in operation. One such system applied to a twin-firetube boiler uses a single burner fitted to one firetube, the combustion gases then passing through the second firetube before entering the

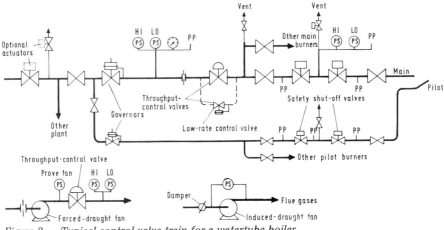

Figure 3   Typical control-valve train for a watertube boiler

tube passes. A number of boilers have also been constructed with three firetubes.

Typical burner designs are shown in *Figure 4*. In *Figure 4a* the gas inlet is on the burner axis and the gas is discharged through ports into the combustion air downstream of a stabilizing baffle. There are several variations of this arrangement, some of which impart rotation to the combustion air. Ignition of these burners is by a pilot burner or of the main burner at a low flow rate. For dual-fuel application an oil gun is inserted on the axis inside the gas-supply tube and the gas ports are then disposed round the periphery of the stabilizer. This is shown in *Figure 4b*.

The ratings of these burners for shell boilers range from 1 MW to 15 MW. The gas pressures are normally between 1.5 and 5 kPa gauge (15 and 50 mbar). Normal piped supply pressure is used for the smaller ratings and gas boosters are used for burners requiring the higher pressures. The combustion-air pressures are normally in the range 0.2—1 kPa gauge (2—10 mbar) but also up to 3 kPa gauge. The air pressure has to allow for the pressure drop across the boilers.

Many boilers are fitted with dual-fuel burners of the type shown in *Figure 4c*. In this design gas is injected radially into the air stream, passing through the annulus around the rotary-cup oil atomizer. This burner is also used on some small watertube boilers. The ratings range between 2 MW and 50 MW. A normal gas pressure is 4 kPa gauge (40 mbar) and the air pressure 0.8—1 kPa in excess of the boiler pressure drop.

A typical controls-layout for these shell-boiler burners is shown in *Figure 5*.

Shell and coil-type boilers have been designed for the heating of thermal fluids for many process applications, as an alternative to hot water or steam. Thermal fluids are normally low-viscosity mineral oils that are chemically stable at the higher operating temperatures.

Low-water-content Coil Boilers

This Section relates to burners for boilers in which the stored water capacity is extremely low. These are usually termed steam generators, flash or coil-type boilers; they are usually of the coil type where

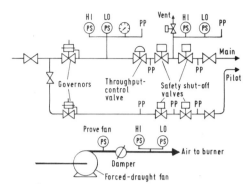

Figure 5   Typical control-valve train for a shell boiler

the combustion chamber is the cylinder enclosed by the coils and the combustion gases flow in a multipass manner along successive layers of coils. The output range covered by the principal types is from 200 to 3000 kg/h steam, or the equivalent output of hot water.

The burners resemble register burners in that the air-supply passages are integrated into the boiler casing and the gas-burner components are also integrated or inserted from outside. The geometry of the air and gas ports produces a short wide flame to fill the combustion chamber. Some boilers can be fired with either gas or oil by changing the burner components but they are not dual-fuel in the customary sense.

The burner controls include conventional flame-safeguard equipment, and when this method is used for steam production careful attention is required to proportioning the fuel input to suit the water input and steam pressure. Because the water is evaporated at the same rate as the steam is produced the fuel rate must be accurately controlled to meet this. The burner turn-down must, therefore, provide the low input needed for low steam flows and also prevent widely fluctuating steam pressures, which would otherwise result from burner cycling. To meet rapid-starting requirements the pre-purge air flows are minimized. The controls are otherwise similar to those used on package-boiler burners of equivalent ratings.

Multiple Firetube Boilers

For small steam needs in the range 45 to 400 kg/h, multiple firetube boilers are frequently used. In these boilers firetubes up to 50 mm diameter are fired by individual gas burners.

The boilers contain from a few to over a hundred firetubes and although the majority have vertical firetubes mounted in a vertical shell there are also horizontal-boiler versions. The burners are supplied with gas at normal low pressure. Combustion air is introduced partially by the entrainment effect of these burners and also by the natural-draught effect of the vertical firetubes, or

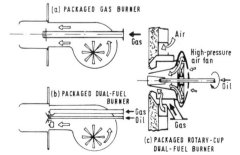

Figure 4   Typical packaged burners for shell boilers

by flue-induced or fan-assisted draught in the case of horizontally tubed boilers.

The gas-burner controls are to some extent different from those used for other boilers owing to the numerous separate burner flames. The majority of these boilers have not been provided with comprehensive flame safeguards, partly because of the numerous small flames and partly because the compact and high-strength combustion chambers formed by the firetubes make this design more resistant to pressure rises. They have, therefore, been manually ignited with the heat input controlled by steam pressure arranged to give either the maximum rating or a low flame rate set to a level below standing heat losses. Control systems are currently being developed to provide semi-automatic or fully-automatic operation as with other larger boilers.

Sectional Cast-iron Boilers

Sectional boilers designed principally for solid-fuel firing are also fired by oil fuels and gas. The majority are for hot-water service, but some have been used for the generation of low-pressure steam. The sizes range up to several thousand kW. Such boilers have been gas-fired by forced-draught package burners and also by natural-draught burners.

Designs of forced-draught package burners were shown previously in *Figure 4*. In the burner sizes applicable to these boilers the gas inlet is usually on the axis of the burner head and there are several techniques used for flame stabilization and flame shape generation. The simplest method is by release of the gas downstream of a stabilizing device in the air stream. More complex designs include means of providing progressive combustion by the admittance of graded quantities of air into the gas stream. Although the results of applying these different types to boilers do not give very significant differences in thermal performance, some give quieter boiler/burner combinations than others. The air and gas pressures for these burners are usually about 0.5–1 kPa gauge (5 to 10 mbar). As the boilers have been designed for natural-draught operation with solid fuel, there is little flow resistance to the combustion products and there is negligible combustion-chamber pressure. Any effect of chimney draught on the boiler is prevented either by the use of a draught stabilizer or by allowing the free entry of dilution air into the flue immediately following the flue exit from the boiler.

These burners are normally used in a simple on–off manner, as the heat storage in the boiler is usually sufficient to avoid excessive temperature hunting. The burner controls are usually fully automatic and provide either interrupted pilot or low main-flame starting. Electronic flame safeguards are incorporated together with protection against loss of, or undue reduction of, combustion air flow.

Typical controls are as shown previously in *Figure 5*. For maximizing thermal efficiency, it is recommended that the excess-air levels are maintained at not more than 25%, and the boiler doors require to be plated over or provided with relief panels to obtain an effective seal and to afford protection against any excessive pressure rises in the combustion chamber or flueways.

Natural-draught burner systems are also used for these boilers. In such installations the chimney draught is reduced to a low constant level by the use of a draught stabilizer and this draught induces the combustion products through the boiler. The burner unit is fitted to the boiler with purpose-designed casings so that combustion air passes to the burner whilst avoiding the ingress of unnecessary excess air. These firing systems require careful matching to the installation and regular servicing to avoid lowering the thermal efficiency owing to changes such as those caused by random chimney conditions or the disturbance of boiler deposits formed prior to gas firing.

The controls for natural-draught burner systems usually include a continuous pilot for ignition of the main flame in response to signals from temperature and time-switch controls.

Purpose-designed Gas Boilers

A range of cast-iron sectional boilers has been specifically designed for gas firing; for hot-water heating systems they are produced in unit sizes up to 1 MW but when used in multiples of units, termed modular boilers, the size range is limited only by the total number of boilers that can be accommodated.

In these boilers either aerated burners or neat-gas burners are provided in manifolds across the boiler base, combustion air entering from below and around the boiler sides. A compact combustion chamber leads the combustion gases to narrow-channel heat-exchange sections provided with baffles or extended surfaces. The boiler height provides adequate draught and a combined draught diverter and dilution air-entry assembly is integrated into the boiler construction. Because of the specific design advantages, the resulting thermal efficiency is normally some 75–80% without recourse to forced draught. The burners are almost silent in operation.

The main burners are normally provided with on–off control and low-rated continuous pilots. Cascade control systems are now being used on modular boiler installations. Flame safeguards are thermoelectric in the smaller sizes and electronic in the larger sizes, and continuous pilots are normally used for ignition.

Melting Furnaces (Ch.8.1)

The following Sections relate to burner equipment for smelting furnaces used in the ferrous and non-ferrous industries and also in non-metal melting processes. The order of presentation is generally from the largest to the smallest type of plant, including similar or allied types where appropriate. This approach has been used so that the similarities or comparisons of methods of firing similar types of plant can be recognized.

Blast Furnaces

The largest smelting furnaces in use are blast furnaces for iron production. Although the primary fuel is coke-oven coke, other supplementary or partial replacement fuels are used. Gas is sometimes used and is injected into the tuyeres. There is therefore no burner as such and as injected the gas mixes with the air stream and burns in the general conflagration in the furnace. A typical gas injection rate is 850 MJ/h ($\approx$ 235 kW) in a 50 t/h furnace. An allied use for gas is in heating the 'Cowper Stoves' which provide, on a regenerator principle, preheated combustion air to the blast furnaces. These tall cylindrical furnaces use forced-draught nozzle-mixing burners to give long flames extending up to five to ten metres through the furnace.

Cupolas and Shaft Furnaces

These furnaces, as used for melting iron and for calcining lime and flints, are normally fired with coke which is either mixed with the charge or included in layers in the shaft. Partial or full replacement by gas is being used and also, in purpose-designed gas furnaces, for melting copper. This latter process is normally carried out in reverberatory furnaces.

Gas-burner applications are basically similar in that burners are fired through the sidewalls in the lower section of the shaft. There are differences in techniques as, for example, in iron-melting cupolas providing partial replacement of coke the burners are positioned at a carefully chosen height above the air tuyeres whereas in copper melting they include burners at hearth level.

Burners for these duties are normally nozzle-mixing but with combustion-tunnel variations to accommodate flame-impingement aspects. Because the pressures in the shafts can be considerable and also change very quickly according to charge variations it is necessary to pay close attention to gas and air control equipment. Similarly the air/gas ratio control can be critical in order that the quality of the charge shall be maintained or modified.

Reverberatory Melting Furnaces

Gas firing has been applied to many types of reverberatory furnaces in several industries. Some of these are briefly described.

Open-hearth Steel Furnaces. Complete gas firing and mixed firing with oil are both used to a considerable extent in many countries. A system originating in the USSR is in current operation in the UK: in this the burner unit includes a cracking section where the main gas flow is partially cracked to give some free carbon which in the main furnace produces a highly radiant flame. The cracking burner comprises a small furnace in which some of the gas is burned to provide a local temperature of around 800°C. The main gas flow is injected in a jet stream through the centre of this furnace, and the resultant gas contains the mixture of raw gas, cracked gas and products of combustion.

When gas is used for mixed firing with oil the latter provides the radiant component in the flame. Flame momentum makes an important contribution to performance, and with oil the high-pressure steam used for atomization also provides this need. Natural gas at up to 700 kPa gauge (7 bar) has been used in place of steam for atomizing, and although the total momentum is lessened, the increased flame temperature due to absence of steam is a compensating component giving an overall equivalent performance. Various gases are used with or without steam, with satisfactory results. Burners are water-jacketed shells up to more than 3 m in length and 0.3 m dia. containing pipes carrying the fuels and steam, projecting through the furnace endwall and angled downwards to fire over the bath of metal.

Burners at each end cycle on and off in unison with the regenerator reversals. Much development has been aimed at burner design to maximize furnace performance, and nozzles include sonic-exit designs, annuli and other variations.

Control of these burners is usually relatively straightforward and equipment is necessarily robust to withstand furnace-shop conditions. Furnace temperatures greater than 1700°C and high-intensity and high-velocity combustion demand careful attention to obtain good performance.

Glass Tanks. The firing of glass tanks is analogous to that of steel-melting furnaces although less demanding in that the furnace temperatures are lower (around 1500°C) and the furnaces are lower-rated. Nevertheless continuous campaigns of up to 7 years without break, and the characteristics of the refractories needed to withstand the molten glass and feedstock, require careful burner application. It is customary to inject gas at up to 50 kPa gauge (0.5 bar) or somewhat higher into the air streams from the regenerators. Different positions are used according to tank size and shape. Side-port firing is typical, in which gas flows converge from either side of an air inlet.

Burner gas velocities, jet divergence, vertical and horizontal firing angles are all interrelated, and in general the aim is to produce a turbulent highly-radiant flame of moderate velocity, avoiding direct impingement on either glass or roof and with the completion of the flame zone before the regenerator inlet ports. Burner designs vary from simple single orifices to complex multiple jets, some including air injection.

Miscellaneous Reverberatory Furnaces. Gas-fired furnaces from fifty t capacity for some metals to 0.1 t capacity for others are used for melting copper and copper-based alloys, aluminium and iron, for casting, refining and recovery purposes throughout industry. As a generalization, the burner equipment is usually of the nozzle-mix type using combustion-air pressures less than 10 kPa gauge (0.1 bar) and gas pressures less than 1.5 kPa gauge (15 mbar). Medium-velocity turbulent combustion provides convective heat transfer which contributes the majority of heat transfer to the furnace charge and furnace walls.

Fuel conservation needs may well see a resurgence of interest in heat recovery from waste gases, and there is much scope for saving either by the adoption of burners which include recuperators (Ch.16.2) to provide air preheat or by separate stack recuperators.

Crucible Furnaces

For metals which are unsuitable for open-flame melting, for example magnesium alloys and some brass and aluminium alloys, and also where small quantities of other metals are needed for casting, the crucible furnaces used are normally fired with gas by burners directed between the crucible and the furnace lining so that the combustion products circulate round the crucible, commencing at the base and ascending to top flues or vents. Provided that there is sufficient momentum to obtain rapid circulation, the type of burner does not greatly affect performance but careful selection and positioning is needed to prevent local overheating. Burners of the premixed gas and air type have been used to a large extent, since the flame characteristics are well suited to this furnace design and the simplicity of the burner results in low maintenance requirements in normal running, particularly if a crucible fractures and releases molten metal. Both

nozzle-mixing burners and integral fan package burners are also used, the latter being usually applied to melting of aluminium and zinc diecasting alloys.

Other firing methods have also been used, for example in metal crucible or pot furnaces. A widespread application is in molten-salt pots used for heat treatment and descaling. There is particular need to avoid local hotspots when using metal crucibles, and surface-combustion radiant panels, radiant-cup and flat-flame burners, and multiple tangential-firing burners have been used in a variety of designs. The efficiency of these systems is greatly dependent on the attention given to design, the cost benefit of the more complex systems, and their ability to withstand operating environments.

Bath Furnaces

Metal baths other than crucible shaped, used for galvanizing, lead and salt baths and operating between temperatures of 300—750°C are fired in a variety of methods. Uniformity of heat transfer is the primary need, to optimize plant size and the economic life of the vessel The customary approach has for many years been to distribute the heat input by using a large number of small burners or by installing flat-flame burners or radiant burners. *Figure 6* illustrates these variations.

There has recently been a radical change in approach by using high-velocity burners to provide high-velocity combustion gases which, when combined with a matched furnace design, promote a rapid recirculation of gases round the vessel or furnace being heated. The system is now used in many furnace types and in connection with metal bath furnaces has given successful results in galvanizing baths. In addition to providing uniformity of heating, the reduction in the number of burners facilitates the economic inclusion of automatic burner-operating controls. Such plant can therefore be made to start and stop by pushbutton activation.

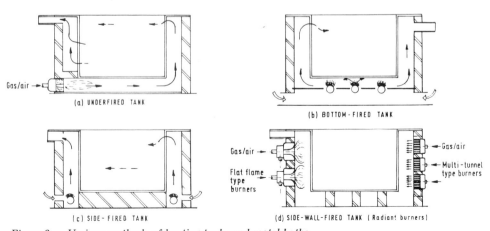

Figure 6   Various methods of heating tanks and metal baths

General Heating Furnaces: Direct-fired (Ch.8.1)

The firing equipment for this class of furnaces is described by referring to the names customarily used in the industries concerned. The ambiguity in the terminology between industries prevents identification by simpler means such as reverberatory furnace or oven furnace as these and other terms are very mixed in practice. The burner systems described are for direct-fired furnaces, that is where the combustion gases are present in the working chamber. Indirect-fired systems, where the combustion products are kept from the work load, are included in the next Section. In general the presentation is in the order of decreasing plant size.

Reheating Furnaces: Steel Production

Burners for these large reverberatory furnaces reheating up to 250 t/h of steel slabs are usually of the nozzle-mixing type and are fired through arches, sidewalls and endwalls so as to give the required zone heating. Individual burner ratings are from 1 to 3.5 MW and use air at up to 8 kPa gauge (80 mbar) and gas at low pressure.

It is necessary to avoid flame impingement, and modern furnaces have been fitted with flat- or wall-flame burners so that a high burner input can be introduced through roof sections whilst at the same time minimizing roof height above the work load. They are controlled in zones and because of the importance of obtaining the correct heating schedule the control equipment is very comprehensive.

Soaking Pits: Steel Production

Soaking pits for steel ingots weighing up to twenty t each, up to ten in a pit, are fired by large nozzle-mix burners. Air is preheated by recuperators or regenerators for the larger pits. A recent innovation is the use of high-velocity burners which promote higher rates of circulation of gases, giving more uniform and faster heating. An example is a pit formerly heated by two burners and now fitted with only one burner. The performance, in general terms, has been improved 20–30%.

Continuous Billet Furnaces: Steel Production

The firing of these furnaces, which cover a wide range of sizes, is similar to that for the larger reheating furnaces but with the size correspondingly reduced. Uniformity of heating is very important and it is customary to use a large number of relatively small input burners, for example fifteen burners in the endwall of a furnace for heating billets c. 6 m (twenty feet) in length. Modern practice is to use higher velocities in conjunction with lower roofs.

Continuous Chamber Kilns: Brick Firing

A kiln may typically consist of 16 chambers each holding between 15 000 and 25 000 bricks, two or three of the chambers being fired at any one time.

The burners used fall into three broad classifications. The simplest employs relatively high gas pressure at about 100 kPa gauge (1 bar); this induces some of the air for combustion into the burner tube which protrudes a short distance into the firing chamber. This protrusion is adjustable. The remainder of the combustion air enters the firing chamber from the chambers where the bricks are being cooled, and this air is thus preheated. Air-blast burners can be used in a similar manner and employ gas at a lower pressure than the injector types described above, and also use low-pressure air. In these techniques the flames simulate the original coal-firing flames. The local heat release is however reduced and as a consequence heating is more even, resulting in lower rejects. In a third method used with gas, trolley-mounted air-blast burners firing at low level are employed. With all the methods employed the burners are moved from chamber to chamber as firing progresses around the kiln, thus minimizing the total number of burners required.

Tunnel Kilns: Brick and Refractories

Large kilns for brick and refractory firing, up to 3 m in internal height and width and up to 100 m long, require firing systems which give even heating in various temperature zones with particular attention to uniformity across the load. This is normally obtained by using a large number of small input burners firing between the cars and/or through spaces arranged within the charge at low level. The burners are usually partial premix, the remainder of the air to complete combustion being obtained from the flow along the kiln. Full premix is also used in some instances. These kilns can also be roof-fired in a manner similar to continuous kilns, and also by burners firing along the kiln between the walls and the ware.

Smaller types of tunnel kilns for the production of glazed tiles, refractory bricks and sanitary ware are similarly fired.

Beehive Kilns: Heavy Clay

These kilns, which are up to 7 m in diameter, as used for firing clayware such as roof tiles, drain pipes etc., were originally designed for solid-fuel firing and are gas fired using several different burner arrangements.

The method which is nearest to the solid-fuel system uses nozzle-mix burners firing vertically up the sidewalls, the combustion products then being drawn through the charge to the central bottom flue outlet. Other methods have been used with the aim of providing a recirculation effect within the ware. One method uses a central roof-firing burner with separate air jets to promote circulation and for additional air for use in the drying part of the cycle. Another method that has been tried comprises recirculation ducts, with a single high-velocity burner used to obtain complete recirculation throughout the ware.

Intermittent Kilns: Ceramics

For batch firing of ceramic ware, open-fired furnaces with pull-out hearths are normally fitted with numerous small burners firing along sidewalls in the space between the wall and the ware. Top-hat furnaces with burners firing vertically from the fixed base section are fired in a similar manner. The burners are either nozzle-mix or premix.

A recent development which is particularly applicable to these kilns is the use of high-velocity self-recuperative burners. A smaller number of burners will provide the same or improved furnace performance, and fuel savings of up to 40% can be obtained.

Stress-relieving Furnaces: Steel

For the heat treatment of fabrications, stress-relieving furnaces require close attention to temperature uniformity in the working range 600–700°C. Within this temperature range radiant heat transfer is limited and there is therefore a need to maximize convective heat transfer. Some furnaces employ forced circulation, using heat-resisting recirculating fans in the same manner as forced-convection ovens.

For large fabrications such as tanks and pressure vessels, the recirculation effect provided from large high-velocity burners is used to promote uniform heating and the furnace enclosure may be purpose-assembled around the work. In the case of vessels with access to the interior, the burner or burners may fire into the interior and the outside is covered with insulating material.

Annealing Furnaces: Iron and Steel

For the various annealing treatments required for steel castings and fabrications, spheroidal iron,

alloy steels, etc., the batch-type furnaces normally used are fired either under the crown or up the sidewalls with burners of nozzle-mix or premix types. Large furnaces have burners distributed in banks at various heights to aid temperature uniformity.

A modern technique is to use high-velocity burners which cause a high degree of recirculation within the furnace, thus resulting in an even temperature distribution.

General Heating Furnaces: Indirect-fired (Ch.8.1)

Some furnace treatments require the absence of combustion products in the work chamber. Most require a special atmosphere, e.g. as in steel carburizing treatments. Examples of combustion systems are described.

Radiant-tube Furnaces

The most frequent method of achieving indirect heating is by using radiant tubes, in which the gas is burned inside heat-resisting tubes and the combustion products are released outside the furnace. Burners for these radiant tubes are shown in *Figure 7*. It will be seen in *Figure 7a* that a simple arrangement consists of a low-pressure burner firing directly into a tube, the heated portion of which is in the furnace space. *Figure 7b* shows a burner firing inside the inner tube of a single-ended tube where the combustion products return along the annulus between the tubes. *Figure 7c* shows a burner firing inside a recirculating tube where the combustion gases are recirculated by the entrainment effect of the burner gases. The burner units in the last two systems include recuperators (Ch.16.2) which preheat the combustion air[4]. This arrangement enables thermal efficiencies of 70% to be achieved at a radiant-tube temperature of 1000°C. Burners of the recuperative type are also being applied to ceramic radiant tubes for operation up to 1500°C.

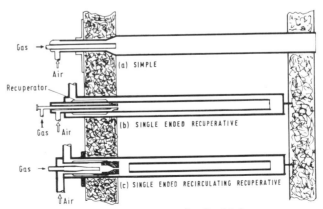

Figure 7   Three common types of radiant tubes

Muffle Furnaces

The largest type of batch muffle furnaces is used in the vitreous enamelling industry, where muffles built up from panels are up to 5 m long. Small muffles are either in one piece or in segments and are used for various treatments in air or special atmospheres. Heating is usually effected by means of a number of burners equally spaced along one or both sides of the furnace. The burners are normally of either the low-pressure aerated type or the air-blast type.

Continuous muffle furnaces are also used where the work can be passed through tubes or box sections that are fired externally by siting them in a furnace enclosure. These furnaces are fired in a simple manner by low-pressure burners spaced at intervals along the furnace, or by air-blast burners suitably dispersed to provide even heating; simple firing methods are normally used for process temperatures up to 900°C and the more complex systems for temperatures up to 1700°C, using ceramic tubes or sections for the higher temperatures.

Although the work space in most of the furnaces is usually of small cross-section there are large plants with cross-sections greater than 1 square metre. Burner applications are specific to individual furnaces.

Drying, Curing and Calcining: Direct-fired

Burners used for these processes can be conveniently categorized into those used for heating air which is then used as the medium for the process, and those where the heat from the burner is used more directly, as in radiant sources or where the flame is used for flame contact or high-source-temperature convection heating. As in furnace firing, these methods can be used directly, i.e. where combustion products are in contact with the work as described in this Section, and also indirectly as in the next Section. Presentation is again generally in decreasing order of plant size, although this factor has little relevance to the firing principles.

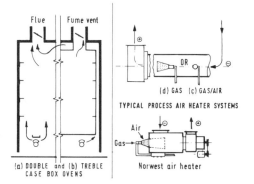

Flue | Fume vent

(d) GAS   (c) GAS/AIR

TYPICAL PROCESS AIR HEATER SYSTEMS

Air
Gas

Norwest air heater

(a) DOUBLE and (b) TREBLE CASE BOX OVENS

Figure 8   Typical low-temperature ovens and process-air heaters

Natural- and Forced-convection: Drying and Curing Ovens

The majority of small ovens are fired by burners fitted in the oven base in a similar manner to food-cooking ovens. The burners are usually of aerated bar type fitted with thermoelectric flame safeguards in the smaller sizes and electronic equipment in the larger sizes[5].

Most forced-convection continuous or batch ovens and driers have a heater unit either appended to or integral with the main working space. The burner and combustion section may then be either before or after the main circulating fan. Where it is before the fan and the combustion-chamber pressure is below atmospheric, a suction burner is often used. This is a low-pressure burner taking combustion air from outside the heater unit; it is simple and effective, and can be used for most applications. Two systems are shown in *Figure 8*. A combustion-chamber negative pressure of 200 Pa gauge (2 mbar) is adequate with a low-pressure gas supply.

Where the burner system is on the outlet of the circulating fan and there is a positive pressure in the combustion chamber, a pressure burner is required. In *Figure 8c* a forced-draught premix ribbon burner, or where there is adequate excess air a neat-gas burner as in *Figure 8d*, is located in the main circulating air duct. The air/gas mixture pressure is several mbar higher than the local pressure in the duct. Alternatively a nozzle-mix or premix burner is fired into a combustion chamber or into a duct. By using forced-draught burners the flame size is minimized and the momentum of the combustion products may also aid mixing with the main circulation. The space taken up by the combustion section is therefore minimized, albeit at higher burner-system costs.

The controls for burners for these ovens and driers usually require a two-level heat input or a proportioning system. Many processes will accommodate a maximum and standing-loss two-level system, but where the work is sensitive to small temperature fluctuations a proportioning system is necessary. To minimize the possible concentration of unburned air/gas mixture in the oven space, the burner systems should include fast-acting flame-safeguard equipment and pressure- and/or flow-switch interlocks with the oven circulation systems. To further safeguard operation and to take account of possible rises in the oven pressure, relief panels[6] should be incorporated.

Radiant-burner and flame-contact drying

For drying and or curing of continuous web material, e.g. paper webs, textile materials or coatings or wires, radiant burners of the type shown in *Figure 9* may be used in preference to, or in conjunction with, convective drying. They are also used for conveyorized systems where the work is suspended from or carried on a conveyor system.

Such continuous systems may alternatively utilize flame-contact heating. Ribbon-flame burners or multiples of separate burners are directed onto the moving work. By these means a very high local heat flux can be obtained and the heat may be precisely zoned. The burners may not need to be enclosed to suit some treatments. Attention must be given to interlocking starting and stopping, and process-temperature control is usually more difficult with these concentrated heat sources than with oven treatments.

Drying, Curing and Calcining: Indirect-fired

Although direct-fired processes are satisfactory for most materials, there are exceptions where the products of combustion are incompatible with the work material. Examples are some paint finishes, curing silicone rubber, the drying of some chemicals, and in isolated instances where condensation of water vapour on the surface of cold work may occur and cause undesirable surface effects.

Natural- and Forced-convection: Drying and Curing Ovens

A typical natural-draught indirect-fired oven comprises a separately ventilated casing which is heated externally by burners firing under the bottom; the combustion products pass up the sides within an insulated exterior casing.

Forced-convection ovens or driers are fitted with or include external or integral indirect air heaters, i.e. heat exchangers that transfer heat from the combustion gases to the air being circulated. These heaters may have a single flame firing into a combustion chamber followed by a bank of heat-exchange tubes, on the same principle as space-heating air heaters, or heat-exchange tubes separately fired. The burners for these follow the same patterns as in boiler firing but there are also many purpose-designed variations.

The controls required are similar to those fitted to space-heating air heaters and boilers. It is necessary to study the combination of the air heater and the application to identify the control requirements.

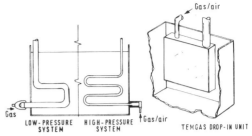

Figure 10   Various types of immersion heaters

The heat storage effect of the heater must be taken into account in deciding whether on—off, high—low or proportional heat-input control is required.

Radiant Indirect Drying

Because most requirements can be met with methods already described there is little demand for these systems. An example where the effect is achieved by using radiant sources is in paint drying, where the work passes between radiant panels. The combustion products ascend immediately in front of the panels and are ventilated without coming into contact with the work. Where complete separation is required it is possible to use radiant tubes at either high or low temperatures. Tubes for furnace applications or simpler systems can be purpose-designed to suit specific applications.

Tank and Vat Heating

Tanks and vessels containing liquids for washing, degreasing, coating, pickling, etc. are heated by external gas firing in the form of burners firing under or along the sides of the vessels. This simple form of heating can be applied to many processes, but unless it is used in a relatively complex manner by jacketing the vessels and providing disciplined travel to the combustion gases the thermal efficiency will be low, with consequent fuel wastage. For example, a simple underfired tank may effectively use only 40% or less of the heat input.

There are however many instances where the nature of the liquid being heated requires a completely open vessel interior, and for these external heating must be used.

Where space can be provided in the vessel, gas-fired immersion tubes can be used in a similar manner to steam-heating coils or tube banks. By this means, thermal efficiencies up to 80% can be obtained, which is higher than many steam-heating applications. For applications where there is ample space for these immersion tubes, low-pressure burners with flue effect obtained from the tube installation can be used simply and effectively. Where space is restricted, as for example when they must not exceed the space taken up by steam coils, it is necessary to use either pressure burners

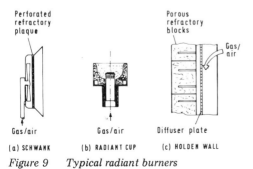

Figure 9   Typical radiant burners

or fan suction at the tube exhaust, or a combination of both. Various types of tank-heating burners are shown in *Figure 10*.

Space Heating

Gas-fired space heaters are mainly comprised in five categories: black emitters, luminous panels, direct-fired air heaters, indirect-fired air heaters, and make-up air heaters.

Working Flame Burners

For certain heating applications and processes it may be desirable or necessary to use burners whose flame impinges directly onto the work to provide the required heating. Generally such working flames are used when controlled localized or spot heating is needed, and heat recovery from the products of combustion is rarely attempted.

Torches

Probably the best known appliance is the ubiquitous air-gas or oxy-gas torch. These simple hand-held torches are used for a wide variety of applications such as soldering and brazing, metal cutting, hardening, jewellery manufacture and glass working. Flame sizes may be obtained ranging from the large torch down to a pin-sized flame.

Automatic Machines

When working flames are used on automatic machines, it is normal to use multi-hole burner heads fed with a premix of gas-air or gas-oxygen, or in some instances with a gas/air/oxygen mix with several burner heads fed from the same source.

Such automatic machines may be fitted with high—low burner control and moving burner heads, and are popularly used on automatic brazing machines, glass-working machines, particularly those concerned with lamp and flask manufacture, and flame-hardening machines.

3.2 Liquefied petroleum gases

Between the permanent hydrocarbon gases such as methane and ethane and the liquid hydrocarbons containing five or more carbon atoms is a range of hydrocarbons containing three or four carbon atoms per molecule such as propane and butane. All are normally gaseous at ambient temperature and pressure but can be liquefied at moderate pressures for the purpose of storage and transportation: one volume of liquid will give rise to approximately 250 volumes of gas. These pure compounds and their mixtures are collectively known as the Liquefied Petroleum Gases (LPG).

References p 38

Large bulk storage for LPG at refineries and major industrial sites is sometimes refrigerated, but more commonly consists of large spherical pressure vessels at ambient temperature. Fixed bulk storage at commercial or industrial sites is normally in cylindrical tanks with dished ends, and may range in capacity from 1 to 100 t. Light industrial applications and the domestic market are covered by refillable cylinders ranging in capacity from 0.5 to 50 kg. Disposable cartridges used largely for leisure applications may contain up to 0.5 kg of product.

Transportation in bulk as a liquid may be by pipeline or ocean-going ships, followed by rail tankers each taking from 10 to 90 t of LPG and frequently travelling as high speed 'liner' trains, or by road tankers which may travel anywhere with loads of 3 to 30 t.

There are potential hazards related to the handling and transport of LPG since any escape of liquid, by its rapid vaporization, can cause severe frost burns if in contact with the skin. Also such an escape would produce large volumes of flammable vapour which, when mixed with air and ignited, may give rise to explosions. The flammability limits for propane are 2.2 to 10.0% by volume in air with a minimum ignition temperature about $500°C$, while the limits for butane are 1.8 to 9.0% in air with minimum ignition temperature about $450°C$. Unlike natural and manufactured gases, LPG vapour is heavier than air and therefore may accumulate in drains, cellars and low ground rather than dispersing freely. This high specific gravity goes some way towards compensating (e.g. in the Wobbe index) for the high calorific value ($93 MJ/m^3$ for commercial propane and $122 MJ/m^3$ for commercial butane) when selecting a burner injector. Although LPG may be used in specially designed burners in the liquid state it is far more commonly utilized directly as a gas or as an LPG/air mixture. Its combustion characteristics most closely resemble those of natural gas and many of the domestic and commercial applications such as cooking, water heating, space heating and refrigeration are shared with natural and manufactured gases.

LPG is used as a feedstock for catalytic steam reforming to produce hydrogen, for blending with air as a substitute for natural gas, and as a chemical feedstock (App.B5) for ethylene, butadiene and acetyl chemicals. LPG is used widely by industry in many heating applications such as heat treatment of metals and metal plate cutting and profiling. The heavy-clay industries manufacturing bricks, pipes and tiles are frequently beyond the piped gas systems and so provide major loads for LPG firing. Agricultural applications for the intensive rearing of chickens, turkeys, ducks and piglets together with grain and grass drying, tobacco curing and greenhouse heating form a further rural load. The main automotive use of LPG is as a fuel for spark-ignition internal-combustion engines. In addition to private vehicles, the fuel is used for industrial fork-lift trucks and more extensively for fuelling refrigeration systems on refrigerated transport vehicles and containers. On building and civil engineering sites advantage is taken of the mobility of the LPG storage units to provide personal amenities such as space heating, hot water and cooking as well as the engineering services such as metal cutting, brazing and soldering, bitumen melting, drying and floodlighting. In recent years LPG has been used in increasing quantities in the leisure trade. Camping, caravanning and boating provide a growing demand for cooking, heating and lighting appliances based on refillable containers or disposable cartridges.

There are many publications, guides and regulations related to all aspects of LPG storage, transportation and utilization which are based upon experience and safe practice. In the UK a number of Home Office Codes of Practice, the Highly Flammable Liquids and LPG Regulations, and British Standard Specifications apply to the product itself, its handling, its utilization and the equipment related to the LPG industry. The L.P. Gas Industry Technical Association is seeking to bring these safety needs together with its booklet 'Introduction to LPG', its 'Emergency Procedures', a guide to handling LPG incidents, and its wide range of Codes of Practice:

No.1.   Installation and Maintenance of Bulk LPG Storage at Consumers Premises.
No.2.   Safe Handling and Transport of LPG in Bulk by Road.
No.3.   Prevention or Control of Fire involving LPG.
No.4.   Safe Operation of Bitumen Boilers and Hand Tools with Propane.

No.5.   Filling of LPG Containers from Consumers Bulk Storage.
No.7.   Storage of Full or Empty LPG Cylinders at Stockists.
No.9.   LPG—Air Plants.
No.10.   Safe Handling of LPG in Mobile Tanks and Mobile Equipment.
No.11   LPG as a Fuel for Motor Vehicles.

References

1   BS 4947:1973. *Test Gases for Gas Appliances*, British Standards Instn, London
2   Coles, K. F. and Wilbraham, K. J. 'New developments in radiant and tunnel burners', *Instn Gas Engrs Publn* 953, London, Nov. 1974
3   *Interim Code of Practice for Large Gas and Dual Fuel Boilers*, Parts 1—4, 1970—71, British Gas, London
4   Bryan, L. J., Masters, J. and Webb, R. J. 'Applications of recuperative burners in gas-fired furnaces', *Instn Gas Engrs Publn* 952, London, Nov. 1974
5   *Standards for Automatic Gas Burners*, 2nd edn, British Gas, London, 1970
6   *Evaporating and Other Ovens*, Department of Employment Factory Inspectorate Health and Safety at Work Booklet 46, HMSO, London, 1974

General Reading

(A)   Priestly, J. *Industrial Gas Heating*, Ernest Benn, London, 1973

(B)   *Gas Engineers Handbook 1965*, The Industrial Press, New York

Chapter 4

Oil firing

Preparation for Burning

Broad Classification of Oil Fuels

Petroleum oil fuels are classified mainly according to their viscosity and this characteristic is of considerable importance in their handling and burning. Pour point or pumpability are also important when considering storage, especially of the heavier grades. Sulphur, vanadium and other impurities are more significant in relation to the process or plant in which the fuel is to be used. The commercial classification of petroleum fuels is dealt with in App.A9 and tests and their significance in App.D2.

Liquid fuels derived from coal — commonly referred to as coal-tar fuels or CTF — are dealt with in Apps.B11 and D4 and in the book *Coal Tar Fuels*[1]. The BS 1469[2] designations of CTF numbers (CTF 50, 100 and so on) indicate the minimum temperatures in degrees Fahrenheit at which the fuels are sufficiently fluid for use in an atomizing burner.

In this Chapter particular reference to CTF is made only when there is a distinct and significant difference, as the handling and combustion of CTF are similar in many respects to those of petroleum oil fuels. The term 'oil fuels' includes CTF as well as fuels derived from petroleum.

Main Characteristics of Oil Fuels

Petroleum oil fuels normally have a density relative to water of less than 1.0 and water and sludge may therefore be removed periodically by means of a simple valve in the bottom of the storage tank, at the lowest point.

Because the relative density of CTF is greater than 1.0, special arrangements are required for the removal of water and sludge. Precautions are also necessary to minimize thermal cycling of CTF in storage and to minimize corrosion of copper and copper-bearing alloys. Jointing materials and pump packings for use with CTF should not contain rubber or bituminous materials. Generally, pump wear and erosion of valves etc. are more severe with CTF than with petroleum oil fuels.

Petroleum oil fuels should NEVER be mixed with CTF if separation of insoluble material is to be avoided.

The low-temperature characteristics of all oil fuels are important in connection with storage and handling, and in all circumstances the fuel must remain sufficiently fluid to permit flow under gravity in storage tanks and ready pumping in fuel lines. Minimum storage and handling temperatures are contained[3] in BS 799: Part 4 (also see *Table 1*), but for fuels derived from certain waxy crudes suppliers may recommend higher minimum temperatures.

The viscosity/temperature characteristics of normal petroleum fuels are indicated in *Figure 1*. From a knowledge of the viscosity of a fuel at a fixed standard temperature (normally $82.2°C$,

180°F), a suitable operating temperature may be determined to give the required viscosity at the burner.

Selection of Oil-fuel Grade

A number of factors, in addition to net fuel cost, require careful consideration in selecting the most appropriate grade of fuel for any particular application.

Effect on the Product. In some cases the material or product being heated may be sensitive to some impurity; e.g. vanadium may be undesirable in certain specialized glass-making processes.

Effect on the Plant. The method of operation of the plant may make certain fuels unsuitable; e.g. where it is necessary to operate plant at low temperature from time to time, a distillate fuel of very low sulphur content may be desirable in order to minimize low-temperature corrosion if other methods are impracticable.

Heat-transfer Characteristics. Although techniques are established to produce flames of high emissivity when using gases or distillate fuels, high radiant heat transfer rates are usually more readily obtained using residual oil fuels of high carbon/hydrogen ratio.

Minimum Burner Firing Rate, and Control System. Individual burners of low heat input or burners with sophisticated controls are more readily fired on distillate fuels. Account should be taken also of the length of any shut-down periods and the sensitivity of burner equipment to viscosity changes, especially if on—off control is used. Residual grades of fuel should always be considered for larger plant of, say, 600 kW heat input or greater.

Clean-air Requirements. The normal method of ensuring correct ground-level concentrations of sulphur dioxide or other pollutants is by the provision of a chimney of adequate height. There may be circumstances, however, where such a chimney cannot be installed or where the use of a lower-sulphur distillate fuel is desirable for some specific local reason.

Delivery and Storage

Detailed guidance on various aspects of delivery and storage is contained in the appropriate parts of BS Code of Practice 3002[4] and also in publications issued by fuel suppliers. Certain points are, however, of particular importance and these are dealt with below.

Reception Facilities. Oil may be delivered by road, rail, water or pipeline and discharged into main storage. Correct supervision of this operation is essential. Generally a representative of the buyer should be present to accept delivery and his prime responsibility is to ensure that the storage tank has sufficient space (or ullage) to accept the full amount

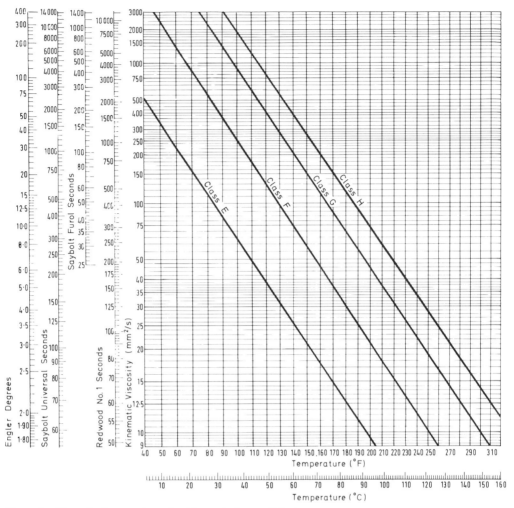

Figure 1   Viscosity/temperature chart for petroleum oil fuels
Class E: Gas oil, BS 2869: 1970
Class F: Light fuel oil, BS 2869
Class G: Medium fuel oil, BS 2869
Class H: Heavy fuel oil, BS 2869
Reproduced by permission of the British Standards Institution, London

to be delivered, especially where high delivery rates are used. Accurate measurement of the tank contents is vital to this operation.

For road delivery vehicles there must be safe access to the site, adequate hard standing, and room for the largest vehicle likely to be used to manoeuvre safely. Properly planned sidings or berthing facilities must be installed for rail and water deliveries. Precautions should be taken to prevent movement of delivery vehicles or removal of rail tank cars from sidings whilst hoses are still connected. In discharging rail tank cars, particular attention should be paid to oil preheating, the correct sequence of operation of valves, and pipework drainage.

Connections should be of the correct size, fitted with non-ferrous dust caps and properly maintained. They should be at a convenient height — normally about 1 metre above ground level — and with easy access. Each storage tank should have a separate filling pipe which should be clearly labelled with the appropriate oil-fuel grade. Ideally the storage tank, and especially the vent pipe, should be within sight of the person effecting the discharge.

All filling pipes should be as short as possible and free from sharp bends. They should be self-draining into the storage tank if possible, but otherwise should be fitted with gate valves close to the hose connections and should be lagged and

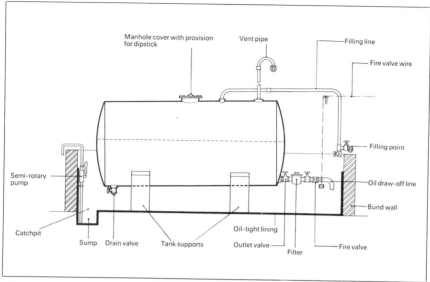

Figure 2   Storage tank for distillate grades of petroleum oil fuels. Courtesy Shell Marketing Ltd. (Industrial Oil Fuels. Delivery, Storage and Handling, p 11)

traced.   The tracing should be thermostatically controlled to prevent overheating.

Storage. Storage should preferably be sufficient for at least three weeks' supply at maximum off-take, and each tank should be capable of accepting full loads.

All oil fuels contain traces of water. This should be drained off periodically under continuous supervision, preferably just before a delivery is made. Drain valves (cast steel or malleable) should be blanked off or securely locked when not in use. Regular drainage of water minimizes tank corrosion and sludge formation.

Any catch-pit around a storage tank should be oil-tight at all times and facilities for removing rainwater from the pit should be carefully controlled. For small and medium-sized industrial installations, the effective capacity of any catch-pit should be at least 10% greater than the capacity of the storage tanks contained within it. This may be reduced for large installations and tank farms such as those operated at oil terminals to which special safety codes[5] apply.

Below-ground steel tanks should be housed in properly constructed brick or concrete chambers. Buried tanks require special protection. Prefabricated and ceramic-lined reinforced concrete tanks may also be used, and in these cases specialist guidance should be obtained.

A vent pipe must be fitted to the highest point of every storage tank. It should be of a bore equal to or greater than the diameter of the tank-filling pipe and never less than 50 mm diameter. If a simple vent pipe rises to an abnormal height, excessive pressure could result in the tank in the event of an overfill.   A pressure-relief device must be fitted to the bottom of such a vent pipe to prevent this.   Every storage tank must have an adequate manhole in an accessible position. Typical storage-tank layouts and fittings are shown in *Figures 2* and *3*.

Heating for Storage and Pumping.   Distillate grades of oil may be stored, handled and atomized at ambient temperature; but heavier grades require heating, at various stages, to facilitate flow and to ensure proper atomization.   The minimum temperatures for storage and for outflow from storage and handling are indicated in *Table 1*, for standard grades of petroleum oil fuels and for CTF.

Heating coils or elements should be spaced evenly over the bottom of the storage tank to minimize cold spots.   The oil draw-off point should always be above any coils, elements or thermostats to ensure that these are always completely immersed in oil. The use of an outflow heater only without at least some background tank heating is not good practice, especially in long horizontal cylindrical storage tanks.   The storage temperatures indicated below facilitate oil flow under gravity within the tank.   The minimum outflow and handling temperatures are designed to facilitate pumping in the oil handling system or ring main — see under *Oil Handling Systems*.

Heating may be by electricity, steam or a liquid-phase system.   For each form there are additional specific requirements that are referred to in the appropriate British Standards, or other standards.

Safety.   Tank cleaning and tank repairs should not be undertaken without proper precautions. Tanks which have contained oil fuels should be well ventilated before entry by properly trained

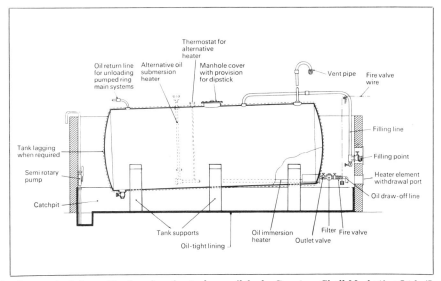

Figure 3   Storage tank for residual grades of petroleum oil fuels. Courtesy Shell Marketing Ltd. (Industrial Oil Fuels. Delivery, Storage and Handling, *p 11*)

Table 1   Minimum storage and handling temperatures for oil fuels

Class of oil fuel	Minimum temperature for storage ($^{\circ}$C)	Minimum temperature for outflow from storage and handling ($^{\circ}$C)
Kerosine, Class C, BS 2869: 1970	Atmospheric	Atmospheric
Gas oil, Class D, BS 2869	Atmospheric	Atmospheric
Light fuel oil, Class E, BS 2869	10	10
Medium fuel oil, Class F, BS 2869	25	30
Heavy fuel oil, Class G, BS 2869	35	45
CTF 50	Atmospheric	5
CTF 100	35	35
CTF 200	15	30
CTF 250	45	60

personnel. Purging to free from gas must be carried out before any work is undertaken. Even if a tank is certified gas-free, hot work may produce sufficient vapour from any residue or scale to create a flammable mixture with air.

In the event of a fire near a storage tank it is desirable that oil from the tank should be prevented from reaching the fire. An automatic fire valve should be inserted in the oil fuel line as close to the tank as possible. A temperature-sensitive element to actuate the valve should be positioned close to the firing equipment and well above floor level.

Valves may be actuated pneumatically, electrically or by gravity. They should close in the event of failure of the actuating medium and a warning device should be included to indicate that resetting is necessary. When a valve is closed by the action of a falling weight, a short free fall is desirable before the weight begins to close the valve, in order to overcome any sticking in the open position.

Fire valves should preferably be of glandless construction. A manual release should be installed on all fire valves so that regular checking is facilitated.

Handling from Storage to Burner

Filtration.   Before oil is permitted to flow in any part of a system, it must be filtered to prevent blockage of small pipes or orifices or damage to pumps or other equipment. Filtration must be adequate to ensure protection, but not so fine as to cause excessive restriction to flow. Filters should be constructed and fitted so that they may readily be cleaned and in such a way as to prevent any trapped solid matter passing to the downstream side of the filtering element. Detailed construction features for filters are contained in various parts of BS 799.

Filtration should normally be considered in two stages.

1. First-stage filtration or straining to provide protection for pumps or similar equipment which may handle the oil at storage temperature or at a temperature lower than the final atomizing temperature.

2. Finer filtration to protect valves, atomizers and similar equipment which incorporate small orifices. Second-stage filtration is sometimes included as part of the burner assembly.

The effective filtering area of any filter must always be sufficient to make frequent cleaning unnecessary. For fully or partly automatic burner equipment, the minimum period between cleaning for distillate and light fuels should be three months and for medium and heavy fuels one month. Shorter periods may well be required however when a system is first commissioned. Adequate filtration area also ensures low pressure loss in the system, although this depends also upon the viscosity of the oil at the point of filtration. The loss across a clean second stage filter should not exceed 14 kN/m² (kPa). Provided that proper protection is ensured for sensitive components, filtration should not be finer than the BS 799 limits indicated in *Table 2*.

Paper, fibre and ceramic filters are generally unsuitable for heated grades of fuel. First-stage filters for these grades should be placed in the draw-off line as near the tank as possible and should incorporate a filtering medium with a mesh equivalent to circular holes of between 0.75 and 2.5 mm diameter.

Oil-handling Systems. Oil temperature is of considerable importance. Distillate grades of oil fuels are usually handled at atmospheric temperature except in areas subject to severe weather conditions where temperatures below the cloud point of the fuel may occur, or where the plant is only used intermittently and 100% reliability is essential (e.g. an exposed peak-lopping gas turbine). In such cases some heating may be necessary to ensure fluidity at all times. Maximum cloud points for winter and summer qualities of gas oil (Class D) are indicated in BS 2869. Where offtake rates are low it should be remembered that summer-quality fuel may remain in the tank into the winter period unless deliveries are properly regulated.

In order to reduce their viscosity, heavier grades of fuel are handled at or above the minimum temperatures indicated in *Table 1*. This improves flow control and pumping and reduces frictional losses in the system. Further heating is normally required at some stage, using line heaters, in order to raise the oil temperature to one suitable for atomizing.

There is wide variety in the systems used to convey oil from storage tanks to the firing equipment. Each installation requires individual consideration, but there are some general principles and requirements, outlined below, which apply to all systems.

1. Oil must be cut off in the event of a fire (see under *Safety*).

2. Oil must be correctly filtered (see under *Filtration*).

3. Essential items like pumps should be duplicated, especially if continuous operation is important. The system should have facilities to isolate and bypass vital components so that they may readily be removed for servicing or replacement. Pumps, valves, filters and steam and electric line heaters can often be built into compact units.

4. The system should be constructed of materials that will ensure long and trouble-free use. All parts

Table 2   Maximum degree of filtration of oil fuels

Class of oil fuel	Approximate viscosity at filtration temperature		Mesh number (BS 410)
	Centistokes (mm²/s)	Seconds Redwood No.1	
C, D and CTF 50	—	—	240
E, F and G	Less than 24	Less than 100	170
	24 — 60	100 — 250	120
	60 — 290	250 — 1200	85
	290 — 490	1200 — 2000	44
	490 — 860	2000 — 3500	22
	More than 860	More than 3500	8
CTF 100, 200 and 250	Less than 490	Less than 2000	44
	490—860	2000—3500	22
	More than 860	More than 3500	8

References p 55

of burning or handling equipment which are in contact with oil (especially hot oil) should be made of corrosion- and erosion-resistant materials. Galvanized metals, low-grade copper and zinc alloys, lead and zinc should be avoided. Copper and aluminium should not be used for filter screens, bellows or diaphragms (unless these are of substantial thickness) nor should they be used for element tubes for electric heaters.

Thermoplastic materials are generally unsuitable, but nylon and PTFE are satisfactory for valve seatings, seals etc, as are some oil-resistant synthetic rubbers. In case of doubt manufacturers should be consulted on the precise suitability of their products.

Copper and some copper alloys are generally unsuitable for use with coal-tar fuels.

5. The system should be designed for straightforward filling, and for draining if necessary. Provision must be made for venting air at the highest point of the system (and at any other point where air may be trapped) during filling. The draining of oil-handling systems is facilitated if there is a continuous slope back to the pumps. It is useful if valves and cross-connecting pipework are incorporated, to allow oil from both flow and return lines to gravitate to the pump suction (with the air vent open). Oil can then be pumped back into the main storage tank. The simplified hot-oil ringmain system shown in *Figure 4* incorporates this facility.

6. Oil temperature must be maintained throughout the system under the most adverse conditions, such as when oil flow is stopped. Tracing is normally required in order to maintain the oil above its minimum handling temperature, and proper lagging of oil lines and fittings (especially filter and valve bodies) is essential for all heated grades of fuel. The heat input from tracing must be adequate for the worst conditions but not so high as to overheat the oil under normal conditions. For systems operating at or near atomizing temperature, thermostatic control of the tracing is essential. The most common methods of tracing are:

External or internal steam or hot water pipes.

External or internal electric heating tapes or cables.

For steam-boiler firing, some electric heating and tracing may be necessary if a steam-heated system is used, in order to ensure start-up after a prolonged shut down.

Heat inputs from various forms of tracing and the insulating properties of lagging materials may be determined from data supplied by manufacturers. Approximate heat losses from pipelines may be determined from *Table 3*.

7. Oil must be brought to the correct viscosity before atomization. This is done by means of a line heater (which may be incorporated in the actual burner). The atomizing viscosity required should be specified by the burner manufacturer; and from this and a knowledge of the viscosity of the oil at a standard temperature, the correct atomizing temperature may be determined from *Figure 1*. Line-heater ratings should be based on the maximum likely oil offtake rate (assuming that no line heating is required for recirculated oil) and the maximum temperature rise. An approximate mean specific heat of 2.0 kJ/kg °C may be assumed for all residual grades of oil fuel.

8. The system should have adequate capacity to minimize pressure losses and pumping costs. This may be facilitated by the use of a ring-main system in conjunction with pressure control as shown in *Figure 4*.

Each burner in a system should be supplied by a sub-circulating loop or by a separate branch line. Pressure conditions are maintained approximately constant at each offtake point by placing a pressure-regulating valve after the last offtake point and by circulating 1½ to 3 times the total maximum offtake of oil. The bore of the ring main should be such that pressure variations are not excessive for the equipment served. Pressure conditions should be calculated at each burner offtake point for all likely conditions of operation. If the burner equipment to be used is especially sensitive to pressure variation, individual pressure-regulating valves may be required.

9. The system should be designed so that it may be restarted without difficulty after a period of prolonged shutdown in cold weather. Particular attention should be paid to correct tracing and lagging of all pipework but especially to lengths of

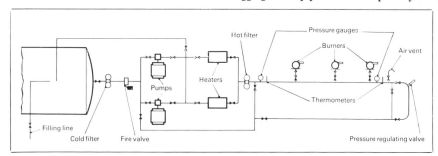

Figure 4   Simplified hot-oil ring-main layout. Courtesy Shell Marketing Ltd. (Industrial Oil Fuels. Delivery, Storage and Handling, *p 27*)

Table 3 Heat-transfer coefficients for exposed horizontal bare steel pipes

Nominal internal diameter (mm)	Heat transfer coefficient $(W/m^2{}^\circ C)$ Temperature differences $(^\circ C)$				
	30	55	85	110	140
15	12.5	14.4	16.3	17.9	19.7
50	10.9	12.6	14.2	15.8	17.5
100	10.1	11.7	13.2	15.0	16.3
200	9.5	11.0	12.4	13.8	15.4

Notes (a) For freely exposed pipes (e.g. in open air) multiply values by 1.5
(b) For well sheltered pipes (e.g. in ducts) multiply values by 0.8

pipework in which oil circulation cannot be maintained, even if the pipework is indoors. Valves, filter bodies etc should also receive special attention.

Oil Burners

Combustion of Oil Fuels

In oil firing, especially of industrial boilers[6], it is essential to match combustion equipment properly to the plant to be fired.

The characteristics of an oil flame are affected mainly by the composition of the fuel and the rate at which fuel and air are mixed and subsequently burned in the combustion zone. Good mixing of fuel and air is assisted by fine atomization of the oil, and particle sizes between 0.03 and 0.15 mm diameter are required for reasonable combustion in most burners. The uniformity of particle size is as important as the average size, though oil sprays are nevertheless sometimes described according to their average surface area characteristics. (Sauter mean diameter.)

The simplest-structured hydrocarbon molecules vaporize most readily. The temperature required to vaporize the more complex ones generally exceeds $400^\circ C$, at which temperature thermal cracking begins, producing heavy residues as well as simpler molecules and carbon. Oxidation of hydrocarbons without cracking produces characteristic blue flames of low emissivity, whereas combustion of carbon particles produces flames of high emissivity. Oil fuel is usually introduced into the combustion zone at a temperature well below that at which vaporization begins in order to avoid cracking and to ensure uninterrupted oil flow. Heat is required quickly to initiate vaporization once the fuel leaves the atomizer and the process of vaporization is accelerated in any case if the fuel is finely and uniformly atomized.

The momentum of the fuel and its atomizing medium causes entrainment of air or products of combustion or both. If secondary (combustion) air is introduced remote from the entry of the fuel jet or with unsuitable velocity and direction, the jet may preferentially entrain products of combustion. This will increase the heat input to the

fuel and its rate of vaporization. Because oxidation will be slow, however, some at least of the fuel will crack and carbon will be formed. On the other hand if highly preheated air is introduced into the combustion zone in such a way that rapid mixing with the fuel occurs, vaporization will also be rapid and a high proportion of the fuel will be oxidized as hydrocarbon vapour rather than as carbon.

Most oil combustion systems consist of all or some of the following:

1. Means of pressurizing the oil and preheating it to reduce its viscosity when necessary, i.e. delivering it to the burner in a suitable condition.

2. Means of supplying atomizing air or steam at a suitable pressure (when required) together with combustion air at a suitable pressure and temperature.

3. A burner — a means of mixing oil and air in order to promote controlled combustion. A burner in which combustion of the fuel is preceded by a preparatory phase in which it is divided into small droplets is known as an atomizing oil burner. This is the type in general industrial use and it is dealt with below in greater detail.

4. Means of ignition which must be rapid, reliable and safe, and means must also be provided for monitoring the oil flame so that if it fails to be properly established or is extinguished the burner is safely shut down.

5. Means of sustaining combustion after ignition. Combustion is assisted by the use of refractory material, especially at the root of the flame, and by partial recirculation of hot products of combustion. A shaped refractory block or burner quarl can be used to raise the temperature at the root of the flame and also to direct combustion air in a suitable way.

6. Means of controlling the starting, stopping and rate of firing (when required) of the oil burner in a safe and reliable manner and of maintaining air and oil supplied to the burner in suitable proportions over its working range. Controls may be required for safety purposes and for matching burner operation to heat demand.

Oil burners may be classified according to:

1. Their capacity (e.g. litres/hour, gallons/minute) or heat output (kW etc).

2. Type of control system used (manual, fully automatic, etc).

3. Type and grade of fuel which may be burned (gas oil, CTF etc).

4. Method of atomization used (pressure-jet, steam atomized etc).

5. Other special features.

Atomization

Atomizing oil burners may be broadly classified according to the method of atomization used.

Pressure-jet Atomizers. In these, oil is supplied under pressure (500 kPa gauge and upwards) and at a suitable viscosity to a conical chamber in which rotation or swirl is imparted to the oil before it leaves the final orifice. On leaving the final orifice, the thin film of oil produced breaks up into droplets. The oil spray produced by a pressure-jet nozzle has little capacity for entraining air and the design of the air director to promote oil—air mixing and to stabilize the flame front is most important.

1. Simplex pressure-jet. This is the simplest type, consisting of a burner body with single atomizing nozzle. As oil throughput varies approximately with the square root of the oil pressure, and as droplet size increases with reduction in oil pressure, the range of oil throughputs for satisfactory operation is limited. This type of burner is often used for simple on—off operation.

2. Spill return pressure-jet. This has flow and return oil passages and turn-down is effected by varying the return oil flow. Good atomization is obtained over a greater range by continuing to maintain a high degree of swirl in the nozzle at lower firing rates.

3. Duplex pressure-jet. This has a single swirl chamber and final orifice but independent flow and swirl passages for both main and pilot. Greater flexibility in firing rate is thus obtained whilst maintaining satisfactory atomization.

4. Duple pressure-jet. This incorporates two independent concentric pressure-jet atomizers, for main and pilot. As with the duplex atomizer, oil is supplied at low flow via the pilot passage only but via pilot and main at high flow.

Two-fluid Atomizers. In these, oil droplets are produced by promoting high rates of shear in the oil by the use of an additional fluid such as air or steam. Some form of metering device is required in the oil-handling system to regulate the oil supply to the atomizer. Two-fluid atomizers are generally more tolerant of shortcomings in atomizer surface

finish and in filtration. A turn-down ratio (high fire:low fire) of 4:1 and greater is easily obtained.

1. High-pressure air (or steam) atomizer. Oil is fed at a controlled rate to the nozzle where it meets a stream of air or steam and is atomized. The atomizing air or steam is usually at a pressure of 100 kPa gauge or higher. Steam should be at least dry-saturated or preferably slightly super-heated, and atomizing air should be free from water and oil.

2. Medium-pressure air atomizer. This is similar to the HPA type, but compressed air is usually employed for atomization at a pressure between 7 and 100 kPa gauge. Air used for atomizing represents under 5% of the total air required for stoichiometric combustion.

3. Low-pressure air atomizer. This is similar to the MPA type but atomizing air is usually supplied by a fan at a lower pressure, between 3 and 10 kPa gauge. About 25—30% of the total air requirement is used for atomizing, which may be completed in several stages.

4. Spinning-cup or rotary atomizer. Oil is fed at a controlled rate to the inside surface of a conical cup rotating at 4000 to 6000 rpm (min^{-1}). As a thin film of oil leaves the outer edge of the cup under centrifugal force it is atomized by the impact of a stream of air supplied through an outer concentric nozzle. To maintain an even distribution of oil in the cup, particular attention should be paid to the correct oil viscosity recommended by the manufacturer. The cup lip must be maintained in good condition.

5. Emulsifying atomizer. Oil at a suitable viscosity is metered and mixed with a controlled proportion of air at the inlet to a compressor, from which the mixture emerges as an emulsion. This is conveyed to a nozzle where atomization is effected, normally by using air as atomizing medium. Alternatively air or steam may be used to emulsify the oil within the atomizer body before it emerges from the final orifice.

6. Steam or air assisted pressure-jet. Air or steam may be used to improve the atomization of a pressure-jet atomizer and to maintain it at low flow rates. In this way, the range of satisfactory firing rates of the burner may be increased.

There are many variations of these basic types and these are usually well described and illustrated in manufacturers' literature. Oil may also be atomized ultrasonically. Schematic sections of several types of atomizer are shown in *Figure 5*.

In some firing equipment, atomization may be deliberately debased in order for example to produce long and highly radiant flames. In some high-temperature kilns a simple drip-feed system may suffice where oil vaporization is completed quickly in the presence of preheated air, and where any carbon formed completes its combustion towards the bottom of the kiln, thus assisting uniformity of heating.

High-duty Combustors. By taking particular care with atomization, and with control of the quantity and direction of secondary air, the production of hot combustion gases of very low oxygen content and low stack solids (carbon) content is possible. Oil may also be burned sub-stoichiometrically so that combustion can be completed in two stages[7]. High-duty combustors or gasifiers of this sort employ advanced recirculation and combustion-chamber cooling techniques. In all cases the atomizer and the combustion chamber must be regarded as an integral unit. This type of equipment finds application in the firing of those kilns and furnaces where a neutral or reducing atmosphere is essential, where even slight carbon contamination of direct-fired products is unacceptable, and where a gaseous flame of low emissivity is needed.

Air and Oxygen Requirements

One kilogram of oil fuel requires between 14 and 15 kilograms of air for complete combustion (between 10.6 and 11.2 m^3/kg for heavy fuel oil and gas oil respectively). Well designed oil-firing

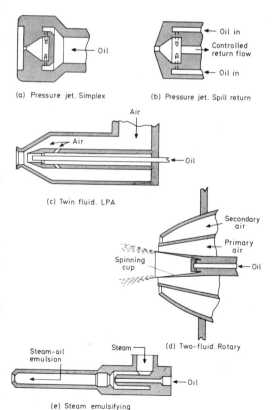

(a) Pressure jet. Simplex

(b) Pressure jet. Spill return

(c) Twin fluid. LPA

(d) Two-fluid. Rotary

(e) Steam emulsifying

Figure 5   Methods of oil atomization

equipment should be able to operate satisfactorily, over its range of firing rates, with an excess of air of not more than 20% and in some instances much lower. Low excess air is desirable (as for equipment designed for other fuels) in order to:

1. minimize the amount of heat carried out in the waste gases (chimney or stack losses);

2. maintain high flame temperatures when required; and

3. minimize the formation of sulphur trioxide (Ch.20.2).

Low-temperature corrosion rates in large water-tube boilers are markedly reduced when they are fired with excess-air levels down to 1–2%. Large well-maintained plant should be capable of operating at these levels without the excessive production of stack solids (carbon) in the waste gases, but this requires a high standard of equipment design and maintenance.

Waste Gases. Oil fuels (consisting mainly of carbon and hydrogen) yield on complete combustion carbon dioxide, water vapour, nitrogen, oxides of sulphur, oxides of nitrogen and any excess air. For petroleum fuels the amount of carbon dioxide present in the dry waste gases from stoichiometric combustion is normally between 15% (for distillate fuels) and 16% (for heavy residual fuels), and for coal-tar fuels about 18%. As with other fuels, measurement of the carbon dioxide content is a useful indication of the excess air employed and hence thermal efficiency. Care is necessary however not to use this criterion where air is used deliberately (as in a drier, Ch.9.1) to reduce the overall waste-gas temperature. Using recirculated waste gases for dilution rather than fresh cold air leads to more efficient operation in many instances.

Stack Solids. Although the solids burden of waste gases is normally increased by operation with very low excess-air levels, other factors are equally important.

1. The uniformity of atomization. An important variable has been shown to be the number of large droplets produced by an atomizer. The mean diameter of the fuel spray is not especially significant.

2. The intensity of combustion (heat release per unit volume of the combustion chamber). Increase in combustion intensity increases stack solids production, and packaged boilers operated at MCR at intensities of 2 MW/m^3 or more are especially critical.

3. The effect of combustion-air swirl. There is a rapid increase in solids burden with increasing degree of swirl beyond an optimum value. The air mass flow for combustion in an average oil burner is about 17–18 times the fuel mass flow rate and it supplies most of the energy required for mixing. The actual velocities of oil and air are

broadly similar, and mixing depends very much upon the relative directions of fuel and air streams close to the atomizer.

4. Flame chilling. Impingement of burning droplets on relatively cold surfaces (e.g. the wall of a shell-boiler furnace tube) can result in a marked increase in stack solids. The use of a large amount of cold excess air also has a chilling effect.

Various legal limits exist throughout the world to control the emission of stack solids (Ch.19.3). Limits for oil-fired plant in the UK where the material being heated does not contribute to the emission are between 0.2 and 0.4% of the weight of fuel burned, depending on the plant size[8]. The more stringent requirement of 0.2% applies to large plant. These limits, based on actual test results and proper scientific evaluation by an official Working Party on Grit and Dust, are realistic and can be met by well-designed equipment.

In considering stack-solids performance, the particular combination of burner and boiler or furnace must be stated together with the operating conditions (firing rate, excess air etc.) if the results are to be meaningful. It is desirable that the combustion equipment for oil firing should be designed and developed as an integral part of the plant and not independently.

No reliable overall correlation exists between the carbon content of waste gases and smoke numbers, and unless comparative results are available for particular burners and firing conditions it is unwise to rely much upon simple optical methods. For oil fuels, if the carbon content of the waste gases alone is determined, the ash content of the fuel (mainly sodium and vanadium) should be added. 0.04 wt% of the fuel (400 ppm w/w) is a typical value for a residual oil fuel.

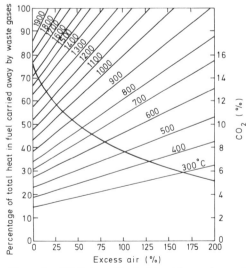

Figure 6   Oil firing. Typical heat losses in waste gases. Courtesy Shell Marketing Ltd. (Industrial Oil Fuel Manual, 6.4)

Preheated Air.   Waste-gas heat losses are dependent upon the amount of excess air used for combustion and also upon the temperature at which the gases leave the zone in which heat transfer takes place. *Figure 6* indicates losses for heavy fuel oil. Reduction of excess air or the use of recirculated waste gases to reduce temperatures and increase mass gas flow may increase thermal efficiency but the reduction of waste-gas temperatures may not always be desirable or even practicable because the process (e.g. metal melting) is a high-temperature one. In such cases heat may be recovered from the waste gases by using recuperators or regenerators (Chapter 16.2).

In addition to the recovery of heat in this way there may be additional benefits for oil-fuel combustion. The speed of combustion is increased and the flame temperature raised, leading to more rapid heat transfer. The amount of excess air necessary to ensure complete combustion may also be reduced.

But there are certain limitations to using preheated air. In particular, overheating of fuel passing through the burner must be avoided to prevent the vaporization of light components or cracking of the oil. The temperature of air passing through a burner body is usually limited to about 250°C. Where it is desirable to use air preheated to a higher temperature it is normal to use this as secondary air and to use only a small amount of cold air or air at lower temperature for atomizing.

The air supply to a burner should be regulated before it is preheated so that any variation in preheat will not affect the correct air/fuel ratio. This will cater also for the progressive reduction at the end of a cycle in the temperature of preheated air passing through a regenerator. Where heat is recovered by preheating combustion air, particular attention should be paid to possible leakages in the heat-exchange units.

Oxygen.   The advent of tonnage oxygen plants and large bulk storage in recent years has greatly increased the use of oxygen for combustion[9]. Oxygen enrichment of combustion air to increase flame temperatures by reducing the proportion of nitrogen used in the combustion process is of particular value to operators of high-temperature furnaces. The dissociation which results when using oxygen limits the theoretical flame temperature which may be achieved (Ch.8.1). However, the proportion of total heat rendered latent through dissociation becomes available as heat is transferred from the flame, the temperature of which is thus maintained by re-association. The percentage of delayed heat available in this way increases to over 50% when using 100% oxygen for combustion with oil (*Figure 7*). But the reduced waste gas volume which results from oxy/fuel firing may be significant where some heat transfer is effected by convection, and accelerated combustion may also markedly reduce flame emissivity.

If care is not exercised, the use of oxygen in burners may result in the rapid deterioration of any part of a burner which becomes overheated. A standard oil burner cannot be changed from air to operation with oxygen without radical redesign. Most oxy/fuel burners are water-cooled, and in this respect oil firing has distinct advantages as the burner size and hence the water-cooling requirements and consequent heat losses are considerably less than for oxy/gas burners of equivalent heating capacity.

In the supply of oxygen to the combustion equipment or combustion zone, strict safety precautions must be observed. Specialist guidance should be sought. Degreasing is essential, and the use of non-ferrous materials is advisable where high oxygen velocities in pipelines or burners are likely to occur. A well defined safe sequence of operations for starting up and shutting down oxy/fuel burners (including emergency shut-downs) must be developed and clearly indicated to operators. Oil and oxygen should not be permitted to mix, except in a controlled manner in the combustion zone.

Selection of Firing Equipment

The selection of the most suitable firing equipment for any particular installation depends upon a number of factors.

1. Grade of oil which it is desired to use. For environmental reasons (e.g. an existing chimney of inadequate height) or because of the sensitivity of the product being heated, constraints may be imposed on the grade of oil fuel which may be used (see *Selection of Oil-fuel Grade*). This in turn may restrict the choice of oil-firing equipment.

2. Type of plant to be fired and desired heattransfer method. There must be adequate combustion volume of the correct shape to match the desired flame shape and size. Flame impingement must be avoided, and due regard paid to the man-

ner in which secondary air will be introduced, taking into account the furnace pressure. From the outset the primary heat-transfer method — by radiation or by convection — must be decided and a burner selected which will give appropriate flame characteristics.

3. Maximum heat input required and range of load variation. The burner selected must be capable of meeting the maximum heat demand but also have sufficient flexibility to meet any variation without deterioration in performance. The ability to meet short-term variations in demand resulting from the nature of the heating process is mainly a function of the control system used, but the burner must be able to match this. If for example continuous operation with a wide variation in firing rate is demanded, then a simplex pressure-jet burner would be inappropriate. If however the heat demand is properly met by simple on—off operation with a fairly constant overall demand then such a burner would be suitable. With large plant with a multiplicity of burners (e.g. watertube boilers), load variation may be met by varying the number of burners fired, thus maintaining their individual performance.

In general, the efficiency of plant falls markedly when it is operated for long periods at low load, and the initial selection of plant is equally as important as the selection of burner to be used. For space-heating loads the installation of two or more boilers has much to commend it so that, for summer operation only, one boiler is used — at high load and efficiency and with reduced risk of low-temperature corrosion.

4. Combustion air. The pressure and temperature of combustion air which may be available or which may be necessary in order to obtain high furnace temperatures should be considered. Where air is highly preheated it may not be possible to pass this through the burner, and the use of LPA burners may be precluded in favour of MPA, steam-atomized burners or pressure-jet burners.

5. Equipment costs. In choosing the firing equipment both initial and running costs must be considered. At a time of rising fuel costs, well-designed equipment that will operate over a wide range at high thermal efficiency has an obvious premium advantage which should not be lost sight of in considering initial costs. The cost of oil preheating, electricity for fans and motors and charges for steam must also be taken into account.

The dependability and cost of burner maintenance services should be examined before equipment is ordered. Operator training and spares stocks are obviously simplified by limiting the choice of new equipment types and sizes to those already in use where this is technically feasible.

6. Other site considerations. Each site requires separate detailed consideration, but the adequacy and suitability of the following should not be overlooked:

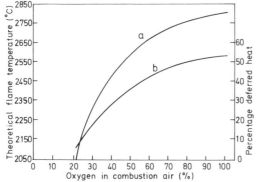

Figure 7   Oxy/oil combustion. Theoretical flame temperatures and percentage deferred heat
a: Theoretical flame temperature (°C)
b: Deferred heat (%)
Courtesy Shell Marketing Ltd. (Industrial Oil Fuel Manual, 6.5)

References p 55

(a)   Electricity supply for fans etc.
(b)   Steam supply for atomizing.
(c)   Space for burner erection and maintenance.
(d)   Chimney and flues.   Draught available.
      Liability to low-temperature deposition.
(e)   Burner noise control.

Operation and Maintenance

Burner Controls and Sequence of Operation

In any oil-firing system, provision must be made for operation in a safe and satisfactory manner. Safety is more important than is generally recognized.   Industrial oil fuels can be handled safely in bulk without any particularly stringent safety requirements; indeed in bulk they are very difficult to ignite.   But in a finely divided or partially vaporized state, mixed with air, and with a possible source of ignition, there is a danger of accidental ignition or explosion unless proper controls are used to prevent this happening.

Controls are used also to perform a second main function — to match the heat output of the firing equipment to the overall quantity and pattern of heat required.   Each control system should be considered individually, account being taken of the temperature sensitivity and heat capacity of the material being heated, the acceptable temperature swing, and the heat capacity of the furnace.   In addition, the ability of the burner proposed to respond to and operate satisfactorily with the pattern of operation required must also be examined.

A suitable sequence of operations may in its simplest form be under manual control, i.e. an operator in full-time attendance making observations and responding by opening and closing valves etc.   At the other extreme, the sequence of operations may be controlled by sophisticated equipment operating fully automatically.   Whatever control system is used, the safe sequence of operations is much the same, though certain items may be rearranged within the overall sequence.   Such a sequence is included in BS 799, Part 4[3] for burners atomizing over 36 l/h, the main operations of which are given below in abbreviated form.

Light-up sequence

1.   Ensure correct conditions for atomization. Oil at correct temperature and pressure.

2.   Purge the furnace or check for combustibles (or both).   No flammable fuel/air mixture must be present before a source of ignition is introduced.

3.   Set air flow for selected firing rate.

4.   Start means of ignition (and prove presence of ignition source when required).

5.   Turn on atomizing medium (when necessary) and turn on and set oil flow to start-up rate. (Normally 'low fire').

6.   Ensure that flame has been established.

7.   Stop ignitor.   To precede item 6 where the detector cannot discriminate between the main flame and the ignition source.

8.   Continue to monitor oil flame.   Oil is shut off if the flame fails.

Running conditions

9.   Adjust oil, air and atomizing medium (when necessary) to meet load demand.

10.   Continue to monitor oil flame and presence of combustion air.

Shut-down sequence

11.   Shut off oil supply to burner.

12.   Purge the burner (where applicable).   To prevent oil in the burner carbonizing, vaporizing or leaking.

13.   Purge the furnace (post-purge operation, where applicable).   To dilute to a safe level any flammable mixture remaining in the furnace.   The extent of purge depends on the furnace volume and subsequent gas passages etc.

14.   Complete required shut-down procedure.

For safety reasons a number of operations may need to be completed within certain time limits; e.g. oil must be shut off immediately if the main flame fails.   Reference should be made to the appropriate British Standard, or other standard applicable, for details of these safety timings.   A great deal of other useful information and notes on the various operations are included in these standards, which also contain definitions of many of the terms used in connection with controls and control sequences.

When oil-firing equipment is first commissioned or after the rectification of any fault, additional sequences and precautions may be necessary.   In any case, particular care should be exercised.

The resetting and repeated operation of a burner which fails to ignite, without proper investigation of the cause, is a practice to be deplored.

Dealing with Operational Difficulties

Storage.   An accumulation of water or the development of a sludge or oil/water emulsion can best be avoided by frequent and careful water drainage (see *Delivery and Storage*) and by the storage of fuels at the correct temperature.   If large amounts of sludge or emulsion do form however, filter blockage and flame failure may result.   If this material cannot be drained off, the affected tank must be cleaned after the removal of any good oil. The work should normally be undertaken by a specialist contractor who will take responsibility not only for the cleaning but also for the subsequent safe and correct disposal of the material removed.   The value in dealing with such situations of duplicated and adequate storage facilities is obvious.

Handling. Irregular oil flow or pressure drop in oil-handling systems may result from such things as vapour-locking or carbonization of the fuel owing to local overheating, filter blockage or oil chilling. Inspection of the fuel-handling system should be systematic and extend to all parts, but especially to sections where oil circulation cannot be maintained on shut-down ('dead legs') and sections which run close to hot surfaces.

Burner Operation. A failure in burner operation needs also to be investigated in a systematic way, and this is greatly facilitated by the use of fault-finding charts and operating instructions provided by the burner manufacturer.

The measurement, at key points, of oil temperature and pressure and the pressure of any atomizing air or steam is valuable in fault tracing.

For proper combustion, correct matching of oil and air quantities over the whole firing range, uniform and fine atomization, and proper oil and air mixing are normally essential. There should be no visible smoke at the chimney, particularly when using distillate fuels.

Poor combustion (flame instability, smoking or carbon deposition) may be caused by:

1. Flame chilling. This may be due to:
(*a*) Impingement on a relatively cold surface (e.g. a boiler furnace tube) of oil droplets before combustion is completed. A poorly-aligned burner or a displaced or damaged quarl or air director may be the cause of this.
(*b*) An excess of cold air in one part of the flame envelope caused by maldistribution of air around the atomizer.

2. Poor atomization and the production of oversized droplets. This may be due to:
(*a*) Incorrect oil viscosity resulting from too high or too low an oil temperature. It is important to measure the oil temperature close to the atomizer and over the whole range of firing conditions. The actual average viscosity of fuel being delivered and not the specified maximum should be used in determining the correct atomizing temperature. Tank heaters, outflow heaters or line heaters may be accidentally bypassed or may be otherwise inoperative owing to such things as: an inadequate steam supply; failure of condensate removal; an electricity supply failure; inadequate air venting of the heater oil space; a defective or badly positioned thermostat.
(*b*) Low oil pressure or low atomizing-medium pressure. Valves, pumps and compressors may need servicing or a blockage may have occurred in the handling system (see *Handling*).
(*c*) Mechanical defects in the atomizer, which may be incorrectly assembled or damaged. The grooves, orifices and edges of pressure-jet atomizers are particularly sensitive to damage and even slight blemishes may result in a marked reduction in performance. Worn or damaged jets should be replaced.

3. Inadequate combustion-air supply or restriction to flow arising from:
(*a*) Choked air passages or damaged refractory within the burner.
(*b*) Dirty air filters or silencers. Fibres and dust may have collected on the fan intake guard, or the fan blades themselves.
(*c*) Unserviced pressure-reduction valves or overfull drain vessels.
(*d*) Restricted air supply into the boilerhouse or other building in which a fan is housed.

Corrosion. Both low-temperature and high-temperature corrosion may result from oil firing, particularly when heavier grades are used. The possibility of corrosion is best considered and dealt with when the plant is designed, and the temptation to cut capital cost and risk corrosion should be resisted.

Corrosion is briefly dealt with elsewhere (Ch.20.1 and 20.2). Some operational difficulties may be alleviated by simple practical steps however. It is important to consider at the outset the extent to which a striving to attain high thermal efficiency is aggravating the position. A practical compromise may need to be struck between efficiency and corrosion.

Low-temperature corrosion and fouling caused by the presence of sulphur trioxide in the waste gases occurs when surfaces in which the gases are in contact fall below their sulphuric acid dew-point temperature (between about 115 and 150°C or 240 and 300° F for most industrial boilers)[10]. Maximum corrosion rates normally occur at 20--30°C below the acid dewpoint. At temperatures below the water dewpoint corrosion may be very severe. A reduction in excess-air levels and improvement in lagging and cladding of hot surfaces to minimize heat losses are obvious steps to be taken. Particular cold spots on which condensation is occurring (e.g. a badly positioned return-water inlet to a hot-water boiler) should receive special attention. Problems may be alleviated by various recirculation, premixing and preheating techniques.

The formation of sulphur trioxide may be reduced by operating with excess air at very low levels, reducing flame temperatures, and other combustion techniques; but the scope for this is limited in most existing plants. A reduction in the sulphur content of the fuel is not likely to be effective in reducing low-temperature corrosion unless a value of well under 1% is achieved. This would normally require a distillate grade of oil. If condensation cannot be prevented, it may be necessary as a last resort to neutralize the sulphur trioxide using ammonia, or to use some other proprietary additive to suppress its formation or to neutralize it. Comparative tests with and without the additive should be conducted under carefully controlled conditions.

High-temperature corrosion and fouling result from the deposition of vanadium and sodium compounds on surfaces usually in excess of 550—600°C (1000—1100° F). The mechanism of deposition and subsequent possible corrosion is a complex one, but usually the more volatile sodium compounds adhere first to surfaces and produce a rise in temperature and sticky surface conditions. These in turn may cause a rapid build-up of further deposits of particles with higher melting point.

Proper settling of fuel in storage, correct storage temperatures and regular draining-off of water assist the removal of any water-soluble sodium salts present in the fuel. Poor atomization tends to increase deposition rates and in conjunction with high excess air encourages the formation of low-melting-point oxides. Flame impingement and hot spots should also be avoided. As with low-temperature problems, additive treatment may be used to combat high-temperature corrosion.

Deposits, and Precautions to be taken when Cleaning. The deposits found in the gas passages of furnaces and boilers fired with residual petroleum fuels consist mainly of carbon, fuel-oil ash and sulphates. Carbon deposits may be minimized by attention to combustion conditions, and sulphates by eliminating any acid-dewpoint condition, but the ash (mainly vanadium and sodium) will accumulate as a fine dust and must be removed periodically.

Whenever reasonably possible, deposits should be removed without entering the flue-ways, by using probes, compressed air or industrial vacuum-cleaning equipment. Care should be taken not to disperse dust into the surrounding atmosphere. When it is necessary to enter the plant, approved masks (to BS 2091B) should be worn. Goggles or a full face mask must be worn to protect the eyes. A boiler suit (close-fitting at wrists, ankles and neck) and gloves should be used to prevent the dust coming into prolonged contact with the skin. A bath should be taken at the end of each working period and protective clothing should be thoroughly washed or discarded.

If adequate precautions are not taken during cleaning then symptoms of vanadium poisoning may occur. These may include irritation of the skin and eyes, a dry cough, wheezing, shortness of breath and a greenish-black discolouration of the tongue. The symptoms may all be delayed by up to 24 hours after exposure. If they occur, proper medical attention should be sought immediately.

Waste oil. Waste oil is sometimes used for burning and this may lead to operating difficulties. Waste oil normally consists of sump drainings and may therefore contain quantities of carbon and finely divided metals and be contaminated also by motor spirit, diesel fuel or kerosine. As well as producing undesirable emissions of lead to the atmosphere, the burning of waste oil containing lead may lead to serious boiler-fouling problems. The presence of traces of motor spirit may well produce an explosion risk from vapour emitted from any storage tank. The dumping of waste oil and recovered spillage into normal oil tanks by unauthorized persons may also be a source of difficulty. Securely bolted manhole covers and a sharp look-out for signs of spillage around manholes are the best methods of tackling this problem. When waste oil must be used, ample provision must be made for preheating because oil viscosity may vary widely. The burner used should also be relatively insensitive to viscosity and other variations.

Waste oil storage requires special consideration.

1. Adequate settling time must be allowed (not less than 24 h) and provision made for removing sludge from the tank periodically. This is all facilitated by the use of two tanks used and refilled alternately.

2. Each tank should have a vent pipe carried to at least 5 m above ground level. Open-topped tanks should not be used.

3. Filtration and fill and vent pipe sizes must be adequate.

4. Where both oil fuel and waste oil are used they must be stored in quite separate and properly labelled tanks.

Maintenance

Storage and Handling Equipment. A clear diagram of the oil storage and handling system, protected from dirt and moisture, should be prominently displayed. Operating instructions for pumps, heaters etc. and a record of routine maintenance should also be available for study and inspection. For every system an individual maintenance schedule should be developed which should include the checking of the following essential items.

1.   Cleanliness of the delivery point and the use of dust caps on filling-pipe ends.

2.   Tank identification and oil-grade labels.

3.   Drainage of water from the oil storage tanks.

4.   Security of the bunded area.

5.   Securely locked and blanked off tank drain valves.

6.   Condition of the tank vent pipes, to ensure that they are clear, especially after an overfill has occurred.

7.   Correctness of tank contents' gauges, by comparison with dipstick readings, to prevent overfilling.

8.   Correct functioning of fire valves.

9.   Condition of filters. Cleaning them and draining off any water.

10.   Condition of all tracing and lagging, making good any defects.

11.   Condition of steam traps.

12.   Operation of oil heaters, ensuring that they are properly vented and that controls are functioning correctly.

13. Leakage from any part of the system. Making good any defects.

14. Condition of oil interceptors fitted to surface water drains.

Repairs to oil storage tanks and oil lines involving hot working should only be undertaken with special precautions (see *Delivery and Storage — Safety*).

Oil-firing Equipment. The cost of maintenance is usually far less than the cost of a factory shutdown caused by the failure of a boiler, drier or furnace, even if this occurs only rarely. Oil-firing equipment and controls should be regularly maintained by a properly trained engineer familiar with the equipment, or by the manufacturer or other competent organization. The opportunity should be taken during maintenance to check the thermal efficiency of the plant if this is not otherwise done regularly. A copy of the manufacturer's recommendations and operating instructions should be kept with the plant.

Maintenance is greatly facilitated if, when a plant layout is designed, adequate access and space for servicing are provided, together with good illumination.

If a systematic maintenance schedule is not supplied by the manufacturer, a schedule should be developed to include the following:

1. Isolating the burner electrically and removing fuses before work is started.

2. Cleaning all filters.

3. Checking any burner heaters, ensuring that they are full and that all controls are functioning correctly.

4. Checking motors, belts and other moving parts for wear.

5. Checking brushes, commutators and switches for wear or loose connections.

6. Checking and cleaning all fans and air ducts.

7. Checking air director vanes, quarls etc. for damage.

8. Lubricating bearings, linkages etc., using the correct grade of oil or grease.

9. Checking for refractory damage within the combustion chamber.

10. Checking for air and waste-gas leakage.

11. Examining the atomizer for wear or damage and replacing when necessary.

12. Cleaning the atomizer using only gas oil (or CTF 50 where CTF is fired) and soft tools. It is particularly important that wire should not be used to clear pressure-jet atomizer nozzles, which should be soaked first to loosen any deposits.

13. Checking all oil pipes, valves etc. for leakage and making good any defects.

14. Checking all tracing and lagging and making good any defects.

15. General cleaning of the burner.

16. On restarting the burner, checking oil and atomizing-medium pressures, and oil temperature at the burner.

During any servicing or maintenance work, prolonged contact with oil fuels, especially distillate fuels, should be avoided. Oil fuels can impair the protective properties of the skin by removing the natural greases, leaving it open to attack by bacteria and to irritation by dirt and abrasive particles. A suitable barrier cream should be used or gloves may be worn where possible, though this precaution is of little value if the insides of the gloves are allowed to become contaminated with oil.

References

1 Huxtable, W. H. (Ed.), *Coal Tar Fuels. Their production, properties and application.* 2nd edn, Association of Tar Distillers (now the British Tar Industry Association), 1961

2 BS 1469. *Coal Tar Fuels*, British Standards Institution, 1962

3* BS 799. *Oil Burning Equipment*, British Standards Institution.
Part 3: 1970. Automatic and semi-automatic atomising burners up to 36 litres per hour and associated equipment
Part 4: 1972. Atomising burners over 36 litres per hour and associated equipment for single-burner and multi-burner installations
Part 5: Oil storage tanks

4* CP 3002. *Oil Firing*, British Standards Institution.
Part 1: 1961. Installations burning Class D fuel oil and CTF 50
Part 3: 1965. Installations burning preheated fuels Class E, F and G fuel oils and CTF 100 to 250

5 *Model Code of Safe Practice in the Petroleum Industry*, The Institute of Petroleum.
Part II: 1973. Marketing Safety Code

6 Tipler, W., 'Combustion equipment for oil-fired industrial boilers', Third Liquid Fuels Conference, The Institute of Fuel, 1966, Paper 6

7 Wheeler, W. H., 'Developments in the application of two-stage combustion', Third Liquid Fuels Conference, The Institute of Fuel, 1966, Paper 7

* These Standards are undergoing extensive revision at the time of writing and care should be taken to refer only to the most up-to-date Standard (listed in the BSI Yearbook) and the appropriate part of each

8   Memorandum 'Grit and dust', HMSO, London, 1967

9   Bagge, L. P., *Development of the Shell Toroidal Oxygen—Liquid Fuel Burner*, The Institute of Petroleum, 1962

10   Barker, K. and Carpenter, P. R., 'Progress made towards achievement of lower exit gas temperatures (Industrial and space heating boilers)', Third Liquid Fuels Conference, The Institute of Fuel, 1966, Paper 9

Chapter 5

Fluidized-bed combustion

In fluidized-bed combustion, any fuel, e.g. coal, oil or gas, is dispersed and burned in a fluidized bed of inert particles. In most applications the temperature of the bed is maintained in the range 750—1000°C so that combustion of the fuel is substantially completed but particle sintering is prevented. The gaseous combustion products leave the bed at its operating temperature, removing about 50% of the heat generated. The remainder of the heat released is available for direct transmission to heat-transfer surfaces immersed within the bed: in boiler applications these comprise a bank of steam-raising tubes. The heat transfer to immersed surfaces is uniformly high, in comparison with the variation of radiative heat transfer through a conventional combustion chamber. Consequently less heat-transfer surface is required for a given output and a boiler system occupies a smaller volume. The low combustion temperature reduces corrosion and fouling of heat-transfer surfaces, allows a lower emission of nitrogen oxides and alkalis, and also permits control of sulphur dioxide emission by direct addition of limestone or dolomite to the fluidized bed.

History

Combustion of coal in a high-temperature fluidized bed with clinker formation was developed in France, as the Ignifluid Boiler. This is described in Chapter 6. The first tests to demonstrate that coal and other fuels could be successfully burnt in fluidized beds at temperatures below 1000°C (i.e. without clinkering) were carried out during the mid 1950s. Several small experimental combustors were built and operated between 1964 and 1969 in the UK, USA, Czechoslovakia and elsewhere. A book describing this early work is available[1].

By 1970 three coal-fired pilot-scale units were in operation in the UK under the sponsorship of the National Coal Board (NCB). These were (1) a 1 MW combustor at the Coal Research Establishment (CRE) to obtain data for atmospheric-pressure watertube boilers, (2) a 2.5 MW vertical shell boiler at the British Coal Utilisation Research Association (BCURA), *Figure 1*, and (3) a 2 MW combustor, operating at pressures of up to 600 kPa (6 bar), at BCURA to obtain data for combined-cycle power generation[2]. The programme was supported by extensive work on smaller combustors and on cold models, and by design studies. Experience of over 17 000 hours operation of fluidized combustors has been amassed at BCURA and CRE[3]. The NCB is also developing fluidized combustion for the incineration of colliery waste slurries.

British Petroleum (BP) began development of oil-fired fluidized-bed combustors in 1969, and under the sponsorship of BP the BCURA shell boiler was converted to oil firing in 1971[4]. In 1971 the exploitation of British expertise in fluidized combustion became the responsibility of the National Research Development Corporation (NRDC), and in 1974 a joint company, Combustion Systems Ltd (CSL) was formed by BP, NCB and NRDC to promote the technique.

In the USA there are several approaches to fluidized combustion. The Office of Coal Research (OCR) has a programme for developing atmospheric-pressure fluidized combustion for steam generation. Under OCR sponsorship, Pope, Evans and Robbins (PER) developed a modular concept for industrial boilers, and by 1968 they had built and operated a 1.5 MW module[5]. In 1972 PER were given a contract by OCR to design, build and operate a 30 MW$_e$ fluidized-bed boiler. The Environmental Protection Agency (EPA) has an extensive design and feasibility exercise on fluidized combustion. The EPA has sponsored research programmes on almost every experimental combustor and pilot plant in the USA and UK, particularly to study sulphur retention by addition of limestone or dolomite, and is now sponsoring a 2 MW pressurized combustor with sulphur retention and dolomite regeneration. Westinghouse have carried out a series of comparative design studies of atmospheric and pressurized combustors for EPA[6]. The EPA also has a programme for the development of a pressurized fluidized-bed incinerator for domestic refuse, incorporating power generation by a gas turbine. Fluidized-bed incinerators for industrial wastes and sewage sludge have been built in the USA.

Small development programmes for fluidized-bed boilers have been carried out in other countries, particularly Czechoslovakia and India, where the process has many potential advantages for local fuels.

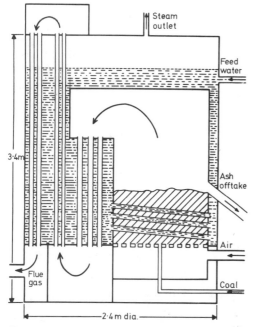

Figure 1   BCURA vertical shell boiler

References p 63

Characteristics of Fluidized Combustion

Properties of Fluidized Beds

In fluidized combustion the fuel is burnt within a fluidized bed of mineral matter, e.g. coal ash, silica sand, or limestone for sulphur retention. The passage of the air for combustion through the bed maintains the particles in a violently fluidized state. The turbulent fluidized bed is an ideal environment for combustion because of the rapid heat and mass transfer rates: the fluidizing air is heated to bed temperature within a few millimetres of entry; solid fuel is rapidly distributed through the bed; the convective-heat-transfer rates to immersed boiler surfaces are high. The pressure drop across the bed is that required to support the weight of the particles; a typical value is 10 Pa (1 mm of water gauge) per 1 mm of bed depth.

Combustion in Fluidized Beds

Relevant Factors

The amount of fuel that can be burnt in a fluidized bed is determined by the air supply rate. and hence by the fluidizing velocity and operating pressure. Velocities of 0.3 to 4 m/s measured at bed temperature have been used, giving heat-release rates of 0.2 to 3 MW/m^2 of bed area at atmospheric pressure. The efficiency of combustion is determined by many operating factors, including fuel type, excess-air level, bed temperature, fluidizing velocity, bed height, and uniformity of fuel distribution. Volume (bed + freeboard) heat release rates for coal are ≈ 2 MW/m^3 (5—8 pressurized).

Combustion of Coal

Most research and development on fluidized combustion has been directed to coal-fired power-station boilers[1]. Ash from the coal is conveniently used as the bed material, and since ash particles can be as large as the largest coal particles, the coal is crushed so that the ash produced is of a size appropriate to the fluidizing velocity. Typical coal sizes are from 1.6 mm top size for 0.5 m/s to 6.4 mm top size for 4 m/s. The bed is normally operated between 750 and 950°C. Lower temperatures may be used at low load, but give lower combustion efficiencies. There is a risk of ash sintering above 950°C, because burning coal particles may be 100°C or more above bed temperature. Provided that the coal is uniformly distributed over the bed area, the volatiles and carbon monoxide produced are burnt mostly within the bed, and their combustion is completed within less than 1 m above the bed surface. Less than 1% of coal in the bed is required to sustain combustion. The coal feed will contain a proportion of fine coal, and there will inevitably be some elutriation (carryover by the flue gases) of unburnt fine particles. In order to minimize this the coal is injected, usually pneumatically, near the bottom of a bed at least 0.5 m deep, and preferably deeper. Also a freeboard of 2

to 4 m is allowed above the bed for burnout of the fines. If the loss of combustibles in the elutriated fines is unacceptably high it can be reduced by either refiring to the main bed or feeding to a separate bed. The fines are readily separated from the gas using cyclones (they are considerably coarser than pulverized-fuel fly-ash). Fines refiring to the bed is the simpler method, but many passes through the bed may be required to achieve a high combustion efficiency, particularly at high fluidizing velocities. A high recycle rate of ash may build up in an efficient refiring system, and if the ash is cooled before each refiring, as is likely, a large amount of heat may be removed from the bed.

In a recent new approach to coal firing for industrial appliances, a method of burning uncrushed coal sized up to 50 mm or more has been developed at the UK National Coal Board's Coal Research Establishment. This uses the fluid-like property of the bed to 'float' the coal by buoyancy in a bed of high-density refractory. Fluidizing velocities are within the usual range of 0.3 to 4 m/s. The method is particularly suitable for 'washed' coals. On combustion in a fluidized bed, these generally produce an ash which is fine and/or friable and is thus readily elutriated from the refractory bed. The use of uncrushed coal reduces elutriation of unburnt coal fines, and allows bed height and overall height to be reduced.

Combustion of Oil

Interest in combustion of oil in fluidized beds is due to their capability of burning the entire range of fuel oils from gas oil to vacuum residues and of reducing the emission of sulphur dioxide by addition of limestone (as discussed later)[4]. Because oil volatilization occurs rapidly and lateral gas mixing in fluidized beds is poor, the distributor plate system for oil and air tends to be more complicated than that for coal firing. With a suitable system, however, even heavy oils can be burnt efficiently in a bed of refractory (e.g. sand) particles.

Combustion of Gas

Interest in combustion of gas is due to the high heat-transfer characteristics in fluidized beds in comparison with radiation from a gas flame. As with oil firing, the poor lateral mixing of gas in the bed makes it essential that the gas should be uniformly distributed. This is most readily achieved by mixing the gas with the fluidizing air in the air plenum chamber below the distributor. The distributor must be designed to prevent any possibility of burn-back from the bed into the plenum chamber. With a premixed gas mixture, combustion in the bed is extremely intense. Complete combustion can be achieved with low excess-air levels in beds of refractory only 50 mm or less deep at fluidizing velocities up to and exceeding the normal flame velocity[7].

Combustion of Waste Materials

Fluidized beds may be used for incineration of waste materials, with or without heat recovery; e.g. oil-refinery and coal-preparation waste sludges, municipal garbage, and sewage sludge. The usual requirement is the production of a dry inert solid for disposal, and high throughput rather than combustion efficiency is the most important requirement. Aqueous sludges are best sprayed on to the bed surface in order to obtain some evaporation in the freeboard and to reduce the risk of sticking bed particles together.

Heat Transfer in Fluidized Beds

Convection

The high heat-transfer rate to surfaces immersed within fluidized-combustion beds, e.g. boiler tubes or metal objects for heat treatment, represents a major advantage of the system. Good convective heat transfer is created by the turbulence of the bed, which constantly replaces the cooled particles against the immersed surfaces with particles at the bulk bed temperature. Since the convective-heat-transfer mechanism is effectively conduction across a gas film between the hot particles and the surface, the heat-transfer coefficient is determined by the thickness of the layer and the mean gas conductivity. Thus the coefficient increases with finer particles, owing to the thinner gas film, and higher bed and surface temperatures owing to increased gas conductivity. Provided that the bed is turbulent, fluidizing velocity is not important except in so far as it determines the particle size for the bed. Convective heat transfer coefficients are usually within the range 150 to 400 W/m^2 K. (See also App.C3.)

Radiation

The importance of radiative heat transfer increases rapidly with increasing bed temperature; it may account for 20 to 50% of the total. The radiation is from the particles around the immersed surfaces, and the mean temperature of these particles may be somewhat below that of the bed.

Operation at Elevated Pressure

Pressurized combustion is being developed primarily for power generation[2]. The hot high-pressure combustion gases can be expanded through a gas turbine which drives the air compressor and generates power. Fluidized combustion offers a practical method of coal-firing an open-cycle gas turbine. With other coal-combustion methods the fly-ash is abrasive and the gas contains unacceptable compounds, particularly vaporized alkali-metal salts. In earlier coal-fired gas turbines this resulted in excessive erosion and/or deposition in the turbine. Because of the low operating temperature in fluidized combustion, however, the ash is friable and most of the alkali-metal salts are retained in the ash.

Operation at elevated pressure increases proportionately the air supply rate at constant fluidizing velocity and thus allows the combustion rate also to be increased in proportion. The area of heat-transfer surface in the bed must also be increased in proportion to operating pressure, and the bed is consequently deeper than in combustion at atmospheric pressure. Pressure *per se* has no significant effect on heat-transfer characteristics in the bed or on combustion efficiency, except in so far as the latter is improved by the deeper beds.

Reduction of Sulphur Dioxide Emission

A major advantage of fluidized combustion is that sulphur dioxide emission can be reduced simply by adding limestone or dolomite to the bed. Most of the research on this has been with coal firing[8], but tests with oil have shown similar effectiveness[4]. The theoretical additive requirement is 3.15 kg of $CaCO_3$ or 5.75 kg of $CaCO_3.MgCO_3$ (the $MgCO_3$ component of dolomite does not take part in the reaction) per kg of sulphur, i.e. additive feed rates of 3.15 or 5.75% of the fuel feed rate per 1% sulphur in the fuel. In practice, under the most favourable operating conditions of low fluidizing velocity, efficient fines recycle and a temperature of about 800 to 850°C, the theoretical addition retains about 80% of the sulphur and double the theoretical rate retains about 95%. The retention efficiency falls rapidly at higher or lower temperatures, at higher fluidizing velocity, and without refiring of unreacted additive fines. Limestones and dolomites from different sources vary greatly in their effectiveness in reducing sulphur dioxide emission because of differences in the pore structure produced on calcination.

In a pressurized combustor the higher carbon dioxide partial pressure prevents calcination of calcium carbonate. With limestone a porous structure is thus not formed and reaction to calcium sulphate takes place only at the surface, giving low effectiveness, although abrasion may slowly expose unreacted calcium carbonate. However, the $MgCO_3$ component of dolomite is calcined, and this produces a satisfactory pore structure in some dolomites. Thus even though only half of the dolomite is effective in reducing sulphur dioxide emission, the additive rate required is lower than for limestone when operating at elevated pressure.

Control of Other Pollutants

Nitrogen Oxides

The low combustion temperature in fluidized beds reduces the formation of nitrogen oxides from all fuels. Most of the NO_x produced in the bed is derived from the nitrogen in the fuel. For combustion of coal containing about 1% N, the NO_x concentration is in the range 50 to 200 ppm v/v in the pressurized combustor and somewhat higher at atmospheric pressure. With residual-oil firing, NO_x concentrations are typically 100 ppm, and with gas, values as low as 20 ppm have been measured.

Particulates from Coal Combustion

The fly-ash from fluidized combustion is not so fine as from pulverized-coal combustion, and because it has not been formed by solidification of molten ash it is less abrasive. For some applications there is a good chance that emission standards can be met by cyclone separators only.

Alkali-metal Salts from Coal Ash

The emission of alkali-metal salts, which may condense in boilers and on gas-turbine blades, causing fouling and corrosion, is substantially lower than in conventional combustion systems because of the low combustion temperature.

Vanadium from Oil

The emission of vanadium from oil is substantially reduced without additives: typically, less than 2% of the vanadium present is released in the gas.

Design and Operating Considerations

Selection of Fluidizing Velocity

The fluidizing velocity is the most important design factor and must be optimized. Operation at low fluidizing velocity gives high heat-transfer rates, good combustion efficiency in shallow beds, good sulphur retention, but a large bed area. Operation at high velocity gives a smaller bed area but reduces efficiency of combustion and sulphur retention. Deeper beds may be needed to pack the heat-transfer surface into the smaller bed area, and additional heat-transfer surface is required because of reduced heat-transfer rates.

Start-up

Coal starts to burn in a fluidized bed at about 500°C; oil and gas burn in the bed from about 650°C. For start-up, the entire bed or a start-up compartment must be heated to the appropriate temperature before oil or coal is fed to the bed. The simplest heating method is to direct a high-intensity oil or gas flame on to the bed surface while it is gently fluidized. It is desirable to use the minimum air for fluidization during start-up in order to reduce heat loss.

An alternative procedure, particularly suitable for gas-fired appliances, is to fluidize from cold with an air/gas mixture. The gas initially burns above the bed and then combustion moves back into the bed as it heats above 600°C. For large particles, the air flow needed to fluidize when cold may give a hot-gas velocity above the bed in excess of the flame velocity. In this situation, or as an alternative method for finer particles, the air/gas supply may be set at a velocity such that the bed is static but would be fluidized if it were hot. The gas burns above the static bed, which is progressively fluidized as it becomes heated throughout its depth.

In some experimental test-plants the bed has been heated by using a hot fluidizing gas, but this is not likely to be used commercially because it complicates the design of the air distributor.

Control and Turn-down

In order to achieve good combustion efficiency and sulphur retention it is preferable that the bed temperature should be within the range 750 to 950°C. For applications in which the bed contains fixed heat-transfer surface, particularly boilers, the heat-transfer rate can only be reduced by a reduction in bed temperature. (The heat-transfer coefficient is independent of fluidizing velocity.) Depending on the maximum operating temperature, the output from a fluidized-combustion boiler unit can only be reduced to 70 or 80% of full load before the temperature becomes too low to support combustion when the fuel- and air-feed rates are further reduced. In order to achieve the turn-down ratio of 5:1 or more required in most boilers, it is necessary to reduce the area of heat-transfer surface during operation. There are two methods of achieving this. The first is to divide the bed into cells or compartments and to fluidize the appropriate number to meet the load. The second is to reduce the effective heat-transfer surface area as the output is reduced. This may be achieved automatically since tubes near the surface of the bed will be progressively uncovered by the contraction of the bed as the fluidizing velocity is reduced. Alternatively, particularly for the deeper beds of pressurized combustors, the bed depth may be controlled by withdrawal or replacement of bed material.

Combustors without heat-transfer surfaces, e.g. incinerators and hot-gas generators, will usually be operated with excess air to remove the heat generated in the bed, and the excess-air level may be used to control bed temperature. The only restriction on turn-down is that a fluidized bed must be maintained.

Applications of Fluidized Combustion

Power Generation

Most of the research and development on fluidized combustion have been directed towards its use in power generation, and many design and costing studies have been carried out. Most of the designs involve several separate fluidized beds in order to achieve control and turn-down. For example, a design for a 660 MW$_e$ mixed steam/gas-turbine cycle plant[6], prepared by the Westinghouse Electric Corporation for the U.S. Environmental Protection Agency, involves four vertical pressure vessels each containing four beds, as shown in *Figure 2*.

The design studies indicate that an atmospheric-pressure combustor with a conventional steam cycle would give savings of about 10% in boiler capital-cost and power-generation cost in comparison with a conventional plant, both for 140 and 660 MW$_e$ units. For a pressurized combustor with a mixed

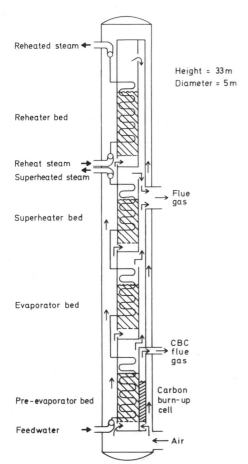

Reheated steam ←

Height = 33 m
Diameter = 5 m

Reheater bed

Reheat steam →
Superheated steam

Flue
gas

Superheater bed

Evaporator bed

CBC
flue
gas

Carbon
burn-up
cell

Pre-evaporator bed

Feedwater →

← Air

Figure 2   Westinghouse pressurized combustor concept

steam-cycle and open-cycle gas turbine, the gas turbine driving the air compressor and producing about 25% of the power, the designs indicate slightly increased savings at 140 MW and savings of about 20% in boiler capital-cost and 10 to 14% in power-generation cost at 660 MW. Complex cycles involving both open- and closed-cycle gas turbines have also been considered[3], particularly for sites where availability of cooling water is limited.

Industrial Boilers

Boilers for industrial steam-raising or heating, of either the watertube or shell type, can be fired by fluidized combustion. Shell boilers could be either vertical, as shown in *Figure 1*, or horizontal. Fluidized combustion should allow a reduction in the size of boiler for a given output.

Other Applications

Fluidized combustion could be used to provide heat for many industrial processes. One potential application is the generation of a hot gas for drying. Another application could be the provision of heat for coal gasification, possibly by combustion of low-grade ungasified residues.

The good heat-transfer properties of fluidized beds can be utilized in the heat treatment of metal objects. A gas-fired fluidized combustion unit is manufactured for high-temperature heat treatment[7].

The ability of fluidized combustion to burn variable and low-grade fuels makes it suitable for incineration applications, both with and without heat recovery. Units have been operated to burn sewage sludge, domestic garbage, oil-refinery waste and aqueous slurries.

References

1 Skinner, D. G. *The Fluidised Combustion of Coal*, Mills & Boon, London, 1971
2 Hoy, H. R. and Roberts, A. G. *Amer. Inst. Chem. Engrs Symposium Series*, Vol.68, No.126, 225
3 Locke, H. B. *J. Inst. Fuel* 1974, Vol.47, 190
4 Locke, H. B., Lunn, H. G. and Roberts A. G. *3rd Int. Conf. on Fluidized-bed Combustion*, Session V, Paper 5, U.S. Environmental Protection Agency, 1972
5 Bishop, J. *2nd Int. Conf. on Fluidized-bed Combustion*, Session IV, Paper 4, U.S. EPA, 1970
6 Westinghouse Electric Corporation *'Evaluation of the Fluidized-bed Combustion Process'* U.S. EPA, CPA 70—9, 1971
7 Elliott, D. E. *3rd Int. Conf. on Fluidized-bed Combustion*, Session IV, Paper 1, U.S. EPA, 1972
8 Highley, J. and Wright, S. J. *3rd Int. Conf. on Fluidized-bed Combustion*, Session I, Paper 4, U.S. EPA, 1972

General Reading

Skinner, D. G. *The Fluidised Combustion of Coal*, Mills & Boon, London, 1971

Chapter 6

Solid-fuel firing

6.1 Fixed-bed firing

Factors in selecting stokers
A note on principles
Types of stoker
Coal handling and ash removal
Selecting a stoker

6.2 Pulverized-coal firing

Coal preparation
Supply of coal and air to the furnace
Burners
Combustion and behaviour of mineral matter
Possibilities for achieving more intense combustion
Recapitulation

6.3 The cyclone furnace

Types of cyclone furnace
Some comparisons with dry-bottom p.f. boilers
Suitability of coals

6.4 Garbage (town refuse) incineration

Heat value in refuse
Reasons for incineration
Cell-type incinerators
Typical grate systems in modern use
Parameters of combustion (residue standards)
Gas cooling and cleaning
Incineration without a grate
Heat recovery
Problems encountered in modern practice
Capital and running costs

6.1 Fixed-bed firing

Historically, all methods available for the combustion of coal unless in suspension are derived from the burning of a pile of coal lumps, either on the ground or raised on a simple open grate, built up and replenished by hand. The open domestic fire is the principal remaining example of this procedure, justified now only on aesthetic and physiological grounds rather than on its economic and technical advantages. In fact there is no other justification, especially in these days when it is realized that fuel should never be used wantonly, for hand-firing any combustion appliance, and no consideration is given to hand-firing in this Chapter.

It was a logical progression from the simple hand-fired natural-draught fire, first to attempt some control over the amount of air supplied to the bed and its distribution, and secondly to develop ways of transferring the coal mechanically into the burning bed and of removing the ash from the combustion zone so that the bed could continue to burn steadily without clogging. The earliest attempt to use forced draught appear to have been made by Gordon in 1825, and mechanical stokers first appeared about the same time[1], as can be seen from *Table 1.*

It was found possible to obtain a much better combustion efficiency with these mechanical stokers and to maintain and control the rate of heat release to the required level, regulating it up and down as required. Mechanical stokers established themselves in watertube boilers before the start of this century, their use becoming more widespread and extending down to smaller sizes of plant over the next half century. In the UK, stoker-fired watertube boilers for power generation were only superseded by pulverized-fuel firing after the second world war; almost at the same time, stokers were being widely and rapidly introduced into the shell-boiler field. Developments in stoker firing were still being made up to the 1960s when, in the UK as in many other countries, oil-firing made such

inroads into the firing of shell-type and other industrial boilers that there was little incentive for further work or development on coal-firing equipment.

There is a tendency these days to view the firing of any solid fuel in a fixed bed as out-of-date but this is a superficial judgement. In the smaller boiler sizes, as for example in the domestic and small commercial range, there is as yet no established alternative to burning the fuel in some form of fixed bed. Even in larger sizes, through at least the size of units covered by shell boilers, stokers incorporating fixed fuel beds are capable of giving economically as high a level of performance and reliability as any other combustion system and should be taken into consideration when deciding how such boilers should be fired.

A major subject of comparison must be between stoker firing and pulverized-coal firing. (The comparison with fluidized-bed combustion is discussed later, where the argument is developed that fluidized-bed combustion can be considered as a development of fixed-bed firing.)

The principal arguments for the use of stoker-firing rather than p.f. firing are that a stoker can:
1. burn direct a coal as delivered without the need of further grinding;
2. achieve in most cases a low level of grit and dust emission, sufficient to satisfy air pollution standards, using a simple cyclone or other inertial-type collector whereas pulverized-fuel firing generally needs an electroprecipitator to meet similar standards;
3. be readily banked to maintain some heat input even when not operating and can often be hand-fired in emergency. Stokers are easy to understand and operate. Controls are simple.

Against these arguments, a stoker:
1. generally contains relatively bulky moving parts, often in contact with the burning coal, leading to high maintenance;
2. takes up a proportion of the furnace volume. This, together with the limitations in the heat-release rates per unit area of grate that are inherent

Table 1   Development of mechanical stokers

1816	Mechanical coal thrower	:	John Gregson	(inventor)
1816	Underfeed stoker	:	John Hawkins & Emerson Davison	(patentees)
1819	Revolving grate	:	William Brunton	(inventor)
1822	Sprinkler stoker	:	John Stanley	(inventor)
1833	Underfeed stoker (first practical design)	:	Richard Holman	(patentee)
1834	Horizontal perforated cylinder stoker	:	J. G. Bodmer	(patentee)
1841	Chain stoker	:	John Jukes	(patentee)
1847	'Drunken screw' stoker	:	J. G. Bodmer	(patentee)
1860	Improved mechanical stoker	:	John Jukes	(patentee)
1874	Mechanical stoker	:	John West	(patentee)
1882	Babcock & Wilcox installed first mechanical stoker in watertube boiler (Hamilton Palace Colliery)			

in fixed beds (see App.C4) means that the output rates are low for a given size of furnace (see *Table 4*).

Arguments have been put forward that it is possible to burn a wider range of coals and higher-ash coals in pulverized form: also that the rate of response of a pulverized-fuel flame is much faster than can be achieved from a grate. Such sweeping statements are hard to prove or deny but evidence will be presented later in this Chapter which tends to refute these claims, at least as *general* conclusions.

Factors in Selecting Stokers

It will be useful, before reviewing the various types of stoker that are available, to list first the factors that have to be considered in selecting the most suitable stoker for a specific application, i.e. coal type, factors affecting costs, ease of operation, and environmental considerations.

Coal Type

No stoker will burn all types of coal and it is necessary to know the range of rank, size grading and ash of the type that it is intended to burn. The rank of a coal is a classification in which all coals of similar rank can be expected to behave in a similar, predictable manner. *Table 2* gives a simplified version of the National Coal Board Grouping as used in the UK. It is based on the volatile matter of the

coals expressed on a dry-ash-free basis and the caking properties of the clean coal. It should be noted that high-rank coals of high carbon content have low NCB Rank numbers (see also App.A6 Tables 7 and 9, App.D3).

The caking properties are important in combustion, as the swelling that takes place and the clinker that is formed — which can vary from large fused masses to light friable powders — largely determine the completeness of combustion in fixed-bed systems and also whether or not the ash is discharged satisfactorily away from the combustion region. Many methods of defining caking properties are available, related to the strength of the coke formed or the extent of swelling taking place. The most useful and widely adopted test is probably the Gray—King assay where a coke sample is prepared under standard conditions and the swelling defined by visual comparison against standards.

Figure 1 shows the general relation between caking and standard volatile matter. It will be seen that where the volatile matter exceeds 30%, coals with widely different caking properties can be found at any given volatile matter; this is reflected in the NCB classification as shown in *Table 2*.

The size of the coal is important. In general, the larger-sized coal is more expensive though not always the easiest to handle or burn. In mechanical stokers, the important differentiation is between graded coal, having a limitation on both maximum

Table 2   National Coal Board (UK) coal classification

NCB* rank	Type of coal*	Caking properties	Gray—King coke type	Volatile matter (daf, %)	Typical calorific value (MJ/kg,daf)
Wood	—	—	—	85—90	19.8
Peat	—	—	—	65—70	22.0
Lignite (brown coal)	—	—	A	55—60	26.7
900	High volatile	Non-caking	A—B	>30	32.5
800	High volatile	Very weakly caking	C—D	>30	
700	High volatile	Weakly caking	E—G	>30	
600	High volatile	Medium caking	G_1—G_4	>30	
500	High volatile	Strongly caking	G_5—G_8	>30	
400	High volatile	Very strongly caking	>G_8	>30	35.0
301	Medium volatile	Strongly caking	>G_6	20—30	36.4
300	Medium volatile	Non-to-weakly caking	A—G	20—30	
206	Low volatile	Non caking	A—D	10—20	
204	Low volatile	Strongly caking	G_5—G_8	17½—20	36.9
203	Low volatile	Medium caking	G_1—G_4	15½—17½	
202	Low volatile	Weakly caking	C—G	14—15½	
201	Low volatile	Non-caking	A—B	9½—14	36.4
100	Anthracite	Non-caking	A	Less than 9½	36.4

*   *For details of the Gray—King method, for information on the 'International' and other systems for classifying coals, and for further information on coal properties and characterization, see refs (A)—(C)*

and minimum particle sizes, and 'smalls' which have a specified upper size only. The grades of sized coals available in the UK are shown in *Table 3*: singles and doubles are the most readily available. The 'smalls' available for stoker firing have generally a top size of 25 to 50 mm. Handling and combustion of these coals is affected by the amount of fine coal (less than 3 mm) present. *Figure 2*[2] shows for example the effect of the combination of moisture and fine material on the flow of coal from a bunker and illustrates how, even with coals of high fines content, handling can be achieved if the free moisture is low. Alternatively, special handling plant for these difficult fuels is available (see later).

Coal may be 'washed' or otherwise treated, i.e. put through a cleaning process which removes much of the loose dirt. Washed coals have obvious advantages of leaving less ash for handling or disposal. In a few appliances high ash may prevent a good burnout of the coal, although too low an ash level can give trouble with the grate bars since ash often acts as an insulating layer between them and the burning bed.

Unwashed coal has the big advantage of a lower cost, but the high (up to 20%) and (often more important) variable ash limits the stokers on which it can be used. There is also a larger cost for handling equipment.

In general, it is the amount of ash formed that is important, not its composition. The composition of the ash may, however, be significant if:

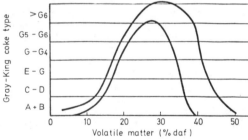

Figure 1   Relation between caking and volatile matter

Table 3   Sizes of graded coal marketed by NCB in UK

	Permitted range of sizes (mm)*	
Name of Group	Upper limit	Lower limit
Large Cobbles	150–200	75–130
Cobbles	100–130	50–75
Trebles	64–90	40–50
Doubles	45–57	25–40
Singles	25–40	13–25
Peas	13–20	6–13
Grains	6–11	3–6

*   *Round-hole screens*

References p 91

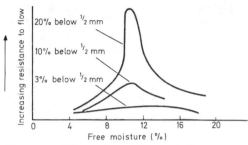

Figure 2   Effect of moisture and fines content on discharge of small coal from bunkers[2]*. Courtesy Inst. Fuel, London*

1. the ash-fusion temperature is abnormally low so that a semi-fluid ash layer can form as a blanket over — and even in — the grate;
2. the chlorine content of the coal (a useful criterion of the amount of alkali salts present) is above 0.3–0.4% when in some circumstances deposits can form in the combustion chamber and smoke tubes. There are ways of reducing this deposit formation by the use of baffles, recirculation of flue gas, or injection of compressed air or steam at the back of the grate[3,4] etc. The main effect of all these procedures is to lower the gas temperature, particularly at the point where it first impinges on metal surfaces;
3. the sulphur content is so high that sulphuric acid can condense on metal surfaces below about 180°C, corroding them. The conditions under which this can occur have been extensively studied; corrosion from this cause is very rare in shell boilers but can occur[5,6] in watertube boilers owing to the different furnace configuration. Methods of mitigating this trouble other than using a fuel of lower sulphur content are available[7].

Factors influencing Costs

Both capital and running costs need to be considered. The capital cost of the boiler and stoker combined may be only half or less of the total cost of the installation including coal and ash handling. Even so, the cost of the former units (which in this case at least should be considered together) is important. The most important factor affecting cost is the overall size of the boiler, if of the shell type. With watertube boilers the heat-transfer area (length of steam tubing) is also important.

A major disadvantage of a stoker-fired unit as compared with boilers fired with p.f., oil or gas flames is that the stoker masks part of the furnace enclosure that would otherwise be available for heat transfer. In a shell boiler, a third of the available heat-receiving area may be shielded in this way. Some typical heat-release rates are given in *Table 4*, equivalent values for oil- and gas-firing being also given. The tendency in modern designs of stokers will be to increase the heat release rates in coal-fired plant towards that obtainable with the other fuels.

It is also necessary to balance the higher initial cost of the additional complexity of a stoker designed to give both a high flexibility and high automaticity against the savings in the range of coals that can be burnt on the stoker and in the cost of the labour that might otherwise be needed at regular intervals.   While the maintenance of some stokers is relatively high, particularly where metal parts are likely to come into contact with the hot bed, the more sophisticated designs are not necessarily associated with higher maintenance costs.

Ease of Operation

It is necessary to consider what performance a stoker will be called upon to achieve. All stokers will operate over a range of loads, though in most cases with some loss of efficiency as the load decreases. Often however a turndown ratio of 3:1 from full to the lowest load at which the plant is required to operate continuously is sufficient, especially if one can also 'bank' the boiler — that is to say, so operate the stoker that the coal on it stays ignited but does not give out appreciable heat. It is then possible to regain load rapidly as required (see e.g. the discussion on chain-grate stoker). Also, lower heat outputs can be attained by on—off operation.

Most modern stokers may be fitted with relatively simple combustion-control systems to regulate:
1. the coal or the air fed to the grate as the steam demand changes, and
2. the air or coal to maintain the required air/fuel ratio.
In many cases the furnace pressure is also balanced to keep it (usually) slightly below atmospheric to minimize inleakage of air.

The dependence of the thermal efficiency of a boiler on the excess-air rate is illustrated in *Figure 3* and discussed later, and the discussion on chaingrate stokers illustrates the high level of performance that it is possible to maintain by such relatively simple control systems.

Much of the manual effort, dirt and high maintenance associated with coal-fired stokers is due to the coal- and ash-handling plant. If it is desired to have a modern, clean and largely automatic plant, it is necessary to consider the installation as a whole.

Environmental Considerations

There is no reason why modern stokers should not meet present Clean Air legislation and (as required) such standards as may be set in the future. All modern stokers are designed not to make smoke under normal operation. Some smoke emission during light-up from cold occurs however with many stokers (and is allowed for in Clean Air legislation) while some stokers, particularly those in which occasional disruption of the bed (either manually or as part of the mechanical operation) is needed to break up clinker, may give intermittent smoke in operation.

All stokers emit some grit or dust particles into the gas stream, the quantity increasing with the more highly rated appliances. A typical carryover (chain-grate stoker) operating at MCR (maximum continuous rating) is 0.5—1% of the fuel fired (say 450—900 mg/m^3 (0.2—0.4 grains/ft^3)). The particle size of this emission, moreover, is such that a large proportion of it can be caught by a simple, and hence reasonably inexpensive, inertial collector such as a cyclone or multi-cyclone unit[8].

A Note on Principles

Some notes on the fundamental considerations of fixed-bed firing are given in Appendix C4. It is useful, here, however to summarize what has to be achieved in a successful design of stoker by considering first a small bed of coal in a cylindrical pot on a grate with combustion air passing up through the bed. The coal may be ignited at the top or the bottom. If lit at the bottom, the air flow helps the flame front to travel through the bed, but volatiles released by heat above the flame front are entrained by the flue gases without passing through a hot zone

Table 4   Typical heat-release rates

	Fixed-bed	Oil	Gas
Shell boilers			
per unit volume of furnace tube (kW/m^3)	300—400	600—1000	500—750
per unit area of total furnace tube surface (kW/m^2)	90—100	180—210	150—170
Watertube boilers			
per unit volume of furnace chamber (kW/m^3)	300—350	450—500	450—500
per unit area of projected watertube surface (kW/m^2)	200—220	200—220	200—220

in the bed. If the bed is lit at the top, the ignition plane has to travel down against the direction of air flow. Ignition is thus more difficult to establish, though once achieved the subsequent spread of the combustion will proceed irrespective of the air velocity. Increasing the air velocity will lead to more of the coke residue above the flame front burning out simultaneously with the travel of the ignition plane through the bed. The volatiles have to pass through the burning coke above the ignition plane and are more likely to be consumed. In either case, once the volatiles have been mostly released, there is a period when the devolatilized coke particles in the bed burn away. Another broad generalization, common to almost all fixed-bed systems, is that for a major part of the combustion process the principal product is carbon monoxide and not the dioxide and this has to be burnt at or close to the bed surface. A successful stoker has therefore to be designed to ensure:
1. stable ignition,
2. means for burning the volatiles released,
3. means for burning CO at or above the bed, and
4. means for feeding fresh coal to the bed, ensuring sufficient time/space in the combustion zone to allow the coke residue to burn off before the ash is removed.

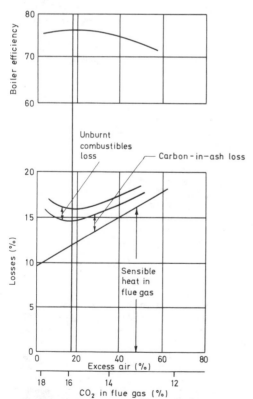

Figure 3   Influence of excess air on combustion efficiency (chain-grate stoker, bituminous coal). Courtesy Inst. Fuel, London

References p 91

Stable ignition requires that the incoming coal is heated to ignition temperature by the already burning fuel, either by direct contact or by radiation, e.g. from a refractory arch over the bed at the ignition end.

Combustion of both gases and volatiles above the bed requires excess air, air turbulence and adequate gas temperature. There is often sufficient air present either by leakage through the bed, particularly along the stoker walls or through holes in the bed, or if the furnace chamber is under suction by inleakage through the front of the furnace tube (shell boilers) or holes in the casing (watertube boilers). It is for this reason that a balanced draught is used on many stokers. The need for balanced draught is reduced in modern, better-sealed boilers where inleakage can be controlled. Moreover it is better to inject secondary air (needed, it can be argued, on all stokers) at a velocity and in a position where it will mix with and increase turbulence in all the gases leaving the bed; good examples are described later under spreader stokers and Vekos-type stokers.

It is reasonable to expect a modern stoker to burn over 90% of the combustible in the coal with an excess air of not more than 40%, giving a dry-flue-gas loss less than 15%; though a lower performance than this might be acceptable in special circumstances if capital cost and simplicity of design and operation are paramount considerations.

There is an optimum excess air, which will vary from stoker to stoker and is dependent on the boiler (amount of convection surface), at which the sum of the heat losses in unburnt fuel and flue gas is a minimum. *Figure 3* shows a typical example for a chain-grate stoker in a modern shell boiler where, at an excess air of less than 20%, over 92% of the fuel is burnt. Typically, such an installation should operate at a boiler efficiency of about 80%: though a boiler efficiency of say 75% must still be considered good for coal-fired shell boilers in general.

In order to convert the simple bed of coal used as an illustration so far into a continuously operating appliance, it is necessary to develop some way to replenish the coal as it is burnt away — and it is the way in which this is done that determines in a large measure the design of the stoker and the range of conditions under which it can be used.

Broadly speaking, there are three alternatives. First, new coal can be fed in through the side of the bed either along the whole height of the bed or, say, by pushing it into the bottom or top half only. In all cases, it will be necessary for the burnt or burning coke to be moved out of the way, either by using the force of the incoming green coal, by mechanical means, or by the use of gravity. New coal can, as another alternative, be dropped onto or fed to the top of the bed. Finally, one can consider forcing new coal up into the bottom of the bed, pushing the burnt or burning coal upwards and over the top of the bed.

A number of designs of stoker involving the above principles will now be described.

Types of Stoker

Gravity-feed Stokers

Only anthracite and coke are sufficiently free flowing throughout the combustion process to be used
in stokers which rely entirely on gravity to ensure
the flow of fuel. Such stokers are therefore in
general restricted to the smaller applications (say
up to 750 kW) where their simplicity and cheapness
offset the higher cost of the fuel.

In the most common form (of which *Figure 4* is
an example) the fuel contained in a hopper above
the combustion region is allowed to flow down an
inclined grate, usually water-cooled, on which it is
burnt. The bed is held in position by the ash which
collects at the bottom of the grate. This ash is removed, manually or automatically, at regular intervals by some form of ram or scraper such that more
fuel feeds onto the grate.

Alternatively, the fuel may be fed into a static
bed located in either a vertical or horizontal cylindrical cooled chamber where it is allowed to burn
until a substantial clinker has built up which is then
removed manually from the bed. Under suitable
conditions this clinker removal need only be carried
out after at least 8 h or so, though this depends on
the nature and ash of the fuel. In particular, as a
clinker rather than a powdery ash is required if it is
to be removed manually, the ash-fusion temperature should not exceed 1300°C.

Combustion air is supplied through the grate
and, in most cases, secondary air is also supplied to
burn up the combustibles, particularly the carbon
monoxide, released from the bed. Fairly simple
but effective combustion control systems are available in which the quantity of combustion air (and
hence the rate of combustion of fuel and the heat

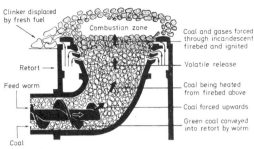

Figure 5   Underfeed stoker

release) is controlled by regulating the forced-
draught fan from an indicator of heat requirement.

A good review of alternative forms of gravity-
feed boilers will be found in reference 9.

Underfeed Stokers

These are the simplest and cheapest class of stoker
for burning coal. In underfeed stokers (*Figure 5*)
the green (raw) coal is forced up into the bottom
of a burning bed with the result that the burning
coal is moved up and eventually falls away sideways, burning away on the outer edges of the bed
from where the ash or clinker is normally removed
periodically and manually. Combustion air is supplied from a forced-draught fan through tuyeres in
the upper part of the bed container.

While stokers are used in sizes up to 5 MW, about
85% of those in the UK fall within the size range
200–1000 kW, where their low capital and running
costs are most appreciated. A full-time attendant
is not needed whilst the simplicity of the appliance
means that maintenance is low and does not need
a highly skilled engineer.

On the debit size, underfeed stokers are selective as to the fuels they can burn, can give smoky
combustion, and have not yet been made fully
automatic. Underfeed stokers to burn coke (though
abrasion, particularly of the screw, is a serious problem) and anthracite (using water-cooled tuyeres)
have been built. Most underfeed stokers, however,
are designed to burn a coal that:
1. is weakly caking (NCB Rank No.201/202 or
700 to 900) so that the coke falls away from the
centre of the bed and does not form a large mass,
2. is low in ash (less than 10%) to reduce de-ashing,
3. has a clinkering temperature above about 1200°C
(Rank 700/900) or 1350°C (Rank 201/202) to prevent clinker forming in the retort. The clinkering
temperature must not, however, be so high that a
powdery ash rather than a clinker is formed along
the sides of the grate, and
4. is preferably sized 25 to 40 mm depending on
the size of the screw, though small coal may be
burnt if it does not contain excessive fines.

The air supply is regulated by the fan speed or,
more normally, by a damper in the forced-air duct.
By linking this to the screw feeding in the coal, correct coal/air rates can be maintained. Turn-down

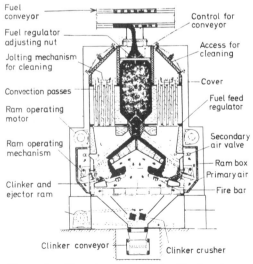

*Figure 4   Magazine or gravity-feed boilers (cour-
tesy Trianco Ltd)*

to about 20% of full load can be achieved either by simple on—off controls, or by fully modulating systems including the control of air/fuel ratio. Underfeed stokers should not however be operated for long periods at low loads as this can lead to burning back, deterioration of the casting, and clinkering and coking in the grate.

Secondary air is now a common feature to control smoke emission, particularly during on—off operation or de-ashing. Automatic de-ashing has and is being investigated, and it is no doubt possible to devise a fully automatic underfeed stoker, though at the cost of some complexity which may nullify the basic advantages — simplicity and low cost — of this widely used design.

Reference 10 — though dated 1954 — reviews in depth the basic principles of the small underfeed stoker.

Overfeed Combustion Units with Static Beds

These include all units consisting of a stationary grate to which fuel is fed by falling through a drop-tube, possibly with some distributor such as a cone to attempt to feed the coal evenly over the bed. In the Vekos system, for example, secondary air fed in an annulus round the coal drop-tube is made to swirl to aid distribution. This design of stoker is generally part of an integrated boiler unit, as shown in *Figure 6*[11]. As in the form of gravity-feed stoker discussed earlier, the ash builds up a clinker over the bed. In the earlier designs as illustrated, this is removed manually at intervals of 8 h or more depending on the amount of ash. With a low-ash coal (2 to 3% is required to protect the grate bars) manual de-ashing can be reduced to once a day. The more modern designs incorporate automatic ash removal. Other designs, such as the 'Segor', developed in France, use a deeper fuel bed rather like a simple gas producer, the ash being drawn out of the bottom of the bed by a revolving scraper system. It is claimed that the heating the coal is subjected to as it drops on to the bed reduces its tendency to swell. The 'Segor' design also utilizes another device: recirculating part of the flue-gas back through the bed. This procedure

lowers the bed temperature (and may have other effects on the reactions in the bed) which helps to reduce clinkering. (Flue-gas recirculation can also reduce the volatilization of alkali salts from deep fuel beds, so reducing deposit formation on the boiler-tubes in watertube boilers[5]. Flue-gas recirculation has also been found useful in reducing deposit formation in shell boilers, but here the benefit is due more to lowering the gas temperature than to changes in the chemical reactions in the bed[3].

In all these designs (as in any stoker where the fuel is dropped or spread on to the bed) the carry-over of particulate matter from the grate is high. This can be minimized by the use of a graded fuel — and the smaller graded fuels should certainly be used in these appliances. Even so it is usually necessary, as in the Vekos system, to collect at least the larger carryover particles in a cyclone or similar device and to refire them.

Spreader Stokers

These stokers also feed the coal onto the grate by distributing it onto the top of the bed through the furnace space. However, for the first time mentioned in this review of different stokers, the coal on the bed is moved horizontally across the area of the grate so that the burnt ash is all brought to one point where it is discharged from the bed.

There have been many variants of spreader stokers developed over the years; the design described here is the latest. The fuel is sprinkled from one end of the rectangular bed across the bed surface by means of a rotating impeller. The finer coal particles burn in suspension while the larger pieces fall on the grate to form the bed. The grate is in the form of a continuous chain of links or bars, running over driving sprockets at the front end (from which the coal is fed) and over a skidplate or freely turning spindle at the other end. The grate revolves in such a way that the burning coal moves towards the front of the appliance, so that the ash is discharged at the same end as the coal is fed in. The convenience of this in fitting the stoker into a boiler, particularly of the shell type, will be obvious. Primary air for combustion is fed upward through the bed and secondary air above the bed. In the design illustrated the latter is injected at high pressure at the rear of the grate to reduce the carryover of fly-ash into the convection passes.

Spreader stokers have been used in watertube boilers up to about 80 MW but their main application has been and is likely to continue to be in shell boilers. They can burn a wide range of fuels, and have the advantage of being able to cope with sudden changes in fuel quality. In general, they burn:
1. all ranks of coal except anthracite,
2. preferably a small size-graded coal (singles), though smalls have been successfully used, and
3. coals with a wide range of ash, though carryover is reduced with a low-ash coal.

In fact, a high carryover of particulate matter, associated with a tendency for unburnt volatiles to

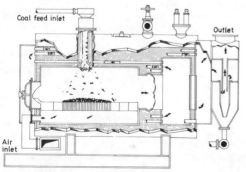

Coal feed inlet

Outlet

Air inlet

Figure 6   Vekos-Powermaster boiler with manual ash removal (courtesy Parkinson Cowan GWB Ltd)

escape, has been the main argument against spreader stokers in the past. In watertube practice grit re-firing is used, and the latest designs for use in shell boilers as illustrated, using a low-ash coal, are claimed to keep this carryover to an acceptable level.

The Coking Stoker

There have been various forms of grate in which coal is pushed horizontally onto the grate by a flat reciprocating ram. The incoming coal pushes the fuel already on the grate across it, assisted by reci-procating fire bars which move forward together taking the bed with them, and return in two stages, alternate sets of bars then moving together, in sequence.

In the coking stoker, the fuel is first deposited on a coking plate where the volatile matter is par-tially released to burn above the bed. It is then pushed by the incoming coal along and down onto grate bars as described above, where the coke burns, the ash being discharged over the end of the grate. In the earlier designs, the ram and coking plate were set fairly high relative to the grate giving a deep fuel bed at the start of the grate. But this design was largely superseded by the 'low ram' type where the drop from the coking plate to the reciprocating grate is much smaller, giving a fuel bed more suitable for burning smalls and a design which permits a wider ram (and hence a more even coal feed) in the shell boiler configuration.

Coking stokers operate best[12]:
1. with slightly caking coals, with which large coke masses do not form but the coke is sufficiently held together for the loss of 'riddlings' through the fire bars (both a loss of fuel and a material which has to be manually removed) to be minimized;
2. with singles and doubles though smalls have also been widely used. If smalls are used, it is necessary for them to be sufficiently wet for the coal to hold itself as a 'ball' when squeezed in the hand: other-wise the loss due to riddlings will be excessive; and
3. with sufficent ash to prevent overheating of the grate bars (e.g. 8% with cast-iron bars).

Control may be exercised by a simple time mechanism, or full automatic control can be app-lied, varying the grate speed or the length of travel

of the feed ram. A maximum burning rate of about 1.5 MW/m² (0.5 M Btu/ft² h) can be obtained.

Chain-grate Stokers

In stokers of the chain-grate type*, a moving grate — similar to that used in spreader stokers (described above) but turning so as to move the bed away from the coal feed end — transports the coal across the grate. A typical design is shown in *Figure 7*. The coal is fed onto the grate under gravity from a bunker, the bed depth being controlled by a guillo-tine. It is then carried through the furnace, the residual ash and clinker being discharged over the far end. To maintain ignition at the top of the bed, an ignition arch of refractory is used to radiate heat down on the bed. As the coal moves across the grate, the ignition front burns downward until, somewhere about two-thirds of the way along the grate, it reaches the bottom. The bed above the ignition plane consists of burning coke.

Chain grates have been built in all sizes up to units for use in watertube boilers of over 70 MW. Probable future applications will however be mostly limited to shell boilers.

Chain grates, in spite of a relatively high capital cost and (sometimes) high maintenance, have the advantages that they can[13,14]:
1. burn small coals very efficiently if the coal is wet enough, and
2. burn all ranks of coal except anthracite (which is both hard to ignite and has a large grit carryover).

The chain-grate stoker for use in shell boilers was the subject of considerable research and deve-lopment in the early 1960s. In addition to research investigations, the principal developments may be summarized as follows:
1. burning rate was found to be limited in practice by the amount of combustion air that could be fed through the grate before it began to carry an un-acceptable and rapidly increasing amount of mate-rial away from the bed. The primary combustion air is fed by a forced-draught fan from the front end into a windbox under the grate. Without baf-fles, most of this air passed along the windbox and up the back of the grate, thereby bypassing the regions of the bed where it was most needed. A system of undergrate baffles[15], shown in *Figure 7*, corrected this distribution and enabled the burning rate to be raised from about 1.5 MW/m² to over 2.0 MW/m²;
2. simple automatic control systems have been de-veloped and tested[16]. In these, either (*a*) the air supply and the coal supply are regulated separately by the steam demand to give a preset air-to-fuel ratio, or (*b*) the air supply is regulated to match

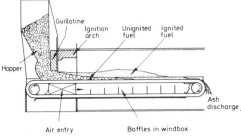

* The design of stoker mostly used in watertube boilers is more correctly termed a 'travelling-grate' stoker insofar as the grate is driven by separate driving chains down each side, rather than as in the true 'chain-grate' stoker by the grate links them-selves intermeshing with the driving sprockets

Figure 7   Chain-grate stoker[15].   Courtesy Inst. Fuel, London

the steam demand and the coal supply is regulated by the air supply. Only one other combustion control is required: to maintain a balanced furnace pressure usually slightly below atmospheric. It was found possible to maintain boiler efficiency within 2% over a turn-down ratio of 3:1 and 7% over a turn-down ratio of 8:1. It was also found that the stoker (and boiler) responded rapidly to sudden changes in load: for example changes amounting to 20—25% of full load were achieved with no significant upset in combustion conditions; and

3. it has been shown possible to build a chain-grate stoker-fired shell boiler to burn the normal run of industrial small coals automatically and at high efficiency over a range of loads of at least 8:1 [17]. The boiler efficiency—load curve that can be obtained is shown in *Figure 8*. The response to sudden load changes is as good as likely to be required and, should the boiler not be required overnight, it can be left unattended at low load or banked; it will then come on line automatically at a preset time as required[18,19].

Methods for avoiding deposits when burning high-chlorine and high-sulphur coals on a chain-grate stoker have already been referred to.

Vibrating Grates

These are somewhat similar in operation to chain-grate stokers except that the coal bed is vibrated over a water-cooled hearth rather than carried on a continuous mattress. Grates of this principle have been used in watertube boilers in the USA and in Germany: the German grates were generally sloping to encourage flow along the surface. A grate developed but never marketed in the UK[20] was horizontal and of a size and lightness suitable for use in a shell boiler. The advantages claimed for a vibrating grate as compared with a chaingrate stoker are:
1. lower capital cost and maintenance,
2. an increase in heat-transfer surface through the water-cooled hearth, and
3. an ability to burn very low- and very high-ash coals.

The Ignifluid System[21]

This system, developed in France for use in watertube boilers, accepts the fact that to increase the

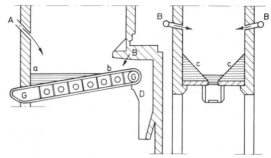

Figure 9   *'Ignifluid' furnace*
A: coal feed
a—b: bed surface
B: secondary air
c: static fuel layer protecting walls
D: clinker hopper
G—G: narrow chain-grate stoker with compartmented wind box

rate of heat release beyond that possible on, for example, a chain-grate stoker, it is necessary to increase the air rate through the bed until the particles become suspended over the grate, moving and mixing throughout the bed volume. This turbulent motion results in a high rate of chemical reaction — hence the first use of the Ignifluid system was to burn unreactive anthracite duff.

The principle of the Ignifluid combustion unit is shown in *Figure 9*. The sloping sides of the bed region cause the velocity to decrease up through the bed. The temperature of combustion is such that the ash particles clinker together in suitably-sized lumps to fall on to the narrow chain-grate at the bottom of the bed. They are then carried by this grate up through the shallow end of the bed to the point of discharge. As with any fluidized-bed system, carryover is large: the smaller particles in the bed, either fed in with the coal or formed by burnout or breakdown of particles, are elutriated by the air velocity required to fluidize the larger particles in the bed. A grit-recycling system is therefore an integral part of the system.

The most suitable fuels for this method of combustion are those graded up to about 13 mm and with only low caking properties. High ash is not a disadvantage.

Fluidized Combustion

New developments are most likely to be based on this system, and applications to the small industrial, shell and watertube boilers have been developed (as at mid 1974) and should soon become commercially available. Fluidized combustion is discussed more fully in this book (see Chapter 5) and all that need be pointed out here is that this system is, in fact, a logical progression from such fixed-bed stokers as the chain-grate. In order to reduce the size and hence cost of the overall boiler, the combustion intensity of the unit has to be increased and this

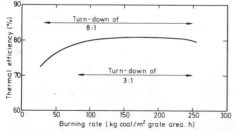

Figure 8   Efficiency—load curve for Economic boiler with chain-grate stoker[16]. Courtesy Inst. Fuel, London

can only be achieved (except by such, at present, unrealistic means as oxygen enrichment) by increasing the air rate through the bed. The effect of this on the stability of the bed is shown in *Figure 10*. Schemes to increase air rate while still holding the bed in place by, for example, various forms of water-cooled grids over the bed have been proposed and even tested, for example at BCURA. But there is logically a strong case for accepting that the use of these increased velocities will lead to fluidized beds, where advantage can be taken of the enhanced combustion intensity and heat-transfer rates to surfaces immersed in the bed; and where problems such as carryover are designed for.

Hence, the maximum combustion rates of 1.5 MW/m^2, obtained on such appliances as the coking stoker and chain-grate stoker, and increased to 2.0 MW/m^2 on the latest forms of chain-grate stoker can, by the use of fluidized beds, be increased to around 3.0 MW/m^2 (burning rate $\approx$ 360, see *Figure 10*).

The Future

There is now a bewildering number of alternative designs of stokers for burning coal in solid beds. The production of some of these and the development of new and improved stokers virtually ceased during the 1960s, and it is probable that fewer will become widely marketed in the new era of world-wide interest in coal burning from the mid 1970s onward, these concentrating on giving coal combustion a new and modern look. An attempt to compare the existing stokers of greatest interest is given at the end of this chapter; but it is likely that the development of new stokers, particularly those based on fluidized combustion, may change the position dramatically over the next few years.

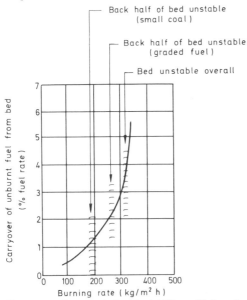

Figure 10   Increase in bed instability with burning rate (chain-grate stoker)

Coal Handling and Ash Removal

The capital cost of storage bunkers, conveyors and ash hoppers can be considerable, while the labour and maintenance costs — and the frustration — that can arise from frequent blockage in bunkers or elevators can become one of the major factors making a coal-fired boiler house unsatisfactory. Moreover, with labour costs rising, it is no longer acceptable to rely on manual coal feeding and ash removal. It is worth giving the handling side of the plant sufficient attention to ensure that the coal is fed to the boiler and the ash removed with minimum manual effort, and maximum cleanliness and reliability.

Coal bunkers should be designed to give a reliable discharge. Alternatively, if coal is stored in a pile on the ground, there are devices such as twin-screw open feeders which will lead the coal towards a centre point where it can be transferred to a feeder[22].

The coal may be transported from the storage hopper to the stoker by screw feeder[23], bucket elevator, belt conveyor or scraper conveyors of various types[24]. Screw elevators working at low speed (about 20 rpm) have been found a cheap and satisfactory way of moving small coals. Devices such as cutting away the top third of the casing and fitting a box section have been shown to reduce the risk of jamming. Bucket elevators tend to become choked if wet small coal is being handled, though there are special bucket designs which will prevent this.

Ash removal to the front of the boiler is common. On chain-grates and similar stokers, for example, a scraper conveyor is generally fitted below the grate to draw the ash to the front of the boiler. Ash handling out of the boiler house to a storage hopper can be carried out using conventional handling gear, though it is expensive. One compromise is to store the ash in a sealed bin near the boiler which can be removed at regular intervals. It is possible to move ash in a slow-moving screw-elevator of fairly heavy mild-steel construction, without excessive abrasion[22].

Selecting a Stoker

To summarize this Section, it is worth considering at least the main stokers described in terms of the factors listed earlier for consideration in selecting a stoker: i.e. coal, costs, ease of operation and environmental considerations.

Coal

No stoker will burn all coals, and it is necessary to decide first the range of coal properties to be used (bearing in mind cost and availability) and then to see what stokers they can be used on. *Figure 11* shows the ranks of coals suitable for the four main types of stoker and *Figure 12* gives similar information for fines content, ash and moisture. Fluidized-bed combustion stokers need not be selective as to rank or ash level of coal, though a washed sized coal will be used on many commercial and small industrial applications. *Figures 11* and *12* are only indicative of the general position, and later developments could well increase the range of fuels that different appliances can use. For example, the use of white-iron grate bars on chain-grate stokers allows the use of low-ash washed coals (down to 1—2%), whereas it was for a long time argued that an ash level greater than 7% was needed to protect the grate bars from overheating.

As has been said earlier, the optimum moisture content when burning smalls on chain-grate stokers or low rams is that at which the coal will 'ball' in the hand. A more scientific rule is that the optimum free moisture (that is, moisture beyond the air-dried condition) should be 1—1½% for every 10% of the coal sized less than 3 mm.

Though coals covering the range of properties shown in *Figures 11* and *12* can be burned satisfactorily, there is a penalty to be paid by way of a falling-off in boiler performance as the ash and fines increase. Systematic experiments on both chain-grate[13] and coking stokers[12] have shown that boiler efficiency can fall 0.2—0.4% per 10% increase in material below 3 mm with a low-ash coal and up to three times this for an untreated coal. No equivalent reduction in efficiency with increasing ash occurs in fluidized-bed combustion units.

Costs

It is not possible to be very quantitative on costs. Obviously, the cost of the stoker is proportional to

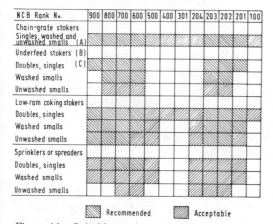

Figure 11   Suitable coal ranks for stoker firing
A: Not doubles
B: Only the 900 coals of coke type B, and 201 coals of volatile (d.a.f.) 13.5—14.0% coke type C considered for underfeed stokers; rank 203 coals are less suitable for space heating and hot-water boilers where long running without attention is required.
C: Doubles only for use on underfeed stokers burning more than 135 kg/h; singles only for stokers burning more than 9 kg/h
Courtesy The Macmillan Press Ltd, Basingstoke (Industrial Fuels, Ed. P. C. Ball, 1971)

References p 91

Stoker	Washed or unwashed coal	Fines (%<3mm) 20 40 60	Ash (%) 5 10 15 20	Moisture (%) 5 10
Chain-grate stoker	Washed			
	Unwashed			
Underfeed stoker	Washed			
	Unwashed			
Low-ram coking stoker	Washed			
	Unwashed			
Sprinkler/spreader stokers	Washed			
	Unwashed			

Recommended   Acceptable

Figure 12   Suitable fines, ash, and moisture levels for stoker firing.   Courtesy The Macmillan Press Ltd, Basingstoke (Industrial Fuels, *Ed. P.C. Ball, 1971*)

its complexity: the chain-grate is in general more expensive than for example the underfeed stoker. However, it is more realistic to consider the overall cost of manufacturing steam or hot water (say in p/1000 kg steam), taking into account the relevant interest and depreciation rates; the cost of the boiler (if new), stoker, coal- and ash-handling plant and also operating costs: fuel (which represents 70—90% of the operating costs), labour, maintenance, and running costs (power requirements etc.). Except where a boiler of adequate capacity exists, most economy in capital cost is obtained by using a boiler of the smallest dimensions to give the required output. The heat release per unit of furnace volume is therefore of crucial importance: the higher rates shown in *Table 4* relate to the chain-grate and spreader stokers and any increase over such rates is likely to be limited to these designs and newer developments such as fluidized combustion.

Ease of Operation

Ease of operation is in general only gained at a cost which has to be offset against the tangible gains in performance or output that can result and the less tangible benefits in amenity, lack of dirt and dust etc. Simple, effective automatic-control systems are available for all stokers and can always be justified in terms of improved performance. Automatic ash removal out of the boiler — not readily obtained with underfeed stokers and designs with static beds (though even here ash removal is possible) — is increasingly attractive as labour becomes more expensive.

Amenity

It may be assumed that no stoker now marketed will give rise to smoke production except possibly during light-up. The amount of grit and dust leaving the bed will vary with the design of stoker. But it may be assumed that a well operated chain-grate stoker, for example, will not have a carryover from the bed in excess of 1.5% of the fuel fired: equivalent to a concentration in the flue gas of about 1350 mg/m^3 (0.6 grains/ft^3 at NTP). It is then possible to keep within standards of 900—450 mg/m³ (0.4–0.2 grains/ft³) with grit collectors of only 30% and 60% efficiency respectively. Emission rates from the bed that are higher than this may occur on some stokers, particularly with overloading, dry fuel or bad distribution; but even so, it will be seen that it should be possible to meet emission limits such as are at present in force or contemplated, with inertial grit collectors of the cyclone type.

6.2 Pulverized-coal firing

The essential feature of pulverized-coal firing is that throughout the combustion the fuel is suspended as a cloud of small particles in the combustion air. Because they follow more or less the same path as the combustion air, fuel-particle residence times are short compared with those in fixed-bed appliances. To ensure complete burnout, and to enable the coal to be transported to the furnace in suspension, a small particle size is necessary, and it is customary to use a size consist in which between 60 and 90% passes a 75 μm aperture sieve.

The most important use of pulverized-coal firing at the present day is in steam-raising for electricity generation, but another significant use is for the firing of cement kilns. Metallurgical furnaces and shell boilers contribute small amounts to the total usage.

The reasons for the widespread adoption of pulverized-coal firing, as opposed to grate-firing, for electricity generation include the suitability of the coal feeding and firing arrangements for use in very large boilers; the strongly radiating flame resulting from the combustion of a cloud of coal particles; the suitability for operation with swelling or caking coals, and high-ash coals; and the minimization of smoke emission, which is normally due to incomplete combustion.

A pulverized-coal-fired watertube boiler incorporates the following features (illustrated in *Figure 13*): coal storage and extraction from storage; drying and pulverization; coal and air supply to the burners; ignition, combustion and radiant-cooling zones in the furnace; means for extracting the larger ash particles which separate from the gas in the furnace; on-line tube-cleaning facilities ('soot-blowers'); tube-banks for convective heat transfer; economizer; air heater; fly-ash removal equipment; fans; and chimney. Because of the relatively high capital and operating costs of some of the ancillary equipment, e.g. mills and gas-cleaning plant, the use of pulverized-coal firing for small-scale heating requirements as in shell-boiler firing has not been widely adopted.

Particular attention must be paid to certain aspects of plant design. For example, in the choice of pulverizing equipment it is necessary to consider the physical and chemical properties of the coal supplied and also the proposed design of furnace

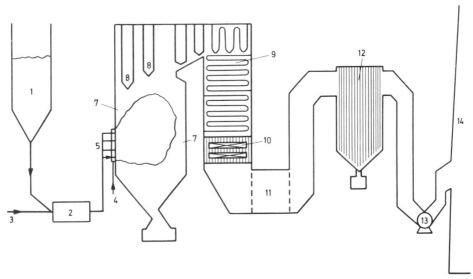

Figure 13 *Essential features of a pulverized-coal-fired water-tube boiler plant*

1	*Coal bunker*	8	*Radiant superheater panels (platens)*
2	*Feeder and mill*	9	*Convective superheater tube banks*
3	*Heated primary-air supply*	10	*Economizer*
4	*Heated secondary-air supply*	11	*Air heater*
5	*Burners*	12	*Electroprecipitator*
6	*Ash discharge*	13	*Induced-draught fan*
7	*Furnace wall tubes*	14	*Chimney*

(see Ch. 8). Other critical aspects of design are the choice of burners to ensure flame stability and avoidance of flame impingement on furnace walls; burner—furnace arrangement to ensure full utilization of furnace volume and good mixing of fuel and combustion air; and provision of sufficient radiant-heat-transfer surface area to ensure adequate cooling of the exit gases before they encounter the convective-heat-transfer surfaces.

In the following discussion, attention will be focussed on these points. For more detailed information on the design of the furnace aspect the reader is referred to Chapter 8.1 of this book and to other publications[25,26]. Gas cleaning is dealt with briefly in Chapter 19.3.

Coal Preparation

The fineness to which the coal must be ground depends on the ease with which it may be ignited, and on the reactivity of the char left after expulsion of the volatile portion. In general, the higher the rank of the coal the more difficult it is to ignite, and the longer the time required to complete combustion of the char (App.C2). These requirements can to some extent be met by the selection of suitable burners and furnace design, but for the good combustion of anthracite a very fine particle size is necessary. Dolezal[26] recommends 92% under 90 μm in size, compared with 65—72% under 90 μm for a low-rank bituminous coal and 40—58% for lignite. The CEGB[25] recommend 80—85% under 75 μm in size for high-rank coals and 70—75% for most other coals.

The moisture content of the coal supplied is usually greater than is permissible if adhesion of the ground coal to surfaces it meets is to be avoided. It is usual to preheat the air supply to the mills to between 250 and 350°C, so that the coal/air mixture after pulverization is at not more than 100°C. All the moisture content of the coal enters the furnace as vapour or as residual water in the coal, unless a system of indirect firing is adopted. In this system, the coal is separated from the air by a cyclone followed by an electroprecipitator or bag filter, and enters an intermediate storage bin from which it is removed at the required rate by a suitable feeder. Direct firing is simpler, safer, and more often used in modern power stations. In some installations, the hot air and the coal are passed through a preliminary drying duct before passing to the mill. This effects a certain amount of heat transfer and reduces the risk of fires in the mill. Coal fed to the mills should not contain lumps larger than about 20 mm diameter.

The types of pulverizer in general use are shown in *Figure 14*. They are the ball, tube or rod mill (low speed); the ring-roll type (medium speed); the ball-and-track type (medium speed); and the impact mill (high speed). In the tube mill the coal is ground by the tumbling of balls or rods in a slowly rotating cylindrical drum with a horizontal axis. The space above the charge is swept by a stream of air, and

this entrains particles which are small enough not to fall back into the charge. The ring-roll type grinds the coal between a roller and a rotating bowl or table upon which it runs, and the ball-and-track type resembles a ball race into which the coal particles fall, to be nipped between the rotating balls and their bearing surfaces. The impact mills are essentially different, in that the coal particles are directed into the path of hammers or pins attached to a disc rotating at high speed. Comminution results from the impact and not from a crushing action as in the other types.

In all types of mill, the major running expenses are power consumption and replacement of worn grinding parts. Both these items depend on physical properties of the coal — its hardness and abrasiveness, and it is recommended that expert advice be sought in the selection of equipment, materials of construction, etc., to give the cheapest possible operation consistent with the fineness required. Typical power consumptions for an average coal are shown in *Table 5*. Repair and maintenance costs are generally of the same order as power costs.

The milling of coal produces a powder containing particles of a wide range of sizes. In some plants a classifier is installed to return the largest particles to the mill. In others the action of the sweeping air is adequate to ensure the absence of oversize particles. It is convenient for some purposes to express the size distribution of the pulverized coal by a simple mathematical relation, and one commonly used is the Rosin—Rammler equation:

$$R = 100 \exp\left[-(x/x')^n\right]$$

where R is the weight percentage of particles greater than x in size, and x' and n are adjustable constants. The constant x' is a measure of the overall fineness of the powder, and n is a measure of the spread of sizes. A high value of n indicates a closely-sized powder. Typical values for pulverized coal[27] are $n = 1.2$, x' between 30 and 70 μm.

Supply of Coal and Air to the Furnace

A large front-wall-fired water-tube boiler may have as many as forty eight burners, and eight mills each of which supplies six burners. These six burners will be symmetrically arranged about the centre-line of the furnace so that when the mill is shut down to reduce the boiler output, both sides of the furnace are cooled by the same amount (*Figure 15*).

Preheated secondary air is usually supplied to the burners through wind-boxes, each supplying one or more burners. It is desirable to ensure that each burner is supplied with the correct quantity of air in relation to the coal supply to that burner, in order that the flame shall be stable, have a high combustion intensity, and be free from extensive zones of reducing atmosphere that can lead to slagging in the furnace. Correct proportioning of

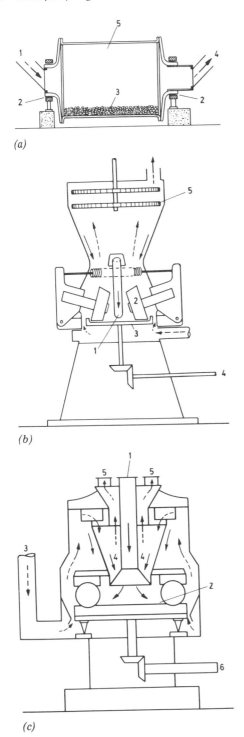

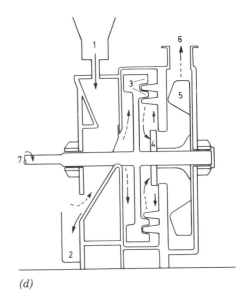

Figure 14 Types of mill used for preparing pulverized coal
$— — →$ *Air flow,* $→$ *Coal flow,* $--→$ *Air and coal flow*

(a)   *Tube mill*
1   *Inlet for coal and preheated sweeping air*
2   *Bearings*
3   *Charge of grinding balls*
4   *Outlet for airborne pulverized coal*
5   *Liner*

(b)   *Ring-roll-type mill*
1   *Coal inlet chute*
2   *Spring-loaded rollers*
3   *Rotating table*
4   *Drive shaft*
5   *Rotating classifier vanes*

(c)   *Ball-and-track-type mill*
1   *Coal inlet*
2   *Grinding zone*
3   *Air inlet*
4   *Return of oversize particles*
5   *Twin coal/air outlets*
6   *Drive shaft*

(d)   *Impact mill*
1   *Coal feeder*
2   *Receptacle for iron, stones etc.*
3   *Pulverizing hammers*
4   *Classifier bars*
5   *Fan*
6   *Coal/air outlet*
7   *Drive shaft*

fuel and air to each burner is facilitated by ensuring that the various burners supplied by one mill receive, as far as possible, an equal supply of coal. The resistances to flow offered by the different primary-air pipes from each mill should not differ greatly, and the layout of the bifurcation points should not be such as to cause coal to feed preferentially into any of the burners. Instruments

Table 5 Typical power consumption of coal-pulverizing mills of 15 tons/h capacity (kW h per ton of coal milled). Courtesy CEGB and Pergamon Press (ref.H)

Mill type	Mill	Feeder	Primary-air fan	Exhauster fan	Total
Ball mill (suction)	12.0	0.2	—	10.6	22.8
Ball mill (pressure)	13.1	0.2	5.5	—	18.8
Medium-speed mill (suction)	7.6	0.2	—	11.9	19.7
Medium-speed mill (pressure)	8.3	0.2	8.0	—	16.5
High-speed* (suction) mill	19.5	—	—	—	19.5

* Motor also drives feeder and fan

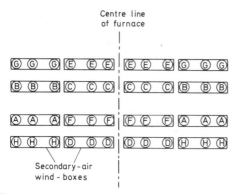

Figure 15 Arrangement of coal and air supplies to a wall-fired boiler furnace. (The burners are lettered to denote the mill supplying them)

to meter the flow of pulverized coal in pipelines are under development[28], but no reliable commercial instrument is yet available.

The concentration of coal in the primary air is typically in the region of 0.4—0.5 kg/m³, corresponding to a coal:air mass ratio of about 1:2. The primary air constitutes about 20—25% of the total combustion air, which is in turn about 120% of the stoichiometric quantity. When burners are not in use, some secondary air may be admitted through them to keep them cool. Primary-air velocities in the supply pipes should be at least 25 m/s, to avoid the deposition of coal in horizontal runs of pipe. It is found that this is well in excess of the minimum velocity required to prevent flashback of the flame into the feed-pipe.

Burners

The function of the burners is to introduce the fuel and the combustion air into the furnace in such a way that a stable flame-front is formed some distance from the burner. If the flame is too close to the burner, overheating and slagging of the burner may occur. The burners have to ensure rapid and efficient mixing of fuel and air (App.C8), and to direct the incoming gases at a suitable velocity so

that the full volume of the furnace chamber is utilized, without any 'dead' zones, and without impingement of the flame on the furnace wall. There must also be provision for the introduction of oil or gas burners when lighting-up the furnace.

The choice of burner must be made in relation to the shape of the furnace and the arrangement of the burners in it. Certain combinations of burner and furnace are usually chosen, depending on the rank of the coal to be burned. Low- and medium-rank coals for electricity generation are usually burnt either in wall-fired boilers using turbulent-flow burners (*Figure 16a*), in which the primary and secondary streams are admitted, with some swirl, through concentric annular openings, or in corner-fired boilers (*Figure 16b*), where the primary and secondary streams enter through alternate ports arranged vertically at the corners of the furnace, and directed so as to generate a vortex in the chamber. Low-volatile coals are mostly burnt in down-fired furnaces; the primary stream is directed downwards from the top of the furnace, and the secondary air is introduced as horizontal jets which impinge onto the primary jets (*Figure 16c*).

We shall consider briefly features of each of these burner/furnace combinations, including the way in which flame stability is achieved with each.

Turbulent-flow Burners

These burners are similar in general construction to those used for oil-firing. Extensive studies of the associated flow-patterns by the International Flame Research Foundation at Ijmuiden, and by other organizations, have shown that the swirling of the jets results in a widened angle of spread, and a recirculation vortex may be formed on the axis if the swirl is strong enough. Stable combustion with these burners is dependent on the fuel being sufficiently easily ignited for the flame to be well established within the limits of an outer or inner recirculation eddy. Burning coal is brought back towards the burner and is mixed with the freshly entering primary stream. Radiation of heat back from the main flame zone onto the coal jet as it enters the furnace also contributes significantly to the stabilization of ignition. At very low coal rates, and until combustion is well established, the use of

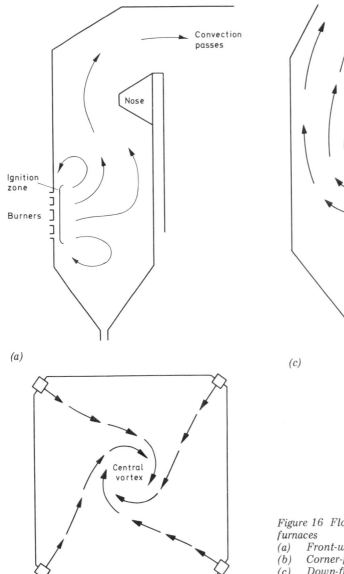

(a)

(c)

(b)

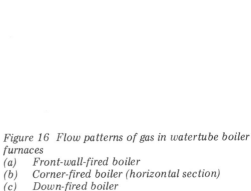

Figure 16   Flow patterns of gas in watertube boiler furnaces
(a)   Front-wall-fired boiler
(b)   Corner-fired boiler (horizontal section)
(c)   Down-fired boiler

an auxiliary burner firing an easily ignited fuel — gas or a light oil — is advisable.

The advantage of this type of burner is generally considered to be the more rapid flame development, giving more intense combustion and therefore requiring a smaller furnace volume to complete combustion. On the other hand, front-wall-firing with turbulent-flow burners suffers from a drawback already referred to: namely that if the air:fuel ratios to individual burners are unequal, either because of deficiencies in design or obstructions developing in the burners or feed-pipes, strongly reducing zones can be created in the flame which can lead to slag deposition on the furnace walls. The slag may run down and partly obstruct the air-opening of the burner, which aggravates the unbalance.

Corner Burners

The primary and secondary air enter the furnace as a number of parallel jets, and the initial coarse-scale mixing of fuel and air is governed by the velocities and positioning of the jets. Mixing continues, with combustion, as the gases rotate in the central vortex. The individual fuel jets entrain some hot furnace gases and are also heated by radiation as they approach the vortex, but ignition is finally ensured by their merging with the zone of intense combustion associated with the vortex.

Because the flame does not develop close to the walls or burner, any departure from ideal fuel:air ratio at the burners does not have such serious consequences as with wall-fired boilers. The incoming streams are subjected to thorough mixing owing to the rotation of the central core of gas.

The burners can be tilted in a vertical plane, to fire in any direction between 30° above and 30° below the horizontal. This has the effect of moving the position of the zone of maximum temperature up or down, and it is used to control the amount of steam superheat, and the furnace outlet temperature.

Down-firing Boiler

This type of firing arrangement, generally used for anthracite and semi-anthracite, aims to delay the mixing with secondary air until the coal/primary air mixture has been heated to a temperature of 800–900°C, which is required for ignition of these high-rank coals. The coal is heated by radiation from the main flame zone, and by the entrainment of recirculated combustion products, as shown in *Figure 16c*. The secondary-air ports are spaced out along the length of the developing flame so that it will not be quenched by too rapid dilution with the relatively cool air (at about 250°C).

Parallel-flow (DSIR) Burner

Most of the shell boilers employing pulverized-coal firing use the parallel-flow burner developed by the Fuel Research Station[29], illustrated in *Figure 17*. The entering fuel and air form a single wide jet comprising a large number of primary-air streams embedded in an enveloping secondary-air stream. This type of burner is claimed to give a stable flame over a 6:1 range of heat inputs, although 4 or 5:1 is a range more frequently obtained in operating plant.

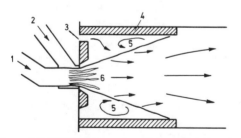

Figure 17 DSIR burner installed in the furnace tube of a shell boiler

1	Coal and primary air
2	Secondary air
3	Tertiary air
4	Refractory sleeve
5	Recirculating gas eddy
6	Ignition zone

Cement-kiln Burners

The requirements of rotary kilns for cement manufacture differ from those of boilers. A fairly long flame is needed, and flame stability is not really a problem because of the high temperature prevailing inside the kiln. A very simple form of burner usually suffices, in which the fuel and primary air enter through a central pipe and the remainder of the combustion air is drawn into the kiln around this pipe, after passing through a heat exchanger in which it receives heat from the cement clinker leaving the kiln.

Combustion and Behaviour of Mineral Matter

The process of combustion in a pulverized-coal flame is best illustrated by considering the history of a single particle of coal in the flame (see also App.C2). As the particle is heated it decomposes with the evolution of inflammable gases. The temperature range over which this occurs, and the quantity of gas evolved, depend on the nature of the coal. For a typical low-rank bituminous coal, decomposition begins at about 300°C and the rate of gas evolution increases as the temperature rises. When the coal is heated in a fraction of a second to 1000°C or higher, the quantities of gas and tars evolved may be 50–100% greater than the quantities evolved during the standard proximate analysis with relatively slow heating. In other words, the coal is volatile to a greater extent when heated rapidly.

The char residue left after the gas has been evolved is highly reactive, and burns away in the surrounding air. Combustion of the evolved gases and of the residual char may overlap, provided oxygen can gain access to the surface of the char particle. The reactivity of the char is a function of the coal rank, in general decreasing as the rank (i.e. % C) increases[30]. In large pulverized-coal-fired boilers, however, the reactivity of the char is not the only factor controlling the rate of burnout, because at the temperatures reached in these furnaces the chemical reaction rate can be greater than the rate at which oxygen will diffuse through a stagnant gas film surrounding the particle, to reach the particle surface. The combustion rate is therefore limited by the rate of gaseous diffusion, and this depends on particle diameter (to which it is inversely proportional), on the oxygen concentration outside the zone of diffusion, and to a small extent on the temperature. Diffusion rates increase only slightly with increasing temperature. For rapid burnout, therefore, efficient gas mixing is needed to ensure that no particles are starved of oxygen. The particles should also all have a sufficiently long residence time in the high-temperature zone, where combustion rates are not limited by the chemical reactivity. This reactivity exhibits a temperature-dependence expressible in the form of an Arrhenius equation, and is therefore strongly influenced by changes in temperature.

Measurements of the burning times in air of coal particles representative of the largest particles that occur in pulverized coal show that complete combustion takes less than one second[31]. Combustion is slower in a flame, however, because the supply of oxygen is limited. Nevertheless, in a well operated, large power-station boiler, the carbon-in-ash content does not exceed 2%, and the carbon monoxide content of the flue gases is negligible, representing over 99.5% combustion efficiency in a boiler with a mean gas residence time of about two seconds.

During combustion, especially in large boiler furnaces and in cement kilns, where flame temperatures may reach 1600°C, the mineral matter in the coal may soften and even fuse into spherical droplets. Solidified globules of ash can be seen in microscope examination of fly-ash. If molten or sticky ash particles are caused to strike a water tube, adhesion may occur, and it is therefore important to ensure that the combustion products are adequately cooled before they meet the convective-heat-transfer surfaces of the boiler. The furnace outlet temperature should strictly depend on the ash-fusion characteristics of the coal fired, but practical experience has shown that these may vary so widely that design calculations should be based on a maximum furnace exit temperature of between 1050 and 1100°C, in which range all but a few coals will give trouble-free running. Some coals, particularly those with a high chlorine content, which is usually associated with a high content of volatile alkali-metal salts, produce a more fusible ash than may be indicated by a laboratory ash-fusion test. When this ash forms a deposit on superheater tubes at metal temperatures above 600°C corrosion of the tubes may result (see Ch.20.1). These coals may therefore impose a more serious limitation on boiler output.

Because the heat-transfer area is in practice limited to the walls of the boiler, and the wall area does not increase so rapidly with size as the furnace volume, the position is now being reached in the design of watertube boilers where the furnace size is governed more by heat-transfer considerations than by the space needed for combustion.

The furnace exit temperature tends to rise if heat-transfer rates in the furnace are reduced for any reason. This may occur owing to an accumulation of dust on the tubes lining the wall of the furnace, and frequent soot-blowing reduces the risk of slagging from this cause.

The effective volume occupied by the flame and the position of the main heat release zone are believed to have an important influence on the overall heat-transfer rate in the furnace[32]. Calculation of heat transfer in this system is, however, a complex and poorly understood subject, and there is undoubtedly a need for further research in this field, including further efforts to apply the Hottel—Cohen zone method for calculating the radiant heat exchange[33—35].

Possibilities for achieving more Intense Combustion

Because gas diffusion is a limiting factor to the combustion rate in most practical furnaces, there are prospects of increasing combustion rates by using various ways of assisting the access of oxygen to the particle surface.

The rate of diffusion is inversely proportional to the particle diameter, so that more intense combustion can be obtained by grinding the coal very finely. This has the effect of increasing the temperature at which diffusion control takes over from chemical rate control, and results in a hotter flame with a higher volumetric heat release rate. However, the cost of grinding coal very finely is not considered to be compensated for by the reduction in furnace size made possible. It has already been pointed out that in most modern large boilers the size is governed more by the area of heat-transfer surface required than by the volume necessary for combustion, and for intense combustion to be economically attractive, some alternative way of utilizing the energy release appears to be called for.

An alternative approach is to try to effect savings on coal grinding costs by developing combustion systems which will rapidly burn relatively coarsely ground coal. One such appliance is the cyclone combustor in which the larger of the particles are thrown against and adhere to a sticky layer of slag on the wall, where they are scrubbed by the gases. The relative movement of gas and solid increases the rate of transport of oxygen to the particle surface, and the particle is held by the slag until combustion is completed. The cyclone combustor has been successfully developed into a working furnace, and is dealt with in Ch.6.3.

Another possibility is pulsating combustion, in which a periodic displacement of the combustion air results in relative motion between particles and air which again increases the rate of access of oxygen to the particle surface. The subject has been discussed by Reynst[36], but the principle has not yet been successfully applied to commercial heating furnaces. One difficulty is the very high acceleration rates needed to produce significant velocity differences between the gas and the particles.

The fluidized-bed combustor, described and illustrated in Chapter 5, is yet another system in which higher mass-transfer rates are obtained by relative motion between particles and gas. This process has also several other interesting features which probably outweigh the one just mentioned.

Recapitulation

Pulverized-coal firing is especially suitable for very large boilers and cement kilns. A wide range of coals can be burnt, but coal preparation and gas cleaning are major cost items.

70% < 75 μm is required for bituminous coals, 80% for anthracite. Coal dried with air at 250—350°C for milling in low-, medium- or high-speed mills. Main milling costs are power and replacement of worn parts. Wear depends on fineness of grinding, and on hardness and abrasiveness of coal.

Coal and air distribution to burners must be uniform to avoid local reducing zones which encourage slagging. Burners for wall-fired boilers are concentric-jet, turbulent flow type, giving short, intense flame. Those for corner-fired boilers are parallel jet type, with primary and secondary ports arranged in a vertical line. For down-fired boilers, primary ports are in furnace roof and secondary ports in side-wall.

Combustion rate generally is controlled by eddy-mixing and burnout rate by diffusion of oxygen to particle surface, but that for anthracite is chemically controlled. Ash-softening temperatures are reached in flames, and gases must be cooled to $1050-1100°C$ to avoid slag deposition on super-heater tubes.

6.3 The cyclone furnace

As its name implies, fuel and combustion air enter the cyclone furnace with a strong swirling motion, the air velocities and fuel size being such that some, or most, of the fuel particles are flung to the walls. The high combustion intensity for which the furnace is designed results in a temperature which is sufficient to melt the ash. The liquid-ash (slag) film which forms on the walls of the furnace restrains the movement of coal particles reaching the wall surface; the resulting relative motion between the air and the particles causes rapid combustion and enables high combustion intensities to be obtained. The liquid slag runs down the furnace walls to the outlet, where it is usually quenched in some form of water bath and granulates to form easily-handled particles.

The main application is for steam generation. Large numbers of cyclone-fired boilers were built between 1940 and 1970 in the USA and W. Germany, and to a less extent in the UK, France and Eastern Europe. At one stage in the early 1960s, over 30% of new boiler capacity in W. Germany and over 20% in the USA was cyclone-fired. In recent years, the decline of the coal

industry in Europe has resulted in a dramatic reduction in all types of coal-fired boilers, and many cyclone furnaces have been converted to oil-firing. Similarly, in the USA cyclone-fired boilers are no longer built — partly because of a lack of suitable coals in some areas and partly because the application of cyclone furnaces to extremely large units (e.g. greater than 1000 MW) has not proved as economical as expected. The situation may alter, however, as the general energy pattern changes.

Types of Cyclone Furnace

There are two main types — horizontal and vertical cyclone furnaces — each of which can be further subclassified.

Horizontal Cyclone Furnace

Figure 18 illustrates diagrammatically the working of one type of horizontal cyclone furnace. Primary air and coal are admitted at one end of the cyclone chamber and enter it with a swirling motion; secondary air (at about 100 m/s) enters at the periphery, tangentially, to maintain the swirl. The wall of the chamber is built up from water tubes, studded on the inner face and covered with a suitable plastic refractory (often silicon carbide). The gas-outlet throat has the form of a re-entrant cone, which reverses the axial motion of the gases and forces them radially inward before they leave the furnace.

Slag deposited on the furnace wall dissolves away the initial refractory coating to a depth where the cooling effect of the tubes prevents further action. The cyclone furnace is then lined with a layer of slag, which reaches an equilibrium thickness for a given furnace heat-release rate. At this thickness (which is determined by the almost-constant tube temperature, by slag conductivity and by flame temperature), the surface temperature of the layer is above the flow temperature of the slag, so that additional deposits run down the walls to the outlet.

In the cyclones manufactured by the Babcock & Wilcox organization the coal and primary air enter either

(a)   along the axis of the cyclone through a small burner unit which imparts a swirl to the coal/primary-air mixture. Coal is precrushed to pass through a 6.4 mm screen and most of it is flung to the walls. This type of cyclone is known as the axially-fired cyclone and is illustrated in *Figure 18*. It is used almost exclusively in the USA. In the boiler application shown in *Figure 19*, the gases from the cyclone pass through the re-entrant throat into a secondary chamber, and then through a tube bundle (known as a slag screen) before entering the radiation (tertiary) chamber of the boiler. In larger and more modern installations, such as that shown in *Figure 20*, the axis is truly horizontal and the slag screen is omitted.

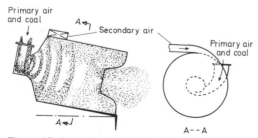

Figure 18 Simplified representation of a cyclone furnace

(*b*) through ports below the secondary air nozzles in a direction either approximately tangential to, or across a chord of, the cyclone body. This type of cyclone is known either as the tangentially-fired or the secant-fired cyclone, respectively, and is used almost exclusively in Europe. It was developed from the axially-fired cyclone in order to burn German bituminous coals, which usually have a lower volatile matter yield than the corresponding U.S. or British coals normally used for power generation. It was found that these German coals had to be crushed to a finer size (e.g. 30 to 55% less than 90 μm depending on the volatile matter), to obtain adequate ignition, and therefore had to be injected nearer the walls of the cyclone to obtain high combustion efficiency.

The horizontal cyclone is designed to have a heat-release rate (heat input rate from coal/internal wall surface area of cyclone) of about 2.4 MW/m^2 at full load. The combustion intensity (per unit volume) varies, therefore, with the diameter of the furnace, being about 8 MW/m^3 in a 2000 mm diameter cyclone (heat input 30 MW) and about 2 MW/m^3 in a 4500 mm diameter cyclone (the largest made — heat input 200 MW). In installations where the secondary chamber is clearly defined by a slag screen, the combustion intensity based on the cyclone volume plus the secondary-furnace volume is in the range 1.5 MW/m^3–2 MW/m^3.

Vertical Cyclone Furnace

Whilst the horizontal cyclone furnace represented a radical break-away in boiler design, the vertical cyclone was a logical development from the corner-fired p.f. (pulverized coal) furnace. Several manufacturers have installed vertical cyclone-fired boilers in Europe and Russia. One design is shown in *Figure 21:* fuel and air are admitted through a number of burner units installed in the sloping roof of the furnace and angled to produce a swirling motion. In another

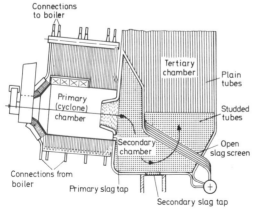

Figure 19 Cyclone furnace applied in firing a watertube boiler

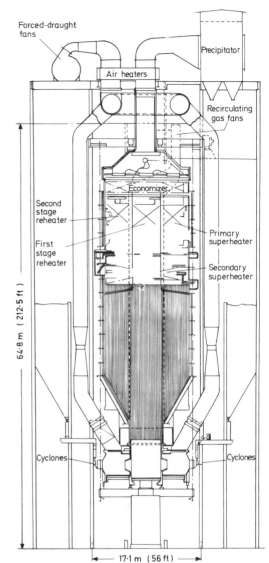

Figure 20   Large multi-cyclone-fired boiler. Courtesy The Engineer, *Feb.19, 1960, p 293*

design the burner units are spaced around the walls of the cyclone.

Partly because of its historical development, more of the fuel is burned in suspension than in the horizontal cyclone, the molten ash particles reaching the walls by agglomeration. The burner units usually attempt to provide intimate mixing of the coal/primary-air and secondary-air streams and hence coals with very low volatile matter yield (e.g. <5%) have been burned successfully. The coal is ground to, typically, 55–75% less than 90 μm depending on the volatile matter yield. This is finer than in the horizontal cyclone but coarser than in conventional p.f. furnaces.

The largest vertical cyclone furnaces are fitted to boilers producing 350—450 tonne/h steam and are of 9000—9800 mm diameter. The combustion intensity is about 0.6 MW/m³. Smaller units operate at higher combustion intensities; for example, a furnace producing 100 tonne/h steam would have a combustion intensity of about 1 MW/m³.

Some Comparisons with Dry-Bottom P.F. Boilers

Although the advantages/disadvantages of cyclone-fired boilers change in emphasis depending on local considerations, the following generalizations are relevant:

(1) Because of higher combustion intensities, the volume occupied by a cyclone-fired boiler is slightly less than its dry-bottom counterpart.

(2) Most of the ash is removed as a granulated (quenched) material which is dense, fairly uniform in size distribution and more easily handled than p.f. ash. However, ash which escapes from the cyclone furnace is in the form of fine re-solidified particles and conventional dust-cleaning equipment is usually necessary to meet environmental considerations. Such equipment is not required to operate at such high efficiency as in its dry-bottom counterpart.

Up to 85% of the ash in horizontal cyclone furnaces and 70—80% in vertical cyclones is removed as slag. If the ash from the dust-cleaning equipment is refired into the furnace —

as is often the case — then virtually all the ash is removed as slag.

(3) The ancillary power consumption is generally the same as for p.f.-fired boilers. Coal preparation costs are substantially lower for the axially-fired horizontal cyclone and slightly lower for the vertical cyclone. But these are counterbalanced by the higher fan-power costs of the cyclone furnace, and particularly of the horizontal cyclone furnace.

(4) Provided that the ash in the fuel can be converted into slag, a wider range of fuels can be burned in cyclone-fired than in p.f.-fired boilers. Details of coal, and other fuel, suitability are given in the next Section.

(5) Load range. As the load on the boiler is reduced, the slag becomes more viscous. In horizontal cyclone-fired boilers the normal full load range is usually met by shutting down individual cyclones. In vertical cyclone-fired boilers (which invariably have only one cyclone per boiler) low loads are achieved by increasing the excess air at a certain point on the load curve, and operating the furnace as a dry-bottom unit.

(6) Furnace explosions ('puffs') are virtually unknown in cyclone-fired boilers; the heat capacity of the slag-coated walls means that ignition is safely and instantaneously re-established after short interruptions to the coal feed.

(7) Because of the higher combustion temperatures in cyclone furnaces, the emission of oxides of nitrogen with the flue gases tends to be higher than in dry-bottom units.

Suitability of Coals

The primary consideration deciding the suitability of a solid fuel for use in any slagging system is that the slag must flow freely under the prevailing temperatures in the system. These temperatures depend upon: (a) the slagging characteristics of the ash; (b) the ash level and moisture content of the fuel; (c) the type or rank of the coal (low-volatile, high-volatile; coking etc.) and the size grading; (d) the temperature of the combustion air; (e) the amount of excess air; (f) the combustion efficiency; and (g) the combustion intensity. Slag flow is also affected by the amount of heat lost by the slag between the cyclone and the slag outlet.

The viscosity of the slag at any temperature depends mainly on its chemical composition; the combustion efficiency (and also ignition stability) depends on the furnace design and on the temperature in the furnace, the volatile yield and the size distribution of the fuel.

Slagging Characteristics

Methods used to estimate the flow characteristics of a slag or ash are:

(a) Determination of the ash-fusion characteristics, particularly the hemisphere and flow points

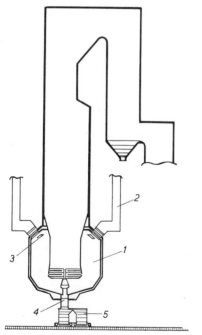

Figure 21   Vertical cyclone-fired boiler.   Based on Dürr Kohlenstaubfeurungen (Figure 11), Dürrwerke A. G., Ratingen, W. Germany

of the ash obtained in the Leitz heating micro-scope. This method is widely used by boiler operators in Europe.

(*b*)   Determination of the viscosity—temperature characteristics of the slag. Ideally these should be obtained by direct measurement in a viscometer, but this is time-consuming and the characteristics are usually predicted from the chemical composi-tion of the ash or slag — sometimes in conjunction with the ash-fusion characteristics.

It is generally accepted that for satisfactory tapping of slag, the temperature in the furnace should be sufficient to maintain the slag below 25 N s/m^2 (250 poise).

Ash Level

Ash level, *per se*, is not a significant factor. However, as it rises above about 20%, the amount of heat required to raise the temperature of the ash increases rapidly, with a consequent lowering of the gas temperature in the cyclone. Thus a 10%-ash fuel lowers the theoretical gas temperature by approximately $17°C$, whereas a 50%-ash fuel would lower the temperature by $170°C$. It is generally accepted that 50% is the maximum level of ash that can be handled satisfactorily, but this must depend to a large extent on other fuel characteristics.

The minimum level of ash is that required to maintain a slag film on the cyclone walls. Coals with 5—7% ash are regularly burned in some cyclone furnaces.

Moisture Content

When the coal is milled in a direct-fired system (i.e. when the coal is conveyed in the milling gases) all the moisture enters the furnace and thus lowers the furnace temperature. Depend-ing on other factors, coals with up to 20% moist-ure have been burned in this way, although opera-tion of slag-tap boilers is more sensitive to moisture content than to ash.

With indirect firing (bin-and-feeder), the mil-ling gases can be admitted to the boiler down-stream of the cyclone furnace. There are then no limits on moisture content other than economic ones.

Volatile Matter

Cyclone furnaces are capable of burning low-volatile fuels — this was one of the reasons for their wide acceptance in Germany. Depending on the design of furnace, fuels with volatile matter as low as 3—5% have been burned, although finer grinding is necessary to compensate for the in-ferior ignition qualities.

Type or Rank of Coal

Both anthracites and brown coals are burned in cyclone furnaces if the other characteristics are suitable. Brown-coal ashes often produce slagging

problems because of the high gypsum content (CaO only acts as a flux for the other ash com-ponents in small amounts). In such cases it is usual to flux the brown-coal ash by adding sand or blending with bituminous coal.

Use of Fuels other than Coal

The cyclone furnace can burn almost any kind of fuel, and although coal forms by far the largest source of supply, local economics can lead to utilization (temporarily or permanently) of other fuels, such as wood, petroleum coke, coal chars, gas and oil. In some installations the fly-ash from other types of boiler, which can contain sufficient carbon to class it as a fuel, is injected into cyclone-fired boilers.

6.4 Garbage (town refuse) incineration

Our way of life is reflected in the contents of our dustbins. A trend towards central heating plus Smoke Control Orders is reflected by progressively less ash and cinders and a higher proportion of paper, packaging and kitchen waste being rejected unburnt by the average household. Another factor is that more consumer goods, e.g. television sets, washers, radios, cars etc., are being sold; these, with their built-in obsolescence, invariably end up on the refuse heap after ever shorter intervals. The charac-teristics of domestic refuse have undergone con-siderable changes over the past 40 years, as illu-strated in *Table 6*.

Heat Value in Refuse

The calorific value of town refuse (*Table 7*) is rising slowly. There was a fairly rapid rise in the 1950—1960 period when there was a large-scale replace-ment of the traditional open fire by more efficient central heating.

A smaller proportion of dust and cinders with a corresponding increase in paper content, together with a reduction in vegetable matter owing to the impact of frozen and pre-packed foods with plastics-coated paper containers of high calorific value, has boosted the heat value. Gross calorific values have exceeded 9.3 MJ/kg (4000 Btu/lb) in many towns for a few years now and indications are that this is still increasing. We might possibly see 11.6 MJ/kg exceeded, at least in dry weather, in the next ten years. The weather can have a significant influence upon the moisture content of refuse, and hence its calorific value. Incineration-plant designers take this factor into account by catering for an overall range of varying calorific values. Heavy thunder-storms can reduce the calorific value of a typical day's refuse by approximately 2.3 MJ/kg. Plastics in general contribute greatly to the rising trends. One important influence will be the containers of liquids. If throw-away glass bottles were replaced

Table 6   Refuse analysis 1934—1972 (UK)

	1934	1964	Year 1967	1970	1972
Component of refuse (%):					
Fine dust and small cinder	66.7	37.3	22.0	14.4	11.5
Large cinder content	8.3	8.5	3.5	3.5	2.0
Vegetable and putrescible matter	6.1	8.9	20.0	24.8	21.5
Paper, cardboard	7.7	31.2	33.5	33.9	41.5
Metal	3.3	4.8	6.8	7.2	7.4
Rags, bagging and textiles	1.5	1.1	2.5	2.1	1.9
Glass (bottles and cullet)	3.2	7.2	8.1	9.0	9.4
Unclassified (not included above)	3.2	1.0	3.6	5.0	4.8
Density of refuse:					
(cwt/cu. yd)	5.76	3.73	2.62	2.49	2.20
(kg/m^3)	381	247	173	165	146
Average weight of refuse per household					
(lb/week)	40.1	34.7	33.5	33.0	33.5
(kg/week)	18.3	15.8	15.3	15.0	15.3

Table 7   Typical heat values and moisture contents of refuse components (1974, UK)

Sample	Moisture content (%)	Calorific value as sampled (MJ/kg)
Fine dust	9.8	6.44
Cinder	5.6	27.40
Vegetable matter	73.5	4.77
Paper	9.6	17.21
Textiles	15.2	13.54
Wood pieces	10.7	13.51
Plastics and rubber	5.5	18.59

by plastics or there is a swing back to returnable bottles, and/or if glass milk bottles were replaced by throw-away plastics or coated paper containers, there would be a considerable rise in calorific value. If all these alterations occurred together, then 12.8 MJ/kg could conceivably be reached. Even today the heat content of 3 t of refuse is equal to that in one tonne of reasonably good coal. The amount of plastics in refuse is still very low, 1—1.5%, and, even allowing for all the possibilities listed above, is unlikely to exceed 3% in future years. Only part of this consists of PVC or other plastics containing chlorine, and although hydrogen chloride is given off in the products of combustion the amount is unlikely to reach a nuisance level.

A typical refuse tonnage collected weekly in the UK is around 6 t (tonne)/1000 population. In a town of average size, say 60 000 persons, the amount of weekly refuse produced would be around 360 t.

References p 91

Reasons for Incineration

The reason for the growth of incineration compared with other methods of disposal is primarily the lack of available tipping sites for crude or partially treated refuse. Controlled tipping demands large holes in the ground at economical distances from the source of refuse; material is deposited in layers less than 2 m deep, each layer is covered by at least 200 mm of earth or fine top cover, and no refuse is left uncovered for more than 24 hours.

Pulverizing can be employed to reduce the volume of crude refuse but the total weight to be dumped remains the same. Composting has been tried and is still carried out in some towns. It attracts rather more interest in developing countries where refuse contains more vegetable matter with less packaging than in the UK.

An incinerator handling towns refuse produces a residue of less than 10% of the original volume and 45% of its original weight, giving an increase in tip life of at least 10 times that of crude refuse. Often this is a conservative estimate since contractors may find uses for the sterile residue if it is graded and all ferrous scrap is removed.

Cell-type Incinerators

Municipal incineration plants have been in use in the UK since the beginning of the century. The early units were all cell types into which refuse was batch charged through a top aperture and the ashes raked out by hand from below the fixed cast-iron grate. Plants of this type are still in use today. The reluctance of operatives to work in close contact with refuse has led to the development of automatic furnaces requiring the minimum of contact between personnel and waste, and these designs facilitate the stricter control over waste gas emission required today.

Cell or batch incinerators are now used mainly by industrial concerns for disposing of relatively small quantities of combustible trade waste such as wood, cardboard and paper. Cell incinerators are also commonly used at municipal incineration plants for disposing of bulky items such as furniture and animal carcasses.

Typical Grate Systems in Modern Use

Essentially these systems (*Figure 22*) have an inclined mechanical grate, usually activated by hydraulic rams, which conveys the refuse through the furnace and supports it whilst it burns. The materials of construction of the moving firebars vary from grey cast iron to alloy steels.

Grates vary in design between manufacturers, and there are conflicting opinions on the degree of agitation needed to break up and stir dense agglomerations of refuse during the process. Often more than one grate is installed, maybe two or three in series, to allow control of speed and extent of agitation. In some instances the first or dead grate which conveys refuse from the feed chute into the furnace is replaced by a push feeder which has a more positive action and can isolate the feed chute from the furnace in the event of a burn-back in the chute. The lower portions of the feed chute are either air- or water-cooled. Combustion takes place in three stages: drying, ignition and release of volatiles, and final burnout. All systems endeavour to produce complete burnout of the material before discharging it off the end of the grate. A number of excellent grate systems are offered by manufacturers; selection would be impossible and invidious. A typical selection of types would be rocking (*Figure 23*), reciprocating, chain-grate (*Figure 24*), roller-grate, rotary, oscillating and reverse cascade.

The smallest commercial moving-grate incinerators offered at present burn around 2 t/h of refuse with a calorific value of 11.6 MJ/kg.

Refuse is normally quite a high-volatile fuel, and early designs failed to appreciate that over 50% of the air supply was needed as secondary air and

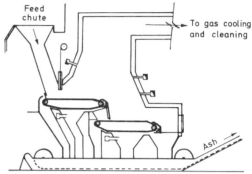

Figure 23   Typical travelling-grate incinerator

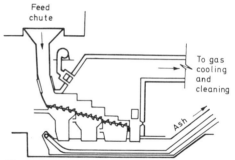

Figure 24   Typical rocking-grate incinerator

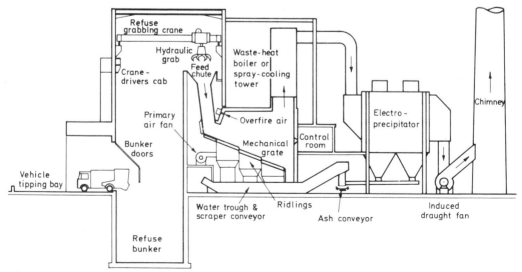

Figure 22   Diagrammatic layout of a typical British refuse incineration plant

at high velocity to encourage turbulence and mixing over the firebed. Experience has shown that the burning rate on these moving grates should be at most 27—37 kg/m² h (60—82 lb/ft² h) for furnace sizes from 15 to 75 MW (50 to 250 million Btu/h) heat output from refuse, assuming a calorific value around 11.6 MJ/kg. Where the furnace is refractory-walled, the volume above the grate should be equivalent to 175—235 kW/m³ h (17 000—23 000 Btu/ft³ h) over the same range of sizes. Where the furnace walls are fully water-cooled as with an integral boiler these furnace heat-capacity figures can be trebled. Although it is essential to obtain as complete a burnout of the refuse as possible, complete combustion of all smoke and volatiles requires excess air to be present and the gases leaving the firebed to be maintained at a minimum of 850°C for at least 0.5 s. With a refractory-walled furnace the maximum temperature is usually limited to 1000—1100°C if relatively inexpensive materials are to be used for most of the furnace structure, so that it is usually difficult to operate with less than 100% excess air. The larger plants of municipal scale can achieve all these conditions with one large furnace chamber.

A typical present-day refuse requires around 4.4 standard m³ of air per kg to meet combustion requirements and also needs around 100% excess air to prevent the furnace temperature rising above 1100°C. To allow a margin for future rises in calorific value, the total air-supply to the furnace should be designed for 6.2 m³/kg, equivalent to 12.8 MJ/kg. Alternatively it may be acceptable gradually to reduce the burning rate as calorific value rises in future years, to suit limited grate areas and air supply. In view of the rising tonnages of refuse however this could be false economy, and design standards today should be based on refuse of at least 11.6 MJ/kg, the grate area and furnace volume being designed for these higher calorific values. (The foregoing assumes no direct recovery of paper.)

Parameters of Combustion (Residue Standards)

The quality of the residue determines the ease of ultimate disposal. Ferrous metals should be recovered by electromagnet, baled and sold for scrap. The removal of tin cans considerably reduces the bulk of residue and gives a valuable return which is credited against operating costs. Reclaimed ferrous metals represent approximately 2.5% by weight of the crude refuse.

Burnout

The first criterion of a good residue should be low carbon content, since in the absence of any other test low carbon must mean low organics. In earlier days 5% carbon in residue, tested by weight loss on ignition, was often specified; but later experience suggests this to be too severe a test, particularly

since heating to high temperature may decompose carbonates, hydroxides, certain strongly bound hydrates and some other susceptible inorganic substances, which are then reported as 'carbon'. A better test for carbon is the BS 1016 test for ultimate analysis of fuels which measures actual carbon dioxide produced from carbon, and the crude 5% test criterion should be relaxed to say 7%, which experience shows is still an excellent burned-out residue.

Putrescibles

Various tests have been postulated to measure the putrescibles. In the Düsseldorf method, alkali-soluble matter is oxidized with permanganate and it is normally stated that if the test answer is less than 0.3% the residue is unlikely to give trouble. This test is unwieldy, takes a lot of apparatus and a long time to carry out, and the production of very small quantities of carbon dioxide for absorption in a heavy weighing-tube train does not tend to accuracy. Also values above 0.3% have rarely been obtained even with residues which to the eye seem objectionable! This method cannot really be recommended.

Gas Cooling and Cleaning

If the waste gases are to be cleaned efficiently, they must first be cooled since multiple cyclone arrestors and electroprecipitators both have upper limiting temperatures for effective operation.

Multiple cyclones can operate at high temperatures, say 700°C, if heat-resisting steels are used, but the cost is thereby increased, and the more usual materials limit the gas temperatures to around 400°C.

An electroprecipitator has the same materials limitation, and the collecting effect is also hindered at high temperatures, so that manufacturers prefer entering gases to be at 250—300°C. With a refractory-lined incinerator, the temperature of gas will be 900—1100°C when leaving the combustion chamber, so that considerable cooling is necessary. Yet in the UK it seems essential, to meet clean-air regulations, that the electroprecipitator be used for the larger plants.

The usual method of cooling for the fairly large incinerators discussed comprises spraying water at high pressure through atomizing nozzles into the gases, usually in separate cooling towers. Although this is cheaper than adding large volumes of additional cold air, it is by no means cheap if town water is used: about 2270 litre (500 gal) of water are needed for each tonne of refuse. The water evaporated into the gases reduces their temperature to 300°C and their volume by approximately 40%.

If a boiler plant is installed, then this cools the gases without adding to their mass; the cooling-tower and water-spray equipment is no longer needed, and the size of the precipitator and induced-draught fan and the chimney diameter can all be reduced, since the volume reduction approaches 60%.

Incineration Without a Grate

Pilot plants are in operation in which refuse is incinerated by less conventional methods than that of a fixed or mechanical grate. If proved successful and developed to acceptable commercial standards, it is quite feasible that in due course one or more of these systems may be favoured.

Pyrolysis is a process which involves heating a batch of refuse in a vertical refractory retort or in a rotary kiln to a temperature within the range of $500-1000°C$ in an oxygen-free atmosphere. At the lower temperatures, oil is produced together with a combustible gas having a calorific value of some $15-19$ MJ/m^3. At higher temperatures, the oil vaporizes or cracks and only gas is produced. All the gas produced is at present used to heat the process.

In the USA a 1000 ton/day plant is under construction for the city of Baltimore.

High temperature gasification systems for solid-waste disposal are also being experimented with in the USA. Air is heated to $1100°C$ and blown through a vertical refractory gasifier which reduces the refuse to a combustible gas and a liquefied slag. The gases produced contain no free oxygen and consist mainly of carbon monoxide.

Another system developing in the UK at present is the fluidized-bed furnace. The fluidized bed comprises a bed of sand held in a fluidized suspension by upward-flowing preheated air. Refuse or sewage cake is fed into the bed and incinerated in the highly turbulent sand. Ash is drawn off at low level in the furnace. Sand carried over with the gas is reintroduced with the refuse charge. Capital and operating costs are claimed to be less than those for conventional incinerators but more work is needed on design problems (as at 1975).

Heat Recovery

Many large Continental plants have watertube boilers which generate high-pressure superheated steam. This course is also adopted in the large Greater London Council plant at Edmonton.

The h.p. steam is used to generate electrical power, using turbines. Surplus power above that needed to run the plant is supplied to the Grid System. Only very large incineration plants can show a reasonable financial return on power generation. But if steam can be used either in the incineration plant or can be sold to a neighbouring factory, industrial estate or district-heating scheme, then the boiler need only operate to supply the highest pressure requirement of the connected loads.

A financial advantage is possible if fairly low pressures, below 20 bar (2 MPa, or 300 psi) or if hot water at up to $180°C$ ($350°F$) is required. This allows the use of smoke-tube boilers, using all the background experience of steelworks waste-heat practice where slag and dust fouling problems are similar. Such boilers are reasonably cheap and contradict the oft-made statement that waste-heat recovery equipment is too expensive to install. Considering the throughput, often considerable, the change from a simple incinerator with (in practice) zero efficiency to a modern plant with 65% thermal efficiency should offset much of these costs. It has been estimated that domestic refuse in a UK community could provide 20% of its heat requirements. In these times of energy shortage, it should moreover be a national necessity.

Problems of Heat Recovery

Slags and Corrosion. Refuse gives off corrosive gases and also low-melting-point materials which leave the grate as tiny molten slag droplets. The combustion chamber must be designed to ensure that when these molten droplets approach any water-cooled surface they are below their melting point, so that only solid material and not a semi-liquid slag impinges onto a tube. Experiments on existing refuse incinerators fitted with boilers have shown that the critical temperatures for slags are above $500°C$. This means that while no trouble is likely to be experienced with hot water-heating or steam-evaporating tube surfaces, even at high pressures, there can be considerable difficulties with superheater tubes if attempts are made to obtain high superheats.

Many early plants ran into great difficulties with fouling, by trying to use conventional water-tube boilers with tube nests above the grates. There must be a large 'empty' space above the grate with water-cooled or refractory walls only. This on the one hand allows adequate time, temperature and turbulence for completing the combustion of volatile vapours and smoke and destroying odours, and on the other hand allows cooling of the gas cloud by radiation to permit most of the molten slag particles to solidify.

At the rear end of the boiler plant, low-temperature corrosion can occur, owing to the well known sulphur trioxide/sulphuric acid dew-point problems and also to the rather less well known hydrochloric acid corrosion. Although refuse contains some sulphur, the sulphur trioxide corrosion problems are not so serious as with fuel-oil-fired plant. This is because the refuse usually contains only 0.2—0.5% sulphur and also certain alkaline materials, some of which pass through the plant as grit and dust and absorb and neutralize a part of the sulphur troxide in the flue gases.

Problems encountered in Modern Practice

Apart from the difficulties with heat recovery, the main problem in the design of a continuous refuse-disposal plant is to ensure its ability to cope with the arduous duty of handling a material so unpredictable as refuse. The grabbing crane feeding the plant must be a robust heavy-duty type. Conveyors removing the residue must be capable of withstanding wear beyond that of normal materials handled in industry.

With spray cooling, water atomization must be good and impingement on refractory brickwork avoided to prevent costly repair bills. Hot refractories and cold water do not mix with impunity. If fine atomization is not achieved there is a danger of water carryover into the gas-cleaning plant which can cause severe problems with dust blockages.

In the interests of efficiency and high plant morale it is essential that the disposal plant be kept clean and well maintained. Laxity in normal attention leads to an unnecessarily high level of remedial work just to keep a plant running.

Planned maintenance and good housekeeping are very important, and emphasize the necessity of good management.

Capital and Running Costs

Early installations were built at very moderate prices, but with the current economic situation plant capital costs are rising rapidly.

In the UK, a single-stream plant in 1974 cost around £90 000 per 1 tonne of hourly capacity installed. A twin-stream unit was £105 000 or more per t/h. This level of price would purchase a basic plant with high-quality gas cleaning. Operating costs have also suffered at the hands of inflation, and these in 1974 were running at £6—£8 per tonne of refuse incinerated, including amortization.

References

1 Carter, E. F. *Dictionary of Inventions and Discoveries*, Fred Muller, 1966

2 Thurlow, G. G. and Wall, A. G. *J. Inst. Fuel* 1962, 35, 455

3 Crane, W. M. and Rolfe, T. J. K. *J. Inst. Fuel* 1968, 41, 426

4 *BCURA Information Circulars*, No. 207, 1966; 345, 1968; and 352, 1968 (obtainable from Institute of Fuel, London)

5 Thurlow, G. G. *Proc. Instn mech. Engrs* 1954, 168, 571

6 Gunn , D. C. *J. Inst. Fuel* 1959, 32, 123

7 The Boiler Availability Committee *Cleaning of Modern Watertube Boiler Plant, Bull.XII MC/334*, 1963 (Sep.)

8 Ref.(G)

9 Bertrand, P. *Journées de la Combustion des Combustibles solides et pulverises*, Inst. Français des Combustibles et de l'Energie, Paris, 1957 (Dec.), p 471

10 Smith, W. D. *BCURA Mon. Bull.* 1954, 18, 189

11 Anon. *National Coal Board Tech. Studies* No.7, 1965 (Aug.)

12 Murray, M. V. *J. Inst. Fuel* 1957, 30, 276

13 MacDonald, E. J. and Murray, M. V. *J. Inst. Fuel* 1953, 25, 308

14 MacDonald, E. J. and Murray, M. V. *J. Inst. Fuel* 1955, 28, 479

15 Rolfe, T. J. K. *J. Inst. Fuel* 1961, 34, 481

16 Rolfe, T. J. K. *J. Inst. Fuel* 1962, 35, 532

17 Thurlow, G. G. *J. Inst. Fuel* 1962, 35, 516

18 Barker, M. H. *J. Inst. Fuel* 1965, 38, 311

19 Rolfe, T. J. K. *Wks Engng* 1968, 63 (Sep.), 6—8, 10

20 Ling, M. D. and Rolfe, T. J. K. *BCURA Inf. Circ.* No. 355, 1968

21 Svoboda, M. *Journées de la Combustion des Combustibles solides et pulverises*, Institut Français des Combustibles et de l'Energie, Paris, 1957 pp 317—321

22 Harris, M., Rivett, W. L. and Thurlow, G. G. *J. Inst. Fuel* 1962, 35, 523

23 Anon. *National Coal Board Tech. Studies* No. 8, 1965 (Dec.)

24 Anon. *National Coal Board Tech. Studies* No. 9, 1966 (Ap.)

25 Ref.(H)

26 Ref.(I)

27 Ref.(J), p 249

28 Somerset, R., Prasher, C. L. and Parkinson, M. J. 'Recent Advances in Boiler Instrumentation', *Papers 1—3 of Conf.* at Central Electricity Research Laboratories, CEGB, Leatherhead, 1969 (Oct.)

29 Hurley, T. F. and Cook, R. *J. Inst. Fuel* 1938, 11, 195

30 Field, M. A. *Combustion & Flame* 1970, 14, 237

31 Ref.(J), p 354

32 Godridge, A. M. *J. Inst. Fuel* 1967, 40, 300

33 Hottel, H. C. and Cohen, E. S. *Am. Inst. chem. Engrs J.* 1958, 4, 3

34 Pieri, G., Sarofim, A. F. and Hottel, H. C. *J. Inst. Fuel* 1973, **46**, 321

35 Anson, D., Godridge, A. M. and Hammond, E. G. *J. Inst. Fuel* 1974, 47, 83

36 Reynst, F. H. *Pulsating Combustion* (Ed. M. W. Thring), Pergamon Press, Oxford, 1961, pp 13—30, 75—111

General Reading

6.1

(A) Francis, W. *Coal: Its Formation and Composition*, 2nd edn, Arnold, London, 1961

(B) Van Krevelen, D. W. *Coal: Typology, Chemistry, Physics and Constitution*, Elsevier, Amsterdam, 1961

(C) Lowry, H. H. (Ed.) *Chemistry of Coal Utilization*, Suppl. Vol., Wiley, New York, 1963

For a summary and comparison of the above three books, see I.G.C. Dryden and D. W. van Krevelen, *BCURA Mon. Bull.* 1963 (Dec.), 27, 529—546

(D) General background information relevant to coal handling: Richards, J. C. *BCURA Mon. Bull.* 1969 (Jan.), 33, 3 (Pressure in bunkers), and 1969 (April), 33, 80 (Flow properties)

(E)   Brown, R. L. and Richards, J. C., *Principles of Powder Mechanics*, Pergamon Press, Oxford, 1970

(F)   The Boiler Availability Committee, *Cleaning of Modern Watertube Boiler Plant, Bull. XII MC/334*, 1963 (Sep.)

(G)   Jackson, R. *Mechanical Equipment for removing Grit and Dust from Gases*, BCURA, Leatherhead, UK, 1963

6.2

(H)   Central Electricity Generating Board (UK), *Modern Power Station Practice*, Vol.2, 2nd edn, Pergamon Press, Oxford, 1971

(I)   Dolezal R. *Large Boiler Furnaces* (English translation), Elsevier, Amsterdam, 1967

(J)   Field, M. A., Gill, D. W., Morgan, B. B. and Hawksley, P. G. W. *Combustion of Pulverised Coal*, BCURA, Leatherhead, UK, 1967

6.3

(K)   *BCURA Mon. Bull.* Reviews Nos. 139 (1954); 196 (1960); 213 (1964). See also ref.(C)

6.4

(L)   Department of the Environment (UK), 'Refuse Disposal', *Report of the Working Party on Refuse Disposal*, HMSO, 1971

(M)   Perry, R. H. and Chilton, C. H. *Chemical Engineers' Handbook* 5th edn, McGraw-Hill, 1973

(N)   *ASME Proc. of 1970 National Incinerator Conference*, Cincinnati, Ohio

(O)   Conference Papers, *The Incineration of Municipal and Industrial Wastes*, Institute of Fuel (UK) 1969

Chapter 7

Electrical heating fundamentals

7.1 Resistance heating

Direct resistance heating
Indirect resistance heating

7.2 Arc heating

Range of applications
Electrode devices
Electrodeless devices

7.3 Induction heating

Penetration depth (skin effect)
Power factor
General advantages
Billet heating
Economics of induction through-heating
Induction hardening

7.4 Dielectric and microwave heating

Range of applications
Nature of high-frequency heating
Radiofrequency heating
Microwave heating

7.5 Radiant heating

Infrared radiation
Types of infrared heaters
Comparison of heater types
Efficiency
Benefits of radiant heating

7.6 Methods of accommodating a varying load

Heating techniques
Characteristics of some electric heating techniques
Energy-demand control

7.7 Lighting

Achieving good lighting with energy economy
Lighting measurement
Addendum

This Chapter is intended to indicate the fundamental principles underlying the various systems of electric heating and lighting as an introduction to the second Section (Electrical) of Chapter 8 on Furnaces, where the development of the various systems will be described. Such a clearcut division however is not always easy, and under a given topic the reader should ascertain what is said in both these Chapters.

Connection to Power Supply. Alternating voltage supply is now usual, but polyphase a.c. supply is commonly used for larger motors and heat-treatment equipment. This leads to advantages in both generation and transmission and the flow of power is more uniform than with single-phase a.c. Motors can have more rugged design and more uniform torque. A three-phase system is perhaps the most usual apart from single-phase. Current and voltage in the separate phases operate individually. With balanced polyphase systems power supplied is equal in all phases. Switching on and off, and overcoming mechanical inertia, are both made easier by their use.

7.1 Resistance heating

The passage of an electric current through a conducting material leads to the generation of heat in the material owing to its electrical resistance. The power developed is I^2R (watt) where I is the current (amp) and R the resistance (ohm). The resistance of the conductor is proportional to its length and inversely proportional to its area:

$$R = \rho \frac{l}{A}$$

where ρ is the resistivity.

The value of ρ differs over about three orders of magnitude even for electrically conducting materials, and also varies with temperature. The resistivity normally increases with temperature for most metals, but may decrease for non-metallic conductors. When materials exhibit a uniform variation of resistivity with temperature the relation is:

$$\rho_T = \rho_0 (1 + aT)$$

where ρ_0 is the resistivity at T_0, ρ_T that at T ($>T_0$) and a is the temperature coefficient of resistivity.

Resistance heating may be used where the material to be heated itself carries the current (direct resistance heating) and the heat is developed within the material, or where the heat is developed in a separate conductor or heating element (indirect resistance heating) and the workpiece receives the energy by a combination of conduction, convection and radiation.

Direct Resistance Heating

For efficient heating by direct resistance a high percentage of the total power available must be dissipated within the workpiece, i.e. the resistance of the workpiece should be high compared with the supply circuit. The resistance of the workpiece should be matched to the supply so that the maximum current is obtained at the rated voltage. Since the resistivity of most commonly used ferrous and non-ferrous metals is within an order of magnitude of the resistivity of the copper conductors used in the supply circuit, this limits direct resistance heating to relatively long workpieces and in practice a ratio of length to width of at least 6:1 is normally required. Uniformity of cross-section is also important since any constrictions in the current path will lead to high local values of current density and local overheating.

The high current and low voltage across the workpiece necessitate the use of step-down transformers. As well as minimizing the losses in the supply circuit, the reactance must also be considered. The contacts used to transfer the current from the source to the workpiece carry several thousand amps and should be of low resistance to avoid power losses and excessive contact heating, be usable many times, and be easy to apply to the material in order to achieve a high throughput. If the contact is allowed to become too hot, then apart from power losses welding may occur; but on the other hand a cool contact will result in cold ends in the workpiece. A number of contact arrangements are possible[1]: these include multiple axial and radial contacts, end contacts with variable pressure, the use of shims between the contact and workpiece, and auxiliary contacts supplying currents across the ends at right angles to the main axial current flow. For continuously annealing wire, electrical connection may be made to baths of low-melting-point metals, or pulleys with insulated segments may be used to reduce sparking.

Direct resistance heating has found its widest application in the through-heating of long rods, billets of ferrous metals prior to forging[1] and continuous annealing of ferrous and non-ferrous wire[2]. Typical heating times for 100 mm square billets are about 3 minutes; wire can be continuously annealed at speeds in excess of 1500 m/min.

Indirect Resistance Heating

Electric resistance heating is used for heating solids, liquids and gases by contact, natural and forced convection, and radiation in almost every conceivable area of industry[3].

Electrical heating elements should have a high resistivity in order that the dimensions may be consistent with a compact design and mechanical strength while keeping the current to acceptable values; there should ideally be only a small variation of resistivity with temperature in order to avoid control and operational problems. The material of the element must have a melting point higher than the operating temperature required, which in turn is higher than the maximum temperature of the furnace or workpiece, and must have high resistance to corrosion by oxidation and other chemical effects. The element should also have good mechanical properties from the point of view of strength at its operating temperature and ductility, at reasonable cost. The heating element itself should be designed so that the terminations are relatively cold.

The materials available include metals and non-metals which have some but not all of the above characteristics, and the final choice must be made on the basis of temperature requirement, operating environment and cost. *Table 1* lists some of the more commonly used materials.

The most commonly encountered class of resistance material is the metal alloys based on alloys of nickel and chromium, commonly available as

Table 1   Materials used   for   resistance-heating elements

Material	Maximum operating temperature in dry air ($^\circ$C)	Resistivity (10^{-12} ohm m at 20°C)
Nickel-based alloys*		
80 Ni/20 Cr	1200	110
80 Ni/20 Cr + Al	1250	
60 Ni/15 Cr/ 25 Fe	1100	110
50 Ni/18 Cr/ 32 Fe	1075	110
37 Ni/18 Cr/ 33 Fe/2 Si	1050	
54 Ni/46 Cu	400	49
Iron-based alloys*		
72 Fe/22 Cr/ 4 Al	1050	139
72 Fe/22 Cr/ 4 Al + Co	1375	145
Refractory metals		
Platinum	1400	11
Molybdenum	1750	5.7
Tantalum	2500	12.5
Tungsten	1800	5.5
10% rhodium/ 90% platinum	1600	19.2
40% rhodium/ 60% platinum	1750	17.4
Non-metals		
Silicon carbide	1600	10^5
Molybdenum disilicide	Up to 1800	40
Graphite	3000	10^3

* Approximate compositions only

wire and strip. The most versatile alloy is 80% nickel 20% chromium, which is capable of operation up to 1200°C in normal oxidizing conditions. The chrome-rich oxide provides a protective layer adhering to the surface and reducing further oxidation and it is resistant to scaling by repeated expansion and contraction during thermal cycling. The addition of a small percentage of aluminium extends the operating temperature to 1250°C. When the element is exposed to low partial pressures of oxygen the chromium is preferentially oxidized, which results in an increase in volume and damage to the oxide layer. The light-green oxide formed is known as green rot. If the atmosphere is also carburizing, chromium carbide forms along grain boundaries and further hastens deterioration. A similar form of corrosion occurs in the presence of sulphur.

The addition of iron to the nickel—chrome alloy results in lowering the maximum operating temperature and the resistance to oxidation and to cyclic deterioration. At the same time the cost is reduced and at lower temperatures the resistance to carburizing atmospheres is greatly improved. The addition of silicon further improves the resistance to carburization and sulphur attack. The nickel—chrome alloys are also capable of operation in endothermic and exothermic reducing atmospheres.

Nickel—copper alloys are used where a high degree of flexibility and only a relatively low operating temperature are required. The nickel—chrome alloys are used for wire, strip and cast-heating elements in ovens and furnaces, mineral insulated and cartridge heaters, and wire in woven flexible heating elements.

The ferritic iron—chromium—aluminium alloys are capable of operating at temperatures up to 1050°C in a wide range of atmospheres, and with the addition of cobalt operation up to 1375°C in dry air is possible. The aluminium forms a protective layer of aluminium oxide which because of the high affinity of aluminium for oxygen is stable even in carburizing or other reducing gases.

Platinum—rhodium alloys are capable of operation in air up to 1750°C. Molybdenum is capable of operating at temperatures in excess of 1750°C in reducing atmospheres and is used in hydrogen muffle furnaces for sintering metallic and ceramic materials. Tungsten and tantalum which oxidize rapidly in air are used in vacuum furnaces[4]. Tungsten is also used in high-intensity heat-lamps enclosed in a vacuum or protective atmosphere for heating by radiation. Source temperatures up to 3000 K are obtainable by using a quartz envelope with a halogen gas.

Of the non-metals, silicon carbide is extensively used in furnaces up to about 1600°C in oxidizing atmospheres for heat treatment of metals and glass. Operation in reducing atmospheres is possible at lower temperatures. Unlike metallic heating elements, silicon carbide has a negative temperature coefficient over part of its operating range, is brittle, and the resistance varies[5] during its life.

Molybdenum disilicide can be used in oxidizing and reducing atmospheres up to 1800°C and is used in the glass industry. The heating elements are fragile and have a high positive temperature coefficient of resistivity[6]. Graphite can be used in non-oxidizing atmospheres or *in vacuo* at temperatures up to 3000°C.

7.2 Arc heating

Range of Applications

Improved combustion stability and fuel utilization[7]: by injecting arc-heated gases into combustion chambers, it is possible to improve flame stability and to obtain net energy yields from otherwise unburnable fuels.

Plasma cutting[8]: plasma jets can carry out a very wide range of metal-cutting operations faster, more cleanly and more cheaply than conventional gas-cutting methods.

Metal refining[9]: by remelting metals in whatever atmosphere the process requires, very highly refined products can be obtained with a wide range of otherwise inaccessible qualities, e.g. extremely high nitrogen content.

Plasma chemical synthesis[10]: the use of very high temperatures makes possible a whole range of chemical process routes — this field is under rapid development at the present time.

Plasma particle-treatments and spraying[11]: refractory particles can be spheroidized by melting and vaporized in a plasma. By melting particles in a plasma gun, refractory coatings can be sprayed onto a wide range of materials.

Plasma melting of refractories[10,11]: plasma heaters can be used to melt refractory materials in a wide range of states of aggregation and in the atmosphere of choice.

Electric arcs are familiar principally because of their widescale use in arc welding. The essential characteristic of an arc is that it is composed of gas raised to such a high temperature (usually in excess of 6000 K) that it becomes highly ionized, i.e. free electrons are present which turn the gas into a sufficiently good electrical conductor for it to be maintained at high temperature by ohmic heating. Gas in this highly ionized state is usually referred to as a plasma. The extremely high temperatures attained in arcs permit heat-transfer rates to surfaces

to be obtained which are orders of magnitude higher than those obtained using, for example, combustion sources; thus whereas oxyacetylene flames can give heat-transfer rates up to about 50 kW/cm^2, plasma jets will give values in excess of 250 kW/cm^2. Moreover, the use of a plasma permits complete choice of atmosphere for processing at very high temperature, e.g. reducing, oxidizing and inert atmospheres. Plasma techniques also make it possible to conduct gas-phase processes at very high temperatures which would otherwise be impossible.

There are two essentially different types of arc-heating device — those with electrodes and those without, the latter depending upon energy transfer to the plasma by induction.

Electrode Devices[8]

Considering the electrode-based device first, *Figure 1*, the arc is struck between two electrodes and may be sustained either by a d.c. or an a.c. voltage; in the latter case the anode and cathode interchange positions each half cycle. There are three different regions of the arc which determine its stability and efficiency of utilization: the cathode region, the long column, and the anode region. Each region has a potential drop associated with it. The greater part of the power represented by the anode and cathode falls is transferred directly to the local electrode, whereas the power in the long column is liberated in the free gas region. Thus, if the purpose is solely to heat the gas, the electrode falls represent heat losses and the system should be designed in such a way that the arc is as long as possible, thereby minimizing the fraction of the power dissipated at the electrodes. For example, a 20 A arc drawn in free air between carbon electrodes has anode and cathode falls of about 10 V and a field in the long column of the order of 15 V/cm; thus an arc 2 cm long would run at a potential difference of 50 V, 40% of the power being dissipated in the electrode regions. Increasing the length to 10 cm would reduce the fractional electrode dissipation to about 10%. This is clearly important in gas-heating applications, e.g. for augmenting flames using electrical energy and heating gas for high-temperature chemical synthesis. If, however, the intention is to heat an electrically conducting solid by means of an arc it will usually be possible to make the workpiece one of the electrodes in which case energy dissipated at the workpiece electrode will not be wasted: e.g. plasma cutting, arc-melting and refining operations. In this case, the problem regarding the efficiency of utilization of energy is to ensure adequate transfer of heat from the gases in the long column to the workpiece. This is usually achieved by blowing the gas at the heated surface — the blast can be generated either by an externally applied gas flow (as in plasma cutting), or by electromagnetic pumping created by the arc contraction around the other electrode (as in the arc furnace).

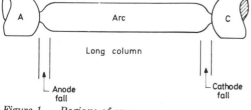

Figure 1   Regions of an arc

References p 113

The relative magnitudes of anode and cathode falls are strongly influenced by the gas, the current and the electrode material. Therefore it is not possible to generalize as to which polarity to make the workpiece in order to maximize energy efficiency; in any case there is a large unknown factor introduced by the fact that vapour emissions from the electrode regions will enter the long column, influencing its characteristics and thereby the convective heat-transfer processes. The reason behind

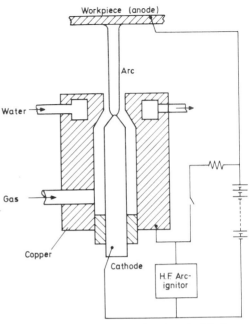

Figure 3   *Transferred arc showing electrical connections*

the choice of polarity will often derive from considerations of arc stability. Since the principal current carriers in the arc are electrons, the free emission of these from the cathode is of key importance in producing a stable discharge. If a cathode spot is created on a molten surface it will tend to move about wildly because of the great constriction of the discharge in its immediate vicinity — this wild movement of the cathode can lead to arc extinction. Extinction can also result if the cathode is subject to chemical attack — chemical attack on both electrodes, but particularly the cathode, is a limiting factor in many potential and actual applications of arc heating.

In both applications, either heating a gas or heating a surface, electrode erosion is a very important consideration. Anodes in particular suffer from thermal erosion processes because, unlike the cathode which is cooled by electron evaporation from its surface, the anode is heated by electron condensation; in these processes the counterpart to latent heat of vaporization is the work function, i.e. the energy expressed in volts necessary to remove an electron from the electrode. There are various techniques to reduce this problem, the most important being water-cooling of the electrode, rapid movement of the arc over the anode surface using swirling flows and/or magnetic rotation of the arc, and transpiration-cooled electrodes. *Figures 2* (gas heating) and *3* (surface heating) show embodiments of the more important points discussed above.

Figure 2: In the simplest case no attempt is made either to protect the cathode or move the anode

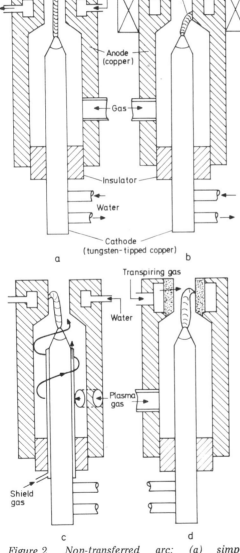

Figure 2   *Non-transferred arc:* (a) *simple,* (b) *magnetic swirl,* (c) *gas swirl,* (d) *transpiration cooled*

spot over the anode surface. Cases (b) and (c) using magnetic swirl and gas swirl respectively lighten the local heat loading of the anode by moving the anode spot at high speed, and with the applied magnetic field much more uniform gas heating is achieved at the expense of consuming extra energy in the magnetic coil (probably 5—10% of total). In case (c) a means of protecting the cathode from chemical attack is also shown. In oxygen-containing or other aggressive atmospheres protection of this kind is usually essential for long cathode life — however, lifetimes of several hours can be obtained by using zirconium cathodes in air, for example. Case (d) shows the anode (nozzle) being cooled by a transpiration flow of gas. Probably the most relevant development is the application of a magnetic field, because the uniform plasma-heating and rapid mixing with downstream additions make it very suitable for chemical synthesis and flame augmentation, both of which usually demand homogeneity for efficient operation.

Figure 3: The transferred arc which is widely used for plasma cutting operations is usually of very simple construction. The key to its operation is efficient cooling of the nozzle. To this end various elaborations such as coating the inner surface by a swirling film of water or replacing the nozzle by a porous water-transpiration-cooled plug are possible. Also shown are the power supply and high frequency arc-ignition source.

The devices discussed so far are d.c. However, a.c. operation is also possible, with saving in the capital costs and energy losses associated with rectifying equipment. The electric arc furnace, discussed elsewhere in this volume, is a good example

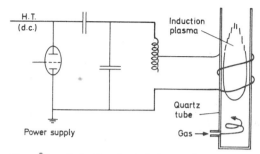

Figure 5   *Induction-heated plasma showing oscillator*

of a three-phase arc heater for applications in the multi-megawatt range. *Figure 4* shows a d.c.-stabilized three-phase system that overcomes the restrike instability encountered in pure a.c. systems. However, the non-uniform instantaneous power input associated with a.c., particularly single phase, could well disqualify its use in a large number of arc-heating applications.

Typically, the efficiency of the transformer/rectifier would be 90% and it is unlikely that much improvement will be made in this direction. The major inefficiencies occur in the arc heater itself in most cases. Typically, one would expect torch efficiencies of about 50—80%, great care being needed to ensure that the plasma's energy losses are minimized by running with long arcs and by keeping the arc away from the interior walls of the torch.

Electrodeless Devices[12]

Electrodeless discharge heaters are typified by the illustration in *Figure 5*. The plasma is contained within a duct usually made from an electrical insulator, e.g. quartz. Other constructions are possible, e.g. a water-cooled copper duct, segmented vertically to prevent the flow of currents around the cylinder. A high-frequency coil, connected to an oscillator, is wrapped around the duct. In the bulk of work carried out so far, operational frequencies have been in the radiofrequency range 0.5—7 MHz, the tube has been of 5—25 cm size, and the power in the range 20—1000 kW. With the high-frequency coil energized by the oscillator, the discharge is usually ignited by inserting a conducting rod into the duct. Once ignited, the discharge draws power from the oscillator circuit by inductive coupling from the encircling coil — the plasma acts as the secondary winding of a transformer, the coil as the primary. Herein lies one of the principal advantages of the induction plasma — no electrodes are required, which relieves problems of electrode erosion and product contamination by electrode materials. Relatively large discharges have been produced using these devices, e.g. up to about 25 cm in diameter. When running at radiofrequency the conversion of mains power to high frequency is about 60%, the coupling efficiency

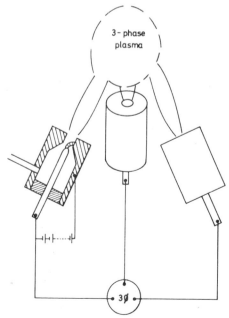

Figure 4   *D.C.-stabilized three-phase plasma*

References p 113

being of the same order; i.e. the overall efficiency is about 40%. However, significant improvements in the efficiency and stability of this type of discharge are to be expected when running at lower frequency (<4 kHz) and large diameters (>30 cm); at the lower frequencies efficient solid-state inverters can be used, and the larger plasmas associated with the lower frequency permit better filling of the coil, thereby lowering the loaded Q of the load circuit and consequently raising the coupling efficiency. In addition, recent studies using small additions of easily-ionized metals, such as potassium, have shown a plausible way of obtaining a plasma of properties compatible with the lower frequencies and larger diameters that are required for efficient operation.

This area of technology is undergoing rapid development at present. If the plasma-stability problem of using low frequencies can be solved economically, overall plasma-heating efficiencies similar to and perhaps even better than those obtained with electrode-based plasmas can be obtained (i.e. from 50 to 80%).

7.3 Induction heating

The phenomenon of induction heating is probably best known as the undesirable heating that occurs in transformer cores if they are not satisfactorily laminated. It is, however, widely and increasingly used for the through-heating, heat treatment and melting of metals, for which it has important advantages. An induction-heating coil surrounding a conducting workpiece is fed with alternating current and the resulting magnetic flux couples with the workpiece. In accordance with Lenz's law, current is induced in the workpiece in such a direction that it attenuates the field of the coil. It is this current, flowing through the resistance of the workpiece, which generates heat.

Penetration Depth (Skin Effect)

If the current in the workpiece is considered to flow in elemental concentric paths, it is seen that the effect of the field from the outer paths will be to attenuate the inner currents. Thus, the current density decays from the surface inwards in accordance with the expression:

$$I_x = I_0 e^{-x/p} \tag{1}$$

where I_x is current density at distance x below the surface, I_0 is surface current-density, and p is penetration depth.

Penetration depth (mm) is given by the expression:

$$p = \frac{5}{\pi} (10^5 \, \rho/f\mu)^{1/2} \tag{2}$$

where ρ is the resistivity of the material ($\mu\Omega$ m), μ is the magnetic permeability (units H/m, but here taken relative to a vacuum), and f is frequency (Hz).

At the penetration depth, the current has fallen to 38.6% of its surface value. Approximately 90% of the induced heat is developed within the penetration depth. For good electrical efficiency the penetration depth should not exceed half the workpiece radius (r). Thus, for a given workpiece diameter and material, a minimum frequency for efficient heating can be calculated:

$$f \nless (10/\pi)^2 \times 10^5 \rho/r^2\mu \nless 1.013 \times 10^6 \, \rho/r^2\mu$$

The smaller the workpiece the more rapidly does the required frequency increase. For example, when $\mu = 1$ we have:

Metal	Temp., t ($^\circ$C)	ρ ($\mu\Omega$ m) at t	r (mm)	f
Steel	1200	$\approx$1.0	630	50 Hz
			31.7	1 kHz
			10	10 kHz
Aluminium	450	0.09	9.5	1 kHz
			3.0	10 kHz

Values of resistivity and their temperature coefficients may be obtained (e.g.) in ref.(N).

With the exception of steel, most metals that require heating are non-magnetic, so that $\mu = 1$. For both through-heating and surface hardening, steel is non-magnetic at the final required temperature. Below the magnetic change point — about 770°C — the value of μ for steel depends upon the flux density to which it is subjected. In practice values are likely to lie between 1 and 40.

For convenience, the entire electromagnetic spectrum is indicated diagrammatically in *Figure 6*. This enables the various frequencies referred to in this Section, 7.4 and 7.5 (for heating) and 7.7 (for lighting) to be placed in perspective.

Power Factor

The resultant current in the workpiece lags 45° in phase behind the induced voltage so that, for a solid workpiece, the power factor (power/current $\times$ voltage) at the coil cannot be better than cos 45° = $1/\sqrt{2}$ = 0.707. Because of the leakage flux in the gap between the coil and the workpiece, however, power factor values in practice lie between 0.1 and 0.5 in the frequency range 50 Hz–10 kHz and are influenced by frequency, coil clearance, and resistivity of the load. Capacitors are normally connected across the induction coil to improve the power factor.

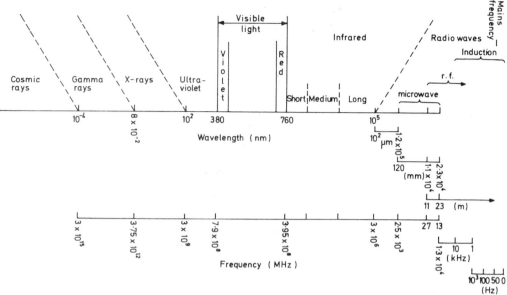

Figure 6 The electromagnetic spectrum (scale not constant overall)

General Advantages

Because heat is developed within the metal itself, heating can be extremely rapid. This in turn leads to metallurgical advantages, makes flow-line techniques possible, and economizes floor space. It is possible to limit heating to selected areas of a component. The environment around an induction-heating furnace is excellent.

Billet Heating

In a typical billet heater, a column of billets progresses on heat-resisting skid rails through a horizontal coil. A pusher feeds a cold billet into the coil at set intervals and ejects a heated billet to a discharge chute. The coil is wound from copper tubing, is water-cooled, and has a lining of thermal insulation.

This type of heater is widely used for heating steel for forging, and temperature is maintained consistently by accurately controlling the billet feed rate and the power to the coil. Frequencies of 1 kHz–10 kHz are used, and power utilization in the region of 320 kW h/t (1150 MJ/t) in terms of medium-frequency power is achieved.

A similar type of heater is used for non-ferrous extrusions, usually working at mains frequency and using either a single or a three-phase coil. In this case the temperature of the billet is measured before ejection, usually by thermocouple probe.

Typical energy consumption figures are:

Aluminium	(450°C)	260–280 kW h/t
Brass	(750°C)	175–210 kW h/t
Copper	(850°C)	240–250 kW h/t

Billets of mild steel, stainless steel and titanium alloy are also induction-heated for extrusion. Diameters can be as large as 355 mm (14 in), so that mains frequency can be used. These large billets are often heated in a vertical position, one billet to a coil. The coils are tapped to accommodate various lengths. Power control is achieved by on/off contactor switching, initiated by a signal from a radiation pyrometer. Other applications include the continuous heating of bar and tube for rolling, heating for soldering, and stress relieving.

Economics of Induction Through-heating

Besides capital and running cost, the overall profitability of the process must be analysed. For steel, the main areas of advantage are reduced production of scrap because of the excellent temperature control and reduced loss of metal (particularly applicable to alloy steels) owing to elimination of scale. Moreover there is no surface decarburization or grain growth.

With non-ferrous metals, the excellent temperature control and freedom from surface contamination is significant. Generally, induction heating gives a greatly increased operational efficiency, start-up is virtually immediate, only a small floor area is required, and the workshop environment is excellent.

Induction Hardening

The hardening process makes use of the skin effect to limit heating and hardening to surface areas only, leaving the centres ductile, ...d requires self-hardening carbon or alloy steels. High power densities are used, up to 25 W/mm², for shallow depths, depending on frequency. For deeper hardening, longer heating times are required to permit more heat conduction from the surface inwards.

There are three main categories of induction hardening:

Single Shot

The heating coil (or inductor) is often a single turn and is designed to embrace the area to be hardened. In some cases, e.g. cams and crankshaft bearings, the inductor is split, being closed onto contact faces. Very short heating times of 1—5 s are usual. For quenching, the workpiece is either lowered into the quenchant or the process occurs *in situ* through a specially designed inductor.

Progressive Hardening

Where a large area, e.g. an axle shaft or a lathe bed, is to be hardened, the single-shot process would require a very high power rating to obtain the required power density. In this case, an inductor is progressed along the work, followed by a quench spray.

Semi-Progressive Hardening

This technique has recently been applied to crankshafts and axles, where high production rates are required. A 'hairpin-type' inductor is arranged axially over the area to be treated to cover the whole length. The component is rotated beneath the inductor and, when hot, is rapidly transferred into the quenchant. While the power requirement is no less than for the single-shot process, this technique permits rapid transfer from inductor to quenchant.

7.4 Dielectric and microwave heating

Range of Applications

Drying: high-frequency electric fields have the capability of rapid uniform drying of dielectric materials, e.g. paper, fibreglass, wood, laminated coatings, textiles (cloth and fibre), ceramics, foundry cores, etc.

Curing, fixation and welding: the rapid volumetric heating achievable with high-frequency fields makes the method suitable for curing synthetic rubber, inks, glues, resins, the fixation of dyes and the welding of plastics.

Food: a wide range of applications, e.g. cooking of food just before serving, sterilization in packages, sterilization of dairy produce, defrosting meat and fish, biscuit cooking, bread baking, deinfestation.

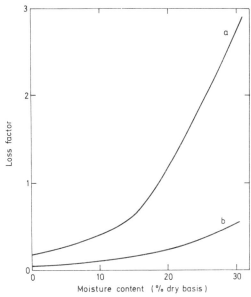

Figure 7   Loss-factor of paper as a function of moisture content with field (a) parallel and (b) perpendicular to web

Plasmas: both high- and low-pressure plasmas can be produced in high-frequency fields with absence of electrode contamination.

This list is indicative rather than exhaustive.

Nature of High-frequency Heating

When an electric field is applied to a dielectric the material becomes polarized, i.e. there is a separation of positive and negative charges on a microscopic scale. If the field reverses direction, the polarization charges reverse their positions. Thus if the field oscillates, the polarization charges oscillate, giving rise to an alternating current. In different materials the motion of the charges is resisted to different degrees and gives rise to local heating, the effect being particularly strong in those with permanent dipoles (e.g. water). The rate of heating per unit volume (W/m³) is given by:

$$2\pi f \epsilon_0 \epsilon_r'' E^2 \tag{3}$$

where E = local peak field strength (V/m), ϵ_0 = permittivity of free space (8.85×10^{-12} F/m), f = frequency (Hz), and ϵ_r'' = dielectric loss factor.

The loss-factor is a property of the material and varies widely from one material to another. It is also a function of the electric field direction in many cases, and always of the degree of compaction and moisture content. *Figure 7* shows the effect of moisture content and field orientation on ϵ_r'' for paper at radiofrequency. The very high loss-factor associated with the wetter areas of materials means that high-frequency drying tends to be selective, the most power going to the wetter areas, thereby leading to uniform moisture profiles.

A very important property for all applications of the technique of high-frequency fields is their ability to heat throughout the volume of a material at a high rate without the limitations set by maximum permissible temperatures, as with conduction, convection and infrared heating. In practice a limitation to the maximum thickness of material that can be processed uniformly is set by the penetration depth δ (m):

$$\delta = \frac{(2)^{1/2}c}{4\pi} \bigg/ f \left\{ \epsilon_r' \left[\left(1 + \left(\frac{\epsilon_r''}{\epsilon_r'} \right)^2 \right)^{1/2} - 1 \right] \right\}^{1/2} \quad (4)$$

where ϵ_r' is the dielectric constant of the material relative to free space, c is the velocity of light in free space, and $(2)^{1/2}c/4\pi = 3.4 \times 10^7$.

It is unusual for penetration depth to be a limitation at radiofrequency, but it could be at microwave frequency: e.g. at 27.2 MHz the penetration depth in seasoned beech is 4.6 m, whereas at 2450 MHz the skin depth is only 4 cm.

When considering the use of high-frequency equipment it is necessary to calculate the field strengths for the required heating rate from equation (3), and the penetration depth from equation (4). Too high a field strength will lead to breakdown, and a penetration depth small compared with the minimum dimension of the workpiece will lead to non-uniform heating (except when using a parallel plate applicator with a plate separation small compared with the wavelength within the material).

Because of interference with telecommunications, only certain frequencies are permitted for high-frequency heating. The most useful are 13.6, 27.2, 896 and 2450 MHz — a code has been laid down in ref.13. The lower two frequencies are in the radiofrequency range and correspond to free-space wavelengths of about 22 and 11 m respectively. These comparatively long wavelengths mean that in the vast majority of cases applicators can be designed as capacitors (the various geometries are discussed later). The higher two frequencies, which are in the microwave range, correspond to wavelengths of about 0.33 and 0.12 m. At these short wavelengths it is necessary to contain the energy either within an enclosure (waveguide or resonant cavity), or by means of a specially constructed transmission line, to avoid excessive radiative losses. In continuous heating systems it is necessary to provide inlet and outlet apertures for the processed material. Since apertures can lead to radiation, their size is limited and care is needed in their design, particularly with microwaves where there is a hazard problem (see ref.14 for code of practice). Aperture height is rarely a limit in practice on the size of products that can be processed.

Radiofrequency Heating[15,16]

Radiofrequency heating is based upon energy transfer in a resonant circuit energized by inductive (transformer) coupling from a power supply oscillating at the specified frequency, as shown in *Figure 8.* Therefore it is necessary (*a*) that the natural frequency of the load circuit should be made equal to that of the power supply within close tolerance and (*b*) that the two inductive loops forming the transformer link between the circuits should be set close to the value for critical coupling.

Typically, radiofrequency power can be generated with an efficiency of about 65–75%. However, there is a further power loss in the applicator itself which can be quantified in terms of the Q-values of the load circuit with and without the workpiece inserted within the capacitor:

$$\eta_{overall} = \eta_{RF} \left(1 - \frac{Q_l}{Q_u} \right) \quad (5)$$

(the subscripts l and u stand for loaded and unloaded respectively). Thus to maximize the overall efficiency it is necessary to minimize Q_l and maximize Q_u. If the material to be heated is very lossy, e.g. a piece of wet wood, Q_l will be so very much lower than Q_u that no special precautions are needed to obtain high efficiencies. However, in many cases this is not so, e.g. partially dried paper and textile fibres; in this case it is important to make Q_l as low as possible, for example by orientating the field in the direction in which the loss factor is a maximum (i.e. parallel to the surface in the case of paper) or by compaction as in the case of fibres. It also helps to aim at the highest possible value of Q_u, by making use of a good electrical conductor for all current-carrying members — copper will give a significantly higher value of Q_u than stainless steel, for example.

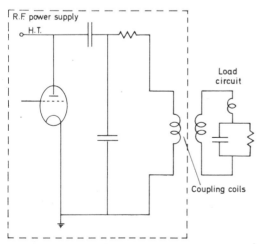

Figure 8   Schematic of r.f. power supply and load circuit

References p 113

Applicators are made in basically three different geometries, *Figure 9*: (a) through-field, (b) stray-field, (c) staggered through-field. Each has its own characteristic field direction to help in matching to the directional properties of the material.

The through-field geometry is particularly suited to thick materials. The stray-field geometry, with its field parallel to the platten, is suitable for thin materials and is particularly to be favoured if the loss factor is greatest for fields in the plane of the surface. The staggered through-field geometry is useful either when the through-field configuration would lead to an unacceptably large capacitance, making tuning impossible, or when a fairly thick mat of material is to be processed which has a maximum loss factor for fields in the plane of the mat.

An advantage of these resonant systems is that very little power is drawn if there is no dielectric in the applicator, even when the power supply is turned on. Moreover, there is a self-adjustment property that is very useful: if appropriately set up, the power will adjust automatically to the load. For example, if a wet material passes through, the power drawn will rise and fall in sympathy with the instantaneous moisture content present between

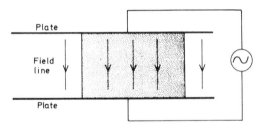

a. Through field

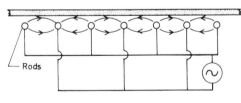

b. Stray field

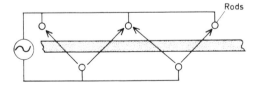

c. Staggered through field

Figure 9   The three basic r.f. applicator geometries showing field lines and location of product (a) through-field, (b) stray-field, (c) staggered through-field

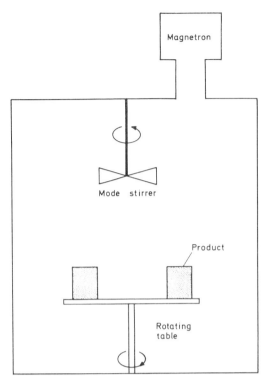

Figure 10   Multimode microwave cavity — field uniformity achieved by means of a rotating table and mode stirrer

the electrodes thereby leading to a uniformly dried product in the direction of product-motion as well as over its cross-section.

Microwave Heating[16—18]

Microwave devices can be based upon either resonant or non-resonant systems. An example of the resonant is the domestic cooker which is simply a metal cavity into which microwaves are transmitted from a magnetron, as in *Figure 10*. Because of the small wavelength, standing wave patterns are set up within the cavity which can lead to non-uniform heating, the r.m.s. electric field strength being a strong function of position. There are two ways of dealing with this situation. On the one hand, either the charge can be rotated within the cavity, or a rotating vane can be inserted to cause the field pattern to vary with time, the objective in both cases being to produce uniform heating by giving a uniform time-averaged exposure to the electric field for each part of the charge. On the other hand, the non-homogeneity of the field can be turned to advantage by matching the field distribution to the shape of the charge. Thus in a cylindrical cavity it is possible to set up a mode with a maximum field strength along the axis which drops off rapidly with radial distance; such a configuration is particularly suitable for high power-density, with very efficient coupling (up to 90% applicator efficiency) into cylindrically shaped charges. An example is a rope of material or a fluid contained in a pipe, passing along the axis (*Figure 11*). Other configurations are also possible.

Resonant microwave systems are subject to the same considerations as regards the loaded and unloaded Q (see equation (5)) as resonant radio-frequency systems. Therefore for high efficiency proper matching is necessary. However, even if no special measures are taken multimode ovens of the type shown in *Figure 10* will usually give overall efficiencies in excess of 50%.

The short wavelength of microwaves allows them to be used in non-resonant systems also. In this type of system the microwaves are transmitted (ideally without any back reflection) down a waveguide within which the charge is placed. One possible geometry is shown in *Figure 12*. As it proceeds, the wave is attenuated both by the charge and by losses to the walls of the guide. The power applied to the material varies exponentially with distance (z) from the magnetron:

$$P = P_0 \, e^{-2az} \qquad (6)$$

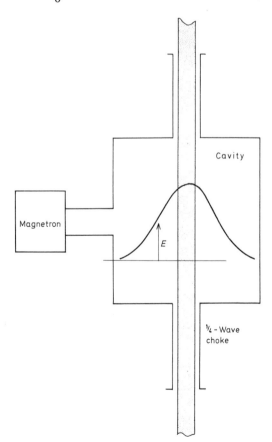

Figure 11   Single-mode microwave cavity showing radial distribution of the electric field

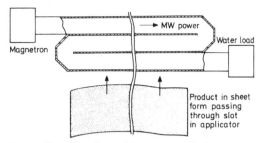

Figure 12   Meander microwave applicator — sectioned plan view

where a is the attenuation constant for the loaded waveguide. This attenuation can lead to non-uniform heating. The usual way around this problem is to use a waveguide that turns back on itself — if the attenuation length is much greater than the width of the charge, uniform heating will be achieved. There is always a certain amount of power reflection from the end pieces of the guide, which give rise to a residual superimposed standing-wave pattern; the deleterious effects of this upon uniform heating can be obviated by, for example, passing the charge through on the skew, or staggering the phase of the residual standing wave pattern by suitable choice of length of limb for the waveguide. It is usual to incorporate a water load at the termination of the waveguide to guard against power being reflected back to the magnetron and causing damage.

7.5 Radiant heating

All radiation can be transformed into heat, but the infrared waveband has the maximum heating effect, and infrared heating is simply an alternative name for radiant heating. Its advantages include precise control or variation of temperature, short heating and cooling times, rapid heat transfer, the ability to focus the heat and the absence of pollution from products of combustion.

Infrared Radiation (cf. App.C11)

All bodies at temperatures above absolute zero are continually emitting radiant energy in the form of electromagnetic waves, which pass through space and are partly absorbed by any other bodies they strike. Some of the radiant energy reaching a body may be reflected from its surface, some absorbed and some may be transmitted through it. That absorbed is converted into heat; and there is a net transfer of heat by radiation from any body to another which is cooler. The rate at which a body absorbs radiant energy is governed by its emissivity and the waveband over which the radiation is spread. Emissivity is defined as the amount of radiation emitted or absorbed by a body compared with that of a black body under identical conditions. A black body is one which theoretically absorbs all of the available radiation at whatever wavelength. The distribution of energy radiated among wavelengths depends on the temperature of the source; the wavelength of peak emission becomes greater as the temperature is reduced, i.e. $\lambda_m\theta$ = constant, where θ is absolute temperature.

Infrared radiation lies just beyond the red end of the visible part of the spectrum, extending to the radio waveband (see *Figure 6*). Being one of the major portions of the total, the infrared waveband is divided arbitrarily into three parts, namely the long (3300—5000 nm), medium (2500—3300 nm) and short (1200—2500 nm) wavelengths.

The total amount of energy radiated away from a heater is proportional to the fourth power of the absolute temperature of its surface. Clearly, there is an advantage in operating a heater at as high a temperature as practicable. Although many heaters operate at black heat, others are designed to run at red heat or even higher temperatures — the short-wave region which has the maximum heating effect of all types of radiation.

Types of Infrared Heaters

Reflector Heat Lamps

These are similar to incandescent lamps, but there are some differences in construction and filament design; used for process-heating up to about 300°C. Tungsten filament normally operates around 2400 K, peak wavelength 1200 nm. Ratings up to a few hundred watt at 240 V. Glass envelope, parabolic in shape, resistant to thermal shock; screw cap, internally coated reflector. Average life 5000 h.

Quartz Tubular Heaters

A tungsten or Kanthal spiral element is supported in a clear quartz tubular envelope. Short-wave types normally operate around 2400 K, peak wavelength 1200 nm, and medium-wave types around 1200 K, 2400 nm. Ratings, from a few hundred watt to 20 kW region, for process-heating up to about 1300°C (nearly 1600 K) at the short-wave end. Very resistant to thermal shock, but neither type should be subjected to excessive vibration or mechanical shock. Short-wave emitters have an average life of 5000 h; medium-wave units should last for several years.

Metal-sheathed Element Heaters

Tubular sheath materials such as stainless steel, Inconel, etc., contain a spiral element of nickel—chrome embedded in mineral insulation (magnesium oxide). Usually mounted in anodized aluminium reflector housings. Operating temperature 700—750°C, i.e. 1000 K, peak wavelength around 3000 nm. Normally for 240 V supply. Various sizes, shapes, and ratings are available.

Ceramic Radiators

A coiled heating-element of a nickel—chrome alloy is fused into a ceramic material to form a tubular or trough heater. Glazed surface gives protection against oxidation and corrosion. Life expectancy up to 10 000 h. Integral or external reflectors can be used. Operating temperatures from 300°C to 700°C, i.e. up to $\approx$ 1000 K, peak wavelength around 4000 nm. Unit ratings up to 1 kW are typical.

Comparison of Heater Types

Reflector Heat Lamps

Respond quickly to switching both in heating and cooling, and therefore suitable for use with materials on conveyors without risk of overheating during stoppages. Radiant efficiencies high and convection losses low, but heat output limited to about 10 kW/m² by bulb size and the need for spacing. Allowance should be made for fragility.

Quartz Tubular Heaters

Capable of high-intensity radiation up to 80 kW/m². Gold-plated or air-cooled anodized-aluminium reflectors are used, depending on ratings. Higher intensities are possible by staggering the heaters in two planes. Quick response to switching is due mainly to the limited thermal mass. The linear shape gives good coverage of moving materials, an improvement over the localized radiation of a heat lamp.

Metal-sheathed Element Heaters

Have the great advantage of mechanical strength and flexibility. Can be bent to suit the contour of the workpiece and reflector. Standard projection units give intensities up to 40 kW/m² but their comparatively high thermal mass means longer heating and cooling times, necessitating retraction devices to prevent damage to products on conveyors during stoppages.

Ceramic Radiators

Multiple units can give intensities up to 40 kW/m² from narrow-beam projectors. High emission factors are obtainable from ceramic surfaces. Retraction devices may be necessary to protect sensitive goods during heater cooling-time. Unaffected by cold water splashes, but less rugged than metal-sheathed heaters.

Efficiency

As has been pointed out in Chapter 1, the efficiency of a process if often difficult to define; it may contain many factors and may ultimately depend upon a point of view. The thermal efficiency calculated as the ratio of the heat absorbed by the product to the nett heat equivalent of the power input can be as high as 70% — or only 20% if based on average gross heat equivalent. But it is necessary to consider the total efficiency of the process. For example, to dry a product naturally in ambient factory air might take several hours as well as considerable floor space. An infrared oven with conveyor could reduce this time to minutes or even seconds, thus preventing bottlenecks, saving space and reducing handling. Comparison of the thermal efficiencies of the two methods would be invidious even though the first requires (apparently) no added heat.

Benefits of Radiant Heating

Compared with convection ovens infrared systems have the advantage of high rates of heat transfer because of the large temperature differential between the source and the product and the fact that the heat can be focussed. In addition rapid start-up saves time and energy. Installation costs are usually lower, as the heaters can be mounted on rudimentary structures. Electric infrared heaters themselves cannot cause pollution as there are no products of combustion.

Oven-space requirements are minimal and maintenance times are reduced as rapid cooling allows access to heaters within minutes. Flexibility in operation is achieved by selection of heater banks, thus changing the effective length and width of tunnel ovens or web-drying systems. Infrared heating has been found to be competitive with other heating methods both in initial cost and running cost.

7.6 Methods of accommodating a varying load

Varying electrical heating loads arise owing to the characteristics of the heating technique, to the need to observe particular heating requirements, to varying production schedules, and to the unregulated and simultaneous application of more than one heating load.

Apart from facilities required to accommodate characteristic process-load variations, there can be a strong financial incentive to install control equipment to minimize the charges of electricity supply undertakings.

Heating Techniques

Industrial-heating techniques can be broadly divided into those used in the metal industries and those in the non-metal industries.

Heating in the Metal Industries

The most important electrical heating processes are: arc melting; induction melting; indirect resistance melting; induction heating for hot working; indirect resistance heating for hot working; plasma-heating processes; indirect resistance heating for stress relieving and preheating; direct resistance heating of billets; induction surface-hardening.

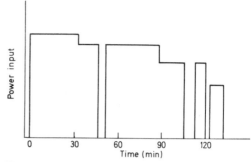

Figure 13 Typical power input programme for an arc furnace

Heating in the Non-metal Industries

The most important electrical heating processes, by industry, are:

Food Industry. Indirect resistance heating; dielectric heating; microwave heating; infrared heating.

Paper, Print and Textile Industries. Indirect resistance heating; infrared heating; dielectric moisture profiling; microwave drying; ultraviolet curing.

Rubber and Plastics Industries. Indirect resistance heating; dielectric curing; microwave curing; infrared heating.

Glass and Ceramics Industries. Direct resistance melting; indirect resistance heating; infrared heating; induction heating; dielectric heating.

Characteristics of some Electric Heating Techniques

Arc Melting

In arc melting, arcs are struck between the material constituting the charge and vertical electrodes. The electrodes are raised and lowered by means of motors, or hydraulic rams, to control the amount of power released in the furnace.

Depending on the amount and type of charge material, furnace size and power rating a given installation will exhibit an optimum power-input programme (*Figure 13*).

Induction Surface Heating

Method of Control. Solid-state inverters are increasingly being adopted for medium frequency (1—10 kHz) power sources. Power control is simple and inexpensive. Maintenance of predetermined power output under varying load conditions is achieved by using a closed-loop control system associated with the thyristor firing circuits.

Direct Resistance Heating

Heat is produced by passing a high current from a low-voltage source through the material to be heated. Direct resistance heating (*Figure 14*) is used to heat billets prior to forging or rolling. Heating times vary between 45 and 75 s, depending on billet size. Close control and short heating cycles enable direct resistance heaters to be integrated readily into an automatically controlled production line.

Method of Control. A radiation pyrometer can be used to initiate control of energy input. To ensure correct heating, complex control equipment is sometimes required. One approach, using computer control, involves weighing the workpiece (billet) before heating to determine the total heat required. The units of energy supplied are integrated and the heater is switched off automatically when the correct amount of energy has been delivered. Precise matching of energy input to each individual billet is thereby assured. On crackable steels, where immediate operation on full voltage may introduce non-uniform cross-section heating, the voltage should be reduced to slow down the heating rate and reduce temperature differences.

Heating in the Non-metal Industries

In the non-metal industries, the most prevalent heating techniques are indirect resistance, dielectric and microwave heating. These are usually applied continuously and their electrical load characteristics tend to bear a direct relation to the rate of throughput of the product being heated.

Energy-demand Control

It is not the function of an energy-demand control system directly to reduce the total amount of electrical energy consumed, although in practice this is likely to happen. The primary function is to stabilize the rate at which electrical energy is consumed and thereby to reduce the demand charge which is levied by electricity undertakings. To use electricity at the lowest cost, it should be taken at a constant rate. Practical considerations make this impossible, but the maximum demand can be reduced by adopting one of several possible control philosophies. These can be summarized as follows:
(1)   The temporary disconnection (shedding) of selected loads.

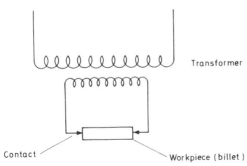

Figure 14   Diagrammatic representation of direct resistance heating

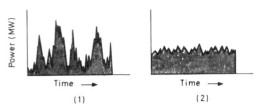

Figure 15   (1) Example of daily demand curve with unregulated use of electricity. (2) Example of daily demand curve with regulated use of electricity. The total energy consumption is approximately the same in both examples. The maximum chargeable demand in case (2) is, however, lower

(2)   Staggering the use of electrical heating processes to avoid the simultaneous coincidence of individual peak electrical loads.
(3)   The redeployment of selected electrical loads to off-peak periods.
(4)   The use of a computer based, resource optimizing facility to minimize overall production costs, including those arising from electricity-demand charges.

Demand charges, as distinct from energy (kW h) charges, arise as a consequence of the investment in plant which electricity undertakings have to make to meet the demand created by customers. It is therefore of mutual benefit to suppliers and users of electricity that the maximum energy be provided by the minimum plant.

Planned Diversity

In almost all industrial plants, some diversity exists in the use of items of production plant, including electrical heating processes, and planned advantage should be taken of this built-in characteristic. Work should be programmed to avoid the simultaneous coincidence of individual maximum demands created by different electrical heating processes. An examination of the possibilities, to reduce demand, might begin with a review of the daily programme of use of the larger heating processes. It will often be found that certain items are in use for a proportion of the day, and if these can be programmed so that their individual maximum demands occur consecutively rather than simultaneously, then a reduction in demand will result. The effects of programmed loading can be seen in *Figure 15*.

Programming Strategies

Programming strategies are used, for example, in steelworks where melting is carried out in arc furnaces. By staggering the operating cycles of furnaces so that they are never simultaneously at full load, substantial reductions in chargeable demand are achieved. Similarly, electrical heating techniques in general offer opportunities for savings. Even in small works it may be possible to transfer certain heating loads to off-peak periods, thereby achieving significant financial savings.

Where processes are capable of interruption without danger or loss, then a load alarm can be installed to warn the management as the electrical demand approaches a predetermined level, and the appropriate action can be taken. Some instruments are provided with an auxiliary relay by means of which selected electrical heating loads can be automatically switched-off (shed) to avoid exceeding the planned maximum demand.

Where, for purposes of maintenance, electrical heating facilities have to be taken out of service, this should be done when possible at times of peak demand.

Each individual plant using electric heating processes will have a changing, yet unique, maximum demand characteristic. In complex industrial plants a computer is the best means of storing, measuring, calculating and controlling the many variables associated with a comprehensive energy-demand control installation.

7.7 Lighting

All workplaces require artificial lighting to facilitate the efficient and safe performance of tasks, and to provide an environment conducive to the comfort and satisfaction of the occupants. Natural lighting by windows and rooflights may make a contribution to the lighting, but large areas cannot be lighted satisfactorily by side windows alone. Furthermore, every window is a potential source of heat loss in winter, and of sky glare and excessive solar heat gain in summer. Last but not least, in many areas both inside and outside activity must continue for 24 hours a day.

By good design of the lighting system, with consideration of the roles to be played by electric and natural lighting, and with proper assessment of the heat gains from the lighting equipment (and the heat gains and losses in the building) an energy-effective system of illumination may be designed. Choice of the types of lamps, luminaires, installa-

tion method and maintenance practices selected will all affect the energy consumption needed to maintain a required standard of illumination, and the cost in use will be significantly affected by these same factors.

Benefits of Good Lighting

The quantity of illumination (the illuminance) and various quality aspects of the lighting must enable the full benefits of good lighting to be achieved, for only then will the potential economic benefits together with safety and staff welfare be fully realized. Restriction of lighting level may well be a false economy, since far greater consumption of energy (and greater cost) may result owing to inefficient operation of plant and a lower standard of work performance by operatives. The diagram below summarizes the lighting/benefit relation.

The list of benefits given in the diagram is far from exhaustive. For example, better lighting may enable a better standard of routine maintenance in a plant to be achieved, resulting in fewer breakdowns and interruptions of production; or, in the event of a breakdown, a return to production may be achieved more quickly. The term 'better work' includes quality of performance; when tasks are heavily vision-dependent (inspection, laboratory work etc) the return on lighting investment may be very great. Indirect and direct benefits will improve staff relations. This can result in reduced labour turnover (with a reduction in the attendant heavy costs), reduced absences for sickness, better time-keeping and above all fewer accidents[19,20].

The relation of productivity to lighting has been clearly established[21]. In general, output tends to increase with better lighting, until the limit of physical dexterity or capability of the operator is reached (*Figure 16*). This applies to 'operator-paced' tasks; with 'machine-paced' tasks, the improvement in productivity arises from reduced down-time, shorter intervals between operations, greater consistency, accuracy or vigilance in performance.

The discrepancy between the visual performance of older and younger workers is greater at lower

illuminances. Also, there is often economic and energy-management justification for providing illuminance higher than that which will just satisfy the needs of the eye as regards acuity, for if the subject feels satisfaction with his environment, his performance may be enhanced because of psychological effects.

Standards of Lighting

Theoretically it is possible to conduct experiments to discover the illuminance at which a task will be performed well. In practice it is simpler to refer to a standard embodying the extensive experience of thousands of previous schemes. Such a standard (and the only one generally available and universally respected in the UK) is the IES Code[22]. This provides reliable recommendations as to illuminance, and also gives guidance on the limiting glare index which will be appropriate to the task and location, as well as helping in the choice of the light source in terms of colour-appearance and colour-rendering properties. A short document which expounds the essential principles in simple language is available[23].

Achieving Good Lighting with Energy Economy

Efficient use of energy for lighting purposes may be studied under such headings as lamp efficacy, luminaire efficiency, installation design, control and maintenance. Attention must also be given to the co-ordination of natural and electric lighting and, in suitable circumstance, to considerations of the principles of Integrated Environmental Design which include the thermal aspect.

Lamp Efficacy

Tungsten-filament incandescent lamps do not produce enough luminous output per watt of energy consumed to justify their use for general lighting

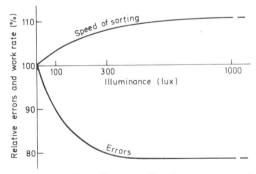

Figure 16   The effect of illuminance on work performance. Workers were asked to sort six easily distinguished types of screws into separate lots. For this simple task the benefits were marginal above about 600 lux

in factories. They should be replaced with higher-efficacy fluorescent tubular lamps or high-pressure discharge lamps. Using these higher-efficacy sources can produce an energy saving of up to 80% for the same luminous flux (lm). With fluorescent-coated lamps, the energy of the radiation first produced is partly converted by fluorescent materials to secondary radiation in the visible region. In discharge lamps, the radiation produced favours predominantly spectral lines in the visible region associated with a particular chemical element.

Typical efficacies for White and Warm White fluorescent tubes are shown in *Table 2*. Tubes with better colour-rendering have somewhat lower efficacies. Typical efficacies for discharge lamps are shown in *Table 3*. In general, the more powerful the discharge lamp, the higher is the efficacy.

Many industrial processes can be performed efficiently under the illumination provided by high-pressure sodium (SON, SONT) lamps, which have colour-rendering properties quite acceptable for tasks involving no critical colour-discrimination. Where better colour-rendering to a reasonable standard is required, mercury-halide (MBI) lamps will often be acceptable. For critical colour tasks, fluorescent tubular lamps of such types as Colour Matching or Artificial Daylight should be employed.

Luminaire Efficiency

Luminaires should be selected for optical performance (cut-off, distribution) and also for a high downward-light/output ratio (within the limits imposed by control of glare). Durable reflector materials having low rates of deterioration in reflecting power should be selected. For discharge-lamp luminaires in high-bay applications, glass refractor-reflectors have much to recommend them. Fluorescent luminaires should be selected from types having reflector surfaces and prismatic enclosures which do not tend to turn yellow with time. Bare-tube batten fluorescent-tube luminaires are inherently inefficient in use; although they have a high light/output ratio, they do not make the best use of the light.

Installation Design

In general, it is more efficient to have a small number of high-wattage luminaires rather than a large number of luminaires of smaller power. But, in most cases, this will necessitate higher mounting-heights which in turn increase the cost of maintenance — though the total cost of replacement lamps will be less. It is not economic to have large numbers of bench-lights and machine-mounted lights in an attempt to make good a poor standard of general overhead lighting. But some degree of localizing of the overhead luminaires to the principal areas of activity can lead to significant economy.

Commonly, a general illuminance is selected as being suitable for movement and for the performance of the simpler tasks; this illuminance is

Table 2 Fluorescent tubular lamps (White and Warm White)

Nominal lamp size (W)	Nominal length (mm)	(ft)	Approx. total circuit energy (W*)	Lighting design (lm)	Typical overall efficacy (lm/W)
40	1200	(4)	50	2700	54
65	1500	(5)	80	4650	58
80	1500	(5)	95	5700	60
85	1800	(6)	95	6100	64
85	2400	(8)	100	6700	67
125	2400	(8)	140	8600	61

* For switchstart circuits

Table 3 Discharge lamps

Type	Nominal lamp size (W)	Lamp shape*	Approx. total circuit energy (W)	Lighting design (lm)	Typical overall efficacy (lm/W)
MBF	250	E	275	12 400	45
SON	250	E	290	21 000	72
MBF	400	E	435	21 800	50
MBI	400	E or T	445	28 000	63
SONT	310	T	350	32 000	91
MBI	2000	T	2080	150 000	72
MBF	2000	E	2080	118 000	56
SONT	1000	T	1100	110 000	100

* E: eliptical; T: tubular

boosted by higher-wattage or extra lamps over the critical areas. Diversity of illuminance in the area must not be so excessive as to produce discomfort or confusion. Direct and reflected glare both tend to reduce the 'seeing value' of the illuminance provided; better control of glare improves the contrast-rendering of what is to be seen, and can lead to economies in the selection of the illuminance.

Control

Where the daylight contribution is sufficient to enable the work to be done without recourse to electric lighting, the latter may be switched out. The control of electric lighting by photoelectric switches may reduce wastage during 'daylighting' hours. The system of PSALI (permanent supplementary artificial lighting of interiors) is one system of combining daylighting and electric lighting[24]. Windows should be kept clean to make the best use of daylight.

Décor

Dark surfaces, dark walls and ceilings, and, indeed, dark floor surfaces, all absorb light. A coat of light-coloured paint on all possible surfaces will raise the utilization factor of the lighting installation.

Maintenance

Replacing discharge lamps and fluorescent tubular lamps towards the end of their rated life (usually 7500 h), plus regular cleaning of the lamps and luminaires, can produce an improvement in achieved illuminance of as much as 25%. If the cost of maintenance is not excessive, it may be possible to design the installation for a higher maintenance factor and so reduce the energy used by as much as 20%. Bulk replacement of lamps is almost invariably more economic than sporadic replacement, for discharge lamps and fluorescent tubes may continue to function (though at low output of light) long after their economic life has expired[25].

Exterior Lighting

Many tasks can be performed efficiently and economically out of doors if lighting is provided to enable work to continue in the hours of darkness. For example, storage of large items, weatherproof pallets of goods etc; and the construction or assembly of items too large to be conveniently housed in buildings.

The night security of premises may be enhanced by the provision of a system of security lighting[26]. In these applications, the use of discharge-lamps will enable the system to operate at high efficacy. Suitable design will permit the achievement of effective lighting with minimum energy usage.

The illuminance within an interior should be graded when persons need to pass from the interior to the exterior frequently. By day, the illuminance in the interior should be graded upward towards the exit; by night, it should be graded downward, so that at both times the eye is not subject to too rapid and excessive illuminance change on passing from one area to the other. This advice applies most particularly to buildings with restricted window area.

Integrated Environmental Design (IED)

This is a system in which heat gains within a building (due to solar radiation, waste heat from processes, from the occupants, and from the lighting) can be conserved by good building insulation, and re-used to reduce the total energy requirements of a building[27]. In some cases, little or no space-heating need be provided.

The system often requires window areas to be small to reduce the summer heat gain and winter heat loss. If the building is 'deep' (having large floor area relative to the window area) the use of electric lighting during all the hours of occupancy may be required. Air handling luminaires facilitate the capture of a significant proportion of the waste heat from the lighting system for its extraction or for its re-use in the building — for example, for warming the make-up fresh air. The principle of heat-recovery is employed to warm cooler areas in the building with surplus heat extracted from the warmer areas, leading to overall economy in the use of energy.

Lighting Measurement

Luminous flux radiated by a light source is measured in lumens (lm). The lumen is the SI unit of luminous flux describing the quantity of light emitted by a source or received at a plane. The lumen is derived from the unit of luminous power, the candela (cd). Thus one lumen is the luminous flux emitted within unit solid angle (one steradian) by a small source having a uniform luminous intensity of one candela, so that 1 lm = 1 cd sr, and the total flux in all directions is 4π lm. The statement 'a 100 W incandescent lamp emits approximately 1200 lm' gives no information about the brightness of the lamp, nor of the directional nature of the flux. A fluorescent tubular lamp of the same luminous flux output would be less bright by virtue of its greater surface area.

Illuminance (the quantity of illumination) is measured in lux, where 1 lx = 1 lm/m^2.

Illuminance is readily measured by a light-meter held at the point of measurement. Some typical illuminances in use are given in *Table 4*. The illuminances listed there are 'standard service illuminances' and may need to be increased if the task or its circumstances are particularly demanding, to arrive at the 'final service illuminance'.

Glare cannot be directly measured, but a glare index can be calculated. Glare indices lie in the span 10—30 representing from virtually no glare to an unacceptable level of glare for any application. A variation of 3 points on the glare index scale is discernible.

Visual acuity is the property of the eye to resolve small detail. It is measured as the reciprocal of the angle (in minutes of arc) subtended at the eye by the smallest detail that can be picked out. It varies within limits according to the brightness of the object and the contrast in luminance of the object to its background. Thus, within limits, increasing the illuminance will increase the acuity of the subject.

Luminance is the measurable attribute of brightness. The SI unit of measurement of brightness is the candela per square metre (cd/m^2), but a practical non-SI unit widely used is the apostilb (asb), where 1 asb = 0.318 cd/m^2.

The subjective attribute of brightness is called luminosity, and there are no direct means of measurement. The subjective sensation of brightness

Table 4 Typical illuminances in use (based on 1973 edition of the IES Code)

	(lx, or lm/m^2)
Storage areas and plant rooms with no continuous work	150
Rough work, rough machining, rough assembly	300
Routine work, ordinary offices, control rooms, medium machining and assembly	500
Demanding work, deep-plan offices, drawing-offices, business-machine offices, inspection of medium machining	750
Fine work, colour-discrimination tasks, textile processing, fine machining, fine assembly	1000
Very fine work, hand engraving, inspection of fine machining or assembly	1500
Minute work, inspection of very fine assembly	3000*

* Localized lighting, if necessary supplemented by use of optical aids, e.g. binocular loupes, magnifiers, profile projectors, etc.

of an object depends on the adaptation level of the eye at the time of observation. For example, a candle flame viewed in sunlight seems barely luminous; but if suddenly viewed after a period of total darkness, the flame would seem very bright. The time to adapt from one level of adaptation to another depends on the ratio of the two field luminances. Although vision appears to be virtually instantaneous, speed of vision (i.e. the time taken to see) is partly dependent on the time it takes the eye to change its focus from one distance to another. This process (accommodation) is faster at higher illuminances.

Addendum

To put into perspective the place of lighting in the energy economy of one country (UK), 12% of the average total power output is used primarily for lighting (S. Brain, *The Times*, Nov. 1974).

Typical lighting consumptions (annual average) are: 20–40 kW/km for a well lit motorway, 3–5 kW/km for a normal urban street. For indoor rooms and factories, usual standards are 75 lx (5 W/m²) in corridors and general lighting about 200 lx (say 5–10 W/m²) when lamps are clean, 500–750 lx for a drawing board (perhaps 20–25 W/m² with fluorescent lighting), and 1000–3000 lx (40–120 W/m²) for very demanding work. Individual machines in factories may require additional local lighting. The IES Code for Interior Lighting is the accepted guide.

Tables 2 and *3* on lamp efficacy can also be related approximately to energy efficiency. The eye is most sensitive at around 555 nm wavelength (greenish-yellow); it falls to a low value near the blue and red ends of the spectrum (*Figure 6*) but near the broad maximum at 555 nm it is about 680 lm/W. Most lamps, which cover a moderate spectral range in addition to radiation beyond the visible region, are consequently far less efficient than 680 lm/W. For fluorescent lamps (*Table 2*) however the efficiency on this basis reaches nearly 10% of the theoretical maximum, because they produce a higher ratio of light energy to heat energy than (e.g.) filament lamps. For discharge lamps it spans the range 6–14% of theoretical; the exceptionally high values of efficiency for sodium lighting (SON and SONT) occur because the light emitted is near-monochromatic at wavelengths of 589 and 589.6 nm, near the maximum of eye sensitivity.

References

1   Landis, G. N. and Trackman, J. C. *IEE 6th Biennial Conf. on Process Heating in Industry* 1963, 145–169

2   Mason, D. J., *Iron and Steel Inst. Publ. 113*, 1968, 63–69

3   Ref.(A)

4   Severs, M. J. and Kuyser, W. C. *VIth Int. Congr. on Electroheat, Brighton*, 1968, N316

5   Evans, J. W. and Berry, J. R. *Vth Int. Congr. on Electroheat*, Wiesbaden, 1963, N608

6   Schrewelicus, N. *VIth Int. Congr. on Electroheat*, Brighton, 1968, N615

7   Harrison, A. J. and Weinberg, F. J. *Proc. R. Soc.* 1971, Vol. A321, 95

8   Ref.(B)

9   Ref.(C), Chs.IV.2 and IV.3

10   Ref.(D)

11   Ref.(E)

12   Ref.(F)

13   BS 4809:1972 'Specification for radio interference limits for radiofrequency heating equipment'

14   'Special precautions relating to intense radiofrequency radiation', HMSO, London, 1960

15   Ref.(G)

16   Ref.(C), Ch.IV.5

17   Ref.(H)

18   Ref.(I)

19   Ref.(J)

20   Herbst, C.-H. *Elektrizität* 1968, No. 11, 294

21   *Better Light — Better Performance* (leaflet), Electricity Council, London, 1972

22   Ref.(K)

23   *The Essentials of Good Lighting* (leaflet), Electricity Council, London, 1975

24   *Lighting during Daytime*, Tech. Rep. No. 4, Illuminating Engng Soc., London, 1974

25   Ref.(L)

26   *Security Lighting* (leaflet), Electricity Council, London, 1972

27   *Integrated Design: A Case History* (booklet), Electricity Council, London, 1971

General Reading

(A)   *Electric Resistance Heating*, Electricity Council, London, 1966

(B)   Gross, B., Grycz, B. and Miklossy, K. *Plasma Technology*, Iliffe Books, London, 1968

(C)   *Elektrowärme — Theorie und Praxis*, Verlag W. Girardet, Essen, 1974

(D)   Venugopalan, M. (Ed.) *Reactions under Plasma Conditions*, Vols. 1 and 2, Wiley–Interscience, New York and London, 1971

(E)   Gerdeman, D. A. and Hecht, N. L. *Arc Plasma Technology in Materials Science*, Springer-Verlag, New York, 1972

(F)   Eyring, L. (Ed.) *Advances in High Temperature Chemistry*, Vol. 1, Academic Press, New York and London, 1967, pp 282–292

(G)   Langton, L. L. *R. F. Heating Equipment*, Pitmans, London, 1949

(H)   Puschner, H. *Heating with Microwaves*, Phillips Technical Library, Eindhoven, 1966

(I)   Okress, E. C. (Ed.) *Microwave Power Engineering*, Vols. 1 and 2, Academic Press, New York and London, 1968

(J)   Lyons, S. L. *Management Guide to Modern Industrial Lighting*, Applied Science Publishers, Barking, Essex, 1972

(K) *IES Code for Interior Lighting*, 1973 edn, Illuminating Engng Soc., London

(L) *Depreciation and Maintenance of Interior Lighting*, Tech. Rep. No. 9, Illuminating Engng Soc., London, 1967

(M) *Integrated Design: A Case History* (booklet), Electricity Council, London, 1968

(N) Electrical Engineer's Reference Book (Ed. M. G. Say), Butterworths, London (current edition)

Use of heat

Chapter 8

Furnaces: principles of design and use

8.1 Gas-, oil- and pulverized-fuel-fired furnaces

Heat transfer in furnace
Heat transfer to stock surface
Heat transfer within the stock; heating times
Fuels and combustion
Combustion and flame temperature
Furnace atmospheres
Modes of heat application
Methods of material handling
Aerodynamic aspects
Fuel consumption and heat economy
Mechanical design and construction
Instrumentation and control
Melting furnaces
Sintering furnaces
Kilns
Furnaces for the glass industry
Soaking pits and reheating furnaces
Heat-treatment furnaces

8.2 Electric furnaces

Melting
Re-melting and refining
Smelting ores and refractories
Heating by indirect resistance
Heating by direct resistance
Induction heating
Electron-beam glow-discharge heating
Dielectric and microwave heating

8.3 Furnaces for high temperatures

Production of high temperatures by chemical flames
Plasma furnaces and other physical methods

8.1 Gas-, oil- and pulverized-fuel-fired furnaces

Almost all manufactured articles go through a process of heating at some stage. The equipment used for heating can be broadly termed a 'furnace'. Because of the diversity of manufacturing processes and raw materials used, the types of furnace required are numerous.

Basically a furnace is a brick-lined chamber, capable of holding/conveying the material to be treated, to which heat is applied by one of various means so as to achieve the required final results. In fuel-fired furnaces, as the name implies, heat is supplied by combustion of fuel and is transmitted by a combination of heating processes. In general, fuel costs are a small fraction of the total manufacturing cost, and in the past fuel economy has often been considered unimportant. However, with increasing costs and decreasing availability of fuels, there is now a considerable financial and a strong conservational incentive to watch fuel-economy aspects.

Although furnaces are put to widely different uses, and conditions may vary very much, the general principles are common to all. These general principles and their applications to design and operation, when not sufficiently covered elsewhere, will be discussed in this Chapter.

Heat Transfer in Furnace

The study of the laws of heat transfer is very important to a furnace designer. A digest of the aspects most relevant to energy efficiency will be found in App.C11. Temperature difference is of prime importance; heat flows from the hotter to the colder body.

Radiative heat transmission in a furnace takes three forms: interchange between two solid bodies at different temperature, interchange between a non-luminous (clear) gas and a solid body, and interchange between luminous (or incandescent) gas and a solid body. Emissivity or absorptivity, the ability to emit (radiate) or absorb radiant heat, ranges for solids from a maximum of 1 (black-body) through a variety of lower values: e.g. brick-work, oxidized steel have emissivities of 0.8, bright surfaces like aluminium have values less than 0.1 (higher if oxidized). Temperature also has an effect on emissivity[1] (*Figure 1*). Because heat transfer by radiation varies as the fourth power of absolute temperature, a slight change in furnace temperature can cause a considerable difference in heating rate when the furnace is mainly heated by this mechanism.

As explained in App.C11, the amount of heat transferred by radiation depends on the relative sizes of the surfaces, their angular relation, and characteristic emissivities; these can usually be combined in a single factor for calculation. The emissivity of gas radiation (App.C11) is seldom more than 0.5. Radiation from luminous flames is even more complex, since it depends on incomplete combustion, and also since the flames may mask the hot surfaces.

In forced convection, whenever there is a change of shape of the boundary or a change in velocity there is increase in heat-transfer coefficient[2], which may be doubled, but this rapidly reverts to normal as the distance from the disturbance increases.

Temperature increase affects convection in two ways, depending on the position of the pump or fan in the circuit. If there is no fan in the hot-fluid circuit the effect is to increase the coefficient of heat transfer by convection. If however there is recirculation due to a hot fan, this is modified by a reduction in mass flow since fans deal with constant volume. Conditions for steady-state conduction do not always exist in furnace practice. During heating of stock, or heating of the furnace walls, besides being conducted from layer to layer heat is also absorbed in the layers thereby increasing their temperature. Heat flow then varies with time (unsteady-state or transient conduction).

Heat Transfer to Stock Surface

In most heating processes, heat is transferred to the stock surface by a combination of convection from the furnace gases and radiation from the

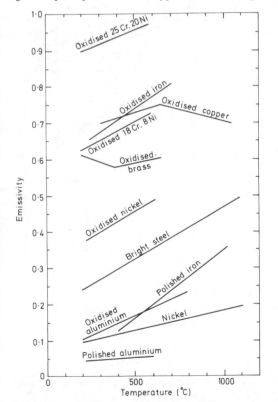

Figure 1   Emissivities of metals and alloys

refractory surfaces and flames; one or other may play the major role. At low temperature (up to about 750°C) forced-convection heating is employed. At higher temperatures, suitable fans are very expensive and recirculation less effective. Radiation is very effective at high temperatures, and plays a prominent role. It is fortunate that most materials with low emissivity require heating to relatively low temperatures, when forced convection can be usefully employed; for materials with high emissivity (e.g. steel), high-temperature heating happens to be required and radiation is doubly effective.

Heat transfer by convection (q_c) to the stock surface area A_s can be written as in App.C11

$$q_c = A_s h_c (T_g - T_s) \tag{1}$$

where T_g = temperature of furnace gases and T_s = stock surface temperature.

Calculation of convective heat transfer coefficients (App.C11) is explained in various books[2-5].

The radiative heat transfer from a wall to stock surface in a furnace is given by:

$$q_w = \sigma A_w F_{ws} (T_w{}^4) \tag{2}$$

(σ = Stefan—Boltzmann constant, A_w = area of walls, F_{ws} = a factor which represents the fraction of radiation leaving walls that is intercepted by stock surface (values can be found in reference books[3,5]), and T_w = temperature of walls).

Flame (or gas) radiation absorbed by the stock surface can be written:

$$a_s q_g = a_s A_s \sigma (\epsilon_g T_g{}^4) \tag{3}$$

(a_s = absorptivity of the stock surface; ϵ_g = emissivity of the gas — for non-luminous flames, it can be calculated easily[3,5]. Typical values for luminous flames may be chosen; T_g = temperature of gases).

The stock itself emits radiation:

$$q_s = \sigma A_s \epsilon_s (T_s{}^4) \tag{4}$$

(ϵ_s = emissivity of the stock surface, and T_s = temperature of the stock surface).

The net heat transfer to the stock surface can, hence, be written as:

$$q = A_s h_c (T_g - T_s) + \sigma A_w F_{ws} (T_w)^4$$
$$+ a_s A_s \sigma (\epsilon_g T_g{}^4) - \sigma A_s \epsilon_s (T_s)^4 \tag{5}$$

The calculations of q from equation (5) are involved and time-consuming, especially for multi-zone furnaces. The above equation can be written in a simplified form, by assuming that heat is transferred to the stock surface solely by radiation from the furnace enclosure at furnace temperature (temperature recorded by thermocouple, T_f):

$$q = f A_s \sigma (T_f{}^4 - T_s{}^4) \tag{6}$$

$$h = \frac{f \sigma (T_f{}^4 - T_s{}^4)}{T_f - T_s} \tag{7}$$

(f is an arbitrary factor such that the above equation gives a stock temperature in good agreement with that measured experimentally; a value of about 0.8 gives reasonably accurate results for steel-reheating furnaces.)

Heat Transfer Within the Stock; Heating Times

Heat is transmitted within the stock by conduction. In calculating the time for a piece of material to heat up, the variable heat flow or transient conduction encountered requires the use of complicated mathematical techniques.

The time taken for a piece of material to heat through depends on its thermal conductivity (k), density (ρ) and specific heat (c_p). These characteristics are combined in a factor known as thermal diffusivity or temperature conductivity:

$$\beta = k/\rho c_p$$

The greater the diffusivity, the quicker the temperature diffuses.

Analytical solutions of transient conduction equations are presented[3-6] in the form of graphs of dimensionless groups of the parameters involved.

In many cases, the actual operating conditions do not correspond to those for which the analytical relations were designed. Such cases can be handled without the use of calculus, by numerical (finite difference) methods, which are discussed here.

Consider one-dimensional transient conduction through a slab of unit cross section (*Figure 2*). Dividing the slab into a number of slices of equal thickness Δx, and temperatures at time zero as indicated, a heat balance on the section ABCD can be written as:

$$\frac{k (T_s - T_1)}{\Delta x} - \frac{k (T_1 - T_2)}{\Delta x} = \frac{\Delta x \rho c_p (T_1{}' - T_1)}{\Delta \theta} \tag{8}$$

($T_1{}'$ is the new temperature at the plane after time $\Delta \theta$). Solving the above for $T_1{}'$ and replacing $k/\rho c_p$ by β

$$T_1{}' = \left\{ \beta \Delta \theta /(\Delta x)^2 \right\} (T_s - 2T_1 + T_2) + T_1 \tag{9}$$

From the above relation, a number of writers have devised calculation methods; the most popular of these is due to Schmidt, who chose x and θ such that $(\Delta x)^2/\beta \Delta \theta = 2$. It is then possible to obtain the required solution by simple graphical construction, which is built up step by step from repeated application of the above equation. This method can also be applied in tabular form[3,4,6]. More accurate results can be obtained by choosing

$(\Delta x)^2/\beta\Delta\theta > 2$, but graphical construction is then not possible.

Reasonably accurate heating/cooling times can be calculated by methods described by Dusinberre[7]. A method following his general approach and including the effect of heat transfer to the stock surface is discussed here. If plane 00 (*Figure 2*) is heated by a warmer fluid with an instantaneous temperature T_f, a similar heat balance on the adjacent half-slice is given by:

$$h(T_f - T_s) - \frac{k}{\Delta x}(T_s - T_1) = \frac{\Delta x \rho c_p (T_s' - T_s)}{\Delta\theta}$$

The above equation can be rewritten as:

$$T_s' = [\Delta\theta/\rho c_p(\Delta x)^2]\,[2h\Delta x T_f + 2kT_1 +$$
$$+ T_s\{((\Delta x)^2\,\rho c_p/\Delta\theta) - 2h\Delta x - 2k\}]$$

Similarly, equation (9) can be rewritten as:

$$T_1' = [\Delta\theta/\rho c_p(\Delta x)^2]\,[kT_s - kT_2 +$$
$$+ T_1\{((\Delta x)^2\,\rho c_p/\Delta\theta) - 2k\}]$$

In the above two equations, to retain a suitable influence of T_s (or T_1) on T_s' (or T_1'), the coefficients of T_s (or T_1) should be positive and hence the time step $(\Delta\theta)$ should be chosen so that:

$$(\Delta x)^2\,\rho c_p/\Delta\theta \geqslant (2h\Delta x + 2k)$$

or

$$(\Delta x)^2\,\rho c_p/\Delta\theta \geqslant 2k$$

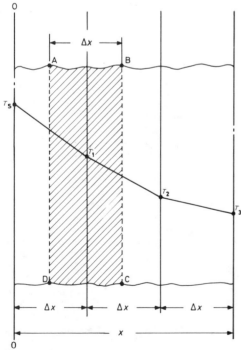

Figure 2   Transient conduction through a slab

References p 147

Table 1		Average time-lag for carbon steel			
Size		Method of heating — time lag (min)			
(in)	(mm)	→O←	→▢←	▢	▢
1	25	0.10	0.10	0.20	0.80
2	50	0.4	0.4	0.80	3.2
3	75	0.9	0.9	1.80	7.2
4	100	1.6	1.6	3.2	12.8
5	125	2.5	2.5	5.0	20.0
6	150	3.6	3.6	7.2	28.8
8	200	6.4	6.4	12.8	51.2
10	250	10.0	10.0	20.0	80.0
12	300	14.4	14.4	28.8	115.2
16	400	25.4	25.4	51.2	204.8
20	500	40.0	40.0	80.0	320.0
32	700	102.0	102.0	204.0	816.0

In using the above approach the value of h can be calculated by use of equation (7), and the temperatures in the previous step can be used to calculate the new temperatures. Calculations made in this manner can be very time-consuming, but the use of digital computers and calculating machines makes the task much easier.

Easy methods of calculating heating times have been developed by various authors. Bloom[8] suggests a method of calculation based on the fact that the centre temperature of a piece of steel follows the surface temperature by a given time lag irrespective of the rate of heating. Average time lags for various sizes and methods of heat application are reproduced in *Table 1*. In order to use this method, the expected furnace temperature, stock surface and centre temperatures can be first plotted against time, and later corrected by making heat balances at various parts of the cycle.

To calculate the heating and cooling times of thin bodies or materials with high diffusivity (e.g. aluminium), the following equation[9,10] can be used:

$$\theta = \{w\rho c_p/3.6\,h\}\,.\ln\{(T_f - T_0)/(T_f - T_\theta)\}$$

(θ = time (h), w = weight of the body (kg/m^2 of exposed surface), ρc_p = specific heat (kJ/kg K), and h = heat-transfer coefficient (W/m^2 K).)

In the above equation the furnace temperature is considered constant, and the only complexity involved is due to the fact that the rate of heat transfer falls as the temperature of the body increases.

Fuels and Combustion

Fuels are a source of energy, which when converted to heat can serve the purpose of raising the temperature of furnaces. Selection of the fuel depends on various factors, but availability is often the sole deciding factor. To avoid dependence on a single fuel, combustion equipment is often chosen in which alternative or combined use of fuels is possible.

All major fuels basically contain carbon, hydrogen etc. in organic or inorganic forms. Combustion is an oxidation reaction producing carbon dioxide, water vapour and sulphur dioxide accompanied by heat release. The total amount of heat obtainable by complete combustion, in air, of unit quantity of a fuel is its calorific value (CV). If the fuel contains hydrogen, water vapour is one of the products of combustion and if condensed the latent heat of condensation is added to the heat of combustion, giving the *gross* calorific value. In general practice, however, combustion gases are not condensed in working parts of the furnace, and deducting this latent heat from the gross calorific value gives the *net* calorific value.

Fuels can be classified in the broadest terms as gaseous, liquid and solid fuels. The selection of the fuel best suited to a particular case necessitates possession of knowledge concerning their more important properties and the equipment needed for their handling and combustion. Liquid (Ch.3.2, 4, App.A2, A5, A9, B11, D2, D4) and gaseous (Ch.3.1, App.A1, A4, A8, B9, B10, D1) fuels can be conveniently handled in pipelines. Solid fuels (Ch.6, App.A3, A6, A10, D3), in general, must be moved in batches, or sometimes transported by conveyors. In pulverized form (Ch.6.2), solid fuels can be blown with air in pipelines.

In the following three subsections, in view of the accounts, cited above, in other parts of this book, only aspects of special relevance to their use in furnaces will be briefly mentioned.

Liquid Fuels

Liquids can be readily stored above or below ground and in out-of-the-way places. Some need no preheating. They are easily transported from storage to furnace and burn without noticeable ash residue. Gas oil, light, medium and heavy fuel oil are the fractions of petroleum most commonly used in industrial furnaces. Much of the coal tar produced is prepared into the standard coal-tar fuels designated CTF 50, 100, 200 etc.

Properties that may affect the choice of liquid fuels are calorific value, sulphur content, and viscosity. Calorific values of all petroleum fractions are similar and higher than those of CTF. Whereas coal-tar fuels usually have a sulphur content less than 1%, petroleum oils (especially heavier fractions) can have as much as 3%. This high sulphur content sets oil at a slight disadvantage compared to CTF, and extra care must be taken in designing some furnace parts. Sulphur oxides can attack ordinary steels at low temperature (acid corrosion), and at high temperatures in reducing conditions it attacks nickel-bearing steels. Most oil fuels contain a small amount of inorganic matter, which on combustion produces a mixture of alkaline sulphates and vanadium pentoxide. These are carried in a molten state in the flue gases and can attack metallic surfaces of equipment such as recuperators.

Viscosities of fuel oils and coal-tar fuels vary over a large range. Whereas light fractions like gas oil have viscosity of about 35 seconds Redwood No.1 at 100°F (or 3.3 cSt* at 37.8°C), the extra-heavy fuel oils have viscosities of about 5000 seconds (or 1250 cSt). Certain of the heavier coal-tar fuels have considerably higher viscosity still. Liquid fuels can be pumped at a viscosity of about 3000 seconds (or 750 cSt), but must be warmed to 100 seconds (or 25 cSt) before they can be atomized for combustion. In general, lighter fuels are more expensive than heavy ones, but when selecting oil fuels the extra costs involved in preheating, pumping, atomizing should be included.

Gaseous Fuels

Of all types of fuel, gases offer the greater number of advantages: ease of transport, smokeless combustion, accurate measurement and hence easy control of temperature and furnace atmosphere. Also, they do not require on-site storage facilities.

Gaseous fuels fall into three groups: (1) hydrocarbons with high calorific values ranging from 37 MJ/m^3 for natural gas with methane as main constituent to a figure about four times as great for gas with butane as main constituent; (2) manufactured (or by-product) gases (App.B9) with little nitrogen or carbon dioxide (16.5—20 MJ/m^3); and (3) lean gases (App.B10) with high inert-gas content (6 MJ/m^3 or less).

Gases with high hydrogen content (e.g. coke-oven gas) burn very rapidly, and with hotter flame than those with a large proportion of inert gases. Natural gas falls between these two extremes. Hydrocarbon gases should not be preheated in recuperators or regenerators, because at high temperature they form soot. Moreover, heat salvage by preheating natural gas is negligible because the weight of the gas is very small compared with that of the air required to burn it.

Rich manufactured gases cannot be preheated for similar reasons. Coke-oven gas generally contains sulphur in the form of hydrogen sulphide, and special care is required in the furnace.

Fuel gases of the third category are commonly preheated and are almost always used with preheated combustion air, in order to attain a high enough flame temperature to achieve the required furnace temperature.

Producer gas, manufactured by blowing air and steam through a deep bed of coal or coke, leaves the producer at a temperature between 500 and 750°C. If this gas is used in the hot raw condition, the benefit of the sensible heat of gas as well as the combustion heat of the tarry vapours is available — adding to the heating value of the gas. The disadvantage is the necessity of providing brick-lined

---

* 1 cSt = 1 mm^2/s

mains, with adequate facility for cleaning and dust-removal. If the furnaces are far from the gas producers, it is advisable to use cooled and clean (de-tarred) gas, which has lower calorific value and burns with less luminosity. Even after gas cleaning, difficulty is sometimes experienced in keeping the gas mains and valves clean, because of small quantities of tar and other chemicals that pass through the cleaning plant.

Another 'lean' fuel, blast-furnace gas, is often used mixed with coke-oven gas.

Pulverized Fuels

Coal is often used in pulverized form suspended in an air stream, and dealt with rather like a gas. Usefulness of pulverized fuels is affected by fineness, moisture content, amount of ash, and ash fusion temperature.

The length of combustion space required depends on the fineness if unburned fuel particles are not to be allowed to escape. In practice fineness is measured by the fraction that goes through a standard 200 B.S. mesh (76 μm) sieve. For industrial furnaces a high fineness factor (90–95%) is recommended. Finer fuel not only makes combustion easier, but also helps to float the ash in the flame and hence carry it out of the furnace.

Though coals with moisture up to about 7% can be ground, they take more energy than dry coals. Powdered coal is hygroscopic and it is normally suggested that it be used as soon after grinding as possible, to avoid compacting. Also damp powdered coal tends to stick to hoppers, feeders etc., thereby interfering with regular feeding.

Another characteristic of powdered coal is its ash, with regard to amount, composition and fusion temperature. The lower the ash level the less is the trouble arising. Most ashes melt below the flame temperature of coal. If the ash has a high fusion temperature, the fraction that melts is small. With a low-fusion-temperature ash almost all will melt into drops, which may become large enough not to float with the flame and may be deposited on the hearth, side walls and the charge. In rolling mills with low angular velocity, the ash deposited on steel may get rolled into the steel.

Another point that must be borne in mind is that an auxiliary plant must be provided for arresting the grit and dust which is inevitably carried away in the exit gases.

Pulverized fuel finds many applications in the metallurgical field for melting, reheating and heat-treatment furnaces. It is very successfully used in cement kilns, when much of the ash becomes incorporated in the cement.

Combustion and Flame Temperature

In most large industrial furnaces the mixing of fuel and air is imperfect and some portions of the air may never come into contact with combustibles. In order to make sure that no fuel remains unburnt, some excess air is nearly always necessary; it is, however, advisable to use the least possible amount of excess air. Whereas for most gaseous fuels complete combustion can be achieved with as little as 10% excess air, liquid fuels require slightly higher amounts, and pulverized fuels may require 20% or more.

The size of furnaces is greatly dependent on the combustion-space requirement. If fuel and air are mixed ahead of the burner (premixed), combustion occurs more rapidly and less combustion volume is required. A great number of variables affect the speed of combustion and it is difficult to predict theoretically the space required for complete combustion. The combustion-space requirement is expressed as amount of heat released in unit time and unit volume (kW/m^3 or Btu/ft^3 s). As a guide, this figure varies from 55 to 670 kW/m^3 (1.5 to 18 Btu/ft^3 s) in most furnaces, with low values used for low/medium-temperature heat treatment furnaces and figures of 185 to 225 kW/m^3 for steel-reheating furnaces.

High rates of heat release can result in high-temperature zones local to the burners, tending to increase wear on refractories. This emphasizes the necessity of selecting the correct burner for the required flame development. Preheat in fuel and air tends to increase the speed of combustion, and hence means a smaller combustion-space requirement.

Much higher combustion rates (3–4 times the above figures) can be obtained with special firing equipment, e.g. radiant tubes. Even higher rates can be achieved in the combustion chambers of forced-convection furnaces, especially when recirculation is used to prevent overheating of the chamber.

It has been mentioned previously that combustion air (and sometimes fuel) are preheated with lean fuels in order to achieve the required flame temperature. Flame temperature can be used as a guide to the maximum furnace temperature that can be achieved. The *It-* (enthalpy—temperature) diagram[4] makes it possible to ascertain the adiabatic combustion temperature at thermodynamic equilibrium. The *It*-diagram was drawn on the basis of a statistical relation that exists between the net calorific value of all industrial fuels, their combustion-air requirement and the combustion-gas volume produced. Another basis is that equal enthalpies of wet combustion products correspond to equal temperatures.

The *It*-diagram covers the range from 100°C to 2500°C and shows the relation between enthalpy and temperature for different air contents of the combustion product gases (in order to take account of excess air used), and also the enthalpy of preheated air. It takes into account the effect of dissociation of carbon dioxide and water vapour above 1500°C, which renders part of the potential heat supplied temporarily latent (recoverable in the furnace as the combustion gases cool). Accuracy is adequate for most purposes, since errors are smaller than those of sampling, analysis, measurement of carbon dioxide, temperature and volume.

To assess the adiabatic flame temperature the total enthalpy of the combustion product gases per unit volume is calculated:

$$I_{total} = \frac{1}{V}\,(C_N + A\,I_{air} + I_f)$$

where I_{total} = total enthalpy above $0°C$ of wet combustion product gases (flame gas) per unit volume of these gases; V = volume of wet combustion product gases (at $0°C$ and 1 atm or at 0.1 MPa) including any excess air, per unit of fuel (V_0 = theoretical or stoichiometric value of V); A = combustion air volume used (at $0°C$ and a standard pressure) including excess air, per unit of fuel (A_0 = theoretical or stoichiometric value of A, and $n = A/A_0$); C_N = net calorific value of the fuel, per unit of fuel; I_{air} = enthalpy of dry air as fed, per unit volume at $0°C$ and a standard pressure; and I_f = enthalpy (sensible heat only) of fuel as fed, per unit of fuel. Unit of fuel here means 1 kg or 1 lb (solid, liquid fuels), 1 m^3 or 1 ft^3 at $0°C$ and a standard pressure (gaseous fuels).

V_0 and A_0 may be obtained from C_N by the following statistical formulae* (SI units):

Solid fuels: $C_N < 23.26$ MJ/kg†
$$A_0 = 0.242\,C_N + 0.5$$

$$V_0 = 0.214\,C_N + 1.65$$

$C_N > 23.26$ MJ/kg†
$$A_0 = 0.242\,C_N + 0.55$$

$$V_0 = 0.240\,C_N + 0.9$$

Liquid fuels: $A_0 = 0.204\,C_N + 2.0$

$$V_0 = 0.266\,C_N$$

Gaseous fuels: Blast-furnace gas
$$A_0 = 0.191\,C_N$$

$$V_0 = 0.152\,C_N + 1$$

Producer gas
$$A_0 = 0.208\,C_N$$

$$V_0 = 0.154\,C_N + 1.11$$

Blast-furnace/coke-oven gas mixtures
$$A_0 = 0.256\,C_N - 0.16$$

$$V_0 = 0.249\,C_N + 0.7$$

---

* Adapted from *Technical Data on Fuel*, 1961, 6th edn, Table 76, p 99

† Of fuel as fired

Rich gases (C_N = 11.2—18.7 MJ/m^3) (town gas etc.)
$$A_0 = 0.262\,C_N - 0.25$$

$$V_0 = 0.324\,C_N - 0.67$$

Hydrogen, carbon monoxide
$$A_0 = 2.38$$

$$V_0 = 2.88$$

Methane (C_N = 35.8 MJ/m^3)
$$A_0 = 9.52$$

$$V_0 = 10.52$$

In general, $A = nA_0$; $V = V_0 + (A - A_0)$.

Excess air (%) = 100 $(n - 1)$. This may be assumed in a design calculation, or deduced from a measurement of the oxygen content:

$$n - 1 \approx 0.95\,\frac{O_2}{21 - O_2}$$

for solid and liquid fuels, or CO_2 content,

$$n = \frac{[CO_2]}{CO_2}$$

of the *dry* flue gas, where $[CO_2]$ is given in the Table below.

C_N may be obtained from reference works (e.g. *Technical Data on Fuel*), or calculated from gross CV (C_G) by the following Table* which also gives the theoretical maximum CO_2.

Fuel		$[CO_2]$ (%)
Wood, peat, lignite $C_N = 1.045\,C_G -$	Wood	20.5
2.326 (±1%)	Lignite	19.5
Bituminous coal $C_N = 0.954\,C_G + 0.465$		18.7
Anthracite $C_N = C_G - 0.745$		19.5
High-temperature coke $C_N = C_G - 0.162$		20.6
Liquid fuels (except creosote) $C_N = 0.750\,C_G + 8.374$		15.8
Blast-furnace gas		24.5
Producer gas from coke		20.5
Producer gas from coal		19.0
Debenzolized coke-oven gas		11.0
Town gas $C_N = 16.8$ MJ/m^3		13.0
$C_N = 20.6$ MJ/m^3		11.0
Methane		11.6

I_{air} can be read from the line for 100% air in combustion product gas at the air-preheat temperature.

---

* Adapted from *Technical Data on Fuel*, 1961, 6th edn, Table 73, p 98, and Table 75, p 99

I_f requires a knowledge of the specific heat of the fuel over the fuel preheat range (e.g. TDF). For a gaseous fuel, I_f will be approximately related to I_{air} and the relative degree of preheat:

$$I_f = B\,I_{air}\left(\frac{t_f}{t_a}\right)$$

(where t_f and t_a are the respective degrees of preheat in °C)

$$= F\,I_{air}$$

Here B has very approximately the values:

Blast-furnace gas	1.03
Producer gas	0.96
Rich gases (C_N 12—18 MJ/m³)	1.15
Methane	1.7

A simplified formula:

$$I_{total} = \frac{1}{V}\,(C_N + (A + F)\,I_{air})$$

may then be used; but for accurate work the respective average specific heats over the temperature range of gas preheating are to be preferred.

The value of I_{total} for the actual combustion product gases can then be calculated. The air percentage in the combustion product gases is given by the expression:

$$100\,(n - 1)/(m + n - 1)$$

where m is the value of V_0/A_0 for the fuel used.

The temperature of combustion products (adiabatic flame temperature) can be read from *Figure 3* (coordinates I_{total} and the appropriate percentage of air in combustion product gases). The SI value of I can if desired be converted to British units to fit the more complete scale.

The loss of heat to the stack can also be read from the It-diagram, using the same air line, if the temperature of the flue gases entering the stack is known. Expressed as a percentage of I_{total} initially calculated, this represents the stack loss. Similar calculations can assess heat used in the furnace, and recovered between the furnace and the stack by (e.g.) recuperators and waste-heat boilers; only the appropriate temperatures at these stages need be known.

References to It-diagram
Rosin, P. O. and Fehling, H. R. *Das It-Diagramm der Verbrennung*, 1929

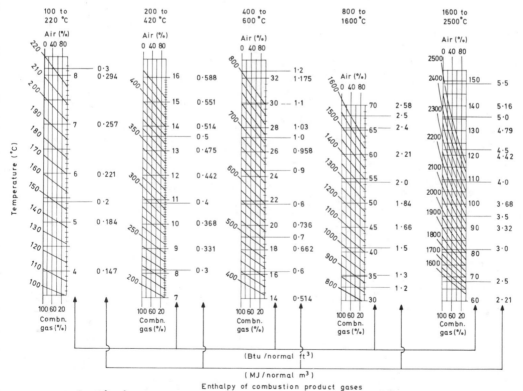

Figure 3   The It-diagram for combustion product gases

Rosin, P. O. and Fehling, H. R. *Le Diagramme It de la Combustion*, 2nd edn, 1935

Kay, H. *J. Inst. Fuel* 1936, Vol.9, 312

Reber, J. W. *J. Inst. Fuel* 1939, Vol.13, 20

Rosin, P. O. and Fehling, H. R. 'The *It*-Diagram for Incomplete and Imperfect Combustion', *J. Inst. Fuel* 1942, Vol.16, 20

Rosin, P. O. 'Total, Recoverable and Returnable Heat in Combustion Gases', *J. Inst. Fuel* 1945, Vol.19, 53

Technical Data on Fuel, 1961, 6th edn
Calorific values: Solid fuels Table 374, pp 297—8; Table 375, p 299. Liquid fuels Tables 349—350, pp 269—270. Gaseous fuels Table 328, p 254; Table 331, p 255.
Specific heats: Solid fuels Table 185, pp 184—5. Liquid fuels Tables 181, 182, pp 271—2. Gaseous fuels Table 331, p 255.

Furnace Atmospheres

Under some conditions, all metals and alloys are affected by the furnace atmosphere. In fuel-fired furnaces, carbon dioxide and excess oxygen may react with the metal heated and cause scaling. Scale can be removed by expensive methods like pickling, but surface finish may alter and dimensional tolerances may be impaired (especially in the case of thin section).

In heating ferrous metals, besides scaling, decarburization is a common problem. Oxidation or decarburization is linked closely with the nature of the furnace atmosphere and to the time and temperature of exposure. Typical furnace atmospheres (resulting from complete or partial combustion or introduction of prepared atmospheres) may contain one or more of a number of gases such as oxygen, carbon dioxide, carbon monoxide, water vapour, hydrogen, nitrogen, methane etc. A list of reactions between the atmosphere-gases, iron, and cementite (Fe_3C) is as follows:

No reaction: N_2 + Fe, H_2 + Fe, N_2 + Fe_3C, Co + Fe_3C.

Oxidizing/reducing (oxidizing to the right):

$$O_2 + 2Fe \rightleftharpoons 2\,FeO$$
$$CO_2 + Fe \rightleftharpoons FeO + CO$$
$$H_2O + Fe \rightleftharpoons FeO + H_2$$

Carburizing (carburizing to the right):

$$CH_4 + 3Fe \rightleftharpoons Fe_3C + 2H_2$$
$$2CO + 3Fe \rightleftharpoons Fe_3C + CO_2$$

Decarburizing (decarburizing to the right):

$$2H_2 + Fe_3C \rightleftharpoons 3Fe + CH_4$$
$$O_2 + 2Fe_3C \rightleftharpoons 6Fe + 2CO$$
$$CO_2 + Fe_3C \rightleftharpoons 3Fe + 2CO$$

Reactions shown above are considered reversible, and at any given temperature come to an equilibrium depending upon the concentration of gaseous reactants and products. It is generally known that a large CO_2/CO ratio will produce scale and result in decarburization. Likewise, the ratio CH_4/H_2, if large, will cause carburization, and if small, decarburization. Similarly, the ratio of H_2O/H_2 will produce scale if large, and if small the action will be reversed. The relation to temperature is shown in *Figure 4*. From the graph showing equilibrium for the system Fe, FeO, H_2, H_2O, the importance of eliminating water vapour in placing the gas on the reducing side can be readily seen. Similarly, reduction or elimination of carbon dioxide is important if a reducing atmosphere is required.

An inert atmosphere of argon can be used for some purposes, but its use is limited because it is very expensive. Most experience has shown nitrogen to be inert, except in nascent form. Under certain conditions it may be decarburizing, especially in the presence of water vapour.

Heat treatment of ferrous alloys is basically carried out in a reducing atmosphere for the following reasons:

1. To maintain a bright surface.
2. To reduce the oxide already formed to the free metallic state.
3. To maintain the carbon content, in conjunction with the above two conditions.
4. To add to the carbon content, i.e. to carburize.

The relative ease with which the above can be performed is in the order given. Item 1 can be achieved by partially burned gas (containing CO, H_2 etc.), whereas a richer gas (containing more CO and H_2) is required for items 2 and 3. The atmosphere gases for item 3, in addition to items 1 and 2, should be higher in CH_4 content and lower in CO_2 and H_2 to prevent decarburization. Care is

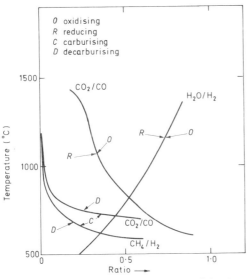

Figure 4   Equilibrium for reactions involving iron

necessary to produce a neutral atmosphere, otherwise carburization may occur if the gas is too rich. Item 4, carburization, is best achieved by precipitation of carbon on steel during the heating-up cycle by means of hydrocarbon gas, after which a gas containing CO_2 is introduced. At temperatures above $870°C$ $(1600°F)$, the reaction $C + CO_2 = 2CO$ is rapid. The CO with the aid of available CH_4 produces carburization. Water vapour should be eliminated in this case.

Figure 5 shows the variation with temperature of the equilibrium constant $(CO)^2/CO_2$ for the reaction $CO_2 + C = 2CO$, $CH_4/(H_2)^2$ for the reaction $CH_4 = 2H_2 + C$, and $[(CO_2)(H_2)]/[(CO)(H_2O)]$ for the reaction $CO_2 + H_2 = CO + H_2O$.

One of the greatest aids to the uniformity of results is the recirculation of the atmosphere gases. Recirculation is also a desirable feature in that it helps to attain uniform temperature and to establish a thoroughly mixed gas in chemical equilibrium. The new gases entering are mixed and heated with the stabilized gases and set in motion.

The iron surfaces of the charge act as a catalyst and the time required for the following reactions to reach an equilibrium is shortened:

$$CH_4 = 2H_2 + C$$
$$CO_2 + C = 2CO$$
$$H_2O + C = CO + H_2$$
$$H_2 + CO_2 = CO + H_2O$$

The desirability of keeping the water vapour content of the original gas low can therefore be readily seen, otherwise increase of H_2O will continue and the gas becomes oxidizing, unless sufficient quantity of new gas is used.

The above gives some idea of the use of controlled atmospheres in the ferrous-metal industry. Oxidation/reduction reactions are similar for other metals and alloys, and information is available in reference books[11],[12] and papers[13]. Commercially available atmosphere-generation plants use LPG, ammonia, town gas or kerosene to prepare atmosphere suitable for any required heat treatment process.

The use of correct controlled atmosphere, though helpful in attaining the properties required by treatment, is not free from disadvantages. Some of these atmospheres contain toxic gases such as carbon monoxide, and care should be taken in their use, especially with regard to leakage. Personnel working close to furnaces using such atmospheres should be advised about the potential hazard, and the usual safety precautions taken.

In using a protective atmosphere for a heat-treatment process the success of the operation depends upon accurate control of the composition. This control requires due attention to purging (replacement of air with the required atmosphere) methods. Many of the prepared-atmosphere gases are explosive if mixed with a range of amounts of air, and such conditions are liable to occur during purging if due care is not taken. A knowledge of flammability range and ignition temperature is essential to the designer/user of furnaces. Such data are available (e.g. ref.4). Explosion can be caused by a spark or a flame. Carbon monoxide, hydrogen and methane are the usual inflammable gases found in protective atmospheres. Out of these methane has the highest ignition temperature $(700°C)$; it can therefore be said that a chamber below $750°C$ should be considered hazardous when such gases are being handled, and approved methods of purging should be followed (see e.g. ref.12).

Some other disadvantages of protective atmospheres are: (1) Carbon monoxide can cause rapid and serious disintegration of refractories containing iron (most firebricks contain iron in some form of

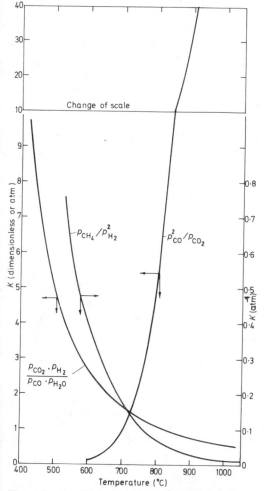

Figure 5   Equilibrium constants for the reactions $CO_2 + H_2 \rightleftharpoons H_2O + CO$, $C + 2H_2 = CH_4$, $2CO + O_2 = 2CO_2$

oxide and it acts as a catalytic agent for the reaction); (2) steel can be damaged by hydrogen absorption in some cases.

The disadvantages mentioned here are not intended of course to lessen the importance of using controlled atmospheres, but to ensure that adequate measures are taken in the design and operation of such furnaces.

Modes of Heat Application

Furnaces will be discussed in this and the following Sections in terms of the method of heat application and material-handling aspects. They may be classified in the former sense as underfired, overfired, sidefired, direct-fired, muffle-type and immersion-type.

Underfired-type. The flame is produced beneath the hearth, and the hot products of combustion pass upwards from the combustion chamber through ports to the heating chamber, from which they are discharged through flue ports (*Figure 6a*).

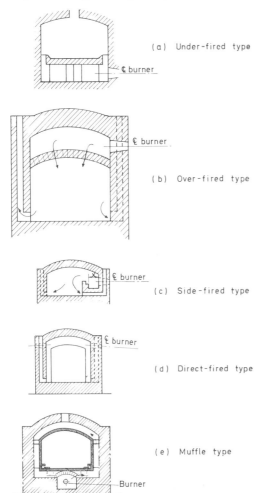

(a)   Under-fired type

(b)   Over-fired type

(c)   Side-fired type

(d)   Direct-fired type

(e)   Muffle type

Figure 6   Modes of heat application

Whereas the temperature in the combustion chamber may be high, the heating chamber is controlled at a required lower temperature. Some of the characteristics of this arrangement are: (1) the material to be heated is protected from direct contact with the flame; (2) excessive scaling is prevented by ensuring complete combustion before the combustion gases enter the heating chamber; (3) a temperature lower than that required for combustion can be achieved in the heating chamber; (4) fuel consumption per unit of stock heated is higher than with direct firing (*a*) because the furnace is larger and (*b*) owing to lower furnace temperatures, heating times are longer; and (5) the hearth is exposed to heat from both sides, and has limited load-bearing capacity and life.

Overfired-type. The combustion chamber is above the heating chamber; the products of combustion enter the heating chamber through a perforated arch (*Figure 6b*), and are usually discharged through ports at hearth level. The purpose of design is the same as for the underfired-type, i.e. complete combustion before the gases reach the heated material, but this type is more efficient because the gases sweep through the charge before leaving the furnace. The perforated arch is made from special tiles, and can be expensive to build and repair.

Sidefired-type. The combustion chamber(s) are on one or both sides of the heating chamber (*Figure 6c*), and take the form of a bridgewall in front of the burner. The bridgewall promotes combustion by creating turbulence, so that fuel and air are intimately mixed when they pass over the wall. Also, the heated material is protected from flame impingement. The bridgewall should extend about 100—150 mm above the centre line of the burner depending on the quantity of fuel burned. A properly designed sidefired furnace need consume no more fuel overall than a direct-fired one unless very small.

Direct-fired-type. This is the simplest form of heat application. As the name implies, fuel is fired into the furnace without any form of baffle. The burners may be located on the sides, ends, or sometimes the roof of the furnace. The flue gases are removed through flues in the roof, sidewalls or hearth (*Figure 6d*).

For direct-fired furnaces with temperatures below about $900°C$ the temperature is not very uniform owing to the large difference between flame and furnace temperatures. For higher temperatures, radiation tends to equalize the temperature and reasonable uniformity can be achieved. For low temperatures, high excess-air can be used to reduce the flame temperature and improve uniformity. The use of high excess-air can cause excessive oxidation and scaling of stock.

In direct-fired furnaces the life of refractories is long even at high temperatures because of the possibility of more robust construction. Disadvantages

are a tendency towards scaling of stock and difficulties in obtaining even temperature-distribution.

Muffle-type. In the above types of furnaces the
products of combustion come into contact with
the material being heated, and this is often
injurious to the material being heated. In such
cases, either the material or the flame is muffled.

Flame and the products of combustion are
muffled by the use of radiant tubes. Alternatively,
the charge can be placed inside a refractory or
metallic muffle, and direct firing applied outside
(*Figure 6e*).

The temperature difference between the heating
chamber and the combustion side depends on the
thermal conductivity (and thickness) of the muffle
material. Needless to say, for a given heat input
the higher the thermal conductivity of the muffle
material the lower will be the combustion-chamber temperature needed, and this will result
in lower flue-gas and wall losses. When selecting
the muffle material, the relative life of various
materials and their cost of replacement should be
weighed against the extra fuel costs involved.

Immersion-type. Salt-bath and molten-metal-bath
furnaces fall into this category; the stock is heated
by immersion. The salt (or metal) is heated in a
pot and neither it nor the material to be heated
comes into contact with the products of combustion. In this type of furnace, material is heated
from all sides and good temperature-uniformity is
achieved. Depending on the liquid heating-mate-
rial, such furnaces are available for a large
temperature range.

Methods of Material Handling

Material Handling Inside the Furnace

Furnaces can be broadly classified into two principal types, according to the method of handling
during the heating process:
1. The in-and-out or batch-type furnace.
2. The continuous furnace.

Batch-type Furnaces. In these the material to be
heated remains in the same position during heating,
and is removed when ready. There are countless
varieties, the simplest being the common box-type
(or fixed-hearth) furnace, in which the stock is
charged by hand or a simple machine with arms.

Certain other batch-type furnaces are briefly
described below.
Bogie (car) type: Usually used for heating heavy or
bulky materials (e.g. annealing of large castings or
pressure vessels). The car or bogie is placed outside the furnace for loading or unloading. The
bogies are generally constructed with flanged
wheels, but sometimes cast-iron balls or steel
rollers are used for heavy loads. Sand seals are
provided on all sides to prevent air infiltration/gas
leakage and to protect the bogie haulage gear from
heat.

Bogie furnaces are rarely underfired. They are
usually direct-fired, but sidefiring is sometimes
employed, especially with low-temperature operation. The flues are usually placed on the sidewalls
of the furnace at hearth level.

Cover or bell type: These are a form of muffle furnace and are extensively used for annealing strip
coils in a controlled atmosphere. Such furnaces
(*Figure 7*) consist of a fixed refractory-covered
base, with a removable metal case (inner cover)
and a removable refractory-lined heating cover or
bell which has integral firing equipment. The
charge is stacked onto the base and the inner cover
placed over it. The inner cover is provided with
adequate sealing arrangements so that a protective
atmosphere can be used. The cooling of coils takes
considerably longer than heating, and continuous
operation is achieved by providing a larger number
of bases and inner covers for this than for heating.

Continuous Furnaces. Continuous furnaces are
used when the furnace forms part of a production
line and when heated stock is required at frequent
and regular intervals. In continuous furnaces the
stock moves while being heated. They were initially introduced on account of their labour-saving
features, fuel economy being an added benefit.

The methods of moving material through continuous furnaces vary greatly with furnace temperature and the size and shape of material to be
handled. Some of the great variety of material-
conveying systems inside the furnaces are: pusher
type, roll-over type, walking-beam/hearth type,
roller-hearth type, chain/belt-conveyor type, shaker-hearth type, continuous-bogie type, rotary-
conveyor (helical motion) type, rotary-hearth type,
and continuous-strand type.

Pusher type: This method of material handling
inside the furnace is very widely used because of
its simplicity and the wide range of temperature
application. The general method of operation is
either to push the pieces through the furnace
directly, where size and shape allow it, or to push
them in trays; the former is preferred. It is seldom
possible to push materials less than 35 mm thick.

Materials with round edges or pieces with a
crooked shape tend to climb or pile on top of one
another. Irregularities in the hearth can contribute
to this. Several ways have been used to reduce
this tendency. The length of furnace is reduced;

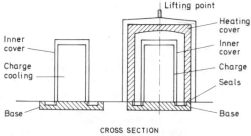

Inner
cover

Charge
cooling

Base

Lifting point

Heating
cover

Inner
cover

Charge

Seals

Base

CROSS SECTION

Figure 7   Portable-cover-type furnace

skids (on which the material is pushed) are made concave so as to cause the pusher force to exert a downward component on the stock. But by far the most popular means has been to construct such furnaces with a sloped hearth. This method also reduces the pusher effort. The disadvantage is high furnace pressure at the charge end owing to buoyancy of combustion products.

Furnaces of this type are usually direct-fired. Such furnaces are very commonly used in steel reheating prior to rolling, and variations are discussed in detail later in this Chapter.

Rollover or rolldown furnaces: These consist basically of an inclined hearth down which material rolls by gravity while being heated. They are provided with a flat hearth at the discharge end to stop the materials rolling out all at once.

The problem is that whereas steel materials roll down successfully up to temperatures of 800–850°C, they become too soft to roll shortly above this temperature. When operating at higher temperatures the operator needs to stand near the furnace in order to rake the billets down from the

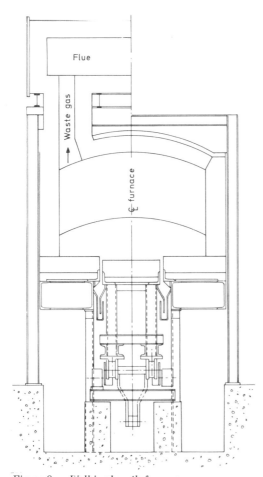

Figure 8   Walking-hearth furnace

spot where they become too soft to roll. Owing to such limitations, few furnaces of this type are built.

Walking-beam/hearth type: This method of conveying stock inside furnaces has become very popular in the last few years. Such furnaces consist of stationary and movable hearths (or beams) positioned alternately along the furnace cross-section (*Figure 8*). The stock generally rests on stationary beams.   In general the moving-beam cycle consists of four motions: up, forward, down and back. In this complete cycle beam(s) lift pieces of stock off the furnace hearth, move forward a short distance and then during downward movement deposit them on the hearth again and return to their normal resting position.   In this continually moving cycle the stock is carried from charge end to discharge end while being heated. This type of furnace is usually direct-fired, though sidefiring is sometimes employed.   Until about 1965, all the walking-beam furnaces constructed in the world consisted of stationary and moving beams of refractory construction with firing from the top only.   The latest development incorporates tubular water-cooled moving and stationary beams, which render both top and bottom firing possible.

With a walking-beam furnace, stock is spaced apart so that it can be heated on three sides — with top-fired furnaces on four sides. This results in faster and more uniform heating.   As stock is not pushed there is no risk of pile-up; and also it is simple to leave a little extra space between batches of stock to indicate changes in grades.   Such furnaces can be easily emptied, as may be desirable for weekend shut-down, visual inspection, or preventive maintenance.

Roller-hearth type: This consists of closely spaced rollers, which when rotated convey the stock from one end of the furnace to the other. In a semi-continuous version, stock is oscillated on the rollers and is discharged when heating is complete. The rollers are driven by means of worm-gear drive, all external to the furnace.   The rollers must revolve fast enough to prevent any noticeable temperature difference between top and bottom; otherwise, warping is inevitable. Also, the rollers sag if they stand still at furnace temperature.

Roller-hearth furnaces are usually direct-fired or radiant tube-fired and are generally available for temperatures up to 1000°C.   Such furnaces have, however, been constructed for heating steel to rolling temperature.

Rollers constitute the largest single cost of this furnace and considerable expertise is required in their design.   One of the earliest rollers consisted of a string of firebrick sleeves (discs), clamped between fixed collars and springs on a water-cooled shaft.   The temperature difference between the outer and inner surfaces of refractory sleeves was excessive, and roller failures were experienced. To overcome this difficulty metallic discs were interposed between refractory discs, so that heat conduction could reduce the temperature gradient in the refractory material, and thereby prevent spalling. Another design consisted of a water-cooled shaft covered with silicon carbide tubing and refractory cement. Silicon carbide can withstand very high temperatures unless it comes into contact with molten iron or steel scale. This type of design was successfully used for heating stainless-steel, which forms little scale. Water cooling of rollers is not desirable, since it complicates the design and also increases heat losses. Improvements in the art of making alloy castings has made possible the use of the rollers shown in *Figure 9*. The main body is a centrifugally cast hollow cylinder to which alloy necks are welded.

Conveyor type: Whereas all the continuous furnaces discussed in this Section can be termed conveyor type, the term is used here to signify chain conveyors, woven-wire belt conveyors, etc.

Chain conveyors come in a variety of designs; in those that have the best chance of success, the chain remains cool. This can be achieved by keeping the chain wholly out of the furnace, or else in grooves under the hearth. Such conveyors will have fingers or attachments (of heat-resisting material) which carry the materials through the furnace. In this case standard non-alloyed link chains can be used and the design can be relatively inexpensive. Also, since the chain remains cold, it can be kept taut with a simple screw adjustment.

Chains that travel through the furnace require careful engineering and operation, especially because they are subjected to repeated heating and cooling; this cyclic change imposes repeated stress which can lead to breakage. Another difficulty in such chains is that linear expansion when heated is a considerable proportion of the pitch of the chain. This can result in the chain getting out of the pitch of the sprocket (unless adequate clearance is given), causing it to ride off the sprocket or skip a tooth with a sudden jerk. Such shock-loading can lead to an eventual breakage.

Mesh belt (woven-wire blanket) conveyors have developed a considerable application for transporting relatively small objects through furnaces. Such conveyors are often driven by rubber-lined cylinders, and avoid the major problems associated with link-chains and sprockets. The expansion is usually taken up by counterweighted idler pulleys.

Shaker-hearth type: Materials can be transported through a furnace by hearth reciprocation, i.e.

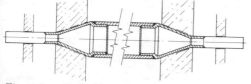

Figure 9   Roller

utilization of inertia and friction. This can be achieved by moving the hearth forward slowly so as to prevent relative movement of stock and hearth, and then pulling it back with such an acceleration at first that the charge slides forward. Alternatively the hearth can be slowly accelerated in forward motion until the required velocity is reached and then suddenly stopped causing the stock to move forward. The hearth can then be retracted slowly. Such furnaces are suitable for heating small flat material.

Continuous-bogie type: This type consists of a number of bogies. The charge is placed on the bogie outside the furnace, which is then pushed in through the end door, each bogie being moved through the furnace by the one behind it. The bogie is withdrawn at the other end, the stock removed, and the empty bogie returned to the charge end by a bogie-transfer mechanism. Sealing between the bogies is not very effective and can cause considerable heat loss.

In the metal industry these furnaces are still used for heating ingots, slabs and irregularly shaped bodies. Such furnaces are usually direct-fired with burners mounted on the sidewalls. The bogie top usually incorporates piers to carry the stock, with burners in two rows, one row firing above and one below the charge in order to reduce heating time.

Such furnaces are more commonly found in the ceramic industry (tunnel kilns). In such cases no scale or slag is formed and the possibility of jamming the conveying equipment is avoided.

Rotary-retort (helical motion) type: This type consists of a horizontal drum of heat-resisting material, inside which there is a screw-thread. The rate of progress of heating stock through the furnace is determined by the pitch of the helix and speed of the drum. The furnace is equipped with a supply hopper on one end, and the material is discharged at the other end. For many pieces (e.g. balls, rollers), this method eliminates pans, trays etc. and produces satisfactory heating.

Furnaces of similar nature are also used for lime burning. In this case the inside of the drum is smooth: but the drum is inclined. The rotation of the drum tends to carry the material in a plane at right angle to the axis of the drum. Material climbs up the side of the arc up to a certain distance (depending on the coefficient of friction) and then drops back in a vertical plane; in consequence the material passes from the top of the kiln to the bottom in a zigzag motion. Such furnaces can be provided with means of adjusting inclination of the drum, and hence the rate of descent of material can be varied.

Rotary-hearth type: The material to be heated is charged on a moving annular hearth, on which it makes one revolution through the hot zone before being removed through the discharge door at or near the starting point (*Figure 10*). Use of this type of furnace is an attractive proposition because the same operator can charge and discharge the stock.   This arrangement satisfactorily handles pieces that lie flat and do not roll about too much. Such furnaces are used for a wide range of operating temperatures up to that of steel rolling. Sizes built vary from very small units to hearths having mean diameter as much as 23 m (and a hearth width of 4 m).

The annular hearth (of refractory construction) in this type of furnace can be rotated on wheels and track or on ball bearings, and is driven through a reduction gear by means of an electric motor. Because of the hearth arrangement and circular shape of the roof, underfiring and overfiring are difficult to achieve. The usual construction is to have direct firing with burners arranged to fire tangentially towards a circle of mean diameter.

Continuous-strand type: Wires, chains and strips can be pulled through the furnaces. In general, such furnaces consist of a decoiler at the charge end and a re-coiler at the discharge end. In the case of wires, many strands can be positioned side by side in the furnace, and these strands can be pulled separately. In case one strand breaks, the wire can be tied to the next strand and pulled out. With strips conveying may be horizontal or vertical.   A continuous operation may be achieved from these by having loops at both ends of the furnace so that the flow of strip is maintained when a new coil is being put into place. For truly continuous operation, much more sophistication may be required including use of a seam welder.

Material Handling Outside the Furnace

Methods of material handling outside the furnace is not strictly a part of the subject of furnace design, but it is a subject with which a furnace engineer must deal, because the choice of furnace type and design of the furnace may be affected by it.

The charging of a furnace depends on the shape, size and weight of the material and the type of fur-

nace.   In batch-type furnaces, if the individual pieces do not weigh more than about 20 kg the handling is done by tongs. Heavier pieces up to about 250 kg may be handled with tongs suspended from monorails, and still heavier pieces are handled by tongs suspended from cranes or by means of electrically driven unit-charging machines or trucks. In some batch furnaces, it is necessary to stack the material outside the furnace. Specially designed charging machines are employed in such cases, and suitable grooves or roller tracks are provided in the furnace hearth.   For continuous furnaces, the material is usually charged onto the conveyor which carries it into the furnace or, with pusher furnaces, on the charging table between the pusher head and the furnace.   In pusher-type furnaces charged from the side, the pieces are either rolled into the furnace on power-driven rollers located inside the furnace or pushed in by a mechanical pusher.

Discharging a furnace, with batch operation, is done in a manner similar to charging.   In continuous furnaces, discharging is often done by gravity, by constructing a slope at the discharge end.   This method is often employed with hardening furnaces where the material must be quickly discharged into the quench tank, and is also common with slab-reheating furnaces of the pusher type.   When discharging billets from continuous pusher-type furnaces, power-driven rollers or conveyors can be employed, though pushers are more commonly used. Special discharge machines are employed with some end-discharged furnaces when layout does not permit normal end-discharge or when it is necessary to avoid marking the surface of the stock. Discharge machines often consist of a number of arms which go into slots in the furnace hearth, pick up the heated stock, and deposit it onto a conveyor outside the furnace.

Handling material to and from a furnace depends on the nature of the pieces or containers and the operation.   In many processes roller or belt conveyors are used to transport the material to the furnace and from the furnace to its next operation.   Selection of furnace type often depends upon the material-handling methods used in the plant, especially in the case of batch furnaces. If cranes and monorails are the general means of transport then vertical furnaces are likely to fit most conveniently; alternatively where conveyors predominate horizontal furnace loading and unloading may be found convenient. In the field of continuous furnaces, however, means of conveying the work through the furnace is not necessarily related to the factory system, but the charge and discharge ends may still be affected.

Aerodynamic Aspects

Flow Pattern

The path taken by the gases flowing in a furnace can have a profound effect on its performance. It

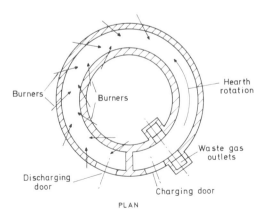

Figure 10   Rotating-hearth furnace

is well known that in furnaces heated by convection the quality of heating depends on uniform flow of gases over the stock, but even with the furnaces heated by radiation the flow pattern is important since the refractory surfaces themselves must receive heat from the gases. Bad distribution of the waste gases can cause damage to the refractories owing to 'spalling by gas impingement (see also App.C6 and C8).

Recourse is frequently had to the use of small Perspex models in which, by distortion of some of the dimensions, one can correct for the difference in size; these operate cold and the effect of buoyancy has to be allowed for.

The most usual way of tracing the flow pattern is to traverse the furnace in all three dimensions with pitot tube(s), but this must be done with the load in position since this changes the flow pattern considerably. But sometimes a smoke test, using a proprietary smoke 'bomb' or even a smouldering oily rag, may show the path of the gases sufficiently.

Furnace Pressure

Fuel-fired furnaces should always be operated under a slight pressure since this tends to even out the flow of gases inside the furnace and prevents ingress of cold air round the doors and other openings. A reasonable pressure to aim for is about 2.5 Pa (0.01 in w.g.), and this should be measured at the hearth level to avoid the buoyancy effect. If the furnace is not kept under positive pressure, it is possible for the gases to short-circuit straight to the flues, thus increasing the heat loss whilst spoiling the heating of the charge.

If the furnace is gas-tight as well as the doors, burner-lighting holes and flue offtakes then a furnace can be operated at high pressure. The furnace, however, must be lighted at atmospheric pressure and sealed (except for the flue) before increasing the flow to full value; the pressure may then be increased to the required level by use of the damper.

It should be emphasized that furnaces using recirculation must be made completely gas-tight, otherwise air will be drawn in at some points and forced out at others.

Maldistribution

Bad distribution of flow can completely ruin the performance of a furnace. It can result from many factors, some in the design and some in the operation of the furnace. Some of these are discussed below.

1. Wrong location of waste-gas offtakes: These should in general be as far from the burners as possible; but their location is not critical with furnaces operating under slight pressure.

2. High-velocity burners (particularly a single burner) create a jet induction effect which can give rise to suction near a furnace opening and allow cold air to enter. This is particularly troublesome in the soaking zone of end-discharged reheating

furnaces. It can, under different circumstances, assist in uniform heating of the furnace by entraining spent furnace gases into the freshly burnt gases issuing from the burner (jet-induced recirculation).

3. Flame impingement on the walls and roof can be caused by wrongly aligned burners, or impingement inside the burner casing of gas or air which is then swirled as it leaves the burner. Non-uniformity in flow of gases as they enter the swirler persists at the outlet despite the swirl and causes deflection of the flame. Badly designed burner blocks can cause such trouble, especially with swirling flames.

4. When air (or fuel) is supplied to several burners by a manifold, it is important that this be large enough to act as a plenum chamber, otherwise the flow to some burners will be excessive and the others may be starved, resulting in a badly heated furnace.

5. A badly located load can spoil the operation of a good furnace by affecting the recirculation of gases.

6. Location of burners: It is a generalization that the top of a furnace looks after itself since heat rises. With batch-type furnaces, in addition to locating the main burners lower in the walls, it is often useful to have a second set of low-capacity high-velocity burners that fire directly towards the load without impinging on it, so creating induction recirculation.

7. It is difficult to keep furnaces with a large turndown ratio under pressure; under these conditions cold air can enter. If the process temperature is not too high, excess air may be used: the percentage of excess air becomes greater as the fuel-flow decreases. This can be achieved by maintaining constant air flow and regulating the fuel flow. With some furnaces steam, or a neutral atmosphere, may be used to augment the gases.

Recirculation

The flow of gases round the charge can sometimes be promoted by careful location of the flues, but recirculation generally requires more positive action, such as the jet-induction effect or the provision of stirrer fans.

Where the furnace is essentially convection-heated (as when heating aluminium), the recirculation is invariably fan-generated, and it is important that the volume recirculated should be adequate. The location of circulation inlets and outlets is then not critical and they can be quite simple. When complicated straighteners or flow-dividers are employed, it is usually an indication of too small a recirculation volume.

The amount recirculated is often based on a simple rule that there should be 30 changes of free-volume (furnace less charge volume) per minute. Another criterion, within the limits of uniformity of the heating, is to calculate the wall heat-losses and to divide this by the heat content per unit volume of recirculation gas.

The volume required for recirculation depends on the heat losses of the system and the difference in temperature between the gases entering and leaving. The uniformity of temperature within the charge may be improved by using a reversing fan or a reverse-flow system. This does not reduce the circulation ratio, but does materially reduce the time required to obtain a given uniformity.

Centrifugal fans cannot be reversed and with them reversing dampers and flues are necessary. These additional flues create higher heat and pressure losses. With axial flow, reversing is quite practicable, but the flow in the reverse direction is only about 70% of that in the normal direction.

If the recirculated gases are directed towards the work through jets, the induction effect may increase the recirculated volume by a factor of 2 to 3. In a recirculation-type furnace the gases should be drawn from the furnace from above the charge with stacked work (billets, tubes etc.) and should be sent into the mass of the work by using jets, which may be simple openings in an inner wall.

Fuel Consumption and Heat Economy

The performance of a furnace installation can be expressed in various ways, the ultimate aim being the determination of the total quantity of fuel required to perform the required heating operation. For the purpose of comparison, however, the fuel consumption of a furnace is often expressed in terms of quantity of fuel, heat or energy required to heat unit quantity of stock.

An accurate balance sheet of heat could be drawn up if all the sources of input and output were known. Sources of heat input are: heat from combustion of fuel; sensible heat of fuel; sensible heat of combustion air; and heat input due to any exothermic reactions within the stock.

The most important forms of heat output are: useful heat (heat content of hot product of operation, and heat required by any endothermic reaction taking place within the charge); and losses.

The losses are: heat losses from furnace walls to surroundings and heat absorbed in raising the temperature of walls; heat loss through doors and openings; by leakage of gases; resulting from incomplete combustion of fuel; loss to water-cooled parts; to conveyors, trays etc.; special losses; and finally heat in flue gases leaving the furnace. (See App.C10).

Heat Input

The net calorific value of solid, liquid and gaseous fuels is available[4]. It can easily be calculated if the composition is known. Solid fuels are not preheated before combustion. To assist atomization, liquid fuels often have to be preheated. Knowing the specific heat and preheat temperature, the sensible heat entering the furnace through liquid fuel can be calculated, but it can almost always be neglected. It is usual to preheat gaseous fuels below a calorific value of about 6 MJ/m^3 in recuperators or regenerators to produce a high enough flame temperature. With rich fuels, combustion air is preheated purely for fuel economy; with lean fuels, primarily to increase the flame temperature. The sensible heat can be assessed by use of the *It*-diagram described earlier. Sensible heat in steam or compressed air used for atomizing liquid fuels should also be accounted for.

In the heating of metals such as steel, oxidation of charge adds to heat input. The heat of oxidation of iron or steel is approximately 5630—7530 kJ/kg, depending on the type of oxide formed, with an average of 6580 kJ/kg. If steel is heated to rolling temperature, 1—4% of the charge[6] is lost by scaling. Although heat input due to oxidation reduces the fuel cost, this will be small compared with the cost of material lost by scaling.

Heat Output

Useful Heat. To determine the heat content of stock, the quantity and temperature of material is used in conjunction with heat content tables or graphs readily available in various books. Account should be taken of preheating prior to charging. Some authors advocate including stock preheat as heat input and the total heat content of product on the output side. If any endothermic chemical reactions take place the heat required should also be included. As an example, the heat absorbed when limestone ($CaCO_3$) used as a flux in cupola operation breaks down into lime (CaO) and carbon dioxide is about 1770 kJ/kg of $CaCO_3$.

Losses. Heat loss from and heat absorption by furnace walls are important in the fuel economy. Furnace walls are usually constructed from more than one layer of refractory, and during the steady state the heat loss by conduction may be calculated as in App.C11.

The coefficient of heat transfer from the outside of the wall to air by radiation and convection may also be calculated as in App.C11, but *Figure 11* gives values of *h* for vertical walls, horizontal walls facing upwards (furnace roofs) and horizontal walls facing downwards, calculated for an emissivity of 0.9.

With known values of T_i (inner wall), T_a (ambient) and an assumed value of *h*, the first estimate of heat loss Q can be made by assuming that T_O (outer wall) = T_a. Successive approximations can then be used to recalculate T_O (=$T_a + Q/h$), h from *Figure 11*, T_O again, and so on. The interface temperatures of the refractories and hence their mean temperatures and the heat stored can then be calculated.

The heat losses discussed above are those incurred during steady uninterrupted operation. In actual practice operating periods alternate with idle periods. During an idle period the heat stored in refractories is gradually dissipated. The exact calculation of heat loss by intermittent operation is complicated, but approximate methods are available. One such method is based on the facts that when a furnace is first shut off the heat loss per unit time will be the same as in the steady state, and that all the heat stored in the refractories will be lost eventually. At the initial rate the time (θ) in which all heat stored would be lost is: $\theta = Q_S/Q$ where Q_S = heat stored per unit surface area of brickwork, and Q = steady-state heat loss per unit area. The total heat lost (Q_O) during the shut-down period (θ_O) is given by:

$$Q_O = Q_S (1 - e^{-\theta_O/\theta})$$

Thring[9] suggests that, if the total period when the furnace is off is less than one third θ, the heat loss may be taken as continuing at the initial rate; if θ_O is more than twice θ the total heat stored may be considered to be lost. The above formula should be used in intermediate cases.

Doors and Openings: For very large openings the heat loss due to radiation may be calculated by evaluating 'black-body' radiation at furnace temperature and ambient temperature and multiplying the difference by emissivity (usually 0.8 for furnace brickwork). For small openings with finite thickness of walls, some of the radiant heat is intercepted by the edges of the opening. Curves drawn by Hottel and Keller[14] give radiation factors for openings of various shapes (see *Table 2*).

Table 2   Factor for radiation through openings

Ratio: Dia. or least width/Wall thickness	Nature of opening		2:1 rectangle	Long slot
	Round	Square		
0.1	0.1	0.1	0.12	0.16
0.5	0.36	0.37	0.45	0.56
1.0	0.52	0.54	0.60	0.68
2.0	0.67	0.69	0.75	0.80
3.0	0.76	0.78	0.81	0.86
4.0	0.80	0.82	0.85	0.89
5.0	0.86	0.88	0.90	0.92

Leakage of Gases: In a fuel-fired furnace there is always a tendency for outward leakage of gases or infiltration of air.

Pressure difference between the inside of a furnace and outside is expended partly in overcoming friction and eddy loss and partly in providing kinetic energy.

The thermal effect of gases escaping from the furnace depends on whether they escape before or after completion of combustion. If they leave a batch furnace after combustion, practically no loss is incurred because it makes no difference to fuel consumption whether the gases escape through doors or stack (unless some form of heat-saving appliance is provided); but if they leave through the discharge door of a continuous furnace, a heat loss occurs, because the same gases could give heat to the charge and leave the furnace at a lower temperature.

Incomplete Combustion: Unburnt fuel passes out of many industrial furnaces, either because of bad burner design or bad operation. Furnaces are often run with little or no excess air in order to reduce oxidation of the charge. With insufficient air supply the products of combustion will contain carbon monoxide and hydrogen in addition to carbon dioxide and water vapour etc. Carbon monoxide is the only combustible constituent of waste gases that can normally be detected, but investigations have shown that approximately the same quantity of hydrogen is almost always present. Knowing the calorific values and determining the volume of carbon monoxide, heat loss due to incomplete combustion can be estimated. Another method of estimating this loss is by the use of Ostwald charts for various fuels. These charts were originally constructed with the assumption that carbon monoxide was the only unburnt gas, but were later modified by Keller to take account of the presence of an equal amount of hydrogen. Knowing the carbon dioxide and oxygen contents of waste gases, the excess (or deficiency of) air can also be determined.

In designing a new furnace, incomplete combustion loss is best estimated on the basis of results from similar furnaces previously constructed.

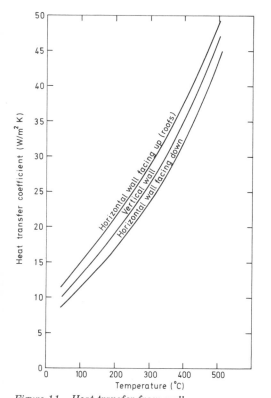

Figure 11   Heat transfer from walls

With the present-day burner designs very good combustion can be achieved, and with proper operation no combustibles should be found in the flue gases. It should be remembered that with almost all fuels as little as 1% CO + 1% H_2 can mean a loss of 5% (or more) of the heating value. Also, combustibles in flue gases can be a potential danger to recuperators and regenerators as combustion can take place in flues owing to addition of dilution air.

Water-cooled Parts: Water-cooled skid pipes and supports are a major source of heat loss in continuous furnaces for metal-heating. Heat is transferred from the surface of the skid pipe (or refractory — if insulated) by radiation, it is conveyed to the pipe interior by conduction, and so to water by forced convection. Calculations of the loss can be made by heat balance including all the processes involved. When calculating radiation from furnace to skid surface, an emissivity of 0.8 gives reasonably accurate results. Heat loss due to bare water-cooled elements can be very high, and can be greatly reduced by using very thin refractory covering.

Conveyors, Trays, etc.: In some furnaces trays or other containers are used for conveying the stock and these normally attain the same temperature as stock. Chain conveyors, on the other hand, especially if thicker than stock (or less exposed) may not attain the same temperature as the stock, and judgement is required in estimating the amount of heat they carry out of the furnace.

Flue Gases Leaving the Furnace: Usually the largest of the losses, and depends upon the amount of flue gases and the exit temperature. The quantity is known depending on fuel and excess air. Their temperature on leaving a batch furnace can be estimated from the heating cycle. With continuous furnaces, it can be calculated from a heat balance or estimated from known practical results.

Estimation of flue-gas loss has been considered one of the foremost problems in fuel consumption calculations. These calculations are simplified by use of the *It*-diagram (described earlier), once the flue-gas volume, excess air content and flue-gas temperature are known.

Special Losses: Apart from the sources of heat loss discussed above, there can be many more such as heat losses from stock projecting out of furnaces, tongs and other charging equipment, exposed surfaces of lead or salt baths. Some may be very small and difficult to calculate, but it is better to use simplified methods than to neglect them altogether.

Heat Balance (and Saving)

A heat balance (App.C10) is a convenient way of assessing the efficiency of a furnace. After carefully evaluating the above details, a heat balance can be drawn up — either in a graphical form by use of the Sankey diagram[12] or by simple tabulation. This can serve not merely as a guide to overall fuel consumption but also as a starting point for attempts to reduce it.

It is well known that the use of better and more insulation on furnace walls helps to reduce losses and thereby effect fuel saving; but in a furnace where such losses are small a reduction may be marginal. For example, a typical heat balance for a 5-zone pusher-furnace for steel reheating is as follows[15]: heat to stock 47%, wall and door losses 5%, water-cooling losses (skids etc.) 13%, waste-gas loss (after air recuperation) 35%. From the above, it is clear that attempts to reduce water-cooling loss by better skid insulation, or increase of air preheat (thereby reducing waste-gas loss) are most likely to result in substantial reductions in fuel consumption. The possibility of reducing heat to stock must also not be ruled out, because reduction in fuel consumption by this means may outweigh a consequent increased energy requirement for rolling, and overall saving may well result.

Thermal Efficiency

This, for a furnace, is one of the most difficult things to define. One way is similar to that for a boiler, i.e. useful heat/total heat input. But whereas the efficiency of a boiler ranges from 60 to 90% (or even higher), that of a furnace on this basis may be as low as 5%. The flaw in this manner of defining the efficiency of a furnace is that it bears no relation to the actual efficiency of the process. Combustion gases give up heat to the charge only so long as they are hotter than the charge. In consequence, in a batch furnace the flue gases must leave the furnace at a temperature in excess of the required final stock temperature. This limits the maximum possible thermal efficiency. In some cases (e.g. annealing) slow heating is necessary to achieve the required properties in the finished product, and this will require low thermal efficiency.

The purpose of finding out the efficiency of a plant is to establish how well the plant is designed or run in comparison with another similar plant. If this view is accepted, then the efficiency of a furnace can better be assessed by measuring the amount of fuel (or energy) required to heat to the required specification unit weight of material.

Mechanical Design and Construction

Furnaces are built up of metal parts and heat-resisting non-metallic materials (refractories) etc. The strength and durability of furnaces are affected by factors such as temperature and its variations, the effects of chemicals and slags, vibrations, etc. Knowledge of the available materials and their properties is essential to a furnace engineer, and much literature and published data are available for reference.

Refractory Parts (see Chapter 17)

Walls, roof and hearth are generally constructed from refractory materials, either in brick form or monolithic construction using castable or plastic (mouldable) materials. Mouldable and castable materials have become popular materials of construction for vertical walls and roofs. They are held in place by metallic or ceramic anchors that are attached to the furnace binding. Rammed plastic materials are often preferred to castables in furnace parts subjected to frequent and severe changes of temperature.

With brick construction, it is important to limit the number of different sizes to simplify building and repairs. The usual standard brick adopted in the UK is 230 × 115 × 75 mm (9 in × 4½ in × 3 in). Whilst a wall is constructed of individual bricks, it will act on heating as a mass and expand and contract with temperature. It is therefore necessary to consider methods of bonding that give the strongest construction and allow for movement. A form of bond used extensively for furnace brickwork is shown in *Figure 12*. All brickwork should be laid with minimum jointing, and fireclay cement should be equal to or better in quality than the brick. High brick walls may have a tendency to bulge inwards and it is good practice to slope these outward either as a whole wall or by increasing the thickness at the base. Building such walls in the form of a vertical arch is another way of decreasing the tendency to collapse.

Sprung arches of brick construction are built either in separate rings or bonded. Bonded arches (*Figure 13*) are stronger but should be built with 'bonders' i.e. bricks 1½ times standard length. The thickness of the arch varies from about 115 mm for a span of up to 1.5 m to 345 mm for a span of about 5.5 m. Where construction permits, it is good practice to support the arch from a 'Skew back' resting in a steel member attached to the furnace binding, so that the arch is supported independently of the furnace walls. Where an arch is built into a wall, as for a door opening, a second arch is often built to allow the lower arch to be replaced without disturbing the wall.

For spans larger than 4–5 m, suspended arches are used. Many proprietary designs of suspended arches are used, consisting of specially shaped refractory blocks suspended either by refractory shapes or by metallic clips. Suspended arches are also made of castable or mouldable refractory held by refractory shapes embedded in the base

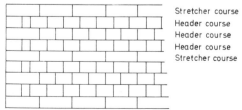

Figure 12   Typical furnace wall construction

Stretcher course
Header course
Header course
Header course
Stretcher course

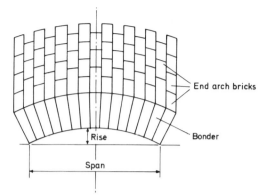

Figure 13   Bonded arch construction

End arch bricks
Bonder
Rise
Span

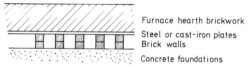

Figure 14   Ventilation of furnace hearths

Furnace hearth brickwork
Steel or cast-iron plates
Brick walls
Concrete foundations

material (the refractory shapes being held by metallic supports), or by embedded supports only. The design and supply of suspended arches is preferably undertaken by companies who specialize in this work, and their advice should be obtained in regard to thermal expansion, method of erection and strength of support.

Most furnace hearths have to withstand high temperature, wear due to movement of product, shock loads, and attacks by scale or slag. Careful selection of refractory material is therefore very important (Chapter 17). When possible the hearth should be ventilated to minimize the amount of heat carried through to the foundation. A method of constructing a ventilated hearth is shown in *Figure 14*. The furnace brickwork is supported on cast-iron or steel plates, these in turn being carried on low brick walls or piers to form air passages. This design, with or without the plates, is also used for ventilating the bottoms or sides of flues.

Substantial foundations are necessary under many furnaces. These should be as simple as possible and reinforced with steel bars. An outline drawing of the foundation should be made to show loads from the furnace, which can be used by civil engineers to decide the thickness of concrete according to the load-bearing capacity of the ground. It is usual to add 50% to the dead load for moving such loads as are encountered with furnaces where the product is pushed over the hearth, and 100% for shock load where there is a risk of the product being dropped, as with ingots in a soaking pit. Pockets should be left in the foundation for grouting the binding members when in their final position. If there is a risk of water penetrating the foundations these should be built with a slight slope to one side, where a sump can be provided.

Metallic Parts

The furnace brickwork/refractory is usually contained by encasing it in steel plate, with structural steel members and tie bolts to prevent distortion of the whole structure when heated. The design is based on well-known theories of structures and strength of materials; it is, however, advisable to allow greater margins of safety than for normal cold structures. It is also advisable to check the deflection under load, particularly for such items as the supports for suspended roofs.

The thickness of steel plating against the refractory will depend upon the size and duty of the furnace. For example, for small heat-treatment furnaces 3 mm plate may be sufficient, but for heavy-duty furnaces as much as 6 or 8 mm plates may be advisable. The size of structural members depends on the size and type of furnace, but these members are usually spaced 1.0—1.5 m apart, and pairs of members are often used to obtain the necessary strength.

Long steel bars are often used as tie bolts to hold the sides and end binding of the furnace. When used, these should be kept as low as possible above the furnace roof. For open-topped furnaces such as soaking pits, it is usual to employ very large steel members placed horizontally near the tops of the vertical side binding, these horizontal members having interlocked corners for adequate strength.

Furnace doors and door surrounds are usually made of cast iron. Small furnace doors may be cast in one piece as a solid ribbed construction; large doors are made of open-type construction (to reduce weight), but very large doors if made of cast iron should be made of panel construction with sections bolted together. Doors should have a generous overlap around the door opening, 100—150 mm being provided for most furnaces, and clamps should be fitted if the door is normally kept closed. Even a simple wedge on the door, sliding against a fitting on the door frame, can be very effective in minimizing air inleakage or gas escape. Another popular method of door tightening is by making use of gravity and having inclined doors.

At the top of the doors, burning of the door frame is frequently quite severe. This effect can be counteracted by water-cooling the top frame or alternatively by making the top part of the frame removable. Sides of doors are also subject to gas escape from furnaces. In extreme cases, therefore, sides and top of door surrounds are made hollow and water circulation is incorporated. In designing door sills, consideration should be given to the fact that stock is sometimes temporarily placed on them.

One of the problems in the use of iron castings subjected to prolonged heating is that of growth, which can be checked to some extent by small additions of alloys. There are many special irons available which are less liable to growth, and have improved mechanical strength and heat-resisting properties compared with ordinary cast iron. These special irons are used extensively for such items as work supports, mechanized parts of furnaces, and for heat exchangers. While more expensive than ordinary iron castings, their use can be fully justified owing to the better life obtained.

Instrumentation and Control (see Chapter 18)

Instruments and controls play a major part in the present-day design and operation of furnaces, especially in the areas of temperature and combustion. It was not always so; only since the instruments began to achieve the necessary reliability and became the unseen eyes of the process have they been accepted. Today it is common practice for the plant and the control scheme to be designed together.

Instruments are available for the measurement of temperature, flow, pressure, level, speed, weight, thickness etc., in fact anything that can be measured and produce a corresponding impulse or signal. They may be classified in three distinct areas: mechanical, which includes hydraulics; pneumatic; and electronic, which may be analogue or digital. Nearly all can be made fully automatic by either open- or closed-loop techniques and, by virtue of feed-forward and feed-back methods, can maintain a constant level of control as required.

The performance of a control system is determined by the nature of the process, the characteristics of the controller, and the magnitude of the disturbances. In furnaces, where flows, temperatures and pressures are interwoven, it is necessary to introduce control for different loops. As the prime requisite in a furnace operation is the control of temperature, the temperature measurement must be related to air and fuel flows which, by control, will regulate the temperature. It is necessary to introduce what is termed cascade control, where the output of the temperature controller will become the desired value of the air and fuel controllers. The pressure within the furnace, which is also a part of the temperature control, also has its own control loop and characteristics. It is therefore possible to devise a control scheme which will cater for most of the plant disturbances and minimize the others.

The field of digital computers and their peripherals opens up an exciting time for the process-control engineer. Systematic study has solved practically all the problems relating to process control, and the control philosophy should be such that the flexibility inherent in the digital technique is totally used. Experience has shown that proper application of the available digital techniques will achieve substantial improvements in throughput and quality of heating, better than an analogue system ever could.

Melting Furnaces

Melting furnaces may be of crucible, hearth or shaft types. The method of operation of these furnaces is dependent on special technical considerations peculiar to each process, and only a general discussion is possible.

Crucible Furnaces

Crucible furnaces are used for melting small batches. The charge is melted in a refractory or metal pot, fired externally by solid fuels, gas or oil. To improve the heat transfer in these, especially before melting, some modern furnaces are designed so that the crucible rotates on an inclined axis. This ensures quick melting and hence improves efficiency. The other main factors influencing the fuel economy of crucible furnaces are the fuel/air ratio, and furnace-pressure control. External waste-heat-recovery appliances are often employed in these furnaces for reducing fuel consumption.

Hearth Furnaces

In hearth furnaces the material or charge, on the floor or hearth of the furnace, is usually heated by convection and radiation from the flames above it. Typical examples of hearth furnaces are the open-hearth furnace for steel making and the glass tank.

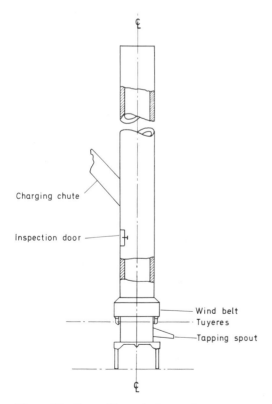

Charging chute

Inspection door

Wind belt
Tuyeres
Tapping spout

Figure 15   General layout of a cupola

References p 147

The operation of the open-hearth furnace for steel making involves various processes. In general it consists of charging and melting a certain percentage of steel scrap; a further quantity of pig-iron is melted by radiative heat-transfer from the flames passing over the surface of the bath. The carbon content of the metal is reduced to a desired figure during and after the melting process, and other impurities are reduced in quantity or removed. It is necessary to attain very high flame temperatures in order to transfer heat to the metal bath. High flame temperatures are achieved in open-hearth steel-melting furnaces by the provision of regenerators for preheating air (and also the fuel when using producer gas or mixed blast-furnace and coke-oven gases). The fuels normally used in these furnaces are oil, producer gas, and mixed blast-furnace/coke-oven gas. Pulverized fuel is avoided owing to the bad effects of coal ash.

The overall fuel consumption in open-hearth furnaces can be kept to a minimum by maintaining a proper fuel/air ratio in all stages of firing, consistent with good combustion; maintaining good furnace-pressure control; and making use of regenerator-reversal equipment to achieve optimum preheat throughout the melting cycle.

Shaft Furnaces

The blast furnace is the most important of the shaft furnaces for smelting metals from their ores. It is also perhaps the most complicated furnace system, as its operation is combined with complex metallurgical operations. Owing to its complexity, it is not further considered here[16]. The cupola may be regarded as a smaller scale and simpler version of the blast furnace.

A cupola (*Figure 15*) consists of a vertical cylindrical refractory-lined steel shell, equipped with wind box and tuyeres at a low level for admission of air. A charging opening is provided at an upper level for introduction of melting stock (pig-iron, scrap etc.), fluxes, and fuel (coke). Holes and spout for removal of molten metal and slag are provided near the bottom (hearth). Cupolas are usually supported on steel columns, and the bottom of the shell is often provided with drop-bottom doors so as to discharge debris at the end of the melt.

The operation of a cupola consists in first setting fire to a bed of coke on the working bottom of the furnace; the coke level is then built up to a level above the tuyeres. When the fire is well alight and the coke at the tuyere level well heated, the air blast is put on and the cupola is charged with alternate layers of metal, coke and flux until full. The metal starts melting within a few minutes, after which the charging rate should be equal to the melting rate in order to keep the furnace full. One of the features of the cupola furnace is that the ascending gases come into contact with the descending melting stock, and an efficient exchange of heat takes place from the hot gases to the melting stock.

The descending fuel replaces that burned from the original coke bed.

The conventional cupola is manufactured in standard sizes, and melting rates depend on the ratio of metal/coke as also on the constituents of the metallic charge (i.e. the quantities of pig-iron, iron scrap, steel scrap). Cupolas charged with pig and scrap iron give higher melting rates than those charged mainly with steel. Cupolas are supplied with a system of tuyeres large enough to provide easy delivery of the combustion air to the coke bed. The total area of the tuyeres is generally of the order of one fourth of the cross-sectional area of the cupola at tuyere level (inside lining), although in practice tuyeres may be found ranging from one tenth to as much as one third of the cupola area. The shape of the tuyeres has little influence on the efficiency of the cupola.

Fuel economy in the cupola depends mainly on the success achieved in utilizing the heat in the coke and in lowering the carbon monoxide content at the top of the shaft. Various methods employed to economize in the use of coke are: (a) use of supplementary fuel, (b) use of more than one row of tuyeres, (c) use of oxygen, and (d) use of hot blast. It has been claimed that as much as 30—50% of coke can be replaced by use of supplementary fuel (e.g. natural gas, oil, tar, etc.), with substantial reduction in the cost of production of cast-iron. The supplementary fuel is usually introduced above tuyere level and sometimes the fuel is completely burnt outside the cupola (with theoretical air) and the products of combustion introduced into the cupola immediately. It is found that with increasing use of supplementary fuel it is not possible to increase the carbon content, and hence this method may not be beneficial when melting a charge rich in steel scrap.

When using two rows of tuyeres, it is essential to control very closely the proportion of blast to each row of tuyeres at all blowing rates. In this case the improvement in operation is not because carbon monoxide in the top gases is lower; but it lies in a spreading of the melting zone, which allows greater time of contact between the molten iron and incandescent coke resulting in higher metal temperature and higher carbon and silicon content.

Use of oxygen in cupolas is more beneficial when injected separately rather than when mixed with the blast. It results in increased metal temperature as well as carbon and silicon content besides reducing the coke requirement per ton of melt. It is often used for raising the metal temperature very quickly, or when starting up the cupola in the morning.

Use of hot blast provides the best means of obtaining the advantages claimed by the above methods, either on its own or in combination with one or more of the methods described above. Hot blast can be provided by use of either an independently fired system or via a recuperator using heat derived from the cupola top-gases. In either case the net result is use of less coke, higher metal temperature, or higher output, depending upon the type of charge and the type of iron required. Government regulations in the UK are such that for cupolas over a certain melting rate, it is necessary to prevent the escape of unburnt gases and hence they must be burned. In such cases, it will be beneficial to burn these gases in a combustion chamber and provide hot blast by use of a recuperator. Such a system needs an additional burner for starting up only.

Another method of effectively reducing the coke requirement is addition of low-melting calcium carbide. It is claimed that it is cheaper to use calcium carbide in comparison with hot blast provided by an oil-fired heater. Calcium carbide is also often used as a corrective to increase the metal temperature.

Sintering Furnaces

The cement kiln is a typical example of a sintering furnace. Kilns are used in the cement industry for calcining lime and are made in various forms. Rotary cement kilns as used in the cement industry take the form of a long cylinder supported on rollers. They vary greatly in diameter and length, from 1.5 × 18 m to 4 × 140 m. In the larger kilns the ratio of length to diameter generally varies between 30 and 40:1. The rotary (drum) is installed at an incline of 3—5°, and generally operates at speeds of 1—2 revs/min. The stock is charged into the kiln at the elevated end and discharged at the lower end, moving countercurrent to the flow of combustion gases derived from fuel burned at the lower end. Such kilns are only charged with about 10% of stone, so that about 90% of the kiln space is filled with flame and hot gases.

Rotary-kiln feed consists invariably of small sized stones (generally not exceeding 50 mm). Such kilns are ideally constituted for heating fine material (e.g. 6—12 mm diameter), since their action causes the fine material to fall (rather than slide) continually through the flame, thus transferring heat by convection to a large exposed surface area. To induce this falling movement, obstructions, usually made from refractories, are built into the linings. Stone sizes smaller than about 6 mm are avoided because they contribute to ring formation in kilns and cause a serious dust-control problem.

Solid, liquid and gaseous fuels are employed in such kilns, but pulverized coal is most widely used, because in practice kilns fired with pulverized fuel are known to give better thermal efficiency. The only disadvantage here is that the problem of kiln-ring formation is much greater, especially with coal having too much ash or ash of an unsuitable nature. Kiln rings not only tend to reduce the output of the rotary kilns, but lead to inefficient combustion because pulverized coal, ash, and stone tend to adhere to these rings.

Rotary kilns have no standard operating temperature, but in the hottest zone the temperature usually ranges between 1250 and 1450°C. The original rotary kilns of short length had very poor thermal efficiencies owing to high exhaust-gas temperatures. Increasing the kiln length (and hence preheating the charge) tends to reduce the exit-gas temperature and improve efficiencies. Further improvement in efficiency is achieved by use of preheaters, heat exchangers and coolers.

The fuel consumptions in the earlier kilns without additional fuel-saving equipment used to be in the region of 10.5 to 14.5 MJ/kg of lime. The fuel consumption of the modern rotary kilns provided with preheaters etc. is about 5.7 to 6.7 MJ/kg.

The other form of kiln used in these calcining industries is the vertical type, which somewhat resembles a blast furnace or cupola. A modern vertical kiln consists of four distinct zones. From top to bottom they are: (1) stone storage, a hopper-shaped zone; (2) preheating zone designed to pre-heat the stone to a temperature just below dissociation temperature; (3) calcining zone where combustion occurs; and (4) cooling and discharge zone, which is constructed like an inverted truncated cone. While stock is cooled in this zone, secondary air for combustion is preheated. In some types of vertical kiln the material is charged in layers alternating with layers of coke. For combustion of coke, air is provided through tuyeres.

The product of such kilns is invariably contaminated with ash. Where a purer product is required, externally fired kilns using producer gas, natural gas or liquid fuels are provided. In such kilns, a series of side ports introduce fuel and primary air into the calcining zone. It is possible to operate an externally fired kiln so as to complete the combustion and give out no excess carbon monoxide.

The size of stone charged into vertical kilns is much larger than with the rotary types, because larger stones tend to produce more voids and improve heat transfer. Fuel consumption in vertical kilns usually ranges from 4.1 to 6.2 MJ/kg.

Kilns

Kilns are used for heating or firing products made by the pottery and refractories industries. In the firing of clayware, three stages of heating are involved — drying, oxidation and finishing (or soaking). In the first stage the moisture is removed by hot air or gas until the ware is dry, in the second the carbonaceous matter in the product is oxidized, and in the final stage a sufficiently uniform temperature is achieved to develop the required degree of vitrification.

The net heat required by many of the materials fired in these industries is very small, since the material gives up a lot of heat during cooling. Older kilns were of batch type in which heat given up during cooling was not utilized; consequently they had very large fuel requirements. One such batch-type equipment is the conventional beehive kiln, which is of hemispherical shape fired by a number of fuel beds around the circumference, with waste gases going down through the ware into

bottom flues. Such kilns are still used for firing pottery and silica bricks. These were later developed into battery kilns in which the air for combustion is preheated by using it to cool the previous charges, while the waste gases go to preheat the following charges.

The kilns that make use of heat given up by the ware in cooling are made in two forms — semi-continuous and continuous types. The semi-continuous kiln consists of a number of chambers connected through suitable openings. In such kilns goods are set and stay stationary, and firing is applied to each chamber in turn thereby moving the preheating, firing and cooling zones. Combustion takes place in the hottest chamber, the combustion air is preheated in the preceding chamber and the gases leaving the hottest chamber preheat the ware in successive chambers, the last of these being connected to the stack. Such kilns give a compact layout.

Tunnel kilns are the continuous types in which preheating, firing and cooling zones are fixed and the ware is moved on bogies operated by an external pusher mechanism. The tunnel kilns are also made circular with an annular moving hearth used instead of cars.

In continuous and semi-continuous kilns, although fuel consumption is very small, control of the firing stages of the batch is very difficult, since the same gases are used to heat different batches at different stages of firing. Where high-grade heating is a requirement, and a variety of batches is to be treated, fuel economy may well have to be sacrificed by the use of batch-type equipment.

In pottery firing the ware is often protected from combustion gases by enclosing it in muffles or refractory containers (saggers). The use of these results in increasing fuel consumption, but in this case the fuel cost is small compared to the value of the ware fired, and fuel economy takes second place.

Most kilns in the heavy clay and pottery industries are coal-fired, but continuous bogie-type tunnel kilns are very often fired by gas, pulverized coal or even oil. The efficiency of kilns, besides being affected by control of air to meet combustion requirement, depends on setting of the ware in the kiln and the firing schedules.

Furnaces for the Glass Industry

Glass-melting Furnaces

The process of glass melting can be considered in three stages.
1. Fusion or melting of the batch, which for most glasses comprises precise proportions of sand, soda ash, limestone and various additions particularly Cullet (waste glass from the production process or bought outside). This stage involves not only conversion of the raw materials into a fluid state, but also chemical reaction: liberation of gaseous components (mainly carbon dioxide, sulphur dioxide) from alkaline salts and alkaline-earth compounds in

the batch, formation of silicates, and mixing of liquid components.

2. Refining: During this stage a high temperature is maintained in the chamber; it comprises removal of gas bubbles from the glass by buoyancy. The rising of gas bubbles stirs the melt, and renders the glass homogeneous.

3. During the refining process the glass has a viscosity of around 10 Pa s (10^2 P). During shaping and gathering etc. viscosities of about $10^2 - 10^3$ Pa s (10^3-10^4P) are required and glass is hence cooled in the third stage to reach the desired viscosity.

Two general types of melting furnace are used in the glass industry — pot furnaces, and tank furnaces. The latter are now more often used, the former being utilized only in special cases.

Pot Furnaces. These are used in plants where a wide variety of glasses is produced in small quantities. As the name implies, the charge is kept in pots, which may be open topped or hooded with a mouth at the front. Pots are hand-made from pot clay and have capacities ranging from 100 to 1500 kg of glass. They are kept in a furnace consisting of a combustion chamber built with high-grade refractory (sillimanite or equivalent), having an arched roof made of silica bricks, and suitably insulated. Pot furnaces are made in rectangular or circular shapes. These furnaces have openings on the side(s) to accommodate one to twenty or more pots.

A typical cycle of a pot furnace comprises hand charging the pots and heating for a period until the charge partly vitrifies; then a second and sometimes a third charge is added. When the charge is fully converted to molten glass, a further period is allowed for refining, after which the fuel flow is reduced (or cut off) to allow the glass to cool down to working temperature. Normally the pots are filled in the evening and the furnace run so that the pots are ready the next morning and are worked out during the day.

Pot furnaces are suitable for firing with any fuel, but operation with open pots requires a careful selection of fuel to keep sulphur content at a minimum. The design of these furnaces lends itself to fuel-saving appliances like recuperators and regenerators.

Pot furnaces, though still used for very small production, are not very efficient and have the following disadvantages.

1. High fuel consumption: (*a*) melting, which takes place gradually inwards from the pot walls, results in practice in longer melting times; (*b*) heat supplied during the melting and refining process is gradually lost during cold stirring.

2. The considerable change in temperature between melting and refining on the one hand and working-out on the other affects the pot material adversely. In spite of all precautions, breakage of pots leading to considerable loss of glass and production is unavoidable.

Day Tanks. This is probably the simplest glass-melting unit in use today. These furnaces are usually rectangular in shape in which a bath of refractory material contains the melt. Above the bath lies the combustion chamber. The operation of these furnaces resembles that of pot furnaces, in that melting and working alternate periodically. Day tanks usually range from 1 to 10 tonne capacity. Whereas the fuel consumption in these may be similar or slightly less than that of the pot furnace, the absence of the fragile pot can be considered an advantage.

Continuous Tanks. These are similar to the day tank. The continuous-tank furnace is a hearth-melting unit, consisting of a bath of refractory material which contains the melt and a combustion chamber which lies above the bath and in which the necessary heat is produced by fuel flames. The raw materials are charged continuously or at short intervals at one end, and after melting, refining, etc. the finished glass is continuously removed for working at the other end. Whereas all the steps in glass production take place in one unit, there are subdivisions which help to control the most important individual stages. These zones of the tank are melting, refining and working. The first two processes are usually carried out in one compartment called the melting end. The third process takes place in a working chamber or 'working end'.

The glass-melting tank furnace most commonly consists of a rectangular bath divided into two separate tanks of unequal size by a double wall built at right angles to the length of the furnace. The double wall is bridged by refractory blocks and the combination of walls and cross block is known as the bridge wall. The two tanks are covered by one continuous crown; the large tank into which the raw materials are fed constitutes the melting end, and the smaller tank is the working end. Glass passes from the melting to the working end through a rectangular covered channel (throat) penetrating the bridge wall. The bottom of the throat is normally on a level with the bottom of the melting end, but sometimes lower. The location of the unheated throat at about tank-bottom level helps to cool the glass to working temperature and also prevents unrefined glass from flowing to the working end.

The modern trend in design of melting tanks is to have separate melting and working ends, so that accurate control of working-end temperature conditions is possible (*Figure 16*).

The tank furnaces are either end-fired (along the length of the furnace) or cross-fired. The type of burner employed depends on the type of fuel used. Although many different fuels may be used, producer gas, town gas, coke-oven gas and oil are commonly employed. Waste-heat recovery equipment in the form of recuperators or regenerators is generally incorporated to preheat the combustion air, and the producer gas where this fuel is used.

The specific fuel consumption of a tank furnace depends on various factors such as furnace load (or glass melted in 24 h per unit area of tank), fuel used, extent of waste-heat recovery, insulation characteristics etc. Automatic control is becoming increasingly important with increasing demand for greater output, better quality of glass and low fuel consumption. The first step in automatic control of a glass tank furnace is the control of gas/air ratio, so that in all stages of firing good combustion is achieved. The other important features are control of furnace pressure, and, where regenerators are employed, control of reversal in order to optimize their use. The following are some of the means of achieving good fuel economy in glass-melting furnaces:

1. Keep the melting-end temperature as low as effectively possible.
2. Reduce the quantity of excess air as far as possible consistent with proper combustion.
3. Keep furnace pressure as low as possible consistent with least sting-out (gas discharge) and no air infiltration.
4. Keep all openings closed as far as possible; also patch up all cracks appearing in the tank, superstructure or regenerator.
5. Use insulation wherever possible.
6. Use cooling only where necessary to improve the life of the refractory.

Glass-annealing Furnaces

The glass-annealing furnaces (lehrs) are used to cool a glass object slowly after forming. When a container leaves the forming machine, it is usually at an average temperature in the region of 700°C. By the time the components reach the lehr, the temperature may have dropped to 500°C or lower. The temperatures in various parts of containers are different, e.g. the outer surface may be cooler owing to the chilling effect of the moulds, and because of differences in thickness and shape some parts will cool more than others. Consequently stresses are set up in the glass. The process of annealing consists in removing these and cooling the object to room temperature without introducing new stresses. Glass has a low thermal conductivity; cooling must therefore be slow enough to avoid excessive temperature-gradients within the container.

The annealing of glass is carried out in the annealing temperature-range, which varies for different glass compositions. The rate of stress removal is greater for higher temperatures; above the upper annealing temperature glass begins to soften under its own weight, and below the lower annealing temperature it is not possible to anneal the glass within a reasonable time.

When designing a lehr, it is necessary to divide the length into three main sections:

1. A number of zones which will allow sufficient time for the glass to be brought to uniform temperature.
2. Slow-cooling zones, where the glass temperature is lowered until it is below annealing temperature.
3. Rapid-cooling zones to reduce the temperature so that the ware may be handled at the discharge end.

Most modern lehrs are of all-metal construction, mineral-wool-insulated where required, using woven-wire mattress-belt conveyors, with air-recirculation/convection heating. This type of lehr comprises a number of zones or sections forming the entire length of the lehr, in which the heating and cooling temperatures may be accurately preset and maintained to give the desired annealing conditions.

The ware is carried on the mattress belt which is either carried on rollers, or more recently on skids placed laterally or longitudinally. The mattress belt is returned to the top conveying side,

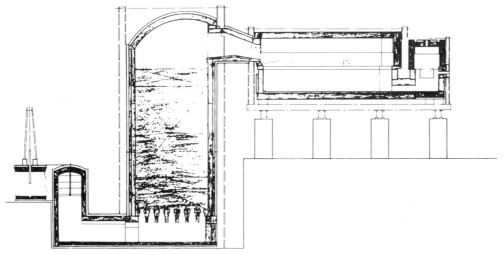

Figure 16   Glass-melting tank (Courtesy Stein Atkinson Stordy Ltd)

References p 147

either inside the lehr tunnel or underneath the main structure, according to the design of the lehr. The former method is preferred since no chilling of the ware takes place from a cold belt; this also tends to give longer life. The method of driving the mattress belt varies considerably in detail. A satisfactory drive depends on the choice and the number of drive drums. To add to the grip, most drive-drums are rubber covered.

In fuel-fired lehrs open gas firing is sometimes incorporated, especially when recirculation is not required, but radiant tubes are often used to ensure a clean heating atmosphere. The design of radiant tubes varies considerably: either they incorporate burners that use both fuel and combustion air at pressure, or they incorporate a burner using fuel only under pressure and induce combustion air around the burner by exhausting the combustion gases from the opposite end of the tube. The second method enables sulphur-bearing gases to be used without danger of contamination of the ware in the event of leakage since the tubes are maintained under suction.

As explained earlier, convection heating is achieved by recirculating the furnace atmosphere by means of fan(s) in each section, designed for upward, downward or lateral flow. Normally upward flow with radiant tubes below the belt is preferred, as this tends to give more uniform temperature.

When using lehrs to anneal articles of different shapes, sizes and thicknesses together, compromise annealing cycles may have to be used. Alternatively different articles may be annealed at different times. In such circumstances, whereas some articles may enter the lehr at higher than annealing temperatures, thereby giving up heat, the others may enter too cool and necessitate heating. This is important to bear in mind when designing or selecting a lehr. Points to remember during the operation of lehrs are:
1. The ware should be charged as soon after forming as possible.

2. The ware should be charged with as small gaps as possible (but not touching each other), and as uniformly as possible. If the charging is not uniform the belt-stretch may be uneven, leading to premature failure.

Soaking Pits and Reheating Furnaces

Depending on the size and shape of the charge and number of items to be heated, either continuous or batch-type designs are used for reheating. It is traditional to use soaking pits for ingot heating. A variety of continuous and batch furnaces are used for heating billets, bloom, slabs, and some of these are discussed here.

Soaking Pits

Soaking pits are used extensively for heating ingots before rolling. The ingots are charged into the pit as soon after stripping from the ingot moulds as possible, sufficient pits being provided to allow a continuous supply of ingots to the rolling mill. Cold ingots are also sometimes heated in soaking pits.

A soaking pit for reheating steel is a rectangular refractory-lined chamber (*Figure 17*) with one or more burners. The pit is provided with a movable cover of arched or suspended construction. To prevent gases from leaking around the cover, various methods of sealing have been developed; one of the latest effective means makes use of ceramic fibres. The pit covers are constructed with integral moving mechanism.

In regular operation the soaking pit does very little heating (except when cold or cool ingots are charged), because the centre of the ingot should be as hot as or hotter than needed, whereas the outside is cool. The idea of having a soaking pit is so to provide the ingot to the mill that it has reasonably uniform temperature from top to bottom and around the circumference. A number of

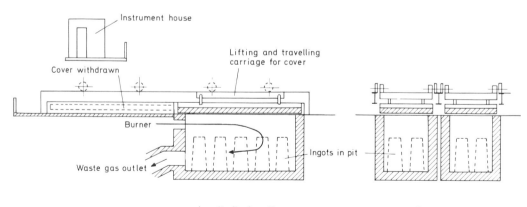

Longitudinal section   Cross-section

Figure 17   Soaking pit

different arrangements have been tried; some of them overheat the top or the bottom of the ingot. Another problem is the drop in temperature between the first ingot drawn and the last, owing to the cooling of the pit each time the lid is opened. A one-way-fired soaking pit, accompanied by a variable-flame burner, overcomes the difficulties mentioned above. With such a burner the flame can be lengthened or shortened without change in the fuel/air ratio, and a uniform temperature can be maintained from back to front of the pit. Continuous control of flame length can be achieved by provision of a centre-zero temperature-controller, which enables a bias to be applied to either end. Another feature of such a burner is that the flame has sufficient momentum at all times to ensure that heat is carried to the bottom of the ingot.

Earlier designs of pits were reversing regenerative types. Almost all the pits designed recently are of the above one-way-fired type with variable-flame burner. The modern pits hold as much as 150 t of steel ingots and are fired by producer gas, coke-oven gas, blast-furnace gas or a mixture of the last, natural gas or oil. Ingots are either charged to stand vertically or lean against the sidewalls. On the bottom of the pit a layer of coke breeze is generally used to absorb slag; spent breeze and slag are removed at regular intervals.

In some cases, continuous-bogie type furnaces are used to preheat ingots before charging into pits, and sometimes also for final heating of ingots.

Aluminium-alloy ingots are also sometimes heated in soaking pits prior to hot rolling, but the design of such pits is different from those for steel heating. Frequently, heating of aluminium ingots is combined with homogenizing, which overcomes brittleness resulting from casting, and involves heating the ingots to a temperature between 400 and over 600°C and soaking for a long period. The temperature is critical; stock temperature needs to be controlled as closely as ±3°C. Most soaking pits for aluminium are indirectly fired by oil, using radiant tubes. High-velocity fans are provided for recirculation of the gases. The recirculated gases are usually drawn through a fan at the base of the pit and pass upwards over the radiant tubes and into the pit above. If the ingots are very long, the volume of recirculation gases must be very large to avoid significant temperature-drop in the gases and hence in the ingots (from top to bottom of the pit).

Whereas the earlier aluminium-soaking pits almost always incorporated oil-fired radiant tubes, maintenance problems with the radiant tubes have led to the development of direct-gas firing, with considerable improvement in thermal efficiency and without deleterious metallurgical effects.

Reheating Furnaces

For steel reheating prior to rolling, pusher furnaces have been the most widely used. Though pusher furnaces had been thought of well over a century ago, most development has taken place in the last 50 years. During this period they have evolved from the single-zone solid-hearth design, capable of heating about 30 t/h of billets up to 100 mm, to modern five-zone designs to heat material up to 300 mm (or more) thick at outputs exceeding 250 t/h.

In the original furnaces (material up to 100 mm) the single-zone end-fired design (*Figure 18a*) was used. The hearth in these furnaces was sloping, which meant air inleakage from the discharge end owing to the chimney effect. To minimize this, the slopes were reduced gradually. These furnaces were generally side- or end-charged; stock was pushed down the hearth by finger pushers and side-discharged by use of push bars.

As the thickness of material to be heated increased and higher outputs were desired, a second zone was added (*Figure 18b*); the sloping hearth was retained. These were also traditionally side-discharged, but as the material to be rolled changed from billets to slabs, the principle of end-discharge from the discharge slope was developed.

A further increase in output and thickness requirement led to the development of furnaces with underfiring, i.e. three-zone (*Figure 18c*) and five-zone (*Figure 18d*) pusher furnaces.

The thickness of stock determines the stage at which underfiring is introduced. It is generally considered that material over 150 mm thick requires underfiring. Below this one- or two-zone top-fired furnaces are adequate; they can give heating

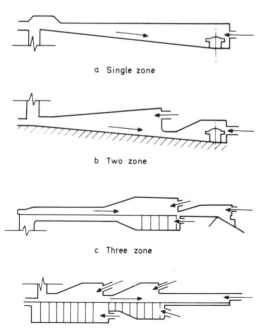

a Single zone

b Two zone

c Three zone

d Five zone

Figure 18   Pusher-type furnaces

rates up to 490 kg/m² h per furnace*. A well designed three-zone pusher furnace can be expected to heat slabs/blooms up to 225 mm thick at rates up to 700 kg/m² h. For even higher production rates and materials thicker than 250 mm, a five-zone pusher furnace must be used, which can give heating rates up to 950 kg/m² h.

In all the variations of pusher furnaces discussed above, the stock moves counterflow to the combustion products, and the furnace temperature increases from the slab entry end to the discharge end. The temperature of combustion products at the charge end (entry to flue system) varies from about 800°C in a conventional two-zone furnace to 1100°C or more in a five-zone furnace. This has led to use of higher air-preheats for five-zone pusher furnaces, but despite this they require more fuel per unit of steel heated.

In the furnaces involving firing below the stock, tubular water- (or steam-) cooled skids carry the stock. The skids cause cold spots on the stock due to contact with a cold surface and the shadow effect. To reduce these cold spots (skid marks), a top-fired soak-zone is provided in these furnaces. This zone may be as much as 30% of the length of the furnace. A considerable amount of research has been done to design skid shapes that reduce the shadow effect.

A further development of pusher furnaces is the opposed-zone reheating furnace (OZRF)[17] (*Figure 19*). High heating rates can be achieved with this furnace, because of the reverse-fired preheat zone (hence high temperature difference/driving force). For a given length such a furnace is capable of achieving 10–15% higher output than a normal five-zone furnace. Owing to the central

---

* To avoid using complicated mathematical techniques to calculate heating time (and hence furnace length), furnace designers, as well as using this criterion, very often decide furnace sizes on this basis also

flue-gas offtake, gases from each zone are separated and good control can be achieved. Flue gases leave such a furnace at a temperature in the region of 1300°C, and hence higher preheats are desirable to achieve required fuel economy.

Walking-beam furnaces eliminate certain traditional limitations of pusher-type furnaces. Some of the disadvantages of pusher furnaces are:
1. Limitation in maximum furnace length (and hence output) owing to the possibility of billets/slabs piling up on one another.
2. The furnace is not self-emptying. This aggravates scale losses during mill delays or interrupted operation. It also introduces problems in access for maintenance of refractories.
3. When water-cooled skids are employed, skid shadows cause appreciable temperature difference between the top and bottom of the charge.
4. Surface defects result from sliding contact between charge and skids (or refractory hearth).

These problems have been substantially solved in walking-beam designs.

Walking-beam furnaces are of two general types — the traditional top-fired walking-beam furnaces, and top- and bottom-fired walking-beam furnaces.

Top-fired walking beam furnaces have stationary and moving beams of refractory construction. The beam refractories are held in castings, and water seals are provided between the moving and stationary beams to prevent leakage. Although this type of furnace offers various advantages over pusher furnaces, they are not without problems. The major disadvantage of this type is choking up of the space between moving and stationary beams owing to scale and to breaking up of the edges of refractory beams.

Top- and bottom-fired walking-beam furnaces overcome the problems associated with pusher furnaces as well as the top-fired variety. In such furnaces the moving and stationary beams are of water-cooled tubular construction and burners are provided both above and below the charge (*Figure*

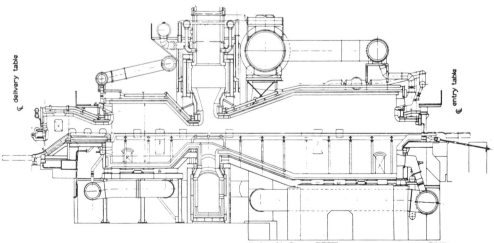

Figure 19   Opposed-zone reheating furnace (Courtesy Stein Atkinson Stordy Ltd)

20). Such furnaces have more skids than the pusher furnaces, but skid marks are less pronounced because the stock does not stay on one set of skids all the time. Also, absence of any pusher vibrations renders possible the use of good skid insulation, thereby reducing water-cooling losses. Perhaps the biggest advantage of such furnaces is their versatility, in that furnaces of this type can handle material in any sequence of length, width or thickness.

Walking-beam furnaces can be designed with two independent sets of beams such that the first section of the furnace is used for slow preheating to 750–800°C, and the final section can be used for fast heating, so that decarburization is minimized. Rapid heating can also be achieved by use of notched-hearth furnaces, the essential feature of which is that the hearth is formed of a series of notches of refractory construction. The stock rests on heat-resisting castings set into the notches, and holes are provided at the rear of the notch for pusher fingers. Every time the billet is pushed from one notch to the next it turns by 90° and a new face is exposed. Continuous turning of the billet gives fast homogeneous heating with minimum scaling. Disadvantages are the short life of the heat-resisting castings, and high fuel consumption because of numerous openings for pusher fingers.

Besides the continuous furnaces discussed above, a variety of batch furnaces are used for steel reheating. The main application of batch furnaces is in the forging industry, for which the weight of the stock varies from a few kg to many tonnes.

Reheating furnaces for non-ferrous metals tend to follow the designs of those used in iron and steelworks, with some modifications. Furnaces dealing with aluminium and copper alloys must operate without damage to the soft metal surface, and hence pusher furnaces are rarely used without provision of chairs or trays for transportation. Furnaces for aluminium alloys incorporate gas recirculation and bogie hearths; chain-conveyor and semi-continuous types are usually employed. For reheating copper alloys, walking-beam, rotating-hearth, or screw-conveyor furnaces are commonly used.

Aluminium and copper alloys lend themselves to rapid heating owing to high diffusivity. Rapid-heating furnaces for heating non-ferrous billets for extrusion take two distinct forms. The first incorporates a series of small high-intensity oil or gas burners, the hot products from which impinge normally on the billets (*Figure 21*). Such furnaces are available for heating billets from 100 mm to about 600 mm diameter. The billets are transported through the furnace on a conveyor or pushed on 'V' shaped rollers constructed from alloy steels and pitched fairly closely along the furnace length. Furnaces of this type are available in recuperative and non-recuperative designs. In the non-recuperative design, the burners are pitched closely throughout the length of the furnace chamber so as to impinge the combustion products onto the billet surface and then envelop it. The furnace is divided into a number of zones along the furnace length, each zone having its own combustion and

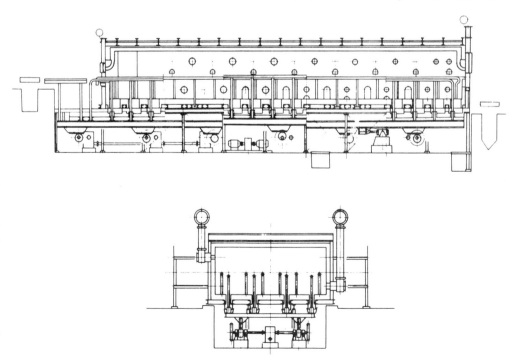

Figure 20   Top- and bottom-fired walking-beam furnace (Courtesy Stein Atkinson Stordy Ltd)

References p 147

temperature-control equipment. Waste gases are discharged through ports in the roof. In the recuperative design, an extra unfired zone is added at the charge end and hot gases from fired zones are drawn down this zone counterflow over the billets and exhausted at the entry end of the furnace by means of a venturi system. Fuel savings of as much as 25% can be achieved by applying this waste-heat recovery system.

When heating aluminium billets in such furnaces, each zone is provided with a pneumatically operated probe device with twin thermocouples, to sense the surface temperature of the billets as they progress through each zone; and by means of appropriate equipment, heat input is controlled. This feature is particularly important since some aluminium alloys are heated to temperatures approaching their melting point. When heating copper alloys, billet temperature may be monitored by using the radiation pyrometer.

The second design of rapid-heating furnace consists of a cylindrical heating chamber in which the products of combustion flow tangentially through slot-shaped burner outlets. High convection heat transfer coefficients are achieved owing to high-velocity recirculating flow over the work surface. The work may be progressed through the furnace in a manner similar to the furnace described above.

Use of such rapid-heating furnaces is also applicable to steel heating, though the billets need to be carried on a skid rail and considerable expertise is required in their design. For steel heating, such furnaces will reduce billet residence times above 800°C and thus minimize scaling and decarburization.

Specific fuel consumption in reheating furnaces may vary over a large range, depending on the material heated, type of furnace, fuel used, extent of waste-heat recovery, etc. Operating the furnaces at or near their design capacity reduces the fuel requirement per unit quantity of charge heated. Also, good operating efficiencies are achieved with good combustion and furnace pressure control so as to minimize air inleakage or gas discharge.

The extent of waste-heat recovery naturally has a considerable effect on fuel economy. Reheating furnaces may be provided with external waste-heat recovery systems in the form of recuperators, regenerators and waste-heat boilers; or in the case of furnaces with stock movement counterflow to the combustion products, recuperative sections may be provided in order to preheat the stock before it enters the main firing zones. Internal recuperation in continuous furnaces usually results in increased furnace length, and only a limited use of such recuperation is therefore possible in pusher furnaces owing to the limitation in pushing lengths. In furnaces incorporating mechanical stock-conveying methods (e.g. walking-beam, roller-hearth, etc.), there is no such limitation and provision of a long preheat section is possible.

Heat-treatment Furnaces

Heat-treatment is a process in which metals are subjected to one or more temperature cycles in order to attain desired properties. In general, in a heat-treatment cycle the metal is heated to a specified temperature in a specified time, held at this temperature for a specified period, and then cooled naturally or at a specified rate in a chosen atmosphere or cooled rapidly in a liquid quenching medium.

The heat-treatment operation is often the last in the manufacturing process, the material value thereby having been enhanced considerably, and it is important that in this last stage the tolerances are not affected. Also, operation in a controlled atmosphere may be required to permit a chosen finish to be obtained. It is therefore important that the charge is handled carefully, especially with non-ferrous metals since they are more susceptible to physical damage.

The general heat-treatment requirements are similar whatever the industry, and therefore furnaces suitable for ferrous metals can be used in the non-ferrous metal industry with little or no modification. Both batch-type (box, bogie, cover-type) and continuous furnaces using various means of conveying the metal through the heating chamber (roller-hearth, walking-beam, pusher, chain-conveyor, shaker-hearth etc.) are employed for heat-treatment operations. Although similar furnaces are used in both industries, the specific mode of heat application depends on the material being heated and the temperature requirements. Materials having low emissivity (e.g. aluminium and its alloys) and low-temperature operation are almost always heated in a furnace incorporating recirculation. For heat-treatment of materials with high emissivity (e.g. ferrous alloys), no recirculation is

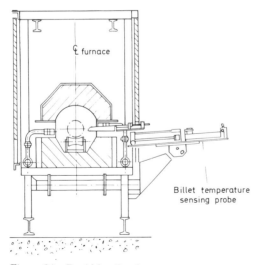

Figure 21   Rapid-heating furnace

incorporated, especially when heating to temperatures in excess of 750°C. For low-temperature operation, however, fan(s) are incorporated in the furnace. High thermal diffusivity combined with reasonably high emissivity make heating of copper, brass etc. fairly simple. In such cases rapid radiant-heating techniques can be employed, except when heating coiled strip and wire below 750°C when recirculation must be incorporated.

In fuel-fired heat-treatment furnaces the choice of fuel is between oil and gas, although some coal-fired units are in operation. Although direct firing is preferred, muffles and radiant tubes are often employed. For heat-treatment of aluminium and its alloys radiant tubes are almost always employed, especially when oil or other sulphur-bearing fuels are to be used, in order to avoid deterioration of surface finish. As the use of muffles and radiant tubes reduces the thermal efficiency, semi-muffle heating should be employed where practicable.

In building a single-purpose furnace there is a wide choice of furnace types; the general principles of design are as discussed previously.

In industries where small quantities of a variety of materials are heat-treated, general-purpose furnaces are used. The furnace designer is then faced with the problem of providing the necessary flexibility in heating, along with temperature uniformity within the limits required as well as fuel economy. It is usual to provide batch furnaces in such cases. As general-purpose furnaces may be used intermittently, the use of refractories of low thermal capacity is preferred since the amount of fuel used for lighting-up becomes an important item. A large turn-down ratio (maximum to minimum firing rate) can be expected in such furnaces because of the variety of heating cycles; hence special attention should be given to choice of burners and control equipment.

References

1   Kay, H., *The Design of Furnaces for Reheating*, Preprints of Rolling Technology Mtg, Metals Society, London, 1974
2   Ref.(A)
3   Ref.(B)
4   Ref.(C), pp 98—109
5   Ref.(D)
6   Ref.(E)
7   Ref.(F)
8   Bloom, F. S., *Iron & Steel Engr*, 1955 (May), pp 64—75
9   Ref.(G)
10   Marsh, K., 'Heat Transfer To Aluminium', *Industrial Heating*, 1954 (Ap.—July), 654—666, 898—910,1120—1132 and 1510—1514
11   Ref.(H)
12   Ref.(I)
13   Fairbank, L. H. and Palethorpe, L. J. W. *Heat Treatment of Metals*, Iron and Steel Inst., London, 1966, pp 57—71
14   Hottel, H. C. and Keller, J. D., *ASME Paper IS55—6*, 1932
15   Kay, H., *Proc. Slab Reheating Conf.* BISRA, London, 1966, pp 31—37
16   *Blast Furnace—Theory and Practice* (Ed. J. H. Strassburger), Gordon & Breach Scientific Pubrs, Inc., 1969
17   Laws, W. R. and Salter, F. M., *Proc. Conf. on Reheating for Hot Working*, Iron and Steel Inst., London, 1967, pp 132—139

General Reading

(A) Shack, A., *Industrial Heat Transfer*, Chapman & Hall, London, 1965
(B) McAdams, W. H., *Heat Transmission*, 3rd edn, McGraw-Hill, London, 1954
(C) Spiers, H. M. (Ed.), *Technical Data on Fuel*, 6th edn, World Power Conf., London, 1961
(D) Perry, R. H. and Chilton, C. H., *Chemical Engineers' Handbook*, 5th edn, McGraw-Hill, New York, 1973
(E) Trinks, W. and Mawhinney, M. H., *Industrial Furnaces*, Vol.1, 5th edn, Wiley New York and London, 1961
(F) Dusinberre, C. M., *Heat Transfer Calculations by Finite Differences*, Industrial Textbook Co., Scranton, Pa. 1961
(G) Thring, M. W., *Science of Flames and Furnaces*, Chapman & Hall, London, 1952
(H) Davies, C., *Calculations in Furnaces Technology*, Pergamon Press, Oxford, 1970
(I) Hotchkiss, A. G. and Webber, H. M., *Protective Atmospheres*, Wiley, New York, and Chapman & Hall, London 1953

8.2 Electric furnaces

Some of the more fundamental aspects of the topics included here may be found in appropriate Sections of Chapter 7.

Melting

Arc Furnace

There are two basic designs, direct and indirect[1]. The most common is the direct type employing three electrodes connected to the secondary winding of a three-phase transformer (*Figure 22*). The three electrodes are made up from machined graphite rods of standard lengths and diameters joined via a threaded socket and tapered nipple. The joints are tied together by a torque wrench to recommended tension values. Each electrode, which is gripped by a pneumatically operated electrical contact clamp, carries heavy current at the supply frequency and its position is regulated independently of the other two by an electro-mechanical or electro-hydraulic hoisting system. Current and voltage signals (voltage measured from each electrode to the earthed bath) are compared with values set for the power control regulator. An increase of current for any one phase over the set value will cause the electrode column to be raised relative to the melting metal, and a decrease, a lowering. The arcs, drawn between the tip of each electrode and the metal scrap arranged directly below, radiate heat directly onto the surrounding scrap-metal charge materials to cause eventual melting (*Figure 23*). As the solid scrap melts to form a liquid pool, the remaining solid — all contained within the refractory-lined steel-cased shell of the furnace — slides into this. Eventually the arcs make contact between the surface of the pool of metal and the tip of each electrode, with additional arcs between the tip and its neighbour. At this stage it is necessary to reduce the applied secondary voltage in order to reduce the length of the arcs and thus avoid burning of the side-wall and roof refractories.

To control power input, and to establish satisfactory arc stability attaining overall efficiencies of 70—75%, a range of secondary voltage tappings and means for adjusting electrode current form a part of the furnace control system[2]. The (a.c.) power of the arc, this being the useful electrical energy directly converted to heat for melting,

requires a circuit having approximately 45% voltage reactance to ensure that as the sinusoidal voltage passes through zero the current-zero occurs a little later to avoid complete extinction of the arc. Furnaces of electrical ratings up to 10—12 MVA have extra reactance added to the furnace circuit, this being in addition to the reactance of the supply line, arc furnace transformer, heavy-current bus bars, flexible cables and electrode bus tubes, to provide a total value of approximately 45% voltage reactance. For furnaces employing still larger electrical ratings, by the correct choice of secondary voltages and the physical configuration of the heavy-current electrical circuit a total percentage reactance is attained for a satisfactory operation. With the larger-capacity furnaces, because of the physical layout of the electrical circuits, the total percentage voltage reactance is higher than that desired for optimum performance; in consequence, special configurations of the flexible cables and copper work attached to the electrode arms are required to give low-reactance-type furnace connections[3,6]. Furnaces with a molten-metal capacity from 5 to 200 t are in common use. The smaller-capacity furnaces are used in iron and steel foundries, those of 60—200 t for mini mills and steel-ingot shops. Furnaces with capacities above 200 t employing powers up to 150 MVA are few owing to the need for special production requirements, power-supply strength (short circuit MVA fault level) and scrap availability[4]. Continuous steel-making furnaces which should be designed to take continuous feed of either shredded scrap or pre-reduced pellets will be required in future; also furnaces of 400—1000 t employing six electrodes directly suspended above the bath and without electrode masts and arms as in the conventional furnace[5]. Such designs enable the optimum value of total percentage voltage reactance to attain the highest average working power factor and effectively to transmit the heat from the six arcs at a greater rate, commensurate with power and molten-bath-capacity ratio, than with three electrodes. Whereas the kilowatt hours per ton energy demand (kW h/t) for melting with the three-electrode direct-arc furnace diminishes with increase of size and power rating, there is an optimum for such upscaling[6]. At this point it becomes necessary either to install more furnaces of optimum size or to consider multi-electrode furnaces together with a completely different concept of mechanical design.

Indirect-arc furnaces employ two electrodes arranged horizontally, each electrode passing through the end face of the refractory-lined metal drum to bring the tips to meet. The arc is formed by withdrawing the electrodes from each other, and the heat radiating onto the metal scrap lying at the bottom of the drum achieves relatively low overall efficiencies of 45—50%. Such furnaces necessarily have small capacities with a maximum molten charge of 500 kg, electrical rating of 250 kVa. They are built for capacities down to 5 kg,

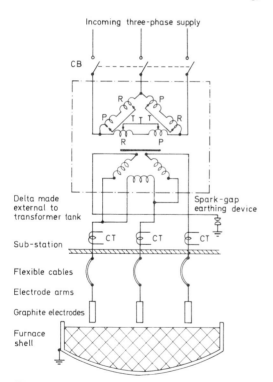

Figure 22   Basic electrical circuit of arc furnace for power ratings up to 10—12 MVA. (Higher power ratings have no current-limiting reactor and usually have 'on-load' tapping changing for selection of the secondary low voltages). Courtesy Electrical Energy (ref.1)

electrical rating 17 kVA, and are generally constructed to allow the drum to be rocked to-and-fro
causing the metal scrap charge to change its position and hence to shorten the total melting time.
By and large, this type of furnace which found use
in the ferrous and non-ferrous foundries has been
replaced by coreless induction furnaces over recent
years[7].

Single-phase direct-arc furnace units, commonly
known as 'baby' arc furnaces, find use in research
and development and pilot production plant. Two
electrodes pass through holes in the refractory-
lined lid of the furnace structure, control is normally by hand, and the arc is struck between the
tips of the electrodes.

Processing times for refining special steels can
be considerably reduced by fitting to the underside
of the arc-furnace-shell bottom a low-frequency
(0.5–1.0 Hz) induction stirrer to move the liquid
metal[8].

Coreless Induction Furnace

This (*Figure 24*) is sometimes known as the crucible induction furnace, the crucible being either
rammed around a metal former or prefired. The
latter form is generally used for small-capacity furnaces from 0.5–50 kg, but rammed linings may be
used where individual melts are to be made and the
lining must be changed to prevent contamination
with the next melt of a different metallurgical analysis. Furnaces up to 40 t are installed, with electrical power input ratings from 5 kVA up to 10

MVA, and furnaces of 200 t and power ratings of
60 MVA are projected. A line frequency of 50 Hz
is used for the larger-capacity furnaces above 1 t,
and frequencies up to 10 kHz for 0.5 kg to 7–8 t
capacity, the frequency diminishing to 150 Hz
with increase in capacity of the furnace[9]. Ferrous
and non-ferrous metals are melted for the production of ingot, but in the majority of cases for cast-

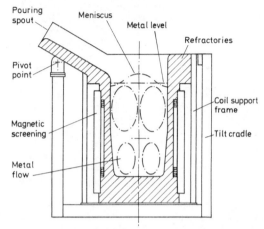

*Figure 24   Coreless induction furnace with mag-
netic-flux screening shield around water-cooled
coil. (The replaceable coil assembly installed in a
'lip axis' parallel tilting cradle). Courtesy British
Cast Iron Res. Assocn*

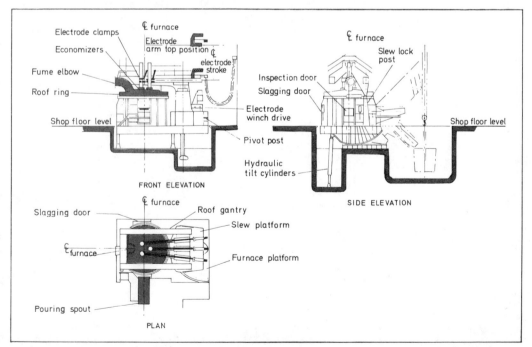

*Figure 23   'Full platform' design of 15—200 t arc furnace of the 'direct arc' type.   Power ratings 7.5 to
56 (nominal), 10—90 MVA Ultra-High Power (UHP). Courtesy GWB Melting Furnaces Ltd*

ing of components in the foundries and pressure die-casting shops. Magnetic laminated-iron shielding packets are normally employed with capacities 250 kg and above; furnaces without shielding are installed, but these are physically much larger for a given capacity than the equivalent with magnetic screening. An improvement in electrical efficiency is also achieved with the screened type, 1—2% over the average overall efficiency of 65—70%. The coil, formed from a water-cooled specially shaped copper section, carries the a.c. and acts as the primary of a transformer; the secondary is the metal charge materials in the crucible. The thickness of the crucible affects the inherent power factor of the furnace in the case of a ram lining; the water-cooled coil also retains a low temperature in the immediately adjacent refractory, thereby extending the life of the lining and freezing off any metal that may penetrate through cracks of the lining. To avoid possible danger to the furnace operator, who may be holding a metal rod when prodding the charge, an earth-leakage device is fitted as standard, which, in some designs, also acts to give a warning of lining deterioration, calling for its replacement.

Power systems for coreless induction furnaces[10] depend upon the frequency of current applied to the coil, and as the furnace is a monophase device it requires connection to the three-phase electricity supply system either as a single-phase load or by a variation of partially, completely or automatically balanced load. Such balancing systems employ, usually, a reactor and capacitor connected in a manner to achieve, at the nominated rated power level, almost equal phase currents on the supply side of the transformer.

To ensure complete balancing throughout the variable power cycle of the furnace load it is necessary to have automatic phase balancing and power-factor control. Switching of the three-phase or single-phase load, whichever is the case, is for preference arranged on the secondary side of the furnace supply transformer, utilizing either single or several stages of series resistance switching to give a 'soft start'. Where furnace loads become appreciable, i.e. 3—4 MVA and above, and where the values of switched capacitor currents become high, it is recommended to employ soft starting to reduce the level of voltage disturbance to the system. Within the low-frequency range but above 50 Hz, frequency-conversion equipment is employed and may be the magnetic type, i.e. 'triplers', giving three or nine times the fundamental supply frequency, or solid-state thyristor converters. The user of such power systems is required to ensure that the equipment is designed with the necessary harmonic filtering, and where necessary voltage-desensitizing equipment, so that the power supply is not disturbed when the furnace plant is in operation. Whilst it is economically an advantage to connect the monophase furnace load directly across two phases of the three-phase supply, in practice and in order to do so, the short-circuit

capacity of the system at the common point of coupling must be sufficiently high to prevent levels of negative phase-sequence-voltage causing overheating of other connected loads, e.g. electric rotating machinery[11].

Coreless furnaces may be employed as bulk-melting units and/or for holding and superheating but because of their relatively high capital cost, the electrical plant representing at least 50% of the total value, it is often more economical to install channel-type induction furnaces[12]. Induction heating of the lining itself allows for heating or melting of non-metallic charges, the heat being conducted from the lining to the charge. Such linings may take the form of steel vessels which in themselves may form a part of a soft vacuum-processing installation or, in certain cases, one under light pressure. The core winding may then be of the dry type, wound around the steel vessel and having, for certain designs, magnetic-flux guides to improve the overall electrical efficiency and to prevent adjacent steelwork becoming heated by the stray magnetic-flux field. Linings of graphite machined from the block may be used for melting exotic non-metallic oxides for which a contamination-free melting condition is required[13,14].

Channel Induction Furnace

The channel or core type induction furnace employs a magnetic laminated core to link the flux from the primary winding to the secondary winding, which takes the form of a loop of the metal being melted contained in its own refractory-lined channel (*Figure 25*). The furnace is employed for the bulk-melting of non-ferrous scrap materials and extensively also in the iron foundries for holding and super-heating irons[15]. The refractory

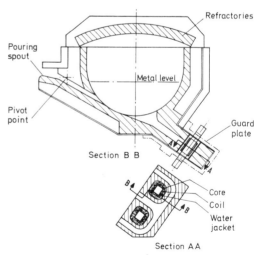

Figure 25 'Shallow bath' chamber with removable roof (twin-loop replaceable inductor units having water-cooled throat, inductor casing and coil shields). Courtesy British Cast Iron Res. Assocn

materials employed for the various metals differ greatly in chemical constitution and grain size (see Ch. 17). Linings may be rammed, or cast *in situ;* because of their importance in directly affecting the continuing use of the furnace, their design, shape and material-specification for each size and design of furnace, and for each specific application for melting the various metals, requires expert advice on the lining required and its installation[16].

The furnace consists of a refractory-lined chamber having a charging-door area and a spout through which the metal is poured out into a ladle or launder. The holding and useful capacity of the chamber is approximately 15—20% less than the rated capacity since molten metal must remain in the inductor-loop channel and throat areas when the furnace is tilted to pour. The furnace bath may take the shape of a rectangle with inductors either vertical and below the chamber or at angles either side of the vertical. The upper chamber, as it is known, may be designed to have a 'shallow' and relatively large surface area for a given bath capacity or to be short with a deep bath. With the shallow bath it is possible to melt light metal scrap, swarf and turnings without undue loss of metal through oxidation. The turbulence present, although not as great as with a coreless furnace, quickly pulls the light scrap under the liquid-metal level.

Furnaces with capacities from 0.5 to 400 t are built, with power ratings from 20 kW to several megawatts.

Vacuum Induction Furnace

Capacities of a few kg up to several tonnes are installed with power ratings from 20 kW at medium frequency to 3—4 MW at mains frequency. The larger-capacity furnaces are usually designed for melting and refining of alloy steels. The smaller furnaces up to 1—2 t working at medium frequency find use in R & D and precision casting plant for the jet-engine and aerospace industries.

The furnace is usually designed for 'lip axis' pouring. Those of smaller capacity are not magnetically shielded, and the whole furnace is housed in a stainless-steel water-cooled tank with a hinged water-cooled lid closing onto an O-ring type seal. Most of these furnaces are designed for batch operation, the feedstock being housed in lock-off chambers to be introduced by the operators in accordance with the melting process. Larger-capacity plants have been built employing a semi-continuous melting process where the casting, having been formed in the vacuum chamber, is transported to a sealed lockable chamber for eventual transmission to the atmosphere. Power sources for these equipments are designed as for air-melting induction furnaces, except that the coil voltage is considerably lower to avoid corona discharge; this results in much higher furnace and capacitor currents.

Resistance Crucible Furnaces

These, mainly used for the light-metal industries, consist of a prefired crucible placed in a steel casing which contains thermal insulation and coil-wire-wound elements supported on refractory brickwork, the whole of which surrounds the outer surface of the crucible. A lift-and-swing-aside refractory-lined lid is normally provided, to minimize thermal losses from the surface of the molten metal and to allow ease of access for charging raw materials and for dispensing the molten metal. An earth leakage detection device to cut off the electric power is normally permitted, in the event of a cracked crucible and spillage of molten metal, to come in contact with the heating elements.

Resistance Immersion Elements

These are employed in baleout, tilting or bottom-poured type furnaces for melting light metals in a temperature range up to $750°C$. The heat is derived from immersion-type sheath elements which may be of the cast-in pattern with ratings of up to 30—35 kW, or mineral-insulated metal-sheathed elements with ratings in the 1½—2 kW range. Depending upon the size of the pot or furnace bath, suitable shapes and numbers of elements are employed to give the necessary heat input[17]. Bulk melting of 'type' metal for the production of stereo plates for the newspaper industry employs baths of 50—60 t nominal capacity with electrical ratings up to 7000 kW. Use of immersion elements in spout areas and small hot-metal holding pots for die-casting machines are typical applications; extensive use is made of the flexibility of this form of electric heating.

Resistance (Submerged Electrodes) for Glass Melting

Bulk melting of glass utilizes either oil or gas firing achieving 25—30% thermal efficiency for the larger furnaces. Electric melting by immersion of electrodes within the glass tank, introduced either through the sidewalls or, for some designs, through the bottom of the tank, is now an alternative. By an arrangement of multi-phase systems and electrode design together with power control through voltage variation at the electrodes, temperature uniformity is achieved while attaining a thermal efficiency of up to 80% on the power supplied. The electrodes are normally of molybdenum, each being held in a water-cooled holder which can be raised or lowered to adjust the current flow[18].

For existing oil- or gas-fired glass tanks, where the desire is for a higher production level, introduction of electric melting to give a greater heat input by installing the necessary electrodes and power equipment has enabled expensive glass-melting plant to meet the higher levels of production[19,20]. For new installations employing electric melting solely, furnaces of special design and shape to take full advantage of electrode position and developing technology are providing improved overall efficiency (20% increase over pot types[21]) and lower costs, together with a substantial improvement in the environment for furnace operators.

Plasma Torch

Torches with electrical ratings of 5—1000 kW operating from a d.c. supply and using argon atmospheres have been installed in melting furnaces for the production of special steel alloys, some of which would have normally been produced in induction vacuum-type melting furnaces. Whilst the capital equipment cost is much lower than for the latter, the operating cost, mainly that of argon, is high. Developments continue, to perfect torches to operate from a.c.[22,23] without an inert gas, and to give a life for the relatively expensive torch which will satisfy the economics of the total melting process. In the area with disturbances to the supply system from a conventional arc-furnace load, the possibility of plasma torches working at megawatt levels commands interest from R & D workers as a potential electric heat source with the advantages of the arc furnace, for metallurgical and steel-making purposes, without the disadvantage of creating objectionable voltage disturbances[24].

Re-melting and Refining

Electroslag Refining (E.S.R.)

Refining high-quality steel alloys by passing strong a.c. or d.c. through an electrode (which has been previously cast in a suitable atmosphere) immersed in molten refractory slag enables droplets of the electrode metal at its tip to descend through the depth of the slag, collect in a water-cooled mould and solidify. The droplets in passing through the slag are refined to form ingots of improved quality and free from inclusions and gas pockets[25]. For certain grades of steel alloys, E.S.R. has replaced the induction vacuum-melting and/or vacuum-consumable electrode arc processes, at lower capital cost. Single or multi-electrodes, as a single-phase load or three-electrode three-phase load, may be employed. Ingots with diameters from a few centimetres to ½ –2 m weighing up to 35—40 t are being produced. Ingots with round, square, rectangular shapes, and with cutouts to form annular rings or other specific cutout shapes, may be produced[26].

Plants exist that allow for continuous feed of electrode stock either as precast sections of the desired diameter/ingot diameter ratio or employ a continuous supply of small electrode (strip or wire), and again that introduce the bulk of the metal in powder or pellets[27,28]. The majority of plants installed use a fixed water-cooled mould which is placed on a water-cooled copper base. Installations employing water-cooled moving moulds allow ingots of greater weight and length

to be produced and take relatively short electrode stock; these require a number of electrodes to produce the desired total ingot weight. Power levels from 100 to 30 000 kW are employed for refining a wide range of alloy steels. Developments of the E.S.R. process are used in the resurfacing of rolls or arbor by casting *in situ*[29] This technique contributes greatly to the energy and cost saving in restoring worn rolls without remelting the complete worn roll, recasting and machining as has been the normal practice. E.S.R. dynamic casting produces various product shapes, i.e. materials for heavy-gauge pressure vessels[30]; and E.S.R. welding fabricates ingots into suitable shapes for rotors for electricity-generating machines[31]. There are many such variations of specially built plant producing engineering components with the minimum of energy and production cost.

Vacuum Arc Furnaces

The example in *Figure 26* uses d.c. power with an input in the range of 1150—1450 kW h/t. Slag removal and separation occurs mainly through reaction of oxidic material in the slag with carbon in the steel as it passes in droplet form through the arc plasma. As in the vacuum induction process, this reaction is facilitated by the pumping away of the carbon monoxide produced. Hydrogen is virtually completely removed; but nitrogen removal and diffusion of high-vapour-pressure metals are not as complete as with vacuum induction since the time in the molten state is much less[32].

Furnaces which use a single consumable electrode are installed to produce ingots up to 1.12 m weighing up to 37 t; these employ electrode currents up to 37 kA with a secondary heavy-current-voltage range of 20 -75[33].

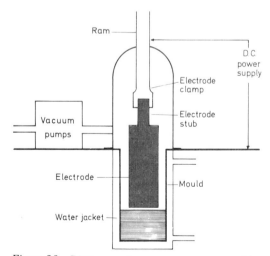

Figure 26   Diagram of vacuum-arc consumable electrode remelting process (showing production of ingot within water-cooled copper crucible). Courtesy The Metals Society, London (ref.32)

Smelting Ores and Refractories

Production of high-heat-intensity sources from an electric arc is applied in many heat/chemical reaction processes in smelting a variety of minerals. In general, such plant is fairly large, employing three-phase a.c. with low secondary voltages (20—200 V) phased to earth[34]. Single-electrode and single-phase plant with a graphite hearth are in use, but for most installations three-electrodes, three-phase is usual; depending on the ore being processed the furnace can have a sealed enclosed roof to permit collection and handling of the gases produced without risk of danger to plant and personnel. The working pressures of such sealed furnaces are important; they are controlled in conjunction with the voltage and current inputs to each of the three-phase power circuits, themselves independently operated, and in conjunction with the rate of feed of charge and fluxing agents. Usually an arc exists, of length dependent upon the material being smelted, and temperatures within the arc area relate to the melting and raw-material feed rates. The arc process termed 'submerged-arc-resistance' provides an electrical load of high average working power-factor. The resistive and reactive voltage drops from the transformer windings through to the tips of the electrodes call for a physical layout of heavy-current interconnections and, where desirable, separately arranged single-phase transformers connected to give a three-phase balanced load in order to minimize electrical losses and achieve the highest working power-factor. Power-consumption levels depend upon the materials being processed but lie in the region of 2000 kW h/t. In general, the graphite-paste Soderberg electrodes used are extended *in situ* by adding paste into thin-wall steel-tube casings added to the electrode column as it moves downwards, the tip wearing away continuously[35,36]. The electrode columns, which may weigh up to 80—90 t, are generally supported by hydraulically operated clamping gear comprising two clamping rings to allow repositioning of the electrode clamps as the electrode moves downwards. Movement of the electrode in answer to the power-control requirement, which may come from a signal either of current or current plus voltage, is provided either through an electromechanical winch system or hydraulic cylinders. The speed lies in the range 5—30 cm/min. Power levels from 2 to 3 MW are used for the smelting of magnesia and alumina, materials used extensively in the manufacture of mineral-insulated electrical cables and heating elements, and levels up to 70 MW for smelting phosphate pellets to produce phosphorus. Furnaces, generally of the non-tilting type for the production of pig iron, calcium carbides, ferrosilicon alloys and slag-refining processes, utilize the basic submerged-arc technology employing very high currents and low voltages, together with furnace-shell shapes and sizes to specific requirements of the process[37].

References p 158

Heating by Indirect Resistance

Heating Elements

Nickel-alloy electrical-resistance materials produced as wire and wound to form 'coil elements' constitute the most common form of electrical-heating element for ovens and furnaces in the range up to 80—100 kW. Higher values of surface load (W/cm^2) may be required, to give higher furnace-chamber temperatures or owing to restrictions of physical space and, in some cases, to strengthen mechanically the elements and prevent premature failure; the element material then takes the form of strip bent into a hairpin shape. For certain applications, elements may take the form of tube or be cast into specific shapes; both utilize high currents and low voltages. Housing of all elements forms an important part of the refractory-brickwork design which, by necessity, has to handle high temperatures and to remain physically intact over several years. When refractory assemblies of the furnace are moved, in the case of lift-off bell-type furnaces, heated hearths and doors, the elements' supporting structures call for special attention, as do the electrical contacts that allow for 'quick release' of the electrical circuit from the fixed portion to the moving part of the structure. Termination of heating elements and the connection to the electrical bus bar and wiring system call for careful design to prevent overheating and subsequent failure. Element life may be expected to be 5—10 years, but the chamber atmosphere can cause rapid deterioration and short life. Metal elements in furnaces operating in the range of 900—1000°C can be affected by a reduction of the protecting oxide film previously built up in air. Selective oxidation of chromium occurs, probably accompanied by carburization. The elements become brittle, a condition known as green rot.

For the lower-temperature ovens and environmental chambers, metal-sheathed elements allow for relatively cheap construction by employing standard element sizes and terminations. Infrared lamps allow ovens and compact heat sources to be built into a heat-processing system[38,39]. A lamp housing may consist of two tubular infrared quartz lamps mounted with a dry-air-cooled polished reflector to give a temperature output level up to 300°C in free air. The requirements of higher working temperatures and of heat density calls for water cooling or forced air to handle the reflections of the heat source from the charge pieces back to the lamp housing. In consequence, with increased working temperatures overall thermal efficiency falls. Infrared heating for forging can provide an easily adaptable heat-processing rig to deal with 'one-off' requirements, with low capital cost and relatively low overall operating costs owing to the rapid fall-off of thermal efficiency at temperatures approaching that used for forging[40].

Operating temperatures are limited by the material employed for the elements. Up to 1150°C nickel- chrome and nickel—chrome—iron are used; up to 1350°C, iron—chromium—aluminium—cobalt. Temperatures of 1400—1600°C require elements of silicon carbide.

Heating elements for high-vacuum heat-treatment furnaces call for materials which are not limiting in the upper range, i.e. above 1150°C. Metals in general use are tungsten, molybdenum and tantalum. Non-metal—graphite elements including graphite cloth and silicon carbide are also used.

Ovens

With designs for natural convection of heat from the elements through the air surrounding the charge pieces, which are stacked inside the oven, operating temperatures up to 200°C are obtained. With fan-assisted convection the operating temperature is raised to higher levels, 300—500°C, while allowing more precise and uniform temperatures in the charge. Ovens may be constructed for batch operation with dimensions permitting easy access and placing of the charge pieces; metal-sheathed elements or wire-wound coil elements assembled into a heater battery with forced-air blow system are used depending on size of oven and kW rating. For applications requiring a continuous flow of work materials, e.g. vitreous enamelling, paint drying and finishing, continuous ovens using radiant heat from metal-sheathed elements or forced-air convection heating systems may be employed. Several heat zones are normally employed to give the desired temperature gradient over the length of the oven, using thermocouples for each zone working with multipoint-recorder controlling instruments. Excess-temperature monitoring thermocouples are used to disconnect the power supply in the event of mal-operation of the normal temperature-control devices. Oven construction and design of heat-circulating systems depend largely on the kind of products that will receive the heat, i.e. shape, mass-density, stacking arrangements and materials. Ageing of light-alloy extrusions for the aluminium industries at 200°C, e.g., require close working temperatures of ±5°C throughout the entire length of the oven which may be 6—15 m long. Drying of paper and like materials not only calls for ovens with well designed heating systems to prevent overheating and combustion of the products, but requires well engineered charge conveyors, reliable and accurate forms of temperature control, heated-air circulation routing, and exhaust-extraction systems to take away water-laden air or fumes emitted as a result of the process.

Ovens constructed for working temperatures up to around 500°C are of similar construction except that thermal insulation materials to deal with the higher possible casing losses are of a different form. Because of the relatively low working temperature of ovens the time taken to reach working temperatures is short. The total operating costs are influenced by the quality of thermal insulation of the casings, and, for forced-air circulation systems, the degree of air loss through circulation. Costs of constructing ovens vary greatly from one manufacturer to another even for a specific design, and reflect in general the quality of engineering represented in the overall operating efficiency, and the performance (temperature uniformity and control).

Environmental Chambers

These enable the low-temperature range (30—150°C) to be achieved within close controlled-temperature limits to meet the requirements of the charge products. Chambers for the pharmaceutical industry require special treatment in regard to materials of construction, uniformity of temperature, and high reliability, since mal-operation is likely to result in costly production losses.

Batch Furnaces

Constructional variations in regard to refractories, thermal insulation, heating-element types and methods of atmosphere circulation, and retorted furnaces vary with the heat-treatment application, type of charge material and size[41]. Access to the heating chamber may involve an opening doorway, lift-off or lift-and-swing-aside lid, drop-down hearth or lift-off bell. Where such furnaces are to be employed with one charge following another through normal day-and-night working, the construction will require thermal losses from the furnace structure to be as low as possible with reasonable thermal-insulation cost. Handling of access-door gear and lifting equipment demands quality engineering, as failure of the component parts can result in breakdown and loss of production. Removable hearths (bogey-hearth-type furnaces) call for designs that minimize hearth-refractory break-up and will stand a reasonable amount of abuse from machines placing the charge into position. For such furnaces, provision of heat seals at points where access is made is reflected in thermal losses value and hence running costs; and where controlled atmospheres are called for the requirement to limit ingress of air is of prime importance. Heating and later cooling of furnace structures directly affects the ability to maintain consistent shape owing to expansion and contraction; in consequence choice of materials of construction in regard to steel specifications, particularly where higher temperatures necessitate alloy steels, is important.

Semi-continuous and Continuous Furnaces

The necessity to work in conjunction with other production facilities and to ensure consistent quality of heat treatment with minimal supervision demands quality furnace engineering. For some applications a heat-treatment line may consist of a charge-delivery system with automatic feed, a shaker hearth to convey components through a gas-seal door gear into the furnace chamber, and a conveyor passing through the furnace out through a gas-seal door into a quenching chamber. For hardening and tempering two continuous furnaces are employed, one for hardening at a higher temperature and one tempering, both with quenching tanks, conveyors and charge-delivery systems. The hearth—conveyor system may be a link or slatted chain moving through the furnace as a continuous belt. Trays on which components have been assembled may require synchronized opening and closing of the door gear to allow the intake and output of one unit. Hearth supporting gear is constructed from nickel—chrome alloys which may be fabricated or built from cast materials. Design for a satisfactory life without cracking or distortion is essential if failure of the entire line is to be avoided. In continuous furnaces such as that for ferritic and pearlitic malleabilizing of iron castings, where quenching systems and mechanical transfer form part of the treatment line, automatic electrical control associated with line movement forms an essential part of the overall furnace plant. Quality of engineering design throughout establishes the overall quality of heat treatment obtained and the efficient use of the furnace plant.

Heat-treatment cycles may require the product, whether component, strip, tube or sheet, to reside at a controlled temperature before being increased or decreased, to a specific time—temperature pattern. In consequence the shape of the furnace structure may in effect increase the chamber length. Cooling at controlled rates is of equal importance; cooling sections may be water-cooled chambers which if badly designed can permit air and water leaks to affect surface finish. All such refinements call for expensive furnace engineering which however can lead to high production levels with lower costs.

Furnaces for Specialized Requirements

The ability to control precisely and provide pure heat without products of combustion permits the design of several forms of electroheating equipment for the more sophisticated heat-processing requirements in the metals industry. The growing use of powder metallurgy makes use of furnaces capable of operating at temperatures in the range up to 1750°C for sintering components connected with tungsten, carbide, sintered metals and ceramics. Such furnaces are of the pusher type and rely on tile hearth construction to effect a reliable high-temperature surface on which are placed boats which carry the charge components. Transporter systems designed for this application call for particularly high hearth loadings to give satisfactory working at temperatures above 1150°C. Such furnaces are completely automatic in charging and discharging, temperature control, protective gas atmospheres and pressure settings.

Reduction of tungstic oxide or ammonium paratungstate to tungsten metal powder requires

controlled-atmosphere tube-heating furnaces designed for infinitely variable pusher-type conveyor systems. Integrated diffusion systems with laminar flow, used for diffusion of silicon and other materials which can be loaded into the furnace as thin slices and carried in quartz or ceramic boats, are used for the production of thyristors. These furnaces consist of a number of tubes and are built in stacks with ratings of 9—11 kW per tube[42].

In electrically heated sealed quench furnaces for gas carburizing and carbonitriding for hardening carbon steels with complete freedom from decarburization, and for carbon restoration in decarburized steels, the furnace has an enclosed oil quench tank and a slow-cooling chamber, allowing either direct quenching or slow cooling without exposing the charge to air. Corrtherm elements, unaffected by carbon deposition, are used.

Hydrogen, having a very high specific heat value and a thermal capacity which rises with temperature, is used for a number of furnace plants of special design and applications in the metallurgical heat-treatment field. The Kepston Hydrogenic furnace process makes full use of hydrogen providing maximum heating rates (over 100°C/min), and cools from 1200 to 300°C in 90 s or less. The furnace is a vacuum- and pressure-tight vessel which can be operated in a vacuum or semi-vacuum conditions with hydrogen or with pressurized hydrogen up to about 400 kPa. At the end of the heating cycle the elements are switched off, with hydrogen pressurized and circulated through a water-cooled heat exchanger. At the end of the cooling cycle the hydrogen is removed by vacuum pump. Used for surface treatment of tools the furnace can also be adapted to handle the most advanced applications in sintering, through hardening, to brazing of a wide range of ferrous metals. In the heat-treatment of die and tool steels the furnace reduces the level of rejects and hence the need for recycling, reduces costly finishing operations, and almost cuts out the human element, in an overall lowering of operating costs.

Ultra-high-powered (UHP) resistance furnaces employing 'low thermal mass' construction and resistance elements bring the possibility of an in-line 'compact heat source' for preheating or superheating of slabs and blooms for rolling[43].

Electroheating equipment constructed *in situ*, for pre-weld heating, stress-relieving of large fabrications built on site, and many applications demanding controlled quantities of 'low heat', is provided by resistance-heating pads, wrap-around cables and tapes[44].

Heating Systems in Vacuum

Recent developments of a unique design of semi-continuous vacuum furnace plant for general-purpose and surface heat-treatment of metals offer the following innovations[45],[46]:
1. A hot zone which combines vacuum with forced-convection heating, thereby benefitting

from the known advantages of both systems for metal processing.

2. A design where the elements and insulation can be opened to atmosphere while hot.

3. An integrated transfer valve/vacuum lock allowing gas and oil quenching chambers to be isolated from the hot zone during processing, thus eliminating contamination.

4. An automatic reactive-gas-controlled system integral with the furnace, to provide the necessary additional gases to perform such surface treatments as carburizing (CarbVac), carbonitriding (CarbniVac) and low-temperature oxy-carbonitriding (TuffVac).

The work to be processed is loaded into the main chamber, treated, and automatically transferred for either oil or gas quenching. The operating costs appear to be lower than those of other furnaces for similar processes; its usage of consumables such as permanent gas and propane is the lowest of any operating furnace. An important economy feature is that it only uses power and protective gases during actual heat treatment.

The gradual development and complexity in design and size of vacuum furnaces[47] has led to a greater use of certain metals whose specifications have in turn been improved. The associated quality of engineering is of prime importance to ensure high utilization of the entire plant, with economic benefits such as low rejection rate and improved value of the end-product, e.g. reduced distortion of tools and dies and increased life. High-temperature front-loading furnaces of the cold-wall construction with built-in cooling coils and efficient radiation shields enable fast heating and cooling rates to be attained with economic use of power. These are suitable for vacuum treatment at temperatures up to $2400°C$ and pressures down to $133 \mu Pa$ (10^{-6} torr).

Diffusion bonding, hot pressing of powdered metals, ceramic and composite materials can be accomplished in specially constructed vacuum hot-press furnaces designed to work at temperatures up to $1650°C$ and at working pressures of 1.33 mPa. The furnace embodies either graphite or molybdenum elements for vacuum or inert-gas operation.

Fluxless Aluminium Brazing

Continuous vacuum furnaces are used in this specialized new field. A considerable reduction of production costs per piece compared with the alternative brazing method, together with much improved technical quality, are claimed for these new designs. A further benefit is the improved environment for operators[48].

Heating by Direct Resistance

The fastener industry which forms wire into a thousand and one component parts that make up devices for the rest of industry receives help from direct-resistance electroheat, by improving the punch life of the stock wire at 'warm temperature' ($200-500°C$) from around 8000 pieces to 25 000 pieces. Such improved production brings savings not only in cost of die-life per piece produced, but also valuable time in changing the punch heads.

Direct-resistance heating of bar stock for forging is attainable from automatic heaters capable of maintaining high levels of production, uniformity of bar temperature and improved forging die-life.

Developments in design of electrical contacts to handle the heavy currents employed with direct-resistance heating bring possibilities of heating medium-sized bar stock section to forging temperatures in just under 1 min[49]. A basic requirement for forging is that the bar stock is at uniform temperature throughout; the problem with an electrical-contact type of heating system is to avoid cold end effects. Whilst through-heating for forging by direct resistance is competitive with induction through-heating, for certain applications the necessity remains to make a good electrical contact, with a satisfactory life/piece-heated ratio and which is quick to replace when necessary. A big advantage is that standby thermal losses are nil and heat losses during process are very low (increasing only over the last 15% of the cycle owing to radiation from the workpiece). As the time cycle is short, there is little adverse effect on the environment. The capital cost is considerably lower than that for an induction heater, and the space requirement on the shop floor is considerably lower than that from resistance-heated refractory-lined furnaces or an equivalent fossil-fuel-fired furnace[50].

Within an 'in-line' production set-up, direct-resistance heating of a relatively small length of the continuous flow of charge (e.g. tube or wire) through roller contacts has proved exceedingly economical. Annealing of wires and rope cables by heavy-current direct-resistance heating through the charge enables a necessary and normally relatively expensive operation to be carried out simply and with little capital expenditure.

Induction Heating

This supplies heat when and where required by inducing alternating currents into the charge piece, placed in the magnetic field produced by a coil supplied with a.c. Hardening of defined surface areas of crankshafts, tappet heads, valve seats and a numerous array of engineering component parts has led to a wide range of electro-mechanical designs and coil configurations being manufactured by this technique. Such equipment must be designed for mass-production, with coils of specific shape to match the workpiece and allow the optimum efficiency. Machines for surface-hardening may treat the whole at one time, i.e. a single-shot treatment, only one part at a time, e.g. tooth-by-tooth hardening of larger cutters and gear wheels, or give continuous progressive treatment of a relatively small portion of the work which moves at constant speed through the effective zone of the inductor. Choice of frequency to harden a given depth (see Ch.7.3) will be different for smooth cylindrical shapes as against complex shapes. Higher frequency inevitably means an increase in inductor voltage, and breakdown of electrical insulation between the inductor and workpiece must be avoided in the design[51].

Through-heating for forging (*Figure 27*) employs coil shapes and lengths dependent upon the shape of billet, slab or bar being heated; electrical efficiencies vary correspondingly. Consequently the mechanical-handling arrangements for conveying billets through coils or placing coils over a slab mounted on a fixed station form an essential part of the entire furnace unit. Such plant has a relatively high capital cost; depending on complexity of the mechanical arrangements, that of the associated electrical plant may represent 60—70% of the total. Success depends largely on high utilization of equipment and minimum loss of charge material through oxidation or other rejection.

The positioning of inductor coils, e.g. wound around the shortest or longest dimension, affects the electrical efficiency as well as the mechanical handling[52–54]. A further arrangement for inducing currents into the workpiece is by the transverse-flux technique in which the magnetic lines of force flow at $90°$ to the surface of the charge. This arrangement allows for heating thin plate or strip charge pieces at 50 Hz with high electrical efficiencies: 78% for low-carbon steel, slab heating; 80% for austenitic steel plates; and 90% for aluminium strip[55]. Developments are in hand for employing the heat produced from induced electric currents arising from 'travelling waves' in magnetic materials.

Power systems may consist of the following:
1. Mains frequency is used for heating non-ferrous billets down to about 75 mm diameter and for steel billets from 200 mm upwards. It avoids the cost and efficiency loss of frequency conversion, but at this frequency a low-voltage, high-current supply is often necessary. Power control is normally by transformer tappings or on/off switching by contactors.
2. Motor alternators comprising a heteropolar alternator driven by an induction motor, both machines mounted on a common shaft, provide medium frequencies (1—10 kHz). They are generally totally enclosed and water-cooled. The output frequency is fixed and typical values are 1, 3, 4 and 8.8—10 kHz. Standard production units have motors wound for 440 V three-phase, 50 Hz supplies, with maximum ratings of 300—350 kW. Larger ratings wound for high voltage supplies are also available. Conversion efficiencies lie between 85% for 1 kHz and 70—75% for 10 kHz.

Figure 27 Induction heating of ferrous and non-ferrous billets for hot-forming, extrusion and rolling. (Designed for automatic loading and delivery of charge and for a 50-Hz 600-kW supply with 1 to 3-phase balancing equipment.) Reproduced by permission of Banyard Metalheat Ltd

Continuous control of output is possible by varying the alternator d.c. field supply.

The thyristor inverter provides an alternative power source in the medium-frequency range 450 Hz—10 kHz. Conversion efficiencies for 1 kHz and 3 kHz can be as high as 95% and 90% respectively and are about 75% at 10 kHz.

Radio-frequency generators comprise a rectifier section — usually solid-state diodes, followed by a triode oscillator feeding a resonant circuit and producing a frequency of some 450 kHz. Control is effected either by varying the d.c. voltage to the valve or by a variable-coupling output transformer.

A range of ratings from 1 kW to about 75 kW is available for surface hardening and through-heating small sections. Conversion efficiency is about 55%. Larger units, 75 kW–1000 kW, are built for continuous-tube-welding applications operating at lower frequencies and somewhat higher efficiencies (60–65%).

Electron-beam Glow-discharge Heating

Production of high-intensity heat sources as used in electron-beam welding, which are defocussed to produce a heat band, and all contained within a vacuum chamber, makes possible efficient heat transfer for annealing strip and wire products[56,57]. The savings in energy are substantial, since no heat is lost other than that taken by the charge materials, and the space required for the heater source is considerably less than for conventional annealing furnaces[49]. In addition, the savings in cost and energy associated with structural materials of the buildings and conventional furnace structures all help to effect a reduction in production cost and an improvement to the environment and social conditions for operators.

Dielectric and Microwave Heating

Most non-metallic materials in industry can receive a controlled dose of heat in precise areas without adjoining parts being damaged by it. Dielectric heating systems built as a part of a process-heat and production machine satisfy many of the mass-production requirements, and, at the same time, increase the range of designs and finishes of the products[58–61].

Some cases, as in drying of paper and board, require uniform moisture levels across the width of the board; dielectric heating (Ch.7) is specially suited to these requirements[62]. Microwave heating for curing rubber moulded products, i.e. tyres, effects considerable reductions in level of rejects. Such industrial heaters use frequencies of 2.45 GHz and 915 MHz. Multiple microwave generators each of 25 kW rating are used for industrial applications. Rigorous control of the design of heating oven and microwave generator is necessary to ensure maximum safety for plant operators[63–65].

References

1 Harrison, W. L. *Electrical Energy* 1958 (July), Parts 1 and 2
2 Ref.(A) pp 464—473
3 McGee, L. *Proc. U.I.E. Congress VII* Warsaw, 1972, Paper N 402
4 Schwabe, W. E. ibid., Warsaw, 1972, Paper N 105
5 Harrison, W. L. *Proc. U.I.E. Congr. VI* Brighton, 1968, Paper N 101
6 Harrison, W. L. *Congrès International Four Electrique A Arc En Aciérie* June 1971, Cannes
7 Harrison, W. L. *Foundry Year Book 1973*, pp 238—242
8 Sundberg, Y. *Proc. U.I.E. Congr. VII* Warsaw, 1972, Paper N 609
9 Harrison, W. L. *Proc. Conf. on Electric Melting and Holding Furnaces in the Iron Founding Industry* BCIRA, 1967, pp 269—276
10 Owen K. T. ibid., BCIRA, 1967, pp 397—408
11 Ref.(B)
12 Child, K. *Proc. Conf. on Electric Melting and Holding Furnaces in the Iron Founding Industry* BCIRA, 1967, pp 331—347
13 Lethen, R. *Proc. U.I.E. Congr. VII* Warsaw, 1972, Paper N 147
14 Kraus, K. ibid., Warsaw, 1972, Paper N 148
15 Calamari, E. *Proc. U.I.E. Congr. VI* Brighton, 1968, Paper N 152
16 Neumann, R. *Proc. U.I.E. Congr. VII* Warsaw, 1972, Paper N 426
17 Barlow, H. D. *Proc. U.I.E. Congr. VI* Brighton, 1968, Paper N 157
18 Schrwelius, N. ibid., Brighton, 1968, Paper N 203
19 Gell, P. A. M. *Proc. U.I.E. Congr. VII* Warsaw, 1972, Paper N 306
20 Reynolds, M. C. ibid., Warsaw, 1972, Paper N 307
21 Scarfe, F. *Proc. U.I.E. Congr. VI* Brighton, 1968, Paper N 201
22 Harry, J. E. ibid., Brighton, 1968, Paper N 123
23 Schoomaker, H. R. P. *Proc. U.I.E. Congr. VII* Warsaw, 1972, Paper N 129
24 Ref.(C)
25 Duckworth, W. E. and Hole, G. *Electroslag Refining* Chapman & Hall, London, 1969, pp 7—16
26 Hole, G. *Electroslag Refining, Iron & Steel Inst./Sheffield Metall. & Engng Assoc. Conf.* 1973 pp 136—144
27 Etienne, M. and Descamps, J. *Proc. U.I.E. Congr. VII* Warsaw, 1972, Paper N 167
28 Dorsch, K. E. *Electroslag Refining, Iron & Steel Inst./Sheffield Metall. & Engng Assoc. Conf.* 1973, pp 145—149
29 Bagshaw, T., Letcher, P. and Crofts, R. ibid., 1973, pp 126—135

30 Ujiie, A., Satos, S. and Nagata, J. ibid., 1973, pp 113–125

31 Paton, B. E. *et al* ibid., 1973, pp 105–112

32 Barraclough *Metals and Materials—Journal of the Metals Soc.* 1974, 8, (3), 192

33 Volokhonsky, L. A. and Pelts, B. B. *Proc. U.I.E. Congr. VII* Warsaw, 1972, Paper N166

34 Baicher, M., Katsevich, L. and Mikulinsy, A. *Proc. U.I.E. Congr. VI* Brighton, 1968, Paper N111

35 Muller, M. B. and Magnussen, T. E. *Proc. U.I.E. Congr. VII* Warsaw, 1972, Paper N404

36 Cavigli, M. D. ibid., Warsaw, 1972, Paper N403

37 Stasi, L. D. and Bigi, S. ibid., Warsaw, 1972, Paper N 407

38 Head, A. *Proc. U.I.E. Congr. VI* Brighton, 1968, Paper N 311

39 Tanaka, Y. and Yamagishi, H. ibid., Brighton, 1968, Paper N 312

40 Gibbs, M. ibid., Brighton, 1968, Papers 317-S & 317-D

41 Lauster, F. *Manuel d'Electrothermie Industrielle* Dunod, Paris, 1968, Ch.4

42 Kortvelyessy, L. *Proc. U.I.E. Congr. VI,* Brighton, 1968, Paper N 313

43 Laws, W. R. and Slater, F. M. *Proc. U.I.E. Congr. VII* Warsaw, 1972, Paper N 422

44 Cooper, P. J. *Proc. U.I.E. Congr. VI* Brighton, 1968, Paper N 324

45 Reynoldson, R. W. *Metals & Materials—Journal of the Metals Soc.* 1973 (Sep.), p 401

46 Reynoldson, R. W. *Metallurgia & Metal Forming* 1973 (Sep), 293

47 Luiten, C. H. *Proc. U.I.E. Congr. VII* Warsaw, 1972, Paper N 226

48 Ruffle, T. W. *Metallurgia & Metal Forming* 1972 (Aug), 265

49 Harrison, W. L. *Energy Conservation in Metal Finishing Industries*, Lecture 3, June, 1974, Technical Conference Organisation

50 Germaine-Bonne, M. *La Metallurgie* 1969 (Oct.), 101, No.10

51 May, E. *Industrial High Frequency Electric Power* Chapman & Hall, London, 1948, pp 197–213

52 Gibbs, M. G. and Harrison, W. L. *Induction Heating for Forging Iron and Steel*, 1962 (April)

53 Peyton, G. J. *Proc. U.I.E. Congr. VII* Warsaw, 1972, Paper N 206

54 Korey, W. J. *Slab Reheating Conf. Proc.* Iron & Steel Inst., 1972, pp 13–19

55 Jackson, W. B. *Proc. U.I.E. Congr. VII* Warsaw, 1972, Paper N 206

56 Dugdale, P. A. *et al. Proc. U.I.E. Congr. VI* Brighton, 1968, Paper N 205

57 Dugdale, P. A. and Thackery, P. A. *Proc. U.I.E. Congr. VII* Warsaw, 1972, Paper N 128

58 Clements, R. G. *Proc. U.I.E. Congr. VI* Brighton, 1968, Paper N 402

59 Carruthers, J. F. S. ibid., Brighton, 1968, Paper N 405

60 Rowe, J. A. ibid., Brighton, 1968, Paper N 406

61 Dvorak, L. and Dormorazek, M. ibid., Brighton, 1968, Paper N 407

62 Jones, P. L., Lawton, J., Sutherland, C. A. O. and Granville, R. *Paper Technology*, 1973 (April)

63 Jolly, J. A. *Proc. U.I.E. Congr. VII* Warsaw, 1972, Paper N 303

64 Guera, M. ibid., Warsaw, 1972, Paper N 310

65 Meisel, M. *Proc. U.I.E. Congr. VI* Brighton, 1968, Paper N 413

General Reading

(A) *Elektrowärme Theorie und Praxis. Union Internationale d'Electrothermie (U.I.E.)* Verlag W. Girardet, Essen, 1974

(B) Electricity Council (UK) *Supplies to Induction Furnaces: Engineering Recommendation*, 1975

(C) Electricity Council (UK) *Supply to Furnaces: Engineering Recommendation P7/2 July 1970*

8.3 Furnaces for high temperatures

A number of techniques are available for the production of high temperatures in furnaces. The best-known present-day method for achieving a high temperature environment is the combustion of a fuel with oxygen, although a number of electrical methods involving plasmas have been developed and these offer great potential.

Production of High Temperatures by Chemical Flames

The use of oxy-fuel combustion based on natural gas, LPG, fuel oils or pulverized coal has many potential applications in which high temperatures, high-intensity combustion or high heat-transfer rates are advantageous. Thus oxy-fuel burners find application in the rapid melting of steel scrap, reheating of metals, tne augmentation of electric arc and induction furnaces, the refining of non-ferrous metals, and in the heating of glass tanks. Other applications are in the fields of cutting, welding and brazing of metals, glass working and in flame reactors.

Thermodynamic Considerations

The thermodynamic principles involved in the production of very-high-temperature gases by chemical means are well known. The maximum flame temperature that can be achieved is given by:

$$T_f = (\Delta H_R - \Delta H_D)/C \tag{10}$$

where ΔH_R is the energy available from the combustion process to give undissociated products, ΔH_D is the energy required to dissociate and ionize the products, and C is the total heat capacity of the products. In most combustion systems the extent of ionization, however, is small and the energy required for dissociation is the major factor. Thus the maximum flame temperature attainable is determined by the extent of dissociation of the combustion products.

To obtain high flame temperatures, therefore, not only must ΔH_R be large, but the combustion

products must be stable molecules to minimize dissociation. A criterion which is conveniently used in assessing the resistance of a molecule to dissociation is the 'bond temperature'. This is defined as the temperature at which the standard free energy change of the dissociation reaction is zero, that is the temperature at which the equilibrium constant for dissociation has the value one. In *Table 3*, the bond energies and bond temperatures of some common combustion product molecules are listed; it is clear that combustion reactions resulting in the formation of carbon monoxide or nitrogen will give the highest temperatures.

The heat-capacity term in equation (10) is also increased by dissociation, and this results in a further reduction of the maximum flame temperature obtainable.

Table 3   Bond energies and bond temperatures of some combustion products. (Based on ref.A)

Dissociation reaction	Bond dissociation energy (kJ)	Bond temperature (K)
$H_2O = OH + H$	485	2880
$CO_2 = CO + O$	527	3320
$N_2 = N + N$	940	7150
$CO = C + O$	1070	7580

Methods used to achieve High Temperatures

Because of the thermodynamic restrictions and because of the inert diluent effect of the nitrogen, few fuels when burnt with air are capable of producing temperatures in excess of 2500 K, as shown in the first part of *Table 4*.

Higher temperatures can be obtained if the reactants are preheated or if the heat of combus-

Table 4   Computed flame temperature (K) at various pressures. (Based on ref.A)

	0.01	0.1	1	10 MPa
Fuels with air				
C_2H_2	2420	2545	2657	2748
C_2N_2	2475	2601	2717	2814
H_2	2301	2388	2450	2489
CH_4	2170	2232	2276	2304
Fuels with O_2				
Kerosine (stoichiometric)	2828	3111	3433	3792
Augmented* kerosine (stoich.)	3041	3426	3908	4518
$H_2 + 0.5\,O_2$	2795	3077	3395	3740
Augmented *$H_2 + 0.5\,O_2$	2987	3363	3834	4427
$C + O_2$	2980	3285	3641	4053
$C_2H_2 + 2.5\,O_2$	3006	3341	3737	4199
$C_2H_2 + 1.5\,O_2$	3076	3431	3849	4324
CH_4	2782	3053	3356	3688
N_2H_4	2763	3026	3312	3606
$2\,Al + 1.5\,O_2$		3800		
Fuels with N_2O and N_2O_4				
$H_2 + N_2O$	2722	2963	3216	3463
$C_2H_2 + 5\,N_2O$	2884	3152	3440	3738
$H_2 + 0.25\,N_2O_4$	2732	2986	3262	3545
$C_2H_2 + 1.25\,N_2O_4$	2913	3210	3546	3917
High-energy systems				
$C_2H_2 + 4\,O_2$	3182	3526	3928	4392
$N_2H_4 + 2\,F_2$	3542	3934	4393	4913
$N_2H_4 + ClF_3$	3137	3444	3786	4142
$C_4N_2 + 1.33\,O_3$	5190	5516	5850	6170

1 MPa = 10 bar

* With heat addition equal to the heat of combustion

tion is increased by the use of oxygen in place of air. For the very highest temperatures a combination of a high-energy fuel and a high-energy oxidant which give combustion products having high bond temperatures must be used. Even so, the highest temperatures reached so far by combustion techniques are only about 5000 to 6000 K.

Use of Pressure. All dissociation reactions result in an increase in the number of moles. Consequently, the dissociation reactions can be reversed by an increase in pressure with a resultant increase in flame temperature. This is most effective in flames (in constant-pressure systems) where there is considerable dissociation, and it has been shown that the maximum effect occurs when dissociation is about 60% complete. The increase in temperature is a logarithmic function of pressure and so there is a much smaller influence at high pressures, as is well illustrated in *Table 4*.

Use of Oxygen. The use of oxygen or oxygen-enrichment of air is a common method of attaining high-temperature flames as well as high-intensity combustion. Using conventional fuels final flame temperatures in the region of 3100 K can be obtained at atmospheric pressure as shown in *Table 4*. With acetylene it can be as high as 3431 K. The highest temperatures can only be reached with fuels which have high heats of combustion and only give products with high bond temperatures. This is best realized when the fuel consists of carbon and nitrogen in the ratio of 1:1. It is clear from *Table 4* that the reactions of C_2H_2 and C_4N_2 to give CO and N_2 fulfil these general requirements and so yield some of the highest known temperatures.

Use of Preheat. The maximum flame temperatures achieved by combustion processes can be further increased by preheating the reactants. In flames where dissociation of the products is small, the flame temperature increases almost rectilinearly with initial temperature. As dissociation becomes more important, however, the flame temperatures become less dependent on initial temperature.

It is clear that a point can be reached where the preheat primarily results in dissociation of the combustion products without increasing their temperature significantly. However, even in such cases the extra energy put into the system as preheat can be recovered by efficient use of the dissociated products; for example, through enhancement of heat-transfer rates.

The use of preheated air has certain attractions. With suitable preheaters, it can be used on a large scale with economic advantage over oxygen enrichment, although much depends on the availability and price of oxygen. Furthermore, preheaters can utilize heat from the combustion products and the use of preheated air does not necessitate the safety precautions necessary with the use of oxygen.

Very High Temperatures. High Energy Fuel—Oxidant Combinations. Although presenting handling and safety problems, high-energy fuel—oxidant combinations are suitable for special applications, e.g. rocketry and high-power gas lasers. If it were possible to choose unstable reactants which burn to give suitably stable combustion products, the highest temperature could be obtained, e.g. 6000 K. Because of the unstable nature of the reactants, preheat could not be used, but high pressures would be advantageous.

Oxy-fuel Burners

In comparison with the more conventional air/fuel flames, oxy-fuel combustion is characterized by higher flame temperatures and higher burning velocities, which greatly influences the burner design. Furthermore, high noise levels are produced with large burners because of the high gas throughputs and the high combustion intensities, and various designs have attempted to overcome this problem.

For small burners of the types used for welding, cutting and glass blowing, premixed flames are used. Premixed burners of this type are usually classified into two groups, namely the plain orifice type and the piloted orifice type. In many instances, where higher heat-release rates are required, several small burner ports may be required to be incorporated in the one burner.

Burners for the combustion of gaseous fuels with oxygen having a capacity greater than 3—6 $m^3 h^{-1}$ are not premixed for safety reasons, and the method of mixing adopted controls the combustion intensity, the flame shape and the burner noise. Commercial burners vary in design from relatively simple systems of concentric tubes to more complex designs in which fuel is injected into or mixes externally with a high-velocity convergent or divergent stream of oxygen. The concentric-tube burner illustrates the simplest form of a post-mixing oxy-gas burner in which oxygen is passed down the central tube and gas down the annulus between the tubes. The provision of the oxygen supply in the central tube is not essential, but it helps to limit excessive oxidation of the furnace charge. Tests have shown that such simple burner designs do not result in efficient mixing and efficient combustion, but such burners are suitable for many applications.

One of the earlier designs of oxy—town gas burner was designed by Babcock and Wilcox Ltd in conjunction with the Scottish Gas Board. In this, the gas and oxygen streams were directed to a point some distance from the burner mouth so that better mixing between fuel and oxidant streams was achieved.

Other oxy—gas burners have been developed by the British Gas Corporation at the Midlands Research Station. Of these, the British Gas Corporation M.R.S. Mark II burner, which uses impinging-jet face-mixing, and the British Gas

References p 165

Corporation Mark III burner which uses coaxial jet-mixing, have been successfully used up to ratings of several MW with both town gas and (with some modification) with natural gas. *Figure 28* illustrates the M.R.S. Mark II burner whose design led to further improvements in the mixing of fuel and oxidant streams; these improvements were a result of the differences in angles of inclination of the ports for the two streams. The Mark III burner (*Figure 29*) may be rated up to 0.6 MW.

A different method of mixing has been adopted in the Shell toroidal burner. This burner (*Figure 30*) involves internal mixing, but the fuel is injected into a high-velocity oxygen stream with sufficient velocity to prevent flashback. The burner was originally developed for the combustion of LPG, but was later used for that of distillate oil and more recently for natural gas. The version used for oil combustion comprises a water-cooled shroud within which oil atomization and flame stabilization occur, the oil being admitted on the outer surface of the flame-stabilizer cone shown in *Figure 30* and atomized by the shearing action of the oxygen stream flowing at sonic velocity through the converging conical orifice. The flame is stabilized near the front face of the water-cooled bluff body and its shape controlled by the position of the bluff body in the outer shroud — a long flame being produced when the bluff body is fully retracted and vice versa. Burners based on flame stabilization by means of a bluff body have also been successfully used for the combustion of pulverized coal.

A number of burner designs for the combustion of distillate and heavy fuel oils based on post-burner mixing have also been devised; these may have capacities up to 5 MW. A typical burner is illustrated in *Figure 31*.

Plasma Furnaces and Other Physical Methods

High-temperature environments in furnaces may be generated by means of plasma devices. The term plasma is generally used to include not only the gas carrying the discharge current but the hot stream of gas which leaves the discharge region and which will contain, because of combination processes, electron densities lower than those normally associated with plasmas. Normally plasma gases are used directly, but in some applications they are mixed with a second gas.

A wide range of processes may be undertaken in a plasma system. These range from the synthesis of endothermic materials, organic syntheses, and the synthesis of pure materials, to the fields of extractive metallurgy and materials processing.

Plasma Arc Systems employing Consumable Electrodes and Unconstricted Arcs

Early work with this type of system was confined to high-intensity arcs which are obtained by passing

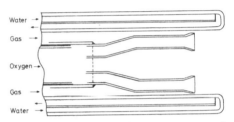

Figure 29   Coaxial jet-mixing oxy–gaseous fuel burner. This particular burner is the British Gas Corporation (MRS) Mark III burner

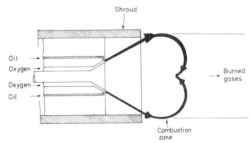

Figure 30   Shell toroidal oxy–fuel burner suitable for distillate fuel-oil combustion

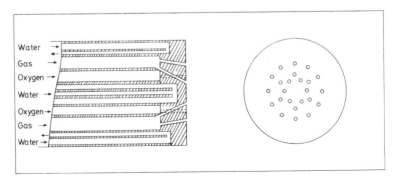

Figure 28   Post-face mixing burner suitable for gaseous fuel–oxygen. This particular arrangement is the British Gas Corporation (MRS) Mark II burner. The design is dependent upon the fuel being burned

a direct current between a cooled metal cathode and a consumable anode. If the current is above a certain critical value, rapid vaporization of the anode materials takes place with the formation of a tail flame of the vaporized material. If a porous anode is used through which gases (or liquids) are passed, attack on the metal anode is considerably reduced and the arc energy is transferred to the transpired materials.

Various forms of d.c. arc plasma systems using consumable electrodes have been proposed or used commercially for synthesis purposes. In particular consumable carbon electrodes may be used in the manufacture of acetylene, for the production of finely divided powders, and for mineral processing. A commercial installation using a d.c. arc operating in the megawatt region has been used for acetylene production by Du Pont for some years although this has now been closed down. Recent important large-scale developments involve the use of three-phase a.c. devices into which three carbon electrodes are continuously fed. Such units may be used for acetylene manufacture and for processing particulate feeds.

Plasma-jet Devices (Constricted Arcs) and Arc Heaters employing Non-consumable Electrodes

One of the best known methods of heating gases to high temperatures is the plasma torch. A typical simple torch consists of a non-consumable cathode (usually thoriated tungsten) and an annular water-cooled copper anode. A gas stream is fed axially or tangentially into the arc chamber, passes through an arc struck between the cathode and anode, and then emerges as a jet which resembles a chemical flame in appearance. The common operating gases are argon, helium or nitrogen.

Numerous designs have appeared in the literature for small d.c. laboratory plasma devices. Typically powers are up to 0.1 MW and thermal efficiencies

are about 60—85%. The gas stream may be injected axially or tangentially as shown in *Figure 32*, and the gas flow causes the arc to pass into the constricted part of the anode. This has the effect of producing higher temperatures than in an unconstricted arc. These axial temperatures are in the range 8000—30 000 K, but there is also a high

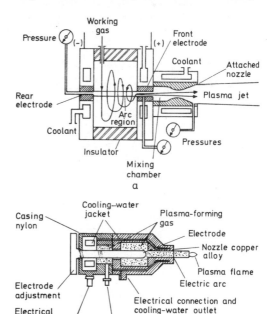

Figure 32   Diagrammatic representation of plasma jet: (a) vortex stabilized; (b) gas-sheath stabilized. Courtesy Instn of Chemical Engineers, London (see ref.B)

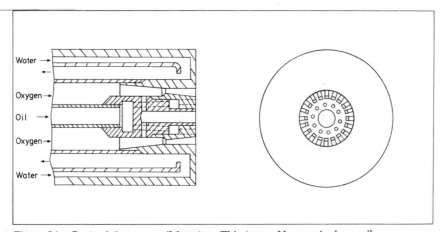

Figure 31   Post-mixing oxy—oil burner. This type of burner is also available for heavy fuel oils in which case it incorporates electrical or steam heating

References p 165

temperature gradient between the core and the outer cooler boundary resulting from the constriction of the plasma column by the solid walls of the anode orifice. Magnetic rotation of the arc may also be used with advantage to minimize electrode erosion and to achieve uniform distribution of energy.

Details of a number of high-power plasma units suitable for furnace applications have been described. In large power units particular attention must be paid to the method of gas injection, so that thermal loads are well distributed and electrode erosion minimized. Expansion of arcs may also be achieved by the rotation of the wall of the containing vessel as in the expanded-arc furnace.

Induction-coupled Plasma Torches

The use of induction heating for the production of thermal plasmas is now widely used following its introduction in 1961 (see also Chapter 8.2). A typical arrangement as used in an industrial unit is shown in *Figure 33*. For chemical syntheses the plasma is commonly enclosed, which may give rise to instabilities. Therefore an important design requirement is the formation of a stable discharge which does not easily become extinguished by added process material. The methods of stabilization may involve sheath stabilization with either single or dual flow or vortex stabilization with either radial or central feedstock feed jets. Vortex stabilization normally involves turbulent flow whilst sheath stabilization may involve laminar

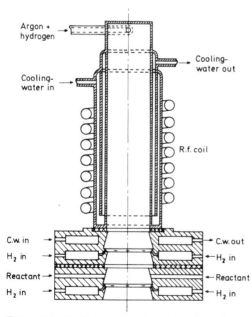

Figure 33   Typical form of an induction-plasma generator. Courtesy Instn of Chemical Engineers, London (see ref.B)

flow. Because induction torches require no electrodes, and since the hot gas can be shielded from the wall by a layer of cold gas, then the plasma may be contained in a silica tube. Thus corrosive or highly pure materials may be processed.

Commercial induction plasma torches of about a megawatt capacity are now available. The frequency of the power unit may be decreased as the power is increased; for large units the frequency may be about 500 kHz compared with 1—10 MHz used for laboratory units.

Electrically Augmented Flames

An electrical discharge offers a relatively simple source of additional energy to a combustion process, provided means can be supplied to distribute the charge reasonably uniformly through the reacting gases (e.g. by magnetic rotation). The main attraction of augmented-flame burners is the increased resistance to blow-off, allowing greatly increased throughputs. A considerable number of laboratory studies of various aspects of this method of producing high temperatures have been reported, but despite a number of attempts, no commercial process based on this technique is currently in operation. However, because of its many advantages it may be expected that some process will be developed using augmented combustion.

Other Physical Methods

Image Furnaces. In these devices the radiation from a high-temperature source, either the sun or some form of electric arc, is focused by means of mirrors on to a target; temperatures in excess of 3000 K can be achieved. The majority of applications are for the determination of the high-temperature properties of materials, e.g. thermal conductivity and expansivity, ablation, etc., although this method has been used in phase studies with refractory compounds and for the synthesis of acetylene from coal.

Radiant-heating Methods. The use of intense bursts of incoherent radiation to activate chemical reactions and to study their kinetics (flash photolysis) is well known, and such techniques can be used for the generation of high temperatures. Thus thin wires or strips of tungsten metal have been vaporized by heating with an intense pulse of light from a capacitor discharge lamp; this method appears capable of wider application.

Lasers are increasingly being used for heating purposes, now that high-power continuous lasers have become available. However their use is generally only justified for the production of high local temperatures, using the ability of the laser beam to be focused into a very small area. Laser heating methods are employed in applications such as nuclear-fusion studies, in micro-machining, and in micro-welding which makes use of the high space resolution of the laser beam.

Electron-beam Heating. This is extensively applied to problems in materials processing such as welding, annealing, melting, machining, evaporation, etc. Normally such operations are carried out *in vacuo* although under the correct conditions the beam can be brought out into the atmosphere. This however results in a general loss of efficiency owing to the scattering of the electron beam by the atmosphere.

Electric Furnaces. Electric heating, either resistance or induction which have been outlined in Chapters 8.2 and 7, offers a simple method of producing moderately high temperatures with good precision of temperature control and ease of operation. Material problems limit the maximum temperature of furnaces to approximately 3700 K for neutral, reducing or vacuum conditions or to around 3100 K for oxidizing conditions.

Exploding Wires. If a large electrical current is passed through a fine wire, the wire explodes, producing for a few microseconds a temperature of about 10^6 K. This technique has applications in producing shock waves, in metal forming, as a light source, and, on a limited scale, in high-temperature preparative chemistry.

General Reading

(A)   Brown, R. L., Everest, D. A., Lewis, J. D. and Williams, A. 'High-temperature processes with special reference to flames and plasmas', *J. Inst. Fuel* 1968, Vol.41, 433

(B)   Williams, A. 'Flame and plasma reactors', *Trans. Instn Chem. Engrs* 1973, Vol.51, 199

Chapter 9

Drying, conditioning and industrial space heating

9.1 Drying and conditioning

Theory of drying
Basic psychrometry of air drying
Performance of driers
Drying plants
Contact driers
Specialized driers
Control of driers

9.2 Heating, cooling and air-conditioning large buildings and spaces

Human comfort
Building construction
Uncontrolled heat gains
Plant capacity
Methods and equipment
Energy economy
Conclusion

9.1 Drying and conditioning

Drying is most commonly the removal of moderate quantities of liquid from a solid by thermal means. Where practicable this is preceded by mechanical methods of moisture removal which have a lower energy requirement. Special cases of drying are:

Evaporation — where heat is used to evaporate large quantities of water from slurries or solutions in such plant as multi-effect evaporators for milk or sodium hydroxide.

Dehydration and freeze drying — removing small to moderate quantities of moisture from materials such as foodstuffs and pharmaceutical products.

Dielectric drying — where high-frequency electric power generates heat within materials (Ch.7), such as foodstuffs, driving out small quantities of moisture.

The main purposes of drying are:

1. To reduce bulk or weight for economic further processing, handling or transport.
2. To produce the material in the desired condition for further processing, handling or marketing.
3. To sterilize or preserve the product.
4. To recover by-products from slurries or solutions.

Minimum energy consumption is usually achieved by drying the material only to the extent necessary, and as rapidly as possible without detriment to the quality of the product.

The theoretical and practical aspects of all forms of drying are covered in detail in such publications as *Chemical Engineers' Handbook*[1]. The present chapter summarizes simplified basic theory and indicates typical types of drying plant.

Conditioning may be described as partial drying or wetting to a controlled end-point—whether of a solid, or of a moist gas, e.g. humidity control or air-conditioning. The principles involved are similar to those used in drying theory (see the following pages and loc. cit.), and discussed to some extent in Chapter 9.2 on space heating.

Theory of Drying

Consider the drying of a wet solid material by a hot air stream. The temperature difference between the air and the material maintains a flow of heat to the material, evaporating the water mainly at the material surface. Simultaneously moisture and/or water vapour migrates from the core of the material to its surface to replace the moisture evaporated.

Drying Rate

The rate at which the material dries is dependent on:

External factors: the drying-air temperature, humidity, velocity and turbulence; the material surface area and thickness.

Internal influences: the nature of the material affects the migration of moisture to the surface by diffusion, capillary flow and flow due to pressure gradients caused by gravity, by shrinkage or by internal vaporization.

The internal movement of moisture is complex and fundamental data on the controlling mechanisms are scanty. Generally the limiting factors on drying rate are (*a*) the rate at which the surface water can diffuse through the static air film at the material surface into the drying air-stream, and (*b*) the maximum moisture gradient through the material which can be tolerated without damaging the material or restricting moisture flow by hardening or shrinking the surface layers.

In practical drying applications it is normal to establish experimentally the most suitable external factors to obtain minimum drying periods and optimum energy utilization.

Effect of Air Temperature on Drying Rate. As the air temperature increases, its potential for holding water vapour increases out of all proportion (exponentially). Higher air temperature also increases the rate of heat transfer to the surface water and the material, resulting in higher evaporation rates, and increasing the driving force assisting moisture or vapour to flow to the material surface.

Highest drying rates will therefore be obtained at the maximum practicable air temperature compatible with the tolerance of the material to be dried, and of course limited by the heating medium available. But high air temperatures are only attainable at the expense of high energy input to the drying system. Drying costs will be high unless measures are taken to lower the air temperature at the drier exhaust to the practical minimum, e.g. by causing the spent air to preheat incoming material.

Effect of Air Humidity (Moisture Content) on Drying Rate. Higher drying rates are also achieved at minimum drying-air humidity. For air at a specific dry-bulb temperature, any increase in humidity reduces its capacity for holding additional water vapour. It can also inhibit the rate at which evaporation takes place.

These effects are most significant at low air temperatures and when the moisture content approaches saturation but become smaller as temperature increases. In practical drying applications, reasonably high humidities are not a significant limiting factor on drier performance. In the interest of economic operation, it is normal to aim at humidities around 80% saturation at the air outlet. This is often achieved by recirculating a proportion of the air, so reducing the intake of fresh air which has to be heated for the same temperature and air velocities within the drier. *Figure 1* indicates the effect on drying rate of variations in air temperature and humidity from laboratory-scale tests on the drying of poplinette[2].

Effect of Air Velocity on Drying Rate. The rate of evaporation from the wetted surface depends on the rate at which vapour can diffuse through a laminar layer at the moist surface and the rate of heat flow to the water. A relatively high-velocity air stream passing over the surface reduces the thickness of the laminar layer and hence increases both heat transfer and evaporation rate.

Turbulence of the air stream and water surface both increase evaporation rate. The direction of air flow relative to the wetted surface can have a significant effect. With tangential air flow the evaporation rate increases as the nth power of the air velocity where n is 0.8. With the air flow normal, values of n as high as 1.4 have been achieved. *Figure 2* shows the results of an experiment drying poplinette with air at $140°C$ through air nozzles normal to the material. The velocity at the nozzles was varied by altering the nozzle discharge areas. Mass flow remained constant.

Drying Periods

It is often useful to carry out drying tests on specific materials with various air temperatures, humidities and velocities to establish the most

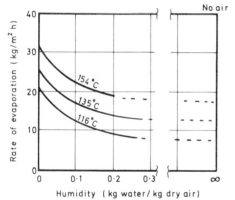

Figure 1   Effect on drying rate of humidity and temperature[2]. Courtesy Textile Institute

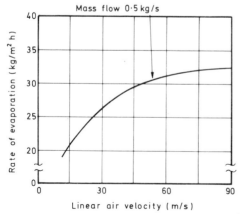

Figure 2   Effect on drying rate of air velocity[3]. Courtesy Shirley Institute, private communication

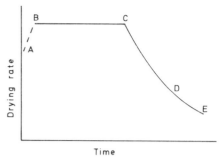

Figure 3   The periods of drying

economic and effective operating conditions. Limited tests can be carried out on full-scale drying plant, but these are often supplemented by laboratory-scale tests.

If the rate of drying is plotted against time, three distinct drying periods emerge as in *Figure 3*.
1. Heating-up Period — AB. Initially part of the heat input goes to heating up the material and raising the sensible heat of the contained moisture. Progressively a greater proportion goes to evaporating the water from the material surface, increasing the drying rate.
2. Constant-rate Period — BC. The drying rate is stabilized at a level equivalent to that from a free water surface for the particular drying-air conditions obtaining. It is limited by the rate of diffusion of vapour through the surface air film and it is in this period that drying-air velocity has its greatest effect. With pure convection drying the surface temperature will remain near wet-bulb temperature. If radiation or conduction heating also occurs the surface temperature will lie somewhere between wet-bulb temperature and the drying-air temperature.
3. Falling-rate Period — CDE. This period is divided into two zones. C to D is the zone of unsaturated surface drying where the moisture is being evaporated from the surface more rapidly than migration from within the material can keep the surface saturated. D to E is the zone where there is no longer an effective wetted surface area. The rate of drying is controlled by the internal liquid flow and the plane of drying moves progressively into the material. The surface temperature rises towards air temperature.

Critical Moisture Content

The moisture content at point C, the end of the constant-rate period, is referred to as the critical moisture content of the material. If the required final moisture content is above this critical moisture content the whole drying process will occur at constant rate after the heating-up period. If the initial moisture content is below the critical value the whole process will be in the falling-rate zone.

A Table of approximate critical moisture contents for a selection of materials is given in

Chemical Engineers' Handbook[1]. Being average moisture contents through the material, the actual values will increase with increased drying rates and material thickness.

Moisture Content — Basis of Expression. The moisture content of either the drying air or the material to be dried is preferably expressed in absolute terms of kg moisture/kg dry substance or as a percentage of the dry material by weight. On this basis equal increments in percentage moisture represent equal changes in weight. Occasionally absolute or percentage moisture contents are quoted on the wet weight of the material. *Figure 4* shows the relation between wet and dry bases and gives an expression to convert from one basis to the other.

Equilibrium or Regain Moisture Content. Any hygroscopic material stored in contact with the atmosphere will tend to reach equilibrium with the moisture content of the atmosphere. This is the basis of traditional methods of crop, fish and leather drying. It is therefore wasteful in energy to dry any product which is to be stored or used in contact with the atmosphere below this equilibrium moisture content. Where materials do not deteriorate rapidly in moist conditions, experience may show that drying can be stopped at much higher moisture contents which may still be acceptable for further processing, storage, transport or marketing. This can result in substantial energy savings and shorter drying times. Materials such as certain foodstuffs which must be dried below equilibrium moisture content for preservation must be subsequently stored in controlled atmospheric conditions, or in sealed containers.

The approximate equilibrium moisture contents of a few typical materials stored in atmospheric conditions of 25°C and 50% saturation (approx. 55% relative humidity) are given in *Table 1*.

Basic Psychrometry of Air Drying

In most drying problems water is the liquid evaporated and the air both supplies the heat for evaporation and carries the vapour away. In order to

Table 1   Equilibrium or regain moisture content (ambient air 25°C; 50% saturation)

Material	Equilibrium moisture content (% dry basis)
Dacron	0.5
Rubber	0.6
Orlon	1.4
Nylon	3.1
Linen cloth	5.1
Paper newsprint	5.3
Cotton cloth	6.0
White bread	6.2
Flour	8.0
Wood (average)	9.3
Soap	10.0
Wool	12.8
Leather	16.0
Absorbent cotton	18.5

carry out any specific calculations on air drying problems it is necessary to have a knowledge of the properties of air and water vapour mixtures over the relevant temperature range.

The relations between most of these properties can be expressed in graphical form by a psychrometric chart, and data derived from such a chart are sufficiently accurate for most practical purposes. A suitable datum for the chart is 0°C and normal atmospheric pressure 1013.25 mbar (101 325 Pa). The chart could show the following properties:

1. Dry-bulb temperature
2. Wet-bulb temperature
3. Moisture content
4. Percentage saturation
5. Total heat of mixture
6. Specific volume of mixture
7. Dewpoint
8. Vapour pressure

Items 1 and 2 above are the most readily measured but given any two of the first six items a point can be located on the chart. The remaining properties can be read off as indicated in *Figure 5*.

The *Saturation Line* gives the maximum weight of water vapour that 1 kg of dry air can carry at any specific dry bulb temperature.

The *Wet-Bulb Temperature* is defined as the dynamic equilibrium temperature reached by a water surface exposed to air when the heat transfer rate from air to water is equal to the rate at which the latent heat of water evaporated is carried back into the air stream. When air at a particular temperature is saturated it cannot take up any further water vapour and there will be no evaporation from the water surface. Dry- and wet-bulb temperatures will then coincide. As the percentage saturation of the air stream is reduced so its

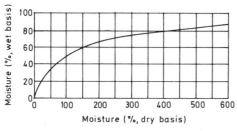

Ww = Wd /(1 + Wd)   Wd = Ww /(1 - Ww)

Ww = kg moisture / kg wet solid

Wd = kg moisture / kg dry solid

Figure 4   Relation between wet weight and dry weight. Adapted from ref.1, Fig.20—10

References p 198

capability for taking up water vapour is increased and the wet-bulb temperature is depressed. The difference between wet- and dry-bulb temperatures is referred to as the wet-bulb depression. It reaches a theoretical maximum when the air stream is dry but in practice this point is never reached.

The *Dewpoint* is the temperature at which air with a given moisture content becomes saturated

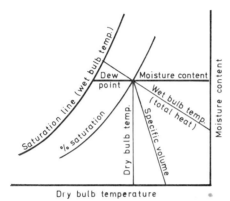

Figure 5   Psychrometric chart (principles)

on cooling. The curve coincides with the saturation line.

The *Total Heat or Enthalpy* of a mixture of air and water vapour, at a point in the chart, is the sensible heat of 1 kg dry air plus the total heat of the weight of water vapour contained, superheated at its partial pressure. If the point considered is on the saturation line the vapour would not be superheated. For simplicity, most low-temperature psychrometric charts show only the total heat of saturated air by lines parallel to the constant wet-bulb-temperature lines. No great error is involved below say 45°C. But since most drier problems involve appreciably higher temperatures, lines of true enthalpy are required. The normal units used are kJ/kg dry air.

Specific Volume as generally used in psychrometric charts is the apparent specific volume, i.e. the volume in m³ of the amount of mixture containing 1 kg dry air. The inter-relation of these functions is shown on the psychrometric chart, *Figure 6*.

Performance of Driers

In a drying operation heat must be supplied to:
1. Raise the temperature of the incoming air.
2. Warm up the material being handled.

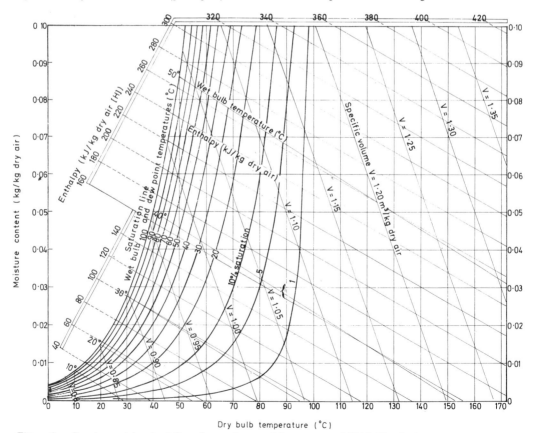

Figure 6   Psychrometric chart (based on a barometric pressure of 1013.25 mbar)

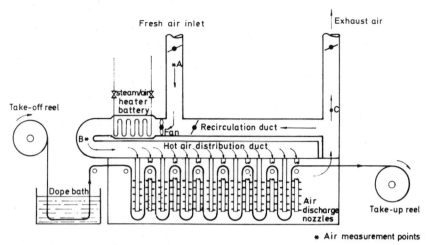

Figure 7   Sketch of festoon paper drier

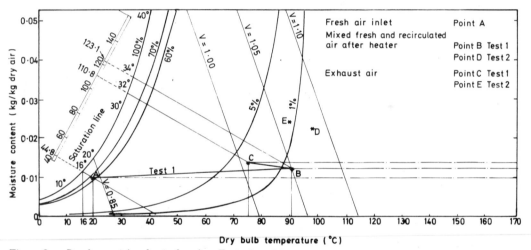

Figure 8   Psychrometric chart showing Tests 1 and 2

3. Heat up and evaporate the water in the incoming material.

4. Compensate for heat losses from the structure, e.g. by convection, radiation and air leakage.

Of these, only 3 is usefully employed. The thermal efficiency of the drying process is given by item 3 divided by the total heat input, the summation of items 1 to 4, expressed as a percentage.

Drier Test

To illustrate the effect of some of the factors previously discussed and also to show the use of the psychrometric chart, consider a test on a simple chamber festoon drier, with air recirculation, for drying coated paper. As indicated in *Figure 7*, the paper flows continuously through a dope bath, then is festooned through the drier. Fresh air is blown through a steam/air heater and then directed to a series of nozzles that impinge heated air on both sides of the loops of paper.

A proportion of the exhaust air from the chamber is recirculated to mix with the fresh-air intake at the heater inlet.

Data in Test 1 represent the 'as found' conditions of a poor-performance installation. Test 2 data represent the same drier after some of the more obvious faults had been corrected. Items marked * were obtained from the psychrometric chart as indicated in *Figure 8* points A, B and C for test 1 and points A, D and E for test 2.

Paper	Units	Test 1	Test 2
Initial moisture content	% dry weight	6.0	6.0
Moisture content after dope bath	% dry weight	23.6	23.8
Moisture content after drier	% dry weight	2.2	5.8
Throughput rate, dry doped paper	kg/h	360	360
Fresh Air at Drier Inlet		(Point A)	(Point A)
Temperature (dry-bulb)	°C	20.0	20.0
Temperature (wet-bulb)	°C	16.0	16.0
*Moisture content	kg/kg dry air	0.0097	0.0097
*Saturation	%	66	66
*Enthalpy (above 0°C)	kJ/kg dry air	44.8	44.8
*Specific volume	m³/kg dry air	0.843	0.843
Air at Drier Exhaust		(Point C)	(Point E)
Temperature (dry-bulb)	°C	75.0	90.0
Temperature (wet-bulb)	°C	32.0	38.5
*Moisture content	kg/kg dry air	0.0136	0.0240
*Saturation	%	3	1
*Enthalpy (above 0°C)	kJ/kg dry air	110.8	153.2
*Specific volume	m³/kg dry air	1.008	1.065
Mixed Recirculated and Fresh Air after Heater Battery		(Point B)	(Point D)
Temperature (dry-bulb)	°C	91.0	98.0
Temperature (wet-bulb)	°C	34.0	38.2
*Moisture content	kg/kg dry air	0.0120	0.0226
*Saturation	%	1	1
*Enthalpy (above 0°C)	kJ/kg dry air	123.1	157.3
*Specific volume	m³/kg dry air	1.050	1.085
Measured air flow rate	m³/h	49 300	49 300
Steam Supply to Heater Battery			
Flow rate	kg/h	863.6	321.7
Pressure	bar gauge	1.86	1.86
Enthalpy (0.97 dry above 0°C)	kJ/kg	2659.8	2659.8
Heat input to air	kJ/h	1 817 000	676 800
Drier Surface Heat Loss	kJ/h	522 700	166 700

Deductions

1. *Air Mass Flow Rate*

(a)	Measured flow rate, hot air duct	m³/s	13.69	13.69
(b)	Specific volume	m³/kg dry air	1.050	1.085
(c)	Mass flow rate = (a/b)	kg dry air/s	13.04	12.62

2. *Determination of Exhaust Recirculation*

By mass balance: each kg air at heater inlet comprises x kg exhaust air plus $(1 - x)$ kg fresh air. By moisture balance:

Test 1: $x(0.0136) + (1 - x)(0.0097) = 1(0.012)$
$$x = 0.59$$

Test 2: $x(0.024) + (1 - x)(0.0097) = 1(0.0226)$
$$x = 0.9$$

Composition of air at heater inlet:

(a)	Recirculated exhaust air	%	59	90
(b)	Fresh make-up air	%	41	10
(c)	Mass flow rate make-up air (1c × 2b)	kg dry air/s	5.347	1.262

3. *Specific Steam Consumption* (steam/kg moisture evaporated)

	kg/kg	11.21	4.96

	Units	Test 1	Test 2

4. *Heat Loss in Exhaust Air.* This is the heat given to the inlet air with inlet moisture content to raise it to exhaust temperature.

		Units	Test 1	Test 2
*(a)	From chart, enthalpy of exhaust air at inlet moisture content	kJ/kg dry air	101.3	117.4
*(b)	Enthalpy of inlet air	kJ/kg dry air	44.8	44.8
(c)	Heat loss in exhaust air per kg (a−b)	kJ/kg dry air	56.5	72.6
(d)	Mass flow rate make-up air (2c)	kg/s	5.347	1.262
(e)	Heat loss (c × d)	kJ/s	302.1	91.6

5. *Heat to Evaporate Moisture*
This is given by the heat content of the quantity of water evaporated from the paper, superheated at its exhaust partial pressure and temperature, less its sensible heat content at drier inlet.

(a) Moisture evaporated

	Units	Test 1	Test 2
Test 1 360 (0.236−0.022) / 3600	kg/s	0.0214	—
Test 2 360 (0.238−0.058) / 3600	kg/s	—	0.0180

		Units	Test 1	Test 2
*(b)	Specific volume of dry air at drier exhaust temp.	m³/kg	0.9884	1.031
(c)	Specific volume of steam at exhaust temperature from steam tables	m³/kg	4.144	2.363
(d)	Mass of water in exhaust air (test data)	kg/kg dry air	0.0136	0.0240
(e)	Volume of moisture in (d) = (c × d)	m³	0.0564	0.0567
(f)	Partial pressure of water vapour 1013.25 × (e)/(b)	mb	57.78	55.73
(g)	Total heat of vapour at partial pressure (f) and exhaust temperature (from steam tables)	kJ/kg	2641.9	2668.3
(h)	Sensible heat at fresh air inlet temperature from steam tables	kJ/kg	83.9	83.9
(i)	Heat added in drier (g−h)	kJ/kg	2558.0	2584.4
(j)	Heat to evaporate moisture (a × i)	kJ/s	54.7	46.5

6. *Heat Balance above 0°C*

		Test 1 kJ/s %	Test 2 kJ/s %
Heat input			
(1)	Heat to circulating air (from steam)	504.7 100	188.0 100
Heat output			
(2)	Heat to evaporate moisture (item 5j)	54.7 10.8	46.5 24.7
(3)	Heat loss by exhaust air (item 4e)	302.1 59.9	91.6 48.8
(4)	Heat loss due to increase in heat content of processed paper over inlet heat content	2.7 0.5	3.6 1.9
(5)	Heat loss from structure (by difference)	145.2 28.8	46.3 24.6
		504.7 100.0	188.0 100.0

7. *Thermal efficiency* (Item 6 (2)/6(1))   %   10.8   24.7

Comments on Test Results and Data. As found the drier had numerous faults resulting in a very low thermal efficiency and an excessive steam consumption of over 11 kg steam per kg moisture evaporated:

(a) Thermal insulation of the structure was poor accounting for almost a third of the heat used.

(b) The initial moisture content of the paper at 6% will be close to the equilibrium or regain value. Heat was being used needlessly in drying the paper to 2% moisture content.

(c) Exhaust air humidity was low.

(d) Although it is not evident from the test data, hot air was escaping from badly closing doors in the drier casing. Heated air was leaking from bad joints in the hot-air duct and passing direct to exhaust without contributing to the drying process. Many air nozzles were choked or distorted, resulting in uneven drying and the need to overdry to compensate.

By increasing the standard of insulation, correcting the faults in (d) and increasing the air recirculation from 59 to 90%, steam consumption was reduced to 37% of the original, for the same material throughput.

References p 198

It will be noted that the largest single loss was that in exhaust air and this was substantially reduced by increasing recirculation and reducing the fresh-air intake to a quarter of the initial rate. At the same time overdrying was largely eliminated. Structural losses were reduced by two-thirds by effective thermal insulation and eliminating hot-air leaks. The final performance at 25% thermal efficiency was still low by good-practice standards, but in this particular case appreciable further improvement was limited by the low temperature at which the drier must operate to avoid damage to the material.

Drying Plants[4]

There is a large range of plants available for drying materials, and in what follows only a brief description is given of the principal types.

Broadly there are two main classifications of driers:

(a) *Convection driers* — sometimes referred to as direct driers because the evaporating medium, usually air or hot gas, impinges on or makes contact with the material being dried and moisture is carried away in the air stream. Examples are: drying rooms, pneumatic driers, spray driers or tunnel driers.

(b) *Contact or conduction driers* — sometimes referred to as indirect driers, because evaporation proceeds because of heat flow through a metal wall or plate on which the wet material rests. Examples are: continuous sheet driers such as paper-machine cylinders, or platen driers.

It is important to note that type (a) may be indirectly heated: air can be heated by steam, hot water or heat-transfer oils in heater batteries, or directly heated by hot gas produced from combustion in a furnace. Indirect heating is used where contamination of the product must not take place, as in fabric driers, whereas direct heating may be acceptable in such cases as fertilizer drying in rotary driers.

Type (b) are nearly always indirectly heated by fluids such as steam, hot water or heat-transfer oils, although in a few cases hot combustion gases can be used.

In addition to the above main classifications, there are two other types of drier, namely radiant-heat driers and dielectric driers. It should be noted that the above classification is a broad one, and indeed some driers are a combination of types, e.g. cylinder or contact driers may be combinations of convection and radiant types. The internally heated cylinder may be fitted with a hood embodying a hot-gas convection system or radiant elements.

Drying plants often constitute a considerable capital investment and the running costs in maintenance and energy consumption can be very high. The decision to install new plant must be based on a thorough economic appraisal, taking all relevant factors into consideration. The possibility of improving the performance of existing driers, however, should never be overlooked for it has been found in practice that the thermal efficiency and output of many driers can be increased.

Driers classified as above take many forms as indicated in the following summary:

Convection driers	*Principal mode of operation*
Drying rooms	Batch, continuous, semi-continuous
Cabinet driers (tray etc.)	Batch
Conveyor driers	Continuous
Tunnel driers (truck)	Continuous, semi-continuous
Rotary driers	Continuous
Vertical cylindrical driers	Continuous
Spray driers	Continuous
Air-swept rotary mills	Continuous
Pneumatic driers	Continuous
Contact driers	
Platen driers	Batch
Cylinder driers	Continuous
Vacuum driers	Batch
Freeze driers	Batch

Specialized driers, including fluid-bed, radiant and high-frequency driers etc.

Drying Rooms

A simple method of drying is to place the material to be dried in a room or chamber through which hot air is circulated. This method is particularly applicable to the drying of large bodies such as building slabs, bricks, wallboard, fibre-board, clothes and foundry cores.

The material to be dried is symmetrically stacked in the drying room so that the maximum surface of each piece is exposed for drying; it may also be stacked on wheeled trucks. Hot air is circulated through ducting arranged in the building, with outlets to discharge the hot drying air or gases uniformly over the surfaces of the material being dried. To obtain maximum thermal efficiency the hot air, after circulating through the room, is drawn off, reheated and recirculated. Humidity is controlled by letting a proportion of moist air escape and admitting new air in its place. For efficient drying it is essential that all material be subjected to uniform hot-gas flow and temperature. *Figure 9* shows an arrangement of a drying room, the operation of which is self-explanatory.

Drying rooms can also be used for continuous or semi-continuous operation, as in the case of the festoon drier for certain types of wallpaper or photographic paper.

Cabinet Driers

These may be used for a very wide range of drying duties such as colours, pigments, chemicals, food products, ceramic ware, textiles, cheeses etc. *Figure 10* shows a common type of cabinet drier arranged for air recirculation. Material may be placed on racks or suspended. With pigments or pastes, the material is contained in trays approximately 4–10 cm deep. In some instances when the material is suitable, e.g. large granules, the trays are preferably perforated to expose additional surface or in some cases, e.g. with fibrous material, the tray may extend the full width of the chamber so that air can be blown through the material (through-circulation).

Conveyor Driers

These comprise a conveyor open or partly open to the atmosphere above, or completely enclosed in a tunnel. The material to be dried passes on the conveyor through the tunnel. The hot drying gases may flow from end to end of the drier, or upwards and downwards through perforations in the conveyor and thus through the material.

Conveyors are made either of woven wire, or of sections of perforated metal attached to link chains on each side of the conveyor. Sometimes the conveyor comprises a fixed perforated plate over which the material is dragged by drag-bars fixed to moving chains on either side of the conveyor. Some conveyor driers comprise multiple conveyors, one over another, the material to be dried being fed on to the top conveyor and passing from conveyor to conveyor down the drier.

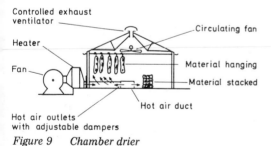

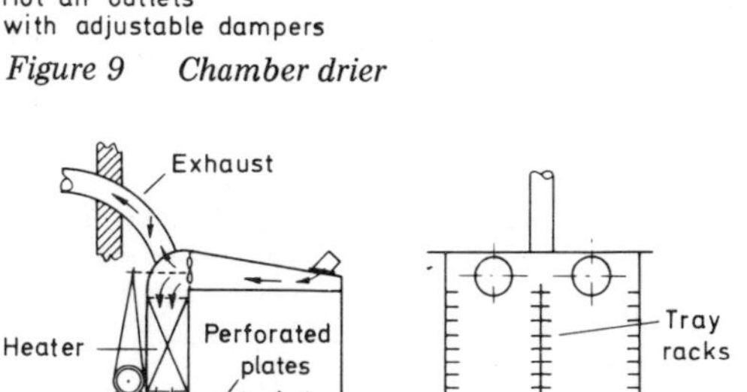

Figure 9   Chamber drier

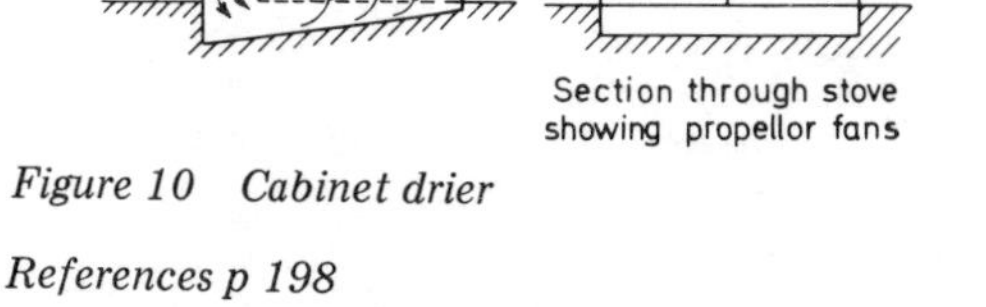

Figure 10   Cabinet drier

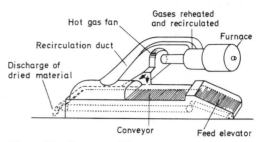

Figure 11   Conveyor drier

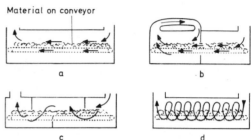

Figure 12   Gas flow through tunnel driers

Figure 11 shows a semi-open conveyor. This differs from those in which the conveyor passes through a tunnel. It is designed to dry goods of high moisture content (70–80%). The material is spread out on the feed elevator in a layer of uniform thickness and discharged onto the conveyor. In the first half of its passage along the drier, hot gases or air pass through the material and are discharged at a high degree of humidity. In the second half the drying medium leaves the material at a much lower degree of humidity, is collected in the hood, and recirculated by the fans through a reheater. In other types of tunnel-conveyor driers the drying medium is circulated over and through material in various ways, as shown in *Figure 12*.

In this Figure, (a) shows the flow above and below the material from end to end of the tunnel; (b) shows the flow above and below the material in the second stage of drying, and recirculated in the first stage; (c) shows hot gases passing up and down through the material as they flow from end to end of the drier; (d) shows hot gases passing upwards through the material, then being drawn down side-ducts in the drier, and forced upwards again through the material in its passage from end to end of the drier.

Tunnel Drier

This is somewhat similar to the conveyor driers, but the material to be dried passes through the tunnel on wheeled trucks. When the material on one truck becomes dry, it is pulled out. The other trucks are then pushed forward, a fresh truck being pushed in at the opposite end. The hot drying gases usually flow from end to end of the tunnel and pass over the material, which is stacked or spread out on trays on the trucks. Alternatively, the hot gases may be circulated as shown in *Figure 13*. This type of drier is suitable for drying large quantities of materials such as bricks, ceramic ware or heavy granular substances.

Simple Rotary Driers

These comprise a horizontal rotating cylinder with a number of longitudinal shelves or lifting flights inside it (*Figure 14*). The smallest driers of this type have cylinders about 2 m long, 0.5 m dia., and the largest 21 m long, and 2.5 m dia. They are used for drying material that has to be turned or tumbled in the hot gas stream to ensure uniformity of drying. Hot gases are generally drawn through the cylinder by a fan. The material to be dried is fed into the cylinder and falls to the bottom. It is picked up by the shelves as the cylinder revolves and is spilled off, through the stream of hot gases. The drum is usually set at a slight inclination towards the outlet. The inclination may vary from 1 in 16 for quick-drying substances to 1 in 30 or 1 in 40 for those that dry slowly. The inclination of the drum, and the shape, width, number and form of the shelves or lifting flights, are determined by experience to produce the best showering effect and rate of feed of material through hot gas. Spiral lifters are sometimes fitted at the feed end of the drier to propel the material well into the cylinder and prevent spillage over the end of the drum.

These driers are used for drying sticky material such as clay, material that is in fairly large lumps, and also for chemicals and a variety of other goods. Where possible the material is fed into the drier in the direction opposite to that of the flow of hot gases.

It is not possible to use the contraflow principle when drying materials that are very light or powdery when dry. Such material is usually passed through the drier in the same direction as the flow of hot gases, and there is an advantage in this, in that the part of the material which dries quickest becomes lighter and is, therefore, carried through the drier by the hot gas stream quicker than the moister material; this automatically assists in uniformity of drying. The rate of feed through the

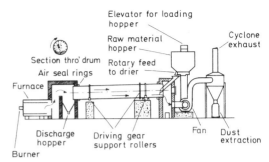

Figure 14   Single-shell rotary drier

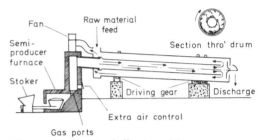

Figure 15   Double-shell rotary drier

drum is controlled by the speed of hot gases, by sloping the cylinder down to the outlet end and by sloping the shelves to feed the material more rapidly to the outlet as necessary.

These driers evaporate water at the rate of about 32—36 kg/m³ of cylinder volume per hour and operate at inlet hot-gas temperatures of 425—820°C and outlet temperatures of 120—230°C.

Rotary Driers: Double-shell

In this type of drier the hot gases pass through a central tube and return through the annular space between the inner and outer drums and are exhausted by an induced-draught fan (*Figure 15*). Lifting shelves or flights are fitted inside the outer drum and also outside the central tube. Hot gases enter the inner tube at a temperature of 540—820°C, and a greater part of the heat is imparted to the material through the surface of the inner tube. The material is usually fed into the annular space at the furnace end of the drier. The lifting shelves or flights pick up the material and shower it on to the shelves on the hot inner tube. The material is carried round about half a revolution on these shelves during which time it is heated by contact with the hot surface of the inner drum, and is then showered off into the outer shell; the operation is repeated over and over again, as the material works its way to the discharge end of the drier.

In another type of double-shell drier, the material (usually light material) is fed into the hot end of the inner tube and passes through the inner tube and back through the annular space between the outer shell and inner tube.

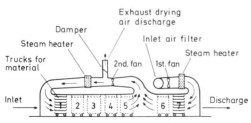

Figure 13   Tunnel drier for wheeled trucks

Double-shell driers normally have a much higher thermal efficiency than single-shell rotary driers.

These driers evaporate 72 kg/m³ h at inlet gas temperatures of 540—820°C. The gases leave the inner tube at approximately 200—260°C and are discharged from the drier at about 65°C.

Special Types of Rotary Driers

There are various proprietary types of high-efficiency rotary driers. One type consists of a horizontal drum with a series of internal channels near the circumference into which hot gases are admitted from a fan. The hot air can pass from these channels through louvres to the inside of the drum, the louvres being so shaped that the material inside the drum cannot spill back through them (*Figure 16*). This drier usually operates at inlet hot gas temperatures of 400—540°C. The material to be dried is fed into the drum, mounts one side of the drum and rolls and tumbles over as it reaches its angle of repose. The hot drying gases are forced through the material as it tumbles over and over.

Another type consists of a horizontal drum with a large number of cross-shaped shelves (*Figure 17*). The material to be dried is continually spilling from one shelf to another and is turned over four times in each revolution of the drum, thus ensuring uniformity of drying. These proprietary types of drier have an evaporative capacity of 95—200 kg/m³ h.

Vertical Cylindrical Driers

Inside a vertical cylindrical casing are a number of concentric rings or shelves. The material to be dried

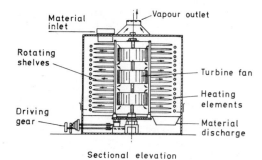

Figure 18   Vertical cylindrical drier

is fed onto the top shelf and is pushed round and turned over at the same time by a revolving rake or scraper. When the material has moved round once, it falls through a gap in the concentric shelf on to the shelf below. Alternatively, the concentric shelves are fixed to a central spindle and revolve, and the material is turned over by a fixed rake. The flow of hot gases is usually outward over one tray and inward over the next.

Figure 18 shows a drier of this type. The material to be dried is fed onto and spread over the top tray. The trays revolve and at every revolution the rake or scraper moves the material through a slot in the tray so that it falls on to the tray below. There are three fans on a central vertical spindle and these draw air inwards over the trays below and discharge it outwards over the trays opposite the fans, so that air passes to and fro over the trays from the bottom to the top of the drier, i.e. in contraflow to the passage of material down the drier.

It will be noted that there are concentric steam coils outside the trays. These serve the purpose of heating and reheating the air in its passage to and fro over the trays. The heating capacity of these steam coils can be adjusted to give the best air-temperature conditions. The fresh material can generally be subjected to a higher temperature than the drier material leaving the drier, and in such cases the capacity of the steam coils would increase the temperature of the air as it flows to and fro and upwards through the drier.

This type of drier is used for materials that have to be turned over during drying, and that need a

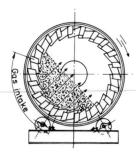

Figure 16   Louvre rotary drier

Figure 17   Cruciform rotary drier

References p 198

moderately long period of drying. They are also used for slurries and pastes as well as for solid materials.

Spray Drier

This is an alternative to the film drier for drying liquid and semi-liquid substances. As its name implies, the substance to be dried is sprayed into a chamber through which pass hot gases.

The total surface area of the many particles in the spray is very large; this, together with the movement of the particles, provides ideal conditions for rapid drying. Drying gases can enter the drier at comparatively high temperatures, because very quick rates of evaporation and heat absorption cause a very rapid fall in temperature, so that the substance being dried does not rise to a harmful temperature. The heavier dried particles fall to the bottom of the chamber, the lighter particles being carried over in the exhaust gases and collected in a dust collector, usually of the filter type.

To obtain satisfactory results with spray driers, it is essential that the substance be sprayed as globules of more or less uniform size, otherwise drying will not be uniform. To prevent drying before the atomized particles are sufficiently dispersed, a cold air duct surrounds the spray or atomizer itself. This type of drier is shown in *Figure 19*.

Spray driers are used for drying milk, eggs, meat, vegetable extracts, other foodstuffs, and a great variety of chemicals. They have high energy requirements, around 5800–8100 kJ/kg water evaporated, and are generally preceded by multiple-effect evaporators with a much lower heat requirement: 830 kJ/kg evaporation for a double-effect evaporator with thermal recompression or 510 kJ/kg evaporation for a modern efficient quadruple-effect evaporator. It is therefore important to condense to the maximum possible concentration, even if the evaporator is relatively inefficient, before spray drying.

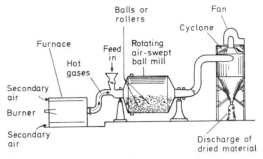

Figure 20   Air-swept mill for pulverizing and drying

Air-swept Rotary Mills

Some materials require to be pulverized and dried, and these two processes can be carried out in one operation.

An example is the preparation of agricultural lime or whiting from chalk. Chalk or carbonate of lime as quarried is partly in lumps three inches or more in size but also contains fines and powder. Agricultural lime can be dried in various types of driers and the large lumps are broken down in the drying process, but air-swept mills are sometimes used.

Another example is the production of pulverized fuel, where it is necessary for the fuel to be pulverized and dried simultaneously.

Figure 20 shows an arrangement of an air-swept rotary ball mill, which pulverizes the material while a continuous stream of hot gas passes through the mill, drying and carrying away the pulverized material. The hot gases are drawn through the mill by an induced-draught fan and the powdered or pulverized material is separated from the hot gases in a cyclone or other type of separating device.

Pneumatic or Flash-type Driers

This type of plant is used for drying materials such as chemicals, clay or sewage sludge in particulate form, with moisture contents ranging from 3% to 900% dry basis. It is suitable for heat-sensitive material, preferably non-abrasive, and can incorporate a facility for recirculation and grinding of coarse material. The plant may be single- or multi-stage depending on the amount and form of the moisture to be removed.

A single-stage plant comprises basically a vertical drying tube some 10 m or more high, a cyclone separator, and an induced-draught fan. Drying air is drawn into the base of the tube through an air heater and travels up the drying tube at sufficiently high velocity to convey the largest particles. The material to be dried is fed into the base of the tube by a screw conveyor, cage, sling or hammer mill (*Figure 21*).

The drying air can be passed through the mill where over half of the total heat transfer may take

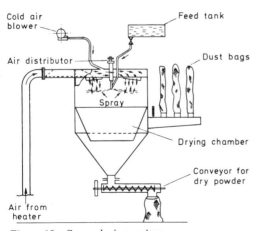

Figure 19   Spray-drying system

place. While the material is being conveyed up the tube by the drying air, drying is effected in a matter of seconds by surface drying. The air and dried material are then separated in the cyclone and the exhaust air may be further cleaned by wet scrubbers, bag filters or electroprecipitators. Total energy requirements for conveying, drying and separation of the material are high, but this type of plant can become attractive when combined with other essential processes, such as grinding and classification.

Contact Driers

Flat-surface Drier

A typical example of a stationary flat-surface drier is seen in the drying floors used in a china clay industry. Driers of this type are heated from below by hot flue gas from a special furnace or by waste gases from a neighbouring furnace or kiln.

In the drying of veneer the stock is dried between two heated plates. The upper plates are lifted off the stock at intervals of time to enable intermittent conveyance of the veneer through the machine to take place. When the two heated plates make contact with the veneer a high rate of evaporation results, and the pressure of the plates prevents distortion.

Film or Roller Drier

This type of drier is used for drying liquids and pasty or pulpy material. Many of these substances exist in a colloidal state in suspension or as emulsions or gels, and include milk, pulped potatoes, yeast, starch, blood, gelatine, glue, chemicals, tannin.

The material is fed or extruded on to a steam-heated revolving cylinder. Drying is completed in one-half or three-fourths of a revolution of the

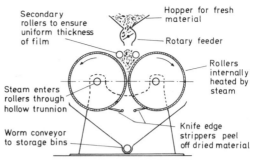

Figure 22   Film drier

steam cylinder. The material is then stripped off in the form of a thin sheet, or breaks up into flakes or powder. For satisfactory operation it is essential that a film of uniform thickness be spread onto the steam-heated drum. The drum is therefore accurately machined and ground, and a roller feeding arrangement may be fitted. Film driers are made with either single or twin rollers (*Figure 22*).

Cylinder Drying: Paper-making

A special application of drum driers is in the manufacture of paper or board. Dewatered pulp is fed onto a series of from one to over 100 rotating cylinders heated internally with low-pressure steam, normally around 100—200 kPa (1--2 bar) gauge. The pulp film passes over and under successive cylinders, being heated first from one side, then the other. In paper-making the bulk of the drying is often effected on a final large cylinder of 3—4 m dia., fitted with a steam-heated hood; heated air at around $110°C$ is blown between cylinder and hood. High-pressure steam may be used on the final cylinder and hood. Typical steam consumptions for paper drying are around 1.75 kg/kg moisture evaporated. A typical heat balance is:

(a)	Heat to heating and evaporating moisture from sheet	64%
(b)	Heat discharged in condensate	18%
(c)	Sensible heat in dried produce	1%
(d)	Balance (radiation losses etc.)	17%
		100%

Vacuum Driers

Driers of this type are expensive, as drying has to be carried out in vessels or chambers sufficiently strong to withstand external pressure, and a condenser and air pump are necessary to maintain the vacuum and draw off evaporated moisture. The majority of vacuum driers are batch driers, as a continuous-feed drier necessitates the incorporation of a seal device to prevent loss of vacuum when the material enters and leaves the drier. The great advantage of this type of drier is that the boiling point of water is very much lowered, as

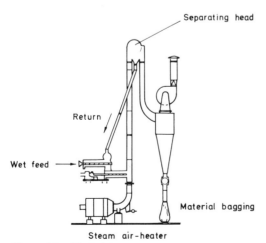

Figure 21   Flash drier

shown by the following figures:

Absolute pressure (mbar or 100 Pa)	130	100	70	40
Boiling point of water (°C)	51	46	39	29

When drying in a vacuum, a high rate of evaporation can be maintained at a low temperature; consequently it is the best method for drying materials that would be harmed by drying at higher temperatures and are difficult to dry owing to a low rate of diffusion of moisture from the centre to the surface of the material. Sugar, chemicals, dyestuffs, rubber, white lead, foodstuffs and explosives are examples.

The material may be heated before it is put into the drier. Inside the drier the heating is by conduction through contact with the hot steam-heated metal surfaces of the drier, and to a lesser extent by radiation. It is sometimes necessary to provide revolving arms or agitators to turn the material over to equalize its temperature; or the material is spread in a thin layer on steam-heated trays in the drier (*Figure 23*).

The thermal efficiency of the vacuum drier is high, with a steam consumption of approximately 1.25 kg/kg of water evaporated. Very economical working can be effected if exhaust steam is available.

Film driers are sometimes incorporated inside vacuum chambers to deal with material which can best be dried by this means and where there are added advantages in vacuum drying.

Freeze Driers

This technique finds its application in the drying of heat-sensitive fluids, e.g. serums and antibiotics and many foodstuffs. The material to be dried is solidly frozen, subjected to a high vacuum, around 100 Pa absolute pressure, and a controlled heat input. In this way the ice crystals sublimate directly to vapour without passing through the liquid phase. This vapour can be collected as ice on the surface of a low-temperature condenser.

The dried product is stored in containers impervious to moisture. Energy requirements are high.

Figure 24 shows the general principles of freeze-drying units.

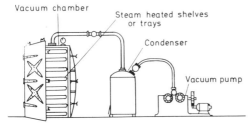

Figure 23   Vacuum drier

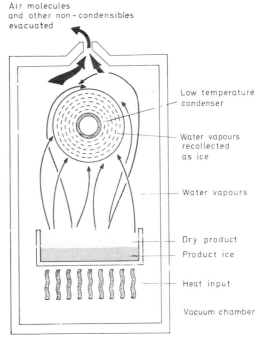

Figure 24   Principles of freeze drying. Courtesy Techmation Ltd

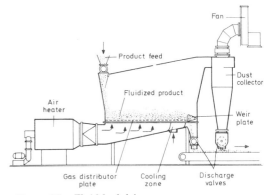

Figure 25   Fluid-bed drier

Specialized Driers

Fluid-bed Driers

A drier of this type is shown in *Figure 25*. Fluid- (or fluidized-) bed units are used for drying chemicals and food products in granular form or other solids such as coal, cement or limestone.

The main element of the drier is the gas-permeable distributor plate which forms the base of the drier chamber. Heated air or gas flows from the air heater through this plate to an induced-draught fan, and is extracted to atmosphere through a dust collector. Gas velocities through the plate are controlled to produce a fluidized state of the product particles as they are fed into the drier and flow along the bed towards the discharge end. Material dried in this fashion is in an ideal state since the maximum surface area is in contact with the hot gases. The depth of the fluidized bed and the retention time of the material in the drier are controlled by adjusting the setting of the weir gate at the discharge end of the drier.

It is possible to fit hot-gas recirculation systems and product-cooling zones to this type of drier. Fuel requirements are approximately 3500—4700 kJ/kg of water removed. Power requirements for fans, feeders etc. are around 0.08 kW h/kg (290 kJ/kg) water removed.

Radiant-heat and Infrared Driers

All hot bodies radiate energy, some of which is within the heating or infrared wavelengths. Wavelengths in the range 2—15 μm are commonly used in drying processes.

A well-known use of infrared radiation is the drying of paints and enamels on mass-produced articles such as motor-car bodies, tin cans etc. The field is now considerably wider and has been extended to such processes as the drying of photographic films and plates, chemicals, drugs etc. Although infrared rays may be produced by heating a surface by any means, in view of the ease of control gas and electricity are generally used.

Electrical infrared generators are normally of two types — lamps and bar generators. Both use a high-resistance wire as the heat-generating element. The wavelength and intensity of heat emitted depend upon the generator temperature: the lower the temperature the longer the wavelength.

Gas-heated generators give longer wavelengths and higher intensity of radiation than do lamp generators. They are of two types — (a) medium-temperature black emitters, and (b) high-temperature incandescent emitters. The medium-temperature range in general has a source temperature of 230—400°C and the high temperature a range of 600—1000°C.

The advantage of very short-wavelength radiation is the high penetration of the waves, especially valuable for drying many types of paint, varnishes, lacquers etc. Heat is generated throughout the depth of the paint film and drying takes place evenly without the formation of a hard surface skin before the inner layers are dry. In some applications the high penetration obtained is a disadvantage, for example in drying transparent strip etc., where some of the radiation passes right through the strip and thus does not generate heat within it.

In such cases, long-wave bar generators or gas generators are to be preferred. The application of infrared methods to drying problems is not as simple a matter as it would appear at first sight. Among other things, the choice of wavelength and the intensity of radiation are of particular importance and require careful consideration, not only on

theoretical grounds but on the basis of empirical and experimental data. Pilot-scale work in a laboratory is often necessary before embarking on the setting-up of large-scale equipment.

Infrared drying plants are usually of the continuous type, and some form of conveyor is used for carrying the articles through the radiation zone. For articles sprayed with paint, this is often a simple chain conveyor from which the painted parts are hung. For drying chemicals, however, continuous or vibrating-deck type conveyors are used.

While it is not normally considered good practice to use high-grade energy for drying, the system can often be justified by taking into account the short drying cycle, often only seconds, the low material temperature, the high thermal efficiency, the low thermal lag and the ease of control (see also Ch. 7, 8.2).

High-frequency Driers

When a material is placed between metal plates to which a high-frequency voltage is applied, heat is generated in the material as a result of the strong alternating electrostatic field which is produced; the material acts as the dielectric of a capacitor.

A high-frequency voltage of the order of 2—100 MHz is produced by an electronic generator; the best frequency depends upon many factors such as the area of the material to be heated, the shape of the material and hence of the electrodes, and the power required. Too high a frequency leads to unequal voltage distribution and local overheating. 15 kV is normal; higher voltages require additional precautions to avoid corona effects and arcing.

The chief feature of high-frequency drying is the uniformity of heating and the fact that even a wet material can be heated at the centre as rapidly as at the surface. This form of drying is used in the drying of artificial silk yarn, rayon cord, vegetables, sponge rubber, paper coatings, carpet backings, etc., and also in the removal of water from slurries and heat-sensitive solutions, as in the manufacture of penicillin.

About 50% of the electrical input is lost in the high-frequency generator but radiation and structure heat-storage losses can be relatively low. Actual gross power consumptions vary considerably with the physical properties and dimensions of the material being dried, but figures of around 2½ kW h/kg water removed are quoted (see also Ch. 7, 8.2).

Control of Driers

Moisture Measurement and Drier Control

One of the most significant factors affecting performance of a drier is the final moisture content of the material. As the moisture content approaches the equilibrium or natural regain, i.e. towards the end of the falling rate period, the rate of drying normally becomes very low indeed. In many instances material is overdried, which means that the throughput of the drier is much less than it need be and thermal efficiency is impaired. Overdrying is very wasteful of energy and it is not surprising that on-stream measurement of final moisture as a means of drier control has received a great deal of attention.

Apart from absolute determination of moisture by oven drying, desiccation etc., moisture content may be inferred by measuring some physical property of the solid that is dependent on moisture content. The principal methods are summarized below.

Electrical Capacitance. The dielectric constant for water is much higher than for most materials, approximately 80 compared with a range of 1 to 5, and therefore a change in moisture content has a considerable effect on the total dielectric constant. The simplest way of making use of this effect in practice is to pass the material between the plates of a capacitor and to measure the change in capacity by a bridge-circuit technique. The change in capacity will depend on the moisture content of the material and the system can be calibrated to read moisture content. The disadvantages in practice are possible uneven distribution of moisture in the material, interference by electrolysis, and the packing density of the material, all of which affect accuracy. The technique has however been used to infer the moisture content of paper, sand, tobacco and agricultural products.

Infrared Absorption. Infrared radiation is absorbed by water at certain wavelengths and therefore the amount of attenuation when radiation is transmitted through a moist material is a measure of the moisture content. In practice water absorption wavelengths of 1.4—1.93 μm are used. High moisture contents necessitate the less powerful water absorption wavelength of 1.4 μm, whereas for materials of low moisture contents the longer wavelength is required. With the reflectance instrument, reflected radiation is measured at two wavelengths and the ratio of these two is used to infer the moisture content of the material.

The infrared technique is substantially unaffected by colour, temperature, composition or particle size and is consequently finding increasing application for sands, yarns, grains, chemicals, plastics etc. Accuracies better than ± 0.2% have been achieved.

Microwave Absorption. Microwaves are high-frequency electromagnetic waves of 1—10 cm length which are strongly absorbed by water but not by dry materials. Microwave attenuation is therefore a measure of moisture content, increasing attenuation implying increased moisture content. With this method, electrolysis affects the measurement less than with the capacitance system. Accuracy is affected however by leakage, reflectance, particle size and packing density, and automatic correction for changes in material temperature is normally required. Extruders and compactors have been successfully used for preparing material for measurement, although with sheet materials the web is passed through a slot in the waveguide so that width as well as thickness is presented to the radiation. The method is suitable for sands, grain, clays, textiles, paper etc and accuracies of ± 0.25% can be achieved.

Electrical Conductance. Temperature correction or measurement at a fixed temperature is necessary. The conductance of most dry materials is considerably lower than that of water and therefore the presence of water will significantly affect total conductance. In practice a sample of the material is placed in a cell with two electrodes and the conductance of the material between them is measured: e.g. in concrete one electrode may be a stud placed flush with the floor of the mixer and the mixer body acts as the second electrode. Other electrodes are rollers, needles or thin blades suited for sheet materials. Variation in moisture distribution through material and changes in packing density can affect the accuracy considerably. In the case of on-stream measurement with granular materials the material is unpacked. Clearly the presence of free ions can produce considerable errors and to some extent efficiency of contact between material and electrodes affects the results. The method is only suitable for measurements below water-saturation level but it has been used widely for grain, seeds and material in baled form. It is quite unsuitable for materials of high conductivity, manufactured chemicals, and for certain minerals such as coal or sand.

Automatic Control by Temperature Difference

A method of drier control developed and patented by the SIRA Institute of Great Britain[5] is worth special mention. This technique is based on the fact that during drying the temperature of a material is depressed below the dry-bulb temperature of the drying environment. If the material is saturated, its temperature will be approximately that of the wet-bulb temperature, whilst if less than saturated the material temperature is not depressed as much. The SIRA technique measures the difference in temperature between material and wet bulb, and this difference is related to the moisture content of the given material. Consequently this method is suitable for measuring moisture contents less than that required to saturate the material. With many hygroscopic materials a temperature difference of 9°C will show that the material is adequately dried for storage.

In the falling-rate drying zone the material temperature and the wet-bulb temperature of the environment together determine the equilibrium relative humidity (ERH) of the material and hence moisture content. Relative humidity is a property of the moist gas whereas ERH is a property of the moist material. It can be shown that over a wide range of drying conditions the temperature difference is uniquely related to the ERH of the material, and in particular when the material temperature is 8°C above the wet-bulb temperature the ERH is 60%. The ERH is measured by taking two temperatures—the wet-bulb temperature of the drying air and the temperature of the material—and subtracting one from the other. The relation between this temperature difference and ERH is shown in *Figure 26* over the range 0—100% and plotted for a wet-bulb temperature of 46°C, although the relation is not materially affected by variations in wet-bulb temperature. In practice it is necessary to measure the temperature of the material near the end of the drying process and the wet-bulb temperature of the air in that vicinity.

Where contact is permissible inexpensive thermocouples or thermistors can be used, but where contact is not permissible, SIRA have developed a special instrument embodying a radiation pyrometer which measures the material temperature and the temperature of a rotating wet wick.

The method is capable of wide application and has many advantages over established methods, being independent of colour, particle size, packing density and material composition. Control of the drier can be accomplished either by varying the throughput rate of the drier or by adjusting the heat input, in accordance with the temperature difference.

With batch driers, two probes can be used to indicate when moisture content has been reduced to the desired level and drying can be terminated. It is claimed that moisture content of powder in fluid-bed driers can be controlled to within ± 0.2%.

Other Factors affecting Thermal Efficiency

Heat Insulation of Enclosure. The heat loss from the drier structure varies enormously, typically 4—30% of the heat input, and in practice the insulation standards are often low. Good insulation is important particularly in high-temperature driers. Where insulation is enclosed by an outer metallic casing it is important to ensure that the effectiveness of insulation is not affected by heat conduction paths which effectively 'by-pass' the insulation material.

Air Infiltration or Leakage. As a general rule contact driers are more efficient thermally than air or convection driers. This is because in the operation of convection driers heat loss in the exhaust air is inevitable, and the exhaust is often considerably below saturation moisture, in many instances much more than it need be. Additionally however, cold-air inleakage into and hot-air leakage from drier body is a common malady; such leakage usually represents a direct loss of heat. On the inlet side of heater batteries air inleakage often results in lowering of hot-air temperature, whilst on the discharge or pressure side of batteries, loss of air reduces the air volume passing to the drier air-distribution system and therefore reduces air velocity.

Mal-distribution of Moisture. If moisture is unevenly distributed, e.g. in a web or sheet material, the drier speed must be governed to suit the wet-test section. It is important to ensure that moisture distribution is as uniform as practicable, and the condition of predrying plant should always be questioned. Unevenness due to distortion of mangle rolls in a textile drier will adversely affect drier performance. Uneven drying-air distribution, often caused by lint blocking air nozzles within the drier, can have a similar effect.

Condition of Heating Equipment. Many driers are steam heated, and it is of the utmost importance

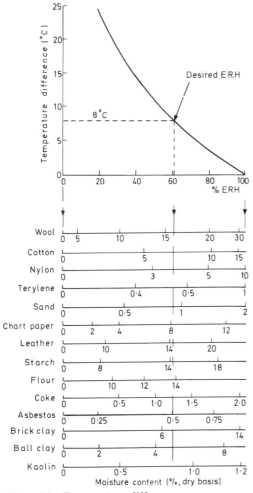

Figure 26 Temperature difference versus percentage moisture content for various materials. Adapted from ref.5

References p 198

to ensure that heater batteries where used are kept clean and free of dust, lints, etc. The output of heater batteries can be seriously affected in this way. Likewise blocking up or linting up of air nozzles can materially reduce output.

Trapping and Air-venting of Steam-heated Equipment. Often the output of heater batteries is materially reduced owing to bad steam-trapping and faulty condensate-removal systems. In cylinder driers, broken syphon pipes or buckets cause serious water-logging of cylinders. This is particularly serious where low or modest steam pressures are in use.

Condensate Losses. The heat loss due to the discharge of condensate from steam-heated driers can vary from 16 to 30% of the gross heat input. If condensate is not required for boiler feed or other uses this loss is serious. The use of flash steam in preheater batteries or cylinders or for space heating should always be considered, condensate being returned to the boiler feed or process hot-water systems. There is no doubt that the use of liquid-phase heating, either by heat-transfer oils or high-pressure hot water, can often show a marked economy and should always be considered.

Recirculation. In convection driers a high degree of controlled recirculation is often necessary to ensure high thermal efficiency. Even where the drier has been designed with this in view, indifferent control of dampers can seriously reduce recirculation and result in high exhaust-air loss, owing to low humidity.

Drier Loading. In addition to controlling production rate, drier loading and evenness of loading, on loose-material machines, have a very significant effect on efficiency of operation. The optimum loading for a continuous drier is usually the maximum throughput rate which can be effectively dried with the heat input, air temperatures, velocities and maximum exit humidities attainable with the installed plant. The temperature in the final drying stages may be limited by the tolerance of the material to be dried. A small margin of drying capacity may, however, be reserved to cater for variations in initial moisture content. At any loading below this optimum level, fixed losses such as exit-air losses, surface losses, air leakage and condensate from heater batteries increase in proportion to the heat usefully employed in the actual drying operation. The effect of unevenness of loading on production rate and efficiency has already been mentioned.

For batch driers the same principles of reducing the significance of fixed losses apply. The upper limit of loading for a particular plant may be fixed by the point at which the packing or stacking of the material starts to limit the flow of heat to the material for evaporation and the removal of the moisture evaporated from the material. This limit can also apply in some continuous machines for (say) slab or palleted materials.

Control Instruments

Many drying plants are sadly lacking in instrumentation and control equipment. Too often it is left to an experienced operator to decide by appearance or feel when a product is sufficiently dry. Control of the many variables may also be manual and arbitrary. This inevitably leads to needless energy consumption, lower-than-attainable production rates, and sometimes an unnecessarily high proportion of damaged or incorrectly processed material.

All drying plant should have a basic minimum of indicating instruments to warn of abnormal conditions and, as a minimum, periodic measurements should be carried out to determine degree of drying, efficiency of operation and the setting of controls for optimum results. For small plants, portable and laboratory equipment could be used.

If energy consumption is appreciable, the material to be dried is subject to variation in moisture content, or optimum production rates are required, more sophisticated permanent instrumentation and various degrees of automatic control can usually be economically justified.

Temperature Indicators and Controls. Most drying plants are fitted with temperature indicators to show material and/or drying-medium temperature. Too low a drying-medium temperature after furnace or heater battery can lead to low production rates and poor thermal efficiency. Too high a temperature can result in damage to the material during the falling-rate drying period, or if a continuous drier is stopped. In the extreme, damage can occur to drier refractory or steelwork. Too high an exhaust gas temperature can result in recoverable heat going to waste with resultant unnecessarily high energy consumption.

Associated with temperature indication, thermostatic control of heater batteries or furnaces to maintain the required gas or air temperature is commonly practised. If a continuous drier is divided into zones, either individual zone-control should be practised, or control should be limited to zones covering the falling-rate drying period with an overriding high-temperature heat-inlet cut-off to provide for possible stoppage.

Moisture and Humidity Instruments. An important measurement for most continuous driers is the humidity of exhaust air. In order to conserve energy this should be kept as high as practicable. For continuous driers with recirculation, the readings obtained can be used to reset the recirculation dampers, or if conditions do not remain stable, to operate automatic damper controls.

For temperatures up to about 90°C, the moisture content of exhaust air can be measured by simple wet- and dry-bulb thermometers, the wet-bulb depression being indicated as % saturation or % relative humidity. Above 90°C, more sophisticated instruments are used. These can be based on direct measurements such as condensing the vapour from a measured quantity of gas, or inferential, based on the variation of electrical conductivity or resistance with the presence of moisture.

Material Moisture Content. To obtain maximum production and minimum energy consumption, the final moisture content of the material after drying must not be lower than is absolutely necessary. Types of instrument to measure this moisture content have already been discussed and these can be used to operate drier speed controls or to control heat input to the drier so as to maintain the required moisture content of the material. Where continuous indicating or control is not justified, periodic laboratory checks of final moisture content should be carried out to ensure that over-drying does not occur.

9.2 Heating, cooling and air-conditioning large buildings and spaces

Among the principal desires of primitive man, the wish to keep warm was by no means the least and, in the present decade, no less than a third of the fuel resources of the UK are devoted to this and allied aims. Of this total, as much as half of the fuel consumed relates to domestic premises (say 25% conversion compared with industrial say 55%). In consequence, the overall average efficiency on conversion cannot be much more than 40%. Assuming potential savings of 10—15% divided equally among the four heads better insulation, improved heating operation and combustion efficiency, and conservation techniques, the potential for economy in this area could easily represent 15% of the total national consumption — a staggering indictment of the low priority given to research into the subject. In world-wide terms and substituting the wish to keep cool as appropriate in tropical climates, there is no reason to believe that the proportions are very different.

Recent work at the UK Building Research Station[6] has shown that the energy consumed in providing a satisfactory working environment in a modern office building is allocated as shown in *Figure 27*. It will be noted that, both in the sense of energy use on site and of dissipation of national resources to produce site energy, consumption for operation of plant auxiliaries exceeds that for the production of heating or cooling. No similar field data appear to exist for industrial enclosures but there is no reason to believe that the ratio between primary and auxiliary consumption differs greatly in such cases.

Inhabitants of the United Kingdom enjoy a temperate maritime climate as is illustrated in *Figures 28 and 29*[7] which are alternative forms of weather frequency diagrams based upon meteorological

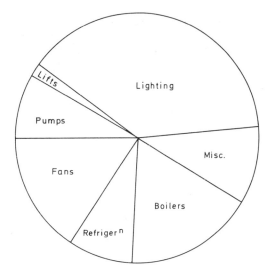

On-site energy

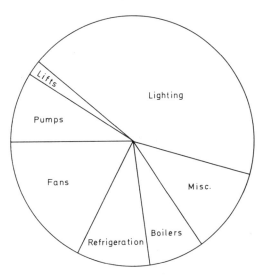

Hydrocarbon energy

Figure 27 Annual energy consumption of the various systems which serve a modern building. (Note the high proportion attributable to auxiliaries)

data. Advantage should be taken of the comparative consistency of external conditions by designing both buildings and their associated heating or cooling plant such that the former act to temper the effect of climate variations and thus permit design of the latter to be based upon economic norms rather than extremes.

In 1972, the Royal Institute of British Architects instituted a project aimed at examination of the concept of 'long life/loose fit/low energy' buildings and many technical and semi-technical papers have been published in pursuit of this ideal[8]. There are innumerable facets to this investigation and the majority has yet to be pursued to finality: as an example of but one, *Figure 30* shows the result of an investigation into economic false ceiling depth for concealment of ventilation duct-work. Here, the annual cost of energy consumed in fan power etc. is related to the annual cost of the required construction in search of an optimum.

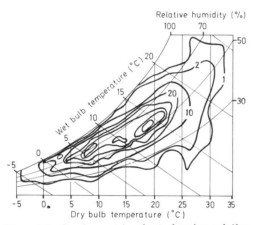

Figure 28   Psychrometric chart showing relative frequency of occurrence of outside conditions at Kew at 1300 hours over a 10-year period

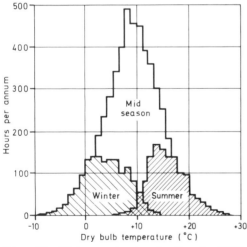

Figure 29   Histogram showing, for Croydon (UK), the average annual seasonal temperatures recorded. (Note that winter overlaps summer and that mid seasons encompass both).   From Integrated Environment in Building Design, *Ed. A. F. C. Sherratt,* Applied Science Publishers 1974, *by permission Construction Industry Conference Centre Limited.*

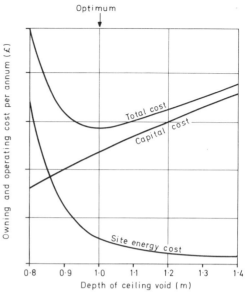

Figure 30   Plot showing how capital and running costs react to varying the depth of a false ceiling concealing ventilation ductwork.   (Note that an optimum exists)

Human Comfort

The criteria which determine whether an enclosure is comfortable or not are many and the interrelation between them is complex: there is no single index which may be used to describe or measure all the variables.   On the one hand there are the human physiological reactions which differ with level of activity, age, sex, race and degree of acclimatization, and on the other are the characteristics of the enclosure in terms of shape, spaciousness, cleanliness and colour.   The general human requirement is that a work place should be comfortably warm, quiet and well lit but the degree of tolerance with which deviation from these ideals is accepted can be quite wide.

Thermal Comfort

The more important of the factors which contribute to human comfort in the thermal sense are dry-bulb temperature, hot or cold radiation, sensation of air movement, air quality and relative humidity.   There is no absolute ideal value for any of these in view of personal idiosyncrasies, ranges of clothing weight and levels of activity: further, each interacts with the remainder.

Dry-bulb Temperature.   Statutory requirements for dry-bulb temperature, as incorporated in the Factories Act[9] and the Shops, Offices and Railway Premises Act[10], are to the effect that 15.5°C must be attained within one hour of the commencement of occupation.   The energy crisis notwithstanding, this requirement is sub-normal and present practice suggests that, for other than the heaviest work, an air temperature of not less than 19°C is required in a factory and 20°C in an office and then only provided that other criteria are met[11].

The instrument used to measure dry-bulb air temperature is the common mercury-in-glass thermometer but best results are achieved if an aspirated or sling-type instrument is used, since a static bulb and scale mounted on a wall or column will be subjected to contact and other effects.

Radiation. This aspect of human comfort, which may be either hot or cold, is best illustrated by example. By merely raising or lowering the dry-bulb temperature it would not be possible to produce a comfort condition in an uninsulated corrugated-iron shed, either in winter in the UK or in summer in a tropical climate. The human body reacts to radiation and, within an enclosure, senses the surface temperature of the surrounding walls, ceiling and floor.

The scale of measurement used for radiation is the mean radiant temperature, which is an area-weighted average of all the room surface temperatures viewed by an occupant, and it has been shown that if this is more than 8°C above or 5°C [12] below the dry-bulb air temperature, comfort cannot be achieved.

The instrument used to measure mean radiant temperature, indirectly, is the globe thermometer which consists of a simple dry-bulb instrument, the bulb of which is mounted in the centre of a 150 mm-diameter blackened metal globe.

Air Movement. Excessive air movement gives rise to complaints of draughts and, conversely, inadequate air movement leads to complaints of a heavy stuffy atmosphere. Standards are variable, dependent upon the level of activity, the weight of clothing and the part of the body affected: the neck is a notably sensitive area. The temperature difference between the moving air stream and the ambient temperature in the enclosure is also an important consideration.

German standards[13] require that, for a moving air-stream temperature of 20°C or below, the maximum air speed should be no more than 0.1 m/s. For higher temperatures greater air speeds are permitted, for example 0.4 m/s at 25°C, but practice in the UK suggests that movement should be limited to 0.2 m/s at that temperature.

Air Quality. An enclosure must be provided with a supply of outdoor air in such quantity as to suppress odours, dissipate contaminants and maintain an atmosphere of freshness. For industrial applications with fume-dispersal problems, each case must be treated on its merits and a wealth of practical data is available in technical manuals.

In non-fume producing enclosures a supply of outdoor air at a rate of 6 litre/s per occupant is the accepted minimum standard[14]. Where smoking is permitted, an increased air supply is required and a rate of 12 litre/s per person has been proposed. Much published work exists in this field and *Table 2* summarizes current practice[15].

Humidity. Relative humidity, which is a measure of the amount of moisture contained in the air at a given temperature, has little effect upon subjective comfort, provided that it is contained within the

Table 2   Recommended outdoor air supply rates (litre/s per person). Based on *IHVE Guide*

| | | Outdoor air supply (litre/s) | | |
| | | Recommended | Minimum (take greater of two) | |
Type of space	Smoking	Per person	Per person	Per m² floor area
Factories * †	None			0.8
Offices (open plan)	Some	8	5	1.3
Shops and supermarkets	None			3.0
Factories * †	Some			1.3
Laboratories †	Some	12	8	—
Offices (cellular)	Heavy			1.3
Cafeteria † ‡	Some			—
Conference rooms	Some	18	12	—
Dining rooms †	Heavy			—

* See statutory requirements and bye-laws
† Rate of extract may override supply demand
‡ Queueing capacity may be greater than seating

References p 198

limits of 40 to 70%. Below 40% complaints of dry throats and of static electricity discharges may arise and, at the other end of the scale, there may be problems of condensation upon cold building surfaces, single glazing etc.

In industrial atmospheres there are statutory requirements that humidity does not exceed certain limits and the Factory Act lists the wet- and dry-bulb temperatures which must not be exceeded in certain circumstances[16].

The instrument used to measure relative humidity is, commonly, a mercury-in-glass thermometer, the bulb of which is covered by a damp wick and thus senses evaporation rate. For a given coincidence of wet- and dry-bulb temperatures, the relative humidity may be read from a chart.

Thermal Indices

Over the years, some thirty or forty different indices have been produced in attempts to combine the requirements for temperature, radiation, air movement and humidity into one finite scale. Equivalent[17], Effective and other temperature scales have had their fashion and been abandoned, owing to shortcomings in ability to represent the influence of one or more of the variables.

The index which has remained acceptable for forty years is that of Resultant Temperature[18] which, in the Dry Scale, reflects all but humidity and in the Wet Scale includes that aspect also, albeit less successfully. Another index currently in use is the Environmental Temperature[19] scale but this is a measure of building performance rather than a comfort criterion. Arising from their respective mathematical derivations, these two scales have similar values for average occupancy of an average building at about 20°C.

Other Aspects

There are, of course, other aspects of human comfort that must be considered: the effect of heat stress; the differing requirements of transient and continuous occupancy; the discomfort which can arise from a cold floor and that from a vertical air-temperature gradient. Reference to an appropriate textbook or technical manual will provide further detail[20].

A timely reminder that some 23 million working days are lost in each year because of industrial accidents has drawn attention[21] to previous investigations into the relation between environmental comfort and accident frequency. To this loss must be added those arising from illness, impaired manual skills, loss of concentration and other causes of error, all of which may derive from unsuitable working conditions. Taking account of air movement and weight of clothing, *Figure 31* shows the relation between body-heat production — a measure of activity — and the desirable Resultant Temperature.

Building Construction

Although the purpose of providing space-heating equipment is to ensure the comfort of the occupants of an enclosure, it is the thermal characteristics of the materials used to create that enclosure which are the principal determinants of the type and scale of apparatus required. The prospect looms before us of an energy ration and whether this be per occupant, as it should be in order to encourage economic planning, or per square metre of floor space, which is meaningless, the necessity for detailed consideration of the suitability of building materials will become paramount. If the UK emulates Edward Lear's Jumblies and goes into the North Sea era in thermal sieves, then we shall deservedly sink!

Properties of Materials

The materials used in building construction have, in the present context, three important properties: thermal conductivity, mass (density) and porosity to air pressure and water vapour. Each of these properties contributes to the adequacy of a material to act as a thermal barrier, to delay internal response to external temperature change, and to admit or exclude undesirable weather influences.

Thermal Transmission. The thermal conductivities of building materials are tabulated in standard works of reference[22], the data there representing steady-state conditions. *Table 3* lists a number of typical values expressed for convenience in terms of thermal resistivity per unit thickness, this being the reciprocal of conductivity; also shown are the corresponding values for specific mass, or bulk density.

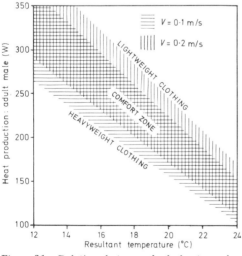

Figure 31 Relation between body-heat production (see Table 8) and optimum Resultant Temperature. (Note the effect of clothing weight and air movement)

The transfer of heat through an element of real building construction, inside to outside or vice versa, depends not only upon the material or materials used but also upon heat transfer through the air layers immediately adjacent to the inside and outside surfaces and through any intermediate air cavities in the construction. Values of the thermal resistance offered by a selection of such surface films and cavities are listed in *Table 4*.

The resistance to heat flow for any part of a building construction is the sum of the individual resistances. Hence, for a 260 mm cavity-brick wall:

$$R_w = R_{si} + R_{so} + R_{sc} + l(R_1)$$

where R_w = thermal resistance of wall, R_{si} = surface resistance, inside, R_{so} = surface resistance, outside, R_{sc} = resistance of cavity, l = thickness of brickwork, and R_1 = resistivity of brickwork, all in coherent SI units. This calculated property of

Table 3 Approximate values for resistivity and specific mass of materials used in building construction and insulation

Material	Properties	
	Specific mass (dry) (kg/m³)	Thermal resistivity (m °C/W)
Construction		
Asbestos cement (sheet)	1550	2.2
Asphalte	2250	0.8
Brick (dry)	1700	1.6
Brick (exposed to weather)	1700	1.2
Clay tiles	1900	1.2
Concrete (dry)	2000	1.1
Concrete (exposed to weather)	2000	0.8
Glass (sheet)	2500	1.0
Plaster	1300	2.2
Plaster board	950	6.3
Stone (marble)	2700	0.5
Timber (softwood)	600	7.7
Insulation		
Asbestos board	700	9.0
Fibreboard	380	16.7
Glass fibre quilt	80	25.0
Lightweight concrete	600	5.6
Mineral wood (mat)	130	27.8
Polystyrene (expanded)	15	25.0
Vermiculite (granules)	100	15.4
Wood wool	600	9.1

Table 4 Approximate values for surface resistances and for structural cavities. Based on *IHVE Guide*

Site of resistance	Thermal resistance (m² °C/W)
Inside (R_{si})	
Wall	0.12
Ceiling or roof (up)	0.11
Ceiling or floor (down)	0.15
Outside (R_{so}) *	
Wall	0.06
Roof	0.05
Cavities (unventilated)	
Vertical	0.18
Horizontal	0.21
Cavities (ventilated)	
Corrugated to flat	0.16
Pitched roof void	0.18

* Varies with exposure etc. Consult *IHVE Guide*, Book A, for precise details

a construction is more usually expressed as a thermal transmittance or U-value which is the reciprocal of the thermal resistance and equivalent to a heat transfer coefficient. For the example shown, $R_w = 0.622$ m² °C/W and $U = 1.6$ W/m² °C; and a .ew other representative figures are shown in the second column of *Table 5*. As may be seen from *Table 5*, if high-resistance (insulating) materials are added the U-value will become smaller, as shown in the third column of *Table 5*.

Specific Mass. The mass of the components of a building structure affects the manner in which internal conditions will vary as the outside temperature fluctuates and, indeed, as internal sources of heating or cooling add to or subtract from the results of external changes. The building structure therefore can act as a flywheel to smooth the effects of both external and internal influences, provided that it has adequate mass and that the internal surfaces have characteristics which allow absorption and rejection of heat. Analysis of problems in this field of transient heat flow is exceedingly complex and a full treatment requires the use of a computer as illustrated in *Figure 32*. The (UK) Building Research Station[23] has done much work in this field and their publications provide tentative solutions to routine problems.

Porosity. It will be noted from *Table 3* that moisture content has a significant effect upon the reaction of brickwork and concrete[24] to heat flow. Such results stem from the effect of driving rain upon the external surfaces of structural components, and no less importance must be placed upon

Table 5   Approximate *U*-values for typical materials used in building construction, with and without insulation

Construction	Thermal transmittance (*U*) (W/m² °C)	
	Basic	Insulated
Walls		
Solid brick, 220 mm	2.3	
Insulated with plaster board		2.0
Cavity brick, 260 mm	1.6	
Insulated cavity		0.7
Asbestos cement sheeting	5.3	
Insulated with glass fibre		1.1
Roofs		
Concrete and asphalte	3.4	
Lightweight screed at top		1.8
Asphalte on woodwool slabs	0.9	
Insulated with glass fibre		0.6
Cavity asbestos cement deck	1.5	
Insulated with glass fibre		0.7
Tiles on battens, flat ceiling	1.5	
Insulated with glass fibre		0.3
Corrugated steel sheet	6.7	
With insulated ceiling		2.0
Floors		
Concrete, on ground (large)	0.11	
Concrete, intermediate	2.5	
Insulated with woodwool		1.8
Timber, suspended (large)	0.14	
Timber, intermediate	4.00	
With plaster ceiling		1.5
Glazing		
In metal frame, single	5.6	
Double		3.2
Rooflight, single	6.6	
With laylight under		3.0

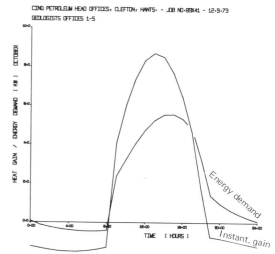

CINO PETROLEUM HEAD OFFICES, CLEFTON, HANTS. - JOB NO.BBX41 - 12·9·73
GEOLOGISTS OFFICES 1-5

Figure 32   Computer analysis of building energy demand and instantaneous heat gain. (Note the effect of the structure in lowing the peak demand and delaying the effect upon the enclosure)

the effects of wind be this either occasionally high or persistent at a lower force.

Most building materials will permit the diffusion of water vapour through them and, where a temperature gradient exists within the material, the probability of condensation exists despite the provision of a vapour barrier. This condensation may be either surface or interstitial. Generally, the problem may be lessened by attention to thermal insulation, and *U*-values below about 1.0 W/m² °C will produce satisfactory results except in extreme atmospheric conditions. It is not possible to prevent some winter condensation internally on single glazing if humidity within an enclosure is held at comfort level.

Similarly, most building construction will allow air infiltration, either through joints between sheets of relatively impervious material such as glass, asbestos, cement or curtain walling or through other less obvious routes. American research[25] upon a nine-storey building showed an air leakage rate of 10 litre/s m² of outside wall area, three quarters of which was through the supposedly solid masonry component of the structure.

Statutory Requirements. Regulations published under the Thermal Insulation (Industrial Buildings) Act[26] require that the *U*-value of a roof of an industrial building shall not exceed 1.7 W/m² °C. The Building Regulations[27], applying to England and Wales, and the Scottish Regulations[28], prescribe levels of insulation for domestic premises but these are extremely low by European standards[29]: there are new draft regulations currently in circulation for comment, but whilst these promise some improvement they still relate only to domestic premises. In defence of the England and Wales regulations, however, it must be noted that these are

Table 6   Tentative proposals for specification of insulation standards[32]. Courtesy IHVE, London

Criterion (maxima)	Heavy structures			Light structures		
	Zone and external temperature			Zone and external temperature		
	I ($-4°$C)	II ($-1°$C)	III ($+2°$C)	I ($-4°$C)	II ($-1°$C)	III ($+2°$C)
Area-weighted *U*-value (W/m^2 °C)	0.5	0.6	0.6	0.4	0.5	0.6
Volume-weighted value (W/m^3 °C)	0.7	0.9	0.9	0.5	0.7	0.9

Table 7   Variation in temperature criteria with *U*-value

Mean thermal transmittance (*U*-value) W/m^2 °C	Temperature values			
	For t_{ai} = 19°C		For t_{res} = 19°C	
	t_{ri}	t_{res}	t_{ri}	t_{ai}
0.4	18.1	18.6	18.5	19.5
0.6	17.6	18.3	18.3	19.7
1.0	16.7	17.9	17.8	20.2
1.6	15.3	17.2	16.9	21.1
2.5 *	13.2	16.1	15.5	22.5
4.0 *	9.7	14.4	12.8	25.5
6.0 *	4.2	11.6	8.0	30.0

t_{ai} = dry bulb air temperature
t_{ri} = mean radiant temperature
t_{res} = resultant temperature
* = difference between air and mean radiant temperature in excess of +8°C or −5°C

published under the aegis of the Public Health Act[30], which has no power to regulate in the interests of energy conservation: the new Health and Safety at Work Act[31], does however contain such power and the publication of appropriate orders will almost certainly introduce added stringency.

It has recently been suggested[32] that the UK should be divided into temperature zones for the purposes of insulation legislation and that, in addition to the imposition of standards that vary from zone to zone, note should be taken of structural mass in addition. *Table 6* shows the proposed criteria which in addition to a maximum area-weighted value also include an alternative datum related to the volume.

Other Aspects

It has previously been shown that, for comfort conditions, the mean radiant temperature of the

surfaces making up an enclosure is significant. This criterion is a function of the level of insulation provided and hence of the mean *U*-value. For an assumed external temperature of 0°C and nominal air movement at 0.1 m/s, *Table 7* illustrates how various temperature scales react one with the other for various levels of insulation. It will be noted that, for mean *U*-values in excess of about 1.6 W/m^2 °C, the differences between air and mean radiant temperature are such that comfort conditions are unlikely to be achieved.

Uncontrolled Heat Gains

Of primary importance during the summer when their presence may often lead to severe discomfort[33], uncontrolled heat gains to an enclosure may be an embarrassment to occupants during the two long mid-seasons and only tolerable during winter. Insufficient account has been taken of such gains in the past and too little attention paid to the possibility of retrieval of at least some part of the available energy for other uses. In the UK, a century or more of cheap fuel and comparatively scarce capital investment has led to an attitude of avoidance of the obvious: the energy problems of 1973 may help to restore perspective and emphasize the necessity for a more rational approach.

Sources

Uncontrolled heat gains may arise from a number of sources, some permanent, some predictable but impermanent, and some entirely adventitious. Examples in these three categories are (1) gains from process equipment (furnaces, machines etc), (2) lighting and metabolic heat from personnel, and (3) solar radiation and infrequent unseasonal outside temperatures.

Process Equipment. Heat gains under this heading are not susceptible to other than general comment. Reference texts quantify the particular[34], but in general, it must be emphasized that the first law of thermodynamics applies and that heat energy applied to furnaces or prime movers is largely dissipated into the enclosure where they are sited via casing radiation, end-product or motive power. Losses via fume dispersal may be considered inevitable but, in energy conservation terms, they may be both inexcusable and uneconomic.

Lighting, Metabolism. Heat gains from lighting, during the current vogue of high service illuminance, may be considerable at up to 30 W/m^2 of floor area served where a particularly high level task illuminance is demanded. Choice of the type of light source and the imposition of a regular system of maintenance and lamp replacement can reduce this heat gain considerably, by 25% in many instances and by 10% in the average case. Technical texts and research bulletins provide both fundamental data and case studies[35],[36] (see also Ch. 7.7).

Metabolic heat output from personnel depends upon the level of activity as shown in *Table 8*, and is divided into sensible and latent heat output[37]. In an office, a department store or a theatre, this component may be among the most important but in an industrial application it may be insignificant.

Solar Radiation. The amount of solar radiation falling upon a building is primarily a function of latitude, season, mean daily sunshine and building orientation: re-radiation from ground surfaces and adjacent structures must also be considered. A variety of publications is available which present these data in tabular or graphical form for worldwide conditions[38]. More important than the incidence of solar radiation on a building, however, is the proportion of the resultant energy which actually enters the occupied space: this depends upon the mass of the various building elements and upon their colour etc. in the sense that light surfaces reflect (re-radiate) whilst dark surfaces absorb.

In the case of solid materials, concrete roofs, masonry walls and the like, an appreciable time lag will occur and a heat input which might otherwise

Table 8 Body activity and heat production (metabolic rate) in an ambient of 20°C d.b. Based on *IHVE Guide*

Level of activity	Heat production of adult male (W)		
	Sensible	Latent	Total
Seated at rest	90	25	115
Light work	100	40	140
Slow walk	110	50	160
Bench work	130	105	235
Medium work	140	125	265
Heavy work	190	250	440

Table 9 Relative efficiency of shading devices in excluding solar radiation (plain single glazing = O)

Shading device	Relative exclusion efficiency (%)
External miniature louvres	83
External canvas blind	82
External louvred sunbreaker	82
External open-weave blind	71
Gold heat-reflecting glass	66
Mid-pane venetian blind	63
Internal linen blind	61
Internal cotton curtain	46
Internal venetian blind	39
Heat-absorbing glass	33
Internal open-weave blind	18
Plain double glazing	16

be a considerable nuisance does not reach the occupied space for some hours after peak solar radiation and thus, perhaps, after working hours. The magnitude of this heat input may be decreased, and the time lag increased further, by insulation of the structure.

Where components of the building envelope have little mass however, the penetration of solar radiation is almost instantaneous: glazing and un-insulated sheet walls or roofs are typical examples. Not all of the incident radiation passes through these materials of course, some being reflected, some being absorbed and some being re-reflected.

For a normal building where the bulk of the structure is opaque and insulated and has reasonable mass, windows and rooflights are the principal routes by which solar radiation reaches the occupied space. In this particular context, the use of clear double-glazing alone does little to reduce the ingress. External shading devices, in the form of structural shields or fins, and external blinds are the most effective barriers[39]. The use of heat-reflecting glass in lieu of plain sheet or venetian blinds mounted between the panes of double windows is effective, but internal blinds of any type are less efficient. *Table 9* shows, in order of merit, the relative effectiveness of a variety of solar exclusion devices.

Effects

Uncontrolled heat input does not necessarily act instantaneously upon the temperature of an enclosure, owing to the delay introduced by the flywheel effect of the structure (mentioned previously). In this sense, however, it is not only the external structure which must be considered but also internal elements, partition walls, intermediate floors and the surface finishes to each. Generally, structural mass is the primary consideration in determining whether a structure is 'heavy' or 'light' but a carpet on a heavy concrete floor or large areas of pin-up board on heavy brick walls in a school classroom can produce what is effectively a 'light' building[40].

Analyses. Theoretical analyses and accompanying field work have shown that the effects of structural influence upon temperature change can be predicted with some exactitude, and computer programs exist which provide quick and convenient results. *Figure 33* shows a plot produced by such a program[41] to illustrate the rise in internal temperature which would occur in a given enclosure due to certain specified levels of occupancy, machine loading, lighting and solar radiation: the Figure illustrates a typical October day but, of course, any other month could be chosen.

Plant Capacity

The installed capacity of a space-heating, ventilation or air-conditioning plant will depend upon a number of variables, among them being the calculated peak load, the thermal characteristics of the building served, the type of system envisaged, and the proposed mode of operation. Misjudgement by the designer of any one of these variables may result in a plant being too large and thus uneconomic in capital cost and energy consumption, or too small and thus inadequate to meet the required duty as normally operated. Four principal considerations have been mentioned and a fifth should have been included with them: it is however a *sine qua non* that the level of maintenance attention

provided to such plants will be meagre in the extreme and the designer must therefore provide for this contingency.

Calculated Peak Load. For a single enclosure, the calculated peak load for a plant will apply to both local and central equipment but, for a large building containing a number of enclosures, the calculated total for the whole will not necessarily be the sum of the local peaks[42]: diversity due to timing and orientation will intervene. Thus, it is appropriate for local plant to have capacity to meet reasonably expected maximum and minimum meteorological impositions.

Building Characteristics. It has already been explained that the mass of a building enclosure will affect the reaction time to both external and internal influences. In the sense of plant capacity, this thermal inertia is taken into account when considering not only maximum and minimum external temperatures but the statistical duration of such conditions as evidenced by meteorological records[43].

System Type. Some types of space-heating system have more ability to deal with occasional situations of extreme external weather conditions than others. For instance, a hot-water plant designed by tradition to operate normally at a mean water temperature of say $75\degree C$ may be boosted in extreme weather conditions to operate at a mean water temperature of say $85\degree C$ without danger. A steam system or one fired directly by gas or electricity has no such inherent overload capacity — unless it is too large for normal operation.

Conversely, some types of space-heating system have the ability to respond more quickly than others to fluctuating demands for capacity to meet normally less extreme variations in external weather conditions or to initial demands on start-up. Warm air and direct-fired gas or electric radiant systems provide quick response whereas hot-water plants have a slower reaction time.

Operational Mode. Systems for space heating and air conditioning may be operated continuously or intermittently over a 24-hour day and/or a 7-day week. There are endless permutations. If operated intermittently, a pre-occupation boost of preheating or precooling may be arranged either for a fixed time period or for a number of hours determined by the season or the external temperature. The mode of operation is pre-empted at the design stage and cannot be changed to suit the whim of an occupier although, of course, it may be tuned and refined to suit specific working conditions.

Combination of Variables. The effect of various combinations of the variables is complex and the variety wide. *Table 10* shows how building mass and plant overload capacity must be considered when choosing winter external design temperature. *Table 11* illustrates that, for alternative rates of

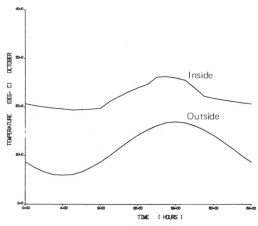

CINO PETROLEUM OFFICES, CLIFTON, HANTS. - JOB NO-BBK41 - 12-9-73
GEOLOGISTS OFFICES 1-5

Inside

Outside

Figure 33 Computer analysis of variation in room temperature with normal diurnal outside variations and internal heat gains. (Note fluctuations with occupancy)

system response, preheat time must be varied for building mass and for the ratio between the plant size and design load[44].

It has been noted in a previous paragraph and illustrated in *Figure 27* that the energy required to operate system auxiliaries often exceeds that required for the source of heating and cooling. Some degree of intermittent use of plant is therefore an advantage and in this context three general rules emerge:

1. Since heavyweight structures cool slowly and do not warm quickly, intermittent use of plant is less effective with these than with lightweight structures.
2. For intermittent use, plant with quick response characteristics is the most suitable.
3. Since intermittent use requires some length of preheat period and some degree of plant excess capacity, in inverse proportion, the relative cost of capital and energy will be the deciding factor.

Methods and Equipment

Space-heating methods, irrespective of the fuel to be used, may be grouped in a variety of ways, including the following:

By location of the energy-consuming apparatus, either actually within the space served or remote therefrom in a plant house.

By heat-distribution medium, whether the fuel source, steam, high-temperature water, low-temperature water, or warm air.

By heating method, either radiant or convective or some combination of the two.

By actual type of terminal heat emitter.

Similarly, ventilation and air-conditioning methods may be grouped into parallel classifications dependent upon plant location, distribution methods and terminal devices. Such groupings are entirely artificial but serve to illustrate the infinite variety of combinations which must be considered when selecting the most suitable plant for a given purpose.

Table 10   Choice of winter external design temperature. Based on *IHVE Guide*

Significant items		Winter design temperature (°C)
Building type	Plant overload capacity	
Heavy	Good	−1
	Poor	−4
Light	Good	−3
	Poor	−5

Table 11   Intermittent heating. Variation in preheating time required with plant type and building mass. Based on *IHVE Guide*

	Preheating time (h)			
	Quick-response plant		Slow-response plant	
Plant size ratio	Heavy building	Light building	Heavy building	Light building
1.2	V. long	7.0	V. long	Long
1.5	6.5	3.0	6.5	4.0
2.0	3.3	0.9	4.4	2.4
2.5	1.8	0.6	3.4	1.9
3.0	1.1	0.4	2.8	1.7

In-situ Energy Consumption

The classic example of raw-energy consumption at the point of use was the domestic coal fire and, in industrial terms, the coal or coke stove. In modern terms, the direct-fired warm-air heater and the high-temperature radiant panel or tube have supervened in the heating sense. In ventilation, the roof ventilator fan operates locally and the room-type air-conditioner provides for summer comfort.

In-situ type energy consumers offer the advantages of relatively low capital cost, high fuel/heat conversion efficiency with no distribution losses, and provide quick response to thermostatic and other control. They have the disadvantage of occupying space and requiring maintenance within the enclosure served; marginal fire risks are incurred at all times and especially in combustible areas.

In terms of capacity to meet peak loads, each item of equipment must be sized for that purpose, both in terms of energy consumption and of heat emission. It follows therefore that each will be oversized for the average load whereas, with a central energy conversion plant, the principal equipment may take account of demand diversity.

Warm-air Heaters. Warm-air heaters of the non-storage type consist of a fuel burner, a combustion chamber, some form of extended heat transfer surface, a secondary fan for air distribution and a flue for dispersal of combustion products: the fuel supply may be liquid or gaseous. A special case exists where electricity is the energy source and thus neither burner nor flue are required, but this refined fuel is decreasingly used for other than the smallest-scale applications in this category. For storage-type heaters, however, where off-peak power is available for misuse at a competitive rate, a warm-air heater would consist of heating elements, a refractory or other energy-storage core, a heavily insulated casing, and a secondary fan for air distribution.

Radiant Heaters. Radiant heaters of the panel type are small in dimension and thus must operate at a comparatively high temperature to be effective: they are normally supplied by gas or electricity. Gas-fired units have application in industrial enclosures of a type where fire risks are minimal owing to the nature of the process, but the units must be fixed well away from occupants to minimize discomfort due to intensity of radiation: flueless-type heaters are not recommended in any circumstances.

Radiant heaters of the tubular type are a comparative innovation. They consist of a number of gas-fired combustion chambers serving long lengths of radiant tube which are fitted with top-mounted polished reflector shields and thus provide a wide area of coverage at black heat. Combustion products are exhausted to atmosphere through a suction pump or fan and the whole apparatus may be controlled thermostatically or by other means[45].

Energy Distribution

For *in-situ* consumption of raw energy, it follows that there must be an entry point for fuel and site storage as appropriate. In electrical terms, for reasons of economics, the supply would be at high potential and thence, via transformers, to a distribution system at normal voltage. With gas, supply to a large complex might require pressure-boosting equipment in addition to a meter house and, with oil, site storage to meet a peak winter demand lasting 14 days without replenishment would be desirable. Whilst solid fuel undoubtedly requires the maximum of site storage space, that necessary for the receipt and processing of other fuels is by no means negligible. In broad terms, the more refined the fuel supply, the smaller but the more sophisticated and more costly is the primary site-storage/processing unit.

With the *in-situ* energy consumption for space heating etc. referred to in previous paragraphs, distribution would be by cable, by gas main or by pumped oil-circulation from the site energy-receipt centre. In other circumstances, however, central generation of a secondary energy medium would be required.

Steam. For space heating purposes, unless the generation plant exists or is required to serve process or other needs, steam is an unsatisfactory medium. It has a high maintenance demand in terms of equipment servicing (which is seldom, if ever, provided) and it suffers from inherent disadvantages in terms of flexibility, temperature control and unseasonable consistency of heat loss. The wise industrialist equipping at the present time therefore, unless faced with an incontestible demand for process-steam use (and they are few) or a surplus of exhaust or geothermal steam, will consider other space-heating media.

High-temperature Water. Capable as it is of flexibility for temperature control, of modulation for flow quantity, of high heat transfer potential and of ease of generation at a central plant, high- ($>$ 120°C) or medium- (100–120°C) temperature hot water is a well understood medium for energy distribution. It is particularly suitable for service to large enclosures and widely dispersed sites.

Low-temperature Water. For local distribution reticulations, low- ($<$ 100°C) temperature hot water is the most suitable medium. The whole system operates at atmospheric pressure, the maximum temperature of the medium is only marginally dangerous, and plant so operated is well within the competence of the low grade of skill commonly available for maintenance. Further, with respect to heat transfer to local warm-air heaters, low-pressure hot water provides an adequate differential and is able to convey comparatively large energy requirements in pipes of quite small diameter.

Warm Air. Heat distribution from a central source by warm air suffers from the disadvantage that, for equal heat-carrying capacity, air ducts are immensely large in comparison with water pipes, in the ratio of up to 100 to 1 in area. Ideally therefore, air handling plant should be sited as closely as practicable to the enclosure served and energy in the form of heating or cooling conveyed to the plant via water in pipes.

Heating Method

In an earlier paragraph, mention was made of the fact that the purpose of a space-heating design was generally to provide comfort for the occupants and not heat to the building. In this respect there are many cogent arguments for the provision of heat supply by radiant rather than convective means. As has been demonstrated, the comfort equation requires that the mean radiant temperature of an enclosure be as important to comfort as is the air temperature.

Radiant Heating. Since radiation is emitted from a source which is warm relative to the immediately surrounding enclosure, it follows that a temperature difference exists. Hence, convection currents will arise and thus a theoretically radiant source becomes a provider of both radiation and convection in some proportion.

In practical terms, radiant heating is provided by those means which permit heat exchange by at least 60% radiation and no more than 40% convection. Examples are heated floors or walls and medium- or high-temperature radiant panels, tubes or strips mounted at high level in the enclosure.

Convective Heating. As a corollary to the case of radiant heating, convective output from a source varies from the total to the adventitious. About 70% of the output of the common hot-water 'radiator', for instance, is by convection[46]. A true convector however will be an appliance insulated to prevent any radiation or one which provides the majority of output by forced (fan-assisted) air movement.

Comparisons. To understand the effect upon comfort of radiant and convective heat sources requires considerable study, beyond the scope of the present text. It must suffice to say that, for an equivalent environmental temperature, considerably more energy input is required of a convective system than of a radiant system. Reference should be made to the appropriate technical reference manuals[47].

Terminal Apparatus

Terminal heating or cooling apparatus may best be studied by reference to technical manuals, textbooks or trade literature. The following notes serve only to illustrate the wide variety of equipment that is available.

Static emitters. Relying upon heat output by radiation or by natural convection, the extreme case of the static emitter is the warmed floor, ceiling or wall. Such an effect may be produced by embedded hot-water pipes or heating cables or by the use of wallpaper treated to act as a low-voltage induction heater. Such an approach provides aesthetic advantages but the response to control is generally slow.

Hot-water radiators, commonly now made of pressed steel rather than cast iron, are much used, as are natural convectors which take the form of finned tubes mounted within a sheet metal or timber casing. These items are best suited to use within small rooms and may be fitted with valves for hand or automatic control[48].

In large industrial enclosures, radiant panels or strips are often used. The principle in each case is that metal plates with insulated backs are fixed to hot-water pipes to provide extended surface. In the case of panels, the plate will be about 2.5 m by 1.5 m and the pipe a sinuous coil whereas, in the case of strips, the plate will be about 300 mm wide by any length required and the pipe a straight run. Such emitters may be mounted on walls or overhead and are most useful in very large buildings when high-temperature water is supplied to them.

Radiant panels may be used in cooling applications with a chilled-water circulation, but much design care is needed if condensation on the cold surfaces is to be avoided[49]

Fan-powered Emitters. One advantage of applying fan power to an emitter is that, size for size, a much greater output results from forced rather than natural convection. A small fan heater or cooler is, in basic principle, little more than a natural convector provided with a powered air supply.

In an industrial enclosure, large floor-mounted blast heaters may be used, similar in many ways to the direct-fired units previously described but incorporating a steam or hot-water battery in place of the burner and combustion chamber etc. Similarly, roof- or wall-mounted units may be used, but these are commonly smaller in size and arranged to discharge either horizontally or vertically downwards. Care must be taken in the application of any such units to avoid too high a discharge-air temperature since this will lead inevitably to an excessive temperature gradient, floor to ceiling or roof, within the enclosure and thence to unnecessary heat build-up and loss through the structure[50].

Automatic control of fan-powered heaters, at the simplest, may be merely an on-off thermostatic switch to the fan motor. Sophistication may lead to fan-speed control and, within limits, to control of the mean water temperature at the heat exchanger.

Fan-powered heaters may be connected through sheet-metal ductwork to outside air and thus act as simple unit ventilation plants. Dampers for adjustment and choice of either outside air or recirculation of room air are often provided, as is some simple form of air filter.

Where cooling is required, units similar to the heaters here described may be used, chilled water being supplied in lieu of the heating medium. Design requirements are however more stringent in such cases.

Energy Economy

In the case of new buildings, it is possible to incorporate the concept of energy conservation into the structural and system designs without disproportionate additional cost. An existing building and an existing plant however may require major modification in order to reduce heat or power input. Normally, such modifications will be self-financing in the long term but this cannot always be predicted. It is important to consider therefore whether the aim is to conserve energy for its own sake or as a fiscal transaction.

Conservation

Consumption of energy may be reduced by applying well-known insulation and heat-recovery techniques, changing the mode of plant operation, improving the level of maintenance or by reducing the standards of provision for human comfort.

Insulation and Heat Recovery. In the sense that insulation prevents energy escape and thus allows re-use, it could be considered the most simple and effective method of recovery. In most buildings however a state of imbalance exists where some areas have an excess and others a deficit of heat or cooling; further, ventilation air must be exhausted either from processes or merely to provide a comfortable freshness.

Heat-recovery techniques may be used therefore to re-cycle energy from areas of surplus to areas of shortage and to transfer energy from exhaust air to that necessarily introduced as replacement. The so-called thermal wheel, as illustrated in *Figure 34*, is a good example of available equipment but other more complex systems exist[51].

Mode of Plant Operation. Provided that a space-heating or air-conditioning plant is suitably designed, significant operational savings may result from intermittent operation. The sophistication of automatic controls increases continually, and devices now exist which enable intermittent plant operation to be monitored by an anticipatory weather-sensitive device; there is evidence that the capital cost of this can be recovered in two years from fuel savings.

Maintenance. The standard of maintenance provided for building installations is execrable world-wide and is nowhere much worse than in the UK. Lack of facilities for operative training and indifference on the part of management are contributory factors. For effective results, it is necessary that between 5 and 10% of the current plant replacement cost be allocated each year to maintenance, the tolerance being a function of the quality of the installation.

Unusual Heat Sources

When energy supplies are short, it is usual to consider unusual sources and the present time is no exception. There is no doubt but that non-critical energy can be used to advantage, but capital costs may be high.

Solar Energy. A potential for the use of solar energy in the UK does exist, but whether this is worth commercial exploitation remains in doubt. One development[52] suggests that, in good weather conditions, 50 litre of hot water at $50°C$ may be produced each day per square metre of absorber area: such a prediction should not be ignored.

Heat Pumps. The heat pump, which in simple terms may be thought of as a thermal transformer stepping up low-grade (i.e. low-temperature) heat

Table 12 Annual availability of free cooling. From *Integrated Environment in Building Design*, Ed. A. F. C. Sherratt, Applied Science Publishers 1974, by permission Construction Industry Conference Centre Limited

Month	Hours below 13°C		
	Day	Night	Total
January	310	430	740
February	280	390	670
March	270	440	710
April	220	370	590
May	170	280	450
June	80	150	230
July	30	70	100
August	30	70	100
September	80	150	230
October	170	340	510
November	250	420	670
December	270	460	730

to a useful level, is a well proven device which can have a theoretical 'efficiency' of as much as 500% and will often produce about half that advantage in practice. The equipment consists of a conventional refrigeration machine in reverse, which provides for use the condenser heat normally discarded and rejects the evaporator cooling normally used. Small units are available in production, these usually being basic air coolers which are reversed in cycle for winter use: larger plants have been built but most have been of an experimental nature[53-55] The principal difficulties lie in the capital cost of reaching and using the base heat source; handling very large air quantities; filtering river water; and laying immense ground coils or pumping sewage for collecting the low-grade heat.

Free Cooling. It is not always necessary to use mechanical refrigeration plant consuming energy to produce a cooling effect. In the maritime climate of the UK, there are many hours of the year when the outside temperature is such that air may be used directly through a conditioning plant for cooling purposes[7]. *Table 12* lists, month by month for Kew, the hours when the outside air is below $13°C$. Air-conditioning plants designed, admittedly at some extra cost, to take advantage of this situation will use less annual energy than those arranged for a more conventional operational mode or for the heat-recovery cycle over-advertised to promote electricity sales.

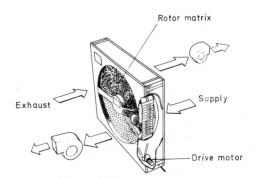

Figure 34 Thermal heat-recovery wheel consisting of rotating heat-transfer cage through which extracted and incoming air pass alternately. (Note simplicity of the device)

Conclusion
The purpose of providing space-heating and air-conditioning plant is to afford comfort to the occupants of a building and not to service the building

as such. This aspect should never be overlooked and, in consequence, unused areas should not have energy wasted upon them.

New buildings and plant can be so designed as to match each other and produce an end-product which conserves energy as a whole. Existing buildings and plant may be poorly matched and require much expenditure to tune. In many cases it may be more economic to adjust the building to suit the plant rather than the reverse. A bad building cannot provide a comfortable working environment and will inevitably lead to wastage of energy.

References

1   Ref.(A), pp 20-4 to 20-58
2   Wadsworth, P. *J. Textile Inst.* 1960, Vol. 51, No.9, 558, Fig.5
3   *Shirley Inst. Bull.* 1956, Vol. 29, No.3, 127, Fig.1
4   *The Efficient Use of Fuel*, 2nd edn, HMSO, London, 1958, Ch.25
5   Harbert, F. C. (SIRA Institute) *Ind. Process Heating* 1973 (June), pp 6—7
6   Milbank, N. O., Dowdall, J. P. and Slater, A. *J. IHVE* 1971, Vol. 39, 145
7   Martin, P. L. 'Air Conditioning—arguments for and against', *RIBA/IHVE Conf.*, Univ. Loughborough, 1974
8   Gordon, Alex. 'Future Office Design, Energy Implications', *IHVE/PSA Conf.*, London, 1974
9   *(UK) Factories Act*, 1961, Sections 3 and 4
10 *(UK) Offices, Shops and Railway Premises Act*, 1963, Sections 6 and 7
11 Ref.(B), Table A9.1
12 Ref.(B), p A1.6
13 *DIN 1946*. Ventilation Plant (VDI Ventilation Rules), 1960
14 *Bulletin No. 19*, Dept of Employment(UK), 'Ventilation of Buildings'
15 Ref.(B), Table A9.24
16 *(UK) Factories Act*, 1961, Section 68
17 Houghton, F. C. and Yaglou, C. P. *Trans. ASHVE* 1923, Vol. 29, 361
18 Missenard, A. *Chauffage et Ventilation* 1935, Vol. 12, 347
19 Danter, E. *J. IHVE* 1974, Vol. 41, 231
20 Faber, O. and Kell, J. R. *Heating and Air Conditioning of Buildings*, 5th edn, Architectural Press, 1973
21 Billington, N. S. *HVRA Newsletter* 1974, Vol. 1, No. 4
22 Ref.(B), Table A3.23
23 Loudon, A. G. 'Summertime Temperatures in Buildings', *IHVE/BRS Symposium* 1968 (BRS Current Paper 47/68)
24 Loudon, A. G. *J. IHVE* 1968, Vol. 36, 167
25 Tamura, G. T. and Wilson A. G. *Trans. ASHRAE* 1966, Vol. 72, 180
26 (UK) Thermal Insulation (Industrial Buildings) Act, 1957
27 (UK) Building Regulations, 1965
28 Building Standards (Scotland) (Consolidation) Regulations, 1971
29 Cibula, E. 'International Comparison of Building Regulations — Thermal Insulation', BRS Current Paper 33/70
30 *(UK) Public Health Act*, 1961
31 *(UK) Health and Safety at Work Act*, 1974
32 Billington, N. S. *J. IHVE* 1974, Vol. 42, 63
33 Wise, A. F. E. 'Designing Offices against Traffic Noise', 1973, BRS Current Paper 6/73
34 Ref.(B), Section A7
35 *IES Code for Interior Lighting*, 1973
36 *IES Tech. Rep. No. 9*, 1967
37 Ref. (B), Table A1.1, Figs. A1.1 and A1.2
38 Petherbridge, P. 'Sunpath Diagrams and Overlays for Solar Heat Gain', 1965, BRS Current Paper 39/65
39 Petherbridge, P. 'Transmission Characteristics of Window Glasses and Sun Controls', BRS Res. Paper, 1972
40 Down, P. G. *Heating and Cooling Load Calculations*, Pergamon Press, 1969
41 Down, P. G. and Curtis, D. M. *RIBA/IHVE Conf.*, Univ. Loughborough, 1974
42 Ref.(B), Table A9.5
43 Jamieson, H. C. *J. IHVE* 1955, Vol. 22, 465
44 Billington, N. S., Colthorpe, K. J. and Shorter, D. N. 'Intermittent Heating', *HVRA Lab. Rep. 26*, 1964
45 Taylor, F. M. H. 'Nor-Ray-Vac Gas Heaters', 1970, *Tech. Manual NRV 14*
46 Peach, J., *J. IHVE* 1971, Vol. 39, 239
47 Ref.(C), Tables B1.1 and 2
48 Fitzgerald, D. and Redmond, J. *HVRA Lab. Rep. 52*, 1969
49 Jamieson H. C. *J. IHVE* 1959, Vol. 27, 245
50 Ref.(C), Table B1.3
51 Hardy, A. C. and Mitchell, H. G. *J. IHVE* 1970, Vol. 38, 71
52 *Energy Report* (Microinfo Ltd), 1974, Vol.1, No.7, 'Sunstor solar water heaters'
53 Sumner, J. A. *J. Instn Mech. Engrs* 1947, Vol. 156, 21
54 Montagnon, P. E. and Ruckley, A. L., *J. Inst. Fuel* 1954, Vol. 27, 170
55 Kell, J. R. and Martin, P. L. *J. IHVE* 1963, Vol. 30, 353

General Reading

(A) Perry, R. H. and Chilton, C. H. (Editors), *Chemical Engineers' Handbook*, 5th edn, McGraw-Hill, New York, 1973
(B) *IHVE Guide*, Book A, Instn of Heating and Ventilating Engrs, London, 1971
(C) *IHVE Guide*, Book B, 1971
(D) Billington, N. S. *Thermal Properties of Buildings*, Cleaver-Hume Press, 1952
(E) Faber, O. and Kell, J. R. *Heating and Air Conditioning of Buildings*, 5th edn, Architectural Press, 1973
(F) Down, P. G. *Heating and Cooling Load Calculations*, Pergamon Press, 1969

Chapter 10

Boiler plant and auxiliaries

10.1 Sectional boilers

10.2 Firetube boilers

Nomenclature
Fuel usage in firetube boilers
Firetube boiler design
Optimum thermal efficiency
Design limitations

10.3 Watertube boilers

Power-station boiler types
Industrial field
Superheaters and reheaters
Economizers

10.4 Boiler auxiliary plant

Boiler mountings
Boiler feed pumps
Fans
Air heaters

10.5 Maintenance; boiler gas-side deposits

Need for removal of deposits
Methods of removal
Effectiveness of cleaning

10.6 Standards and operational problems

Standards
Operational problems
A brief note on boiler efficiency

10.7 Use of steam

General properties
Distribution, storage and process requirements of steam
Power from steam

Boiler ratings are commonly expressed as evaporation (steam) in t/h from and at (F and A) 100°C. Alternatively the rate of heat output in the steam is given as kW or MW.

10.1 Sectional boilers

The advantages of sectional boilers lie in the relatively small component part involved. Since the life of a central-heating boiler is almost always less than that of the building which it serves, the question of boiler replacement must be considered at an early stage in the design of a building and its services. Sectional boilers allow access for replacements to be had by doors and passages of normal dimensions. This gives greater flexibility in boilerhouse positioning.

Most sectional boilers can be extended, so that if the load which the boiler serves is increased, further sections can be added and the output is also increased. A further advantage is that if a section fails for any reason it may be replaced at reduced cost and inconvenience compared with complete boiler replacement.

Sectional boilers are usually constructed of cast iron and are exclusively used for central heating and hot-water supply. The sections are joined by using hollow nipples which provide the water interconnection. These nipples may be screwed left- and right-hand so that they hold the sections together themselves. Alternatively, they are conical or spherical fitting into tapered holes, the sections then being secured in position by bolts and nuts or tie rods. In the case of screwed nipples the water joint is made on the section faces and a joint ring or gasket is used between those faces. When barrel nipples are used, the water joint is made between the outer wall of the nipple and the tapered bore of the hole. Each intersection joint may have two or three nipples and, in some designs, each section is made in two halves so that there are four nipples for each joint.

The most common arrangement of sections is horizontal, extending from front to back. In this arrangement the combustion-chamber and flue travel may also be horizontal, or the flue travel may be vertical. When the design incorporates horizontal travel, the heating surfaces and flueways are included within the profile of the section. Where vertical travel is used, the gases pass upwards between sections.

In some smaller designs, the sections are arranged vertically, when the gas flow is also vertical and the flueways are within the profile of the section.

The most usual layout used in sectional boilers is that of a 3-pass boiler, where the combustion chamber (comparable with the furnace tube in a shell boiler) forms the first pass, the second pass (or first convection zone) transfers the gases from the rear of the boiler to the front, and the third pass (or second convection zone) transfers the gases from the front of the boiler to the rear again, whence they are discharged to the chimney (*Figure 1*). Usually the combustion chamber and flueways are all contained within a common section, but in some designs the combustion chamber is separated from the convection zone. In these cases, different sections are used for each purpose.

The combustion chamber may either be open at the base, or provided with a waterway bottom. If it is open, then all forms of combustion equipment may be applied. Waterway-base boilers are more usually designed for use with liquid and gaseous fuels only. Where waterways are not present below the combustion chamber, attention must be given to the hearth design, not only to limit the heat losses but also to avoid damaging the boiler-house floor. This is particularly important where water-proofed structures are involved.

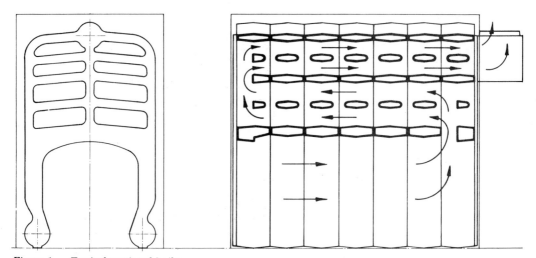

Figure 1   Typical sectional boiler

References p 247

The requirements for the design and manufacture of cast-iron sectional boilers are contained in a British Standard Specification[1]. This used to limit this type of boiler to a maximum working pressure of 410 kPa (4.1 bar), a maximum operating temperature of 116°C, and steam boilers to a maximum working pressure of 100 kPa (1.0 bar). However, this has now been revised so that these arbitrary pressure and temperature levels are avoided and maximum permissible pressures are related to the ultimate hydraulic strength of the boiler, incorporating an appropriate factor of safety. The maximum permissible operating temperatures are then related to the maximum pressures so obtained by reference to the equivalent steam temperature, allowing a suitable anti-flash margin.

Previously, sectional boilers were rated at a standard output proportional to the total area of heating surface contained (13.8 kW/m^2). The current method of rating is based on a measured test output, carried out under standard conditions, and declared at a minimum level of efficiency. This method results in specific ratings ranging from 13 kW/m^2 for small all-fuel natural-draught boilers to 40 kW/m^2 in large pressurized combustion boilers.

Sectional boilers intended for use with solid, liquid or gaseous fuels are usually designed to operate with a negative pressure within the combustion chamber and flueways. The resistance to gas flow, and the entry of some or all of the combustion air, is overcome either by the draught created by the chimney or by an induced-draught fan. With mechanical firing equipment, a forced-draught fan is very often provided but it is still usual for the system to be so arranged that the pressure within the combustion chamber and flueways is negative.

With the development of oil and gas burners capable of operating satisfactorily against a back pressure, the advantages of operating a boiler with positive pressure conditions in the combustion chamber have become apparent. Because of the higher heat-release rates obtainable, the combustion-chamber volumes can be reduced, and the higher heat transfer rates possible enable the primary heating-surface area also to be reduced (*Figure 2*).

The usual method of imposing the back-pressure condition is to increase the resistance of the flueways to flue-gas travel. This is achieved by increasing the flue-gas velocity, either by reduction in cross-sectional area or by the introduction of baffles. The increase in velocity also increases the secondary transfer rate, thus enabling the necessary secondary heating-surface area also to be reduced. By this means the total size of a boiler of given output is reduced. By designing sections which completely enclose the combustion chamber, and by developing sealing systems which can be satisfactorily applied to the inter—section joints, the principle of over-pressure combustion has been applied with advantage to sectional boilers. This type of sectional boiler is usually suitable for firing by oil and by gas through forced-draught burners. The characteristics, especially the size differences, of a typical natural-draught boiler and a typical pressurized-combustion boiler are contrasted in *Table 1*.

When boilers operate with pressurized combustion chambers the chimney is not required to develop draught to overcome boiler resistance, but it is usual to design the chimney to overcome its own resistance, so that the pressure at the flue outlet from the boiler is approximately equal to that of the atmosphere.

Sectional boilers intended solely for use with gaseous fuels, burned through atmospheric burners, are usually based on a vertical gas flow, although the sectional arrangement is commonly horizontal.

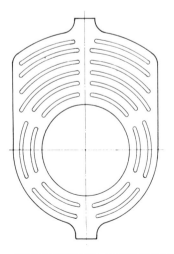
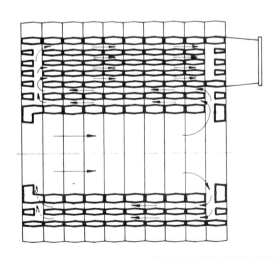

Figure 2   Pressurized-combustion sectional boiler

Table 1   Boiler comparison. Courtesy Instn of Heating & Ventilating Engrs (J. M. Riley, *JIHVE*, Oct. 1970)

	Natural-draught boiler	Pressurized-combustion boiler
Rating (kW)	1471	1466
Overall dimensions		
Length (excl. smokehood and burner) (mm)	5055	1854
Height (mm)	2336	1778
Width (mm)	1955	1346
Weight (boiler only) (kg)	17480	6096
Water content (litre)	3633	600
Total heating-surface area (m^2)	106	36
Mean heat transfer rate (kW/m^2)	13.88	40.38
Combustion chamber volume (m^3)	7.845*	0.595
Mean heat release rate (kW/m^3)	250†	3081‡
Output/floor area (kW/m^2)	148.8	587.0
Output/overall volume occupied (kW/m^3)	63.6	330.0
Output/weight (W/kg)	84.0	240.0
Specific electrical power requirements§ (W/kW)	2.5	6.4
Combustion chamber pressure (Pa gauge)	−13	+1245
(mbar)	(−0.13)	(+12.45)
Chimney draught required (Pa gauge)	50	NIL
(mbar)	(0.50)	NIL

* An allowance of 5% has been made for refractories
† At 75% efficiency
‡ At 80% efficiency
§ Burning oil not requiring preheat

In some designs, the sections are arranged from side to side, rather than from front to rear. Sectional boilers of this type are usually designed to operate with a down-draught diverter placed at the outlet of the boiler. This is to protect the combustion chamber from the effects of down-draught in the main chimney and also from excessive up-draught. The boiler has to be able to operate independently of any draught created by the chimney, that is, the resistance of the boiler to the passage of flue gases through it must be overcome by the buoyancy of those gases within the boiler itself (*Figure 3*).

Since iron sectional boilers are made up of a number of separate waterway elements, the water flow through each section is of utmost importance to ensure long and satisfactory service. The general principle employed to make sure that the water flow through each section is constant is to design each boiler section to have the same resistance to water flow. If, then, the external connections are arranged diagonally across the complete boiler, parallel and equal flow through each section is achieved.

Exceptions to this system are found in some vertically arranged sectional boilers, in which series flow achieves equal distribution, and some horizontal arrangements where internal distribution pipes or manifolds are provided.

In sectional boilers operating without forced circulation, the external system circulation has very little effect upon the internal velocity of the water. Within the sections a general thermo-syphonic circulation is set up and the sections are designed to encourage this. In addition, there is a system of small sub-syphons, forming a small, rapid pattern adjacent to the heat-exchange surfaces. In this type of internal circulatory system the heat transfer to the water will take place at a constant rate, irrespective of the external circulatory velocity.

Where the heat transfer rates exceed those normally associated with natural circulation, forced circulation can be incorporated by including a circulating pump in the boiler circuit.

It is important in all types of boiler to maintain efficiency levels by minimizing the random infiltration of air. With sectional boilers, because of

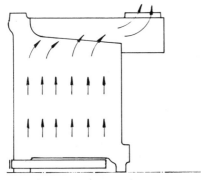

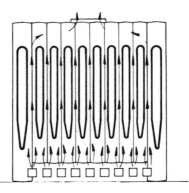

Figure 3   Atmospheric gas-fired boiler

References p 247

their form of construction, this is particularly important and care should be taken to ensure that all joints between sections and the boiler-house floor or plinth are adequately sealed against air inleakage.

Because a very large proportion of sectional boilers is used in hot-water central-heating systems, steps must be taken to minimize the incidence of gas-side corrosion (see Ch.20.1, 2). Corrosion of the heating surfaces of sectional boilers is manifest in three general ways. First, wastage of the metal causes premature failure. Secondly, the deposition of the hard, closely bonded insulating layer of corrosion products on the heating surfaces considerably reduces the heat transfer through those surfaces. Thirdly, the setting up of crystalline growth between sections may produce stress patterns damaging to the structure of the boiler.

10.2 Firetube boilers

A boiler in which all or most of the heat is transferred from hot gases flowing inside tubes to water surrounding them is classified as a firetube boiler. The term shell boiler is also in popular usage to describe this type of plant, this latter name being derived from the fact that water and steam are contained within a pressure vessel or shell which also contains the heating tubes.

In general, firetube boilers are ruggedly and simply constructed and, with modern designs, offer high efficiencies at a capital cost generally lower than that of an equivalent capacity watertube boiler.

Nomenclature

Owing to the confusion that often exists when describing the different regions of firetube boiler plant, it is useful to define the terminology in common use.

Shell.   The pressure-tight vessel containing the water and steam together with the heating tubes.

Firetube, Furnace tube, or Flue. The large-diameter tube or tubes contained within the shell into which the primary heat source (solid-fuel grate, oil burner, gas burner etc.) fires. Heat transfer within this tube is both by radiation and convection.

Combustion chamber.   The chamber at the rear end of the firetube into which the hot gases pass. The gases are usually reversed in the combustion chamber for further traverses of the boiler. This name is somewhat anomalous as combustion should have occurred before this region.

Side flue, Bottom flue. The convective heat-transfer sections of such boilers as Cornish and Lancashire. The flueways are large to facilitate access to the rear end of the boiler, and are contained within the brickwork setting at the sides and bottom of the shell.

Smoke tubes, Fire tubes. A large number of small-diameter tubes (usually 50–75 mm) which provide the convective-heat-transfer section of Economic and package-type boilers. Unlike the side and bottom flues of the Lancashire boiler, the smoke-tubes are contained within the shell.

Transfer box or Reversal chamber. A chamber situated at either end of the boiler for reversing the hot gases for subsequent passes through the boiler shell.

Smoke box. The chamber which collects the gases leaving the final pass through the boiler before discharge to the chimney or stack.

Superheater tubes.   A tube bank occasionally installed in the combustion chamber through which the saturated steam from the shell is passed before finally leaving the boiler. This further heating of the steam which is no longer in contact with water, allows its temperature to be raised above the saturated-steam temperature at the boiler operating pressure. The extra heating surface leads to an increase in overall thermal efficiency.

Economizer. An extension of the total boiler heat-transfer surface area, often situated between the smoke box and the stack. Its purpose is to preheat the boiler feedwater before entry to the boiler; its use increases the overall thermal efficiency of the plant.

Air heater. An extension of the total boiler heat-transfer surface area through which the air for combustion is preheated before mixing with the fuel. If used it is installed between the smoke box and stack, and has the effect of increasing overall thermal efficiency.

Stack, Chimney or Flue. The means by which the combustion products, after leaving the boiler, are discharged to the atmosphere.

Fuel Usage in Firetube Boilers

Until the 1950s, coal was for economic reasons the primary boiler fuel available in the UK, and with the subsequent history this country is a convenient basis for illustrating the technical alternatives available in many industrialized and developing countries.

At that time oil began to compete with coal. This change-over was originally initiated by the relative economics of the two fuels but was undoubtedly given further impetus by the introduction of the packaged compact boiler, which offered high efficiency, low price, and fully automatic operation and which occupied only about half the space of an equivalent coal-fired boiler. Coal-burning boilers were of necessity large as their furnaces were required to accommodate grates, and coal- and ash-handling gear was also large and expensive. As a result, for comparable output the cost of a coal-fired shell boiler was considerably greater than that of its oil- or gas-fired counterpart.

These facts, combined with the competitive price of fuel oil at that time, ensured that most industrial firetube boilers even of the larger capacities were of the packaged compact variety, designed specifically for oil firing. This remained so until the early 1970s, when natural gas became available in sufficient quantities to be considered as an industrial boiler fuel, and fuel oils became both uncertain in supply and expensive. As a result industrial consumers and boilermakers have considered gas-firing such plant, and this has led to some fundamental changes in burner/boiler design, especially in features relating to the differences in flame emissivity and heat-transfer properties in the combustion of the two fuels. Most modern boilers are still of the compact package type, and can usually be dual-fuel fired with oil or gas; in some cases change-over between the two fuels can be achieved without interruption to the steam supply. This facility for on-line change-over has popularized dual-fuel firing of boilers, because the customer may then use whichever fuel is economically advantageous.

At the present time the high cost of fuel oil and the limited availability of natural gas for this market have led to considerable renewed interest in solid-fuel firing. With grate-firing techniques, the compact package boilers which have dominated the new boiler market for the last decade cannot be coal fired without prohibitive downrating and considerable technical difficulties. However, a step in the direction of increasing the flexibility of fuel usage has been the introduction of the multifuel boiler which has provision for coal as well as gas and oil firing, the basic design being based upon a coal-fired package three-pass configuration.

As coal firing has been extensively dealt with in previous editions of *The Efficient Use of Fuel* and other publications (also in matters of detail in Chapters 5 and 6 of this volume), it is considered unnecessary to dwell unduly on this particular aspect. Emphasis in this Section is therefore placed upon firing with liquid and gaseous fuels. In respect of firing detail these have appeared in Chapters 4 and 3 respectively.

With conventional coal and stoker firing, maximum furnace heat-release rates are normally limtted to below 1 MW/m^3 based on total furnace tube volume. Exceeding this rate can give rise to smoke emission, and to excessive temperature of the combustion products entering the firetubes leading to fusion in the tube ends of ash constituents and thus to a need for frequent boiler cleaning. Moreover the fuel/air mixture within and above a coal grate tends to be very stratified, so that subsequent burning out of the flame requires considerable furnace length and volume unless the design incorporates devices for promoting turbulence.

The Vekos Powermaster, shown in Figure 6 of Chapter 6 under stokers, is a package unit of 3-pass design, having either one or two furnace tubes, depending upon rating. Originally of dry-back construction, it is now available with either a dry- or wet-back design. With this boiler the fuel-firing system is an integral part of the boiler and boiler design. A coal-feed device is fitted to which fuel may be delivered either mechanically by means of a screw conveyor from an overhead fuel bunker, or pneumatically from a coal storage area located at or below ground level. The fuel can be delivered from any direction, giving complete flexibility of boilerhouse design. With the fixed-grate system, the unit is de-ashed manually. An automatic grate consisting of two sets of grate bars, one above the other, forming a continuous support for the firebed is now available as standard on the single-flue models. The upper grate bars can be given a sideways movement by means of an electrically driven fly-wheel and clutch mechanism to bring the spacings in the upper and lower grates into line, so that the ash falls through the grate on to a screw conveyor placed beneath which carries the ash to the front of the boiler for discharge through a rotary valve. De-ashing is needed for only a few minutes per 8—10 h. Any unburnt carbon, grit and dust are collected and returned for refiring. With the Vekos pressurized combustion system, burning rates of 343 kg/m^2 h and furnace heat-release rates of about 1.5 MW/m^3 are achieved. The units have a turn-down ratio of about 3:1.

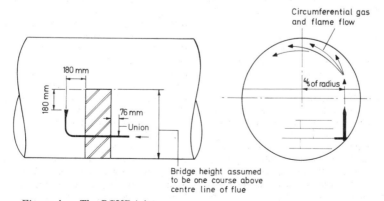

Figure 4   The BCURA jet

A tendency to high combustion-chamber temperatures sometimes experienced can be overcome by means of the BCURA jet. This jet, a rectangular orifice 30.2 × 2.4 mm formed from flattened pipe, is mounted in an off-set position behind the bridge wall, long side parallel to the longitudinal axis of the furnace tube. The arrangement of the jet is illustrated in *Figure 4*. It can be used with either steam or compressed air at pressures of 35—105 kPa (gauge), and increases turbulence by imparting a swirling motion. A greater heat-release rate and complete combustion within the furnace tube are achieved, which causes reduction in combustion-chamber temperatures of not less than 110°C, these temperatures being maintained below the critical 925—955°C level. Correct positioning and maintenance of the jet are however essential for safe operation and optimum efficiency. The jet also prevents grit accumulating on the floor of the furnace tube behind the bridge wall, thus freeing this heat-transfer surface.

Fluidized combustion (Ch.5), especially in its pressurized form, should permit coal firing rates for *watertube* boilers (Ch.10.3) up to 5—8 MW/m³ to be obtained.

Atomized fuel oil and air are more readily mixed by their mode of introduction, and ash substances in the products of combustion, though by no means negligible, are present in much smaller amounts than with coal. Heat-release rates in the range 2—3 MW/m³ are therefore common with oil firing. An even higher rate is possible, but difficulties may then be encountered with stack-solids emission; the appearance of about 0.4% or more of the fuel fired as stack solids is not permitted by the Clean Air Regulations 1971 for the class of boiler under consideration. Clearly the size of furnace needed for oil firing is considerably less than for the combustion of an equivalent amount of coal. However, it must be borne in mind that sufficient heat-absorption area must be provided to cool the exit gases to an acceptable temperature before they enter the combustion chamber and subsequently pass into the smoke tubes. This prevents the danger of ash fusion and of structural troubles in the tube plate and tube ends, such as distortion, tube-seat leakage and ligament cracking.

Gaseous fuels are even more readily mixed with air and burned than fuel oil, and contain only trace impurities. As a result, furnace heat-release rates can be in excess of 3 MW/m³ (300 000 Btu/ft³ h) since there is no possibility of ash fusion. However, again sufficient heat absorption must occur in the furnace to cool the gases to a temperature not detrimental to the rear tube plate. Also the comparatively low emissivity of the gas flame when compared with a fuel-oil flame means that convective heat transfer must play a greater part within the confines of the furnace tube in order to offset the lower radiative heat-transfer rate. There are two methods by which this may be achieved. First, the burner can be designed so that the combustion products leaving the burner nozzle are discharged with a swirl motion down the furnace tube. This ensures that the hot gases scrub the heat-transfer surfaces with increased velocity, and as a result heat is transferred faster; but care must be taken in order to prevent the flame itself impinging directly onto the furnace tube, as local overheating would result. Secondly, the combustion-product flow along the tube can be disrupted by the addition of a refractory target ring, wall or block. The effect of the obstruction is to increase turbulence within the furnace and hence improve the heat-transfer rate. In addition, the refractory inserts themselves are heated to red heat and radiate to the cooler heat-transfer surfaces. This in part alleviates potential problems caused by the low emissivity of natural gas flames, but it has the disadvantage of increasing the resistance to gas flow through the boiler so that greater mechanical power is required to overcome the increased pressure drop.

The increased furnace heat-release rates characteristic of modern compact package boilers have meant increased thermal stressing of the furnace, leading to extensive use of corrugated furnace tubes. The effect of these corrugations is threefold. First, they allow thermal expansions and contractions to be taken up without undue stress; secondly, they provide a greater heat-transfer area per unit volume and hence permit a lower heat flux; and finally there is increased turbulence at the boundary layer.

Firetube Boiler Design

Cornish and Lancashire Boilers

The single-internal-flue shell boiler (Cornish) and double-internal-flue shell boiler (Lancashire) were for many years the mainstay of the shell-boiler industry. For new installations they have now been entirely replaced by Economic and package boilers, but many are still in service after some 40 years. The general arrangements of the two types of boiler are shown in *Figures 5* and *6*. In both arrangements the boiler shell is situated within a brickwork setting which contains the side and bottom flues. These boilers were originally built for coal firing, either by hand or with mechanical stokers and utilizing either natural or forced draught. The small convective heating surface characteristic of this design severely limited efficiencies, to little over 60%. However, if these surfaces were supplemented by a superheater, economizer, and/or air heater, efficiencies of up to 75% were realized.

Owing to the low firing intensities, and consequent low heat fluxes across the metal heating surfaces, these boilers give exceptionally long service, even with inadequately treated water. However, they are bulky and the brickwork setting requires constant maintenance to ensure that the gases do

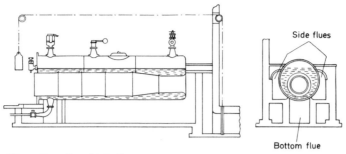

Figure 5   Cornish boiler

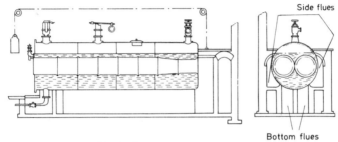

Figure 6   Lancashire boiler

not short-circuit the convective-heat-transfer sections.

In recent years, techniques for firing this type of boiler with liquid and gaseous fuels have been developed. In addition to increasing the basic efficiency to 75%, an uprating in excess of 50% can be achieved. These considerable improvements are obtained by using a forced-draught burner and recirculating a proportion of the combustion products through the furnace tube before eventual discharge to the stack.

The Economic Boiler

This type of boiler is essentially self-contained and requires no brickwork setting. Combustion takes place within a cylindrical furnace inside the shell, the gases then reversing in a combustion chamber at the rear of the boiler, travelling forward through a nest of smoketubes contained within the shell and below the water surface to a smokebox at the front end of the boiler. The smoketubes are normally about 50–75 mm in diameter. Gases pass directly from the smokebox to the stack. This arrangement (*Figure 7*) where the gases traverse the boiler shell twice is called a two-pass system. If the smokebox is replaced by a transfer box where the gases are again reversed and travel from front to back in a second set of smoke tubes to a smoke box at the rear of the boiler, we have the now more usual three-pass design. Some manufacturers have introduced a third set of smoke tubes to give a four-pass arrangement. For obvious

reasons this type of boiler is also commonly called a multitubular shell boiler.

The combustion chamber can be outside the boiler shell and refractory-lined (designated dryback), within the shell and entirely water-cooled (wet-back) or a combination of the two (semi wetback). These three basic configurations are illustrated in *Figure 8*.

The use of a large number of small-diameter smoketubes breaks the gas stream into a number of small elements, increases the gas velocity over the convective-heat-transfer surface area, and increases the area available for such heat transfer. The combined effect of this is to improve significantly both the rate and efficiency of convective heat transfer and thus to improve the overall thermal efficiency of the boiler, while enabling shorter

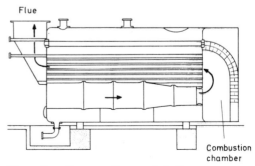

Figure 7   *Two-pass Economic boiler*

furnace tubes to be used. However, the smoke-tubes provide an increased resistance to gas flow through the boiler, which can only be overcome by use of forced- and/or induced-draught fans.

Tubes of special patterns promoting swirl etc. have better heat-transfer properties than the more conventional straight tubes owing to the extra turbulence created by the tube shape. As a result, one pass of such tubes may, in some instances, be used in place of two passes of conventional straight tubes.

Increased heat-transfer rates and efficiencies can also be obtained by fitting conventional straight smoke tubes with flow retarders. This use of twisted metal strips or spiral coils within the tubes promotes gas turbulence, and in addition the inserts act as direct emitters of radiant heat. Efficiency increases with a coal-fired boiler of up to 7% have been observed after fitting retarders. This increase in efficiency is accompanied by an increase in draught loss of, typically, 0.5—1.25 mbar (50 to 125 Pa).

For similar steam outputs an Economic boiler occupies only about half the length of a Lancashire boiler, and with the absence of brickwork setting it is easier and cheaper to install. When siting an Economic boiler, a space at least equivalent to the length of smoketubes must be left clear in front of the boiler to allow for the removal of tubes should retubing be necessary.

Economic boilers are equally suited to coal, oil, or gas firing without major changes to the boiler structure, and in all cases efficiencies of 75% are readily realized without the use of superheaters, economizers or air heaters. The Scotch Marine boiler, which has for many years found a wide variety of applications on land as well as at sea, is basically a wet-back Economic boiler with a short shell of comparatively large diameter accommodating up to four furnace tubes. As with other types of Economic boilers it is readily adapted for coal, oil or gas firing. Watertube boilers operating at high pressures have now displaced the Scotch Marine as a main power unit on most ships.

The Package Firetube Boiler

The name package derives from the fact that the boiler is engineered and built, inclusive of all ancillary equipment such as control systems, feed pumps, air fans, fuel-burning equipment etc. at the boilermakers' works. The boiler is delivered to the customers' premises requiring only connections to water, fuel and electrical supplies before it is ready to be fired. By this means complete boiler installations can be achieved in 48 hours. Most package boilers are variations on the Economic design and are available with 2, 3 or in some cases 4 passes, with wet-, dry- or semi wet-back configurations. The Vekos unit (Figure 6 of Ch.6 and p 205) is a special case. Many modern boilers are compact package units designed to take full advantage of the very high furnace-heat-release rates possible with oil or gas firing. This type of boiler is characterized by its small size, low water content and high efficiency, usually at least 80% with oil firing and marginally lower with gas firing; it is not suitable for coal firing by conventional techniques. The comparatively small furnace volume means that the convective-heat-transfer sections of the boiler play a more important role in the heat-transfer process.

The trend towards compact, high-intensity boilers with the consequent reduction of heat transfer surface area within the furnace tube has meant a considerable increase in the heat flux across the metal of the furnace and rear tube plate. If these heat fluxes are to be maintained without damage to the boiler structure, it is essential that there is no resistance to heat flow in the form of water-side scale. It is therefore of paramount importance with boilers of this type that the most rigorous attention is paid to feedwater treatment; quite small deposits of scale will result in excessive metal temperatures and subsequent boiler failure. It is across the rear tube plate, particularly when firing with gas, that the heat flux is most critical, and in this region extra precautions are also necessary on the gas side. The turbulence created by the reversal

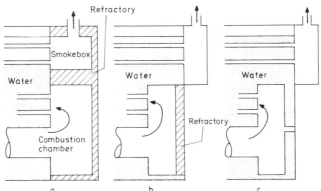

Figure 8   Economic boiler: (a) dry-back, (b) semi wet-back, (c) wet-back

of the gases in the combustion chamber, and their entry into the smoketubes, gives rise to very high convective-heat-transfer coefficients in this region. At this point the path along which the heat flows offers a higher thermal resistance than that for the corresponding path across the flat tube plate. This is illustrated in *Figure 9a* which shows a smoke-tube fitted in the conventional manner. The gas-side surface of the tube end is at a higher tempera-ture than that of the plate itself, and it is in this region that cracks and leakages can occur. The amount by which the tube end-temperature exceeds the tube plate temperature is clearly dependent on the relative values of resistance offered by the two thermal-conduction paths, and to ensure that this difference is kept to a minimum the smoketubes should be refitted as shown in *Figure 9b*. The extra cooling of the combustion-chamber region pro-vided by a wet-back design may make this con-figuration more suitable for high-intensity firing.

Rear-tube-plate and tube-end metal tempera-tures can also be elevated to harmful levels by indifferent or poor burner-combustion perfor-mance; this may result in combustible gases and oxygen persisting in the combustion chamber so that combustion continues as these gases traverse the smoketubes. This further (often localized) liberation of heat causes the damage. Therefore in the already hot and vulnerable region of the com-bustion chamber it is imperative to avoid delayed energy release; with modern compact boilers, the design of the burner and its subsequent commis-sioning must ensure that the combustion process is confined strictly to the furnace tube. Recent evi-dence has shown that comparatively small amounts of carbon monoxide can, if allowed to burn out within the combustion chamber and smoketubes, cause substantial damage to these compact boilers irrespective of the fuel being fired.

A new approach to boiler design, aimed at reducing or eliminating the thermal stressing in the combustion-chamber region, is the development of a three-pass arrangement with a closed, reverse-flow furnace. By this means two passes are accom-plished within the furnace tube before the gases enter a single pass of smoke tubes, and hence the

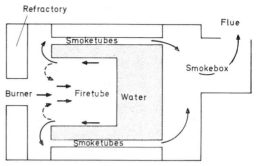

Figure 10   Reverse-flow furnace configuration for three-pass shell boiler

temperature of gases entering these tubes is con-siderably reduced. The principle of operation of this boiler is shown in *Figure 10*. A further advan-tage of the closed-furnace principle is that the expansion of the flame into the furnace will cause an injector effect and hence recirculation of a pro-portion of gases; therefore in reality more than two passes are achieved within the furnace. Effi-ciencies from this type of boiler are similar to those obtained from the more conventional boilers based on the Economic design.

Other Designs

The types of boiler so far considered cover the majority of installations in current use. There are however further variations of the basic firetube design (such as the vertical firetube boiler and the locomotive boiler) which have now, in most instances, been superseded. For this reason these designs are not discussed but details can be obtained from other sources[2].

Optimum Thermal Efficiency

The purpose of the heat-transfer-surface area within a boiler is to extract heat from hot combustion products and to transfer this heat to the water. The efficiency with which this operation is carried out is limited and is subject to a law of diminishing returns. For example, the addition of an extra pass of smoketubes to extract low-grade heat from already cool gases may improve the overall effi-ciency by only 2 or 3%. However, this will involve additional capital cost and additional running costs in upgrading the induced- and/or forced-draught system to overcome an increased draught loss. In addition to this economic limit on efficiency, there is a lower limit on exit-gas temperature (and hence an upper limit on efficiency) below which it is not possible to operate without danger of gas-side cor-rosion caused by the condensation of dilute sul-phuric acid from the combustion products. This is a particular problem when burning fuels with a high sulphur content, such as certain heavy fuel oils. In general, an exit-gas temperature of $55\,^{\circ}C$ above the saturated-steam temperature at the operating pres-sure is considered to be a good compromise from

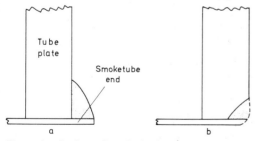

Figure 9   Package firetube boiler: (a) conventional welded smoketube end, (b) flush welded and ground-back tube end

the heat transfer point of view, leading to high efficiency and trouble-free operation. It is possible to reduce this temperature by the use of economizers and air heaters, but then the danger of corrosion should never be overlooked. In addition, with boilers operating at low load, exit-gas temperatures are lower than those at the maximum rating, and although they may operate satisfactorily at maximum output they may be prone to corrosion problems if required to operate for long periods at reduced output.

Design Limitations

The use of a large shell to contain the water and steam imposes certain inherent limitations to the duties which this type of boiler can perform.

The ideal shell design to resist the internal steam pressure would be a sphere, but practical considerations lead to the use of basically cylindrical designs as a good compromise. In a shape of this type the longitudinal forces are considerably greater than those around the girth, and diagonal stays, through-bolts and stay tubes have to be used to give added strength to the flat end-plates. In the longitudinal direction the strength required to resist lateral bursting is proportional to the product of the pressure and diameter, and hence high pressures and large diameters require very thick shell plates. As a result there is an economic limit on the pressures that can be achieved with this design. This practical ceiling is often considered to be an operating pressure of 1.7 MPa, 17 bar (or 250 psi), and for pressures above this watertube boilers may show a capital-cost advantage. But they are made up to 3 MPa (see Ch.10.3).

The difficulties in manufacturing and transporting the very large shells required for high-output boilers also impose an upper limit on the capacity of firetube boilers. At the present time capacities up to 18 MW are obtainable, although once again a watertube design may show cost advantages at these high ratings.

10.3 Watertube boilers

Pressures up to 1.7—3.1 MPa (cf. Ch.10.2) can be catered for by shell boilers. But in the larger ratings, 18 t/h (11.3 MW) and above, this proves somewhat difficult and in consequence watertube boilers are employed.

Power-station Boiler Types

These units of various types can be supplied to meet all requirements. In general three types are available:

1. Natural circulation, in which the force providing circulation is due to the density difference between the water in the downcomer tubes, or pipes feeding the heated circuits, and the steam-and-water mixture in the heated or steaming tubes.

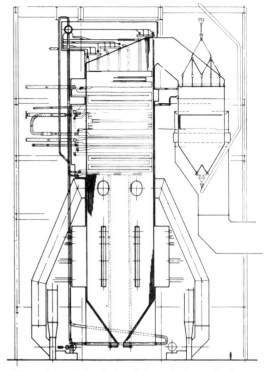

Figure 11   Natural-circulation boiler of tower type — brown-coal fired
Evaporation at MCR: 1134 t/h
Steam pressure at boiler outlet: 12.76 MPa
Steam temperature at boiler outlet: 541°C
Reheater outlet temperature: 541°C

Clearly this density difference is governed by the heating rate and the operating pressure. As the latter increases, the specific density difference between water and steam progressively gets less until at 22.11 MPa (critical pressure) the values are the same and water and steam co-exist and interchange without heat being required since the latent heat of vaporization is zero at critical pressure.

Even though high-head and tower units have been designed to increase the force producing circulation by ensuring the maximum head possible, operating pressures in the boilers of 19.3--19.7 MPa, which means some 17.9—18.3 MPa at the superheater outlet, are the maximum that can be safely used.

Figure 11 gives details of a tower-type unit for a 350 MW$_e$ turbine. The operating details are as stated. It will be observed that despite burning raw brown coal at 67—70% moisture, the furnace is completely water-cooled. It should be particularly noted that all the convection surface in the first upward pass is supported on water-cooled tubes, thus eliminating heat-resisting or other extraneous materials for supporting purposes. Further, the disposition of the live-steam, reheater and economizer heating surfaces in this section are in banks of shallow depth, thus permitting ready cleaning by soot blowers and consequently with maximum availability and accessibility.

In the downward pass before the gases at outlet proceed to an electroprecipitator, two further economizer banks and Ljungstrom-type regenerative airheaters are located. It will be observed that the economizer in this section is cleaned by equipment of shot-cleaning type. The moisture content of the fuel ranges on average from 62 to 71%, but a greater range is possible. For basic design purposes, fuel of 67% moisture with a net CV of 6.35 MJ/kg is used.

Figure 12 indicates the comparison between this unit burning coal with a net CV of 6.35 MJ/kg and a typical 350 MW$_e$ boiler burning a British bituminous coal of 23.25 MJ/kg gross.

2. Forced, and controlled or assisted, circulation. This type of unit employs pumps for ensuring or assisting circulation, and has been increasingly used in the UK, in Europe and in the United States.

With pumps, fully water-cooled furnaces may be used with no undue restriction regarding tube size. Small-diameter tubes with high radiation and convection heat-transfer rates, coupled with greater flexibility in tube arrangement, may thus be employed.

The type of pump now employed is usually of the 'drowned', 'submerged' or 'canned motor' type, with elimination of associated glands, and thus

substantially trouble-free. For units which employ water circulation, the circulation factor may vary from as low as 3 to 20 times the MCR of the boiler: 3—6 with assisted or similar controlled-circulation units in which, with individual tubes or circuits, metering nozzles or control tubes are used; up to 20 for specially highly rated circuits, or to ensure stability and adequate cooling in waste-heat units where extremely erratic and unknown heat fluctuations occur. There are very few examples of the Loeffler principle of steam circulation with all the evaporation taking place in the evaporator drums. With these, the circulation factor (in the 2—3 region) depended on the latent heat, and while good control of steam temperature resulted, the power consumption of the steam pump was some 2% of boiler output, to be contrasted with 0.3—0.5% with forced- and controlled-circulation boilers. Operating pressure imposes some restrictions regarding the above; with water circulatory means, owing to the possibility of cavitation in the suction system, natural circulation limits apply, whereas for steam circulation restriction to a reasonable latent heat is required, the most economic pressure being 13.1—13.8 MPa.

Figure 13 gives details and operating conditions of a forced- and controlled-circulation boiler suitable for 500 MW$_e$ output. The unit is oil fired, using 32 burners in the combustion chamber frontwall. Pressure-jet mechanical atomization at a

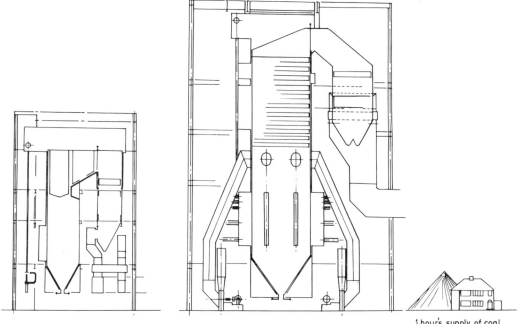

1 hour's supply of coal
600 tons with 67% moisture

Figure 12   Comparison of 350 MW$_e$ *boilers fired with brown coal (67% moisture) and normal British bituminous coal (14% moisture)*

References p 247

maximum oil pressure of 8.27 MPa is employed and the furnace is liberally rated at a heat release of some 0.28 MW_t/m^3 at MCR. Four glandless pumps, of which one is standby, are used to circulate the latent-heat-absorbing surfaces. It will be observed that the only evaporative heating surface is located in the combustion chamber, and particular attention is drawn to the superheater and reheater surfaces. These are extremely extensive since approximately 47% of the total heat generated is absorbed by these sections.

The high-temperature superheater and reheater surface is pendant, followed on the gas side by horizontal banks of low-temperature superheater and reheater surfaces. Additionally, in the live-steam circuit a small amount of radiant superheater surface is incorporated at the top of the furnace in the front wall; this is provided to facilitate turbine temperature-matching for hot restarts. There are no high-temperature uncooled tube supports. In the heat recovery-system an extended surface steel tube economizer and twin vertical-spindle regenerative air heaters are incorporated. Bled steam air-heaters are provided to heat the incoming air, to enable the main air-heater metal temperatures to be maintained at a minimum of $110°C$ to minimize corrosion.

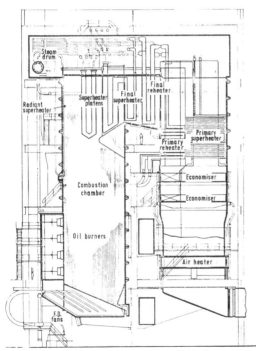

Figure 13   Forced-   and   controlled-circulation boiler — oil fired
Evaporation at MCR: 1610 t/h
Steam pressure at boiler outlet: 16.6 MPa
Steam temperature at boiler outlet: 541°C
Feedwater temperature: 254°C
Reheater outlet temperature: 541°C

3. The O/T or once-through boiler does not in general employ circulating pumps or a steam-and-water drum. The feed pump, which has an appreciably higher pressure than usual, pumps the feedwater not only through the economizer but also through the evaporative circuits (if any) and finally through the superheaters, which are usually both radiant and convective.

In simple terms, two types of this unit are in general use. In the first, a separator is located at or near the end of the evaporative section. From this (as its name implies) the water, which has now concentrated the salt content some 20–25 times, is withdrawn from the circuit either continuously or intermittently and returned to the water-treatment plant; the saturated steam then proceeds to the superheaters before the final plant outlet. In the second type, this form of salt concentration is not provided and the boiler salts are ultimately deposited on the inside of the tube coils as scale. This requires to be removed periodically and is carried out without taking the boilers out of commission by isolating a particular circuit, flushing through with hot treated water, and then recommissioning; the next circuit is then similarly treated, and so on. Obviously this, while apparently successful, is a palliative and does require skill and vigilance during operation.

These two systems are suitable for sub- and supercritical operation — in fact they are the only types suitable for supercritical operation; units have been supplied for pressures in excess of 41 MPa. Although a few years ago a number of power utilities installed steam generators with pressures in the 34 MPa region, the gain expected, in comparison with the difficulties experienced and the increased costs entailed, was found to be unjustified. Consequently there has been a slight recession in pressure to the somewhat more moderate range of 25–27.6 MPa when using supercritical conditions.

In the once-through boiler, a high length-to-diameter ratio is required for flow stability and other purposes, a high mass flow being required. This means that if tube conditions are satisfactory at full load, then at one-third load or thereabouts the flow will probably be insufficient to ensure adequate cooling of the tubes, and unless other safeguards are taken the tubes will overheat, and fail. Such safeguards can be that at all flows up to approximately one-third of full load recirculation or adequate cooling flow is maintained through a blowdown vessel and condenser; or more simply, that at loads up to 50% a recirculating pump may be employed as described in Section 2. This latter pump protects the heating surfaces during startup and low-load conditions and can be manually or automatically shut off as the load increases to a safe value. Alternatively, the surface-metal temperatures of the elements can be used to control the operation of the pump.

Figure 14 is a cross-sectional arrangement of a once-through boiler suitable for 375 MW_e output.

The boiler is of the twin-furnace type with tangential corner firing; the two identical separate furnaces each have an evaporation of 567 t/h. The boiler is arranged for cubical expansion, thus ensuring (in theory at least) freedom from expansion problems. The cross-sectional arrangement of the boiler conforms to the same physical layout as those adopted for natural-, forced- and controlled-circulation units when burning similar fuels; with a completely water-cooled furnace from horizontal meandering-tube circuits.

The pendant platen superheater arrangement in the furnace near the gas outlet is in the steam temperature region before the final outlet temperature; the final superheater is in fact pendant and arranged immediately following this superheater. The next heating surface traversed by the gases is the final reheater section, which again is pendant.

In the horizontal section, the first heating surface is the inlet stage of the reheater, at the inlet of which is the Triflux section. Following this heating surface is the initial section of superheater following the steam separators; the next heating surface swept by the gases is the section of economizer or feed-water-heating surface before the gases pass to Ljungström-type airheaters and then to electroprecipitators, I.D. fans and chimney.

The construction details and tube-supporting arrangements are outside the purpose of this book, but brief reference to the necessary controls is important, in fact vital, to the satisfactory, safe, continuous operation of the unit.

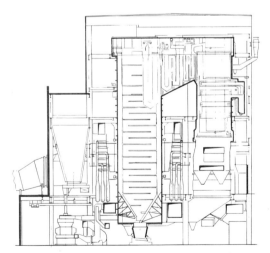

Figure 14   Once-through supercritical boiler —
bituminous-coal fired
Evaporation at MCR: 1134 t/h
Steam pressure at boiler outlet: 25 MPa
Steam temperature at boiler outlet: 599°C
Reheater outlet temperature: 566°C

References p 247

Controls
Feedwater Control. There are three feed pumps, one steam driven, of 100% capacity, for normal service, and the other two standby, being electrically driven, of 50% capacity. Feed-pump control in conjunction with single-seat valves at the inlet of each of the four parallel boiler circuits ensure a delivery pressure sufficient to meet the highest load-demand in the circuits in accordance with the heat absorption. A monitoring of temperature increase and load provide the starting signal for the spare feed pumps. If this temperature continues to rise beyond a safe limit, the pulverized-fuel control is tripped automatically.

Steam-temperature Control. Steam temperature is controlled in two stages by spray desuperheaters, the first set located at the inlet to the radiant superheaters which form the upper furnace wall, and the second set at the inlet to the final superheater outlet. The control system is designed to maintain the steam outlet temperature to within ±5° of 593°C.

Reheater-temperature Control. This is achieved by a Triflux arrangement which is a patented Sulzer system of control. It will be seen from the cross-section that the reheater consists of a high-temperature contraflow pendant section, and a horizontal self-draining section, the last loop of which is the Triflux reheater.

The temperature of the high-pressure steam passing through the inner tube is varied by a spray desuperheater. The reheat steam-temperature control impulses are taken from an intermediate point and at the reheater outlet.

Steam Stop-valve Control. This control is provided to prevent the boiler pressure from falling below a preset value and so protect the turbine from the admission of steam at too low a temperature. The boiler is fired by means of six vertical-spindle mills incorporating classifiers at the mill outlets. The fuel pipes then quadruplicate to burners located in the corners, firing into the two separate furnaces in tangential fashion. Some steam- and reheat-temperature control is obtained not only by altering the angle of tilt of the tangential burners, but also by a degree of differential firing between the two furnaces. Thus further means of steam- and reheat-temperature control are provided for varying fuel conditions and operation, to facilitate setting the general boiler operating parameters. Additionally, provision is made for possible future gas recirculation.

Fuel is East Midlands untreated coal all passing ¾ in B.S. mesh: average moisture 14%, ash 16%, gross CV 22.7 MJ/kg, Hardgrove index 50–55. Ljungström airheaters and the associated induced- and forced-draught fans are of conventional design. Similarly, the electroprecipitators conform with the normal grit emission figures considered satisfactory within the limits required by the Clean Air Act (UK).

General

These examples serve to illustrate the types of watertube boilers in current use, and are all power-station units reasonably typical of such practice in the UK and Australia. The capacities and cycle conditions shown are representative of British (sub-critical with capacities up to 500—660 MW$_e$) and Australian (sub—critical with capacities of 350—500 MW$_e$) equipment. A similar range in capacities but with pronounced use of once-through units prevails in continental Europe. In the U.S. and Japanese power-station fields, with increasing adoption of supercritical pressures of 25—27.5 MPa, the size of units has increased as the requirements of the grid, state or area systems have increased with growth of population and increasing standards of living. Consumption in the USA is some 8500 kW h/head, Sweden 8300, compared with 4200 in the UK. In the large electrical systems of America, where air-conditioning loads in summer are in excess of heating loads in winter, boilers of 1000—1200 MW$_e$ capacity are entirely justified, being within some 9% of total system load, and therefore likely to be economically used.

Industrial Field

In this area the watertube boilers installed in the larger sizes, 225—680 t/h, are clearly influenced by

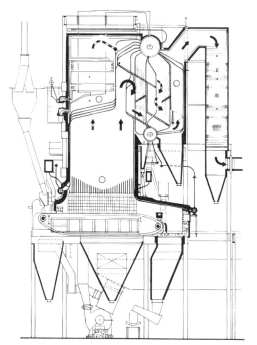

Figure 15   Typical two-drum boiler — industrial and power station, coal fired
Evaporation at MCR:118 t/h
Steam pressure at boiler outlet: 4.38 MPa
Steam temperature at boiler outlet: 491°C

the trends and growth in the power industry. Further, in the past decade and recently, the fuels being generally used are oil and gas and these have dictated the type and design of unit. A typical boiler is frontwall fired with 8 sets of combined oil/gas burners. The furnace is completely water-cooled and the secondary superheater is protected from direct furnace-radiation by a suitable nose. The secondary superheater is pendant, the primary superheater being horizontal and carried on steam-cooled tubes. The economizer, of the bare-tube type, is immediately below the primary super-heater. This section of the convection surfaces as well as the combustion chamber is arranged for water washing. The unit is designed for pressurized combustion, with F.D. fans only.

Large numbers of this size and type of unit are not ordered in the UK owing to the general policy of power generation employed, and consequently they are only installed in large companies with process as well as power loads, such as ICI, Alcan and the oil refineries.

In the range of sizes up to 225 t/h the 2-drum bottom-supported boiler is generally favoured. The operating details stated are reasonably conservative and pressures up to 10.3 MPa and steam temperatures of 510°C can be readily accommodated with this arrangement; although particularly suitable for oil and gas firing, can be arranged for coal and fibrous fuels. An example of a 2-drum coal-fired unit in this range, with operating details, is indicated in *Figure 15.*

Twin travelling grates each 4875 mm wide × 7315 mm long are necessary for the evaporation in question. The furnace is more or less completely water-cooled, with appropriate arches and secon-dary-air ports for ignition purposes which ensure complete combustion. Horizontal secondary and primary convection superheaters are provided in the upward pass from the furnace, these banks of tubes being carried by the initial steam offtake tubes from the steam drum to the superheater inlet header; thus no special material is required for superheater tube supports. The gases then make three passes through the main boiler bank before proceeding to a bare-tube economizer, then to a tubular airheater with hot and cold sections, cyclone-type grit collectors, and finally to I.D. fans and the outlet to the chimney.

Steam-temperature control is by means of a damper located at the gas outlet from the main bank of boiler-heating surface. This damper controls the gas flow over the superheater and main bank, and the gas flow through the by-pass indicated. The stokers used are virtually the largest size that would be recommended, and consequently this unit is perhaps the largest size of coal-fired boiler using mechanical stokers, regardless of type. This particular unit, while illustrating the top range of installation, is typical of stoker-fired installations in general, except of course for overall dimensions.

Other Fuels

In the case of fibrous fuels such as bagasse, palm oil nut residue, wood waste and similar, stoker-fired units of the required capacity are employed. Travelling grates, inclined grates or spreader stokers can be employed and successfully burn these fuels; the spreader unit using mechanical or pneumatic means of fuel distribution has been found extremely successful and well adapted to burn these fuels, with good flexible operation.

For units of moderate output, the boiler illustrated in *Figure 16* is reasonably typical. Oil burners in the sidewall are provided for starting up and emergency operation during mill-plant stoppages and shortages of fuel. A dump grate is shown in a water-cooled furnace.

The gases pass upwards in the combustion chamber, and down through a vertical pendant superheater before making three passes through the main bank of convection surfaces; they then pass to a tubular airheater and finally the plant outlet. Fibrous fuel is admitted to the furnace by pneumatic means; high-pressure pulsating air blows the fuel off a plate, the inclination of which can be altered to suit variations in fuel characteristics and load changes and around which secondary air is admitted. Other primary air is supplied through the grate. In combustion, approximately 50% of the fuel is burned in suspension and 50% on the grate.

For moderate outputs, simple deashing at the front by manual means is usually adequate; for larger units, 27–36 t/h upwards, continuous ash discharge stokers are usually fitted. Pneumatic fuel distribution is still employed, the ashing being mechanically carried out, discharging at the front

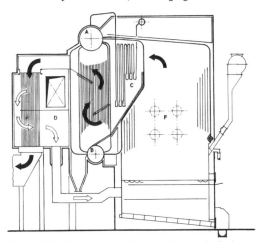

Figure 16   Two-drum boiler for fibrous-fuel plant, spreader-stoker fired
A: steam drum; B: mud drum, C: superheater;
D: air heater; F: oil burners
Evaporation at MCR: 12 t/h
Steam pressure at boiler outlet: 2.0 MPa
Steam temperature at boiler outlet: 300°C

to a suitable means of disposal. Disposition of boiler heating-surfaces, access, deashing etc. are then as in *Figure 16*.

Bagasse, which invariably has a moisture content in the region of 45–50%, requires hot air for its ready combustion and an airheater is normally provided in preference to an economizer. Good combustion of hogged wood and like fuels is ensured by similar means.

In palm-fruit waste, the moisture content is appreciably lower, normally some 15%, and the high oil content makes it readily combustible. In this case, to avoid slagging in the furnace, water-cooling is extremely desirable, even on the smaller units.

A most important point is that in the world's industrial market Package boilers are the largest single sector and represent 75% of the total order value.

These remarks have in general referred to the utilization of waste solid fuel for energy conservation in their associated processes. However, certain liquid residues — tank bottoms, oil sludges, black liquor from paper-making plants — can also be burned to conserve fuel supplies. Additionally, in oil refineries, gasworks, and steel-processing plants combustible gases such as carbon monoxide, methane, blast-furnace gas, coke-oven gas and tail-gas are products and their full utilization can effect appreciable economies in fuel supplies (see also Chapter 19.9). Reverting to waste solid fuel, domestic refuse is dealt with in Chapter 6.4. In the UK there is sufficient heat in garbage to maintain a 2000 MW_e station in full continuous service.

Package Boilers

At the lower range of industrial boilers, 16–45 t/h, the true package boiler, namely, a complete one-piece power unit, is the normal type of watertube boiler supplied.

These have the distinct advantages that they are cheaper than the conventional field-erected boiler, can be pre-tested, checked and all controls can be set before being despatched; coupling-up time at site if well arranged can be a maximum of 2–3 weeks, thus achieving earlier commissioning. Further, correlation of assembly of the complete unit by a single manufacturer should result in the production of a more satisfactory unit than a boiler erected and commissioned on the site.

While relatively clean fuels such as gas and oil are generally used in these units, solid fuels may be burned, but are not recommended. However, the development of fluidized-bed techniques in comparison with conventional combustion equipment indicates possibilities of increasing outputs, so that units thus fired may then become more competitive with gas and oil.

From *Figure 17* it will be seen that in the general interests of compactness and ease of fabrication, a 2-drum natural-circulation boiler is supplied. The boiler is specially designed to comply

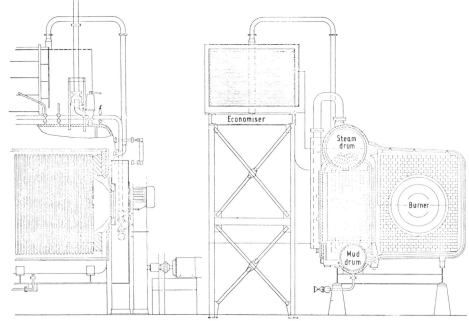

Figure 17   Typical gas- and oil-fired package boiler
Evaporation at MCR: 36 t/h
Steam pressure at boiler outlet: 2.52 MPa
Steam temperature at boiler outlet: 343°C
Feedwater temperature: 88°C

with UK standards for road transport, particularly as far as the cross section is concerned, but without any particularly stringent mandatory restriction regarding length.

Other fuels, liquid and gaseous, can be burned subject to practical problems in handling the fuel and the amounts dictated by the calorific values.

Forced-circulation and once-through boilers are available; they are flexible in operation, not unduly restricted by dimensional limitations, but tend to be somewhat higher in cost than natural-circulation units. They are particularly suitable in the lower output ranges and also in the high and even supercritical pressure regions. However, due regard is required to water conditions and in general they are used in special applications.

Since clearly the head is restricted and associated with the relatively highly-rated combustion chamber, to ensure safety the operating pressure is limited to 6.9—8.3 MPa. If transport changes or making the units in transportable sections permits increased vertical centres, then higher operating pressures are possible. Alternatively, a single-drum unit using outside downcomers is possible, and suitable for any normal operating pressures.

Forced- and controlled-circulation boilers do not have this restriction, being satisfactory at 13.8 MPa (the only problem being adequate head for the circulating pumps), no limit applying to once-through designs. While 45 t/h boilers can be readily supplied as package units, compact boilers

of 73—91 t/h can be supplied as standard designs, and also as package units. Special designs for capacities of 127—136 t/h in three or more sections are available.

All these units of whichever type have low thermal inertias; hence their ability to take load swings without severe pressure fluctuations may be questioned. However, with gas and oil firing, and modern quick and accurate response controls, this presents no difficulty and they are certainly as good, if not better, than similar conventionally designed field-erected boilers.

A typical boiler supplied to the power station of a chemical works for power and space heating etc. is indicated in *Figure 17*. The unit is fired by gas and/or oil; the appropriate decisions regarding the choice are decided by economic reasons.

A single rotary-cup burner is provided, the combustion chamber dimensions being designed and arranged specifically to suit this equipment. The long length of furnace ensures that the heat release rate is reduced to the moderate figure of 0.62 MW/m^3, and that the furnace outlet-gas temperature at entry to the screen before the superheater is reduced to a satisfactory figure for long continuous operation with good availability.

The combustion chamber is completely water-cooled with the exception of the burner wall, from tubes either in tangent construction (tubes touching each other) or from tubes to which fins are attached, namely membrane-wall construction.

The gases from the combustion chamber pass through the screen mentioned before traversing the superheater, then through the main bank of heating surface, before reaching the gas outlet at the front of the boiler. The gas, in fact, makes a complete 'U' type flow.

From the boiler outlet the gases proceed to further heat-recovery systems employing plain or gilled-tube economizers or airheaters, depending on the plant requirements and efficiency necessary. In the illustration a large bare-tube economizer is incorporated.

The shape of the combustion chamber is more or less tailored to a single-burner system; thus if the burner has to be extinguished or removed for any purpose, output falls extremely rapidly, and clearly this is not acceptable in a single-boiler installation supplying power and process demands. However, in most single-burner installations the design provides two 100% sprayers, thus ensuring complete reliability. All normal safeguards required by good operation, complete burner-management and safety interlocks, are built into these package units to exactly the same standards as apply in field-erected boilers.

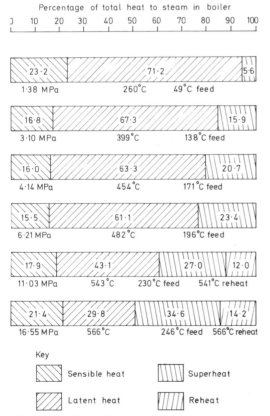

Percentage of total heat to steam in boiler

Key

Sensible heat

Latent heat

Superheat

Reheat

Figure 18   Chart illustrating heat distribution in respective section of boiler plant

References p 247

Superheaters and Reheaters

Reference has already been made to these two most important, in fact critical, sections of heating surface in modern boilers. In *Figure 18* the increasing percentages of heat absorption in the superheater with different pressure and temperature conditions from low-pressure-industrial to British-power-station practice are illustrated.

In the latter conditions some 45—50% of the total heat is absorbed in the steam through the medium of these heating surfaces. To ensure that they do not become excessive in extent, and to maintain reasonable temperature differences, in particular at the gas outlet from these surfaces and the economizer, these superheaters are located nearer and nearer to the combustion chamber. On the large units, *Figures 11, 13* and *14*, they are located immediately at the furnace outlet, with no intervening evaporative or convection-heating surface. More particularly, on *Figures 13* and *14*, platen sections of superheater heating-surface are located in the combustion chamber itself close to the gas outlet.

Platen heating surface, absorbing heat by direct radiation in the convection chamber and convectionally from gases, is distinct from the radiant-heating surface located in the front wall of *Figure 13*. In *Figure 14* with supercritical conditions, appreciable amounts of superheater surface are present in the meandering-tube circuits in the combustion-chamber walls. A detailed study of these circuits would indicate that practically all the combustion-chamber heating surface above the burner belt consists of radiant superheater panels.

If the steam temperature is over 535°C and reheat is included, even on sub-critical units a small amount of radiant-heating surface is of benefit in enabling the following three conditions to be met:
1. assisting turbine hot-restart conditions,
2. assistance in maintaining reasonable temperature differences throughout the unit, and
3. appreciably reducing the rising steam-temperature characteristic with the load normally present with solely convection-type superheaters.

With supercritical units having one, and perhaps two, stages of reheat, increasing amounts of radiant-heating surface have to be included.

The purpose of this sub-section is not to consider the thermodynamic aspects and economic advantages of high-pressure steam with high steam and reheat temperatures, but to indicate how they are integrated into modern boilers, the reasons, the problems which arise, the general characteristics and experience with the use of the necessary plant.

For turbine and similar plants, steam temperatures over 425°C require controlling to perhaps 7—10°C to ensure that the superheater tube-metal temperature does not exceed the maximum design figure, and that the turbine-blading design temperature is not exceeded (prejudicing design stresses and life and also running tolerances); for reasons of efficiency a constant temperature is most desirable.

The normal characteristic of a convection superheater is that the final steam temperature rises with increasing load. Thus means to control or maintain constant are necessary, although by locating the superheater nearer and nearer to the furnace outlet and with the platen surface receiving heat by both radiation and convection, the increase is appreciably reduced. Nevertheless, with steam temperatures of 510—565°C, control to within 3—5°C is essential.

Apart from the normal steam temperature-characteristic, appreciable variations in steam temperature arise from changes in fuel quality; in particular varying ash, moisture and VM, together with such factors as combustion-chamber fouling, faulty or inoperative soot blowers, and other operational changes or difficulties.

Several methods of steam-temperature control may be employed. Some systems are universally applicable, suitable for small low-pressure and even high-pressure power-station units. For example, gas bypassing or variation in gas flow by gas-biasing dampers is a very effective system, particularly as the bypass and dampers can be located in low-temperature gases with minimum hazards and maintenance problems. Such an arrangement is indicated in *Figures 13* and *15*, and can be readily accommodated by the plant shown in *Figure 16*.

A further universal system of steam-temperature control is by using recirculation of inert gases. In this method, to increase the steam temperature, gases withdrawn from the outlet of the unit are reintroduced into the furnace, and, by blanketing one or more walls, they reduce the radiation heat pick-up in the furnace; this, associated with the greater mass gas flow over the superheater, has the effect of elevating the steam temperature. The converse applies if this inert gas is introduced immediately at the superheater gas inlet; it then tempers the gas temperature through the superheater, reduces the gas—steam temperature difference, and consequently reduces the superheat. Provision for the addition of this system of steam-temperature control is shown in *Figures 13* and *14*.

A frequently used system of steam-temperature control employs desuperheaters. The superheater is divided into two or more sections; interposed between them is heating surface supplied with water from the boiler.

Reduction in the temperature of steam leaving the initial superheater section is controlled to maintain a constant final outlet steam temperature. In low-pressure units the heating surface is contained in vessels similar to feedwater heaters. These are not satisfactory for high-pressure high-temperature installations; thus when this system of control is used, for example with reheaters, the actual heating surfaces (i.e. steam coolers) are arranged in the reheater cross-over pipes (see *Figure 11*). This system is suitable for high-pressure live-steam industrial plants, where the

feedwater is perhaps largely make-up and in consequence sprays would not be suitable.

But for power-station use the most usual method of steam-temperature control is by means of spray injection at one or more points in the superheater system. Clearly, with the excellent water in these units, this is entirely admissible. This method is simple and efficacious and for live steam is thermodynamically correct; it is not suitable for installations where doubts exist regarding the purity of the injection water. It is thermodynamically inadmissible as a reheater control, being used only in an emergency since it reduces turbine efficiency and disturbs the balance of steam flows.

Altering the heat input or heat pattern in the combustion chamber is a further system of steam-temperature control. The burners can tilt upwards or downwards; the number at a particular level may be changed, thus varying heat input and pattern, and this can be combined with twin or divided furnaces.

Mention has been made of platen-type superheaters. This term is given to superheater surfaces, normally pendant, located at the top of the combustion chamber immediately before the gas outlet but where combustion is still being completed. For ease of arrangement and support, the heat transfer by convection being negligible, the tubes in each platen are arranged to touch and are only spaced to ensure and maintain alignment. The pitching of these platens across the width of the boiler will usually vary from 500 to 1500 mm. Heat transfer is by direct radiation from the furnace, gas radiation between the platens themselves, and a small amount of convection.

Further pendant superheaters are located at the furnace outlet but normally protected from direct radiation; they receive heat principally by convection, but also a reasonable measure by radiation from the gas layers between the pendants. These pendant sections are pitched transversely to the gas flow at some 500 to 300 mm, the pitching of the tubes in the direction of the gas flow being normal to obtain reasonably good heat transfer by convection. This heating surface is then followed in the gas stream by either further pendant superheater surface or pendant reheater surface, still with normal pitching in the direction of the gas flow to ensure good convective heat transfer but with transverse pitches varying from 300 to 150 mm; the decrease depends on the gas- and steam-temperature conditions and the particular fuel undergoing combustion. Next usually follow horizontal superheater and reheater surfaces, located at perhaps 100 mm horizontal pitch, depending on the actual tube size used for this heating surface. Platen superheaters having extremely onerous operating conditions are normally located in the middle of the superheat region so that the materials required are not too unusual. In order to minimize the use of heat-resisting or similar materials, the complete elements are simply spaced for alignment and normally carried by their own tubes. Movement of these platens hanging in the furnace may be a swing of 100—150 mm but could be 300—400 mm, this movement being claimed by a number of sources to help in maintaining the elements clean.

Pendant heating surface elements also in general carry themselves without support. In view of the tube pitchings in the direction of gas flow in this heating surface, supports can be minimized by using the actual tubes for spacing, but adequate attention must be given to maintain and ensure alignment. Thus in the event supports are minimal and the use of expensive heat-resistant or similar material is avoided. For example, in the horizontal convection sections of the superheater and reheater the gas temperature is reduced and supports need only be of moderate quality to give good life; and in general this heating surface is carried on its own steam tubes so that simple ear or similar supports can be arranged, providing low loading conditions and complete freedom for expansion.

For good availability, individual sections or banks of heating surface should not be more than 12—16 tubes deep. This facilitates cleaning by soot blowers or other means, assists inspection, and ensures that the heat pick-up or temperature change on the steam side for each section of bank is reasonable. This last feature can facilitate controls and avoid undue temperature changes in large wide boilers.

Close definition of material requirements for superheaters and reheaters is beyond the scope of this Chapter. Suffice it to say that for steam temperatures of about $425°C$ uncontrolled or $440°C$ controlled, good quality mild steel is satisfactory. For temperatures above this figure to $495—510°C$, 1% Cr + 0.5% Mo is suitable, and for temperatures to $535—550°C$ 2.25% Cr + 1% Mo is satisfactory. If steam temperatures $550°C$ upwards are required, suitable stainless-steel alloys are available, but these are best avoided if oil or similar fuel is employed.

Economizers

This section of a watertube boiler plant was formerly regarded as heating surface arranged to extract or reduce the exit gas temperature so as to improve the overall efficiency.

In modern watertube boilers, however, economizers not only fulfil the important function of heat-recovery surface, but are in general an integral part of the boiler unit, particularly of power-station units. With the economizer as an integral section, the fullest possible advantage of the temperature difference between water and gas can be utilized, and with drum-type units using natural circulation (*Figure 11*) and forced- and controlled-circulation (*Figure 13*), steaming or near-steaming economizers discharge directly to the drum. It will be appreciated that if the feedwater enters the drum at or near saturation temperature, then the disturbance produced in the main water contents

is minimal, and better water-level control is obtained. In the case of *Figure 14*, the economizer section is by virtue of the supercritical pressure all the sensible-heating surface provided before the commencement of superheating, since there are no latent-heat or evaporative sections. Apart from a convective section of economizer heating surface for temperature-difference reasons, the final section of economizer or sensible-heat absorbing surface is located in the lower combustion-chamber walls.

Considering large industrial units and (*Figure 16*) smaller power-station and industrial units similar reasoning applies. For the smaller industrial units an economizer is added solely for conserving energy, and *Figure 17* is typical. With special fuels, the processes in question may decide the temperature of the feedwater and this, associated with the combustion characteristics of the fuel, may well dictate that the best form of heat recovery is an airheater.

In general five types of economizer surface are in use:

1. Cast-iron gilled tubes, in which the heating surface is composed of cast-iron tubes which have integrally cast fins. The finning may be such that with parallel tubes the fins form parallel gas passages or, alternatively, diamond-type fins with staggered tubes. This arrangement has no particular restriction regarding fuels, is limited to a working pressure of approximately 3.1 MPa, can accept water temperatures as low as $55—71°C$, but $115°C$ is recommended.
2. Cast-iron gills on steel tubes. No particular limitation on fuels except that, as with cast-iron gills, there is a possibility of high-temperature blockage if stoker fired with fuels containing phosphate. There is no particular limitation in pressure, and feed temperatures of $71°C$ upwards can be accepted.
3. Plain tube can be used with most fuels with no restriction regarding pressure. However, it is imperative the feedwater is maintained at approximately $150°C$ when using fuels containing sulphur. Plain tube economizers still represent the best solution for most of the difficult fuels, but have the disadvantage of occupying more space than their steel-finned counterparts.
4. Steel fins on steel tubes. No limitations apply in operating pressure. The configuration of the fins may be such as to produce parallel or staggered gas passages, with varying pitching and fin heights depending on the fuel. As with plain tubes (and for the same reason) a minimum feed temperature of $150°C$ is required with sulphur-bearing fuels. Consideration must also be given to the maximum gas temperature presented to the economizer, which, in view of the limitation of the tip temperature of the fin, should not be more than $600—650°C$.
5. Welded spiral-fin tubes. The remarks under item 4 are equally applicable, with the added

References p 247

limitation that it should be restricted to clean fuels, and is therefore particularly suitable for nuclear applications.

10.4 Boiler auxiliary plant

Sections 10.1–10.3 generally deal with the main components of steam boilers. Although the items discussed in this Section are termed auxiliary plant they are just as essential in the integrated whole of a boiler plant. Some are required for proper running of a boiler and are incorporated in the basic design, whilst others are needed to comply with boiler laws and standards.

Boiler Mountings

Boiler Supports

With weights of boilers varying between 5 t for the small package units to total weights in excess of 5000 t for large power-generation units their support poses a considerable design problem. Two main methods have been adopted: the base-mounted, where the boiler sits directly on a plinth, and the top-supported where the boiler is hung from steelwork which is independent of the main boiler house structure.

There is a very wide range of base-supported boilers, such as firetube and package units. The boilers are invariably fabricated in the manufacturer's works and transported to site, which limits the physical size. Works-fabricated boilers are capable of evaporations up to approximately 150 t/h, but site-erected base-mounted boilers may exceed this figure, dependent on the design. On all marine applications, the base-mounted boiler is utilized as total stability can be assured: the boiler rests on cross bearers and other specially designed steelwork within the hull of the ship. Where seismic conditions prevail base-supported boilers are not normally recommended, since top-supported units absorb shock waves much more readily.

The top-supported boilers are erected on site, individual boiler members being fabricated at the maker's works. On the larger power-generating boilers evaporations of up to 2000 t/h have been installed; the steam drum on this size of unit would be approximately 60 m above ground level.

Safety Valves

It is recommended in both the British and International Standards that each boiler be fitted with at least two safety valves. Each valve should be capable of discharging the total peak evaporation of the boiler. In addition to these, safety valves should also be fitted to both superheaters and reheaters where incorporated in the boiler design.

Safety valves can be classified into two main types, one the direct-operated valve (including assisted valves), and the other the pilot-operated valve. Each type is manufactured in a variety of patterns to suit most applications. It is advisable to install safety valves in the vertical position and close to a comparatively rigid section, e.g. directly mounted on the boiler shell or drum. By installing them in this manner the escape pipework can be more easily designed, and also the shock load created by the valves when they blow will be absorbed. The set pressure is very important; the valve must not continuously open and shut, (known as 'feathering') as this will cause both damage to the valves and also waste of fuel.

Before commercial operation, insurance companies will require all safety valves to be floated; also, periodic testing is required to ensure correct operation.

High and Low Alarms

The British Boiler Regulations require that a steam boiler be fitted with a high and low (level) alarm which is clearly audible to an operator. Provision should be made for isolating the actuating chamber from the boiler, and means should be provided for testing the alarm and blowing out accumulated deposits.

The most common type of alarm is the float or displacer-operated type, which may be fitted inside the steam drum or boiler shell or externally in a separate chamber. This type of alarm consists of a float (two if an internal type) attached to a lever arm; as the water rises or falls, the float activates alarm switches. In more recent times the thermostatically operated, or the differential-pressure-switch-operated alarms are most commonly used. The former is located externally to the boiler in a separate chamber, and consists of two rods, one immersed in water, the other in steam. Its principle relies on the expansion or contraction of the rods to activate the alarm switches. The expansions or contractions are caused by the water level rising or falling; for example if the water level started to rise, the rod immersed in the steam would be cooled by the rising water and contract. The differential-pressure-switch-operated alarm is also located outside the boiler and relies on a very large sensitive diaphragm. A condenser is provided at the steam drum to maintain a fixed head of water on the steam-space side whilst the water side is subjected to varying heads. For equilibrium, both sides of the diaphragm would be at the same pressure, i.e. the boiler pressure. As the water level rises or falls the differential pressure at the diaphragm is changed, causing it to move and activating the alarms.

Water Gauges

Each boiler should be fitted with at least two independent means of indicating the water level, with the following exceptions:

1. On boilers having a design pressure of about 6 MPa and above, two independent, compensated,

manometric, remote water-level indicators may be used in place of one of the water gauges. Each remote indicator should be independently connected to the drum.

2. For boilers of less than 150 kg/h evaporation, one gauge is acceptable.

3. Once-through boilers do not require any water gauges.

Gauges should be so placed that the water level is easily read and so arranged that the visible part of the transparent face is at least 50 mm above the lowest permissible water level. There are three main types of gauge, the tubular glass type, the through-vision or reflex type, and the type with independent circular ports. The gauge should ideally be fitted directly to the boiler and should include a drain cock or valve with suitable drain piping.

Blowdown Valve

British Standards require that all boilers be fitted with a blowdown valve or cock placed at, or as near as practicable to, the lowest point of the boiler. Blowdown may be continuously or intermittently operated; the former is used to deconcentrate boiler solids and the offtake is then often located near the surface of the water in the shell or top steam drum. Depending on water quality and percentage of continuous blowdown, some form of

heat recovery can be obtained by fitting a flash vessel in the discharge pipework. The intermittent blowdown is used for the systematic removal of boiler sludge and for boiler emptying. As it is located at the bottom of the unit, excessive use will result in waste of fuel; it is therefore advisable to devise a programme of operation which limits the blow to once per shift. The position of blowdown valves should be such that they can be operated with a clear view of the water-level indicator. Boiler protection is the main concern; it is therefore advisable that only one boiler be blown down at any one time. Each outlet connection should have an isolating valve, and if the boilers are on a range system non-return valves should also be incorporated.

Boiler Feed Pumps

Pumps for boiler feed application cover a wide range of duties from the small on—off or modulating controls on package boilers to the complex feed controls on 660 MW_e boilers. The design capacity of a pumping installation should be equal to the total steam output of the station including any continuous blowdown and spray-water requirements for desuperheaters. As it is always advisable to have sufficient standby capacity, it is common to have two pumps each of 100% capacity, one electrically driven, the second (steam-turbine driven) acting as the standby unit. On larger boiler installations it is usual to have three pumps per boiler, each pump having a 65% MCR.

In determining the discharge pressure of a feed pump, all system resistances must be overcome, i.e. losses through pipework, valves, feed heaters, economizers and boiler static head. The total system resistance is then added to the maximum boiler pressure, namely the drum safety-valve set pressure. It is advisable to add a design margin of 2½% to the discharge pressure to cover variations in pump speeds.

Feed-pump pressure and capacity can be represented in a graphical form, the characteristic curve of the pump; a typical curve is shown in *Figure 19*. As the load is reduced from the duty point to zero, the pressure rises to a maximum, known as the closed-valve or no-load pressure. It is advisable to limit this rise in pressure to 15—20%; this will give a good falling characteristic and a reasonable design pressure for the external feed pipework. When the curve is very flat, the pump could be discharging either side of the duty point with very small pressure changes. Fluctuations in discharge flow could then possibly set up water hammer in the pipework. When two or more pumps are to work in parallel each pump should have a similar characteristic curve to the others; this will minimize load hogging by any one pump.

The third important factor in the design of a feed pump is the suction condition. Before a pump can discharge the required quantity it must first be possible to get the quantity into the pump. It

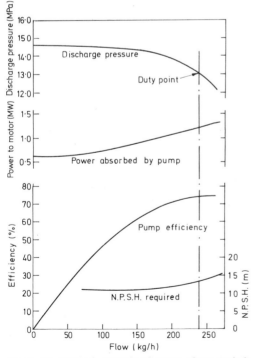

Figure 19 Typical centrifugal-pump characteristic curve

References p 247

cannot be sucked or pulled but must be pushed into the impeller. The equivalent head in metres of liquid must be determined; this is known as the net positive suction head, i.e. N.P.S.H. It is always measured over and above the vapour pressure of the liquid at the design feed temperature. Two types of N.P.S.H. are to be considered; one, termed available N.P.S.H., is determined by the plant-design engineer; the second, termed required N.P.S.H., is specified by the pump designer and is based on test data. To assure full capacity and to prevent vaporization in the entrance to the impeller the available N.P.S.H. must be equal to or greater than the required N.P.S.H.[3] Typical N.P.S.H. for centrifugal pumps are shown in *Figure 20*.

It is difficult to state categorically which design of pump would be most suited to any specific plant requirements, as design temperature and site conditions need to be known. The following is a short resumé of the pump designs in most common use, together with their application.

Steam-driven Vertical Direct-acting Pump (V.D.A.). These are manufactured in a range of sizes up to 100 000 kg/h, are most suitable for boiler pressures up to some 2 MPa, and consequently are mostly used on shell or package boiler installations. The pump is of the reciprocating type which displaces water by the reciprocating motion of a piston; over the years the design has proved to be very reliable.

Centrifugal Pumps. In these pumps the water flows by reason of the centrifugal force imparted to the water by the rotation of one or more impellers. This basic design is the most commonly used as it can be arranged to suit most design duties at very competitive prices. The overall efficiency of this type of pump can vary between 65% for flows of 20 000 kg/h up to 85% for flows of 400 000 kg/h. To meet the many varying duties, the construction

of the centrifugal pump can be changed by increasing the number of impellers or by increasing pump speed. The most common design is the multi-stage ring-section type with external tie-bolts; these are suitable for high pressure and temperature and are very competitive. Their main disadvantage is that external pipework must be dismantled before any pump maintenance can be carried out. The barrel-casing pump has a similar range of duties to the ring-section pump and is also similar in construction, except that the ring section is placed inside a casing of a circular barrel. This gives one major advantage, in that there is only one high-pressure joint and the pump can also be left in place during maintenance without any disruption to the external pipework. The third common type is the horizontal split-casing pump; these again can be designed for most duties. This type of pump is easily maintained on site, as the casing parts on the horizontal axis. With the advent of large boiler complexes, the availability, capital costs and power consumption of pumps have become of prime importance. Pumps for these duties consist of short rigid rotors, run at high rotation speeds, and tend to be of the barrel-casing construction.

To protect pumps, continuous bleed or automatic leak-off devices should be incorporated, the discharges from these being led back to the feed tank. Points to be avoided during normal running of the pumps are attempting to control the pump with the suction valve, and continuously running the pump against a closed valve; these practices cause cavitation and damage to the pumps.

Fans

Fans are basically another form of pump, the medium being either gas or air. They consist of bladed rotors or impellers which do the work and a housing which directs the gas discharged from the rotors. The fan power required is dependent on volume of gas, system resistance and efficiencies of fan and motors. Power may be expressed as shaft power, input power to motor terminals or theoretical power computed by thermodynamic methods. The most vital figures are power input to the shaft and theoretical horsepower. The shaft power is evolved using volume rate of flow V (m^3/s), differential pressure P (Pa), and fan efficiency f; power (W) = 100 PVk/f. k (dimensionless) depends on the pressure ratio and is ≈ 0.95. Theoretical power is calculated on the fan pressure—volume cycle, using differential pressure as measured by a manometer.

Fan performance can be altered by changing one or many design variables. There are essentially two different fan types, centrifugal and axial-flow. The centrifugal fan is the most common type in use, its principle is to accelerate the gas radially outwards in the rotor from the heel to the tip of the blades. The axial fan accelerates the gas parallel to the fan axis. The axial-flow fan is usually found where maximum flow with minimum

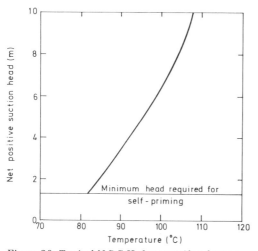

Figure 20 Typical N.P.S.H. for centrifugal pumps

space available are the main objects, as in marine applications.

Each fan has a distinctive characteristic which is a complicated function of relations best expressed in a graphical form (as with pumps for liquids); these graphs are termed the fan characteristic curves. *Figures 21* and *22* give typical curves for forward- and backward-curved blades. The main contributing factor to the curve shape of the graph is the design of the blade which basically falls into three shapes, (*a*) forward-curved, (*b*) flat blade, and (*c*) backward-curved blades. For a given fan it is possible to change the characteristic by varying the fan speed, the nature of the graphical curve itself remaining relatively unchanged. The performance of the fan at different speeds can be related by the following rules:

1. Capacity is directly proportional to speed.
2. Head is directly proportional to speed squared.
3. Power input is directly proportional to speed cubed.

In theory any type of fan can be designed to meet the requirements of volume and pressure, still maintaining maximum efficiency. However, for boiler requirements certain blade types are more suited to particular applications. The backward-curved blades give a steep head-characteristic and therefore a high-speed fan which is more suitable for forced-draught applications. The forward-curved blades are commonly used for induced-draught fans as the characteristic curves are flatter and therefore able to handle larger volumes. The flat blades are used in induced-draught fans where the gases have a heavy dust content.

It is important to have sufficient fan control to meet any load condition imposed on the boiler.

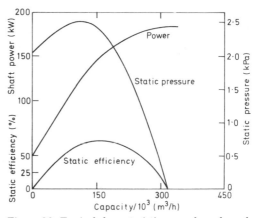

Figure 22 Typical characteristic curve for a forced-draught fan with backward-curved blades. Air temperature 40°C; rpm (min⁻¹) 1000

The most elementary control is the damper, which acts as a throttle valve; it is very cheap to install and easy to operate. But dampers can be wasteful as the fans must generate more energy to overcome throttling. The variable-speed control is the most efficient method of controlling output, and the types of drive used are magnetic-coupled motors, hydraulic coupling, variable-speed motors, and variable-speed steam turbines. Another form is the inlet-vane control, which although more expensive to install than the damper is considerably more efficient.

Air Heaters

The air heater in a boiler system is a means of reclaiming heat which otherwise would be lost. The incoming air to the boiler is generally preheated by the products of combustion, but other heating media may be utilized, such as steam or an independently fired furnace. Preheated air will increase the speed of combustion at all loads, improving the extent at low loads. The efficiency of the boiler will be increased by approximately 2% for every 60°C rise in air temperature; thus air temperatures in excess of 150°C up to a maximum of 315°C can result in fuel savings of between 5 and 10%[4]. Operation with minimum excess air is essential to achieve optimum efficiency and to minimize sulphur trioxide formation. The minimum metal temperatures that can be acceptable for air heaters while avoiding serious corrosion are usually 130—140°C.

Before an optimized design of air heater can be determined, consideration should be given to space limitations, fuel and air temperatures. Another consideration is corrosion, which is liable to occur within the air heater, the main contributing factors to this being metal temperatures that fall below design limits, the sulphur content and the moisture

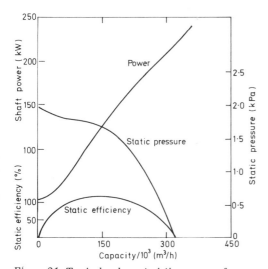

Figure 21 Typical characteristic curve for an induced-draught fan with forward-curved blades. Gas temperature 350°C; rpm (min⁻¹) 1000

content of the fuel. Effective control of corrosion can be achieved by bypassing a portion, or all, of the cool air to increase metal temperatures; this is very effective at light loads when the danger of reaching the dewpoint may be considerable. Similarly, increasing the spacings of the front row of tubes will effect cooling by decreasing the mass flowrate.

The air heater may be classified according to its principle of operation, (*a*) recuperative where heat is transferred directly from the hot gases on one side of a surface to the air on the other and (*b*) regenerative where the heat is transferred indirectly by an intermediate storage medium. The most common recuperative types in use are the tubular and plate air heaters. There are horizontal, vertical and parallel tubular types. They consist of straight tubes expanded into a reinforced casing, and can be so arranged that either the gas or air can pass through the tubes. Thermal expansion of the tubes is accommodated by fixing one end and permitting the other to move freely within the duct-work. It is common to fit a flexible bellows-type expansion joint to permit movement and make a suitable seal within the duct-work. The plate type incorporates thin flat parallel plates spaced with alternate wide and narrow spacings to match the mass-flow ratio of gas to air. The gas flows through the wide-spaced passages in contra-flow to the air which flows through the narrow passages. Strategically placed soot-blowers are incorporated in the design of these air heaters to keep all passages clear by dispelling dust.

Other recuperative air heaters are the steam-coil type, which are basically steam heat-exchangers used where gas—air heaters are not practicable. There are also separately fired air heaters, which usually consist of a refractory furnace with a tubular heater arranged for a number of gas passes. These provide hot atmospheric or pressurized air for industrial use.

The most common of the regenerative types is the rotary air heater. This is circular in construction and can be manufactured in a number of standard diameters. The centre shaft may be mounted vertically or horizontally. Steel plates, which are the storage medium, are placed in segmental groupings around the shaft. As the shaft is rotated, the plates pass through the gas sector obtaining heat and then (via a 'turnstile') through the air sector heating the air.

Their advantages are: they are compact, they can be produced in standard designs and at competitive prices, they can be operated with a low flue-gas exit temperature, and the cold end can be readily replaced; and their disadvantages, air in-leakage, and they have to be rotated mechanically.

Typically 4—10% of leaving-gas weight is air which has leaked across cold end seals (differential 57—73 kPa). The main drawback of this is increased F.D. and I.D. fan size and power consumptions. Boiler efficiency is not greatly affected.

Conclusions

High efficiency can be achieved using modern fuel technology and instruments, and is compatible with safety, good boiler availability with minimum maintenance, and minimal atmospheric pollution. Low excess air tends to reduce SO_3 (acid smuts) but to increase NO_x emission.

10.5 Maintenance; boiler gas-side deposits

Solid and liquid fuels and waste materials contain various proportions of non-combustible intrinsic and extraneous matter. The extent to which these are conveyed into the boiler flueways depends on the method of firing and various operating conditions. With pulverized solid fuel for example, a high proportion of the ash becomes gas-borne. With the coking type of solid fuel stoker, very little ash is carried into the gas stream if the operation and fuel characteristics are satisfactory.

The combustion process is never complete; small quantities of unburnt fuel pass into the flue-ways together with the ash and dust to form the basis of boiler deposits (Ch.20.1). The term 'soot removal' is a misnomer, since the term should refer to all deposits in boilers.

Need for Removal of Deposits

The deposits are likely to contain inorganic compounds formed from various elements present in the fuel, i.e. vanadium, sodium and sulphur. They can result in high-temperature corrosion, agglomeration of deposits, and low-temperature corrosion.

Where deposits contain or are contaminated by corrosive compounds and the boiler is taken out of service for cleaning, it is essential that this material is removed as soon as possible. Severe corrosion can occur if a boiler is taken off line and allowed to stand cool and dirty for a number of weeks. After normal boiler cleaning has been completed, or before if cleaning is delayed, the introduction of a neutralizing spray wash possibly using lime water may be beneficial, particularly with cast-iron sectional and shell-type boilers.

In general terms there are three main criteria which should be considered in relation to boiler cleaning, namely safety, availability and efficiency.

Safety

Deposits on the flue-gas side, unless of a corrosive nature, do not usually constitute a mechanical hazard since they protect the boiler metal from the heat source; but some can be toxic to boiler cleaners.

Availability

In the design of any boiler, the maximum continuous steam output is related to the fuel fired and heat-absorption capacity. It follows therefore that any reduction in heat transfer due to deposits will progressively reduce the steam output and/or steam temperature until the boiler unit cannot meet the minimum requirement. The return to normal performance can be achieved by either cleaning the surfaces while the boiler is on line or by taking the boiler out of service. The period of operation between shut-downs for cleaning is significantly extended by soot blowing and other means; but eventually the boiler has to be taken out of service for thorough off-line cleaning.

The increased use of liquid and gaseous fuels has reduced the necessity for soot blowing on many industrial plants comprising shell-type boilers. The boiler availability on such boilers is improved owing to the reduction in boiler deposits; in addition, modern boilers of the multipass economic type are designed for high flue-gas velocities throughout the boiler and therefore the tendency for solids to drop out of suspension is reduced. The installation of soot-blowing equipment on gas- or oil-fired shell-type boilers is now seldom justified, and as no grit arrestors are normally installed, its use could lead to severe local nuisances.

Efficiency

Deposits in the boiler flue passages retard the transfer of heat to the metal and the effect on the overall boiler thermal efficiency depends on the nature and thickness of the deposits and also their position in the flue system. Deposits in the high-temperature zone have a more adverse effect than those in the convection zone. With shell-type boilers there is a tendency for deposits to occur in smokeboxes where there is a change in direction of gas flow, and these do not affect the thermal efficiency. It is not always appreciated that normal boiler deposits are almost as good an insulating material as asbestos.

Methods of Removal

Boiler deposits may be removed while the boiler is on line or during a complete shut-down. The on-line method is generally used on the large water-tube boiler installations where maximum boiler efficiency and availability are of prime importance. Shell-type boiler plant burning solid fuel or waste materials may incorporate on-line cleaning. The majority of industrial boiler plants with shell-type boilers use the off-line method of cleaning, either at weekends when the factory is shut down or by using a standby boiler.

Off-line Cleaning

With all boilers, whether of the large power-station or small industrial-user type, a complete shut-down enables the unit to be physically examined and cleaned. The loose deposits are generally removed by hand brush or mechanical sweeper in conjunction with an industrial vacuum cleaner. *Figure 23* shows an air-operated brush which gives vibratory-tube scrubbing and vacuum removal of deposits. The relatively common practice of manual cleaning while the induced-draught fan is in operation and the chimney or main-flue dampers are fully open may provide a more congenial atmosphere in the boiler house but this can result in serious accumulations in a main flue which may be seldom if ever available for inspection. In addition it can cause a local public nuisance, particularly if grit arrestors are not installed or in operation.

Strongly bonded deposits may be loosened using a compressed-air percussion lance, rotary gear-type cleaner or expandable scrapers. In using the latter two methods care should be taken to ensure that loss of metal is minimized, and their use should be limited for this reason. The gear-type cleaner is an abrasive wheel driven electrically by cable, mounted on the end of a long flexible tube, and resembles somewhat a large dental scaling drill.

It is essential that adequate protective clothing be provided for persons cleaning out boiler fire-side deposits. This should include overalls, face mask, protective eye glasses and gloves. Abrasion of the skin while boiler-cleaning should be reported and examined by a competent person.

On-line Cleaning

Removal of deposits (e.g. by lance, scraper etc.) from heating surfaces while the boiler is steaming is usually accomplished by reducing the firing rate and air supply to give approximately 70% steam output, but a higher suction is maintained by the induced-draught fan in order to minimize the risk of blowbacks during cleaning.

Steam Soot-blowers

Soot blowing is most commonly carried out using dry saturated or slightly superheated steam at boiler pressure. The use of dry saturated steam generally has a better impact effect in removing deposits than superheated steam, but precautions must be taken to preclude slugs of water from entering the boiler flueways.

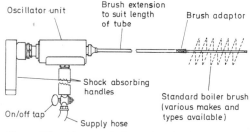

Oscillator unit

Brush extension to suit length of tube

Brush adaptor

Shock absorbing handles

On/off tap

Supply hose

Standard boiler brush (various makes and types available)

Figure 23   Air-operated   tube-cleaning   brush. Courtesy Airnesco Products Ltd

References p 247

The installation of manually operated soot blowers is normally limited to shell-type boiler plants and to small watertube boilers. Electrically controlled and sometimes remotely operated soot blowers are an essential feature on most large watertube boiler installations including marine boilers. The soot-blower nozzles are located at strategic positions in the flue-gas passages and the frequency of use depends on the extent and nature of the deposits. As a general guide, a complete sequence of soot blowing may be carried out once per 8-hour shift with a varying duration of blow-time.

Single high-pressure steam nozzles are (*Figure 24*) usually located some 1 m or more away from the boiler metal in order to minimize erosion. Multi-jets at somewhat lower steam pressures may sometimes be situated less than 0.2 m from the boiler metal.

Sootblowers used in high-temperature zones, for example in superheater tube banks and (seldom) combustion chambers on watertube boilers, are generally of the twin opposing-nozzle type and are fully traversing and retractable when not in use. Many include facilities for rotation to give better coverage (*Figure 25*).

In lower-temperature zones, the nozzles are usually non-retractable, although lateral movement to give a wide area of cleaning may be possible. Single- or multi-jet equipment may be used. With a single-return-pass economic boiler one or two retractable nozzles are located in the rear wall facing the return tubes. On double-return economic boilers, traversable multi-jet nozzles are usually installed in the front smokebox (*Figure 26*).

Stationary- or traversing-element soot blowers of the multi-jet non-retractable type may be used in the convection tube banks and economizer sections of watertube boilers. Where a number of soot-blower units are installed, their use is generally in sequence from the combustion chamber to the

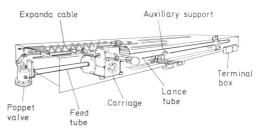

Figure 25   Long retractable and rotating soot blower. Courtesy Diamond Power

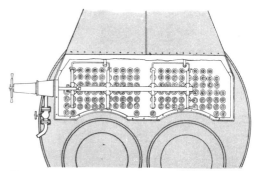

Figure 26   Multijet soot blower in boiler smoke-box

economizer zone. This ensures that deposits are not only loosened but are encouraged to terminate at the grit-arrestor plant or suction-removal point. Grit-arrestor equipment should be adequately sized to allow for heavy dust concentrations which prevail at the time of soot blowing. The correct sequencing and timing of each soot blower is usually accomplished by automatic control, but on small boiler units this may be done manually. The amount of steam used as a percentage of the total steam generated may vary from less than 0.5% for automatically operated soot blowers to 1.5% or more on manually controlled equipment.

Water Lancing

Single-nozzle water lances of the portable type are sometimes used in conjunction with a soot-blowing system where fused and bonded deposits or 'bird nesting' occur in the high-temperature zone of watertube boilers. These deposits cannot generally be removed by steam, and therefore water suitably boosted in pressure is used. The principle applied is simply that of creating thermal stresses in the deposits which then crack and break away from the boiler surfaces. Stubborn deposits may require repeated chilling and heating before eventually being removed. Care should be exercised in the use of water lances to ensure that no water reaches refractory brickwork, tube headers, superheater supports and other non-watercooled parts of the boiler. Water lancing should be kept to a minimum and not carried out so frequently as soot blowing.

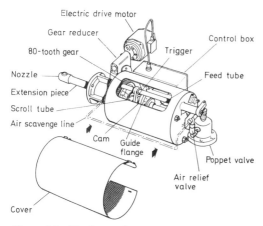

Figure 24   Single-nozzle   retractable   blower. Courtesy Diamond Power

Compressed-air Soot Blowers

Compressed air at approximately 3.5 MPa (35 bar) may be used as an alternative to steam for soot blowing, although from a cost aspect compressed air is usually more expensive. In addition, compressed air even at the higher pressures has not proved to be as effective as steam. The type of nozzles, their location and soot-blowing procedure are basically the same as those used for steam blowers.

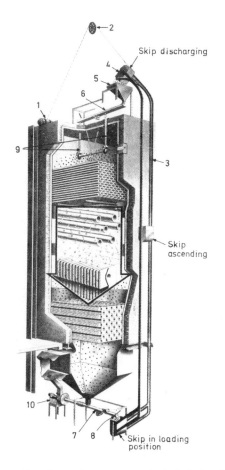

Figure 27   General arrangement of shot-cleaning system

1   *Hoist*
2   *Pulley wheel*
3   *Skip hoist*
4   *Skip*
5   *Shot distribution hopper*
6   *Shot inlet pipe*
7   *Shot outlet pipe*
8   *Automatic valve*
9   *Shot inlet pipes and distributors*
10 *Soot exhauster fan*

Shot Cleaning

This method of on-line cleaning is restricted to large watertube boilers of the high-head type when the superheater, economizer and air-heater zones are installed directly above each other. The system is fully automatic and can be continuous or intermittent in operation. In essence (*Figure 27*), cast-iron shot enters the upper boiler zone via distributors and falls progressively through the lower zones to a hopper in the base of the combustion chamber. The collected shot is then transferred to a skip conveyor where it is elevated to the upper hopper to be recycled.

The quantity of shot entering the boiler is automatically controlled to give the required cleaning effect. The lighter deposits loosened may become gas-borne and pass out to the grit-arrestor plant, whereas the heavier particles fall into the collector at the base of the combustion chamber with the spent shot. A vibrating-pipe system separates the deposits from the shot; the former is exhausted by a fan and returned into the flue system prior to the grit arrestor.

Water Washing

Economizers installed on many watertube boiler plants are of the gilled-tube type and are more difficult to clean with the normal soot blowers, particularly if the deposits are moist. An alternative method is to wash all of the economizer surfaces using spray bars and cold mains water. The economizer is bypassed on the flue-gas side when washing and the process is continued until clear water of neutral pH is being discharged. This process cannot usually be applied other than to economizers which are specifically designed for this operation.

Effectiveness of Cleaning

Boiler cleaning however carried out is expensive, and its merit must be measured in terms of greater boiler availability and a higher overall operating thermal efficiency between major shut-downs.

The additional cost of on-line cleaning can be justified where no standby boilers are installed and maximum boiler availability is consequently essential. In other cases where standby boilers are installed, the improvement in thermal efficiency is the main justification for cleaning costs. Excessive or ineffective cleaning is a waste of energy and will result in high maintenance costs. It is therefore essential that the economics of boiler cleaning are examined for each plant in the light of its effectiveness and return against expenditure. There are frequent instances where shell-type boilers are shut down and cleaned every weekend, incurring high labour costs which are seldom justified by subsequent fuel saving.

Cleaning routines established for coal firing have been perpetuated after conversion to oil-firing, where less frequent cleaning should be adequate. It is generally recommended that the frequency of smoketube cleaning be determined by a maximum of say $20°C$ rise in flue-gas temperature from the cleaned conditions for a given firing rate. If oil-fired, the fall off in thermal efficiency due to fouling in this case would be approximately 1%. By plotting boiler exit flue-gas temperatures daily at a set firing rate the desired frequency of cleaning can be established. Should the result of such an exercise indicate weekly cleaning of, for example, an oil-fired economic-type boiler, then the combustion performance of the oil burner should be examined rather than resort to such frequent cleaning.

When new boilers are commissioned, the initial flue-gas outlet conditions (carbon dioxide and temperature) at various firing rates throughout the turn-down range provide a useful reference for the effectiveness of future cleaning.

10.6 Standards and operational problems

Standards

Standards used in Boiler Design and Erection

Most industrial countries have drawn up Codes and Standards of Construction for those involved in construction and use of boiler plant. Most Codes require inspection and testing by some inspecting authority or body. No uniformity exists between standards, although there may be similarity; an intending purchaser should ensure that plant is designed and manufactured to meet standards applicable in the country in which the plant is installed, which may well not be exactly the same as in the country of manufacture.

In the UK, boiler makers are not obliged by law to follow any particular standard; however, the Factories Act places upon the owners of boilers responsibility for their sound construction and their strength, and purchasers can thus ensure that boilers are constructed to appropriate standards, usually to British Standards. These are formulated by Committees drawn from all bodies interested in boiler design, manufacture and use, and thus represent current good practice. For most plant of this nature, compliance with British Standards requires inspection and testing during all stages by an Inspecting Authority approved by the purchaser. Should an imported boiler be constructed to another standard, e.g. ASME (American Society of Mechanical Engineers), it should be verified that the design and construction will be acceptable within the provisions of the Factories Act.

Most other industrialized countries have their own Standards and Codes for design and construction, and in these cases the Inspecting Authority may be a State organization or a private body.

Again, it should be ensured that plant conforms to the Standards applicable to the country in which the boiler is installed, and that designs are so approved before manufacture. Drawings should thus be submitted to the appropriate Certifying Authority, and should indicate the particular Standards applied in the design and the extent of inspection and testing by a recognized Inspection Authority. Documents required before a plant is allowed to operate will include material and weld-test certificates, radiographic reports etc., before a final certificate of completion and test is issued.

The important details are:

The selection of and agreement on appropriate standards with the Certifying Authority in the country of installation.

The selection of an Inspecting Authority to meet requirements of the standards during construction, this Authority to be specified when the order is placed.

Consultation with the Inspecting Authority to ensure that adequate inspections and tests are made at specific stages, and to ensure correct documentation.

The Associated Offices Technical Committee in the UK is an association of Inspecting Authorities which is represented on many British Standards Committees and has associations with the International Standards Organization, the Comité Européen des Normes, and the Committee of European Inspecting Organizations. This is the kind of body to be consulted for inspections and tests during design, manufacture and construction.

Standards used in Operational Proving of Boilers

Many of the above remarks apply also to these standards, in that each country may have its own standard for operational proving. It is essential that early agreement should be reached on which standard (British Standard, ASME Code, DIN (German) or otherwise) shall be applied. It is the manufacturer's responsibility to prove to the purchaser or his agent that the boiler fulfils guarantees set down during the ordering of the boiler; the standards applied to any tests undertaken to prove boiler efficiency and running conditions are nominated as a basis for agreement on how these tests are to be carried out, and what calculation methods are to be used. Perusal of BS 2885 'Code for Acceptance Tests for Steam Generating Units' or ASME Power Test Code PTC 4.1 'Steam Generating Units' will indicate their scope, and a comparison will prove interesting. Cross-referencing to other codes used in testing other parts of the plant (e.g. dust testing) is made.

Insurance Requirements

Insurance of boiler plant had its origins in the 19th century and was the result of a number of serious incidents involving plant under steam pressure. The emphasis was on the prevention of accidents through adequate inspections before insurance cover was given. The inspection service is still a major part of the insurance system. Insurance in the UK can be split into five classes:

'Inspected Classes' Risks which cover the material damage to boilers etc. Linked with this is a periodical inspection which is mandatory by law.
'Contingency' Risks, which could include contractors 'All Risks' insurance for general erection.
'Consequential Loss' insurance, which covers losses in production or increased costs owing to plant failure.
'Works Damage' insurance in respect of damage to products etc. owing to plant failure.
Miscellaneous Risks.

Consultation with a member firm of the Associated Offices Technical Committee is recommended for specific plants.

Insurance abroad can be arranged under an individual policy, which would be subject to inspections complying with Government regulations of the country involved. Statutory inspection regulations apply in most countries, and *Table* 2 gives an indication of some of these. All material-damage insurances are usually dealt with under a combined 'machinery breakdown' policy in which there is no general warranty that inspection must be carried out. A reputable insurer should be consulted at an early stage in ordering.

Operational Problems

To anyone with an interest in the running of boiler plant from whatever aspect, it will be evident that plant availability is of prime importance. Unscheduled outage of any boiler would almost certainly involve stoppage of complete power units or process plants representing a large capital investment, and in certain cases may have dangerous repercussions. It will also be evident that, given a boiler unit correctly designed for its service, maximum availability can best be promoted by a correct and intelligent mode of operation of the unit, with due appreciation of the type of problem that might occur in a unit firing a particular type of fuel, or having a particular type of service loading.

Basic objectives for operators, then, given a well designed already commissioned plant with correctly

Table 2   Overseas boiler examination requirements

Country	Government Inspection Requirements	Inspecting Body
Argentina	None	Purchaser arranges
Australia	Annual inspection	Purchaser arranges
Austria	Inspection every 3 years	Purchaser arranges
Belgium	Annual inspection	Purchaser arranges
Bermuda	Periodic	Purchaser arranges
Brazil	Periodic	IRB
Canada	Periodic	Purchaser arranges
Denmark	Inspection every 5 years	
France	Inspection every 10 years	Purchaser arranges
West Germany	Annual inspection	TUV
Holland	Inspection every 2 years	ITEB
India	Annual inspection	Purchaser arranges
Italy	Periodic	ANCC
Luxembourg	Annual inspection	GAPAVE
Mexico	None	
New Zealand	Annual inspection	Purchaser arranges
Pakistan	Annual inspection	Province Boiler Inspector
Singapore	Annual inspection	Purchaser arranges
South Africa	Annual inspection	Purchaser arranges
Spain	Periodic	
Sweden	Periodic	SSU
Venezuela	Periodic	

IRB — Instituto do Resseguros do Brasil
TUV — Technisclur Uberwachungs
ITEB — Inspektie Taxatie en Expertise Bureau
ANCC — Association Nazionale per il Controllo della Combustione
GAPAVE — Groupement des Associations de Propriétaires d'Appareils à Vapeur et Electriques
SSU — Swedish Steam Users Association

Usually Government Inspectors undertake inspection work, and/or inspect the log book to ensure that an 'authorized' engineer has undertaken periodic inspections

References p 247

functioning water treatment and auxiliaries, are the following:

To ensure that the combustion is both stable and safe.

To ensure the efficient steaming of the boiler, i.e. to obtain the most efficient fuel/air ratio, assessing this from oxygen and carbon dioxide contents in flue gases.

To avoid smoke and grit emissions and corrosive conditions.

To avoid flame impingement and excessive metal temperatures in furnaces.

To avoid damage to burners both in and out of use.

Combustion Equipment Problems

Coal (see also Chapter 6). The regulation of the combustion process involves the control and distribution of air and fuel up to and into the boiler furnace and this involves milling plant, which prepares the fuel, burners, which mix and ignite the fuel and air and sustain combustion, and air-control equipment. The objective in general is to burn the fuel with a minimum of excess air. Limits to achieving this in the case of pulverized coal are the need:

To cool combustion products below the softening temperature of the ash before leaving the combustion chamber, thus avoiding tube fouling.

To avoid deposition of partially burned coal on wall tubes, causing corrosion by combustion in pockets of a reducing atmosphere.

To avoid acid corrosion of gas passes at the lower temperatures.

The primary air stream is critical, in that it has to be sufficient to carry the p.f. particles in suspension and to burn off volatiles. The ideal mixture is thus dependent upon p.f. fineness (a function of mechanical condition in the mill), and involves both volume flow and velocity of the air, as well as the temperatures.

Owing to their proximity to the furnace, air-damper blades may become distorted by radiant heat. Vanes may sometimes be left partially open for cooling, but this is undesirable since p.f.-pipe and joint failures may result, because hot gases may be drawn in, and unwanted excess air may be admitted to the furnace.

An incorrect primary/secondary air ratio can create a vortex in the primary-air/fuel mixture, and can draw in hot combustion gases. The burner nozzle would then be prone to damage by overheating and a local combustion-air deficiency could ensue.

Difficulties with front-wall firing may ensue if:

The primary-air diffuser is not withdrawn when a burner is taken out of service (causing the diffuser to overheat).

The primary air flows and p.f. burner become unbalanced across burners served by one mill group.

(Flame impingement, badly formed flames and a higher unburnt content of grits may ensue.)

The secondary-air vane openings chosen are unsuitable, giving flame impingement on the side walls if too small, and on the front wall if too large.

The burner quarl becomes defective owing to collapsing refractory. This would cause the secondary-air inlet to be irregular in contour and area. During erection or repair of burner quarls, strict dimensional accuracy is essential.

Burners should be inspected and slagging conditions etc. observed at every outage, for whatever reason.

Fuel pipes should be inspected; these are prone to erosion, mainly at bends and at the coal nozzles into the furnace registers. The quartz content of the coal tends to enhance its abrasive properties.

Oil. The combustion of oil fuels is in some ways more complex than that of coal, as the supply of oxygen for combustion is much more critical. Too little oxygen would present stack emission problems, and too much would give the risk of unacceptable corrosion rates. Operational success depends upon the burner/combustion chamber combination and is measured by the increase of unburnt products in the flue gas with decrease in the oil/air ratio towards stoichiometric.

Good mixing in a flame is largely dependent upon design features, and the only real efforts the operator can make, assuming that the burner manufacturers' settings are adhered to, are to keep the burner components clean with oilways free and orifice, swirl and distribution plates free from scratches. Air controls are usually simple, and settings of various vanes will be provided by the manufacturer during the commissioning stage, the control provided by the operator being to see that casing leakage is kept to a minimum. Typical results of air inleakage in terms of excess oxygen in boiler flue gases are shown in *Table 3*.

In view of the critical nature of the combustion, care should be taken in the siting of oxygen-meter probes in gas outlets. Ideally, several probes should be used, preferably with instantaneous multi-point sampling (this would even out velocity and chemical composition changes in the dust cross-section).

In oil burners used to support p.f. ignition:

High windbox pressure required for good oil burning may not be compatible with best p.f. ignition. Slagging of burners is a danger. This is the object of placing an oil burner down the centre. Any slag formed in front of a sprayer plant could deflect back burning oil into the support tube, and if this should result in burn-through of the tube, could ultimately result in a windbox fire.

Flame-failure devices may also be obscured by slag.

Gas. The combustion of natural gas in a furnace probably presents fewer difficulties operationally than that of most other fuels. There are, however,

Table 3   Air inleakage in terms of excess oxygen (typical). Courtesy CEGB and Pergamon Press (*Modern Power Station Practice*, 2nd edn, Vol.7, 1971, p44)

| | Combustion-chamber pressure (in w.g.*) | | |
| | −1.0 | −0.1 | −0.01 |
Size of aperture	Percent excess oxygen leaking in		
0	0.5	0.5	0.5
50 ft × $^1/_{32}$ in (15.25 m × 0.8 mm)	0.57	0.52	0.51
50 ft × $^1/_{16}$ in (15.25 m × 1.6 mm)	0.64	0.54	0.51
50 ft × $^1/_4$ in (15.25 m × 6.4 mm)	1.07	0.68	0.55
Inspection door: 4 in × 2 in (102 mm × 51 mm)	0.72	0.51	0.50
Access door: 18 in × 12 in (0.45 m × 0.3 m)	1.28	0.75	0.58

* 1 in w.g. = 249 Pa

certain factors to which strict attention should be paid to avoid unsafe running.

Two phases are necessary in lighting and operating a gas burner: preparation, resulting in safe lighting of a burner; and supervisory phase, i.e. observation of correct running.

Even with automatic systems, featuring flame-monitoring devices, operational factors can be overlooked which might result in unsafe running. Thus:

1. Check that shut-off valves are tightly closed before start-up. These should automatically close in an emergency, and ought therefore to be extremely reliable.

2. Ensure the absence of combustibles in the combustion chamber before light-up. Usually no positive check can be made for this, and an adequate purge time should be allowed, related to the amount of gas which could possibly accumulate in the combustion chamber. Particularly important during start-up following a flame failure.

3. Trial for ignition. The duration of this period should relate to the rate of potential combustion energy build-up in the furnace.

4. Pilot flame. If a flame-detection device proves the presence of a pilot flame and allows the main gas supply to flow, it should sense the pilot flame at a point where it cannot fail to ignite the main flame.

5. The function of safety shut-off valves is particularly important. Tolerable leakage rates are very low, as a small leak for a very long period could result in a considerable build-up of gas. Automatic valves should be backed up by manually operated valves; plug cocks with large lapped seats are satisfactory in this instance. Regular functional checks of all safety systems, including line vent valves, are important.

Mixed-firing and Alternative Fuels.   Problems involved in dual-firing systems are largely those of design, and many of the basic problems outlined

above are also applicable.   The basic reason for using dual-fuel systems on a boiler is in general to use some combustible material which is a by-product of a process on a particular plant. For instance, bagasse on a sugar plant and blast-furnace gas on a steel plant are both used in conjunction with some other more normal fuel in the boilers. The problems with this type of fuel lie in their variable calorific values and possibly their unreliable supply which may vary with plant processes. With bagasse, variable moisture content and consistency make the running of the boiler more of a manual operation.   Similar problems can be experienced with blast-furnace gas, where pressure fluctuations can be difficult.   The moisture content of blast-furnace gas is affected by the method of cleaning the gas.   The maintenance of clean, dry gas to all burners running on waste gases, and the routine maintenance of safety cut-off systems, is all-important.

Waste Heat.   Present temperatures available from gas turbines can in general be withstood by metals used in boiler construction, and in many cases the normally disastrous occurrence of a dry boiler can be withstood.   The largest problem operationally stems from vanadium and other deposits in the gas passes; this depends upon the gas-turbine fuel.

Gas-side Problems

On coal-fired boilers, several kinds of deposit may be formed: slag type, alkali-matrix type (bonded), and the phosphate type, only obtained from coals with a high phosphate content (see Chapter 20). Alkali compounds and phosphates are volatilized and condensed on cooler metal surfaces; the alkalis react with sulphur dioxide and sulphur trioxide forming sulphates which bond to the metal and pick up fly-ash on their molten surfaces and phosphates behave similarly.

No universal solution for the problem has been arrived at, but methods such as gas recirculation, humidification of combustion air, the use of additives to the fuel, and treatment of heating surfaces, can ameliorate the position. The addition of cuprous nitrochloride to coal gives a more friable slag, more easily removed and handled. But the correct application and maintenance of suitable air/fuel ratios is the best way of cutting down these problems to a minimum.

In oil firing, gas-side deposition is due in general to the presence of vanadium, sodium and sulphur in the oil. The vanadium and sodium compounds formed during combustion interact to form sodium vanadates, low-melting materials which form on the metal surfaces. Again, particular attention to combustion control is important. Solid additives to fuels, comprising compounds of magnesium, calcium, aluminium and silicon, are also used to avoid deposits.

High-temperature corrosion may also occur, particularly at metal temperatures above 640°C. In coal-fired installations, high-sulphur coals containing an appreciable amount of chlorine give the most severe corrosion problems. Tube temperatures should be kept below 560°C in most cases to avoid this type of corrosion.

Vanadium-containing fuel oils cause great corrosion, accelerated at above 640°C owing to the fluxing action of molten vanadium complexes. As with coal, additives can be used in the fuel to form higher-melting-point compounds. Magnesium oxide, dolomite, calcium, aluminium and silicon have the desired effect; their selection depends upon the temperature of operation, the composition of the oil, and the resulting deposit formation.

The routine maintenance of cleanliness by onload sootblowing and water-washing of low-temperature metal areas should not be overlooked.

Water-side Problems

Steam quality and purity are now of vital importance in many process plants, as well as in turbine use, where deposits on blades and erosion must be avoided.
1. Carryover in steam is normally taken out by the internal equipment in the boiler drum, e.g. baffles and separators; but sudden reductions or increases in load can seriously change their effectiveness.
2. Priming: caused by a very high water level, high steaming rate or low boiler pressure; slugs of water are present in the steam flow. Steam may carry less than 1% by volume of water, but this would represent more than 50% by weight, and could cause serious damage to plant. The condition can be identified by very high water levels or surging in the gauge glass. Additional factors are alkalinity of boiler water, high total solids, and organic matter and sludge.
3. Foaming is basically due to wrong boiler-water conditions, when the foam layer on the drum water is thickened so much as to be carried into the steam separators.

Carryover may be influenced by variation in class of coal burnt in stoker-fired boilers, as a result of a variation in boiler circulation. Other water-side problems involve the deposition of various salts on the tube internals. Correct water treatment will avoid this. Acid-cleaning, alkaline-boil-outs etc. will remove contaminants and a certain amount of debris from the inside of a boiler after such work as re-expansion of tubes has been undertaken; weld beads, mill scale and other debris must be removed by a steam or water purge.

Problems with Boiler Auxiliaries

Problems likely to arise with boiler auxiliaries are those associated with any type of moving plant, and the possibilities are too numerous to mention. However, it is important to point out that for the continuous, safe operation of a boiler its auxiliaries should be either foolproof or duplicated, to ensure safety. A decision must be made as to the basis of maintenance. With maintenance only at breakdown, there must be duplication of spares and maintenance staff. Preventive maintenance depends upon much greater planning and control. Factors other than economic are to be weighed when the decision is taken as to which method is to be used.

Problems with Control Systems

The main problem to be encountered, given a correctly designed and commissioned system, is an environmental one. The primary elements and transmitters etc. of a control system must be kept clean and dry and may need to be protected against undue heat. The important objective with the control system is to set up the system with a realistic approach to the ability of the unit being controlled and its response and duty.

A Brief Note on Boiler Efficiency

The bulk of the above objectives are necessarily concerned directly with the efficiency of the unit. Optimum settings can be established, by heat-balance tests, for boiler operation with minimum losses. However, these settings are subject to certain variable factors, such as boiler load, nature of the fuel, the state of fouling, the condition of combustion equipment, etc. Practical problems involved in the continuous optimization of boiler plant stem from the quality of fuel being burnt, and the difficulty of measuring small changes in boiler efficiency quickly and accurately.

A quick method of checking efficiencies on oil- and coal-fired boilers is outlined below.

Five sources of loss are usually taken into account:

Hydrogen and moisture content in fuel
Dry flue-gas loss
Unburnt gases
Carbon in residue
Radiation and unaccounted losses

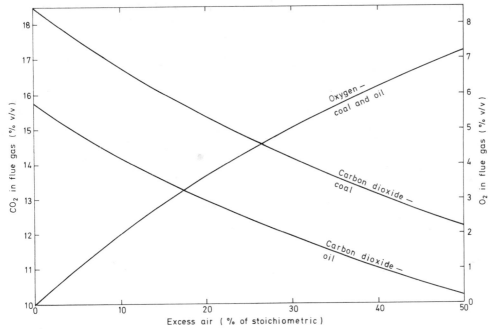

Figure 28   Excess air vs flue-gas composition for coal and oil

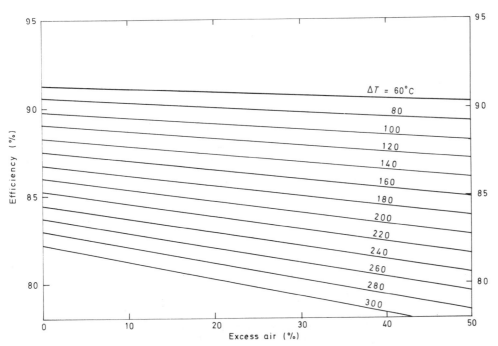

Figure 29   Efficiency of boilers at various values of ΔT vs excess air

The first two account for the major part of losses from a boiler. The other three account for about 1% of the losses, and have here been estimated from past efficiency trials.

The only losses under direct control of the operator are dry flue-gas composition and unburnt gas. Other losses are dependent on the fuel burnt and the boiler design. Thus the efficient operation of the boiler is mainly dependent upon minimizing the excess air supplied (i.e. lowering the air/fuel ratio) but also maintaining sufficient excess air to ensure the complete combustion of carbon monoxide. Reference may be made to *Figures 28* and *29*. The information required is:

(a)   Volume % of carbon dioxide in flue gas at economizer outlet.
(b)   Air-inlet temperature at F.D. fan inlet.
(c)   Gas outlet temperature at airheater outlet (or at economizer outlet if no airheater is fitted).
(d) $\Delta T = (c) - (b)$

(N.B. If a steam airheater is fitted and uses steam from an auxiliary boiler, the air-temperature rise across the heater must be deducted from ΔT.)

From *Figure 28*, read off the excess air value for a given O_2 or CO_2. Transfer this excess air value to *Figure 29*. Find the point of intersection between this and the temperature differential ΔT. Thus e.g. in a coal-fired unit, $O_2 = 5\%$ and $\Delta T = 100°C$, excess air = 30%; therefore the efficiency is 88.8%.

This method gives the boiler efficiency to within ±1.5% of that determined using the methods in BS 2885 'Code for acceptance tests for steam generating units'.

10.7 Use of steam

General Properties

Steam is a simple and relatively cheap form of energy carrier. It can give up its heat as such where required or, in the form of work, in a prime mover.

Steam Formation

When heat is added to water the temperature rises until boiling point is reached, which at normal atmospheric pressure is 100°C; under higher pressure this occurs above 100°C and conversely if the water is at a pressure below atmospheric it will boil below 100°C.

The heat which is not recorded by a thermometer but which is absorbed at boiling point during conversion to steam is termed latent heat or heat of vaporization. The units of heat which are apparently lost during the conversion at constant temperature have been employed in performing a certain amount of work, i.e. in changing liquid to gas and separating the water molecules.

As long as water remains liquid it cannot be colder than 0°C (except under great pressure).

Steam and water heat calculations are, therefore, based on the convention that water at 0°C has no heat. The standard unit of heat is the joule (J). The amount of heat required to raise the temperature of one kilogram of water through one degree Celsius (°C) is 4186 joules or 4.186 kilojoules, i.e. the specific heat capacity of water. To raise the temperature of 1 kg of water from melting ice to normal boiling point requires 418.6 kJ. When water at boiling point is further heated the temperature remains constant but some of the water turns into steam; to convert 1 kg of water boiling at 100°C into 1 kg of steam at 100°C we must add 2257 kJ, i.e. approximately 5½ times as much heat as would have heated that kilogram from freezing point to boiling point.

In order to convert steam back into liquid water this heat must be abstracted. It is by giving up its latent heat when condensing that steam can be made to heat materials. Because the latent heat is needed to force the molecules apart, the amount of latent heat involved in this change of physical state is greater at low pressures than at high pressures. Steam itself obeys (with deviations) the normal gas laws. But if the pressure is sufficiently high the volume of the steam becomes the same as the volume of water from which it was produced. At this pressure, called the critical pressure, which is about 22 MPa (220 bar), the latent heat becomes zero.

Thus the total heat needed to convert water initially at 0°C into steam at any specified pressure is the sensible heat needed to bring it to the boil plus the latent heat needed to convert the boiling water into steam.

When the transparent vapour (steam) is at a temperature corresponding to the liquid boiling point appropriate to its pressure it is said to be saturated steam. But steam may contain water either as drops or as an impalpable mist — the visible steam of common parlance; the water may be due to effervescence in the boiler or to partial condensation. Saturated steam in this condition is often referred to as wet steam, but steam that contains neither superheat nor water drops nor mist is called dry saturated steam. Because properties of steam cannot be expressed by simple relations, tables have been compiled giving them over a wide range of pressures and temperatures, and a copy of these steam tables is necessary in any calculations involving the use of steam[5].

When the steam becomes superheated it obeys the usual gas laws moderately closely.

Advantages of Steam

Steam possesses many outstanding qualities, e.g.:
(a)   It gives up its heat at constant temperature when it condenses. This property is very important. It simplifies plant design because counterflow arrangements are seldom needed. It gives complete and simple control of heating operations

and enables a heating operation to be exactly repeated.

(*b*)   It has a very high heat content. This allows a comparatively small steam pipe to carry a great amount of heat. For example, at 1.4 MPa (14 bar) a 200 mm pipe can carry (as steam) all the heat obtained from burning 2½ tonnes of coal an hour. A hot gas or hot liquid can carry heat in quantity if it is at a very high temperature which may be quite unsuitable for the process. Steam contains about 25 times as much heat as the same weight of air or flue-gas at the same temperature. Whereas air or flue-gas must fall in temperature to give up any heat, steam at (for example) atmospheric pressure can give up five-sixths of its heat without any drop of temperature.

(*c*)   The necessary water is usually relatively cheap and plentiful. To compete with steam the competing substance must be one that can be vaporized and condensed at a temperature approximating to that of boiling water, and have better physical properties. No other substance approaches water in its heat-carrying capacity. Any other substances that might have a claim are expensive or inflammable, or possess undesirable taste or odour.

(*d*)   It is clean, and in principle odourless and tasteless. This is important where the steam comes into contact with the processed material, e.g. in most food industries, textile treatment, dye works, bleacheries, laundries etc. Care must be taken to ensure that certain boiler feedwater additives such as filming amines are not carried over in the steam (see Ch.11).

(*e*)   Its heat can often be abstracted in successive stages by lowering the pressure. When steam is used for evaporating liquids containing water most of the original steam heat is contained in the outgoing vapour. The heat in this vapour can often be used for another heating or evaporating operation, and in appropriate circumstances this process can be repeated five or six times giving great economy (i.e. multiple-effect evaporation[6]).

(*f*)   It can first generate power and then be used for heating. This is perhaps the most important property of steam. Wherever steam can be generated at reasonably high pressure and used for process at moderately low pressure, and where power is required, some or all of the power can be produced by using the highest practicable boiler pressure, reducing the pressure by passing this steam through an engine or turbine, and exhausting at the pressure required for the process.

(*g*)   It can be easily controlled and readily distributed. Only a simple valve is needed to control the flow; this can be hand-operated or automatically controlled by temperature or pressure. Steam, especially if superheated, can be sent quite long distances in pipes. Provided the pipes are well lagged and the pressure-drop needed can be tolerated, it can be transmitted economically for more than 1 km in favourable circumstances.

Steam was described by Sir Oliver Lyle as 'Industry's most wonderful, flexible and adaptable tool'.

There are three important and fundamental rules in relation to the use of steam.

1. For heating and process: remember that with reduced pressure the boiling point of water decreases and the latent heat of steam increases.

2. For power: use the highest practicable initial pressure and temperature, and use the lowest practicable exhaust or back pressure.

3. For any purpose: never permit steam to expand from one pressure to a lower pressure without getting some useful result from the expansion.

Pressure

Normal atmospheric pressure is 100 kPa (1 bar). The pressure normally read from a pressure gauge is called the gauge pressure; the real or absolute pressure is the gauge pressure plus atmospheric pressure. So a gauge pressure of 5 bar is an absolute pressure of 6 bar, and of zero is 1 bar (100 kPa, 1000 mbar) absolute. This is true unless the barometric pressure is abnormally high or abnormally low.

The ordinary pressure gauge is rather unreliable for measuring pressures below atmospheric, although widely so used. In such cases we speak of measuring vacuum, i.e. the amount *below* atmospheric pressure or a negative pressure, frequently expressed for industrial purposes as — millibar (mbar).

Note.   The use of SI units is growing rapidly throughout the world; but in the decade 1970—1980 superseded units continue to be used, and dual usage (even on gauge calibrations) will be common. Often, even individual sections of an industry such as the boiler industry may temporarily adopt divergent conventions. Pressure units are a particular problem in the 1970s, and final adoption of a single standard will simplify thought and calculation immensely. The basic (coherent) SI unit is the Newton per square metre or Pascal (N/m^2 or Pa). The convenient approved derivatives are 10^{-3} or mPa, 10^3 or kPa and 10^6 or MPa. But the bar (10^5 Pa) has been much used meanwhile because it is about the average value of atmospheric pressure; particularly in interim steam tables, previously in atm which was 1.013 bar. Similarly the mbar (100 Pa) is used in meteorology and still in some industrial vacuum measurements. kgf/cm^2 or at (98.07 kPa), lbf/in^2 or psi (6.895 kPa) and other superseded units such as dyn/cm^2 (0.1 Pa), torr or mmHg (133 Pa) and in w.g. (249 Pa) cause needless complication and are best avoided.

Dryness Fraction

It has already been mentioned that steam when formed may contain moisture in suspension. Further application of heat to this wet steam induces the particles still present to evaporate, but so long as steam is in contact with the water surface from which it was formed it remains saturated. If (e.g.) 0.9 kg/kg is the dryness fraction, the steam contains 10% of suspended moisture; hence instead of steam at atmospheric pressure containing 2257 kJ of latent heat it will actually contain only $0.9 \times 2257 = 2031$ kJ. It is as well to note here that many impossible results have been claimed for boilers through the neglect to estimate the quality of the steam obtained (the dryness). If the steam is assumed to be dry and 10% of moisture is in fact present, the evaporative efficiency of the boiler will be exaggerated. Suppose that a boiler were to supply fully dry steam at a pressure of 7 bar absolute corresponding to a temperature of $165°C$ from feedwater at $15°C$, the total heat of evaporation would be $2066 + (697 - 63) = 2700$ kJ/kg (63 kJ/kg in the feedwater). If however there were 10% of suspended moisture, the total heat expended per kg of wet steam would be $(0.9 \times 2066) + (697 - 63) = 2493$ kJ. Therefore the gross weight of water which would appear to be evaporated would be relatively $2700/2493 = 1.08$ or 8% more than the quantity had the steam been dry.

The wetness of steam is determined usually by the throttling or wire-drawing calorimeter. High-pressure steam is wire-drawn through a small nozzle into a chamber at lower pressure, the droplets of water evaporate under the reduced pressure (almost atmospheric), and dry steam is produced which becomes superheated. The temperature of superheat is the measure of the quantity of moisture originally carried in the steam. The dryness fraction of the steam is expressed by the formula

$$f = \frac{(2676 - h) + 1.93\,(t - 100)}{L}$$

where f = dryness fraction, h = sensible heat of water corresponding to boiler pressure, t = temperature of steam after the orifice, and L = latent heat corresponding to boiler pressure.

The superheating effect is very small at the lower pressures and can only evaporate a small amount of wetness; thus there is no guarantee that all the moisture in wetter samples has been evaporated. The throttling calorimeter therefore cannot be regarded as a precision instrument. Sometimes a separating calorimeter is used or both types of calorimeter may be used in series, the separator being usually before the throttling orifice. The results are then calculable from steam tables, and the combination gives the advantages of both types and is not limited in range.

Superheated Steam

As long as steam is in the presence of water it is impossible to raise its temperature above the appropriate boiling point. If however heat is added to steam that is not in contact with water the temperature will rise; it is then superheated and will expand if at constant pressure. If the steam is confined in a closed vessel the extra molecular energy added in superheating will result in an increase of pressure.

From the moment steam leaves the boiler whether it is saturated or superheated there is an immediate cooling and if no reserve of heat has been imparted in the form of superheat the steam begins to condense as the result of loss of heat. The particular advantage of superheating lies in this reserve, because the whole of the superheat must be dissipated before condensation can begin. Superheating thus serves two purposes: (1) the avoidance of condensation and (2) the increased work available from the superheated steam on account of the additional temperature when it is used in a prime mover.

If heat is applied at constant pressure to 1 kg of dry saturated steam to superheat it from t_1 to $t_2°C$ the quantity required is found by multiplying the temperature rise by the mean specific heat of steam over that particular range of temperature. If superheat is added solely for the prevention of condensation or in conjunction with process steam supply, the following points should be noted. Steam used for process heating or boiling is usually employed in one of two ways, in a calandria type of vessel such as a jacketed pan or by passage directly through a steam coil. If steam is used in a steam jacket, the latent heat at the prevailing pressure is surrendered to the heating surface and condensation is deposited there. The mass of steam within the jacket is at a uniform temperature, again corresponding to the prevailing pressure. But if superheated steam be employed, immediately it enters the jacket it is desuperheated through coming into contact with the condensate; the superheated steam entering is reduced to saturated steam which then surrenders its latent heat. Thus the heating surface is maintained at the temperature of the condensed steam in either case.

The situation in the steam coil is different. Superheated steam entering the coil does not come into contact with condensate and is not desuperheated. Upon entry the superheated steam (as a dry gas) imparts heat to a dry heating surface and in this early stage the superheated steam gives an appreciably lower rate of heat transfer than would saturated steam at the same pressure. This is because superheated steam behaves as a gas until it has lost all its superheat. In other words, although the superheat is indicated by a relatively high temperature the additional quantity of heat it carries is slight, and if the superheated steam is flowing along a heat-transmitting surface this slight quantity of superheat is the total quantity of heat it will surrender despite a considerable fall of superheat temperature in the distance of flow necessary for saturation to supervene; the main surrender of latent heat then begins. Where the steam space is large and the steam path short, as in some calandrias, superheated steam can sometimes be used without any lowered heat transfer provided the degree of superheat is not too great.

Thus in neither case does superheat offer (with rare exceptions) any advantage over saturated steam for heating. Indeed it is usually disadvantageous. When steam is required for process work from a superheated steam supply, after employment of the exhaust from an associated prime mover has first been considered, desuperheating may prove worthwhile.

Superheated steam should not be fed to sterilisers — particularly hospital sterilisers which have short time cycles — because it cannot give up its latent heat until cooled to the appropriate phase-boundary temperature. Furthermore the rate of killing of bacteria depends on the degree of moisture present with the heat. But the quality of steam fed to sterilisers should be as near the dry saturated condition as possible, with a dryness fraction not less than 0.98.

In important contrast, superheat when used for power generation ensures that the steam reaches the prime mover in a dry state; it may then be used expansively without undue condensation. When steam gives up some of its heat energy to do useful work by expansion in an engine or turbine it loses heat, cools, and unless appropriately superheated must partly condense. Superheating reduces the steam consumption of engines by approximately 1% for each $5\frac{1}{2}°$ Celsius of superheat, but the rate is rather greater for the lower degrees of superheat and less for the higher degrees. In other words superheating makes it possible to increase the power which prime movers may develop over what they would produce with saturated steam. This gain in economy may be shown by reference to an example. Suppose we have steam entering at 1.2 MPa or 12 bar gauge pressure, $85°$C superheat, and exhausting at a vacuum of -950 mbar (pressure 5 kPa), the relative steam consumption may be set at 1.0; but with dry saturated steam it becomes 1.158 corresponding to an increase of 15.8% in steam consumption. If the superheat is increased from $85°$C to $170°$C the steam consumption becomes 0.89 or a reduction of 11%. Thus the steam consumption with $170°$C superheat is better than with dry saturated steam by 1.16/0.89 or by 23%; an average of approximately 1% gain for each $7°$C of superheat is obtained. The introduction of superheat for reciprocating engines was the biggest single improvement since James Watt's invention of the separate condenser.

Flash Steam

Much energy is wasted by failure to collect and utilize waste heat. The collection of waste heat and its second or even third use is probably by far the largest heat saver, but having said that, the matter must be approached with deliberation and circumspection; for unless there is a use for the flash steam it is no use collecting it and unless the hot condensate can be pumped there is no use returning it.

Consider the formation of flash steam. The temperature of the hot water leaving a steam trap will be at or just below the water boiling point corresponding to the pressure. If a piece of plant is supplied with steam at 7 bar gauge the condensate will therefore be discharged from the trap at about $170°$C. In this state the water will contain about 721 kJ/kg. If it is discharged into a tank at atmospheric pressure some of it must at once boil off, using up latent heat until the temperature has fallen to $100°$C. It must therefore get rid of $721-422$, i.e. 299 kJ surplus heat; a flash of steam is the inevitable result. As the latent heat of steam at atmospheric pressure is 2257 kJ/kg the amount of flash steam produced will be $299/2257 = 0.133$ kg/kg of condensate. Flash steam can be piped to an evaporator, calorifier or vat or can be used for heating in a contact heater.

In many situations however it will hardly pay to collect flash steam. Suppose a trap is draining a steam pipe at 3.5 bar; apart from the heavy condensate load at the start-up the discharge from the trap may only be 4 or 8 kg/h. If this condensate is allowed to flash to atmosphere 9% of the condensate will be flashed into steam, but unless this flash steam can be collected very cheaply, with a minimum of piping, it clearly will not pay to collect it.

It therefore pays to avoid a loss of flash heat and this can be accomplished in several ways. The water can be passed through a heat exchanger before being discharged at lower pressure, the flash from high-pressure condensate can be piped to a low-pressure main, or the flash steam at atmospheric pressure can be used to heat water.

As flash steam is liberated from overheated water almost instantaneously, the flashing can be very violent. Any flash vessel must be sufficiently large to allow proper separation of the steam from the water without the carryover of a lot of water droplets. Two principal points should be borne in mind; the first is the amount of steam that experience with steam accumulators has shown can safely be liberated from a water service without undue carryover: flash steam (km/m³ h) = 210 × absolute pressure (bar). In practice this formula gives a larger flashing surface than the cross-section of the vessel, and the rule will therefore always give the upper limit for the flash-tank size.

The second point is that the steam velocity in the flash tank must be limited to such a speed that entrainment of water drops is unlikely. This limiting speed can be little better than a guess, but expressed as a low limit it can be said that the flash tank should never be so small that the velocity of the flash steam exceeds 3 m/s.

Boiler blowdown, particularly from high-pressure boilers, contains a great deal of flash heat: e.g. blowdown from a boiler working at 17 bar gauge contains 465 kJ/kg more than it can hold at atmospheric pressure. This will cause 20% of the blowdown to be flashed off as steam. If the blow-down amounts to 5% of the water fed into the boiler, the flash from the blowdown, unless it is recovered and used, represents a loss of 1% of the steam output of the boiler. If the blowdown is intermittent it is recognized that it is difficult to arrange for flash recovery, but if continuous blow-down can be arranged (and this is often desirable) various means are available for recovering the flash heat.

Consider a dilute process liquid at $95°C$ with a following process-temperature of $65°C$: a suitable way of cooling it is to spray it into an empty vessel connected to a condenser and a vacuum pump. The steam tables show that if the vacuum is main-tained at -760 mbar ($+24$ kPa) the liquid will boil at $65°C$; it will therefore flash-off its surplus heat and lower its temperature immediately from $95°C$ to $65°C$. The steam tables also show that the surplus heat will be $398-272$, i.e. 126 kJ/kg. The latent heat of steam at -760 mbar vacuum is 2346 kJ/kg, so $126/2346 = 0.05$ kg of steam/kg of liquor will be flashed off. This may save subsequent evaporation, and if the condenser takes the form of some kind of process heater, all the heat can be recovered.

Distribution, Storage and Process Requirements of Steam

General

Generally speaking a steam pipe should carry steam by the shortest route in the smallest pipe with the least heat loss and the smallest pressure drop that circumstances will allow. Quality of steam in distribution is important. Steam for power should be as hot and as high in pressure as circumstances allow. Process steam however should usually be at as low a pressure as possible and should not be highly superheated. In the design of steam-distribution systems where sizing has been carried out on the basis of a specific pressure drop between source and consumer the velocity also needs to be checked.

Steam leaving a boiler at too high a velocity entrains droplets of water, resulting in wet-steam distribution. Recommended velocities are: steam exit from boiler, 12 m/s; exhaust steam, 18–23 m/s; dry saturated steam, 23–30 m/s; superheated steam 36–48 m/s. Steam should be generated at the highest pressure required by any consumer, plus an addition for pressure drop in the pipeline to the consumer.

In deciding pipe sizes there are, generally speaking, only three main considerations, apart from cost: (1) What pressure drop along the pipe can be permitted? (2) What steam velocity can be tolerated? (3) What is the most suitable stock pipe-size?

If steam is being fed to or coming from a reducing valve there is sometimes a temptation to use small-diameter pipe in order to allow some or all of the pressure drop to take place in the pipe. There are some objections to this; first of all the pressure drop will vary with changes in the rate of flow and this may give considerable operating trouble; secondly, at the very high velocities which accompany a large pressure drop, the steam may quickly erode the pipes, especially at bends; thirdly, steam flowing in a pipe at high velocity (even tenuous vapour at high vacuum) can be extremely noisy.

Except in the case of long runs, it is wasteful of fuel to feed steam at a higher pressure than is necessary and subsequently pass it through a pres-sure-reducing valve. Steam reduced in pressure by this means performs no useful work and in the process becomes slightly superheated.

Metering

It is often essential for costing purposes to meter the steam flow, and meters are also useful for conducting boiler efficiency tests. They can be either of the carrier-ring orifice-plate type with condensing chambers and remote indicator/recorder, or electrically operated with mercury filled manometers instead of condensing chambers, or direct-reading shunt meters.

Generally speaking carrier-ring orifice-meters (Ch.18, Figure 38a) with a remote indicator and recorder of the circular or strip chart type are used; but where individual-outflow steam lines are metered, direct reading shunt meters with integration are usual. Whether the listed steam outflow (excluding auxiliary and soot-blower steam) is metered, or whether individual meters on outflow from the steam header are used, depends on individual requirements for costing steam consumption.

With a dual or triple-dual boiler installation, metering of steam output can normally be achieved by selecting a suitable time and load and running on one boiler at a time. For multi-boiler installations, metering of the steam flow separately off each boiler is usually required.

Reducing Valves

The higher the pressure at which steam is generated the higher is the capacity of the pipeline; or if a certain amount of steam has to be transmitted, the smaller is the pipe needed to carry it. By reducing to a minimum the size of pipe required to carry the steam two things have been achieved — a reduction of the capital cost of the transmission plant and a reduction of the heat losses from it. But steam at high pressure is not so valuable for process work as steam at a lower pressure, because the higher the pressure the lower the latent heat. It may therefore be found that it is better to generate at a higher pressure than the pressure at which the steam is to be used and to reduce the pressure at a certain place in the distribution network.

Pressure reduction of steam is effected by pressure-reducing valves, and it will be seen from calculations that reduction of steam pressure in this way improves the dryness fraction and can superheat the steam.

There are three types of pressure-reducing valves: those that employ an external source of power to operate the main through a pilot or relay valve, those that employ the steam pressure itself to operate the main valve through a pilot valve, and simpler direct-acting valves that are spring- or weight-balanced. The first type are somewhat uncommon in the more normal process industries; they find their place on high-pressure and high-capacity installations and where extreme accuracy of process-control pressures is needed. A direct-acting reducing valve is cheaper, but control cannot be as accurate as with the pilot-operated type; nor should the direct acting valve be expected to shut off steam completely at times of no demand without a noticeable rise in downstream pressure. The correct sizing and installation of pressure-reducing valves and their protection against water, scale and dirt build-up are vital if they are to give accurate and trouble-free service. A safety valve should always be fitted downstream.

Pipe Drainage and Condensate Removal

If a steam pipe is not properly drained the steam in it cannot reach the plant it is feeding in a dry state. If an appreciable amount of water is allowed to collect in a pipe, water-hammer may occur. This condition is caused by the water being given kinetic energy and set in motion at high velocity in the pipeline until stopped, for instance by a pipe bend. The water then gives up its kinetic energy in the form of blows or impact; in many cases excessive stresses can be set up in the pipework system. The dangers from excessive water-hammer have resulted in a number of fatal accidents, as well as extensive damage to plant and equipment. In order to avoid water-hammer and to ensure that steam is reaching the plant in as dry a state as possible it is essential to observe recommendations on pipework gradient, to install steam separators where appropriate, and to fit an adequate number of drain pockets. For drain pockets, a full-size T-piece or other full-bore collecting pot into which the water is certain to fall should be provided; once it is in the pocket a steam trap can then remove it. It is quite ineffective and can be dangerous to screw small pipes into the bottom of the steam main. They seldom or never remove water until it is present in dangerous amounts.

The condensate resulting from the condensation of steam onto a heating surface having been kept to a minimum by supplying the plant with as dry a steam as possible, and having been made to leave the heating surface as quickly as possible, must be drained from the inside of the heating surface so that it does not waterlog the plant and prevent access of steam to the surfaces. This draining must be done in such a way that only the water is removed; steam must not accompany the water and be thus wasted.

Some plants are still fitted with hand-operated drain valves or cocks which rely on an operator for efficient working, but the only way to keep a steam space clear of condensate without waste of steam or risk of waterlogging is by the use of a device which will automatically release the condensate and trap the steam; that device is the steam trap.

Assuming that the right kind of trap has been fitted to the plant in the right way, the condensate may be returned via condensate lines back to the boilerhouse; if it is considered to be too far away from the boilerhouse to justify piping it back, some other use should be considered for it rather than running it to waste. For example, can it be used locally for process purposes, or can it be used for a canteen, or toilets or the like? If it is to be used for boiler feed it is generally collected in a tank called the hot-well, from which the boiler feed-pump draws its supply. A pump will not lift very hot water because the reduction in pressure causes a flash of steam which 'breaks the suction'. In order to keep the hot-well cool enough to permit the pump suction to operate, cold water has often to be run in at the wasteful expense of the heat in the condensate. Particular care should be taken to ensure that a sufficient suction lift or pressure head is available; this is determined by the temperature of the water itself.

Wherever possible condensate should drain by gravity to a receiver. A minimum gradient to ensure flow is 1 in 240. Most steam traps will discharge condensate to a height above the trap outlet, the height depending on the pressure of the steam inside the trap, which drives the water out. It is possible to raise water approximately 1 m for every 10 kPa or 100 mbar pressure gauge inside the trap, but not all condensate can be lifted in this way. Steam traps which drain steam mains should not be allowed to discharge their condensate above their own level, lest dangerous water-hammer be induced at the start-up, when the amount of condensate is heavy and the pressure probably insufficient to lift it. Generally speaking it is not good practice to lift the condensate above the trap outlet, but there are occasions when it is convenient to do so, and apart from the draining of steam mains there is often little risk in practising this. If the trap discharges at low level the condensate needs to be raised by pump or by other means to the high-level tank. Usually electrically driven condensate pumping sets are used for the purpose, but sometimes a lifting or Ogden trap is used. Technically speaking a lifting or return trap is not a trap but a pistonless pump. Inside the body of the pump is a float operating the valve gear; when the float is low, water can run into the trap by gravity through a non-return valve, but when the trap fills, the float trips the valve and allows the steam supply to blow the contents of the trap through a non-return valve to an elevated tank. When the float falls as the trap empties it cuts off the steam supply and opens a release valve which blows off the steam pressure in the trap. The exhaust steam from the trap can be blown into the water in the tank or sent to any other convenient destination. If the exhaust from the trap is merely discharged to atmosphere a lifting trap can be very wasteful. A lifting trap can also be used to remove condensate from a heating surface under vacuum, the exhaust steam being returned to the heating surface. It is a very wasteful device to use in draining a condenser.

Condensate from high-pressure steam lines and at low pressure from equipment should be taken back separately from the condensate-receiver or hot-well, and condensate from live-steam pipes or equipment should not be taken into a pumped-condensate-return main.

Steam Traps

A steam trap could be described as a device which distinguishes between water and steam, automatically opening a valve to allow the water to pass out, but at the same time trapping the steam and preventing its escape. Most steam traps also allow air from a steam-distribution main to escape, but air may also be discharged by means of an air vent or eliminator in much the same way as a steam-trap discharges condensate.

Condensate and air when present in any steam-using plant have the following detrimental effects on the efficient operation of the plant: (*a*) they take up heating space which should otherwise be occupied by steam, and (*b*) they cause a water film on heating surfaces, or a pocket of air, which restricts the transfer of heat.

Steam traps can be classified into three main groups:
1. Mechanical: where the trap operates owing to difference in density of steam.
2. Thermostatic: where the trap operates owing to difference in temperature of steam and water.
3. Thermodynamic: where the trap operates owing to difference in velocity of steam and water.

Steam traps and air vents are required to operate under a number of varying conditions, including varying steam pressure, varying steam load, restricted space (which makes maintenance difficult), and exposed outdoor position where there can be a danger of damage from frost.

It is not good practice to fit a steam trap in the outlet of an item of plant without including one or more items of associated equipment, i.e. strainers (if not already incorporated in the trap itself), sight glasses, and check valves.

Strainer. As a trap is fitted at the lowest point of any run it is liable to become a receptacle for dirt, pipe scale and other extraneous matter which can adversely affect its operation, and a strainer should therefore be fitted before the trap to intercept such materials. The strainer will require frequent and regular cleaning.

Sight Glass. In order to ensure that the trap is operating and discharging condensate it is usual to

fit a sight glass at a distance of about 1 m from the trap outlet. This enables a visual check to be made at all times.

Check Valves. In applications where the condensate is lifted to an overhead condensate main a check valve (non-return valve) should be fitted after the sight glass and before the rising line. This will prevent any back-flow to the trap.

There is a danger that when traps are placed out of doors freezing may take place in winter if the steam is turned off. Thermodynamic or bimetallic types of traps are not affected by freezing up. If however traps such as bucket types are used it is possible to protect them with thermal insulation, and in these cases the trap manufacturer should be consulted.

Air Vents

When steam is shut off and the plant cools down, the steam inside condenses, forms a vacuum and draws air into the plant through all the tiny leaks at joints and bends that are invisible or ignored when the plant is under pressure. Air may also be present in the steam as it passes along the steam mains.

When steam condenses on a heating surface in a piece of plant this air cannot condense, and a blanket or film of air is gradually built up between the body of the steam and the condensate film. Apart from this insulating effect, the air mixed with steam lowers its partial pressure and hence its temperature. Air can be removed by permanently open vents, hand-operated cocks or automatic venting valves; but often no provision whatever is made for this.

In some instances, where several different steam pressures are operating in a factory, the plant working at a high pressure is fitted with permanently open air vents piped into a lower-pressure main. Whilst this gives very satisfactory results in the high-pressure plant it makes the task of properly venting the lower-pressure plant all the more important. Automatic air vents give excellent service, but they must be examined regularly and maintained properly. It is quite useless to fit good air vents unless in the right places, but unfortunately many pieces of plant have in fact no right place. All steam-heated plant should be designed so that the steam is made to follow a definite path to a point remote from the steam inlet; the air will then be driven to this remote point which is the place for the air vent. In this context remote means merely the end of the steam's journey.

The following brief notes list the characteristics of the various types of traps and air vents available.

Ball-float Trap (mechanical, blast discharge)
Advantages: robust; can withstand water-hammer.
Disadvantages: bulky; poor air discharge; can be damaged by frost;
Inverted Bucket Trap (mechanical, blast discharge)

Advantages: robust; can withstand water-hammer.

Disadvantages: bulky; poor air discharge; requires a water seal; drop in steam pressure can 'flash-off' water seal; can be damaged by frost.

Open-top Bucket Trap (mechanical, blast discharge)

Advantages: robust; can withstand water-hammer.

Disadvantages: bulky; poor air discharge; can be damaged by frost.

Balanced-pressure Trap (thermostatic, continuous discharge)

Advantages: small bulk — large capacity; valve open when cold — (a) discharges air freely on start-up, (b) capacity is greatest when required; discharges air adequately and continuously; not affected by varying steam pressure; cannot freeze unless discharging into a vertical rising line.

Disadvantages: element can be affected by water-hammer; slow response.

Liquid-expansion trap (thermostatic, intermittent discharge)

Advantages: discharges condensate at preset temperatures; valve open when cold; not affected by vibration, water-hammer or variations in steam pressure.

Disadvantages: can waterlog steam space; slow response; will not discharge condensate as it forms; not suitable for varying steam load nor sudden variation in steam pressure.

Bi-metallic Trap (thermostatic, continuous discharge)

Advantages: small bulk — large capacity; valve open when cold; can withstand water-hammer and variations in steam pressure; acts as a check valve.

Disadvantages: slow response; can waterlog steam space; not suitable where condensate discharge must be at steam temperature.

Ball-float Trap (thermostatic, continuous discharge)

Advantages: robust; withstands reasonable water-hammer; not affected by wide and sudden fluctuations in steam pressure or load; best type to use where steam supply is thermostatically controlled.

Disadvantages: subject to damage by heavy water-hammer; can be damaged by frost.

Thermodynamic Trap (thermodynamic, blast discharge)

Advantages: works over full range of pressure; can withstand vibration and water-hammer; small bulk — large capacity; unaffected by frost; robust; disc acts as a check valve.

Disadvantages: noisy; can air-bind if pressure builds up too quickly.

Air Vent (manually operated)

Advantages: small bulk; relatively inexpensive; discharges air/steam mixture — can easily be seen.

Disadvantages: depends on manual operation — can be forgotten; small discharge capacity; if left open can be wasteful of steam.

Automatic Air Vent (automatic operation)

Characteristics are similar to the balanced-pressure steam trap.

Loss of Heat from Steam Mains

It is evident that as far as possible heat must be prevented from escaping to places where it is not wanted. Apart from steam leaks (obviously wrong), steam loses heat from the pipes and vessels that contain it to the surrounding air. Of all the methods of saving heat, the curing of leaks and the prevention of heat leakage by lagging with insulating material the hot surface are the easiest and most straightforward. In these days of relatively high fuel costs it is all the more necessary to conserve heat by adequate lagging of the services. Lagging materials vary in price, conductivity, ease of application and length of life, and in all cases economic exercises should be carried out to determine the thickness (App. C11, equations 17, 18) and type of insulation required. This is done on the basis of relating the cost of insulation to the cost of the heat loss from the unlagged main itself to arrive at an economically justifiable thickness of insulation. The use of asbestos for this purpose is no longer recommended. Croccidolite, which is a component of blue asbestos, causes cancer if inhaled. Very significant savings can be made by the insulation of pipes and equipment.

Steam Peaks and Thermal Storage

Maximum efficiency in boiler operation can be obtained only under a reasonably steady load which results in efficient operation of the plant. Excessive fluctuations in the demand for steam militate against efficient boilerhouse operation; there is also a tendency to increase the thermal losses that are inevitable in steam raising. Variations in pressure in the steam system result and often lead in turn to the slowing-down of processing with consequent damage to the product. A sudden peak demand can also result in priming (water carryover) in the boilers owing to sudden drop in pressure. Quite apart from the carryover of impurities with the steam, this in turn leads to a higher moisture content of the steam and condensation in the mains. The cumulative effect is a significant increase in the amount of fuel required per unit of product.

Such a situation may, also, result in an inferior product, e.g. inadequate sterilization of beer in the brewing industry, faulty milk-processing, damage to rubber products, and the like. Everything possible should be done to prevent peaks and valleys in steam demand, more particularly today, where current boiler plant — especially packaged boilers— will not cope with any significant overload of stated capacities. Even where plant is capable of coping with overload conditions there will always be a time lag in response to a change in firing rate. If therefore reasonably steady load conditions cannot be obtained, consideration should be given to some form of thermal storage.

Thermal storage can be applied in one or more of several ways. In a feedwater accumulator the temperature of the boiler feedwater can be topped up by surplus steam not required elsewhere in the plant; but the peak carrying capacity of this method is limited in comparison with a live-steam accumulator. In the latter, surplus steam is stored in water under a gradually rising pressure and regenerated as steam under a falling pressure. In an exhaust-steam storing system, the exhaust steam from say a colliery or rolling-mill engine is stored intermittently and regenerated at a more or less steady rate to meet the needs of a mixed-pressure turbine, thus levelling out the peaks and valleys between the main consumer and the powerhouse.

The thermal-storage boiler which was developed some years ago can also be of assistance in meeting a fluctuating steam demand. Provision is made for meeting the peaks and valleys by substantial and controlled rise and fall of water level. The shells of these boilers are considerably larger in diameter than in normal boilers for a given amount of heating surface, and are designed for pressures higher than those required in the particular factory or steam-using plant. The required working pressure or process pressure is held constant by a reducing valve.

Steam for Process

Four broad groups cover general process heating. These are:
1. Evaporation by means of a heating surface.
2. Raising the temperature of the process substance.
3. Maintaining the temperature of the process substance.
4. Distillation by open steam.
The first group covers calandrias in which the steam for heating is on the outside of the tubes and the liquid to be boiled is inside the tubes; coils or worms where the steam is inside the tube and the boiling material outside; and shells where the steam is inside a cylinder whose exterior heats the material, e.g., ironers, calenders, or jackets heating via the interior surface (boiling pans etc). The second group raises the temperature of the process substance by either a heating surface or by direct contact. In the third group the temperature of the process is maintained either by heating surface or by direct contact; the heat is employed to make good losses through radiation or heat absorbed by endothermic action, or to effect a curing or conditioning of the material. The fourth group is used for fractional distillation or for the deodorizing of fats etc.

Heat transfer is the main problem of groups 1, 2 and 3, but group 4 is different because heat transfer is not the main factor in steam distillation.

In this process open steam is blown into a liquid, and when the liquid has a much higher boiling point than the water, the steam and the process-liquid both exert a partial pressure, and the liquid will boil when the sum of the two equals the external pressure. It follows that substances with high boiling points may be distilled at relatively low temperatures. Processes in which steam distillation is used include stripping of crude benzole from wash oil in the carbonization industries and deodorization of fats in the food industries. It is particularly useful for processing materials that are liable to decomposition or degradation at temperatures near their normal boiling points.

The rate of heat transfer through a heating surface from the steam to the material which is to be heated depends on the following factors: the temperature difference between the steam and the material; the actual temperatures; the area of the heating surface; the thickness and composition of the heating surface; the nature and thickness of heat-resisting films on the heating surface; the movement of the steam, and the movement of material (see App.C11).

Efficient Use of Steam

Steam users should search continuously for ways to save steam. The following simple rules should prove helpful in this respect.
1. Prevent the escape of heat by:
   (a)   Lagging of steam and process pipework.
   (b)   Attending to leaks.
   (c)   Reducing excessive boiler blowdown.
   (d)   Reducing unnecessary safety-valve blows.
   (e)   Ensuring adequate maintenance and operation of steam traps.
   (f)   Avoiding condensate and flash-waste.
   (g)   Preventing the overflow of hot-water tanks.
2. Reduce where possible the net heat required by:
   (h)   Lowering the process temperatures.
   (i)   Partial removal of water mechanically before drying materials.
   (j)   Rapid processing.
   (k)   Full loading of plant.
   (l)   Reducing power loads.
   (m)   Recirculation of drying medium, e.g. air.
   (n)   Reducing amount of reprocessing by a reduction of the amount of spoilt work.
   (o)   Processing at lower water content.
3. Re-use part of the heat by:
   (p)   Multiple-effect evaporation.
   (q)   Flash-steam recovery.
   (r)   Use of waste heat for heating, e.g. air, hot water for toilets, process liquors, etc.
4. Never allow steam at high pressure to expand to a lower pressure without — wherever possible — getting useful energy from the expansion.

Power from Steam

Work done and Power

Work is done when a force overcomes a resistance over a finite distance. Energy is the capacity for doing work and can be looked upon as existing in four forms: potential energy, mechanical energy, thermal (or heat) energy, and electrical energy.

Apart from its heating effect, steam can also perform mechanical work. The work that steam can do depends on the amount by which it can expand; this degree of expansion depends on the machine in which the expansion takes place and the temperature of the steam.

Power is the rate of doing work or the rate of flow or use of energy. A steam engine or turbine is a heat engine which turns heat into work; the measure of the potential of heat for this purpose is temperature. The greatest amount of work can be got out of an 'ideal' engine when the input temperature is as high as possible and the exhaust is as cool as possible.

When steam is allowed to expand against a piston or in the nozzles of a turbine it does work on the piston or the turbine blades and gives up some of its heat energy; as a result it cools down. If it were initially saturated some of it would condense. Condensation in an engine cylinder is very wasteful and condensation in a turbine can cause damage to the blades. Water droplets will wear away the blades if there is more than 13% of moisture at exhaust. On the other hand if the steam were initially highly superheated it might still be superheated at exhaust; this too could be wasteful. The amount of superheat should be adjusted to the amount of possible expansion. The higher the initial pressure and the lower the exhaust pressure, the more superheat should there be.

Reference has already been made to the fact that whenever there is any superheated steam fed to a reciprocating engine the engine cylinder must be lubricated. Cylinders can sometimes be lubricated with graphite, which gets over the difficulty of having oily exhaust steam. Difficulties with cylinder lubrication however preclude the use of very high pressures with their appropriate superheats in reciprocating engines. In such engines the extent to which the steam can expand is limited by the size of the low-pressure cylinder whilst in turbines this problem of oil contamination does not exist however great the superheat. There is practically no limit to the amount by which steam can be allowed to expand in a turbine. Therefore, the turbine can take full advantage of very high pressure and highly superheated steam, and can obtain full benefits from a very-high-vacuum exhaust. It is not however as efficient as a reciprocating engine for powers below about 300 kW. The reciprocating engine is a most reliable, efficient and easily controlled machine and is recommended for small plants.

It is possible to calculate the amount of energy that an engine or turbine can take out of steam by use of the function known as entropy. This amount of energy is called the heat drop and is the difference between the total heat in the inlet steam and the total heat in the outlet steam. In an 'ideal' engine this energy that should be available is called the adiabatic heat drop since this engine, in addition to the ideal cycle, has no heat losses.

The heat that is given to steam in boilers is the difference between the heat in the feedwater and the total heat of the steam leaving the boiler. The adiabatic heat drop expressed as a percentage of the heat put into the steam gives the cycle efficiency for the conditions chosen. The actual proportion of the adiabatic heat drop which a real engine or turbine can use varies between just over 80% for very large machines to below 40% for very small machines. This excludes the condensing loss which can however be used for process work. Although cycles show an increasing efficiency with increased initial pressure, the relation is not practically applicable; a very small turbine working at very high pressure is not so efficient as one of similar power working at moderate pressure. The general principles are: (1) that power should be generated using the highest available initial pressure and the lowest possible exhaust or back pressure; (2) that there must be enough superheat to suit the conditions; and (3) that the larger the machine the more are very high pressures and temperatures justified.

The amount of superheat that should be given to the steam that is feeding an engine or turbine is fairly simply estimated:

(a) In a reciprocating engine exhausting to a condenser there should be enough superheat to ensure dry steam at exhaust.

(b) In a turbine exhausting to a condenser there should be enough superheat to prevent the wetness of steam at exhaust exceeding about 13%.

(c) In a reciprocating engine or turbine exhausting into a process steam main there should be enough superheat to allow the exhaust steam to enter the process main slightly superheated, so that on arrival at its point of use it will be as nearly dry saturated as is possible.

(d) In a reciprocating steam engine exhausting to a process where the steam comes into contact with a product that must not be contaminated by oil there must, generally speaking, be no superheat because superheat calls for cylinder lubrication.

Thermodynamic Principles

Thermodynamics is the study of changes in which energy is involved.

The first law of thermodynamics is that energy can be neither created nor destroyed. A consequence (or one of the formulations) of the second law is that conversion of heat to work is limited by the temperature at which conversion occurs.

The mechanical work that steam can do depends on the amount that it can expand, which depends upon the machine used and the temperature of the steam.

Isothermal and Adiabatic Expansion or Compression

Isothermal. If after any change of volume or pressure the temperature is the same as it was before the change, then $P_1 V_1 = P_2 V_2$.

Adiabatic. Adiabatic expansion or compression is that state where heat neither enters nor leaves the gas, and in an *ideal* process none will be generated in it by friction or turbulence. In the ideal process the gas obeys the law PV^γ = constant, which is the equation to the curve of the theoretical indicator diagram. γ is the ratio of specific heat at constant pressure to that at constant volume.

Isothermal and adiabatic processes represent the 'ideals' with which real processes may usefully be compared. In every real case of expansion or compression some heat enters or escapes from the gas or is generated within it by friction or turbulence.

The expansion curve of a gas may be represented by the law: PV^n = constant (where a suitable value is assigned to n). Thus by substituting n for γ in the adiabatic laws, the areas of the theoretical indicator diagrams and the changes in pressure, temperature and volume can be obtained. For the expansion of superheated steam without gain or loss of heat $n = 1.3$. For wet steam the approximate value of n may be taken as $1.035 + 0.1\,x$ (where x is the dryness fraction at the beginning of the expansion).

Entropy. The entropy of a substance is dependent only on its state or condition and for practical purposes it may be regarded as a number related to the thermal condition of a substance. The integral $\int(\mathrm{d}Q/T)$ is called the entropy change, and the physical significance is that every increase in entropy implies that heat has become less available for doing work. Whenever heat enters or leaves a body there is a change of entropy; this is also the case when heat is generated within a body by the action of friction. It is the one quality of a substance which does not change during reversible adiabatic expansion or compression (e.g. in an 'ideal' engine cycle), although total heat, temperature, pressure and internal energy are all changing. The maximum heat that can then be converted to work, since $Q_1/T_1 = Q_2/T_2$, is $(Q_1 - Q_2)/Q_1 = (T_1 - T_2)/T_1$, and the thermodynamic efficiency is $100\,(T_1 - T_2)/T_1$. The 'adiabatic heat drop' mentioned elsewhere is $(Q_1 - Q_2)$. In a real engine the heat drop is less than this.

As the *changes* of entropy alone are of significance it is measured from an arbitrary zero point. In the case of steam and water it is reckoned from the freezing point of water, at which the entropy is assumed to be zero. Steam Tables give the various values of entropy as well as the more usual properties such as enthalpy.

Temperature/Entropy Diagram. Certain properties of vapour mixtures and superheated vapours may be shown graphically by means of diagrams or charts. Such diagrams show the behaviour of the vapours and have a practical application in the solution of a number of problems. Thus any area on a diagram which has absolute temperature and entropy as its ordinates represents work measured in heat units.

In isothermal expansion or compression, and in evaporation or condensation at constant tempera-

ture, the change in entropy is equal to the heat added or taken away divided by the absolute temperature, whereas there is no change in entropy in adiabatic processes.

Temperature/entropy diagrams are useful for considering engines working on the Rankine Cycle when, for example, taking superheated steam at a temperature T_1, expanding it adiabatically to a temperature T_2 and discharging it in a wet condition to a condenser, the available heat and dryness fraction may be deduced.

Mollier Diagram (Enthalpy/Entropy Diagram). In this diagram the enthalpy and entropy are the ordinates. Whereas the principal value of the temperature/entropy diagram is to represent in graphic form the heat changes as an aid to calculation, the Mollier diagram — in many cases — obviates the need for any calculations at all. Mollier diagrams are usually supplied with Steam Tables. Every point on the diagram represents steam in some definite condition, and direct readings may be made of, e.g., work done (available heat) and dryness fraction. This can be achieved whether the expansion is truly adiabatic or is carried out for example with frictional losses as in a turbine, where the steam will be partially dried by the frictional heat.

The diagram is helpful when high-pressure steam is expanded to produce a proportion of power and exhausted at medium pressure for use in process or other heating. The adiabatic heat drop is read directly from the enthalpy coordinate by following the isentropic line from the point corresponding to the initial pressure and temperature in the Steam Tables to that corresponding to the exhaust (process steam) temperature and pressure. This heat drop is multiplied by the mechanical efficiency of the engine (perhaps 0.65), and the result in kJ/kg divided by 3600 and multiplied by the generator efficiency (≈ 0.95) yields the electrical energy in kW h obtained from each kg of steam used. Returning to the 'exhaust point' on the Mollier diagram, the enthalpy available for use in heating can be read directly, again per kg of steam. A heat/power balance for the system can thus readily be struck; or the two steam pressures may be adjusted by trial and error to match a desired balance.

Steam Saving by Superheating

The principal defect of reciprocating engines is that a good deal of the steam is condensed during the working stroke on the relatively cold cylinder walls and evaporated during the exhaust stroke, thus cooling the walls again and rejecting heat uselessly to the exhaust. If superheated steam is used a good deal of this loss will be avoided. Indeed, the practical saving brought about by superheating is considerably greater than that which would be expected merely on account of the greater heat drop because of the higher temperature. With simple slide valve or piston valve engines in which the steam enters and leaves by the same ports a reduction of about 12% in the steam consumption may be effected by using steam with a superheat of 28°C. With a superheat of 56°C the saving would be something in the region of 20% and could even rise to 30% with a superheat of 112°C.

Although steam temperatures up to 510°C and even higher are used today in large steam turbines, a temperature of about 340°C is as high as is practicable with reciprocating engines on account of difficulties with lubrication, distortion and the like. Therefore unless an engine has been specially designed for high temperatures it is far better to keep on the safe side and limit the temperature to around 260°C which corresponds to a superheat of about 90°C with steam at 7 bar gauge pressure or to about 60°C with steam at 14 bar.

A steam pressure of between 10 and 10.35 bar is recommended for compound engines, though the full rated output under non-condensing conditions can normally be obtained using 8.27—8.96 bar. The non-compound types are for steam pressures of about 5.5—6 bar non-condensing, or for higher pressures when exhausting against a back pressure.

In the case of triple-expansion engines a steam pressure of between 11.7 and 14 bar is called for, although such engines tend to be more economical than compound engines at pressures down to 8.6 bar (under condensing conditions). In some cases however the rated output may not be obtainable at these lower pressures.

Back-Pressure and Pass-Out Engines

Factories which require steam at low pressure for process work or steam heating can produce power very cheaply by generating steam at a higher pressure and passing it first through a back-pressure engine or turbine, the exhaust of which serves for the low-pressure duties. Considerable savings accrue owing to the fact that the amount of fuel required to generate steam at say 10 bar is only 2 or 3% more than that consumed in generating the same quantity of steam at say 1 bar. Thus, by combining power production with process steam or heating it is possible to make effective use of as much as 80% of the heat in the fuel compared with a maximum of around 40% for power alone even in the most efficient of the large public power stations. Furthermore it is interesting to note that the efficiency of back-pressure engines in these circumstances is of little importance because, provided that they produce the required amount of power and that there is use for the whole of the exhaust steam, such things as losses due to leaky pistons or valves or by wire-drawing in the steam ports will not affect the overall efficiency of the combined process, but merely increase the heating value of the exhaust steam.

If the demand for power is insufficient for the engine to deliver the required quantity of back-pressure steam it may be supplemented by live steam from the HP main or the boiler, supplied automatically through a reducing valve. But if there is a surplus of back-pressure steam it will have to be discharged to atmosphere through a relief valve and wasted, unless part of the load on the back-pressure engine can be transferred to be met elsewhere. For the highest economy to be obtained, every back-pressure engine should be kept working at full load, so long as maximum use can be made of its exhaust steam.

In situations where the demand for process steam is considerably less than the demand for power steam, it is better to obtain process steam from a pass-out engine. This is a compound or triple-expansion engine exhausting to a condenser, the process steam being drawn off from a receiver between the cylinders.

Engines v. Turbines for Back-pressure and Pass-out Working

The decision whether to use an engine or a turbine will depend on a number of factors, e.g. the size of the unit, the conditions of the live steam and process steam, the quantity of process steam required, and the cost of the fuel. For sizes up to 400 to 500 kW preference would generally be given to an engine on account of its higher efficiency and lower first cost. If the engine is of the pass-out type, the fact that the process steam has to be abstracted between the cylinders limits the pressure it may have, on account of the necessity of avoiding too great a difference in the work done by the cylinders. A turbine however is not subject to any such restrictions. The pass-out turbine also has the advantage of utilizing a high vacuum more efficiently than an engine and of being able to operate with steam at any pressure and temperature, whereas a pass-out engine should preferably work with saturated steam, and in any case the steam temperature should not exceed 290°C.

Quality of Process Steam

More often than not steam taken from back-pressure or pass-out engines is required to be free from traces of grease or oil; the condition is imperative when the steam is required for process work in which it will come into contact with foodstuffs or delicate materials. A well designed high-speed vertical engine will deliver steam as clean as that from a turbine provided that oil is not used for cylinder lubrication. Long experience has indicated that such engines when fitted with bronze piston rings and working with saturated steam will not require cylinder lubrication, while no oil can be carried into the cylinder by the piston rods if the cylinder glands are separated from the casing by a distance not less than the length of the stroke. But for engines working with superheated steam some cylinder lubrication is necessary. In most cases an oil separator of adequate size will generally remove practically all the oil from the back-pressure or pass-out steam, but the final decision will depend on the particular process concerned and the degree of cleanliness required.

Steam Turbines

Turbines are usually classified according to the steam conditions under which they operate. Thus high-pressure turbines are those which take steam directly from the boilers and where expansion takes place in the turbine down to the pressure in the condenser; this may be regarded as the fundamental class of steam turbine. For turbines of up to say 5 MW capacity steam conditions normally range from 14 to 42 bar pressure and 315–428°C temperature, with vacua of −930 to −980 mbar (+7 to +2 kPa). For turbines of greater output higher steam conditions are used, 62 bar/480°C being usual for turbines of 60 MW capacity, whilst the largest units employ pressures up to 138 bar and temperatures up to 565°C or even 593°C.

Low-pressure turbines are usually fed with steam at 1.4–3.5 bar absolute and exhausted to a condenser. They used to be common as adjuncts to non-condensing reciprocating steam engines.

But turbines may also be classified according to the method of steam expansion:
1. Impulse turbines with steam expansion only in the stationary blades or nozzles.
2. Reaction turbines with steam expansion in both stationary and moving blades.

Back-pressure Turbines

These take steam directly from the boiler and usually exhaust at or above atmospheric pressure. In the smallest sizes, they are used to drive auxiliaries, in which case the exhaust steam may be utilized for feedwater heating; but in larger sizes they are often used where there is a demand for process steam, the exhaust pressure being fixed by the steam temperature required for process. The power from the turbine can then be obtained in an economic manner.

So called 'topping sets' are back-pressure turbines usually exhausting into existing steam mains. When supplied with steam from additional boilers at higher pressures and temperatures and driving its own generators, the topping unit has been used as an economical means of increasing the output of stations formerly using only moderately high steam conditions.

It should be noted that the energy available in adiabatic expansion of steam from 12.4 bar gauge and 55°C superheat to −950 mbar (+5 kPa) vacuum is about twice that available if the expansion is carried down to atmospheric pressure only. Hence the addition of condensing plant approximately doubles the amount of work that can be done per kg of fuel without necessitating any increase in boiler capacity.

Reducing Turbines

It is clear that when using a back-pressure turbine the steam available for process and the power sup-plied by the turbine are directly related. Most installations however require to be able to vary these quantities independently of each other, and to meet this requirement reducing turbines were introduced. These consist of a back-pressure turbine to which a low-pressure turbine is added, the back-pressure turbine being designed to pass the full quantity of steam required for process. As the demand for process steam is reduced, automatic governor gear bypasses the excess steam into the low-pressure turbine and at the same time the supply of high-pressure steam is reduced. As the turbine is also supplied with the usual governor gear to maintain speed approximately constant with varying load, the load and the process steam may be varied independently of one another, and the process steam supply maintained at almost constant pressure.

The economical results which can be obtained when using this plant for both electrical power and process steam have resulted in an increased demand for this type of turbine in recent years. In most cases the heat transferred to process is derived almost entirely from the latent heat of the steam. 28°C superheat at the turbine exhaust is usually desirable, so that the steam will reach the heater or plant in a dry condition but without excess temperature.

A general tendency in recent turbine development has been in the direction of increasing the initial temperature at which steam is supplied to the turbine; this results in increased thermal efficiency.

Steam Pressures and Temperatures

Steam pressures and temperatures in current practice have reached 97–164 bar and 480–565°C or 593°C. The rate of increase in efficiency with increase in pressure becomes less as the pressure increases.

A limit to the practicable operating pressure is set by the difficulty in separating moisture from steam in the boiler drum. At 138 bar gauge the steam density is 13.7% that of the water. At 180 bar gauge the relative density is 24.3%. Temperatures are limited by creep of the metal.

Most modern stations resuperheat the steam from the high-pressure turbine back to the initial temperature. Where fuel costs are high, load-factor high, units large and arranged for one boiler per turbine, the overall gain in station efficiency of 4–5% may justify the extra initial cost of a reheat installation in the majority of cases.

In the industrial field, the trend has been towards higher steam pressures, the initial pressure being determined by the balance of power- and process-steam requirements. Although there have been a number of large installations of between 97 and 124 bar and 496–510°C, general practice among industrial plants is to employ 28–31 bar for the smaller installations and 42–59 bar for those of larger capacity. Steam temperatures seldom exceed 400–455°C.

High-pressure units are smaller for a given capacity, and the initial cost is only slightly greater than for lower pressures. To protect the turbine from excessive temperature, it is necessary to employ closer control of the superheat where high steam temperatures are involved.

General Advantages of Steam Turbines

Compared with other prime movers steam turbines require less floor space, lighter foundations and less attendance, have a lower lubricating-oil consumption with no internal lubrication (the exhaust steam therefore being free from oil), have no reciprocating masses with their resulting vibrations, have uniform torque, and lower maintenance costs. They are capable of operating with higher steam temperature and of expanding to lower exhaust pressure than the reciprocating engine. Their efficiencies may be as good as steam engines for small powers and much better in ranges of greater capacity. Single units can be built of greater capacity than those of any other type of prime mover. Small turbines cost about the same as reciprocating engines, but large-capacity turbines cost much less then corresponding sizes of reciprocating engines, and they can be built in capacities never reached by reciprocating engines.

Effect of Changes in Operating Conditions

The effect on performance of a change in some operating conditions such as pressure, temperature, or exhaust pressure from that for which the turbine is designed cannot be generalized, as it depends upon the other operating conditions. The percentage effect of a change in initial pressure is greater the lower the total enthalpy drop. For small changes within which the engine efficiency remains sensibly unchanged the effect may be taken to be proportional to the enthalpy drop provided exhaust pressure remains constant.

A change in exhaust pressure, particularly of condensing turbines, produces a result that depends upon the volume of steam flow, the area of the lowest-pressure turbine elements, and the exhaust pressure. The effect varies with the output of the turbine and is greater in fractional outputs. For example in a turbine with steam at 28 bar superheated to 425°C (total temperature), the reduction in steam consumption in going from −950 to −980 mbar vacuum (+5 to +3 kPa) may be from 0 to 6.8%, depending upon the congestion of the lowest-pressure turbine elements and on the moisture content in the exhaust steam.

The effect of changes in any of the operating conditions is usually twofold: on steam or heat consumption, and on steam flow, the latter affecting output. Correction values are required for both of the preceding in the case of acceptance tests, when the operating conditions of the test deviate from those specified. These corrections are preferably embodied in the contract; failing this they must be agreed upon before tests are commenced.

Overall Heat Consumptions of Large Power-station Plants

The overall net heat consumption at the switchboard outlet for a 55 bar gauge/485°C/34 mbar absolute (+6.6 kPa) back-pressure plant with four stages of feedwater extraction will vary from approximately 11.6 to 12.2 MJ/kW h, depending upon the size, arrangement of the turbines, quality of the fuel, and the load factor at which the power plant is operated. Since 1 kW h = 3.6 MJ_e, this corresponds to 31 to 29.5% thermodynamic efficiency.

The heat consumption will increase about 6% each time the steam pressure is halved and correspondingly decreased approximately 5% each time the steam pressure is doubled; the change in economy with steam pressure is less on the small plants and greater on the large plants, and tends to decrease for a given percentage pressure rise as the pressure increases. The heat consumption will change approximately by 1% for each 17°C that the initial temperature is changed. On the other hand the heat consumption will be reduced approximately 5% if the steam is resuperheated, at approximately a quarter of the initial pressure, back to the initial temperature, with a loss of not more than 12% of the pressure at which resuperheating takes place. The heat consumption of a plant with 68 mbar (9.32 kPa) absolute back-pressure is approximately 3% greater than that of one with 34 mbar (9.66 kPa) absolute back-pressure.

Large steam plants at high pressures with about 538°C superheat, resuperheated, and with efficient extraction cycles to high vacuum are operating at as little as 10.5 MJ/kW h, and some plants may reach 9.7 MJ/kW h based upon net power sent out (over 37% efficiency).

References

1 British Standard Specification 779:1975
2 Ref.(B)
3 Weldon, R. 'A new type of boiler feed pump for 660 MW_e electricity generating sets in England', *Sulzer Tech. Review* 1972, Vol.54, No.3, 189–198
4 Babcock & Wilcox Ltd, *Steam: Its Generation and Use*, pp 5-8 to 5-15
5 e.g. Ref.(I)
6 Ref.(H), pp11-33 to 11-36

General Reading

(A) *The Efficient Use of Fuel*, 2nd edn, HMSO, London, 1958
(B) Williams, J. N., *Steam Generation*, Allen & Unwin, London, 1969

(C)   Gore, W. H., Gunn, A. D. C. and Horsler, A. G., *Inst. Fuel (Midland Section) Symposium on natural gas firing of boilers*, Aston University, 1972

(D)   Central Electricity Generating Board, *Modern Power Station Practice*, 2nd edn, Vol.7, *Operation and Efficiency*, Pergamon Press, Oxford, 1971

(E)   Lyle, Sir Oliver, *The Efficient Use of Steam*, HMSO, London, 1947

(F)   Northcroft, L. G. and Barber, W. M., *Steam Trapping and Air Venting*, 4th edn, Hutchinson, London, 1968

(G)   Guy, Sir Henry, *Steam Turbines*, Instn Mech. Engrs, London, 1956

(H)   Perry, R. H. and Chilton, C. H., *Chemical Engineers' Handbook*, 5th edn, McGraw-Hill, 1973

(I)   *IHVE Guide*, Book C, paragraphs C2.1, C2.2, C2.3

Chapter 11

Water treatment

The principal uses of water in industry are:

1. For steam raising, the steam being used for electrical or mechanical power generation, heating, or chemical reaction.
2. As a cooling medium.
3. For hot water supply and hot-water heating systems.
4. As a solvent and reactant in chemical processes.
5. As a suspending medium in processes of mechanical separation such as coal washing.
6. For drinking.
7. For laundering, washing and general domestic purposes.

The last four items are beyond the scope of this book although some of the principles and methods of water treatment discussed are applicable to those waters in appropriate instances. The main topic discussed in this chapter therefore is feedwater conditioning for boiler plant, but the treatment of water for cooling purposes and hot-water heating systems is also considered.

Impurities in Natural Waters

All water supplies contain impurities in the form of dissolved gases, and solids in solution or suspension, and although a 'bad' water may contain less than 0.05% of dissolved solids, the cumulative effect of these is considerable. Most industrial water supplies require treatment for one or other of the purposes mentioned above. Such treatment consists essentially either in removing objectionable constituents or at least reducing their concentration to a level at which they are relatively innocuous, or in the addition of substances which suppress the undesirable effects.

The quantities of substances present in waters used for steam raising may sometimes be sufficiently high to be stated in percentages but are more usually stated in weight parts per million, generally written ppm.

The common impurities in natural water supplies are:

Dissolved gases — always present
Carbon dioxide (CO_2) and oxygen (O_2), up to about 90 ppm and 12 ppm respectively, and occasionally hydrogen sulphide (H_2S) up to 4 ppm and ammonia (NH_3).

Suspended solids — sometimes present
Mineral and vegetable detritus such as sand, silt, and decaying plants in varying amounts.

Dissolved organic substances — sometimes present
Materials such as peaty acids and complex coloured substances extracted from soil or plants and proteins.

Living organisms — always present to some extent
Natural water bacteria, soil bacteria and sewage bacteria. Algae, etc.

References p 269

Dissolved salts — always present
The bicarbonates, chlorides, sulphates and nitrates of calcium, magnesium and sodium in varying proportions, together with silica, and sometimes small amounts of iron, manganese and aluminium.

The calcium (Ca) and magnesium (Mg) salts are the cause of hardness in water, and most of the scales and deposits formed from natural waters in boiler plant and cooling plant are largely compounds of calcium and magnesium. (See *Table 1*).

The calcium and magnesium salts may be divided into two groups:

1. The bicarbonates $Ca(HCO_3)_2$ and $Mg(HCO_3)_2$ which, because they are easily decomposed by heat, cause alkaline hardness, also known as temporary hardness or carbonate hardness.
2. The sulphates, chlorides and nitrates $CaCl_2$, $MgCl_2$, $CaSO_4$, $MgSO_4$, $Ca(NO_3)_2$ and $Mg(NO_3)_2$ which are not decomposed by boiling and cause non-alkaline hardness also called non-carbonate or permanent hardness. The nitrates are normally present in very small quantities.

The decomposition of alkaline hardness substances by heat may be represented by the following equations:

$$Ca(HCO_3)_2 \longrightarrow CaCO_3 + CO_2 + H_2O$$

calcium bi-carbonate → precipitated calcium carbonate + carbon dioxide + water

$$Mg(HCO_3)_2 \longrightarrow MgCO_3 + CO_2 + H_2O$$

magnesium bicarbonate → magnesium carbonate (more soluble than $CaCO_3$) + carbon dioxide + water

It should be noted that the CO_2 liberated will give an acid steam condensate.

Types of Water

The chemical composition of water supplies varies widely, but in the UK we can distinguish according to source three classes of water, each of which has characteristic qualities:

(a) Surface drainage from upland areas, i.e. streams, lakes and reservoirs situated in moorland type of country.
(b) Surface drainage from lowland areas, i.e. agricultural countryside.
(c) Underground supplies, i.e. deep springs and wells and water from the lower measures of coal pits.

Moorland waters are usually fairly constant in composition, and are clear except perhaps in times of spate. They may be coloured brown, and are often slightly acid, owing to dissolved carbon dioxide or to weak organic acids derived from peaty material, which makes them corrosive; and, although the hardness (Ca + Mg) is low, they form sufficient scale in boilers to justify treatment. A

Table 1   Composition of typical hard boiler scales

	Sulphate scale (%)	Carbonate and sulphate scale (%)	Carbonate and silicate scale (%)	Analcite scale (%)
$CaCO_3$	6.03	45.6	37.4	—
$Mg(OH)_2$	6.32	9.8	2.52	1.5
$CaSO_4$	83.23	34.7	1.65	2.3
Apatite (Ca phosphate)	—	—	—	40.5
SiO_2	1.02	7.4	29.2	24.8
Calcium silicate	—	—	20.6	—
Analcite (NaAl silicate)	—	—	—	41.7
$Fe_2O_3 + Al_2O_3$	1.56	1.9	2.26	4.3
Water and organic matter	1.54	0.6	2.65	7.1
	99.7	100.0	96.28	97.4

small amount of iron is often present, probably in combination with the organic matter, and as a result of the activities of certain strains of iron bacteria the iron may be abstracted and deposited on the walls of pipes as hydrated oxide.

Lowland surface waters are not often coloured, but may contain fine suspensions of mud which settle only after prolonged periods unless assisted by coagulants. Their composition can vary widely as a result of heavy rainfall, and the hardness, which is present as both temporary and permanent hardness, is usually high enough to form serious amounts of scale and deposit in boilers, economizers, and cooling plant.   Rivers and canals may be contaminated by trade or sewage effluents which can interfere with the normal methods of softening.

Deep well waters are usually fairly constant in composition, though occasionally slow changes may take place as a result of contamination by other waters percolating through faults in the strata. When freshly drawn they are often 'water-white', clear and sparkling, but sometimes a brown opalescence develops on exposure to air, for which small amounts of iron are responsible. This iron is probably in the ferrous condition and is deposited as hydrated ferric oxide by the combined effects of oxidation and the loss of carbon dioxide. Such waters usually contain traces of hydrogen sulphide, easily detectable by odour, and no dissolved oxygen; hence the slight delay in the precipitation while oxygen is being absorbed from the atmosphere. Iron bacteria can also exist in these waters, and the iron may then be deposited on the walls of pipes by their action as well as by air oxidation.

The hardness is often entirely in the form of temporary hardness and sodium bicarbonate may be present as well. The silica content is often high, e.g. about 15—20 ppm.

Effects of Impurities in Water

The most objectionable effect of using raw water in a boiler is the deposition of hard adherent scales on the heating surfaces.   These have a low thermal conductivity estimated at between 1.15 and 3.45 W/m °C, so that the metal is not properly cooled and its temperature rises to the point at which it softens, bulges and splits under pressure, with dangerous results.

The parts of the heating surface most sensitive to this effect are water tubes exposed to radiant heat, or furnace tubes of shell boilers, where rates of heat transfer and water evaporation are high. In tubes receiving heat by convection and conduction, greater thickness of scale can be tolerated before failure takes place.   The direct loss of heat or waste of fuel caused by scale has been estimated at about 2% or even less in watertube boilers, but may be up to 5 or 6% in smoketube boilers where heating surfaces are short.   The more important losses are the cost of repairs, interrupted service of plant during mechanical descaling, and fuel required to start up another boiler in the event of tube failure.

Many waters also deposit solids in the form of soft scale and sludge, at and below boiler working temperatures.   These deposits are found in cooling plant, feedwater heaters and economizers as well as in boilers. A certain amount of sludge can be tolerated in a shell boiler where it settles into the cooler parts, from which it can be removed by opening the blow-down cock momentarily a few times each day; but if the accumulation is not controlled in this way sludge may stick to the heating surfaces in a soft layer, which retards heat flow even if it does not cause failure. This is particularly objectionable in small firetube boilers, because of the difficulty of cleaning between the tubes.   The accumulation of more than small quantities of sludge in watertube boilers should be avoided, because it can accentuate the formation of foam in the boiler drum and carry-over of boiler water into the steam, and because severe corrosion can take place underneath deposits in tubes in which the circulation is sluggish.   There are numerous examples of this type of corrosion in boilers working at pressures above 2.76 MPa (27.6 bar).

The gases carbon dioxide and oxygen which are dissolved in water, together with that carbon dioxide which is liberated when waters containing bicarbonates are heated, can cause corrosion in economizers and boilers, and since they pass out with the steam they reappear in condensate, which is therefore corrosive. Lastly, salts and suspended solids in boiler water can under certain conditions be carried out of the boiler in the steam and deposit in superheaters, steam mains, and turbines.

The aims of water treatment are, therefore, to prevent the formation of hard scale in the boiler and of other deposits in ancillary (pre-boiler) equipment such as economizers, to control the amount of sludge in the boiler, to reduce or eliminate corrosion caused by salts in the boiler water or carbon dioxide in steam, and to avoid the contamination of steam by salts in the boiler water.

Scale

Chemical analysis has shown that the chief constituents of hard boiler scales are calcium sulphate ($CaSO_4$) and calcium and magnesium or complex aluminium silicates, often mixed with calcium carbonate ($CaCO_3$) and magnesium hydroxide ($Mg(OH)_2$), and that soft deposits or sludge are composed of $CaCO_3$ and $Mg(OH)_2$, with only small quantities of other constituents. *Table 1* shows some examples of hard boiler scales.

Scales consisting essentially of calcium sulphate, and those having calcium sulphate as a major constituent, are relatively rare today, because most boiler waters receive some form of preventative treatment (see below). Calcium carbonate often occurs as a minor constituent in scales, because calcium bicarbonate predominates in the supply water and decomposes to the carbonate on heating.

Scales containing appreciable amounts of silica are less common, because silica is not usually a major constituent of water supplies. In waters which are naturally soft, and in those which have been treated to remove hardness — but not silica — this may not apply, and silica or silicates may then form a major part of the scale composition. Thin layers of siliceous scales can cause tube failures. The use of phosphates in feed-water conditioning — or the more recent chelant-type treatments — have usually been successful in preventing the formation of calcium and magnesium scales in boilers. Scales consisting almost entirely of silica can form in modern highly rated boilers. In some cases the position can be controlled by suitable regulation of boiler-water alkalinity and concentration, but for high-pressure boilers it has become standard practice to use a treatment process which reduces the silica input to the boiler.

Analcite — sodium aluminium silicate, and acmite — sodium iron silicate, are scale constituents which contain no calcium or magnesium; they can occur even when the boiler feedwater contains negligible amounts of aluminium or iron. For this reason it is thought that formation of these scales is due to localized concentration of boiler water at areas of high heat input. Even in boilers whose design and loading do not produce such areas, analcite may be formed if excessive amounts of aluminium salts are used in prior clarification or softening processes and the silica content of the boiler water is also relatively high. Although chiefly incident in watertube types, analcite and acmite have been noted in firetube boilers.

Water-treatment Processes

Water-treatment processes for boiler feed preparation are of two kinds: (*a*) those that remove the calcium and magnesium ions which are the main offenders in scale-forming processes, and (*b*) those that also remove all or part of the dissolved solids. There are four processes chiefly used:

1. Precipitation processes, in which chemicals are added to precipitate calcium and magnesium as compounds of low solubility. The lime—soda process is typical of this class, but other precipitating agents such as caustic soda and sodium phosphate can be used when the composition of the raw water permits.
2. Ion-exchange processes, in which the hardness is removed as the water passes through a bed of natural zeolite or synthetic resin and without the formation of any precipitate.
3. Water which is essentially free from dissolved salts can be prepared by evaporation and condensing the steam.
4. Synthetic resins can also be used to remove dissolved solids from a water, the process being called demineralization or de-ionization.

A brief and very general description of these processes is given below, but it must be emphasized that the choice of process depends on the composition of the raw water and the purposes for which the treated water is required, and a decision on such points is best worked out with the help of an expert.

Precipitation Processes

Lime—soda. In the lime—soda or lime—soda/sodium-aluminate or other coagulant process, calcium and magnesium are precipitated as compounds of very low solubility, namely $CaCO_3$ and $Mg(OH)_2$, by the addition of hydrated lime and sodium carbonate, or by hydrated lime alone when the raw water contains sodium bicarbonate. The amounts of chemicals added must be in proportion to the Ca and Mg present in the water and will therefore require adjustment if the composition of the water changes. The process can be worked hot (70—90°C) or cold, the former usually giving better clarity and hardness removal, particularly with turbid waters or waters containing organic matter. The amount of sodium aluminate required for coagulation of the magnesium hydroxide is approximately 10 ppm in hot softening and 20 ppm in the cold, irrespective of the amount of hardness to be removed; it is, however, used only when the raw water contains appreciable amounts of magnesium salts. A small amount of residual hardness is always present in the treated water and amounts to about 10—15 ppm $CaCO_3$ in the hot process and 15—25 ppm in the cold. Aluminium sulphate and modified (activated) silicate may also be used as coagulants.

The plant comprises stock tanks for preparing and storing the mixture of reagents, usually as a thin slurry containing about 50—100 g/l (½ to 1 lb/UK gal) chemicals; an apparatus for adding measured quantities of this slurry to the raw water in proportion to its rate of flow; a means of mixing chemicals and water intimately; a settling tank with a retention time of one hour (for hot softening) or three hours (for cold); and filters. The settled precipitates are removed as a slurry containing about 5% of solids, which is run to waste.

Various types of reaction tank designed to speed up the reaction rate in precipitation softening carried out in the cold have been developed since about 1945. The general principle is to introduce previously formed precipitate — calcium carbonate and magnesium hydroxide — to the water at the point where the softening chemicals are added. This encourages the precipitation of fresh solid on the existing particles, instead of forming new particles. The sludge from this type of plant will contain 15—25% of solids.

Phosphate treatment. The solubility of calcium phosphate is less than that of $CaCO_3$; a precipitation process using sodium phosphates instead of sodium carbonate should thus give a lower residual hardness. This is in fact the case, but phosphates are expensive, and their use is limited in practice to a second stage of softening at high temperatures for the special purpose of treating water for use in boilers working in the higher pressure range, i.e. above 3.45 MPa or 34.5 bar (500 psi).

The procedure is to submit the water to hot lime—soda or lime—soda/sodium-aluminate softening at 70—90°C, and then to pass the treated water containing about 10—15 ppm of residual hardness to a similar plant in which sodium phosphate is added in sufficient amount to precipitate the residual calcium hardness. The water leaving the lime—soda section should contain enough caustic alkali to precipitate magnesium as $Mg(OH)_2$. After settling, the water is passed through filters, and the hardness should be reduced to about 2 ppm.

Ion-exchange Processes

An ion can be considered to be an atom which has either lost or gained electrons, and has thus acquired an electrical charge. Salts dissolved in water separate into their constituent ions; these have some degree of mobility. Thus magnesium sulphate ($MgSO_4$) dissolves to form the two ions Mg^{2+} and SO_4^{2-}. Ions carrying a positive charge are termed cations and include the metallic and hydrogen ions. The anions carry a negative charge, and those chiefly of interest in water treatment are SO_4^{2-}, Cl^-, NO_3^-, HCO_3^-, and CO_3^{2-}.

Interaction between ions in solution underlies a considerable number of chemical reactions, including those of the precipitation processes previously described. The term hardness in fact arose from the tendency of 'hard' water to destroy the detergent properties of natural soaps by precipitating their anions as the calcium or magnesium salts. In addition, there are some solid materials which will exchange ions with those dissolved in water passing through them. This phenomenon was first noticed in relation to certain minerals known as zeolites, which are essentially sodium aluminium silicates.

When water containing hardness is allowed to percolate through a bed of suitably graded zeolite nearly all the calcium and magnesium ions are replaced by sodium and the water is thus softened. Eventually all the sodium in the zeolite is used up and the bed consists essentially of calcium and magnesium zeolite, but this can be reconverted into the sodium zeolite by treatment with a strong solution of brine (NaCl).

Many other cations, e.g. potassium, barium, strontium, and iron, can be exchanged for sodium in the same way to varying extents; sodium in its turn can be displaced by ammonium and hydrogen ions.

Synthetic zeolites have been made that are more efficient in softening than the natural minerals, but these materials have been surpassed since 1935 by substances of quite a different class. The first of these were products obtained by partial dehydration and sulphonation of coal, and were followed by the discovery of Holmes and Adams that resins made by the condensation of phenols and formaldehyde also have good exchange properties. Other types of resins having similar properties have more recently been developed, such as the polystyrene and carboxylic resins.

These processes work best with clean water. Suspended solids in the raw water should be removed by filtration, using coagulants if necessary, otherwise they will clog the pores of the exchange material and reduce its exchange capacity. There are also working losses due to abrasion and carry-over of fine material, so that some fresh material must be added after a year or two, as make-up. These losses vary according to working conditions, and the plant suppliers should be consulted for estimates of losses in any given case.

Base-exchange Softening: Cation Exchange on the Sodium Cycle. This process consists of a cycle of operations in which the raw water is pumped through a bed of exchange material until tests on the issuing water show that the bed is failing to remove calcium and magnesium ions completely. At this point the flow of water is shut off, or diverted to an adjoining fresh bed of material. After a short backwash with water, a strong solution of brine (5—10% NaCl) is pumped or syphoned through the exhausted bed for some 5 or 10 minutes.

During this period the relatively high concentration of sodium ions in the brine replaces the calcium and magnesium ions in the zeolite and reforms the sodium zeolite. The calcium and magnesium ions pass into solution and out to waste when the brine is run off, and the regenerated bed is then washed free from brine, generally by an upward flow of water before more raw water is admitted. The full cycle of operations is softening, backwashing, regeneration, rinsing, softening, and the plant often consists of two or more units or trains so that the regeneration and washing stages do not interfere with the continuity of the supply of softened water: otherwise, extra storage capacity may be required. In some plants all these operations can be set in motion automatically after a measured volume of water has been softened.

The exchanges may be expressed by the following simple equations in which Z represents the anion of the base-exchange material, which may be either a zeolite or a resin:

Softening

1. $\underset{\text{zeolite}}{Na_2Z}$ + $\underset{\substack{\text{temporary}\\\text{or alkaline}\\\text{hardness}}}{Ca(HCO_3)_2}$ = $\underset{\substack{\text{calcium}\\\text{zeolite}}}{CaZ}$ + $\underset{\substack{\text{sodium}\\\text{bicarbonate}}}{2NaHCO_3}$

2. $\underset{}{Na_2Z}$ + $\underset{\substack{\text{permanent}\\\text{or non-}\\\text{alkaline}\\\text{hardness}}}{CaCl_2}$ = $\underset{}{CaZ}$ + $\underset{\substack{\text{sodium}\\\text{chloride}}}{2NaCl}$

Magnesium salts behave in the same way as those of calcium.

Regeneration

$\underset{\substack{\text{strong}\\\text{brine}}}{CaZ}$ + $2NaCl$ = $\underset{\substack{\text{regenerated}\\\text{zeolite}}}{Na_2Z}$ + $CaCl_2$

This method of softening has disadvantages for waters having a high alkaline hardness that are to be used as boiler feed, because the alkaline hardness (calcium and magnesium bicarbonates) is converted to sodium bicarbonate which is decomposed in the boiler to carbon dioxide and caustic soda. For reasons given later a large amount of caustic soda in boiler water may be objectionable, while carbon dioxide in the steam can cause corrosion.

Hydrogen-ion Exchange: Cation Exchange on the Hydrogen Cycle. The most important difference between the zeolites and the carbonaceous or resin-exchange materials is that when the latter are regenerated with acids they can exchange hydrogen ions for sodium, calcium and magnesium ions, so that all the salts in the water passing through the bed are converted to the corresponding acids. Thus bicarbonates become carbonic acid which can be decomposed into carbon dioxide and water by air blowing, whereas chloride, sulphate and nitrate remain as the corresponding mineral acids.

The equations representing these changes are:

(a) $\underset{\substack{\text{carbonaceous}\\\text{exchanger}\\\text{(regenerated}\\\text{with acid)}}}{H_2Z}$ + $Ca(HCO_3)_2$ = CaZ + $\underset{\substack{\text{carbonic}\\\text{acid}\\\text{(in}\\\text{solution)}}}{2H_2CO_3}$

(b) $2H_2CO_3 = 2CO_2 + 2H_2O$
     (air blowing)

(c) H_2Z + $CaCl_2$ = CaZ + $\underset{\substack{\text{hydrochloric}\\\text{acid}}}{2HCl}$

(d) H_2Z + $MgSO_4$ = MgZ + $\underset{\substack{\text{sulphuric}\\\text{acid}}}{H_2SO_4}$

The treated water is not suitable for use at this stage because it is corrosive, but, as explained in the next Section, the acidity can be neutralized. The plant must, of course, be lined to protect it from attack by acid.

Carboxylic resins form a special group which will only exchange H^+ for calcium or magnesium in the alkaline form, i.e. they only undergo equations (a) and (b). For complete hardness removal, they must be followed by a sodium ion exchanger — the 'de-alkalization/base-exchange' process.

Blended Acid and Base Exchange. If part of the water is passed through a hydrogen ion exchange unit and the remainder through a sodium ion exchange unit, the two treated waters may be blended, with the result that not only is the mixture softened but the bicarbonates are destroyed by conversion into carbon dioxide, which is removed by aeration.

By this means the mineral acids from the hydrogen ion exchange unit are neutralized by the sodium bicarbonate in the base exchange water, and there is some reduction in dissolved solids.

If the composition of the raw water is subject to wide variation, then the proportions of acid and salt must be altered to correspond to the proportions of bicarbonate and non-alkaline hardness, in accordance with the variations.

Waters containing calcium alkaline hardness can also be treated by a combined process, in which the bicarbonates are removed by precipitation with lime and after filtration this treated water is passed through a base-exchange plant to remove non-alkaline hardness.

Anion Exchange.   Anions in water such as Cl^- and SO_4^{2-} can also be removed or interchanged by passing the water through another type of resin bed.   Holmes and Adams found that resins made from formaldehyde and metaphenylene or similar substances have the property of absorbing acids. Improved resins with a polystyrene structure have been available since 1949.   The reactive group of the resin structure is commonly an amino group. Weak base anion exchangers have primary, secondary or tertiary amino groups, and strong base anion resins have quaternary amino groups.   Strong base resins undergo the following reactions (where R indicates the resin structure):

Exchange: $R.OH + HCl = R.Cl + H_2O$

Regeneration: $R.Cl + NaOH = R.OH + NaCl$

The strong base materials have the advantage that they will remove weak acids such as carbonic and silicic acids.   Weak base reactions are:

Exchange: $R + HCl = R.HCl$

Regeneration: $R.HCl + NaOH = R + NaCl + H_2O$

The weak base resins will not remove weak acids such as carbonic or silicic, but have the advantage that they can be regenerated with the cheaper alkalies, ammonia or soda ash.   The exchange reactions so far discussed occur with the anion resin in the alkaline form.   Interchange of anions is also possible: for example if an anion resin is put in the salt form by regenerating it with sodium chloride, it can be used to interchange sulphates in water with chloride, viz:

$R.Cl + Na_2SO_4 = R.SO_4 + NaCl$

Demineralization or Deionization.   The preceding equations show that it should be possible — in principle — to prepare water equivalent in quality to distilled water by ion-exchange processes.

Thus, by passing a water supply first through a bed of strong acid cation-exchange resin and then through a bed of strong base anion-exchange resin, all or nearly all of its dissolved solids should be removed.

A simple two-bed system such as this is adequate with certain types of supply water, provided requirements as regards dissolved solids and silica reduction are not exigent.   One method of improving performance is with a three-bed system, the extra bed being a second anion exchanger which will remove the silica and any other anions remaining.

Alternatively, the third bed can be a mixed bed. In this method the cation and anion exchangers are mixed during the treatment process, but are separated by an upward flow of water before the caustic soda and acid regenerating solutions are applied.   After regeneration and rinsing, the materials are re-mixed by means of compressed air. The intermixture of the two types of material enables de-ionization to be very efficient and, as the anion resin is strongly basic, silica is removed.

A third bed allows the use of a weak base anion resin in the second bed in many cases. This is advantageous in that the efficiency and cost of regeneration is improved.   In similar fashion, improvement in working costs can be made by preparing a fourth bed of carboxylic resin before the cation exchanger, if alkaline hardness constitutes an appreciable proportion of the dissolved solids.   The optimum choice between these various trains or systems for a particular set of conditions is a matter for the specialist in this field.   Demineralization is generally considered to be limited to the treatment of waters having dissolved solids of 500 ppm or lower.   By the correct choice of constituent units it is possible to produce water of almost any quality down to 0.5 ppm dissolved solids and 0.05 ppm silica.

Saline or Brackish Waters.   Waters having considerably more than 500 ppm of dissolved solids can now be treated by ion-exchange processes which use specially developed resins.   At present, the main application of these processes is to the provision of potable water, but with the increasing scarcity of good quality water supplies, they may have some future use as preliminary processes to those previously described.

In the processes currently available, cation exchange proceeds as in demineralization, but the anion resin is in the salt form.   Thus, in the Sul Bi Sul process the anion resin is in the sulphate form, and the solution of dilute acids provided by the cation exchanger undergoes such reaction as:

$R_2.SO_4 + HCl = R.HSO_4 + RCl$

If the saline water being treated contains appreciable sulphates, it can be used to regenerate the resin.   The pH of the effluent may require adjustment.   In the opencast mining districts of Pennsylvania, water sources are contaminated by rainfall percolating through from the waste material left after mining — the chief contaminant being calcium sulphate.   At Smith's Township, Penn., a plant has been installed which reduces the dissolved solids from 1500 to 150 ppm.

In the Desal process of Rohm and Haas a bed of special anion material (IRA-68) precedes the cation exchanger, and is in the bicarbonate form.   The bicarbonate is exchanged for other anions (such as Cl^-) in the entering water, and from the resulting solution of bicarbonates the cation unit produces a dilute solution of carbonic acid.   This is passed through a second column of IRA-68 to convert it from the free base to the bicarbonate form.   When the first column of IRA-68 is exhausted, it is regenerated to the free-base form with ammonia, caustic soda, or lime. The flow through the three units is then reversed, so that the second bed of IRA-68 takes over the function of the first, and vice versa. The relevant equations are:

1st anion unit:
$$R - NH.HCO_3 + NaCl = R - NH.Cl + NaHCO_3$$

cation unit:
$$R - COOH + NaHCO_3 = R.COONa + H_2CO_3$$

2nd anion unit:
$$R - N + H_2CO_3 = R - NH.HCO_3$$

regeneration to
free-base form:
$$R.NH.Cl + NH_3 = R - N + NH_4Cl$$

Evaporation. Evaporation is often employed to provide make-up feedwater for boiler plants in which a large proportion of condensate is recoverable from process or turbines and the amount of fresh make-up is therefore small. The raw water to be evaporated should be softened or otherwise treated to prevent scale formation and liberation of carbon dioxide in the evaporators. The evaporators may be single- or multiple-effect though it is not usual to employ more than three effects. The steam for heating may be waste low-grade steam or be taken from low-pressure boilers, or, as in many power stations, be tapped off from one of the turbine stages.

In 'flash' distillation, the evaporated feed is preheated and fed into a series of chambers or 'stages' kept at successively reduced pressures, so that evaporation occurs in each stage. This method is used chiefly for potable water preparation, but also in association with marine boilers. The heat in the distillate vapour is recoverable by heat exchange with the incoming evaporator feed. To secure the most economical working conditions the evaporator must be incorporated into the heat balance of the whole boiler plant in the design stage. The methods of water treatment used for evaporators are similar in principle — but not in detail — to those used for boilers.

The distillate should be nearly as pure as fully demineralized water, but it is important to design and operate an evaporator so that carry-over of the concentrated water is avoided at peak loads.

Typical examples of the effects of various methods of treatment on the composition of three types of raw water are shown in *Table 2*.

Internal Treatment

This is a combined softening and conditioning treatment carried out in the boiler itself, thus avoiding the installation of a softening plant. The alkaline hardness in the raw water is decomposed and precipitated as the water is heated, and the permanent hardness is precipitated in the boiler by the addition of alkali in the form of sodium carbonate, caustic soda, or sodium phosphates. The latter are the most effective precipitants, but because of their cost they are not generally used in the substantial amounts needed when the hardness of the feedwater is high, but for treating feedwater for boilers working at pressures above 1.38 MPa or 13.8 bar (200 psi) or low-hardness feedwaters, especially naturally soft upland waters, they are essential.

Organic materials such as tannins, starches, lignins, synthetic polymers etc. are often used in conjunction with the inorganic precipitants with the object of forming a more mobile sludge and reducing deposits on heating surfaces. Magnesium salts may also be used for the same purpose. Sodium polyphosphates or mixtures of these with tannin help to keep economizers and preheaters cleaner by delaying the precipitation of calcium carbonate from the raw water passing through them.

The main features of the process are:

1. Continuous addition of polyphosphates or mixtures of polyphosphates and tannin to delay precipitation of hardness in economizers, etc.
2. The intermittent addition of precipitating alkalis, alone or mixed with organic materials, to the boiler itself by means of a pump or pressure pot.
3. The continuous addition of catalysed sodium sulphite or hydrazine to feedwater to remove dissolved oxygen.
4. A carefully operated programme of continuous and/or intermittent blowdown to keep the dissolved solids content of the boiler water at a specified concentration and prevent undue accumulation of suspended solids.
5. Chemical control of the process by regular tests on samples of boiler and feedwater.

This treatment has been applied with great success to many boiler plants, having both shell and water-tube boilers. It is usually restricted to feedwaters having a hardness less than 150 ppm calcium carbonate, but has been applied to harder waters under favourable circumstances. The points of addition of chemicals and the precise chemicals or mixtures to be used must be chosen to suit the water and the boiler operating conditions, and the process should not be applied without the guidance of an expert.

Boiler Feedwater Conditioning

There is usually some residual hardness in a softened water and small amounts of Ca and Mg salts are often found in condensate owing to contamination in the process from which the condensate is derived, or to cooling-water leakage in the turbine condenser. In many cases, therefore, small amounts of chemicals are added to the boiler feedwater to counteract scale formation by the residual hardness, and also to adjust the composition of the feed and boiler water to prevent corrosion and caustic cracking. This process of adjustment is generally known as conditioning.

The principles of conditioning against hard sulphate scale formation were worked out by

References p 269

Table 2   Effects of water treatment on composition of treated water (approximate figures)

Treatment	*Parts per million* — Ca	Mg	Na	HCO_3	CO_3	OH	Cl	SO_4	NO_3	SiO_2	D.S. (HCO_3 calculated as CO_3)	Hardness as $CaCO_3$ — Total	Alkaline	pH approx.
Effects of softening on surface water from agricultural area														
Raw water	63	18	25	159	—	—	43	79	16	10	332	232	130	7.5
Hot lime–soda/sodium-aluminate above 70°C	2	1	109	—	36	10	43	79	16	6	301	10	—	10
De-alk./base exchange	1	0.5	76	16	—	—	43	79	16	10	230	5	—	6.5–7.0
Base exchange (cold water)	1	0.5	129	159	—	—	43	79	16	10	357	5	—	6.5–7.5
Acid exchange and base exchange — blended (cold water)	1	0.5	74	12	—	—	43	79	16	10	230	5	—	6.5–7.0
Demineralization (cold water)	nil or very small	—	—	—	—	—	—	—	—	10*	15–20*	nil or very small	—	7
Distillation	—	—	—	—	—	—	—	—	—	—	10	nil or very small	—	5–6.5
Effects of softening on deep well water														
Raw water	48	21	72	348	—	—	48	10	—	17	387	205	205	7
Hot lime–soda/sodium-aluminate above 70°C	2	1	72	—	36	10	48	10	—	12	191	10	—	10
De-alk./base exchange	1	0.5	82	35	—	—	48	10	—	17	179	5	—	6.0–7.0
Base exchange (cold water)	1	0.5	164	348	—	—	48	10	—	17	411	5	—	6.5–7
Acid exchange and base exchange — blended (cold water)	1	0.5	38	12	—	—	48	10	—	17	120	5	—	7
Demineralization (cold water)	nil or very small	—	—	—	—	—	—	—	—	17*	22–27*	nil or very small	—	7
Distillation	—	—	—	—	—	—	—	—	—	—	10	—	—	5–6.5
Effects of softening treatment on moorland water														
Raw water	7	5	8	15	—	—	10	31	—	4–8	76	40	12	6.5–7
Hot lime–soda/sodium-aluminate above 70°C	Unsuitable because raw-water hardness is too low for effective precipitation													
De-alk./base exchange	No advantage over base exchange because HCO_3 is too low													
Base exchange (cold water)	1	0.4	25	15	—	—	10	31	—	4–8	82	5	—	6.5–7
Acid exchange and base exchange — blended (cold water)	No advantage over base exchange because HCO_3 is low													
Demineralization (cold water)	nil or very small	—	—	—	—	—	—	—	—	4–8*	13*	nil or very small	—	7
Distillation	—	—	—	—	—	—	—	—	—	—	10	—	—	5–6.5

* Lower figures are possible, depending on choice of resins and plant design (see p 256)

D.S. = dissolved solids

Dr R. E. Hall in America, following on his studies of the mechanism of calcium sulphate scale deposition. Hall found that by the addition of the right amount of sodium carbonate (Na_2CO_3) to a boiler water containing calcium and sulphate the calcium could be preferentially precipitated as calcium carbonate, which is only slightly soluble in water at boiler temperatures and forms a loose soft sludge or very thin soft scale.

Sodium carbonate decomposes in the boiler, to a degree depending on the working pressure and steaming rate, forming caustic soda (NaOH) which will precipitate residual hardness due to magnesium salts as a sludge of magnesium hydroxide ($Mg(OH)_2$). It is not practicable to give a definite figure for the amount of Na_2CO_3 to be added in any specific case because it depends on the boiler operating conditions and the amount of sodium carbonate or bicarbonate in the softened feed water; for example, the amount of sodium carbonate left in a water softened by the lime—soda/sodium-aluminate process is often sufficient to prevent sulphate scaling in boilers working at pressures up to 1.38 MPa or 13.8 bar (200 psi).

The proof of correct conditioning in this respect is that the water in the boiler shall have a hardness of less than 5 ppm of $CaCO_3$ when tested by the EDTA Test.

At working pressures above 1.38 to 1.73 MPa or 13.8 to 17.3 bar (200 to 250 psi), the decomposition of Na_2CO_3 is usually too great to maintain the required concentration of carbonate in the boiler water, and sodium phosphate is used instead, since the phosphate ion (PO_4^{3-}) is stable at all boiler temperatures, while calcium phosphate is less soluble than $CaCO_3$; it also forms as a sludge. The quantity of phosphate added should exceed the equivalent of the calcium in the feedwater by a small amount so as to build up and maintain a suitable excess of Na_3PO_4 in the boiler water, acting as a temporary reserve against any unexpected intrusion of calcium salts. To avoid the deposition of calcium phosphate in economizers and preheaters the phosphate may be added intermittently at the suction of the boiler feed pump, or injected directly into the boiler drum by means of a force pump or displacement vessel.

Conditioning with phosphate is generally recommended for all boilers working at pressures above 1.38 to 1.73 MPa or 13.8 to 17.3 bar and in those at lower pressures if the silica content of the feedwater is high, for example, above 10 ppm since it also prevents scales due to calcium and magnesium silicates. Besides trisodium phosphate (Na_3PO_4), other phosphates such as disodium hydrogen phosphate (Na_2HPO_4) and sodium hexametaphosphate ($NaPO_3$) — glassy metaphosphate — can be used when there is surplus of alkali in the softened feed water. Organic materials are used with the inorganic precipitants, as in internal treatment.

More recently, chelating compounds such as nitriloacetic acid (NTA) and EDTA, usually in the form of their sodium salts, have been used as conditioning agents. These chemicals form soluble compounds with Ca^{2+} and Mg^{2+}, thus preventing scale and sludge formation. The economics of this method make it expensive for use with feeds having more than 3 ppm total hardness. Its use is also limited to boilers working below about 6.9 MPa or 69 bar (1000 psi), because chelants have some thermal instability; oxygen scavenging is essential to limit further their instability at high temperatures.

Oil in Feedwater

Oil in feedwater may occur either in suspension or emulsified. Its harmful effects are well known; emulsified oil is believed to increase the risk of carryover or priming, while oil in suspension tends to adhere to heating surfaces and, because of its high resistance to heat transfer, results in overheating of the metal, distortion and possibly rupture.

The source of oil in feedwater is usually condensate from steam passing through the lubricated cylinders of engines. Exhaust from reciprocating feed pumps is a very likely cause and in this case it is generally better to recover the heat by means of a coil in the feed tank and reject the condensate rather than to attempt to remove the oil from it.

Steam which has been used in back-pressure engines can be partly freed from oil by means of oil separators situated before the process plant, but the lower-boiling fractions cannot be removed by this means and must therefore be separated from the condensate. There are several ways in which this can be done, with varying degrees of success. Mechanical methods include tanks in which the oil is removed by flotation. The method is suitable only for water containing a considerable quantity of oil. Filtration of the water through wood wool or sand, or even Turkish towelling, will remove much of the suspended oil, but these methods will not remove emulsified oil where the particles are too small to be filtered off and carry an electrical charge causing them to repel each other. Electrical methods can neutralize the charge and allow condensation of the particles into droplets of sufficient size for mechanical separation.

Chemical means include flocculation with a coagulant such as aluminium sulphate in alkaline solution and subsequent filtration.

Oil can be removed during lime—soda/sodium-aluminate softening, but the method is not recommended except where the proportion of condensate containing traces of oil to the total flow through the softener is small; otherwise, softening conditions may be upset.

Corrosion

Important factors affecting corrosion of steel at ambient temperatures — namely, acidity and dissolved oxygen — also come into play with boilers and their feed systems. The alkalis previously mentioned in relation to hardness precipitation therefore serve a double purpose, in that they also neutralize acidity, and their concentration in both feed and boiler waters is adjusted accordingly. Dissolved oxygen can be removed by mechanical deaeration. (This may in any case be desirable to improve the heat-transfer properties of the steam.) In any case, an oxygen scavenger such as catalysed sodium sulphite or hydrazine is normally used, either to remove residual oxygen after mechanical deaeration, or to deal with the entire oxygen content where a deaerator is not available.

The influence of deposit accumulations in localizing corrosion has already been mentioned. Owing to the present-day cost of fuel, even low-pressure plants are returning as much condensate as possible. Without treatment, such condensate attacks the return system, and brings back appreciable amounts of iron and copper oxides to the boilers: these can initiate boiler corrosion and interfere with water circulation in watertube boilers causing tube starvation. To counter these effects, volatile amines are used, and are of two types, neutralizing or filming. Neutralizing amines such as cyclohexylamine volatilize with the steam and condense with it. In condensing, they neutralize acid gases such as carbon dioxide, thus preventing attack at the condensation point or later in the return system; usually the pH of the condensate is thereby adjusted to 8.8—9.3. The amount of neutralizing amine needed is proportionate to that of carbon dioxide present in the steam, and treatment costs therefore rise uneconomically if a high percentage of a make-up is used which releases a large amount of carbon dioxide in the boiler. Filming amines — on condensing with the steam — form a unimolecular film on the metal which is water-repellant, and thus do not permit the usual reactions to occur at the metal surface, i.e. the film is protective. They are most used where the neutralizing type would be uneconomic.

The direct reaction of pure water with steel to produce magnetite (Fe_3O_4) is of increasing importance as boiler temperatures and pressures rise. Under optimum conditions, a magnetite layer of fine, close, texture is produced, which is protective to the metal beneath; the thickness of the magnetite layer is thus stabilized. Under conditions which are still, to some extent, a matter of research and debate, the magnetite layer can become massive, coarse, porous, and non-protective, or may be dissolved away, allowing craters to be gouged out of the metal. The influence of caustic soda in boiler water on these phenomena is reasonably well defined. First, at the optimum concentration, caustic soda improves the magnetite layer and causes breaks to be repaired more quickly. Secondly, excessive amounts of caustic soda can dissolve magnetite and reprecipitate it in a coarse non-protective

form. The optimum concentration of caustic soda appears to decrease as boiler pressure rises. Since other factors such as heat flux, water purity, and boiler design are thought to affect the stability of magnetite films, there is some uncertainty about precise values, but *Table 3* probably reflects present-day trends.

In high-pressure boilers, porous deposits, arising from metallic oxides derived from the feed system or inferior magnetite formation, provide a structure within which concentration of boiler water can occur. In circumstances still not completely explored, soluble contaminants such as nickel salts (derived from corrosion of stainless steel and cupro-nickel elements in the feed cycle) and magnesium salts (derived from condenser leakage) can then concentrate and hydrolyse to the respective acids, producing acid attack. Deciding the appropriate concentration of caustic soda that will be sufficient to prevent acid attack, without dissolving away magnetite or causing caustic gouging, is then a matter of some difficulty.

The congruent phosphate method (see page 261) has been advised by some authorities as a means of avoiding the difficulties of deciding and keeping to the optimum hydroxide concentration in high-pressure boilers.

Anyone requiring values for operational use should consult his boilermaker or a water-treatment specialist for the most recent information on this aspect of boiler control.

Idle Boilers

Unless special precautions are taken, some corrosion is almost inevitable when a boiler is drained and allowed to stand, because some tubes or headers cannot be completely emptied, and the walls of empty drums and tubes are wet. If a period of several months standing is expected, the boiler should be thoroughly dried out with a stream of hot air and then closed up; and trays of silica gel or quicklime should be placed in the drums to keep the atmosphere dry. On the other hand if the boiler is to stand for only a week or two, and may be required at short notice, it can be filled up to the top of the drum with alkaline feedwater to which sodium sulphite has been added at the rate

Table 3   Boiler pressure and hydroxide alkalinity

Boiler pressure	Maximum hydroxide alkalinity
3.1 MPa or 31 bar (450 psi)	150 ppm as $CaCO_3$
34.5 bar (500 psi)	100 ppm as $CaCO_3$
45 bar (650 psi)	80 ppm as $CaCO_3$
55 bar (800 psi)	70 ppm as $CaCO_3$
69 bar (1000 psi)	50 ppm as $CaCO_3$
83 bar (1200 psi)	40 ppm as $CaCO_3$
104 bar (1500 psi)	20 ppm as $CaCO_3$
124 bar (1800 psi)	10 ppm as $CaCO_3$

of 200—300 ppm, samples being tested for sulphite once per week and any losses being made good. When the boiler is required, all that is necessary is to lower the water to normal working level and blow down a little more heavily during the first few days' working. Superheaters also can become corroded during idle periods and, where they can be completely drained, it is advisable to fill them with the same solution as the boiler for short standing periods, and to wash out, drain and dry for long standing periods.

Caustic Cracking or Caustic Embrittlement

This form of attack is intergranular, i.e. a network of fine cracks is produced which are propagated around the boundaries of metal grains. In some cases boiler plate has been so weakened that complete failure and explosion has occurred. Research has shown that two principal factors are involved: a high tensional stress in the steel, and contact of caustic soda with the steel in a concentration of some 100 000 ppm.

Concentrations of this order do not occur in boiler water, and hence some concentrating mechanism is necessary to explain them. In riveted boilers this is provided by slight leakage between overlapping plates and rivets. While such leakage can be remedied by re-caulking all seams inside and outside the boiler, there is a practical difficulty in detecting slight leakage and arranging immediate action. Hence, it is possible for boilers of this type to be exposed to conditions favourable to caustic cracking for uncertain periods, despite seemingly good maintenance. For this reason, most authorities agree that riveted or part riveted boilers should receive throughout their lives one of the chemical treatments described later.

In the case of boilers having solid forged stress-relieved drums, or drums or shells which have been welded and stress-relieved, the stress factor concerned in caustic cracking has been virtually eliminated; also the prospect of a concentrating mechanism due to slight leakage is negligible. Hence these types of boiler are not considered to require chemical precautions against caustic cracking. Some firetube boilers now made according to BS 2790:Part 1:1969 are of welded boiler plate, the individual plates or sections having been previously stress-relieved, but not the completed shell. Residual stress may therefore be present in the seams of such boilers, while weld defects such as undercuts, voids, etc. could be sources of a concentrating mechanism. Thus boilers made in this way cannot be considered to be entirely immune from the possibility of caustic cracking: to date however no failures seem to be known that are due to caustic cracking of this type of boiler.

Sodium Sulphate for Caustic-cracking Prevention.

F. G. Straub and co-workers concluded from their researches (1926—1930) that caustic cracking could be prevented by a sufficiency of sodium sulphate in the boiler water, and the diminution in the number of cases occurring in the USA from 1926 to 1938 in stationary boilers has been attributed to the widespread use of this additive. In the UK it has been standard practice for many years to hold a weight ratio of sodium sulphate to caustic soda in boiler water of 2.5 or above where the sulphate method of control is employed. In many cases the sulphates naturally present in the make-up, together with those contributed by interaction of sodium sulphite with dissolved oxygen, are adequate to keep the ratio above the minimum. Where this is not so, sodium sulphate is added.

Sodium Nitrate for Caustic-cracking Prevention.

Although satisfactory in stationary boilers, sodium sulphate was found to be relatively ineffective in American locomotive boilers. Sodium nitrate, on the other hand, was very successful in reducing the incidence of caustic cracking in locomotive-type boilers. The ASME Boiler and Pressure Vessel Code (1962) states that sodium nitrate — in proportions up to 0.4 parts by weight per part of caustic soda — has been successfully used in this form of protection. In the UK, the general recommendation is that any natural nitrate content should be so augmented that in the boiler water the weight ratio of $NaNO_3$ to total alkalinity expressed as NaOH is 0.4 or greater. Nitrate protection is not usually used outside the pressure range 0.69—4.5 MPa or 6.9—45 bar (100—650 psi) although Berk briefly mentions its use up to 6.2 MPa (900 psi).

Congruent Phosphate for Caustic-cracking Prevention.

This method is essentially a refinement of the co-ordinated phosphate method, which was also known as the 'captive alkalinity' or 'zero caustic' method. Purcell & Whirl introduced the co-ordinated phosphate method in America about 30 years ago. The basis of the method was that acid phosphates are used to react with all caustic soda in the boiler water, thus eliminating the possibility of caustic cracking. The reaction creates trisodium phosphate which has a ratio of 3 sodium atoms to 1 phosphate radical.

Later research showed that evaporation of a saturated solution of trisodium phosphate deposited a solid having only about 2.65 to 2.85 sodium atoms to 1 phosphate radical. Thus some caustic soda would be left in the mother liquor, and under appropriate conditions could initiate caustic cracking.

To meet this difficulty, D. E. Noll introduced the congruent phosphate method in 1964. The ratio of sodium atoms to phosphate radicals is kept at or below 2.6. In this way it is ensured that, even if solid sodium phosphate is deposited, no caustic soda is left in the remaining solution.

For boilers whose construction does not require precautions against caustic cracking, acid phosphates can also be used to control the caustic soda concentration to any value (down to zero) which may be considered an optimum under the particular conditions of operation, for the preservation of the magnetite layer and the underlying steel.

Steam Purity

The purpose of the boiler plant is to make steam that will not cause corrosion or deposits in superheaters, turbines or process plant. Corrosion by carbon dioxide and oxygen in the steam has already been mentioned but the steam can also contain salts as a result of entrainment or carryover of particles of boiler water.

As the steam passes through the superheater these particles dry out and solid salts may be found at various points in the system, e.g. inlet header boxes and inlet tubes of superheaters, outlet header boxes, in steam lines, on steam valve spindles, turbine governor valves, and turbine blades.   The presence of boiler-water salts in condensate from steam traps and process plant, or deposits in process plant, may also indicate carryover.

The presence and amount of salts in steam may be determined by means of special sampling nozzles inserted in the steam lines, the sample being cooled and condensed and the dissolved solids determined either by evaporation to dryness in the laboratory or by measurements of conductivity.

A more accurate measure of salt contamination can now be obtained by using a glass electrode which responds selectively to the sodium-ion concentration in the condensate. In association with a reference electrode, this develops a potential which varies with the sodium-ion concentration and which can be indicated as such on an instrument similar to a pH meter. By comparison with the sodium-ion concentration of the boiler water, the degree of contamination can be calculated.

Sudden rises and falls in steam-condensate conductivity may indicate a tendency to prime, which can be counteracted by lowering the boiler water level or reducing boiler output, whereas a steady increase in conductivity probably indicates continuous carryover, which can usually be diminished by blowing down to reduce the concentration of dissolved salts in the boiler water.

Some superheater failures may be explained on the supposition that when the boiler is shut down the salts on the walls will dissolve in the condensate and drain down, to the lowest parts of the superheater, where they will be concentrated by evaporation when the boiler starts up again and destroy the magnetic oxide layer on the metal, thus exposing it to fresh corrosion. Such a set of circumstances is more likely to occur in superheaters that cannot be completely drained, and it is suggested that it might be a useful precaution to wash out the superheater with condensate whenever salt deposits are observed at the inlet ends.

Experience shows that carryover of boiler water is usually promoted by circumstances such as high water levels, sudden changes in steam load, leakage past baffles in the steam drums, and the concentration of suspended and dissolved solids in the boiler water, or the presence of oil, operating together or independently. The influence of the dissolved solids in boiler water has been widely studied, and it seems to be established that for any given set of steady boiler operating conditions the entrainment of particles of boiler water increases rapidly when the concentration of dissolved solids in the boiler water exceeds certain figures. *Table 4* gives a rough indication of these critical concentrations in boilers working under steady load and at the maximum continuous rating (MCR) specified by the boiler manufacturer.

As a result of carryover, silica can appear on turbine blades along with other suspended or dissolved constituents of boiler water. There is another process whereby silica can be transferred to turbine blades — silica volatilization. F. G. Straub and others established that silica has a definite solubility in steam, which increases with temperature, and with the concentration in the boiler water. It is generally accepted that a safe practical limit for SiO_2 in steam is 0.02 ppm. To keep to this limit, the maxima for boiler water given in *Table 4* should not be exceeded. To avoid inordinate amounts of blowdown, evaporation or demineralization of make-up may be required to comply with either the D.S. or SiO_2 limits of *Table 4*. This matter should be decided at the design stage, and the following formulae can be used to give an indication of probable blowdown requirements (as a percentage of the total feed), in a particular situation:

$$\frac{\text{D.S. of make-up}}{\text{Allowable D.S. in boiler}} \times \text{percentage make-up}$$

$$= \text{percentage blowdown}$$

$$\frac{SiO_2 \text{ of make-up}}{\text{Allowable } SiO_2 \text{ in boiler}} \times \text{percentage make-up}$$

$$= \text{percentage blowdown}$$

The values given by the two formulae will not generally agree, and the larger result represents the real blowdown requirement. Thus, for a 41.5 bar (600 psi) boiler, the maxima from *Table 4* are 2000 ppm, and 40 ppm, respectively, for D.S. and SiO_2. If the other relevant data are:

D.S. of make-up = 50 ppm: SiO_2 of make-up = 7 ppm: percentage make-up = 25%

then blowdown to control D.S. works out to 0.63% and that for silica control to 4.4%. If the position regarding silica is ignored, and the boiler is operated to the D.S. maximum of 2000 ppm, then turbine difficulties are inevitable. The real choice lies between operating the boiler with

4.4% blowdown, or eliminating silica from the make-up by using evaporators or demineralization.

Anti-foams

Foam formation arises in boilers from the presence of traces of such foam formers as soaps, certain compounded lubricating oils, detergents, etc. Apart from carryover, difficulty arises in the working of boilers with foam present because the true water level cannot be distinguished. High water levels then tend to be carried as a precaution against low water and overheating of the boiler, but high levels increase the risk of carryover. Efficient antifoams which will deal with these problems have been available for about twenty years; on addition of an adequate quantity, foam suppression is instantaneous. It is now usual for proprietary chemicals used in water treatment to contain some antifoam to offset minor contamination. Antifoam formulations can also be obtained as separate items for the larger usage required by severe contamination with foam formers.

Disposal of Blowdown

There are four methods of disposing of blowdown:

1. Blowing to waste. If this method is adopted the most rigorous control is necessary to prevent waste of fuel.
2. Passing through a heat exchanger, in which the feedwater is heated, or through a flash vessel from which the steam is passed direct into the feedwater and the heat in the residue is recovered in a heat exchanger. Alternatively, part of the residue from the flash vessel can be returned to a hot-softening plant in which its heat and chemical content are used.
3. Returning a portion of the blowdown direct to the feed so that the chemicals in it are used for conditioning purposes. As a rule the amount of blowdown recoverable in this way is small.

Table 4   Maximum permissible dissolved solids (D.S.) and silica (SiO$_2$) in boiler water

Type of boiler		D.S. (ppm)	SiO$_2$ (ppm)
Lancashire		17 000	—
Packaged firetube		4 500	—
Vertical		4 500	—
Watertube:			
psi	bar		
0—200	0—13.8	4 000	150
201—300	13.8—20.7	3 500	100
301—600	20.7—41.4	3 000—2 000	50—40
601—900	20.7—62.1	2 000—1 400	30—20
901—1100	62.1—75.9	1 400—1 000	20—10
1100—1500	75.9—103.5	1 000— 750	10— 5

4. Using the blowdown from high-pressure boilers working on good quality feedwater as feedwater make-up to low-pressure boilers. For example, in one case the blowdown from 45-bar boilers is used as part feed to boilers working at 13.8 bar.

Boiler water returned to the feed system by methods 2 and 3 does not reduce the dissolved solids in the boiler, and must not be accounted as part of the blowdown for deconcentration.

Continuous blowdown is the most satisfactory method of removing water from the boiler because it can easily be adjusted to maintain the D.S. in the boiler water consistently at the desired level. The method preferred in watertube boilers is to fit a short horizontal length of pipe below the water level in the steam—water drum and at the end opposite the feed entry. In cylindrical boilers an open-ended vertical pipe may be used, but the open end should be more than 51 mm (two inches) above the top of the main flue tubes so that these cannot be exposed through excessive blowdown. An orifice of suitable size or a fine-control valve is fitted in the blowdown line outside the boiler to control the amount discharged, and should be adjusted to maintain the desired D.S. in the boiler water. Continuous blowdown is not really practicable when the amount of water to be removed is less than 108 litre (25 gallons) per hour.

For boilers operating up to about 31 bar (450 psi), automatic blowdown valves are now available. In one type, the interval between blows and the duration of each blow is set by an electrical timer device, the operation of which can be varied over a suitably wide range, either manually, or automatically according to boiler-water conductivity. Other types of blowdown valve are automatically opened when the conductivity of the boiler water reaches the desired maximum level or range, and shut again when some minimum value is indicated. These effects approximate to that of continuous blowdown, and can be arranged to deal with blowdown rates considerably less than the 108 l/h mentioned in relation to continuous blowdown. These valves are therefore particularly useful with smaller low-pressure boilers.

Most boiler waters contain some sludge which must be removed by intermittent operation of the ordinary bottom blowdown valve. This valve should be opened as often as is reasonable, taking into account the total volume of water to be removed, i.e. once per shift, or not less than once per day, so that the concentration of salts in the boiler water is kept fairly steady.

To ensure removal of accumulated sludge, the valve should be opened momentarily several times with a short pause in between so that the disturbance in the neighbourhood of the valve can die down and fresh sludge can move into position ready to be flushed out.

When feedwaters of low D.S. are used, or the rate of evaporation is low, and the maximum permissible D.S. in the boiler water is high, it may be found that this figure is reached only after several weeks' running, and perhaps not at all. In such cases the intermittent blowdown should be operated once per day after the first few days' working.

Modern packaged boilers are fitted with automatic level controllers. It is most important that the float chambers be blown down once a shift, to avoid filling with deposit and so failing to indicate low water.

It is advisable to have a clear set of blowdown instructions issued to each boilerhouse covering all conditions of operation.

Hot-water Heating Systems

Water heating systems used for space heating are theoretically closed circuits requiring little or no make-up water. Difficulties due to scale and corrosion should be comparatively rare, but the following general observations may be of interest:

Make-up should be reduced to an absolute minimum. Deliberate draw-off should be eliminated and accidental losses avoided as much as possible. If there is any doubt in this regard suitable treatment should be introduced.

Converted steam boilers have been used to provide hot water for circulation. One advantage of such 'pressure hot water' systems is that the benefit of high temperature is available without the disadvantages of steam trapping and condensate return. If little make-up is used scale formation will not be appreciable, but as the results of corrosion in the circulation system might be very serious it is usually advisable to treat such systems and in particular to use an oxygen scavenger such as sodium sulphite to remove dissolved oxygen. When a boiler serves a dual purpose for steam raising and hot-water heating, soft water make-up is essential to avoid appreciable amounts of sludge in the circulation.

Hot-water systems for industrial space heating frequently operate at temperatures above $100^\circ C$. This requires that the system be pressurized. Nowadays this is done by keeping a nitrogen blanket above the water level in the boiler, in place of the 'steam cushion' of earlier types — which were essentially converted steam boilers. This blanket is maintained from a nitrogen pressurization vessel which also acts as a reservoir for expansion water. The principles of treatment are the same for this type of system, i.e. the emphasis is on corrosion prevention, since the scaling potential is low unless the system has a high make-up requirement.

Another type of system utilizes steam from a boiler which heats water in a separate pressure vessel, from and to which the hot water is circulated from the heating-system proper. These systems accumulate water owing to the condensation of steam, and this has to be removed —

usually to the feed tank of the steam boilers. Thus, for these systems, treatment for the hot-water system affects conditions in the boilers, because of the transfer of treated water. Specialists should be consulted about the treatment of this type of system if the best results are to be obtained in both system and boilers.

Hot-water Supply Systems

The two water treatment problems in hot-water supply systems are calcium carbonate scale formation and corrosion. Scales due to calcium sulphate or silicate or to magnesium compounds are uncommon. Corrosion may be fairly uniformly spread over the metal surfaces or may take the form of pitting, and the products of corrosion may be deposited on site or carried forward to other parts of the system, where they may choke pipes.

Calcium carbonate scales are caused by the decomposition of calcium bicarbonate, with the evolution of carbon dioxide. Methods available for control are (1) temperature limitation, (2) softening, and (3) addition of inhibitors. Some control may also be obtained by reducing the flow of water at the outlet valve so that the system is maintained under slight pressure. Rise in temperature increases the scale formed, so that any temperature control that can be imposed will reduce the scale problem.

Softening is probably the most satisfactory method of dealing with hard waters for hot-water supply, particularly for domestic use or for certain process purposes. Either precipitation or ion-exchange processes may be used, but if plant is being installed for hot water only, the more convenient, and usually cheaper, ion-exchange plant may be preferred.

Where scale prevention is required but softening is not possible, threshold treatment with glassy phosphates can be used. Applied at the rate of a few ppm, these compounds interfere with the crystallization of calcium carbonate and so induce an apparent super-saturation. Above $82^\circ C$ the glassy phosphates are converted into orthophosphates which do not possess the same property, but if the contact period at high temperature is small a considerable reduction in scaling may nevertheless be expected.

Corrosion in hot-water supply systems can usually be avoided by suitable choice of materials but even with the more resistant materials difficulties are occasionally experienced. Soft waters with pH values under 7.0 (acidic waters) are particularly troublesome with non-ferrous metals. With lead they may pick up sufficient metal to make them dangerously toxic; with copper sufficient metal may go into solution to produce green stains on fittings or fabrics or even to discolour the water. In such cases raising the pH value with an alkali such as caustic soda, hydrated lime or sodium silicate is usually satisfactory. Over-dosing must, however, be avoided.

Both soft and hard waters may be corrosive to ferrous metals, though more difficulty is experienced with the former. Lime softening of corrosive hard waters is often effective; the added lime removes dissolved carbon dioxide as well as hardness. Sodium-exchange-softened waters may be blended with a small proportion of hard water to reduce this corrosiveness. But in some instances the addition of an alkali is also necessary. Alkali additions to corrosive hard waters are also often effective, but simultaneous addition of an inhibitor is usually required to prevent scale deposition which will otherwise be increased by the alkali addition.

Chemical conditioning of corrosive soft waters for use in ferrous-metal systems is more difficult as addition of alkali alters the nature of the corrosion products. Below a pH value of 7.0 corrosion tends to be general and corrosion products are carried away with the water, but in the range of approximately 7.5—9.0 the corrosion products may adhere to the metal in the form of 'tuberculation' and corrosion is more localized. If it is possible to raise the pH value to over 9.0 corrosion control is often successful, and sodium silicate is sometimes effective. Some inhibitors which would otherwise be effective, such as the chromates and various metallic salts, are toxic and cannot be used for domestic or many trade purposes. When the rate of flow is sufficient to ensure a continuous supply of inhibitor to the metal surface the glassy phosphates can be used, although the amount required is more than for scale prevention alone and no protection is afforded to tanks or other large surfaces where lack of turbulence prevents the necessary supply of inhibitor from reaching the metal surface. For tanks, certain proprietary protective coatings, or sprayed aluminium for hot tanks and cement-wash or bitumen for cold tanks, can be used.

General Principles

Boiler Feedwater Treatment —A Checklist

The principles are summarized below, but not all the qualities mentioned are essential in any given case and it cannot be too strongly emphasized that expert advice should be sought when choosing a treatment.

1. *Choice of external treatment to obtain:*

(a) low hardness — to avoid sludge in the boiler;

(b) low bicarbonate — to reduce corrosion by steam;

(c) low dissolved solids — to reduce boiler blowdown; and

(d) low silica and low alumina where necessary.

Objectives (a) and (b) may be reached by external process such as lime—soda, sodium ion exchange, etc. Such processes are generally appropriate for

boilers working below 2.4 MPa or 24 bar (350 psi).

As pressures progress upwards, boilers are worked at increasingly lower dissolved solids (see *Table 4*). Hence objective (c) requires that evaporation or deionization of low-grade supplies be considered.

For objective (d), coagulation and precipitation processes can be useful in reducing silica and alumina. Complete elimination is desirable — certainly for boilers operating at and above 4.2 MPa (600 psi). To accomplish this, strongly acidic and strongly basic resins must be included in the ion-exchange train.

2. *Conditioning treatment*

The number and types of additives which may form part of a conditioning treatment have increased appreciably over the past twenty years. Brief notes on a not all-inclusive list are given below. A programme devised for a particular boiler plant would be unlikely to include all items shown.

(a) Sodium carbonate: to promote zero hardness in low-pressure boilers operating below about 1.38 MPa (200 psi) and so prevent scale. Also to raise alkalinity of feed so as to minimize corrosion. Some external treatment processes provide adequate sodium carbonate in the treated make-up.

(b) Caustic soda: Can be used in place of sodium carbonate in low-pressure boilers as above. Again, sufficient may be provided by external treatment processes. In high-pressure work, the hydroxide in boiler will be controlled along the lines indicated in *Table 3*.

(c) Phosphate: All forms used for scale prevention at boiler pressures above about 1.38 MPa (200 psi). Glassy phosphates can also reduce precipitation of calcium carbonate in hot feed lines. Both glassy and acidic phosphate may be used to eliminate caustic soda from the boiler water.

(d) Chelating agents: Used as an alternative to phosphates as preventatives of scale in boilers. Application limited (by economics) to good-quality feed water.

(e) Antifoams: Used to prevent foam formation in boilers. Proprietary boiler-chemical mixtures often contain an antifoam, but these materials can be obtained separately for individual application to severe cases.

(f) Neutralizing amines: Used to neutralize carbon dioxide in steam condensate and feed lines and so diminish corrosion. Not economic in systems with high make-up of untreated water. Unsuitable where steam comes into direct contact with foods, beverages, or pharmaceutical products.

(g) Sodium sulphite: Used to eliminate dissolved oxygen and so diminish corrosion. Catalysed sodium sulphite is reputed to react 20 to 500 times as fast as the uncompounded material, and this offers more protection to short feed systems.

(h) Hydrazine: Also used to eliminate dissolved oxygen and so diminish corrosion. Has the advantage of not increasing dissolved solids. Reacts slowly at temperatures below about 245°C. Not used where steam processes food or beverages.

(*i*)   Sodium sulphate: Used to prevent caustic cracking in riveted boilers.

(*j*)   Sodium nitrate: Also used to prevent caustic cracking.

(*k*)   Sludge mobilizers: Natural and synthetic organic materials are used to reduce adherence of sludge to boiler metal. Some of these materials have temperature limitations; the advice of the vendors should be followed closely in their use.

(*l*)   Internal treatment: Essentially a form of conditioning treatment used without any prior external treatment process. Its use is confined to firetube and low-pressure watertube boilers. The larger water spaces of firetube boilers tolerate a higher concentration of conditioned sludge. For this reason, feedwaters (to firetube boilers) of up to 150 ppm total hardness are treated by this method, but for the modern packaged firetube boilers the trend is to limit internal treatment to cases where the feedwater hardness is 50 ppm or lower. Similarly, whereas for older low-pressure watertube boilers the limit was about 50 ppm, fcr the modern packaged type feedwater hardness is preferably not above 10 ppm. Internal methods require extra blowdown for sludge removal, but avoid the capital cost of external treatment plant.

3. *Blowdown control*
Control of blowdown to maintain the D.S. or SiO_2 content of the boiler water in the ranges given in *Table 4*. Intermittent blowdown to reduce the amount of sludge in the boiler to the lowest possible amount, without excessive loss of heat.

4. *Testing*
The institution of a schedule of regular testing to observe and control the treatment at all stages.

5. *Idle boilers*
Special addition of alkali and sodium sulphite to protect idle boilers against corrosion.

Control of Water Treatment

To secure satisfactory softening and conditioning requires control by routine testing. Most of the methods of test are simple and rapid, and it is of course important that immediate action should be taken on the results to modify the softening or conditioning chemical charges as required. Testing should be carried out in a clean and well-lighted place, shut off from boilerhouse fumes and dust, and supplied with hot water for cleaning the glassware used.

It is impossible to specify in detail the tests which should be made for every plant owing to the differences in operating conditions, and expert advice should be sought. In addition, the bibliography to this Chapter lists the principal relevant British Standards.

Management of Water Treatment

Whatever methods of treatment are used it should be recognized that the changes that take place in water during preliminary treatment, and in the boiler under the influence of heat and evaporation, are chemical changes which must be controlled at all stages in the process of converting water into steam. It is suggested, therefore, that where the boiler plant is large, or working at high pressure, full responsibility for the supervision and direction of water treatment should be given to a chemist, who should have sufficient knowledge of modern water-treatment practice and of the design and operation of boilers to advise the engineer in charge on the treatment and control tests required, and to explain the reasons for them. In many large works there is scope for such a man in the study and solution of other water-treatment problems, such as the prevention of scale, bacteriological growths and corrosion in cooling systems and condensers, or the treatment and disposal of works effluents. In view of recent legislation, the latter will require much more attention than it has hitherto received.

When a chemist is not available, a consultant can help materially by paying periodical visits, making check tests, and teaching the boiler staff to carry out simple control tests and the action to be taken in the light of the results. Many small boiler plants are successfully controlled in this way after preliminary consultation with an expert. The correctness and efficiency of any form of feedwater treatment can be gauged only by frequent and regular tests done by a competent tester whose results are confirmed at regular intervals by a qualified analyst specializing in this work.

Cooling-water Circuits

Interference with heat transfer in the heat exchangers of cooling systems can occur from four principal causes:

1. The instability of calcium bicarbonate with rising temperature and concentration produces deposits of calcium carbonate.
2. Corrosion products tend to be transferred from other parts of the system to the heat-exchange surfaces.
3. Microbiological organisms may grow *in situ*, or the dead organisms may be transferred as foulants from other parts of the system which are open to atmosphere, e.g. cooling towers, sprays, or sumps.
4. Natural foulants such as clay, silt, fibrous plant material, etc. which are present in the raw water and escape any filtration, trash separation, or other preliminary treatment processes.

Once-through systems suffer the same problems as circulating systems, and the following remarks apply to both types; closed systems are dealt with later.

Calcium Carbonate Deposits

Such deposits may be dealt with by one or more of the following methods.

External Treatment. Lime softening or carboxylic resins will remove most of the bicarbonate hardness, while base exchange on the sodium cycle will remove total hardness almost entirely. The resulting water may be more corrosive, but this is easily dealt with by one of the modern corrosion inhibitors. External methods are not usually employed with the large volumes of make-up associated with major installations.

Acid Addition. This destroys bicarbonate and substitutes a stable salt. If all or nearly all the bicarbonate is eliminated, the resulting water will be corrosive, but again, a corrosion inhibitor can be used to offset this condition. An alternative method is to destroy only such part of the bicarbonates as are expected to prove unstable at the anticipated temperature and concentration, and to use a scale inhibitor (see below) to cater for minor deposition.

Scale Inhibitors. Complex inorganic phosphates, such as the sodium metaphosphates, have the property of modifying the growth of crystals of calcium carbonate in such a way that the crystals do not have the same tendency to form aggregates of deposit. These compounds have therefore been used for forty years or so to restrict formation of calcium deposits, and to allow systems to work at a higher concentration than would otherwise be possible. Other materials, e.g. tannins, are incorporated in proprietary mixtures to augment the effect.

The inorganic phosphates have the disadvantages that, when dissolved, they gradually revert to the orthophosphate, and that heat accelerates the reversion. This limits their effectiveness in cooling systems, since orthophosphates do not inhibit calcium carbonate: on the contrary, they will cause precipitation of calcium as phosphate. During the past decade organic phosphonates and polyol esters have been introduced. The molecular structure of these compounds is such that reversion to orthophosphate does not occur under the conditions pertaining to cooling systems. Consequently, the range of conditions for which scale inhibition is possible has been considerably extended by this class of compounds.

Control of Concentration and pH. Calcium carbonate and bicarbonate are of low solubility, and so cannot be concentrated indefinitely in cooling systems. Since they are alkaline, an increase in their concentration also results in a rise of pH in the cooling water. Langelier defined the relation between the calcium, alkalinity, and pH of saturation in mathematical terms. The pH of saturation (pH_s) is theoretically the maximum possible pH before the water becomes unstable with respect to calcium carbonate. Comparison of the calculated pH_s with the actual pH of a water shows whether the water is supersaturated and likely to create deposit ($pH > pH_s$), or whether further concentration is

possible ($pH_s > pH$). This is the basis of the Langelier or Saturation Index:

$$\text{Saturation Index} = pH \text{ (actual)} - pH_s$$

pH_s can be calculated by the formula:

$$pH_s = A + \frac{(DS)^{1/2}}{100} - (\log CaH + \log M)$$

where DS is dissolved solids (ppm), CaH is calcium hardness (ppm), and M is total alkalinity to methyl orange (ppm). Values for A are dependent on temperature, and are given in *Table 5*.

Calculation of Saturation Index, pH_s

Data:

Calcium hardness (CaH) = 300 ppm as $CaCO_3$
Total alkalinity (M) = 200 ppm as $CaCO_3$
Dissolved solids (DS) = 600 ppm as such
Temperature = 30°C
pH (actual) = 8.2

A = 11.58 at 30°C (from *Table 5*)

CaH = 300 ppm $\therefore \log CaH$ = 2.477
M = 200 ppm $\therefore \log M$ = 2.301

DS = 600 ppm $\therefore (DS)^{1/2}$ = 24.2

and $\dfrac{(DS)^{1/2}}{100}$ = 0.242

Hence,

$$A + \frac{(DS)^{1/2}}{100} - (\log CaH + \log M) =$$

$$11.58 + 0.242 - (2.477 + 2.301)$$

Table 5 Values of A to be used in calculating saturation pH (pH_s) of calcium carbonate

Temperature (°C)	Value of A
0	12.31
10	12.04
20	11.80
30	11.58
40	11.40
50	11.25
60	11.11
70	10.95
80	10.83

Note Interpolate for intermediate temperatures, e.g. for 25°C use 11.69

Hence

$$pH_S = 11.822 - 4.778 = 7.044$$

Saturation Index = pH (actual) − pH_S =

$$8.2 - 7.04 = +1.16$$

Stability Index = $2pH_S$ − pH (actual) =

$$14.08 - 8.2 = 5.88$$

The positive value for the Saturation Index predicts incrustation.

The Stability Index calculated suggests light to moderate amounts of deposit. The accuracy of this type of calculation is such that more than two decimal places should not be quoted in the final result.

When positive, the Index indicates deposit-forming conditions, and when negative that there is a margin of safety. The Saturation Index is not quantitative, i.e. a small positive value does not necessarily mean that the system will have only a small amount of deposit. To overcome this difficulty, Ryznar proposed the Stability Index:

Stability Index = $2\ pH_S$ − pH

Ryznar correlated field results with his Index, and the correlation indicates that as the value of the Stability Index diminishes from about 6.5 to 3.0 the likelihood and severity of scaling increases. With Ryznar Indices above 6.5—7.0, his correlation suggests that deposits will not be a problem but corrosion may occur.

These, and similar calculations, enable an estimate to be formed of the working concentration and pH possible for a particular set of conditions, but the Indices are most useful in checking whether the actual circulating water is likely to deposit scale or not. This method is not always applicable to hard supply waters, since the degree of concentration possible may be small or zero: in practice, it is almost always used in conjunction with scale inhibitors.

Corrosion

Many systems operating with a hard circulating water and a scale inhibitor enjoy low corrosion rates of under 5 mpy (1 mpy = 1 mil or thousandth of an inch per year = 2.5×10^{-2} mm/year). This is because the concentration of calcium bicarbonate is high in such systems. Calcium bicarbonate is believed to act as an inhibitor in near-neutral solutions by reacting with hydroxyl ions which are produced by the corrosion reactions. This gives carbonate ions which precipitate calcium locally and so stifle the corrosion cell.

Soft supply waters — especially in industrial atmospheres from which they absorb acidic gases —

are generally more aggressive, and require stronger forms of corrosion inhibition. All corrosion inhibitors take effect by forming a protective film on the metal. Anodic inhibitors, such as nitrites and chromates, form a protective film at the anode, i.e. the point at which the metal is dissolving. Cathodic inhibitors, notably the zinc ion, form a coating at the cathode of the corrosion cell. The conjoint action of both anodic and cathodic inhibitors has a synergistic effect, giving remarkably low corrosion rates for very low inhibitor concentrations. Each of these combinations has an optimum pH range over which corrosion protection is most effective. For some of the more popular synergistic mixtures, the usually accepted pH range is given below:

	pH *range*
Zinc/organic	7—8.0
Zinc/chromate	6.5—7.2
Zinc/phosphate	6—7.0

Users of proprietary mixtures should accept the pH range specified by the suppliers. Harder water supplies operating within these pH ranges may require acid dosing of the make-up. The use of chromates in any form is somewhat restricted by the effluent problem created by disposal of bleed-off from the system.

An alternative to conventional effluent procedures for dealing with chromates is an ion-exchange process which allows recovery of chromate from the bleed-off for re-use in the system. Even for large systems the economics do not seem promising. The process does not seem to have been used outside the USA.

Microbiological Organisms

Organic growths restrict water flow and impede heat transfer, and it is important that they be dealt with effectively. Until about thirty years ago, chlorine and the hypochlorites were virtually the only biocides to be used in cooling systems. As oxidizing agents, they have the disadvantage that the dose required is proportional to the oxidizable matter present, and the amount needed may therefore be unduly large for polluted make-up waters. Hypochlorites are not usually applied to large systems, while chlorine can only be present transiently, or at relatively low concentration, owing to the stripping action of the air in the cooling tower.

For such reasons, other materials were developed such as sodium pentachlorphenate, chlorophenols, organosulphur or amino compounds, bisthiocyanates, organometallic compounds, etc. These materials have the advantage of longer residence time in cooling water, which extends their potential for killing unwanted organisms. Choosing the biocide most suitable for a particular problem is a specialist area. Because organisms can develop a resistant strain to a particular biocide, it is becoming customary in difficult cases to use two different biocides alternately. Most of these substances are toxic and, owing to their persistence, cannot be used for sterilizing potable-water equipment, etc. Care is needed in handling and feeding chlorine and the hypochlorites also, although feeding methods are usually simpler for the new materials. In all cases, handling precautions should be checked with the suppliers.

Inert Foulants

As mentioned earlier, corrosion products, clay, silt, and other debris may give rise to heat-transfer problems, especially in areas of low water velocity (below 1 m/s). To deal with these, antifoulants have been evolved, usually natural or synthetic organic reagents, which have a dispersing action on these types of deposit. The effect is believed to be that of electrostatic repulsion between colloidal particles of the same electric charge. This prevents these particles forming the larger aggregates which precipitate and may adhere to the pipework, etc. It is important to note that antifoulants are not a replacement or alternative to antiscalants or scale inhibitors, which serve an entirely different purpose. Antifoulants can make an appreciable contribution to plant cleanliness and heat economy when a surface-water supply is used and where the system is subject to airborne contamination with industrial dusts.

Closed Systems

Evaporation from these systems is minimal, but there are usually losses through vent pipes, pump glands, etc. The small amount of oxygen supplied by make-up to remedy these losses is sufficient to maintain any corrosive action which may have been initiated when the system was first filled, if the water used has aggressive tendencies. Corrosion prevention with inhibitors is therefore common practice. The theory that the oxygen in the filling water will be used up in surface rusting, and that thereafter there will be no corrosion, has been found wanting in many cases. Hence it is most desirable to have a specialist's opinion about the need for corrosion prevention when installing a new system.

Scaling may not be a problem, unless the filling and make-up supplies are particularly hard, the amount of make-up is unduly great owing to leakages, repairs and refilling, etc., or the maximum temperature in the system is high. Acid dosing is not usually employed with closed systems, but the other remedies suggested for open systems are applicable.

Biological growths are not usually a problem unless surge tanks etc. are open to atmospheric contamination. Sulphate-reducing bacteria may gain access in this way, when the presence of iron or copper sulphide as a corrosion product is an indication of their presence.

Antifoulants may be useful in minimizing deposits below which the anaerobic environment exists which is needed by sulphate-reducing bacteria. They are also useful in establishing inhibition quickly in systems which have been previously corroded and are subsequently treated with a corrosion inhibitor.

General Reading

Manual on Water, 3rd edition, ASTM Special Technical Publication No.442, 1969, 355 pp

Water Treatment for Industrial and Other Uses, Eskel Nordell Rheinhold Publishing Co.

Noll, D. E., 'Factors that Determine Treatment for High-Pressure Boilers', *Proc. Am. Power Conf.* 1964, Vol.26, 753

Scriven, B. R., 'The Boilermaker's Attitude to Water Treatment', *Chemy Ind.* 1969 (Sep.13), p 1292

Masterson, H. G., Castle, J. E. and Mann, G. M. W., 'Waterside Corrosion of Power Station Boiler Tubes', *Chemy Ind.* 1969 (Sep. 6), 1261

Marcy, V. M. and Halstead, S. L. 'Improved Basis for Co-ordinated Phosphate — pH Control of Boiler Water', *Combustion* 1964, Vol.35, 45

Hoult, E. 'Scales, Corrosion, and Hydrogen Embrittlement in Boiler Plant', *Chem. Proc.* (UK) 1970 (Jan.), Vol.16, 32

Man'kina, N. N. and Romadin, V. V. 'The Effect of Boiler Water Phosphate Treatment Conditions on the Rate of Corrosion Processes', *Thermal Engng* 1969 (Sep.), p 33

Straub, F. G., 'Proven Practices in Chemical Control for Steam Power Plants', ASME, *J. Eng Power* 1967 (July), p 305

Coulter, E. E., Pirsh, E. A. and Wagner, E. J., 'Selective Silica Carryover in Steam', *Trans. ASME* 1956, Vol.78, 869

Jacklin, C., 'Chelating Agents for Boiler Treatment — Research and Actual Use', *Proc. Am. Power Conf.* 1965, Vol.27, 807

Berk, A. A., 'Caustic Cracking in Steam Boilers', *Chemy Ind.* 1953 (Ap.18), 360

Langelier, W. F., 'The Analytical Control of Anti-Corrosion Water Treatment', *J. Amer. Water works Assn* 1936, Vol.28, 1500

Emerson, A. G. D., 'Application of Equilibrium pH to Recirculating Cooling Systems', *J. Soc. chem. Ind.* 1945, Vol.64, 335

Ryznar, J. W., 'A New Index for Determining Amount of Calcium Carbonate Scale Formed by a Water', *J. Amer. Waterworks Assn* 1944, p 472

Farnsworth, N. B. and Robertson, R. S., 'New Non-chromate, Non-phosphate Organic Cooling Water Treatments', 21st Ann. NACE Conf, St. Louis, Miss., March 15th, 1965 (Nalco Chemical Co. Reprint No.151)

Brooke, J. M., 'Inhibitors: New Demands for Corrosion Control', *Hydrocarbon Processing* 1970, Vol.49 (Jan.), 121 and (March) 138

Stafford, A. E., 'Water Treatment Requirements for Open Evaporative Cooling Systems', *Eff. & Water Treat. J.* 1972 (Feb.), Vol.12, 75

British Standards

1170:1968 Methods for treatment of water for marine boilers

1328:1969 Methods of sampling water used in industry

1427:1962 Routine control methods of testing water used in industry

1647:1961 pH scale

2455:1973 Methods of sampling and examining deposits from boilers and associated industrial plant: Part 1. Water-side deposits

2486:1954 Treatment of water for land boilers

2690:1956 Methods of testing water used in industry

2690:Part 1:1964 Copper and iron

2690:Part 2:1965 Dissolved oxygen, hydrazine and sulphite

2690:Part 3:1966 Silica and phosphate

2690:Part 4:1967 Aluminium, calcium, magnesium and fluoride

2690:Part 5:1967 Alkalinity, acidity, pH value and carbon dioxide

2690:Part 6:1968 Chloride and sulphate

2690:Part 7:1968 Nitrite, nitrate, and ammonia (free, saline and albuminoid)

2690:Part 8:1969 Cyclohexylamine, morpholine and long-chain fatty amines

2690:Part 9:1970 Appearance (colour and turbidity), odour, suspended and dissolved solids and electrical conductivity

2690:Part 10:1970 Sodium, potassium and lithium

2690:Part 11:1971 Anionic, cationic, and non-ionic detergents and oil

2690:Part 12:1972 Nickel, zinc, chromate, chromium and manganese

2690:Part 13:1972 Dichromate value (chemical oxygen demand), non-volatile organic carbon, tannins and chlorine

2690:Part 14:1972 Arsenic, lead, and sulphide

2690:Part 15:1974 Free EDTA, total salts of EDTA, polyacrylate and polymethacrylate

Production of mechanical and electrical energy

Chapter 12

Devices for direct production of mechanical energy

12.1 Internal combustion engines

Criteria for selection of the prime mover
Operating costs
The diesel engine
Spark-ignition gas engines
Dual-fuel engines

12.2 The industrial gas turbine

Mechanism
Advantages and disadvantages
Applications
Noise and pollution
Availability and maintenance
Future developments

12.3 Heat recovery in reciprocating engines and gas turbines

Exhaust conditions
Exhaust-gas analyses
Choice of prime mover
Heat-recovery systems
Summary

Chapters 12, 13, 15 and to some extent 16 deal with aspects of the simultaneous production of heat and electrical energy, from somewhat different standpoints. The intelligibility of each by itself therefore requires a degree of overlap, but this has been kept to a minimum except where it supplements the total information presented.

12.1 Internal combustion engines

Internal combustion engines can be grouped into five categories by the fuels used.

Petrol. Very high speed adapted motor vehicle engines, low output (up to 100 hp); used for special standby duty only and so not considered in the context of this book.

Distillate Fuel Oil. BS 2869 Class A_1 A_2 B_1 B_2, ASTM D 975-68 Nos. 1-D, 2-D and 4-D, or D 936-67 Nos.1 and 2. High speed, up to 2000 rpm. These are either adapted lorry/bus engines (up to 500 hp) or light industrial units (up to 4000 hp). Used for standby or the production of shaft power where space is at a premium. Also medium speed, 720–1000 rpm, for the same duties but where space is not so restricted.

Light Residual Fuel Oil. BS 2869 Class F, ASTM D 936-67 No.5 Heavy. Medium speed, 600–1000 rpm, for the continuous production of shaft power using an inexpensive fuel (up to 5000 hp).

Heavy Residual Fuel Oil. BS 2869 Class G or ASTM D 936-67 No.6. Slow speed, 400–600 rpm, for continuous duty on a cheap fuel (up to 30 000 hp).

Gas. These units can either be Spark-Ignition, i.e. a gas/air mixture ignited by a high-tension electric spark (up to 2000 hP), or Dual-Fuel, where the gas/air mixture is ignited by a pilot injection of diesel oil (up to 10 000 hp)

Note For the purpose of the power classes hp refers to brake horse power but has deliberately not been converted to kW (normally 1 hp = 0.746 kW).

Criteria for Selection of the Prime Mover

Basic Economic Criteria

Even when desirable from an engineering standpoint, on-site generation is only feasible when certain economic factors reach significant proportions (see Ch.13, 13.1).

Security. Danger to life; certain dangerous processes (e.g. exothermic); the economics of continuing a process where failure would cause a disproportionate loss in production (e.g. heat treatment); Trades Union requirements for security of work.

Partial Generation. If security is a necessity, by an incremental increase in capital expenditure, some or all of the installed capacity could become a revenue earner by operating in parallel with the local Electricity Supply Board. When costing such a scheme either the incremental capital is charged against generating cost or all capital is charged and an annual credit for security is given to generating costs.

Total Generation: When the cost of energy is a significant factor in the manufacturing cost of the end product. For example, if this is more than 10% then the accruing savings will have a high priority at Board level.

When there is a high utilization factor, so that the capital amortization is reasonable. This usually occurs beyond 40% plant load factor.

When a suitable economic fuel is available locally. When the added security of on-site generation is an additional factor to be considered.

Selection of Unit Types and Sizes

Standby. The criteria are:
Minimum capital.
Minimum space for unit and its fuel.
Minimum attention yet reliable.
Design to cater specially for the site requirements. The high-speed diesel engine on a premium fuel (35 seconds Redwood No.1 Class A) meets these requirements provided that the set is suitable for the duty required.

Ratings. BS 649 gives a sound basis for engine selection with BS 2613 for the alternator. The rating given is continuous with a 10% overload for 1 hour in any 12 without detriment to the intervals between overhauls. 'Standby' or 'Sprint' ratings should be avoided. A 1500 rpm 'standby' rated engine is liable to require a top overhaul after 750 hours operation. The same engine rated to BS 649 requirements would have a 2000/3000 hour period between overhauls.

Most engines now have bmep (brake maximum effective pressure) ratings of 1.4 MPa or more. This is quite satisfactory for its continuous rating but if the site duty requires acceptance of load from no load it must be remembered that such engines will only accept up to about 1 MPa bmep in one initial step. This ability to accept load has to be considered in conjunction with the connected-load characteristics, e.g. the starting characteristics of large motors, thyristor-controlled lights, etc., and the voltage and frequency parameters acceptable to other items of electrical plant.

If frequency and voltage control is of importance, the set builder must be asked to quote the data required in BS 649:1958, Amendment No.4 1970, Clause 5, Governing A1. Alternatively, the user must specify in the call for tenders the acceptable transient and permanent parameters of voltage and frequency, together with full details of load and the timings of their imposition.

Unit Sizes and Fuels. 1500 rpm units are available from 5 kW_e to 2000 kW_e using Class A fuel and generating approximately 15 kW h per UK gallon (3.3 kW h/l) of fuel oil. Heavy fuel oils and gas are not usually suitable for standby duty.

Peak Lopping: Partial Generation (see also Ch. 13.1). Private peak-lopping equipment operated by industry in collaboration with the local Electricity Authorities can significantly improve the system load factor to the benefit of all. Electricity tariffs should be such as to encourage this co-operative effort. If security requires the installation of generation plant, then it is always worthwhile investigating using a set or sets for peak lopping. This converts idle capital into a revenue earner.

Hours of Operation. In industry peak lopping of the maximum demand may be needed usually for 5 days a week, 8—10 hours a day, during the 4 critical winter months, i.e. for less than 1000 hours a year; but always during working hours when simple waste-heat recovery plant could usefully supply heat energy. In practice we find that such units operate between 500 and 2000 hours a year, so that even with 1500 rpm machines top overhauls are not more frequent than every second year.

Equipment. Peak lopping requires certain additional equipment, and this incremental cost should be charged against peak lopping, *not* the total capital employed, in order to obtain a true picture. Ultimately, for convenience, all the capital can be allocated to this duty, but the financial mechanics should be appreciated in the costing presentation by noting that standby security is then 'free'.

If parallel operation with the Electricity Supply Board is not catered for in the generating equipment purchased, peak lopping can then only be carried out by supplying isolated circuits independently. This is not very satisfactory, but in some recent instances this practice has had to be used where standby plant was installed and the

local Board had requested that the site load be reduced to say 60% of normal site requirement. If the standby plant had been designed to be capable of peak lopping the exercise would have been easy to carry out.

Basic Requirements. Investigations into peak lopping must be carried out in conjunction with the local electricity supply authorities. In addition to the essential requirements listed in Ch.13.1, the generating unit should conform to BS 649 (mechanically) and BS 2613 (electrically), the addition of a generating unit on the system must not raise the fault level of the system above the rupturing capacity of the local switchgear, and the generator should operate at a power factor not greater than that of the normal load of the site. The neutral earthing shall conform to the Electricity Supply Regulations.

Continuous Independent Operation. For the production of shaft power the high thermal efficiency/shaft power ratio of the reciprocating engine (*Table 1*) is the first choice for investigation. The analysis of a scheme is then concentrated on the availability of a cheap fuel, and its suitability to the size and speed of unit required.

Operating Costs

Fuel Costs

For the production of shaft power only, and using reciprocating engines, the cost of a viable fuel is between 1/10 and 1/6 the cost of purchased electricity expressed in heat units. This ratio has been established in the USA as 1/10; Holland 1/8; Belgium 1/7; UK 1/8½. The following comparison (*Table 2*) of gross cost of electricity with viable cost of a fuel for generation (no heat recovery) is given as an example (UK, 1975).

Table 3 compares the cost of different fuels. It should be remembered that the additional expenditure involved in heating and treating Class G fuel as compared with Class F is at least 1% of the fuel cost. This, allied to the use of slower speed engines and higher maintenance costs, means that unless Class G fuel is already in use on the site, it is usually not worth considering fuels outside Class F.

With heat recovery, higher fuel prices are viable.

For approximate analyses the following rates of consumption can be used:
(*a*)   600 and 750 rpm.   Class G and F fuels. 4.4 kW h/l
(*b*)   750 rpm. Class A/B fuels. 3.75 kW h/l
(*c*)   1500 rpm. Class A fuel. 3.7 kW h/l
(*d*)   750 rpm. Dual-fuel gas engines. 9.5 kW h/100 MJ
(*e*)   750 rpm. Spark-ignition gas engines. 8 kW h/100 MJ

When considering cheap residual fuels, the contained impurities must be analysed. The usual limiting specification (depending on the type of engine) is:

Table 1   Heat balance in reciprocating engines (full load)

Item	Diesel	Dual-fuel	Spark-ignition
To shaft power	41	39	29
To exhaust	32	34	41
To jacket water	14	13	21
To lubricating oil	4	4	5
To intercooler	6	7	—
To radiator	3	3	4
	100%	100%	100%
Bmep pressure (MPa)	1.4	1.2	0.7

Table 2

Cost of electricity: (p/kW h)	1	1½	2	2½	3
(p/100 MJ)	27.8	41.6	56.5	69.5	83.0
Viable cost of fuel @ between					
1/6 and 1/10	2.8	4.2	5.7	7.0	8.3
(p/100 MJ)	4.6	6.9	9.4	11.6	13.8

Table 3   Comparison of relative fuel costs in UK conditions, Jan.1975, and consumption per energy unit

Fuels Type	BS Class	Secs Redwood No.1	Relative cost/l	Gross CV (MJ/kg)	Gross CV (100 MJ/l)	Relative cost/100 MJ
Gas oil	A/B	35	1.00*	45.5	0.380	1.00†
Residual	F	1000 (245mm²/s)	0.77	43.0	0.410	0.72
	G	3500 (850mm²/s)	0.74	42.6	0.412	0.68

*   4.4 p/l, Jan. 1975
†   11.6 p/100 MJ, Jan. 1975

Table 4   Scheduled maintenance costs as a percentage of fuel and lubricant cost, and overhaul periods

Speed (rpm)	Fuel Class BS	Period to top overhaul (h)	Major period to overhaul (h)	Maintenance costs (% of fuel and lub. costs)
1500	A	3 000	10 000	4
1000	A	4 000	15 000	3½
750	A	10 000	30 000	2
	F	5 000	20 000	3½
	Gas	10 000+	30 000+	2½
600	A	10 000	30 000	2
	F	10 000	30 000	3½
	G	5 000	20 000	4
	Gas	10 000+	30 000+	2½

Sulphur (wt%) 3.5; vanadium metal (ppm w/w) 150; sodium (ppm w/w) 70; vanadium pentoxide (ppm w/w) 270.

The critical factor for exhaust-valve life is vanadium in the presence of sodium.

Lubricating Oil Consumption

The consumption of lubricants is approximately 1–1½% of the full load fuel oil consumption. To this should be added the sump capacity every 2000 hours of running.

Spares and Maintenance

As some guide to this item, the percentages given in *Table 4* (Jan. 1975, UK) allow an average for labour and materials for all scheduled maintenance including major overhauls.

Breakdown Insurance

It is possible to insure the cost of labour and materials for all breakdowns. which are defined as all stoppages requiring non-scheduled maintenance/repair work. Depending on conditions, a usual fee is 1% of the proportion of capital cost for engines required to meet the maximum demand and ½% of the portion of capital cost for standby units. These premiums cover for all claims up to half the value of the engine involved in any one incident, irrespective of the number of incidents.

Staffing

For standby and partial generation, particularly as present practice is for automatic operation of the equipment, existing site staff are utilized and so this cost item is minimal.

In a total generating installation operating 24 hours of the day throughout the year on a fully manned 3-shift basis, the cost would be of the order of 0.15 p/kW h sent out (salaries + 50% overheads 30% plant load factor). With full automation this figure can be reduced to 0.05 p/kW h for a capital cost of about £3½ per kW installed (UK, Jan. 1975).

References p 284

Table 5   Summary of operating costs as percentages of p/kW h (UK, Jan. 1975)

Engine speed rpm	750–1000	600–750	500–600	600–750
Fuel type	BS Class A/B	Class F	Class G	Dual-fuel
Fuel	91.5	82	77	89
Lub. oil	2.5	4	4	4
Spares and labour	2	3	3	2
Staffing	4.0	11.0	16	5
Total*	100	100	100	100

* UK Jan. 1975, 1.27, 0.98, 0.96, 0.96 p/kW h respectively

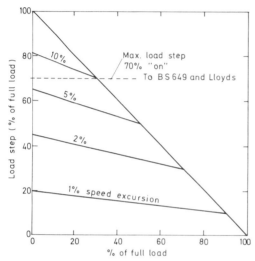

Figure 1   Diesel engines: percentage speed drop when accepting load

1. *Engine at normal operating temperature*
2. *Full load bmep 1.54 MPa max.*
3. *Governor type. R.E. 1100. Max. G.D.F. 0.5*
4. *Fuel-pump stop set at 110%*

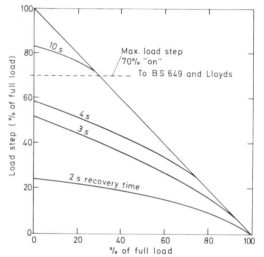

Figure 2   Diesel engines: recovery times when accepting load

1. *Engine at normal operating temperature*
2. *Full load bmep 1.54 MPa max.*
3. *Governor type R.E. 1100. Max. G.D.F. 0.5*
4. *Fuel-pump stop set at 110%*

The figures in *Table 5* summarize in general terms the costs of typical engines with different speeds/fuels. Automation is feasible with engines on Class A/B fuels and dual-fuel but not with residual fuels. It is also worth noting that (UK, Jan. 1975) waste-heat recovery from the exhaust system was valued at 0.20 p/kW h while total heat recovery was worth 0.45 p/kW h.

The Diesel Engine

In order to improve the volumetric efficiency and to increase the bmep, modern compression-ignition engines are supercharged and charge-cooled with compression ratios of between 12:1 and 13:1. Bmeps are now generally over 1.4 MPa. These bmeps are satisfactory in the context of continuous ratings but do affect the unit's capability of accepting load swings. This fact is accepted in the latest Amendments to BS 649 and by Lloyds.

Load Acceptance

Figures 1 and *2* show typical percentage speed drops and recovery times respectively for a 750 rpm diesel engine with various load steps. When the driven unit is an alternator, the speed-dip effect has to be added to the alternator voltage dip associated with the automatic voltage regulator. This aspect is most important where computers or tape-controlled machine tools etc. are on the electrical supply system.

Fuels

The fuel must be heated until the oil has a viscosity of between 60 and 70 Redwood No.1 seconds (13–16 mm²/s) at the injectors. Examples of various fuel temperatures are given in *Table 6*. The limiting factor in the use of a fuel is the contained impurities (see earlier, and Ch.20.1).

Table 6   Residual-fuel temperatures

Viscosity of fuel (Redwood No.1 seconds (or approx. mm²/s)) at 38.5°C	200 (49)	450 (110)	1000 (245)	1500 (370)	3500 (600)
Temperature at fuel rail (°C)	73.5	90	107	113	129
Heater inlet and outlet temperatures (°C)					
Storage tank outflow heater	7–24	11–28	20–36.5	25.5–42.5	32–49
Settling tank outflow heater	24–45	27.5–55	36.5–60	42.5–66.5	49–72.5
Purifier line heater	45–65.5	56–82	62–87.5	66.5–90	72.5–95
Service tank outflow heater	37.5–51.5	37.5–68	49–76.5	49–82	49–87.5
Final line heater	51.5–73.5	73.5–90	76.5–107	82–113	87.5–129
Tank temperatures to be maintained (°C)					
Settling-tank immersion heaters	24	27.5	36.5	42.5	49
Service-tank immersion heaters	37.5	37.5	49	49	49

Spark-ignition Gas Engines

Although efficiency increases with compression ratio, the ultimate figure is 11.7:1 on a methane gas so as to limit the risk of detonation or 'knock'. The knock characteristic varies with the gas used, and so the unit ratings are usually of the order of 700 kPa to 1 MPa bmep.

Detonation

Sir Harry Ricardo observed: 'It is detonation, and detonation alone, that sets a limit, and a relatively early limit, to the power output and economy of a petrol engine; the incidence of detonation, in fact, determines both the weight of air which can be consumed, and the efficiency with which the heat thus liberated can be converted into power'.

This statement is equally true of gas engines, and the following explanation by Ricardo of the phenomenon of detonation is generally accepted: 'Where a combustible mixture of fuel and air is ignited by the passage of the spark, there builds up, slowly at first but with a rapid acceleration, a small nucleus of flame, at first somewhat in the form of a soap bubble; this spreads outwards with ever-increasing rapidity.   As the flame front advances it compresses ahead of it the remaining unburnt mixture, whose temperature is raised by both compression and radiation, until a point is reached when the remaining unburnt charge will

ignite spontaneously, and almost instantaneously, thus setting up a detonation wave which will pass through the burning mixture at an enormously high velocity, such that its impact against the cylinder wall will give rise to a ringing knock, as though it had been struck with a light hammer.'

Detonation causes rough running and lack of power and, if allowed to continue, may damage the engine.

Detonation or knock is dependent on:
Fuel characteristics.
Fuel/air mixture ratio.
Inlet air temperature and pressure.
Compression ratio.
Injection timing.

Fuels

For various hydrocarbon gases it is possible to define the critical compression ratio (CCR) at which detonation starts to occur. This and the net (lower) calorific value are shown in *Table 7*.

The usually available gases are producer gas (net CV 4.5 MJ/m³), town gas (net CV 15), sewage-sludge gas (net CV 18.5–22.5 with 55% methane), natural gas (net CV 35 with 92% methane as from the North Sea). It should be noted that methane is the main constituent of sewage-sludge and natural gases and is the least susceptible to knock. Propane, butane and pentane are minor constituents of

References p 284

natural gases and will increase the tendency to knock; careful on-site testing is required if these heavier hydrocarbons constitute 10% or more of the gas.

Dual-fuel Engines

This type of unit is a compromise between the diesel and the gas engine. It has to be able to start and run as a diesel engine and so have a compression ratio of more than 11.4:1, but less than 11.7:1 so as to limit the risk of detonation when used on a gas/air mixture ignited by a pilot injection of fuel oil. The bmep is usually between 1 MPa and 1.2 MPa.

Load Acceptance

The load-acceptance characteristics of a dual-fuel unit are better than those of a spark-ignition engine but not so good as those of a straight diesel engine. For example, a 1% frequency excursion on load change is usually acceptable. On a diesel engine this limits the load applied/rejected to a step of 35% of the full-load diesel rating (1.4 MPa). With a dual-fuel engine the step is limited to 20% of the full-load dual-fuel rating (1.2 MPa). The recovery times are about the same, approximately 1½ seconds. Associated with the 1% frequency excursions, the load changes will cause a voltage excursion of about 6% which is well within the usually specified +10% to −8% called for in computer (etc.) installations.

Ratings on Different Fuels

As discussed, a dual-fuel engine is usually a modified diesel engine. On a methane gas a rating of 1.3 MPa bmep can be achieved without risk of detonation. If the engine is to conform to BS 649 ratings the unit as a 'dual-fuel engine' will therefore have a 'continuous' rating of 1.2 MPa to allow for the 10% overload on dual-fuel for 1 hour. This does not allow full use to be made of the engine structure. It is preferable to quote a 'continuous gas rating' (in this instance 1.3 MPa bmep). An 'overload' can be taken equivalent to the continuous diesel rating if the electrical equipment is sized correctly. This takes advantage of the capabilities of the engine structure. But it is unusual for a dual-fuel engine to be able to take advantage of the full 1 hour 10% overload of the continuous diesel rating owing to limitations imposed on the injector pump required for dual-fuel operation.

As mentioned briefly in an earlier sub-section, rating is limited by the pre-ignition or knock characteristics of the gas.

Considering the modern charge-cooled supercharged dual-fuel engine, pre-ignition starts at about the following ratings for pure combustible gases (*Table 8*). The rating which can be obtained on a gas mixture is inversely proportional to the weighted sum of the thermal energy of the combustible gases contained in the mixture. Each gas thermal energy can be stated as the percentage volume of that gas multiplied by its lower (net) calorific value, i.e.

$$\frac{100 \times LCV}{\text{Rating on mixture}} = \frac{LCV \times \% \text{ (methane)}}{1.4}$$

$$+ \frac{LCV \times \% \text{ (ethane)}}{1.25}, \text{ etc.}$$

A safety factor, of say 5%, in the rating should be allowed. However, this can be modified if nitrogen or carbon dioxide is present, as these act as 'knock inhibitors' by reducing the velocity of flame propagation.

Noise and Vibration (See Ch. 19.7)

With a standard exhaust silencer on a 750/900 rpm engine the peak is in the 30—32 Hz octave band and is about 110 dbA at 2 m from the end of the pipe.

With a 'residential' type silencer the peak is shifted to 125 Hz and 36 dbA. To obtain the best results it is essential for the inner tubes of the silencer to be designed to cater for the particular firing frequency of the engine so as not to create a resonant frequency in a particular octave band.

If a waste-heat boiler is used the attenuation is greatly improved owing to the mass within the boiler.

On a turbo-charged engine the noise from the air intake louvres is concentrated in the 4000 Hz octave band. If there is residential property within about 400 m of the engine then the air must be

Table 7

Gas	CCR	Net CV (MJ/normal m^3)
Methane	12.6	34
Ethane	12.4	61
Propane	12.2	89
Isobutane	8.0	
Butane (straight chain)	5.5	

Table 8

Gas	Methane	Ethane	Propane	Butane	Hydrogen
Rating (MPa)	1.4	1.25	1.05	0.7	0.5

drawn from within the station, if solidly built, or the air intakes must be ducted and silenced. Passage in air has greater attenuation effect on high frequencies than on low.

The International Standards Organisation recommends a noise rating of 31 NR as unlikely to produce complaints from residential areas. The I.S.O. rating requires that the pressure levels must be determined in the octave bands whose lower frequencies are 32, 63, 125, 250, 500, 1000, 2000, 4000 and 8000 Hz. These sound pressures can then be converted to NR values from Table 1.3 of the I.S.O. document. The significance of the NR value can then be modified depending on circumstances, e.g. duration of noise, existing noise in the area, etc. (See ISO/TC 43 (Secretariat-194) 314 E).

Airborne noise within an average station is between 95 and 105 dbA depending on engine size, speed and weight. A noise-absorbent inner surface improves conditions within the station, while an 11 in (280 mm) cavity wall gives an attenuation of approximately 50 dbA between 100 and 3000 Hz. Above and below this it averages 45 dbA.

H.M.S.O. (UK) have issued a 'Code of practice for reducing the exposure of employed persons to noise' (SBN 11.360887X) 1972. This recommends that for 8 hours continuous exposure of unprotected ears the noise level should not exceed 90 dbA. A useful nomogram gives the maximum recommended time for higher and fluctuating noise levels; for greater or longer exposures ear protectors are recommended.

Transmitted vibration in some installations can be a problem for surrounding buildings. Engines should be isolated in some form or other.

1. For large low-speed units, spring isolators can be used between concrete bases and the engine foundation block. This will absorb 90% of the dynamic forces. Viscose dampers can also be fitted to limit the spring movement. For sites where the soil conditions are less likely to transmit vibration, the building and the foundation block can be isolated one from the other by the use of a resilient mat such as 'Coresil'.

2. For medium- and high-speed engines the engine and alternator unit can be mounted on an underframe which maintains alignment and itself is isolated from the foundation block by anti-vibration pads. This also allows the foundation block to be reduced in size, often giving considerable savings in the cost of the building.

Fire Precautions

In all new installations the head fire officer should be consulted about any particular requirements. The main fire dangers are from fuel oil spilling on to hot exhaust pipes and manifolds, gas leaks igniting, and crank-case explosions.

For fuel oil, bulk storage tanks have to be surrounded by a 'band' capable of retaining the oil from the tank should the tank burst. Pipe routes should not pass over any exhaust manifolds or pipes, and high-pressure injector pipes should be shielded so that fuel spray from a breakage cannot reach any hot surface. On a fire alarm, fuel transfer-pumps should be automatically stopped.

These areas should be protected by automatic sprinkler systems. Ideally daily service tanks should be sited in a separate '4 hour' area and the pipe from the daily service tank to the engine should have a fusible valve that closes on the indication of a fire in an engine.

Gas detectors should be positioned in dual-fuel and spark-ignition stations to give warning of gas leaks. The effect of a severe gas leak into a station, where the aspirated air for the engine is taken from within the station, is to cause the engines to become ungovernable owing to gas being drawn into the engine with the air. This gas/air mixture can explode in the air manifold and so ignite back into the station. This can be prevented by ducting the aspirated air from outside the station. If escaping gas is detected, dual-fuel and gas engines should automatically be changed to straight diesel operation and the main gas valve closed. With long exhaust pipes on dual-fuel engines it is advisable to have gas detectors in the exhaust pipe. Unburnt gas can be carried into the exhaust gas either through the gas valve sticking partially open during the scavenge period or through the misfiring of a cylinder. Gas detectors should also be fitted in the station ventilating system.

Crankcase explosions are very rare and the effects can be minimized by the use of explosion vents and by taking the crankcase breather outside the station.

The usual fire equipment used in stations is a combination of automatic mulsepyre and non-automatic foam or carbon dioxide.

12.2 The industrial gas turbine

Derived from industrial turbine technology which has produced the heavy-duty, long-life gas turbine, and aircraft technology from which the compact, light-weight industrial aero-derivative gas turbine has been developed, this form of smooth rotary power, based on a constant-pressure, continuous-combustion thermodynamic cycle, is today serving the varied power requirements of all sectors of industry. This universal acceptance of the gas turbine for non-aero duties is clearly demonstrated by world-wide sales over the past decade (1964 to 1974), which have amounted to some 50 000 MW_e of new plant.

The industrial gas turbine is, in essence, a larger, heavier unit with conservative temperatures, stresses and speeds, simplified construction, and a wider acceptance of the heavier grades of residual oil fuels. By comparison, the major design criteria of the aero-derivative turbine are maximum power/weight ratio, high thermal efficiency and low installed volume — prime requirements for the original aviation specification. Each type has achieved proven reliability and life in industrial service and, with continual improvements to meet common problems, modern designs are benefitting from the technology each offers. While the industrial machine is available in single, twin and multi-shaft types, the aero-derivative turbine uses either a turbo-prop engine, incorporating its own output drive shaft, or a turbojet (or modified turbofan) turbine which forms the gas generator (engine) section discharging gas into a separate industrial-type free-power turbine, mechanically linked to the driven unit.

Mechanism

Basically this is very simple. A 'clean' fuel, gas or a variety of distillate oils, is burnt with compressed air. Excess compressed air bypasses the flame in a cooling jacket, and is then used to dilute the combustion products to reduce the temperature to 700—950°C, when they are used directly to turn the turbine blades, while expanding. The turbine exhaust gases in a single stage, at about 500°C, still contain much recoverable heat; i.e. the temperature drop in the simplest cycle is thermodynamically inadequate.

Advantages and Disadvantages

For the discerning energy user, the gas turbine provides a vibration-free power unit with higher specific output, lower weight and reduced dimensions than a reciprocating engine of similar power. This installational advantage where space is at a premium or where mobility is required is coupled with rapid starting ability, high reliability and availability, multi-fuel capability, low lubricating-oil consumption, short construction and installation periods, and competitive costs — both in initial equipment outlay and subsequent operation.

Despite its multi-fuel capability, burning gaseous and liquid fuels, the simple-cycle gas turbine has a major disadvantage in its relatively high specific fuel consumption, resulting in thermal efficiencies of only 10 to 27% depending on size and type of plant; this can be increased to 35% if a heat exchanger is added to the cycle. While modern designs are improving these figures, particularly highly developed 'second generation' aero-derivative engines which have thermal efficiencies of around 35% so avoiding the use of costly heat exchangers, this situation can be further offset by utilizing the constant flow of relatively high-temperature, oxygen-rich (around 18% by volume) turbine exhaust for combined-cycle systems, power-recovery cycles, district-heating schemes and numerous total-energy systems (Ch.15.1). In this manner, total plant efficiencies of 80% or even higher can be achieved.

Where a steam turbine is integrated with the gas turbine, exhaust from the latter is fed to a boiler converting it directly, or with supplementary firing, to steam for the steam cycle. For a typical total energy scheme serving a building complex, exhaust heat is recovered for space heating in winter, for absorption type air conditioning equipment in summer, and for hot water supply throughout the year. In a factory complex, the heat may serve numerous factory processes.

Often however the characteristics of the gas turbine in meeting a specific requirement, as for example rapid start ($\approx$30 s) and load-acceptance for short-duration standby, emergency or peak-load operation, outweigh the higher fuel consumption of the simple-cycle machine. Further, when one takes into account the current world situation regarding fuel availability, increasing costs and varying grades, the gas turbines multi-fuel capability to burn, within reason, whatever fuel is available with running changeover and without frequency fluctuation, is commanding additional user assessment in comparison with other forms of power supply. In order to exploit fully the multi-fuel characteristic and achieve lower fuel bills, fuels of lower calorific value derived from gasification or waste-product utilization are proving acceptable to certain types of gas turbine.

Applications (*Figures 3—5*)

In utilizing these distinctive features, electrical power generation accounts for the largest proportion of industrial-gas-turbine sales, as also those of internal combustion engines (Ch.12.1). Duties include standby protection against main supply failure to safeguard the continuous operation of industrial processes, specialized equipment (such as computers), essential public services, and the vital auxiliaries necessary for main-power-station operation. Peak-load applications, often combined with standby duty so giving maximum plant utilization against capital invested, permit industry to avoid purchasing power at high network tariff rates, or combine with utility base-load plant — nuclear or fossil-fired — to give flexibility of supply at high consumer-demand periods.

For base-load applications, maximum fuel economy is achieved when burning low-cost gas or cheaper grades of residual fuel oil.

Mechanical-drive gas turbines account for an equally diverse range of duties. One of the major uses is for the oil and gas industries, where the turbine, driving pump or compressor, is called upon to cope with particularly arduous conditions, ranging from unattended, continuous operation in dusty, high-temperature zones or under arctic conditions on overland pipelines, to the many offshore operations which are growing as the search for new energy sources extends into hostile areas such as the North Sea.

Often, the discharge of high-temperature, non-toxic exhaust gases by the gas turbine is sufficient to justify its application for process use.

Sizes and Types Available

To serve these diverse applications, an extensive variety of gas-turbine power is available supplied either as individual items or as part of a turnkey installation. Sizes range from small, self-contained, manually handled sets of 30 kW, through skid-mounted, package, or mobile equipment in the mid-power range, up to larger sizes of 70 MW, built in packaged form or installed in conventional power stations for all types of intermittent or continuous service for the public utility supply companies. At the upper end of the power scale, sets with individual outputs of 100 MW are now available (*Figure 3*). Principal manufacturers throughout the world, and turbines in production or under development, are listed in the current edition of *Sawyer's Gas Turbine Catalog*[1].

Depending on type of duty, designs are usually available in single, twin or multi-shaft arrangements. Where accurate, close-governing characteristics for close-frequency power generation have to be met, as required by computer services, the single-shaft gas turbine, operating at constant speed and constant load, is preferred. Throughout the range of electrical demands, both the single-shaft and multi-shaft turbine systems, the latter utilizing aero-derivative turbines, are in use.

For mechanical drive or other duties that require rapid turbine response to meet fluctuations of load or rotational speeds, the flexibility of twin or multi-shaft arrangements is favoured.

Special Applications due to Flexibility and Mobility

In addition to conventional installations where the gas turbine and its associated equipment are located internally in orthodox buildings, a high

Figure 3 Artist's impression of a joint American—Swedish design due to enter service during 1975, combining aero technology with industrial technology: 100 MW generating plant, available in 50 or 60 Hz, to meet the demands of electric utilities for larger increments of peak and base-load power. Aero technology and industrial technology from Stal-Laval Turbin AB, a member of the ASEA Group. Photograph by Gas Turbine News

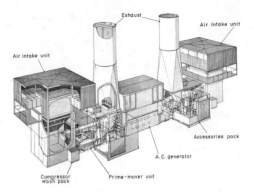

Figure 4 Self-contained packaged generating plant; two aero-derivative gas turbines drive into a central generator developing 40—50 MW. By using one gas turbine only, with the other de-clutched, power can be halved. (Courtesy Rolls-Royce (1971) Ltd)

demand exists for fully packaged, weather-proofed and sound-attenuated plant (*Figure 4*). This method of construction allows a complete set to be built and tested in minimal time under controlled factory conditions before shipment (or airfreighting), either in a single module or multiple modules depending on plant size and output. On arrival at site, packages can be installed rapidly (within a week or so) and with minimum cost, on simple concrete raft foundations. As power demands increase, additional packages may be added, linked to a single control, so enabling growth on the supply network to be met in incremental steps without high advance capital expenditure on larger plant.

The lightness of gas-turbine packaged equipment compared with that of reciprocating-engine-driven plant permits its use in mobile applications either by road, rail or on a barge. This added flexibility is particularly important in meeting intermittent demands for power at different sectors of the distribution network or beyond the reach of transmission lines; coping with natural disasters; or providing blocks of barge-mounted power in city areas where ground space may be restricted.

The gas turbine is ideally suited to applications where space is at a premium; for example on offshore drilling and production platforms serving the oil and gas industries, or where additional or emergency standby power is added to an existing building complex. The ability to fit a gas turbine into 'shoe box' dimensions or restricted spaces has been demonstrated by installations in office areas. In such installations, the weight advantage permits location of the equipment at a high level or on a building rooftop (*Figure 5*).

Noise and Pollution

When operating in a residential area, a further advantage of the gas turbine in conforming to modern living standards is its environmental acceptability in terms of low emission levels (Ch.12.3) —

owing to the high oxygen content in the exhaust —
and acceptable noise levels, attenuated by well
proven intake and exhaust silencing, and by
acoustic enclosures: to 35 NR at 100 m (Ch.19.7)
for stringent residential requirements.   Already
these environmental features, particularly the
exhaust aspect, have had an important bearing on
the selection of combined-cycle plant for a number
of major power stations, principally in the United
States and other densely populated areas governed
by environmental regulations.

Availability and Maintenance

Throughout the world, gas turbines in continuous
operation are regularly achieving reliabilities and
availabilities of 97% or more. Normal inspections,
often made at one year frequency depending on
type of service, can usually be handled with training by site personnel.   Major inspections are
required between 20 000 and 30 000 hours.
Although industrial turbines are designed for
major overhaul on site, aero-derivative units are
usually subjected to gas-generator-section removal
and replacement as a complete assembly. On later
designs, the increasing use of modular construction
allows replacement of individual sections of the
turbine without disturbing the complete engine.
For this purpose, turbines are fitted with boro-
scope inspection devices and other condition-
monitoring devices which give advance warning
of component deterioration. The gradual adoption
of this form of maintenance procedure will in
future eliminate fixed engine-overhaul periods,
thus significantly reducing running costs and
assuring maximum availability.

*Figure 5   Roof-top installation in London with
two 500 kW gas-turbine generating sets supplying
additional power for a major clearing bank: illustrates the installational flexibility. They provide
quiet and vibration-free power without disturbance
to a hospital located next door. (Courtesy Cen-
trax Ltd)*

Future Developments

To consolidate the markets already won for the
gas turbine, the manufacturing industry is pursuing
a continuous development programme to improve
performance, life, and efficiency. In addition to
fuel-system, component, blade-cooling and other
general studies, the possibility of using ceramics
for high-temperature components, so permitting
an increase of cycle temperatures and hence
efficiency with reduction in specific fuel consumption, is likely to reach a productive state by
the early 1980s.

In terms of future requirements, the trend
towards larger, long-life plant of 100 MW size, able
to operate as individual units or in combination
with a steam cycle, so providing higher utilization
efficiency for the utility supply industries, has
already been mentioned. Equally, combined-cycle
plant that gives operational flexibility and higher
fuel utilization is in much demand.

Repowering of existing steam stations, so
extending their useful life span and deferring high
expenditure on new plant, is a further role for the
versatile gas turbine. By adding gas-turbine power
and a heat-recovery steam generator to an existing
steam installation, it is possible to increase the
power output and thermal efficiency of a station,
and to replace existing steam boilers that have
become uneconomic.

Another demonstration of the flexibility
of the gas-turbine system is its recent integration in plant capable of supplying low-cost
peak-load power by the use of compressed air,
stored underground during off-peak periods.
For this duty, the compressor operates separately
from the combustion system and turbine sections
of the gas turbine. During the air-storage phase,
the electrical generator acts as a motor to drive the
compressor which replenishes the storage facility.
At times of peak demand on the grid, pressurized
storage air is fed via the combustion system to the
turbine driving the electrical generator. By operating the gas-turbine components separately, the
normal power of the turbine is increased more than
three-fold.

With the development of the high-temperature
gas-cooled nuclear reactor, the opportunity of
integrating a gas turbine, based on a closed-cycle
system using helium as the circulating medium
taking heat from the nuclear reactor, to produce
electrical power without the intervention of
steam boilers, appears a distinct possibility. A test
plant producing heat and electricity to assess this
system, but with coke-oven gas providing the heat
source, was placed in service in West Germany this
year (1975).

In a further attempt to conserve the earth's
dwindling energy resources, the likelihood of
utilizing solar energy directly, again with a helium
closed-cycle gas turbine, is also receiving
consideration.

12.3 Heat recovery in reciprocating engines and gas turbines

If shaft power is the primary requirement then the reciprocating engine with its high efficiency is the obvious prime mover to use. Waste-heat recovery being of secondary importance, the system should be kept as simple and as inexpensive as possible. If process heat is as significant as shaft power then the ratio of these two factors will indicate the most suitable prime mover to be used. The simple engineering choice will be modified by the availability and cost of fuels.

Exhaust Conditions

Table 9 compares the mass flows per kW and temperatures with load for the three main prime movers used to produce shaft power and recovered heat. This shows that the most stable is the dual-fuel unit; mass flow per kW and temperature are virtually constant because of the necessary throttling in the air manifold. The diesel engine and, to a greater extent, the gas turbine, have exhaust conditions which deteriorate rapidly (from a heat-recovery point of view) with the decrease in load. It should be remembered that the final gas temperature after any heat-recovery equipment should not be allowed to go below the acid dewpoint due to oxides of sulphur, i.e. say 200°C, at the lowest load where operation could continue for any appreciable length of time (see Ch.20.2).

Exhaust-gas Analyses (*Table 10*)

There is sufficient oxygen in the exhaust gases to sustain combustion, using an 'after-burner' in the exhaust stream before the waste-heat boiler. However, this is not acceptable practice with reciprocating engines, owing to burner pulsations that affect the turbo-charger and the overall back pressure of the system. All the heat cannot be extracted from the exhaust gases because of the danger of overcooling the exhaust on low loads. If the dewpoint is reached consistently excessive corrosion will take place of the exhaust stack and even of the heat-recovery boiler. Usually calculations are based on a final gas temperature of 200°C at half

load. This enables about 60% of the available heat to be recovered. This figure can be improved if: (*a*) normal operation for long periods of time can be assured at a higher load; and (*b*) sulphur-contaminated fuels are not used, when a lower final stack temperature can be accepted.

Choice of Prime Mover

In a heat-recovery project the choice of prime mover is often governed by the ratio of steam (or heat) output to shaft power. *Figure 6* gives an example showing the ratio of weight of steam generated to shaft power (kW h). It is essential to analyse the heat and power profiles plotted against time and only to use the in-phase values. Out-of-phase values either require dumping or separate boiler facilities. Ideally, the recovered heat should be slightly less than the demand. Supplementary boilers, or after-burners in gas-turbine installations, supply the balance. In this way maximum thermal efficiency is achieved and the variable boiler pressure allows the heat and power demands to be out-of-phase.

The theoretical ideal choice of prime mover sometimes has to be modified by the availability of a cheap fuel. After-burners and supplementary boilers need not necessarily use the same fuel as the prime mover.

Heat-recovery Systems

Figure 7 shows a simple heat-recovery system on a reciprocating engine. Jacket-water heat is recovered in the form of hot water at $70-85^\circ$C and used direct or via a heat sink, which, if saturated, allows the water to be bypassed to a conventional radiator

Table 10   Exhaust-gas analyses

Exhaust gas (% v/v)	Diesel	Dual-fuel	Gas turbines
N_2	75	84	78
O_2	14	12½	16½
CO_2 and SO_2	8½	3½	3
H_2O	3	3	2½
CH_4	—	½	—

Table 9   Variation of exhaust conditions with load

	Diesel		Dual-fuel		Gas turbines	
Load	Mass flow (kg/kW h)	Temp. ($^\circ$C)	Mass flow (kg/kW h)	Temp. ($^\circ$C)	Mass flow (kg/kW h)	Temp. ($^\circ$C)
Full	8.9	325	7.9	400	31.2	525
¾	9.4	305	8.2	415	39.6	465
½	10.1	280	8.2	405	51.9	420

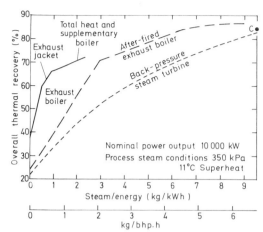

Figure 6   Overall thermal efficiency and ratios of heat to power reciprocating engines, gas and back-pressure steam turbines, with heat recovery
———— *Diesel plant 2.8 MPa, 400°C*
--------- *Steam turbine 2.8 MPa, 405°C*
— — — *Jet gas turbine*

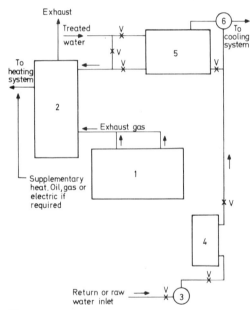

Figure 7   Diesel engine: heat-recovery system
1. *Diesel engine*
2. *Exhaust gas calorifier silencer*
3. *Raw water pump*
4. *Jacket water cooler*
5. *Heat exchanger*
6. *Thermostatic valve*

or cooling tower. Higher temperatures (and pressures) can be used on diesel engines but not on dual-fuel engines owing to their tendency to pre-ignition.

From the heat sink, water is pumped to the exhaust-gas boiler to produce pressurized hot water or low-pressure steam for whatever heat demand is required.

To raise the quality of the heat recoverable from the jacket water, it is possible to pressurize the water jacket and so increase the temperature of the water from the engine to about 110°C, permitting vapour-phase cooling to be applied.

It is essential to conform to the manufacturers' recommendations as the ability to use a vapour-phase or ebullient cooling system depends on the designed thermal rating of the engine. It is not advisable to use this type of heat-recovery system on a dual-fuel engine as the system raises the temperature of the unit, thus increasing the tendency to pre-ignition or 'knock'.

The simplest system in conjunction with boilers is for the exhaust-gas boiler to operate on a closed circuit heating the heat sink, to which the boiler condensate returns. The jacket water heats the make-up water to this heat sink and supplies domestic requirements and low-temperature space heating.

Gas turbine heat-recovery systems, as they can incorporate after-burning so countering the fall in temperature with decreasing load, are very simple and do not need supplementary boiler systems.

Gas turbines are usually fired with a clean fuel (natural gas or gas oil) and so the exhaust can be used for direct drying or some such process.

Summary   *(Table 11)*

Reference

1   *Sawyer's Gas Turbine Catalog.* Gas Turbine Publications Inc., Stamford, Conn., USA (Annual)

General Reading

(A)   Pounder, C. C. (Ed.), *Diesel Engines: Principles and Practice*, George Newnes Ltd, London, revised 1962.
       Diesel Engine Users' Association Papers, 1973 onwards
(B)   Shearer, A. B. 'Selection and Use of Standby and Peak Lopping Generating Equipment', *Diesel Engine Users' Association Paper 362*, June 1974
(C)   Shearer, A. B. 'The Economics of Diesel Engines for Standby and Energy Generation', Institution of Engineers, Ireland, Nov. 1974
(D)   Simmons, R. C. *Gas Turbine Manual*, Temple Press — Hamlyn Group, Feltham, Middx, 1968
(E)   Cohen, H., Rogers, G. F. C. and Saravanamuttoo, H. I. A. *Gas Turbine Theory*, Longmans, London, 1972

Table 11 Comparison of efficiencies of prime movers, recoverable heat and primary fuels that can be used

	Diesel	Dual-fuel with gas	Spark-ignition (gas)	Industrial gas turbines	Industrial condensing steam turbines
Percentage heat to: Shaft power	41	39	29	>20	25
Exhaust	32	34	40	<70	15
	←————60% of these recoverable (high-grade heat)————→				Not recoverable
Water	14	13	21	Nil	43
	←——————All recoverable (low-grade heat)——————→				Partially recoverable
Miscellaneous radiation, etc.	13	14	9	10	17
Sizes available (kW)	Up to 20 000	500 to 5000	20 to 1500	500 to 100 000	250 to 60 000
Fuels usable	Distillate and residual oils	Natural and sewage gases plus distillate oil	Natural, propane, butane and sewage gases	Natural, propane, butane and sewage gases. Distillate oil	Any combustible fuel in boilers
Heat/power ratio (kg steam/kW h)	2.15:1	2.15:1	2.15:1	8.7:1	—
Best overall efficiency	74%	72%	74%	85%	Up to 45%

Notes
1. Process heat is assumed to be steam at 350 kPa/11°C superheat
2. Industrial back-pressure steam turbines with a heat/power ratio of 3.3:1 to 6.7:1 give an overall efficiency up to 75% (see *Figure 6*)
3. Combined-cycle gas turbines (gas turbine with heat recovery to a steam turbine and condenser) give a shaft efficiency of up to 40% but little recoverable heat from the condenser
4. Major CEGB steam stations (up to 500 MW) have a shaft efficiency of 27—30% with recoverable heat from the condensers but at some detriment to shaft efficiency
5. Large gas turbines (100 MW) using premium fuels have a thermal efficiency of 27—30%

Chapter 13

Production of electrical power

13.1 Background principles

13.2 Plant suitable for industry

Generators
Prime movers
Parallel, independent or sectioned operation
Implications of standby requirements
Use of standby sets for peak-lopping or base-load operation
Reliability
Summary

13.3 Magnetohydrodynamic and electrogas-dynamic conversion

Magnetohydrodynamic (MHD) conversion
Electrogasdynamic (EGD) conversion

13.4 Electrochemical energy conversion

Electrochemical cells
Basic principles
Primary cells
Secondary cells
Fuel cells
Metal—air cells

13.5 Thermionic and thermoelectric energy conversion

Thermionic generation
Thermoelectric conversion

The public Electricity Generating Boards in the UK fulfil a statutory function to provide electricity at the lowest possible cost to the consumer. A proportion of the fuel they burn is of very low grade, which could not be effectively used by industry with existing plant. On the other hand, the method of generation is based on the condensing steam-turbine cycle, and the overall power station thermal efficiency is limited, largely by the thermodynamics of the heat—mechanical energy conversion (Ch.10.7), to around 35% and scope for improving this is marginal. Taking account of transmission losses between the power station and consumer, on average, the power delivered to a factory represents approximately 26% of the heat content of the primary fuel burned.

The situation in the UK has developed differently from that in some other countries; because of the plant—population distribution and statutory limitations it is unlikely to change appreciably in the near future. The installation of combined heat-and-power stations (Ch.15.2) and large hydro-electric and tidal schemes to conserve primary energy resources would clearly improve efficiency (at a large capital cost) to a major extent; however the lead time associated with such projects must limit significant change over the next two or three decades. But in some other countries combined heat-and-power systems have existed on a significant scale for decades, leading to efficiencies based on primary energy up to 85%.

From the industrialist's point of view, a supply of electricity is indispensable. In the UK electrical energy represents over half industry's total energy consumption neglecting specialized use of coal in coke ovens. It is the most expensive form of energy used, being at least 2½ to 4 times (as purchased) the cost of the other important forms of energy, coal, oil and natural gas; but this is offset in many applications by a high utilization efficiency, instant control, time and space saving in process work. Minor disruptions in the supply have however demonstrated to the industrialist that a disproportionate cost in lost production can result.

Because of the disadvantages of scale, industry in general cannot compete economically with the public supply in the straight production of electricity. But if a particular industry has both a power and heat requirement and these are in reasonable balance, an economic case can often be made for the combined generation of power and heat at an overall thermal efficiency of between 60 and 80%. In addition to the financial advantage to the industry concerned, this can make much better use of a nation's primary energy resources. This type of system has been traditional in activities such as the paper and chemicals industries and can be applied much more widely. A case for private power generation also exists where substantial quantities of waste heat are available, e.g. either low-grade vapours from evaporators or high-grade waste heat from furnaces.

In recent years there has been a trend in industry to install standby generating plant to safeguard essential production in the event of failure of the public supply. The economics of this are essentially an insurance calculation, but the return on capital expended can sometimes be improved by running the generating plant to provide a proportion of the works power requirements or to limit expensive peak-demand costs on the purchased power supply.

There is no general case for industry to generate its own electricity requirements and each case must be carefully evaluated on its own merits. Where the conditions are favourable, the economic advantage to industry and to the nation can be substantial. In the UK, financially viable schemes for private generation may apply for about a quarter of the electricity at present purchased by industry, potentially saving the nation approximately 10 Mt coal-equivalent p.a. in primary energy usage. Scale is not the sole criterion; economic cases have been established in special circumstances for generating as little as 50—100 kW.

Apart from flow through water turbines (considered for large-scale generation in the neighbourhood of large waterfalls or estuaries and in very small isolated units), and small windmills, oil or gas engines for isolated installations, power is generated mainly via steam turbines (Ch.10.7), gas turbines (Ch.12.2) and reciprocating engines (Ch. 12.1). The basic source of energy for these may be oil, coal, gas, peat, waste fuels such as wood waste or refuse, and increasingly nuclear reactors (Ch.14).

Public supplies are now frequently distributed by a national or international grid, which is fed from both nuclear reactors providing the base load, and conventional fuels via large watertube boilers (Ch.10.3). Steam turbines (Ch.10.7) are the main prime movers, but peak loads may be more flexibly handled by gas turbines or diesel engines. Fluctuations are a major problem; fault detection etc. and load balance call for constant vigilance.

The reason for industrial generation has already been discussed. The source of energy is often integrated with the total energy consumption of the plant served, or may be waste-heat-boiler steam. Distribution, and load balance, can be a problem on a large site.

Further comment on public utility stations must be very restricted; as mentioned earlier they are normally as efficient as is possible within the system chosen at the outset, their operation is complex and specialized, and the principles in respect of electricity generation *per se* have been well documented[1]. The overall efficiency beyond the combustion stage is the product of cycle efficiency (1), turbine efficiency (2), and generator efficiency (3). 2 and 3 cannot be improved much further, and so the limitation is the cycle efficiency. With a back-pressure or pass-out turbine this can approach 100%. The pros and cons of systems incorporating district heating are discussed in Ch.15.2. The case for this must be affected by the system already established in a country and by the relative geographical distribution of fuel, population, and industry. The extent to which heating load has been or still could be reduced by better external insulation (Ch.9.2) is also relevant, as are the relative costs and time scales of this partial alternative and of the adoption of district heating in connection with new, and less attractively existing, power stations. One of the most favourable situations is a new town so planned that heat can be supplied from temporary boilers until the heat load builds up to allow of connection to the power station. This situation requires long-term planning and is likely to occur more often with a developing country.

All this said, avoidable rejection of 50—60% of the heat in the power-station fuel to cooling water or atmosphere remains an undesirable practice in principle, more easily justified by .cost-benefit analysis than by the increasingly attractive alternative of energy accounting.

A modern public utility station can have over 35% efficiency at the generator producing electricity alone or up to 60% delivering heat and power combined. Alternator capacities range from 50 to 500 MW_e increasing in overall efficiency as the size increases. Initial voltages are commonly 25 kV, stepped up for transmission purposes to 132 kV, 275 kV or 400 kV, then down by substations as required, using oil-cooled transformers from which large amounts of heat due to losses (about 0.5%) have to be dissipated.

13.1 Background principles

Electrical power generation is based on Faraday's law of mutual electromagnetic induction: in an electrical circuit moving (e.g. rotating) so that it cuts magnetic lines of force, an e.m.f. is generated that is proportional to the rate at which unit lines of force are cut. The magnetic field is generated by a d.c. excitation current in a primary circuit, produced by rectifying a.c. from a subsidiary generator geared mechanically to the main alternator. According to Lenz's law the direction of the current set up in the subsidiary circuit is such that electromagnetic interaction between the

two circuits tends to hinder the motion setting up the e.m.f. in the secondary circuit. Hence any energy flowing from the secondary, ultimately dissipated by the consumer, derives from the energy of relative motion between the circuits, and can be supplied by a heat engine in order to maintain this motion.

With the secondary circuit open (no load), the self-induction of the primary causes it to act as an automatic choke so that energy transfer from primary to secondary ceases and the load on the prime mover (e.g. heat engine) for maintaining rotation is minimized; the speed control on the latter then reduces the flow of heat (e.g. of steam to a turbine). Thus some measure of flexible control matching the output of electrical load to heat-energy input is inherent. Most generators now deliver a.c.; rectification at source is not justified. With rotating fields and stationary armatures, the moving contacts are at low voltage, handling field or primary current. The principle of three-phase a.c. generators (alternators) may be simply described diagrammatically by an example (*Figure 1*).

While in many industrial plants the necessary d.c. is produced more or less as shown (direct or belt driven), in large stations it may be derived from a separately driven source via excitation bus bars. Owing to the large capacity and high voltage in the secondary, dissipation of the heat due to energy losses requires large quantities of cooling air.

Three-phase alternators are normal; there are three groups of windings on the armature, 60 degrees between their central axes, so that by reversing the middle winding the three phases are obtained 120° apart. In general, polyphase a.c. is preferred for larger systems because of transmission and control advantages, more uniform flow of power, and suitability for motors of rugged design with more uniform torque. Current and voltage in the separate phases operate independently, and are commonly balanced; but they are interconnected in a number of alternative ways for economical operation.

Transmission costs are lower at higher voltage, hence high-voltage transmission is chosen for long distances, when the losses (owing to line capacity, self-induction and resistance) can be minimized. A.C. can be increased and decreased in voltage by the use of transformers; the ratio is determined by the number of turns in the two windings. When the secondary circuit is open, the transformer becomes a 'choke', or reactance coil, on the flow of energy in the primary circuit.

Loads consumed differ considerably in electrical characteristics. Interior lighting (Ch.7.7) requires a small current, usually nowadays from a 240 V a.c. supply. Process heating (Ch.7, 8.2) can take a variety of forms; the simplest is resistance heating at standard voltage, controlled by a periodic on—off mechanism or an a.c. choke rather than by a series resistance which wastes as heat the energy

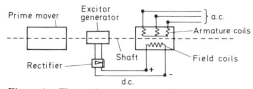

Figure 1   Three-phase a.c. generator

not required. Electrolysis (Ch.8.2) on the other hand requires a large flow of d.c. power at low voltage, and therefore a very large current; conductors of large cross-section are needed to minimize resistance losses. Motors utilize a moderate current, at voltages increasing with the load required, and a polyphase supply and complex sub-circuits for starting and stopping in the larger sizes. Their efficiency increases from 70 to 95% with increasing size (say 5--5000 nominal brake horsepower); it decreases markedly with reduced load, especially in the smaller sizes.

Electric motors are in effect generators run in reverse, converting electrical to mechanical energy. They vary in permissible load, capability and speed, and the starter equipment is complex because the supply voltage and current are connected in a controlled manner to protect the motor itself and to prevent damage to other circuits and controls on the same supply source. Many protective devices are incorporated. There are two main classes of starter, low-voltage (<600 V) and high-voltage (>600 V); they differ in insulation class and robustness of construction. Above 5000 V, switchgear is required. The amount of power handled largely determines the choice of low- or high-voltage starters: very approximately <150 kW and >150 kW respectively.

13.2 Plant suitable for industry

Generators

Generators used in private schemes normally produce a 50 Hz a.c. supply and operate at 440 V, 3.3 kV or 11 kV depending on the size of the set and the distribution voltage employed in the particular factory. Brushless generators are now used almost exclusively. The output current is controlled by means of an electronic voltage regulator which controls the exciter field and maintains the generator output voltage at a preset value.

Protection systems generally comprise overcurrent, earth-fault and reverse-power relays with differential protection on the larger sizes. Undervoltage protection may also be employed to guard against excitation failure.

In the few cases of partial generation operating in parallel with the public supply, induction generators may be used. These are similar to an induction motor, with excitation current, at a low leading power factor, taken from the public supply. The major advantage of this system is simplicity, since no synchronizing gear is required. A disadvantage is that if this is the sole means of generation, excitation current and therefore the whole private generation is lost in the event of failure of the purchased public supply; hence the installation of a small auxiliary petrol-driven exciter is usually called for. An induction generator

is also slightly less efficient than a synchronous alternator and moreover normally requires power-factor correction gear.

But with an independent private generation system, the industrialist is not necessarily tied to the public supply frequency, and generators may be installed operating at say 400 Hz for lighting, or at high frequency for special process requirements such as induction heating or radio-frequency drying.

D.C. power is very occasionally generated privately, but usually this is dictated by the need to supply existing d.c. motor power drives or small loads, such as emergency lighting or battery charging, without the complication of rectifiers. It is unlikely to be an economic proposition for new installations.

Prime Movers

The most common generator prime movers for industrial applications are steam turbines (Ch. 10.7), diesel engines (Ch.12.1) and gas turbines (Ch.12.2); petrol and gas (including fermentation methane) engines (Ch.12.1) may be used on a limited scale. Although regarded by many as obsolete, high-speed steam engines are still used to a limited extent for power generation and for local direct drives.

Steam Turbines

Three basic types of steam turbine are used to generate power as a by-product of process or exhaust steam: condensing, pass-out condensing, and back-pressure. These are manufactured for outputs from 50 kW$_e$ to 30 MW$_e$. Most industrial installations are in the range 1—5 MW$_e$. A reduction in the exhaust pressure effects a greater increase in power generation than an increase in inlet steam pressure. Specific steam consumptions depend on the absolute pressure ratio over the turbine and the size of set considered.

Condensing turbines (*Figure 2*) tend to be high in cost, thermally inefficient — around 10—25% efficiency — and bulky relative to power output. They can generally only be used economically for private generation where steam can be generated from waste heat from some other process, or to utilize low-pressure vapours from evaporators or similar plant. They are never used for total generation schemes, but can, in favourable circumstances, be used to provide a proportion of works electrical power requirements. Inlet steam pressures can range from 5 MPa to slight vacuum.

Pass-out condensing turbines (*Figure 3*) are the most common type for total generation schemes since, within limits, electrical output can be adjusted by altering the proportion of steam passing to the condenser; steam demand in excess of turbine pass-out capacity can be provided through a reducing valve and desuperheater bypassing the turbine. Inlet steam conditions can be in the range 4.5 MPa, 400°C down to 1.75 MPa saturated;

exhaust vacuum is generally 7 kPa absolute or greater. Thermal efficiencies in the range 50–70% can be obtained but may be much lower if more than the minimum steam is passed to the condenser.

Back-pressure turbines (*Figure 4*) are inexpensive, thermally efficient and compact, and usually the most economical proposition for partial generation schemes; but inflexible. Power generation is dependent on the steam flow required to meet process requirements, and the pressure drop over the turbine. The running cost for electrical generation is therefore the marginally additional cost of generating steam at higher pressure and temperature than would be required for process usage only, plus the fuel equivalent of the heat drop across the turbine. Exhaust pressure from the turbine is determined by process steam pressure requirements. Inlet steam conditions therefore depend on the power generation required; seldom more than 4.5 MPa gauge, 400°C, although in large-scale chemical industry pressures may be up to 13 MPa. Thermal efficiencies in the range 75–85% are common, and this is virtually unaltered by varying the back pressure provided that all the steam can be used. Steam can be blown to atmosphere or passed to a dump condenser to increase electrical generation in relation to process steam demand, but if this is practised for more than short-term emergency occasions, the costs involved can cancel out the financial savings of the scheme.

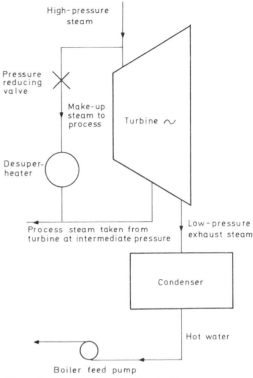

Figure 3   *Pass-out condensing turbine*

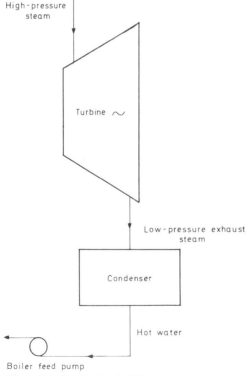

Figure 2   *Condensing turbine*

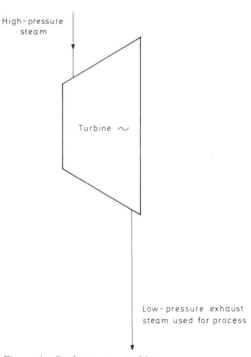

Figure 4   *Back-pressure turbine*

Here the 'insurance' aspects must be considered if the set is running in parallel with public supply. If the public supply fails, it may still pay to waste exhaust steam in order to generate the full electrical capacity of the set to keep as much of the factory in production as possible.

There are many variations of these basic types of turbine to suit special applications, e.g. mixed-pressure turbines accepting or passing out steam at more than one pressure; a back-pressure turbine and condensing or exhaust turbine driving the one alternator through a common gearbox; exhaust or topping turbines.

Modern industrial steam turbines run at high speed, e.g. 10 000—12 000 rpm, and are connected to their generators through reduction gearboxes to give the required generator speed of 1500 or 3000 rpm. Steam turbines are extremely dependable, requiring little maintenance and attention.

Many industrial steam-turbine installations require higher-pressure boiler plant than would be required for process steam only. In an existing works, the cost of replacing the boiler plant may make a power-generation scheme uneconomic. When considering a new factory installation, or if a power-generation installation is timed to coincide with replacement of an existing boiler plant, only the difference in cost between new LP and HP boiler plants need be related to the saving in cost of purchased power, and the scheme is much more likely to be economic.

Diesel Engines (Ch.12.1)

Into this category come engines running on light fuel oil, heavy fuel oil and natural gas. Dual-fuel arrangements can also be obtained, normally using natural gas and light oil.

The common size range of diesel generators is 30 kW to 1000 kW but both smaller and larger sets are readily available. These nominal ratings are reduced in relation to the height of the installation above sea level, ambient temperature above 30°C, and such factors as cooling-water temperature.

There is also a range of engine operating speeds available from high speed, around 1200 to 1800 rpm, to low speeds below 400 rpm. The less expensive high-speed engines, mainly in the smaller sizes and running on light fuel oil, tend to be the most economic proposition for standby duties. The more expensive lower-speed engines, often running on heavy fuel oil, tend to be selected for larger units and continuous operation because of their higher efficiency, greater life and lower maintenance requirements. Larger engines are usually pressure-charged and fitted with air coolers, since this can increase power output by up to 50% and improve efficiency by up to 4%. Most diesel engines operate on the four-stroke cycle, but in the larger low-speed range there are some very dependable two-stroke units.

Specific fuel consumption varies with engine size, design and imposed loading. Turbo-charged medium-speed sets in the 1000 kW$_e$ output range can consume around 0.22 kg—0.25 kg light fuel oil per kW h output at full load, rising to 0.24—0.27 kg/kW h at half load. Performance below half load deteriorates rapidly. Thermal efficiency as a straight electrical generator can vary from as low as 25% at full load in the 30 kW size to as high as 39% at full load for sizes above 1500 kW. Considering the relatively high cost of fuel and maintenance, this cannot compete economically with the purchased power supply unless other factors, such as safeguarding production in event of public supply failure, are also taken into account.

Of the total heat in the fuel supplied, approximately 30% is rejected in the exhaust gases and between a half and two-thirds of this can be readily recovered by heat exchangers such as waste-heat boilers to provide hot water or steam to the factory services. Also, some 18% of the heat output is dissipated by jacket and lubricating-oil cooling water. Nearly all this is recoverable by suitable heat exchangers, but this is not always economic. This potential heat recovery can raise the overall thermal efficiency of the plant to between 40% and 70% provided that full use can be made of the recovered heat. Heat recovery, however, falls rapidly with reducing engine load and, in the case of steam generation, with increasing steam pressure requirement.

On any medium-to-large installations, say of over 500 kW$_e$ installed capacity for continuous duty, it is always worth considering the additional capital cost of full heat recovery against the overall running-cost savings. In some cases, particularly on heavy oil, a combined power and heat scheme can compare favourably with purchasing electricity and generating heat requirements independently.

With all installations, adequate provision of combustion and cooling air is essential. A cooling-water system with cooling towers is generally required (or heat exchangers if river water can be circulated). In most continuously operating installations and many standby plants, consideration has to be given to acoustic treatment of foundations and buildings, or acoustic enclosures.

Petrol Engines (Ch.12.1)

Petrol-engine driven generating plant is available, normally restricted to outputs up to 100 kW. In industrial practice their use is usually limited to small compact portable units of up to 10 kW output for emergency lighting, battery charging duties, or to enable larger plant to be started up after a power failure. High engine and fuel costs preclude their consideration for larger size ranges.

Gas Turbines (Ch.12.2)

For small to medium industrial applications, simple open-cycle gas turbines are generally used, running

References p 308

on either natural gas or light fuel oil. Dual fuel arrangements are readily accommodated and in some cases automatic changeover on load between the two fuels can be provided.

The machine may be of the simple single-shaft type where power turbine, compressor and output drive are on a common shaft; or the twin-shaft type where the compressor and its driving turbine are on one shaft and the power turbine is on the output drive shaft. Both types are suitable for alternator drives but have slightly different power-control characteristics; also they may be developed from engines initially designed for aircraft application, or units specifically designed for land application. The former are claimed to be lighter, more compact and more efficient on partial load while the latter are claimed to be more robust, but there is a large area of common ground. In this country, common unit sizes are 0.5 to 3.5 MW output, but units up to 60 MW can be obtained.

In the small industrial size range electrical-generation efficiency can be as low as 14 to 20% at full load, but larger units can achieve more than 30%. Specific fuel consumptions of a typical 1.25 MW set can vary from 0.51 kg gas oil/kW h at full load to 0.63 kg at half load with an ambient temperature of 15°C, increasing however with higher ambient temperatures, particularly on partial load.

Deducting the heat equivalent of electrical output from total heat input, virtually all the remainder is available for heat recovery. Radiation and oil-cooler losses only amount to 1—2%. The waste gas is clean — nearly all air at over 500°C on full load. If the waste gases were passed through a waste-heat boiler to generate steam, the 1.25 MW set could generate some 4500 kg/h at 1.75 MPa gauge, raising the combined electrical and steam-generation efficiency to around 50%. If additional steam is required, the exhaust gas can be used as preheated combustion air for firing additional fuel in the waste-heat boiler; thus an additional 13 000 kg/h steam at 1.75 MPa could comfortably be raised, increasing the overall thermal efficiency to more than 70% — and the supplementary fuel could be heavy oil.

As an electrical generator alone the gas turbine is therefore less effective than the diesel engine. Its waste-heat-recovery potential is however greater and it is more flexible.

The diesel engine would therefore tend to be used where the electrical demand is high in relation to steam or heat demand, say less than 3 kg steam per kW h. The gas turbine cannot develop its full thermal efficiency potential unless more than 4 kg of steam is required per kW h. These figures compare favourably with a straight back-pressure steam turbine, where with typical medium industrial ranges of inlet and process steam pressures the steam load must exceed 10 kg/kW h.

As an alternative to steam generation, the waste gas may be used directly or through a heat exchanger for many industrial drying applications.

Gas turbines have many advantages over diesel engines for combined power and heat installations. They are low in weight and operate with little vibration, requiring inexpensive foundations. Maintenance costs are low and dependability is high. Also no cooling-water system is usually required, since a simple fan-blown air-cooler is generally sufficient for oil cooling.

The initial cost tends to be higher than diesel engines, although this is generally offset by lower running costs for continuously operating plant. A problem with gas turbines can be noise emission. Even small units in the 1—3 MW range can emit sound pressures around 100 dBA at 1 m (Ch.19.7), though mainly in the high-frequency range (4000 s^{-1}) which can readily be attenuated. The most common approach is to fit acoustic enclosures round the machine itself, or to house each machine in an acoustic cell of solid construction. In either case, silencers are required on air inlets and cooling-air exhaust. The heat-recovery equipment and chimney often provide sufficient attenuation for exhaust-gas sound emission.

Parallel, Independent or Sectioned Operation

When a firm considers installing private generating plant to operate continuously, as opposed to emergency standby equipment, there is the option to operate in parallel with the public electricity supply. This can apply whether it is intended to generate the total works electricity requirements or only a portion of them.

Parallel Operation (Figure 5)

Private generating plant operated in parallel with the public supply is in effect an extension of the national network. As such, the standards of design, installation and operation must satisfy the Electricity Boards that they will in no circumstances endanger the Board's system. Some of the essential requirements are:

(a)   The protection fitted to the private generator must be fast-acting and immediately disconnect the generator circuit from the system on occurrence of a fault. This is usually met by provision of differential protection with over-current and earth-fault back-up.

(b)   There must be no neutral earth connection when running in parallel with the public supply. The system neutral must, however, be adequately earthed on any occasion the private generation system is isolated from the public supply. This requirement is generally met by interlocking between the incoming supply circuit breaker, the generator circuit breaker and the system neutral earthing circuit breaker.

(c)   Adequate synchronizing facilities must be provided, such as manual synchronizing with a back-up check synchronizing relay.

(*d*) When the parallel-operation agreement is on the basis of import only from the grid. a reverse-power relay must be fitted to prevent export into the mains.

(*e*) Major switching operations of the incoming public supply must be effected by or under the control of the Generating Board. On larger private generating plants, particularly when export is permitted, trained staff may be required to be in attendance at all times to carry out switching operations on the Board's instructions.

The chief advantages of parallel operation are:

(i) Any deficiency between private generation and total works demand is automatically compensated by drawing the balance of power requirements from the public supply.

(ii) The frequency of the works electricity supply is locked to and governed by that of the public supply with its closely maintained limits of variation.

(iii) Because of (i) the private generating plant can be run continuously at full load or at its most economic overall load factor, to achieve minimum works costs taking into account purchased and works-generated power costs and waste-heat recovery.

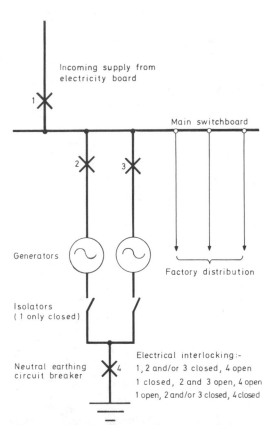

Figure 5 Two generators arranged for running in parallel with the national grid

Incoming supply from electricity board

Main switchboard

Generators

Factory distribution

Isolators
(1 only closed)

Neutral earthing circuit breaker

Electrical interlocking:-
1, 2 and/or 3 closed, 4 open
1 closed, 2 and 3 open, 4 open
1 open, 2 and/or 3 closed, 4 closed

(iv) In event of failure of the works generating plant, the public supply will provide an automatic standby subject to any faulty plant being disconnected from the system. This avoids the capital and maintenance costs of private standby generating plant.

(v) In event of failure of the public supply the private generating plant can still be operated within the limits of its capacity, provided that means of starting essential auxiliary equipment have been installed.

(vi) Where the parallel-running agreement provides for exporting power, private generation surplus to works requirements can be automatically exported to the public supply network. Should this export be substantial, and over long periods, a small unit price for exported energy payable to the industrialist can be negotiated with the public supply authority.

Against these advantages certain costs will have to be weighed. The industrialist will be required to pay the Electricity Board an annual charge related to the standby capacity element of the Board's connected supply and related equipment. This will vary from Board to Board, but is usually based on the installed or normally operated rating of the private generating plant, and may be at a rate of £4.5 to £7 p.a. (1975) per installed kVa. Also if this standby capacity has to be used owing to private plant failure or shutdown for maintenance, the purchased power over this period will be charged at full maximum demand tariff or other special high rates written into the agreement for parallel operation.

If power is exported only minimal rates are paid to the industrialist, usually based on the Generating Board's fuel costs. It is not therefore economic to export power to the public supply except on a very large scheme with preferential agreed tariffs or other very special circumstances.

Independent Operation

In some cases when a firm can generate all its own power requirements from process steam or a similar 'total energy' basis, there may be no need to operate in parallel with the public supply or even have a public electricity supply at all. In such circumstances, independent operation can be the most economical method of power generation and is common practice in the paper-making and chemical industries. Major considerations are the number of generating units required to meet the total power requirements and the extent to which standby plant should be installed to cater for essential maintenance or unprogrammed failure of generating units and essential auxiliaries. This will be considered in more detail under Reliability.

If production is to be shut down every weekend, it may not be economical or practical to generate the small amount of electricity required for essential lighting and maintenance operations over this period. A small public electricity supply could then be retained serving segregated power and lighting circuits, enabling the generating plant to be shut down. If the works is served by a small number of large boiler units, it may also be economic to have a small standby boiler to provide essential heating services over the weekend shut-down period.

Consideration has also to be given to the power requirements for starting up the main boilers and generating plant after the shut-down. This could be met by a small diesel generator or in the case of a gas turbine installation, either a.c.-motor start from the public supply, compressed-air start with a diesel-driven compressor, or d.c. start from a trickle-charged storage battery system.

Sectioned Operation

When only a proportion of the works electricity requirements is to be supplied by private generating plant, sectioned operation is the alternative to parallel operation. In its simplest form the plant to be served from the private generating equipment can be on completely independent circuits from the remainder of the works served by the public supply. Purchased maximum demand and units will be reduced to the extent of private generation. This reduction will, however, be limited to the actual usage of the connected equipment which may be less than the potential generating capacity. The dependability of the independent section will also be less than the remainder of the works served by the public supply.

If it is required to increase the dependability of the private-generation section by using the public supply as a standby, the public and private supplies to the switchgear serving the plant must be connected through interlocked circuit breakers to make it impossible to connect the public supply until the private supply is isolated and vice versa. With such an arrangement, the Electricity Board will generally apply standby charges as in the case of parallel operation.

In general, sectional operation can be lower in first cost than parallel operation, and is often used when private generation is only a small part of total electrical requirements and slightly lower dependability than the public supply can be accepted. As the extent of private generation increases, and if it is desired to retain the public supply as standby, the economic balance moves in favour of parallel operation.

Implications of Standby Requirements

The main reasons for installing generating plant for standby duty only are to provide a supply of electricity to enable production processes to be continued or to prevent damage to or loss of product or manufacturing plant in the event of an interruption of the public electricity supply.

The economics of standby generation are essentially an insurance-type calculation. The cost of probable loss of production or product and all associated costs due to an infrequent and probably short-term failure of public electricity supply must be weighed against the capital, maintenance and running costs of the necessary equipment to minimize such loss.

Factors such as non-productive wages, and external considerations such as the ability of suppliers to continue to supply raw materials or customers to accept supplies over a prolonged failure of public electricity supply have also to be taken into account.

Since there is no guaranteed return from the capital outlay involved and the actual running time of installed plant is unlikely to be more than a few weeks a year, low-cost self-contained generating units that can fit into existing factory space are favoured. This generally means high-speed diesel generators or in special cases gas turbines. Also, it is seldom economic to consider duplicating the total works electrical requirements on a standby basis. The capacity of installed plant is generally limited to that necessary to power certain essential process loads and emergency lighting, which could be at most three quarters of normal requirements and usually much less.

The standby generators may be connected to the main factory switchboard, enabling generated power to be distributed throughout the works. On large installations where there are a number of switchboards, generators feeding into selected switchboards are sometimes employed. Whichever system is adopted however, it is necessary to ensure that the switch on the Electricity Board's supply is opened before starting the standby plant and connecting it to the switchboard, and conversely isolating the standby connections before reconnecting the public supply. The necessary interlocking arrangements must be agreed with the local Electricity Board (*Figure 6*).

The time delay that can be tolerated between a failure of the public supply and the connection of the standby supply largely determines the type and cost of the changeover system selected. There are three basic options:

1. Manual — this is the most common and lowest-cost system. On failure of the public supply the incoming circuit breakers are manually opened, releasing the interlocks to enable the standby-plant circuit breakers to be closed. Distribution circuit breakers are opened, the standby plant is then manually started, the selected generator and distribution circuit breakers closed, and motors etc. throughout the works restarted. Depending on the extent of the installation this operation can take from 10 to 30 minutes, provided that staff are always available to carry out the necessary operations.
2. Semi-automatic — contactors which will drop out on mains failure and not reclose when the standby supply is brought on can be provided on (i) the main supply and (ii) selected non-essential

distribution circuits to limit the remaining connected load to within the capacity of the standby generator. The standby generating plant can also be arranged to start automatically on mains failure. Manual operations are then reduced to closing the standby generator circuit breaker and restarting motors etc. This can be taken a stage further by having automatic changeover contactors between the mains and standby supplies and the switchboards on plant served, reducing manual operation to restarting motors etc.

3. No-break System — this is the most expensive in both capital and running costs and is only used in extreme cases of individual plant items, such as a computer, where any break in electricity supply can result in very high losses. The alternator of the standby generating set is continuously run at operating speed by a mains-driven electric motor. On mains failure, the standby-generator prime mover is automatically started and the alternator connected to the supply via a changeover contactor.

A more recent system ideally suited to computers is a standby supply comprising basically a storage battery with an invertor floated across the load in parallel with the mains supply (*Figure 7*). On the occurrence of a mains failure, the entire load is automatically taken by the battery/invertor,

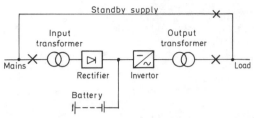

Figure 7   No-break supply using rectifier, invertor and battery

so maintaining an uninterrupted supply to the load. The effective duration of operation of such an arrangement depends on the capacity of the battery in relation to the load, and should it be required to maintain the supply for more than an hour or two it would be necessary to provide suitable back-up by say a diesel generating set. This system gives the very close control of voltage and frequency necessary to avoid malfunctioning of the computer on changeover, and is less costly and more reliable than a no-break diesel set.

In cases where more than one set is installed, whichever scheme is adopted, it is necessary to ensure that the load connected to the first set on start-up is within its capacity. This might be arranged by initially connecting say the lighting load to the first set and waiting until the other sets are synchronized before connecting substantial power loads.

Use of Standby Sets for Peak-lopping or Base-load Operation

If a case can be made out to install standby generating plant, it is always worthwhile considering the possibility of generating a proportion of the factory electrical requirements on a continuous or regular periodic basis before selecting plant or designing the system. While this will certainly involve some additional capital outlay on more robust plant, and more sophisticated control equipment and revenue expenditure on running costs, it will give a positive calculable financial return on the installation. In favourable circumstances, a good economic case can often be established.

No industrial generating plant without waste-heat recovery can compete with the public supply on average cost per unit generated. However, most electricity purchased by industry is on a two-part maximum-demand tariff where the maximum-demand charge per kW is 200–500 times the average unit cost. If a firm can reduce its maximum demand for purchased power by judicious private generation, there can be an overall reduction in total (purchased plus generated) power costs.

On the basis so far considered, this is purely a financial saving to the firm concerned and does not necessarily make the best uses of the nation's primary energy resources. This aspect will however be mentioned later.

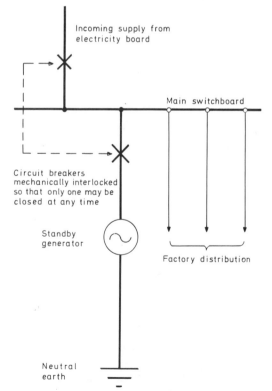

Incoming supply from electricity board

Main switchboard

Circuit breakers mechanically interlocked so that only one may be closed at any time

Standby generator

Factory distribution

Neutral earth

Figure 6   Single generator arranged for standby operation only

References p 308

Basic Plant

While generating sets selected for purely standby operation could be used for partial continuous generation, it will generally be more economic to consider more efficient and more robust units and a more permanent installation to minimize operating and maintenance costs. This could mean for example selecting lower-speed diesel engines, and if the size of the units warrants it using lower cost heavy fuel oil instead of diesel oil. Also if the size of the sets is beyond the range covered by radiator cooling of jacket water, a cooling circuit with cooling towers should be considered instead of a 'once-through' cooling system.

If parallel operation is chosen, and particularly if peak lopping is the selected method of operation, it will generally be worth considering the installation of an additional generating unit to minimize the risk of savings being nullified by failure of the private plant. It is these additional capital costs, over those of purely standby installation, that have to be weighed with running costs against the financial savings on the total works electricity costs.

Methods of Operation

If the works daily, weekly or monthly purchased electrical load curve has one or more substantial peaks, operating the private generating plant to supply these peaks may be the most economic method of operation. If the load curve is relatively level, base-load operation of the private plant may be indicated. If peaks in demand are due to a small number of individual plant items, it may be possible to supply those plant items from the works generating plant on an electrical distribution system separated from the public supply. Base-load operation can be practiced by operation in parallel with the public supply or sectioned operation of part of the total load.

Only a detailed analysis of the particular works load pattern and an economic evaluation of all the possible alternatives, taking into account running costs and possible standby charges, will establish which method of operation will give the greatest financial savings for comparison with the additional capital costs discussed previously.

Waste-heat Recovery

Any scheme for partial power generation can be enhanced if it is possible to recover waste heat in the form of steam, hot water or hot gases for use in process. If the potential for using such waste heat exists, there may be economic grounds for considering more expensive installations, such as higher-pressure boiler plant and steam turbo-alternators or gas turbo-alternators. Such combined heat and power installations have the added advantage of generally making more efficient use of the nation's primary energy resources than purchasing electricity and generating the heat requirements independently. Roughly 90% of privately generated electricity is produced on combined heat-and-power plants.

Reliability

The reliability of any private generation scheme will naturally be compared with that of the public electricity supply. On average, the likelihood of mechanical interruption of the public supply to industry is appreciably less than two hours per year (reliability >99.98%). This very high dependability is achieved by the installation of additional generation, transmission and distribution equipment and excellent standards of maintenance and fault rectification, reflected in the purchase price of electricity produced. But in the UK the crises of recent years have brought home the vulnerability of the public supply to political and industrial action. Failure of the public supply for one working week a year could reduce the reliability to below 98%.

The comparable reliability of private generating plant units depends on the extent of non-availability for generation purposes due to two factors — the time required for essential planned maintenance and overhaul involving shutdown, and unplanned outage due to electrical or mechanical failure in service. Failures in service are unlikely to occur every year and realistic estimates are difficult to obtain. Reliability, taking account of these two factors, can be calculated assuming that the average industrial firm will have a holiday shut-down of two weeks per year during which planned overhauls can be carried out.

For typical prime movers the approximate indications from records are:

Type of unit	Programmed shut-down	Failure (h/year)	Overall reliability
Diesel engine	250 to 680 h/year. This could comprise one shut-down for major overhaul lasting two to three weeks and three to four weekend shut-downs.	200	94.0%
Gas turbine	Varies with type and manufacturer. 4 to 24 h in first year of operation. 60 to 80 h every second year.	0–24	98.7%

	Four weeks every 16 years when set is virtually rebuilt.		
Steam turbine (excluding associated boiler plant)	72—180 h/ year depending on size. Possibly reblading 2 weeks every 10 years	0—24	99.7%

These figures refer to a single unit of each type. Obviously, the effect of a failure in operation can be reduced by having more than one unit which together can meet the maximum total load. Also, the installation of standby units in excess of total required rating will improve overall reliability and can enable maintenance or repairs to be carried out outside the annual shut-down period. The overall reliability of multi-set installations is a simple probability calculation.

When private generating plant is installed purely for standby duty to the public supply, the chances of a period of non-availability coinciding with a public supply failure are remote. A single unit of the required rating is all that need be installed, unless the works switching arrangement dictates the need for multiple units.

But when private generating plant is used for partial generation on a reasonably continuous basis, either for peak lopping or as base-load plant, the financial loss resulting from the failure of a single unit could be substantial. If diesel engines are used, it is usual to install at least two units to provide the required rating. If gas turbines or steam turbines are used, owing to their greater reliability and higher capital cost it is seldom economic to install more than a single unit for small- to medium-sized installations.

If private generating plant is used for total works power generation, the choice will tend to be the more reliable gas or steam turbines with full waste-heat recovery. Some measure of standby is desirable, but the high capital cost of this is difficult to justify. One compromise frequently adopted is to replace the plant before the end of its useful life and retain the older plant as standby. If space is available, this can be more economic than retaining a public supply as standby.

Fuel Supplies

A further major factor in considering the reliability of private generating plant is the availability of fuel. For purely standby generating plant the storage requirements of fuel can be assessed on the basis of the duration of public supply failure against which it is intended to insure, plus a reasonable margin for periodic operation of the plant to keep it in good condition. The more continuously it is intended to operate the generating plant, the more important does it become to maintain adequate stocks of fuel to cover for possible breaks in supply.

With natural gas it is almost certainly uneconomic to consider the installation of storage capacity.

With natural gas on an interruptible tariff, a secondary fuel, such as gas oil, is generally used to supply dual-fuel plant or essential services in the event of interruption of supplies. It would be prudent to maintain stocks of the secondary fuel sufficient to cover the full periods of breaks allowed in the supply agreement: in the event of a supply shortage natural gas customers with only a standby requirement for oil will come at the bottom of the oil companies' supply priorities.

In the past, oil companies have recommended a minimum oil storage of three weeks supply at maximum usage rate or two weeks at maximum usage rate plus one normal delivery, whichever is the greater. For power generation the former is recommended. Many companies are in fact now considering doubling their past oil-storage capacity to cater for possible prolonged interruptions in supply of up to two months. With heated grades of oil, as far as is possible, only one storage tank should be kept up to normal storage temperature and it should be thermally insulated to minimize heat losses. Similar storage is desirable for LPG, but the high capital cost of storage equipment may justify some compromise.

Coal deteriorates slightly in stock and this loss and the additional handling costs, should be budgeted for. A minimum stockpile of around two months' supply at maximum usage rate is not considered unreasonable.

Summary

Since the individual conditions vary so vastly between firms, no conclusive generalizations are possible. The following pointers, however, frequently apply.
1. Generation for standby purposes can give no guaranteed return on the capital expenditure involved and can only be justified on an insurance-type calculation. As a result, the simplest possible installation with lowest-cost equipment consistent with reliability is first choice. In most cases this will be high-speed diesel generators.
2. Peak lopping or base-load partial power generation can be justified on its own merits in very few cases. A sound economic case can often be presented, however, if it has already been decided to install standby equipment and only the additional capital cost has to be considered, and if a substantial proportion of the waste heat from the prime movers can be usefully recovered for process applications, or if generation can be effected by back-pressure steam turbines or by utilizing steam or heat from process which would otherwise be wasted.
3. Total generation can result in the greatest financial savings, but is only applicable in comparatively few firms where there is a relatively

large steam or heat requirement such that the power-generation potential is in reasonable balance with or in excess of works power requirements. Capital costs tend to be high, but in favourable cases, so is the return. With steam-turbine installations, the capital cost is minimized if power generation is considered when process-steam boilers are due for replacement.

Any potential scheme for power generation requires the fullest evaluation of all the many factors in a particular firm, and possible alternatives involved. Such evaluations should take full account of tax on savings, tax allowances, grants etc., and future changes in process steam pressure; they should preferably be evaluated on a discounted-cash-flow basis to establish a realistic payback period and the return on capital expenditure.

13.3 Magnetohydro-dynamic and electrogas-dynamic conversion

Magnetohydrodynamic (MHD) Conversion[2-5]

Basic Principles

A high-velocity, electrically conducting, fluid stream crossing a magnetic field may be regarded as taking the place of the moving conductors of a conventional dynamo. The operation of a MHD generator can be understood from *Figure 8*. If a jet of conducting fluid, velocity u, moves through a magnetic field of flux density B at right angles to u, then an electric field E will exist where

$$E = uB \tag{1}$$

If electrodes are suitably placed in contact with the jet then energy can be extracted into an external load, displacing the equilibrium so that $E \neq uB$. The system is seen to be analogous to

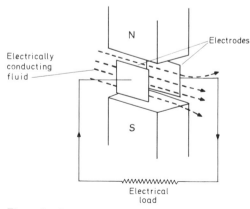

Figure 8   Principle of MHD generator

that of a turbine with electromagnetic braking taking the place of the mechanical braking of the turbine blades. The energy to drive the fluid through the MHD duct may be obtained by means of either a Rankine or a Brayton cycle. Normally the thermodynamic fluid, vapour or gas, is also used in the MHD duct. It is however possible to use a conducting liquid in the MHD duct in which case it is necessary to use a two-phase system in which the liquid is accelerated by a jet of vapour or gas.

Assuming that the working fluid behaves as a normal electrical conductor of conductivity σ we see that the current density j is given by

$$j = \sigma(E - uB) \tag{2}$$

and the electrical power output per unit volume of duct is

$$-jE = \sigma E(uB - E) \tag{3}$$

If the ratio of load resistance to total resistance is K, i.e. $K = E/uB$, then the electrical power generated per unit volume of duct is

$$W = -jE = K(1 - K)\sigma u^2 B^2 \tag{4}$$

This power is obtained from the work done by the moving stream against the body force:

$$jB = -(1 - K)\sigma uB^2 / \text{unit volume of duct} \tag{5}$$

So the work done by the stream

$$-jBu = (1 - K)\sigma u^2 B^2 \tag{6}$$

and the difference between equations (6) and (4) represents ohmic heating in the fluid and is given by

$$(1 - K)^2 \sigma u^2 B^2$$

We see by using this method that the thermal energy of the source must first be converted to kinetic energy; the energy-conversion chain thus involves a mechanical-energy link.

Though the MHD method is identical in principle with the turbo-generator, it has a number of potential advantages at high temperatures: no moving parts, no close tolerances, and high power/volume ratio in large units.

Electrical Conductivity

Since the power output of an MHD generator is directly proportional to the electrical conductivity it is important for this quantity to be as high as possible; this means that the gas must be partially ionized. Ionization may be achieved in a number of ways.

Thermal Ionization.   All hot gases are ionized to some extent; the degree of ionization depends strongly on the ratio of the ionization potential of the gas atoms to the temperature of the gas.

The degree of ionization and hence the electrical conductivity of a gas can be enhanced by several orders of magnitude by adding a small quantity of material of low ionization potential; this is known as seeding.

If x is the atomic fraction of the seed material, then the fractional degree of ionization n_e/n_0 is related to temperature T, pressure p and ionization potential V_i of the seed material:

$$\frac{n_e}{n_0} \propto T^{5/4} \left(\frac{x}{p}\right)^{1/2} \exp\left(-\frac{V_i}{T}\right) \qquad (7)$$

Thus the electrical conductivity is proportional to $(x/p)^{1/2}$ and is very sensitive to temperature because of the exponential term.

Magnetically Induced Ionization. Since the electrons are the most mobile of the current-carrying particles, the energy of the ohmic heating of the fluid is given to the electrons, which are normally maintained at essentially the same temperature as the ions and neutral particles by energy exchange on collision; the rate of energy transfer from the electrons to the atoms and ions is proportional to the temperature difference between the electrons and atoms, and hence, if the currents flowing in the gas are sufficiently great, it is possible to attain a state where the electron temperature is significantly greater than that of the main body of the gas. This is most easily attained in the monatomic gases, since with these, because of the large atom/electron mass ratio and the absence of molecular vibrational excitation, only a small fraction of the energy difference is exchanged on each collision.

Since the dominant ionization process is normally electron—atom collisions, it might be expected that the electron temperature will determine the degree of ionization and hence the electrical conductivity of the gas; experiments have shown this to be the case.

This method of ionization is suitable for use in closed-cycle generators employing monatomic gases but cannot be used with molecular gases and is therefore unsuitable for combustion-product generators.

External Ionization. A variety of possible methods for increasing ionization above the thermal equilibrium limit has been discussed in the literature, including high-voltage electron beams, photo-ionization by lasers, a.c., r.f. and d.c. discharges. The major disadvantage with all these methods is the high energy input required relative to the generator output. Other proposals include the use of thermionic emitting particles, radioactive particles, and frozen-in flow condition. In addition to steady-flow designs, some work has also gone into the use of striated flow or pulsating systems. The Rayleigh—Taylor instability poses a practical difficulty in these methods.

Most work has gone into studies of systems employing either thermal equilibrium ionization, magnetically induced ionization, or liquid metals.

Applications

MHD generation has possible applications as a topping plant for use with conventional steam-generating plant, as a peak power supply, and as a power supply for propulsion of large space vehicles.

There are three main categories of generation system:
1. Fossil-fuelled open-cycle system.
2. Nuclear-heated closed-cycle inert-gas system.
3. Nuclear-heated closed-cycle liquid-metal system.
Most effort to date has gone into the fossil-fuelled open-cycle system, *Figure 9*. The feasibility has been demonstrated, but major problems associated with the constructional and the seed materials remain. In order to achieve a sufficiently high temperature to give an adequate level of thermal ionization of the seed it is necessary either to provide oxygen enrichment, or to preheat the incoming air to 2000 K. The air can be preheated either in an external heater or by means of a recuperator. Both oxygen enrichment and external heating are expensive, and it is economically very desirable to develop a suitable recuperator. A further problem arises from the interaction of the alkali-metal seed and the residual ash in the combustion gases, giving rise to erosion and slag formation in the MHD duct. It is necessary to extract and recycle the seed material, partly in order to avoid pollution and partly because of the cost of the material.

The nuclear system is shown in *Figure 10*. The working fluid is an inert gas seeded with an alkali

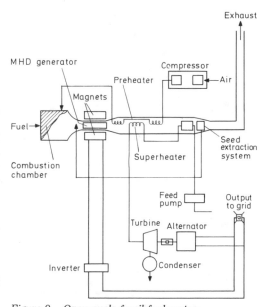

Figure 9   Open-cycle fossil-fuel system

metal. An overall cycle efficiency of 50% appears to be possible. In order to use thermal ionization, a coolant outlet temperature of 2000 K is required; this is unlikely to be achieved so it is necessary to consider the use of a non-equilibrium method of ionization. Magnetically induced ionization has received most attention; however, at present there are major uncertainties in the understanding of the electrothermal instabilities which arise, and it is too early to say whether this system can be operated stably. A further design problem is in the selection of operating pressure; a high pressure (1—2 MPa) is required in order to keep down the size of the reactor core, whereas the magnetically induced ionization increases as the pressure is reduced and requires a value of ≈ 100 kPa to give a reasonable duct volume.

The liquid-metal systems have the advantage of a high electrical conductivity; however, a two-phase system is required in order to drive the

Table 1

	EGD	MHD
Potential gradient	10^7 V/m	10^3—10^4 V/m
Voltage output	$>10^5$ V	10^2—10^4 V
Impedance	High	Low
Power output	>1 kW	>10 MW$_e$
Power scaling	As channel cross-section	As channel volume
Temperature	Room temperature to $1000°$C	High temperature $>1000°$C
Working fluid	Unipolar charged particles in non-ionized gas	Neutral slightly ionized gas
Magnetic field	None	Yes

liquid metal through the duct. A high-temperature vapour is injected into the liquid metal before the latter enters the duct. The vapour is separated from the liquid on leaving the duct. Several methods have been proposed, but in all cases friction and mixing losses are high and the overall system efficiency is low (a few percent). A reactor coolant outlet temperature of 950—1200°C is required. The system is unlikely to emerge as a major base-load power-generation system in the near future.

Electrogasdynamic (EGD) Conversion[6,7]

In an EGD converter, charged particles of one polarity are transported by gas flow against an electric field. The enthalpy of the gas stream is thus converted directly to electrical energy (*Figure 11*). The process is similar to that taking place in a van de Graaf generator, the gas flow taking the place of the moving belt in the latter. *Table 1* compares EGD conversion with that using MHD.

EGD can be used with both Rankine and Brayton cycles, together with fossil fuel, nuclear or solar energy sources. EGD provides a high-voltage, low-current electrical source particularly for power levels in the 1—100 kW range. It is still in the development stage and has not yet reached commercial exploitation.

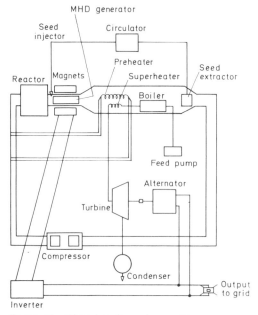

Figure 10   Closed-cycle nuclear system

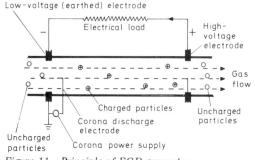

Figure 11   Principle of EGD generator

13.4 Electrochemical energy conversion

Electrochemical Cells

When a fuel is burnt, the chemical reaction involved is non-isothermal and if, as is usual, no work is done, the chemical energy released will appear as heat, i.e. as random energy of the products of combustion. This random energy can only be converted into work by some form of heat engine. In an electrochemical converter the chemical energy released in the reaction by the oxidation of the fuel appears directly as electricity and no random energy stage is involved. Such a converter is called a galvanic cell (*Figure 12a*). In its simplest form it consists of two electrodes separated by an electrolyte. Fuel is provided at the negative electrode, or anode. Electrons flow from the anode to the cathode via the external circuit and the circuit is completed by ions flowing through the electrolyte.

Electrochemical cells can be classified as follows.

1. Primary cells in which the fuel, usually a metal, forms the anode and the cathode provides the oxidant, normally in the form of a metal oxide. During the reaction the electrodes are consumed and the cell cannot be reconstituted.

2. Secondary cells. These cells are similar to primary cells in that the electrodes are consumed during discharge. They differ from primary cells in that they may be reconstituted or recharged by passing a current in the reverse direction through the cell. Secondary cells are also referred to as accumulators or rechargeable cells.

3. Fuel cells. In the fuel cell the electrodes are not changed in the reaction but serve to dispense fuel and oxidant into the cell from external containers. An important advantage of the fuel cell over primary or secondary cells is that the size of the cell is determined only by the power output required and the size of the fuel tanks determines the total energy stored (*Figure 12b*).

4. Semi-fuel cells. In this type of cell atmospheric air provides the oxidant by means of a non-consumable electrode whilst the fuel is provided by a consumable electrode. Metal—air cells fall into this class (*Figure 12c*).

Batteries

All electrochemical cells have a small d.c. output voltage of around one volt. It is usual to combine a number of cells in a series or series-parallel arrangement to increase the output voltage. Such an arrangement is termed a battery.

Basic Principles

The maximum energy available from a chemical reaction is $-\Delta G$, the change in the Gibbs free energy for the reaction. The corresponding ideal voltage for a reversible reaction is

$$V_0 = \frac{-\Delta G}{nF}$$

where n is the number of electrons transferred in the reaction, and F is the Faraday (96 000 C/mol). In practice irreversibilities will reduce the available cell voltage.

Important sources of irreversibility include:

Activation polarization. The loss due to high activation energy for electron transfer at the electrode.

Concentration polarization. Change in concentration near the electrode surface.

Ohmic polarization. Resistive loss in the electrolyte.

These three polarization losses reduce the on-load voltage from the ideal value V_0. In addition there may be losses (specified as a current efficiency) in current due to side reactions, and in the fuel cell inefficiencies in preprocessing of the fuel and in energy requirements by ancillary plant such as circulating pumps.

In order to reduce the effects of polarization, the electrodes are constructed of highly porous material to increase the surface area. Polarization is a particularly serious problem in fuel cells and catalysts are incorporated in the electrodes; the cells may also be operated at high temperature to give improved performance.

In primary and secondary cells the anodes are usually lead, zinc, cadmium or iron, and the cathodes metallic oxides, though chlorides or sulphides may also be used (the electrodes must be electrical conductors). The combined discharge—charge operation of a secondary cell, or oxidation—reduction process (redox), can be represented by

$$M^a + 2\,OH^- \rightleftharpoons M^a(OH)_2 + 2\,e$$

where M^a represents the metal anode. The electrolyte, which makes possible the transport of ions, should have a high electrical conductivity.

Primary Cells

Leclanché (Dry, Zinc—Carbon)

The Leclanché cell employs a zinc anode, which normally acts as the container, and a manganese dioxide cathode. The latter surrounds a carbon rod which forms the positive terminal. The Leclanché cell is cheap but the energy density is

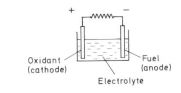

a. PRIMARY OR SECONDARY CELL

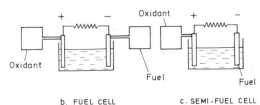

b. FUEL CELL   c. SEMI-FUEL CELL

Figure 12   Fuel-cell comparison

not high and the voltage falls readily under discharge. Recent improvements in design have led to the development of the high power (H.P.) range which have a watt-h capacity from 4 to 18 times that of the standard cell.

Mercury-Oxide Zinc (Ruben—Mallory, Kalium)

These cells use a zinc anode together with a mercuric oxide cathode. The electrolyte is potassium hydroxide. The cell is contained in a steel case. Whilst more expensive than the Leclanché cell, this cell has the advantages of constant voltage output during discharge and long shelf life.

Alkaline Manganese

The electrodes are similar to those used in the standard Leclanché cell; the construction and electrolyte the same as in the mercury-oxide zinc cell. This cell has a long shelf life but the voltage stability under load, though better than that of the Leclanché cell, is inferior to that of the mercury-oxide zinc cell.

Selection of Primary Cells

The Leclanché cell is used where low-cost energy is required, for example for torches, transistor radios and small domestic equipment. For heavier current loading or where intermittent power is required over long periods the alkaline manganese cell is used. The mercury-oxide zinc cell is selected if voltage stability is important or where very long reliable life is required, for example in cardiac pacemakers. The characteristics are summarised in *Table 2*.

Special-purpose Primary Cells

Most requirements for primary cells will be satisfied by one of the three major types described above. There are, however, requirements for which special-purpose cells have been developed: for example sea-water-activated cells for emergency lifeboat-radio use.

Copper chloride—magnesium cells using sea-water as the electrolyte are used to provide lightweight power supplies for applications such as

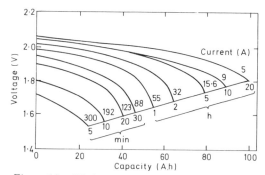

Figure 13   Discharge curves for lead accumulators at various rates of discharge (100 amp hours at 20 h rate)

radiosonde, and other meteorological use. For use in fresh water the separators are impregnated with potassium chloride which is primed by immersion in the water.

Sea-water is also used as the electrolyte in silver chloride—magnesium cells which have a high power density at high discharge rate and are used particularly for torpedo propulsion.

Lead dioxide—magnesium and lead dioxide—zinc cells are also used, the former with sea-water electrolyte and the latter with sulphuric acid electrolyte.

Secondary Cells

Lead—Acid Cell

The positive electrode is lead dioxide, and the negative electrode is made of pure lead in a spongy form. The electrolyte is dilute sulphuric acid. The active materials are mounted on lead grids which both provide mechanical support and act as electrical conductors. The reaction may be written:

$$PbO_2 + 2H_2SO_4 + Pb \overset{\text{Discharge}}{\underset{\text{Charge}}{\rightleftharpoons}} PbSO_4 + 2H_2O$$

During discharge, the concentration of the electrolyte is reduced by the removal of sulphate ions and the production of water. This effect is used to indicate the state of charge by measuring the specific gravity using a hydrometer. The specific gravity for automobile and traction batteries is normally maintained at 1.275—1.285; and 1.200—1.215 for stationary batteries.

The major reasons for the dominant position of the lead—acid cell are:
1. The relatively high nominal voltage, 2.0 V/cell.
2. The versatility in providing high or low currents over a wide range of temperature.
3. The high degree of reversibility: it is capable of hundreds of charge/discharge cycles.
4. The relative cheapness and ease of fabrication of the principal material, lead.

The capacity depends on the rate of discharge. Typical discharge-voltage curves and capacity are shown in *Figure 13* for various rates of discharge.

Table 2   Characteristics of primary cells

Cell	Nominal e.m.f. (V)	Specific energy (W h/kg)	Normal operating temperature (°C)
Leclanché	1.5	49	0 to + 70
Alkaline manganese	1.5	77	−30 to + 70
Mercury-oxide zinc	1.3	100	−30 to + 70

A typical charge/discharge characteristic is shown in *Figure 14*. Drycharged batteries are available and will give up to 75% of their rated capacity on filling after a dry-storage period.

Alkaline Secondary Cells

A number of commercially available cells employ dilute potassium hydroxide as the electrolyte. The most important are nickel—iron (NiFe), nickel—cadmium, silver—zinc and silver—cadmium. The characteristics of these cells together with the lead—acid cell are summarized in *Table 3*.

Nickel—Iron Cells

The positive electrode is nickel hydroxide and the negative electrode is of iron. The active material for the latter is formed into pellets which are contained in a 'pocket' of perforated nickel-plated steel. The nickel hydroxide forming the positive electrode is stacked with alternate layers of nickel flakes in perforated steel tubes. The electrolyte (KOH) has a specific gravity of 1.200; it does not take part in the reaction, and its specific gravity remains almost constant.

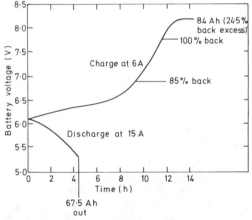

Figure 14   Charge/discharge   characteristic   for lead—acid accumulators (6 V. 90 amp hours at 20 h rate)

The overall reaction is:

$$2 \text{ NiOH} + 2\text{H}_2\text{O} + \text{Fe} \underset{\text{Charge}}{\overset{\text{Discharge}}{\rightleftharpoons}} 2 \text{ Ni(OH)}_2 + \text{Fe(OH)}_2$$

Nickel—iron cells are robust and capable of operation under deep-cycle conditions; they will operate at higher temperatures than lead—acid cells, but have some disadvantages over the latter, e.g. their charge retention and low-temperature performance is poor and they are more expensive.

Nickel—Cadmium Cells

The reaction for the nickel—cadmium cell is similar to that of the nickel—iron cell. The 'plates' or electrodes are either of the pocket type, or employ sintered plates which allow very high rates of discharge. A later development of the sintered-plate type is the hermetically sealed cell which requires no maintenance. Sealed cells must be charged carefully and the recommended charge rate must not be exceeded. Vented cells can be charged at very high rates. They have good cycle life, mechanical robustness, low maintenance requirement and good performance down to -40°C. They are, however, more expensive than the lead—acid cell.

Silver—Zinc Cells

These cells are characterized by high energy density per unit of weight and volume and are capable of a high discharge rate. The discharge voltage of 1.55 V is constant over most of the discharge. They are expensive and have a limited number of recharge cycles ($\approx$100).

Silver—Cadmium Cells

This has a high energy density though somewhat lower than that of the silver—zinc cell. Its advantages over the latter are the increased number of recharge cycles and its availability in hermetically-sealed units. The cost is high, and it is likely to find its use in small sizes or special applications.

Table 3   Characteristics of secondary cells

Cell	Nominal e.m.f. at discharge (V)	Specific power (W/kg)	Specific energy (W h/kg)	Operating tempera-ture (°C)	Operating life	
					(cycles)	(years)
Lead—acid	2.1 − 1.46	1.5—6	4—6	−20 to +25	Hundreds	5—15
Nickel—iron	1.3 − 0.75	1.5—8	6—7	0 to +45	Thousands	10—20
Nickel—cadmium	1.3 − 0.75	1.5—9	3—8	−40 to +45	Thousands	10—20
Silver—zinc	1.55 − 1.1	5—30	16—20	−30 to +85	≈hundred	≈2
Silver—cadmium	1.3 − 0.8	4—14	10—12	−25 to +65	Few hundred	≈3

References p 308

Future Developments in Secondary Cells

There is considerable interest in new types of secondary cell, offering high energy density. Three cells currently under study are sodium—sulphur, lithium—chlorine and zinc—chlorine. The characteristics are given in *Table 4*.

Selection of Secondary Batteries

The major fields of application for large secondary batteries are:
1. Static applications (emergency lighting, telephone exchanges, fire and alarm systems).
2. S.L.I. (automobiles; starting, light and ignition).
3. Traction (industrial trucks, road delivery vehicles, electric cars).

Lead—acid cells are chosen for 1 and 2 and frequently for 3 also. Nickel—iron batteries are also used in traction applications. They are more expensive than an equivalent lead—acid battery but this must be offset against the longer cycle life. A detailed study of the actual service conditions is necessary before the choice can be made. Nickel—cadmium batteries are used for aircraft and diesel-electric locomotive starting and for emergency supplies, and are extensively used to supply small portable electrical equipment ranging from lawn mowers, hedge cutters and electric drills to toothbrushes and electric razors. The high energy density, portability, and lack of maintenance offset the high cost of these batteries. Provision must be made for charging secondary batteries, and the required duty cycle should include adequate charging time.

Fuel Cells

Figure 12b shows a simple schematic diagram of a fuel cell. A practical fuel-cell converter will be a complex system incorporating circulating pumps, fuel-conditioning plant and control gear. In principle, the fuel cell is reversible and can be used as an energy storage device in the same way as a secondary cell. Most effort has gone into the design of fuel cells for the direct conversion of chemical energy to electrical energy, and the fuel processing arrangements are not, in general, reversible. The practical fuel cell is comparable to the mechanical/electrical converters such as the diesel generator set.

Table 4   Secondary cells under development

	Operating temperature (°C)	Specific power (W/kg)	Specific energy (W h/kg)
Sodium—sulphur	300	45	68
Lithium—chlorine	650	18—32	68
Zinc—chlorine	25	18—27	23—34

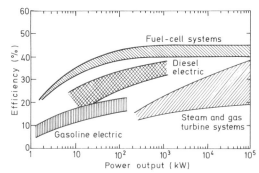

Figure 15   Comparison of power-system efficiencies. Courtesy Pratt & Whitney Aircraft

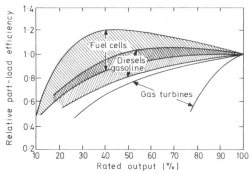

Figure 16   Comparison of part-load efficiencies. Courtesy Pratt & Whitney Aircraft

Fuel cells compare favourably with other conversion systems. They have both a high efficiency and a high part-load efficiency (*Figures 15, 16*). The exhaust is relatively clean, the device contains few moving parts, and it operates silently over a long lifetime with little maintenance.

The earliest and most developed fuel cell, the Hydrox cell, uses hydrogen and oxygen as fuel and oxidant. The Bacon Hydrox cell constructed in 1959 gave a power output of 6 kW at an efficiency of 70% and a current density of 15 A/cm^2 and a cell voltage of 0.8 V. The electrolyte was a 45% KOH solution and the cell was operated at 200°C in order to reduce polarization effects at the electrodes. The electrodes were of sintered nickel. It is possible to operate at lower temperatures ($\approx$25°C) if noble-metal catalysts such as Pt or Pd are used. A low-power, low-temperature (25°C) variant of the Hydrox cell due to the General Electric Company used an exchange resin as the electrolyte.

If atmospheric oxygen is used then it is necessary either to remove the CO_2 by a scrubbing process or to employ an acid electrolyte. Hydrogen gas as a fuel is both expensive and bulky. Much work has gone into the development of cells which will burn natural gas or liquid hydrocarbons. Various combinations of high temperature and heavy catalyst loadings have been tried. The corresponding electrolytes are acids, sulphuric or phosphoric, at a temperature of 150°C, and molten carbonates in the temperature range 350—500°C. An alternative approach to the direct use of hydrocarbon fuels which is currently under study is the conversion to hydrogen in separate plant and the conversion of the hydrogen in a Hydrox cell.

In spite of extensive development programmes the fuel cell is not yet commercially viable. The development of cheap alternative catalysts to the noble metals palladium and platinum has not yet been achieved. The Hydrox system is the exception and in the form of ion-exchange-membrane cells has been used in the Gemini space programme and the Bacon cell in the Apollo programme. A current research programme, Target, by Pratt Whitney Aircraft and a number of gas industries, is aimed at developing a system fuelled by natural gas for use in domestic installations. If successful, Target would offer a single fuel supply for heating, cooking, lighting, etc.

Application of Fuel Cells

Space Vehicles. The major advantage of fuel cells over alternative converters for this application is saving in weight. This, together with the reliability of the system, has resulted in outstanding success for the Hydrox fuel cell. For example, the cells used in the Apollo programme achieved an energy density of 880 W h/kg compared with 110—132 W h/kg of the silver—zinc cells normally used on these missions.

Stationary Power Plants. The Target programme has already been referred to; in addition to domestic, commercial and industrial use, such a system could also be used as an alternative to diesel generator sets for remote site and emergency use.

Transport. The separation of the energy-storage function from the power output have long been

recognized as a considerable advantage for transport applications. A number of experimental vehicles have been constructed. However, such an application still awaits the development of a cheap cell operating on a liquid hydrocarbon fuel.

Metal--Air Cells

In these the fuel is provided by the negative electrode and oxygen from the atmosphere is dispensed at the positive electrode which does not itself take part in the chemical reaction. Zinc, aluminium, magnesium, cadmium and sodium have all been used as the fuels, but development has concentrated on the zinc—air system. Metal—air cells can be made both as primary and as secondary cells. A further variant is the mechanical replacement of the fuel electrode.

Primary metal—air cells have high power/weight and power/volume ratios compared with other primary cells, the voltage remains constant throughout the life, and heavy currents can be drawn. The low-temperature performance is also good. It is necessary to seal these cells from the atmosphere before use and they have a relatively short shelf life, losing 80—90% of the capacity after one year; and on breaking the seal their active life is only a few months.

The characteristics of some commercial zinc—air batteries are shown in *Figure 17*; characteristics of common primary batteries are also given for comparison. Zinc—air primary batteries are used in toys, shavers, recorders, communications equipment, paging systems and hearing aids. Work is in progress on the development of high power zinc—air secondary batteries for electric vehicle propulsion.

13.5 Thermionic and thermoelectric energy conversion

Thermionic Generation

In a thermionic generator, or diode, *Figure 18*, electrons are emitted from the heated cathode (emitter) and collected by a cooler anode (collector). The electrons return to the cathode by means of an external load. Like the thermocouple, the thermionic generator is a heat engine using an electron gas as the working fluid. Its useful features are the high power density (up to several tens of watts per square centimetre) and the high-temperature heat rejection (700—1000°C). The latter is particularly valuable in space applications. Single units may be constructed from a few watts output to several hundred watts. Thermionic generators may be operated in series/parallel connection to give higher power levels.

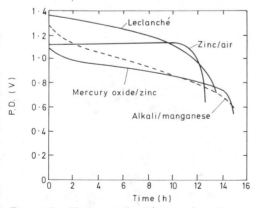

Figure 17 Characteristics of some batteries. (Energy Conversion Ltd)

References p 308

Laboratory generators have been operated at energy efficiencies of 10—18% and with development may be expected to achieve 25%. Lifetimes of greater than 10 000 h have been reported.

Space-charge Neutralization

One of the principal problems in thermionic generators is that of space charge. Electrons leaving the emitter will fill the interelectrode space with a negative charge which will tend to inhibit further emission.

In a vacuum diode in order to withdraw even low current densities, it is necessary to arrange for a very small interelectrode spacing — less than 5 μm. Thermionic generators employing such close spacings have been constructed for low power levels (a few watts) but are not practical for higher power levels, and methods of avoiding the space-charge limitation are required. The method which has received most attention is to provide positive ions in the interelectrode spacing. The simplest source of ions is obtained by filling the generator with a readily ionized vapour such as caesium at a pressure in the range 1—1500 Pa. Various ionizing mechanisms including contact with the hot emitter result in a source of positive ions.

Applications

Space-power Supply. Thermionic generators may be used in association with nuclear (fission reactors or isotope) heat sources. Solar heating is also possible. With nuclear heating the thermionic generators may be placed either inside the reactor (in core) or heat from the reactor may be piped to the generator situated outside. A considerable amount of development work has gone into this application.

Central Power Stations. Thermionic generators may be used as toppers for both nuclear and fossil-fuel-fired power stations. In both cases the economic justification is difficult to make. Problems include life time, reliability and cost.

Small Generators. There may be some application for fossil-fuel-fired generators to provide electrical power in the 100 W range. Integrity of the vacuum envelope is a problem with flame heating.

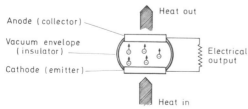

Figure 18   Principle of thermionic generator

Thermoelectric Conversion

Thermoelectric Effects

Seebeck Effect. When the junctions of two dissimilar conductors are maintained at different temperatures an electromotive force V is produced in the circuit. This effect is known as the Seebeck Effect. The Seebeck coefficient a is defined as the open-circuit voltage observed for unit temperature difference; a is also referred to as thermoelectric power:

$$a = \frac{dV}{dT}$$

Typical values for a range from 10 μV/K for metals, 100 μV/K for highly doped semiconductors and 1000 μV/K for lightly doped semiconductors.

The sign convention is such that a is positive if the cold end is positive and negative if the cold end is negative. In n-type semiconductors a is negative and in p-type semiconductors a is positive. Since the Seebeck coefficient for a thermocouple is the difference between the coefficients of the two limbs, pn junctions are normally used in order to take advantage of the summation of the two coefficients.

Peltier Effect. If a current is passed through a junction between two dissimilar conductors there is either a heating or cooling effect at the junction. This is known as the Peltier effect and its magnitude is given by the relation:

$$Q = \pi I$$

where Q is the rate of heat release or absorption, I is the current, and π is the Peltier coefficient.

Peltier coefficients range from 3 V for metals, 30 V for highly doped semiconductors to 300 V for lightly doped semiconductors.

Thompson Effect. Heat is released or absorbed when a current is passed along a conductor if a temperature gradient exists in the conductor. This, the Thompson effect, may be expressed by the relation:

$$Q' = \gamma I \frac{dT}{dx}$$

where Q' is the rate of heat released or absorbed/ unit length, I is the current, dT/dx is the temperature gradient, and γ is the Thompson coefficient.

Thompson coefficients vary from 10 μV/K in metals to 100 μV/K in semiconductors.

Kelvin Relations. The three thermoelectric coefficients are connected by the Kelvin relations:

$$\pi_{12} = a_{12} T$$

$$\frac{da_{12}}{dT} = \frac{\gamma_1 - \gamma_2}{T}$$

where the subscripts 1 and 2 refer to the two limbs of the thermocouple.

Thermoelectric Energy Generation and Refrigeration

The thermocouple is a heat engine converting heat to electrical energy. The ideal efficiency of conversion is less than the Carnot efficiency owing to irreversible effects, namely the Joule loss due to the finite electrical resistance of the conductors and the heat loss by conduction along the conductors. Thermocouples may be reversed by passing a current through them from an external source and will then act as heat pumps or refrigerators.

The advantages of thermoelectric devices include long life, reliability and absence of moving parts. The major disadvantages are low conversion efficiency, around 5% in practical generators with a possible upper limit of perhaps 12–15%; low output voltage; and direct current output. Similarly refrigerators are limited in temperature depression and coefficient of performance by the irreversible losses inherent in the device.

Thermoelectric Generation. It can be shown that the maximum efficiency of a thermoelectric generator is given by

$$\eta = \eta_c \frac{(1 + Z_m T_m)^{1/2} - 1}{(1 + Z_m T_m)^{1/2} + T_c/T_h} \qquad (8)$$

where

$$\eta_c = \frac{T_h - T_c}{T_h}$$

is the Carnot efficiency, T_h is the temperature of the hot junction, T_c is the temperature of the cold junction,

$$T_m = \frac{T_h + T_c}{2} \;,$$

and Z_m is an averaged figure of merit for the thermoelectric materials.

Thermoelectric Refrigeration. The coefficient of performance (c.o.p.) for an optimized thermoelectric refrigerator is

$$\text{c.o.p.} = \frac{T_c}{T_h - T_c} \frac{(1 + ZT_m)^{1/2} - T_h/T_c}{(1 + ZT_m)^{1/2} + 1} \qquad (9)$$

Thermoelectric Materials. Equations (8) and (9) show that it is desirable for the figure of merit Z_m to be as high as possible. Z for a single material is defined as:

$$Z = \frac{a^2 \sigma}{k}$$

where a is the Seebeck coefficient, σ is the electrical conductivity, and k is the thermal conductivity.

Semiconductor materials have higher Z values than metals and possess the additional advantage of, by p- and n-type doping, providing Seebeck coefficients of opposite sign. Z_m is defined as:

$$\frac{(a_1 - a_2)^2}{[(k_1/\sigma_1)^{1/2} + (k_2/\sigma_2)^{1/2}]^2}$$

If $a_1 = a_2$, $\sigma_1 = \sigma_2 = \sigma$, $k_1 = k_2 = k$ then Z_m reduces to

$$a^2 \sigma/k$$

a, σ and k are all dependent on doping level, as shown in *Figure 19*; Z is found to have a maximum at a doping level of about $10^{25}/m^3$; the corresponding value for a is around 200 $\mu V/K$. *Figure 20* shows Z values as a function of temperature for the best thermoelectric materials. No single pair of materials is best over more than a few hundred degrees temperature range. Bismuth telluride-type materials are selected from room temperature up to 200°C. In the intermediate temperature range up to 500–600°C materials based on lead telluride, germanium telluride, silver antimony telluride and lead tin telluride are used. For higher temperatures up to 1000°C, germanium silicon alloys are suitable.

Applications of Thermoelectric Devices

The advantages and disadvantages of thermoelectric convertors have already been referred to. Thermoelectric generators can be used with any heat source, e.g. solar, radioisotope or fossil fuel. They

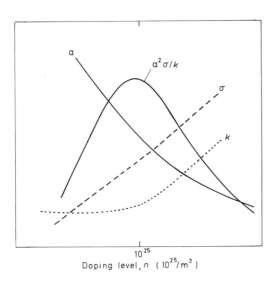

Figure 19 Curves of a, σ, k and a²σ/k against n

References p 308

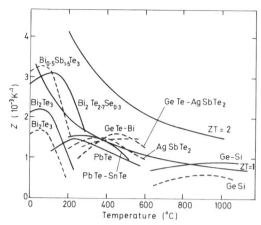

Figure 20 Figure of merit of 'best' materials. Adapted from D. A. Wright, Metallurgical Review, 1970, Vol.15, 148
———— **n** *type*, - - - - - **p** *type*

are particularly suitable for low power (<100 W_e) applications in remote locations. Such applications include power for communication satellites, underwater repeater stations and navigational buoys. Very small ($\approx mW_e$) generators are used in cardiac pacemakers. Thermoelectric refrigerators have applications in vacuum traps, cooling of infrared detectors, and similar uses.

References

1 Central Electricity Generating Board (UK), *Modern Power Station Practice*, Vols.1—8, 2nd edn, Pergamon Press, Oxford, 1970 onwards
2 Proceedings of the International Conferences on Magnetohydrodynamic Power Generation organized jointly by the European Nuclear Energy Agency and from 1966 with the International Atomic Energy Authority
3 Proceedings of the more recent conferences (see ref.2): Paris, 1964; Salzburg, 1966; Warsaw, 1968; Munich, 1971
4 MHD Electrical Power Generation — 1972 Status Report, *Atomic Energy Review* 1972, Vol.10, 301
5 Heywood, J. B. and Womack, G. J., *Open-Cycle MHD Power Generation*, Pergamon Press, Oxford, 1969
6 Musgrove, P. J. and Wilson, A. D. *Advances in Static Electricity* 1970, Vol.1, 360
7 Lawson, M. and von Ohain, H. *J. Engng for Power (Trans. ASME)* 1971 (April), p 203

Further Reading

MHD Generation

(A) Spring, K. H., *Direct Energy Conversion*, Academic Press, London and New York, 1965

(B) Sutton, G. W. and Sherman, A., *Engineering Magnetohydrodynamics*, McGraw-Hill, New York, 1965
(C) Womack, G. J., *MHD Power Generation: Engineering Aspects*, Chapman & Hall, London, 1969

Electrochemical Energy Conversion

(D) Barak, M., *Electrochemical Power Sources*, Institution of Electrical Engineers, London, 1975 (Good general introduction to electrochemical converters)
(E) Collins, D. H. (Ed.), *Power Sources*, Oriel Press (Up-to-date account of recent developments. Contains the proceedings of the International Power Sources Symposium, and is published annually)

Secondary cells

(F) Smith, G., *Storage Batteries*, 2nd edn, Pitman, London, 1971
(G) Falk, S. U. and Salkind, A. J., *Alkaline Storage Batteries*, Wiley, 1969

Fuel cells

(H) Gregory, A. M., *Fuel Cells*, Mills & Boon, London, 1972
(I) Liebhafsky, H. A. and Cairns, E. J., *Fuel Cells and Fuel Batteries*, Wiley, 1968

Metal — air cells

(J) Gregory, D. P., *Metal Air Batteries*, Mills & Boon, London, 1972

Thermionic Generation

Ref.(A)
(K) Sutton, G. W., *Direct Energy Conversion*, McGraw-Hill, New York, 1966

A comprehensive account of developments over the international field is to be found in:
Proc. Int. Conf. on Thermionic Electrical Power Generation, I.E.E. and E.N.C.A., London, Sep.1965, pubd by Instn of Electrical Engrs, London

Proc. Int. Conf. on Thermionic Electrical Power Generation, E.N.E.A. and Euratom, Stresa, May 1968, pubd by Euratom, Stresa, Italy

Thermoelectric Generation

Ref.(A)
Ref.(K)
(L) Heikes, R. R. and Ure, R. W. Jr, *Thermoelectricity: Science and Engineering*, Interscience, New York, 1961
(M) Corliss, W. R. and Harvey, D. G., *Radioisotopic Power Generation*, Prentice-Hall, Englewood Cliffs, N.J., 1964
(N) *Proc. Int. Symposium on Industrial Application of Isotopic Power Generators*, Harwell, UK, Sep.1966, E.N.E.A. and O.E.C.D.

Chapter 14

Nuclear power generation

Introductory Nuclear Physics

The Nuclear Atom

According to Rutherford's model, the atom consists of a central core or nucleus made up of protons and neutrons and surrounded by electrons. Since the protons and electrons are equal in number and carry equal and opposite charges ($+1.602 \times 10^{-19}$ and -1.602×10^{-19} coulomb respectively) and the neutrons are uncharged, the atom as a whole is electrically neutral.

The number of protons in the nucleus is known as the Atomic Number and indicated by the symbol Z, the number of neutrons in the nucleus is indicated by the symbol N, and the total number of nucleons, the collective name for neutrons and protons, is the Mass Number, indicated by the symbol A. It follows that $A = Z + N$.

The electrons around the nuclei are formed into a number of shells in accordance with Pauli's Exclusion Principle, and the configuration of the electrons, particularly those in the outer unfilled shell, determines the chemical behaviour of the atom. A chemical element can therefore be specified by its atomic number Z.

Atoms with the same atomic number (and therefore identical chemical properties) but different numbers of neutrons in their nuclei are known as isotopes. They may have very different nuclear properties. Nuclei with the same number of neutrons but different numbers of protons (and therefore different chemical properties) are known as isotones.

Nuclei can thus be identified unambiguously by their atomic number Z and their mass number A, and are usually indicated by writing the chemical symbol of the element with the mass number as a superscript. For example, the symbol ^{235}U specifies the isotope of uranium which has 235 nucleons and since uranium has 92 protons ($Z = 92$) this isotope has 143 neutrons. Similarly, ^{238}U has 92 protons and by subtraction 146 neutrons.

The atomic number Z is sometimes also indicated as a subscript e.g. $^{235}_{92}U$, $^{238}_{92}U$, but since this is already specified by the chemical symbol the information is strictly redundant.

The radius of the outer electron orbit which can be considered as the atomic radius is of the order of 10^{-10} m, and the radius of the nucleus is given approximately by $R = 1.3 \times 10^{-15} A^{1/3}$ m. The radius of a medium-weight nucleus is thus about 5×10^{-15} m. Since neutrons and protons are approximately 1838 and 1836 times as heavy respectively as electrons, it follows that most of the mass of the atom is concentrated in the very small nuclear volume which consequently has a density between 10^{13} and 10^{14} as great as the density of atoms as a whole (or of liquids and solids with which we are familiar in everyday life).

The standard of mass against which nuclear particles are measured, the atomic mass unit or amu, is equal to one twelfth of the mass of the carbon atom, and on this scale the mass of the proton is 1.007276 amu and of the neutron 1.008665 amu. 1 amu = 1.66×10^{-27} kg.

Electrons are attracted to the nucleus by the well known Coulomb forces of attraction between opposite charges $F = Ze^2/4\pi\epsilon d^2$ and occupy fixed orbits or quantized energy states which may be deduced from the Bohr postulates or from wave-mechanical considerations. The energy required to remove one of the outer or valency electrons from the atom is of the order of a few tens of electron volts, rising to a few tens of kiloelectronvolts (keV) for the inner electrons; the electrons are therefore said to have negative binding energies of these values. Their stability increases as the negative binding energy increases.

Nuclear Stability

The Coulomb forces in the nucleus are repulsive and the nuclei owe their stability to the internucleon or nuclear forces which act between the nucleons. These forces are less well understood than the Coulomb forces, but they are known to be attractive and extremely strong over their very short range, which is generally similar to nuclear dimensions (10^{-15} m). Typical (negative) binding energies of the nucleons within the nucleus as a result of these forces are of the order of 8 or 9 MeV. Nuclear forces appear to be independent of charge, although the Coulomb repulsive forces between protons are, of course, superimposed upon them. Because of their short range, nuclear forces exhibit a 'saturation effect' each nucleon interacting only with its few immediate neighbours. To a first approximation therefore, the binding energy of a nucleus would be expected to be proportional to the number of nucleons present, or, in other words, the binding energy per nucleon would be expected to be constant. This is modified in the case of the lighter nuclei, where the high proportion of nucleons near the surface reduces the average binding energy, since these cannot interact fully with neighbouring nucleons. For the heavier nuclei there is a reduction in binding energy owing to the positive energy associated with the long-range Coulomb forces increasing approximately as Z^2. The variation in the average binding energy per nucleon with mass number is shown in *Figure 1*. Remembering that high negative binding energy corresponds to high stability, we should expect the very heavy nuclei to be unstable and this is illustrated by natural radioactivity of some of these nuclei and by exothermic fission. Exothermic fusion is possible for some of the light nuclei.

For a given mass number, the stability of a nucleon is determined primarily by the ratio of its neutrons to protons (N/Z ratio), which varies from approximately unity for light nuclei to about 1.5 for the heavier nuclei where the effect of the Coulomb forces starts to be significant. With the exception of the very heavy naturally radioactive nuclei ($Z > 82$), all nuclei have at least one stable isotope and may have appreciably more. The largest numbers of stable isotopes and stable isotones are found in the so-called 'magic number nuclei' that is nuclei with 2, 8, 20, 50, 82 or 126 neutrons or protons, which form closed shells in much the same way as electrons form closed shells around the nucleons. Such nuclei usually have exceptional nuclear stability just as atoms with closed electron shells have exceptional chemical stability.

Nuclei formed with an N/Z ratio which does not correspond to stability 'decay' to a more stable nuclear system. This may take place by emission of a neutron or a proton if energy in excess of the binding energy of these particles is available. These are almost instantaneous processes.

Where the difference between the initial and final energy states of the nucleus is less than the nucleon binding energy the N/Z ratio may be changed by the β-decay process, in which a neutron within the nucleus changes into a proton plus an electron and an anti-neutrino $n \rightarrow p + e^- + \nu$, the positron emission process in which a proton changes into a neutron plus a positive electron or positron and a neutrino $p \rightarrow n + e^+ + \nu$, or by the emission of α particles which, because of their high internal binding energy, require less energy than protons or neutrons to escape from the nucleus. These processes are in general not instantaneous, the fraction of nuclei decaying within any time interval being proportional to the number of unstable nuclei present

$$\frac{dN}{dt} = -\lambda N$$

where λ is the decay constant. Integrating, we obtain the well known relation $N_t = N_0 e^{-\lambda t}$ where N_0 is the number of unstable nuclei present at time $t = 0$ and N_t the number at time t. The time for the number of radioactive nuclei to fall to $1/e$ of its initial value, $t = 1/\lambda$, is known as the mean lifetime of the activity and the time to fall to half the initial activity, $t_{1/2} = \ln 2/\lambda$, the half-life.

The Energy Balance in Nuclear Reactions. A nuclear reaction is expressed symbolically as follows: $i + T \rightarrow$ (C.N.) $\rightarrow R + e + Q$, where i is the incident particle or quantum, T the initial or 'target' nucleus, R the residual nucleus, e the particle or quantum emitted, and Q the energy change.

For low-energy reactions it is helpful to consider that the reaction proceeds via an intermediate stage, a compound nucleus C.N. formed by the capture of the incident particle subsequently splitting up into the residual nucleus and emitted particle or quantum. In all nuclear reactions, as in chemical reactions, mass is converted into energy according to Einstein's mass—energy equivalence $E = m_0 c^2$. The amount of energy liberated or absorbed is known as the Q value of the reaction. If Q is positive (energy liberated) the reaction is exothermic or exoergic, if Q is negative (energy absorbed) the reaction is endothermic or endoergic. Thus Q is given by:

Q = (initial mass − final mass) c^2
= final kinetic energy − initial kinetic energy

More specifically

$$Q = c^2 (M_i + M_T - M_R - M_e)$$
$$= (E_R + E_e - E_i - E_T)$$

Here M and E are the masses and kinetic energies of the particles.

Nuclear reactions differ from chemical reactions in that the fraction of mass converted into energy is much higher, typically about $1/10^3$ of the total mass involved in the reaction as compared with about $1/10^9$ for chemical reactions.

Nuclear reactions are often indicated more briefly by the symbols $T(ie)R$, thus the symbols $^7\text{Li(pn)}^7\text{Be}$ would indicate the reaction $^7\text{Li} + \text{p} \rightarrow {^8\text{Be}}^* \rightarrow {^7\text{Be}} + \text{n} - 1.646\,\text{MeV}$. The asterisk added as a superscript to the compound nucleus ^8Be indicates that this is formed in an energetic or excited state. The above reaction is an endothermic reaction with a Q value of $-1.646\,\text{MeV}$.

The probability that a nuclear reaction will take place is known as the reaction cross-section, and this is indicated by the symbol σ which is effectively defined as

$$\sigma = \frac{Y}{nN\theta}$$

where Y is the fractional yield of the reaction, n the number of incident particles, N the number of target nuclei per unit volume of target and θ the thickness of the target material. $N\theta$ therefore corresponds to the concentration of target nuclei per unit area of target presented to the incident beam. It will be noted that dimensionally cross-sections are in units of area and the unit area σ is conventionally expressed in terms of the barn ($10^{-28}\,\text{m}^2$), which approximately equals the physical cross-sectional area associated with a medium-sized nucleus. Classically, the reaction probability is said

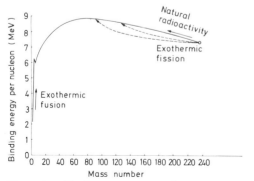

Figure 1   Binding energy per nucleon versus mass number

to be proportional to the area of an imaginary disc representing the nucleus so that cross-sections are more correctly expressed in terms of barns per nucleus. The actual physical size of the target nucleus, however, plays a very minor part in determining the magnitude of the reaction cross-section. For incident positively charged particles, particularly at low energies, the dominant factor is the Coulomb repulsive force between the incident particle and the nucleus which reduces the reaction probability by inhibiting the close approach of the incident particle, the so-called Coulomb potential barrier. No such barrier exists for incident neutrons, and the cross-sections for neutron-induced reactions in general increase as the velocity of the neutrons is reduced — the so called $1/v$ law — reaching values of hundreds or frequently thousands of barns per nucleus for neutrons which have energies of fractions of an electron volt. Charged-particle-induced interactions even for incident particle energies of a few MeV are expressed in millibarns or even microbarns per nucleus. Like whole atoms, nuclei can exist only in certain quantized energy states and for both neutrons and charged particles the probability of capture is enhanced when the energy of the incident particle is just sufficient to produce the compound nucleus in one of its permissible excited states; this is known as resonant capture.

The term capture cross-section σ_c refers to the probability that the incident particles will be captured to form a compound nucleus whilst the reaction cross section σ_{ie} refers to the combined probability that the incident particle will be captured and that the compound nucleus so formed will de-excite by the emission of the specified particle or quantum e. If there are other energetically possible ways in which the compound nucleus can de-excite σ_{ie} will of course be less than σ_c.

More specifically, the capture cross-section for the formation of the compound nucleus, σ_c, is equal to the sum of the reaction cross-sections:

$$\sigma_c = \Sigma\, \sigma_{ie_1} + \sigma_{ie_2} + \sigma_{ie_3} + \sigma_{ie_4} + \dots$$

where e_1, e_2, e_3 and e_4 are the particle or quanta emitted in all energetically possible modes of de-excitation of the compound nucleus.

In most nuclear experiments the target nucleus has effectively zero kinetic energy, i.e. is at rest in the laboratory frame of reference. It is sometimes mathematically convenient to consider that the centre of mass of the reacting system is at rest, in which case the reaction is considered in the so-called *centre-of-mass* frame of reference.

Basic Reactor Physics

The Fission Reaction

On capturing a neutron ^{235}U forms the compound nucleus ^{236}U which is unstable to fission as indicated by the following equation:

$$^{235}U + n \rightarrow\, ^{236}U\,
\begin{array}{l}
\nearrow \text{ fission-product nuclei} \\
\rightarrow 2 \text{ or } 3 \text{ neutrons} \\
\searrow \text{ fission-product nuclei}
\end{array}$$
$$+\ 205\ \text{MeV}$$

The process is exothermic, approximately 205 MeV of energy being emitted. About 180 MeV of this is initially in the form of kinetic energy of the fission-product nuclei and neutrons. This subsequently degenerates to thermal energy as these particles are slowed down in the surrounding material. As indicated earlier, the fraction of mass converted into energy is approximately one thousandth of the total mass present. The excited compound nucleus ^{236}U can be split up in a large number of ways, but asymmetric splitting is most usual to form nuclei of mass number slightly above 90 and about 140 as shown in *Figure 2*. These fission-product nuclei, having in general too many neutrons for stability, are β-active, and about 25 MeV of the energy emitted per fission is associated with the β-activity of these nuclei and the associated γ emission.

The Fission Chain

The most significant feature of the fission reaction is that it is produced by a neutron and in turn produces either two or three neutrons (more rarely one or four), but on average 2.44 neutrons per fission.

Clearly if the so-called fission neutrons strike further ^{235}U nuclei to produce fission in what we like to think of as the 'second generation' of fission processes, the possibility of a self-sustained and rapidly growing chain reaction of the type shown in *Figure 3* exists.

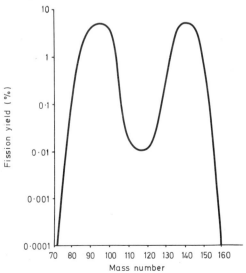

Figure 2   Fission yield versus mass number

References p 339

Perhaps fortunately, it is not so easy to produce a self-sustained chain reaction in natural uranium. The isotopic abundance of fissile ^{235}U is only 0.72%. Natural uranium consists mainly of (99.27%) ^{238}U, with a very small percentage (0.006%) of ^{234}U. On capturing neutrons the most plentiful isotope of uranium, ^{238}U, forms the compound nucleus ^{239}U which then de-excites predominantly by β-decay to form first neptunium and then plutonium:

$$^{238}\text{U} + \text{n} \rightarrow {}^{239}\text{U} \xrightarrow[\;t_{1/2}\,=\,23.5\text{ min}\;]{\beta} {}^{239}\text{Np}$$

$$\xrightarrow[\;t_{1/2}\,=\,2.34\text{ days}\;]{\beta} {}^{239}\text{Pu}$$

This reaction converting ^{238}U into ^{239}Pu, which does not occur in nature, is important because ^{239}Pu itself undergoes fission on capturing slow neutrons; so the reaction provides a method of converting the plentiful isotope ^{238}U into fissile material. It is known as the *breeder reaction* since fissile ^{239}Pu is considered to be *bred* from ^{238}U which is consequently known as *fertile* material.

All reactors convert some ^{238}U into ^{239}Pu since some neutron capture by ^{238}U is inevitable, and the conversion ratio is defined as the amount of ^{239}Pu produced divided by the amount of ^{235}U undergoing fission.

As will be seen later, it is important for the long-term utilization of nuclear power that reactors of high *conversion ratio* should be built. But in natural-uranium reactors neutron capture by ^{238}U is a source of neutron loss competitive with the ^{235}U fission process; and since there are approximately 140 ^{238}U nuclei for each ^{235}U nucleus the fission chain can only be preserved if neutrons are preferentially captured by ^{235}U.

The neutrons at fission share the total kinetic energy available and have an energy spectrum shown in *Figure 4a*, the median energy being about 2 MeV but the energy spectrum extending to about 10 MeV. As shown in *Figure 4b* the ^{235}U fission cross-section obeys a classical $1/v$ law, reaching

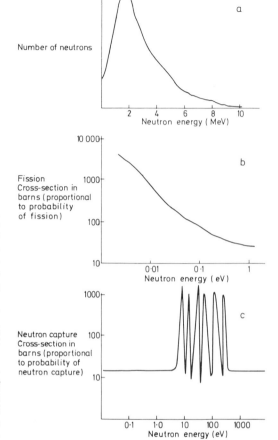

Figure 4   (a) The fission-neutrons energy spectrum. (b) Variation of ^{235}U fission cross-section with energy. (c) Variation of ^{238}U capture cross-section with energy

values of between about 600 barn per nucleus for neutron energies of 0.025 eV. At these lower energies the neutrons are near thermal equilibrium with the surrounding material and are consequently known as thermal neutrons. The cross-sections for the competitive process, capture by ^{238}U, remains comparatively low at about 10 barn/nucleus (*Figure 4c*). In the region of 10 eV to 1 keV, however, this process exhibits strong resonant capture. The problem is therefore to take neutrons in the MeV energy range and reduce their energies to a fraction of an electron-volt, where they will be preferentially captured by ^{235}U, without losing too many of them in the resonance trap of ^{238}U, and this is done by a process known as moderation. A neutron (mass — approximately 1 amu) on striking a nucleus of mass number A will in general undergo an elastic or 'billiard ball' type of collision in which some of the energy is transferred to the scattering (target)

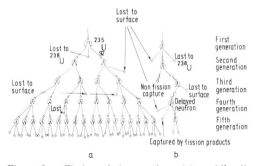

Figure 3   Fission chain reactions (a) rapidly diverging (b) near criticality

nucleus. By applying simple kinetic considerations to this interaction it is easy to show that:

$$\frac{E_{after}}{E_{before}} = \frac{A^2 + 1 + 2A \cos \phi}{(A + 1)^2}$$

where E_{after} and E_{before} are the neutron energies after and before the elastic collision and ϕ is the angle of scatter in the centre-of-mass frame of reference. For a simple head-on-collision in which the neutron rebounds along its previous path but in the opposite direction ($\phi = 180$) the expression simplifies to:

$$\frac{E_{after}}{E_{before}} = \frac{(A - 1)^2}{(A + 1)^2}$$

From this it is clear that nuclei with small A are more efficient as moderators. A neutron could for example lose all its energy in a single head-on collision with a proton. In fact the scattering angle can assume all values from 0 to $180°$, and the average numbers of collisions required to reduce neutrons in the MeV energy range to thermal energies are shown in *Table 1*. In contrast, over 2000 collisions are required with the uranium nucleus to slow neutrons down to thermal energies.

Additional requirements are that the moderator must be readily available in a dense and fairly cheap form and must not itself capture neutrons strongly. The latter requirement eliminates the use of hydrogen in the form of water since hydrogen has a capture cross-section of 0.33 barn for slow neutrons; and helium, being an inert gas, can only be obtained in a concentrated form by liquefaction. Beryllium is in general eliminated on the grounds of toxicity, high cost and scarcity, whilst lithium is chemically very reactive and the ^{6}Li isotope captures neutrons strongly (950 barn/nucleus). By elimination therefore, deuterium in the form of heavy water and carbon in the form of graphite are the only two moderating materials that are acceptable for use with natural uranium. In order that the neutron moderation should take place in a region remote from ^{238}U to reduce capture in

Table 1   Average numbers of collisions for neutron moderation

Element	Average number of collisions required for moderation
Hydrogen	18
Deuterium	25
Helium	43
Lithium	60
Beryllium	86
Boron	95
Carbon	114

the resonance trap, there is the further restriction that the reactor must be heterogeneous in design, that is the uranium and moderator must form separate volumes within the reactor.

As shown in *Figure 3*, it is convenient to consider a fission chain-reaction in terms of successive 'generations' of fission processes. The ratio of fissions in one specified generation to that in the previous generation is known as the multiplication factor K. Clearly if K is greater than unity the chain-reaction will grow and the system is said to be supercritical; if K is equal to unity the reaction will proceed at constant rate and the system is critical. When K is less than unity the reaction rate decreases and the system is said to be subcritical. An associated parameter, the reactivity of the system, is defined as $\Delta K = (K-1)/K$ and clearly ΔK positive corresponds to supercriticality, $\Delta K = 0$ refers to a critical system and ΔK negative to a subcritical system. When ΔK is positive this is sometimes called excess reactivity*.

With careful moderator and fuel design and the choice of reactor materials with low capture cross-section it is possible to construct a supercritical system using natural uranium fuel, as shown in *Figure 3b*. A qualitative estimate of the upper limit of K may be obtained from the following considerations.

Let us assume that in the first generation of neutrons under consideration n fission neutrons are produced with average energy 2.0 MeV. In the moderation process some of these neutrons will be captured by ^{238}U as discussed above, and a fraction p defined as the *resonance escape probability* will escape the ^{238}U resonance trap. We now have np neutrons at thermal energies. Some of these will be captured by the moderator and structural material of the reactor (however carefully these are chosen) and only a fraction f (known as the thermal utilization factor) of the thermal neutrons will be captured by the uranium fuel. The uranium fuel thus captures npf thermal neutrons but not all of these will produce fission. Some are captured by ^{238}U, and about 16% of those captured by ^{235}U do not produce fission. Rather more than 50% of these thermal neutrons do produce fission and each fission process produces an average 2.44 neutrons. We therefore define a *fuel utilization factor* η as the number of fission neutrons produced per thermal neutron captured by the uranium fuel (^{235}U and ^{238}U). Our initial n fission neutrons in generation 1 have therefore produced $npf\eta$ neutrons in the following generation.

Here we have considered the fission of ^{235}U only. ^{238}U is not fissile to thermal neutrons but can undergo fission on capturing neutrons of energy above 1.4 MeV. A small fraction of the fission neutrons will therefore produce fission in ^{238}U whilst being slowed down from their initial fission energy

* Excess reactivity is sometimes also defined as $\Delta k = K - 1$

to 1.4 MeV. This can best be taken into account by multiplying the number of neutrons in the second generation by a factor ϵ, the *fast fission factor*, which is slightly in excess of unity. Our n fission neutrons in the first generation have thence become $npf\eta\epsilon$ neutrons in the second and the multiplication factor becomes $npf\eta\epsilon/n = pf\eta\epsilon = K_\infty$; since we have not considered neutron losses from the surface of the system this refers to an infinitely large system.

For a carefully designed natural-uranium assembly the maximum values of p and f are about 0.9, η is 1.31 and ϵ is 1.03, giving the maximum possible value of $K_\infty = 1.09$.

In fact some neutron loss from the surface is inevitable and this will be proportional to the surface area whereas the number of neutrons produced will be proportional to the volume of the system. The fraction of neutrons lost will therefore be reduced as the size of the system is increased and K therefore also increases with size.

The minimum volume for which a chain reaction is just possible ($K = 1, \Delta K = 0$) is called the critical size of the system. This varies with the shape; since the area/volume ratio is less for a sphere than for a cube the critical volume for a spherical reactor will be less than that of a similar cubic one, and that for a cylindrical reactor between these limits.

A typical critical volume for a cubic graphite-moderated natural-uranium assembly is about 200 m³. The critical size can be reduced by fitting reflectors of light material around the core so that a fraction of the escaping neutrons is reflected back.

Turning to enriched-uranium reactors, in which the percentage of fissile material has been increased beyond 0.72% by adding separated ^{235}U or ^{239}Pu, the proportion of neutrons captured by ^{238}U is reduced, leading to an increase in p and η. The fractional neutron loss which can be tolerated at the surface is therefore increased and the critical size is reduced. It is also possible to tolerate a higher frac-

tional loss of neutrons for the constructional material of the reactor (i.e. a lower value may be accepted for f) so that these may be chosen for their strength and high-temperature performance rather than primarily for their low neutron capture cross-section.

In the extreme case of highly enriched cores (say above about 15% of fissile material) the competitive capture by ^{238}U is so reduced that moderation is unnecessary and a chain reaction becomes possible using the fast (high-energy) fission neutrons.

Reactivity Changes during Operation

So far we have considered only the factors affecting a freshly built 'clean' reactor core. As a result of operation the reactivity is affected by two main additional factors. The first is the reduction of the ^{235}U content of the fuel owing to fission, the so-called *burnup* of the fuel and its partial replacement by the other fissile isotope ^{239}Pu bred from ^{238}U, and the second is the capture of neutrons by the fission-product nuclei which are not of course present initially.

Considering first the ^{235}U burnup and ^{239}Pu production. Not all ^{235}U nuclei undergo fission on capturing neutrons, some ^{236}U is formed, and not all ^{239}Pu nuclei undergo fission, some ^{240}Pu is also formed and even heavier isotopes ^{241}Pu and ^{242}Pu. The reactions are as follows:

$$^{235}\mathrm{U} + \mathrm{n} \nearrow \,^{236}\mathrm{U}$$
$$\searrow \text{ or fission}$$

$$^{238}\mathrm{U} + \mathrm{n} \to \,^{239}\mathrm{U} \xrightarrow{\beta} \,^{239}\mathrm{Np} \xrightarrow{\beta} \,^{239}\mathrm{Pu}$$

$$^{239}\mathrm{Pu} + \mathrm{n} \nearrow \,^{240}\mathrm{Pu} \qquad ^{240}\mathrm{Pu} + \mathrm{n} \to \,^{241}\mathrm{Pu}$$
$$\searrow \text{ or fission}$$

$$^{241}\mathrm{Pu} + \mathrm{n} \nearrow \,^{242}\mathrm{Pu}$$
$$\searrow \text{ or fission}$$

Initially, in a clean reactor containing natural uranium, the increase in reactivity due to ^{239}Pu production is greater than that due to the reduction in the ^{235}U. But since the ^{235}U concentration falls steadily with time, and that of ^{239}Pu approaches asymptotically to an equilibrium value where its rate of production from ^{238}U is equal to its rate of removal owing to fission and ^{240}Pu production, the initial rise in reactivity is followed by a steady fall. *Figure 5* shows the changes in concentration of ^{235}U and ^{239}Pu with reactor operating time for a typical natural-uranium reactor with a thermal-neutron flux of about 10^{13} neutrons/cm² s and also indicates the changes in reactivity. ^{240}Pu captures neutrons quite strongly, so its accumulation leads to a steady fall in the reactivity which however is less than that due to reactor fission-product poisons. The build-up of fissile ^{241}Pu is small and has a negligible effect. With enriched reactors the ^{239}Pu

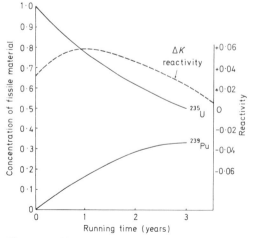

Figure 5   Changes in reactivity and concentration of ^{235}U and ^{239}Pu during reactor operation

build-up and its effect on the reactivity are correspondingly less, so that in the extreme case of a reactor containing 100% ^{235}U the reactivity would fall smoothly with burnup. In all reactors, burnup is greatest where the neutron flux is highest, near the centre of the reactor, and the effects of this can be minimized by 'shuffling', that is interchanging the inner and outer fuel elements.

Fuel elements are usually left in the reactor until about 40% of the initial ^{235}U has undergone fission (40% burnup), but this varies with reactor type.

The second main cause for reactivity change during operation is the accumulation of fission-product nuclei, some of which capture neutrons strongly. Of these, xenon ^{135}Xe is the most serious since it has a capture cross-section of 3.5×10^6 barn/nucleus, about 5000 times as great as that of ^{235}U. Although it is produced directly in only 0.3% of fission processes, it is produced indirectly by the successive β-decay of tellurium ^{135}Te which is produced in 5.6% of fissions. ^{135}Xe is itself unstable and decays first to caesium ^{135}Cs and then to stable barium ^{135}Ba.

$$^{135}\text{Te} \xrightarrow[t_{1/2} = 2 \text{ min}]{\beta} {}^{135}\text{I} \xrightarrow[t_{1/2} = 6.7 \text{ h}]{\beta} {}^{135}\text{Xe}$$

$$^{135}\text{Xe} \xrightarrow[t_{1/2} = 9.2 \text{ h}]{\beta} {}^{135}\text{Cs} \xrightarrow[t_{1/2} = 2 \times 10^4 \text{ yr}]{\beta} {}^{135}\text{Ba}$$

In fact ^{135}Xe is formed from ^{135}Te more quickly than it decays to ^{135}Cs and therefore accumulates in an operating reactor until it reaches a level where its rate of production is equal to its rate of removal by β-decay plus its removal by neutron capture (^{135}Xe + n → ^{136}Xe). ^{136}Xe does not capture neutrons strongly. When the reactor is shut down the level of ^{135}Xe increases since its half life for formation is shorter than the lifetime for decay. This can result in a ^{135}Xe build-up and a reduction in reactivity by as much as 0.1 for a high-flux power reactor, making it impossible to re-start the reactor until the ^{135}Xe had decayed away for a period of perhaps 20 h (*Figure 6a*).

The second main reactor poison, samarium, is less serious in that the capture cross-section for slow neutrons ($\sigma = 10^5$ barn/nucleus) is about 1/30 of that of ^{135}Xe and its effect on reactivity is correspondingly less. It is formed mainly from the decay of the fission product neodymium, ^{149}Nd, which is formed in 1.6% of the fission processes, via the intermediate nucleus promethium:

$$^{149}\text{Nd} \xrightarrow[t_{1/2} = 1.7 \text{ h}]{\beta} {}^{149}\text{Pm} \xrightarrow[t_{1/2} = 4.7 \text{ h}]{\beta} {}^{149}\text{Sm}$$

As with ^{135}Xe, ^{149}Sm is held at a steady level by neutron capture during reactor operation, ^{149}Sm + n = ^{150}Sm, but the process ceases when the reactor is shut down so that the ^{149}Sm level then increases, being fed by the decaying ^{149}Pm. A reactor poisoned by ^{149}Sm would not recover its reactivity as in the xenon case since ^{149}Sm is stable, but as the reduction in reactivity due to samarium poisoning at shut-down is about 0.01 for a typical power reactor (*Figure 6b*) samarium poisoning is unlikely.

Flux Flattening

Since neutrons are lost from the surface of a reactor assembly, in a uniform system the neutron flux, power density and temperature would all be higher at the centre than near the surface. In many reactors the power output is limited by the maximum temperature which the fuel elements will withstand, T_{max}, and for highest efficiency it is necessary for the average fuel temperature T_{av} to approach T_{max} as nearly as possible. This can be done by a procedure known as 'flux flattening', an attempt to obtain a uniform flux throughout the reactor assembly: by varying the distribution of fuel elements within the moderator, or by placing the control rods or additional neutron absorbers in the central regions of higher flux. With enriched reactors a similar result may be obtained by using fuel elements of higher enrichment near the edges.

A more uniform temperature throughout the reactor may also be achieved by increasing coolant flow rates in the central region, as an alternative or additional measure to flux flattening.

Reactor Control

In order to understand the problems of reactor control it is first necessary to examine the fission process and the fission chain-reaction in a little more detail. 99.35% of all the neutrons associated with the fission process are emitted spontaneously at fission, the so-called 'prompt' neutrons; the other 0.65% are emitted by fission-product nuclei which

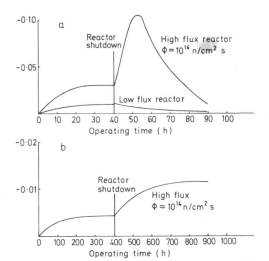

Figure 6   Reactivity changes in ΔK due to fission-product poisons (a) ^{135}Xe and (b) ^{149}Sm

have first undergone β-decay so that they are emitted after the fission process with a delay equal to the half life of the β-decay process of their precursors. It is not clear how many such 'delayed-neutron' emitters there are, maybe as many as thirty, but reactor behaviour can be explained satisfactorily on the assumption that there are six of half lives between 55.7 and 0.23 s produced in the proportions shown in *Table 2*. Of these the 55.7 s activity has been identified as that of ^{87}Br decaying by β-decay to ^{87}Kr which is the neutron emitter.

The rate at which these neutrons are released to the fission chain is given by $\lambda_1 C_1 + \lambda_2 C_2 + \lambda_3 C_3 + \lambda_4 C_4 + \lambda_5 C_5 + \lambda_6 C_6$ where λ is the decay constant and C the concentration of the precursor of the delayed-neutron group under consideration.

The time interval between generations of fission processes produced by the prompt neutrons is determined by the time taken to slow the fission neutrons down to thermal energies (in a thermal reactor) plus the average time for which thermal neutrons diffuse throughout the system before being captured: i.e. neutron lifetime = slowing-down time + diffusion time, $\approx 10^{-3}$ s for a graphite-moderated natural-uranium reactor and 10^{-4} s for an enriched liquid-moderated reactor. With a fast reactor there is no moderation, and the time from one generation of reactions to the next is about 10^{-7} s. The time taken for the actual fission process is negligible compared with these times.

Considering the prompt neutrons alone we have the rate of change in neutron population:

$$\mathrm{d}n/\mathrm{d}t = n\Delta K_p/l \tag{1}$$

where ΔK_p is the reactivity due to prompt neutrons alone. If we include the delayed neutrons the rate of change becomes:

$$\frac{\mathrm{d}n}{\mathrm{d}t} = n\left(\frac{\Delta K_p}{l}\right) + \sum_{i=1}^{6} \lambda_i C_i \tag{2}$$

$$\frac{\mathrm{d}C_i}{\mathrm{d}t} = n\frac{\beta_i}{l} - \lambda_i C_i \tag{3}$$

where β_i = delayed-neutron fraction in group i, and C_i = concentration of delayed-neutron precursors in group i.

Table 2 Delayed-neutron emitters

Isotope	Effective $t_{1/2}$(s)	Effective decay constant λ (s^{-1})	Fraction of total neutrons, β
1	55.7	0.0124	0.000215
2	22.7	0.0305	0.001424
3	6.2	0.1110	0.001274
4	2.3	0.301	0.002568
5	0.61	1.14	0.000748
6	0.23	3.01	0.000273

Let us first examine the situation when ΔK_p is positive. If we neglect the effect of delayed neutrons in (2) then the neutron population increases exponentially:

$$n_t = n_0 e^{(\Delta K_p/l)t} \tag{4}$$

where n_0 is the neutron density initially and n_t is the neutron density after time t. The neutron population will therefore increase by a factor e in a time $t = l/\Delta K_p$ which is known as the *e-folding time* or the *reactor period* l = neutron lifetime.

If we take the modest value of $\Delta K_p = 0.001$ for a natural-uranium reactor ($l = 0.001$ s) we see that after 10 s the neutron population would increase by a factor of e^{10}, i.e. by a factor of over 20 000, if this situation were not corrected during this time.

For an enriched water-moderated reactor ($l = 0.001$ s) multiplication by a factor of 20 000 would occur in 1 s, and for a fast reactor ($l = 10^{-7}$ s) in 10^{-3} s.

The usual mechanism of reactor control is to insert rods of neutron-absorbing material usually containing boron ($\sigma = 4000$ barn/nucleus for ^{10}B) or cadmium ($\sigma = 20\,000$ barn/nucleus for ^{113}Cd) which remove a fraction of the neutrons from the fission chain. Deeper insertion of the rods thus reduces ΔK, extraction increases ΔK. In some water-moderated reactors the reactivity is controlled by varying the level of the moderator thus varying the degree of moderation and of selective neutron capture by the fissile material, or by adding neutron-absorbing solutions.

These are relatively slow processes and could not easily be made to respond to reactor periods of a small fraction of a second. The control of a fast reactor with a reactor period of the order of 10^{-4} s would be even more difficult. In operation therefore the reactivity due to the prompt neutrons alone must be kept to negative values, that is the 'prompt' critical situation must be avoided by insertion of absorbers. The total reactivity in operation is thus limited to less than $\Delta K = 0.0065$.

Under these circumstances the first term in Equation (2) is negative and $\mathrm{d}n/\mathrm{d}t$ only becomes positive when we consider the terms involving the delayed neutrons. The reactor period under these circumstances is governed mainly by the half lives of the precursors to delayed-neutron emitters and is of the order of seconds, allowing the reactor control mechanisms adequate time to correct any unwanted change in power level. If the total reactivity ΔK is much smaller than 0.0065 the power rise is slow and an approximate solution to Equations (2) and (3) is

$$n_t = n_0 e^{\Delta K t/(l + \Sigma(\beta_i/\lambda_i))} \tag{5}$$

which has a period of

$$(l + \Sigma(\beta_i/\lambda_i))/\Delta K$$

where

$$\Sigma \, \beta_i/\lambda_i \approx 85 \text{ ms}$$

Hence the delayed neutrons exhibit a large damping effect and the prompt-neutron lifetime is unimportant. As ΔK increases through prompt-critical the reactor period varies between the limits of Equations (5) and (4).

Very approximate values for the reactor periods as a function of ΔK for the three types of reactors considered are shown in *Table 3*.

Thermal Reactors

The following Sections describe the main characteristics of existing and projected nuclear power-generation systems. Burnup figures quoted are not in the more usual megawatt-days per tonne (MWD/t) which is a guide to the life of a fuel charge and to irradiation, but not to total fuel utilization which depends on the initial enrichment. For comparison purposes the burnup has been converted to percentage utilization of natural uranium, which takes into account the total uranium fed to the enrichment plant and includes burnup of ^{235}U and burnup of ^{238}U after conversion to plutonium.

Complete burnup of all the ^{235}U in natural uranium (enrichment 0.72%) is equivalent to 5500 MWD/t and complete burnup of all the ^{235}U in 4% enriched fuel is equivalent to 30 800 MWD/t. As outlined in the Section on reactor physics, plutonium is produced from neutron capture by ^{238}U and partial burnup of this increases the fuel life. In general the spare-plutonium production rate is higher for low burnups and low enrichments.

Gas-cooled Reactors

The first reactor to achieve criticality, in 1942, was natural-uranium fuelled and graphite moderated. By 1955, with the first of the Calder Hall reactors, the capabilities of this system for large-scale power production had been demonstrated. From then until 1971 a series of power stations based on this design have been brought into operation in Great Britain with a total generating capacity of 5300 MWe.

These reactors have many features in common. The moderator is graphite, the fuel metallic natural uranium clad in Magnox, a magnesium alloy which has given the colloquial name to this reactor type, and the coolant is high-pressure carbon dioxide circulated in a closed circuit through the core. Steam is raised in carbon-dioxide/water heat exchangers and so electricity is generated via conventional steam turbines and alternators. The natural-uranium-metal fuel rods are typically 28 mm in diameter and 1 m long and the Magnox fuel cans, which prevent corrosion of the uranium in the hot carbon dioxide atmosphere and also prevent release of fission products into the coolant, have deep transverse fins to aid heat transfer. Single rods are stacked end-on in vertical coolant channels of about 100 mm diameter and on a square pitch of about 200 mm in the graphite. Between 1600 and 6000 coolant channels exist in the cores of the various reactors. Hence it is possible to remove one complete channel without markedly affecting the criticality of the system and minor control rod movements can cater for this. Consequently all except the earliest of the British Magnox reactors are designed for on-load fuel changing, thus enabling the system to be used at a high load factor since no shutdown time is needed for refuelling. Control is by means of boron steel neutron-absorbing rods.

Each of the four Calder Hall reactors[1] and the four virtually identical Chapelcross reactors owned and run by British Nuclear Fuels Ltd has a thermal power output of 262.5 MW$_t$ and gross/nett electrical outputs of 60.5/50.5 MW$_e$ from 110 t of natural uranium, giving a thermal efficiency of 18.8%. The carbon dioxide coolant has a pressure of 800 kPa (8 bar) with inlet/outlet temperatures of 150/345°C. It is contained by a mild-steel pressure vessel 51 mm thick in the form of a cylinder with hemispherical ends 11.3 m in diameter and 21.6 m high. The active core is 9.4 m diameter and 6.4 m high and the average power density is 0.45 kW$_t$/l (MW/m^3). Four coolant loops circulate the hot gas through heat-exchangers to raise steam for two 30 MW dual-cycle turbines at 320°C/1.55 MPa and 190°C/340 kPa. These reactors are not designed for on-load refuelling, and fuel changing is carried out at approximately 12-monthly intervals under normal operating conditions. As in all the Magnox reactors, uranium utilization is about 0.5%. This limit is set both by loss of multiplication factor and by irradiation damage to the metallic fuel and can. *Figure 7* shows the main features of a Magnox reactor.

The first CEGB nuclear power stations at Berkeley and Bradwell show considerable development from the Calder Hall design both in power output and efficiency. At Berkeley[2], uranium loading is doubled at 231 t but thermal power has increased to 620 MW$_t$ with electrical output 160/151 MW$_e$ and an efficiency of 24.6%. This improvement in performance was brought about by increasing the coolant pressure to 1 MPa and the pressure-vessel thickness to 76 mm. Coolant temperatures are virtually unaltered from the Calder Hall design at 160°C inlet/345°C outlet, but the steam pressure has increased to 2.2 MPa with temperature still at

Table 3 Reactor period as a function of reactivity

Total reac- tivity, ΔK	Prompt reac- tivity, ΔK_p	Neutron lifetime, l (s)		
		10^{-3}	10^{-4}	10^{-7}
		Reactor period (s)		
0.001	−0.0055	60	60	60
0.003	−0.0035	10	10	10
0.0045	−0.0020	2.5	2.5	2.5
0.0065 (Prompt critical)	0.0000	0.8	0.2	0.1
0.01	+0.0035	0.3	0.03	3×10^{-4}

320°C. The pressure vessel is larger, 24.2 m high and 15.2 m in diameter, with active-core 13.1 m in diameter and 7.4 m high. Power density is still similar to Calder Hall at 0.55 kW$_t$/l. Eight coolant circuits and heat exchangers supply steam for two 83 MW$_e$ turbines. The complete station contains two reactors of this type and so has a nett electrical capability of 302 MW$_e$. It has proved very reliable since commencing full-power operation in 1962 and has averaged a load factor of 78% since then.

Further detailed developments to the Magnox reactor contained in a steel pressure vessel took place culminating in the station Sizewell A having two reactors each of 290 MW$_e$ nett electrical output from a core containing 321 t of natural uranium. This is contained in a spherical steel pressure-vessel of 19.4 m diameter and 105 mm thick containing the carbon-dioxide coolant at a pressure of 2 MPa. By raising coolant outlet temperature to 410°C and using a dual steam cycle with temperatures and pressures of 394°C/4.65 MPa and 397°C/2.05 MPa, the efficiency was raised to 30.5%.

To make further increases in efficiency it was necessary to increase coolant pressure and reactor power-output. This would have necessitated even larger and thicker steel pressure-vessels, but the last two Magnox stations (Oldbury and Wylfa) show a very significant change in design. All reactors need very thick concrete shielding around the core against the radiations emitted from the reactor. By including steel prestressing tendons in the concrete and lining it with a thinner steel lining to make the system gastight, the biological shield itself can be made to take the forces due to the gas-coolant pressure. The final Magnox station at Wylfa[3] has two reactors each containing 595 t of natural uranium and generating 590 MW$_e$ nett electrical power at an efficiency of 31.4%. Coolant pressure was raised to 2.8 MPa and the interior of the prestressed concrete pressure-vessel has an internal diameter of 29.3 m. Power density is 0.9 kW$_t$/l on average. Coolant outlet temperature is 414°C and

steam temperature and pressure 401°C/4.68 MPa. Uranium utilization is again about 0.5% of natural uranium, but the low burnup obtainable with metallic fuel gives better spare-plutonium production than the slightly enriched gas-cooled reactors and the light-water reactors. Plutonium production rate from Magnox reactors is about 550 kg/yr for 1000 MW$_e$ electrical output and this may prove very significant in starting a fast-breeder-reactor programme.

The upper limit to the operating temperature (and hence thermal efficiency) of Magnox is set by the need to use metallic fuel, since it is necessary to keep the fuel-centre temperature below the phase-transition temperature of 660°C. Some margin of safety is necessary and in the Wylfa reactors the fuel-centre temperature is limited to 569°C and the maximum can-surface temperature to 451°C. Further increases in operating temperature can only be made by a change of fuel form. The Advanced Gas-cooled Reactor (AGR) is the next generation and shows many changes from the Magnox design. Use of UO$_2$ fuel slightly enriched in ^{235}U and clad in a thin stainless-steel can is the most significant change. Oxide fuel can be taken to much higher temperatures than metallic uranium, thus enabling the gas-outlet and steam temperatures to be raised. A 40 MW$_e$ electrical output AGR has been operational at Windscale since 1963[4] and has provided design and test information for large-power-output systems. Five two-reactor stations of this type are under construction (1974) with a planned nett electrical output of 6182 MW$_e$[5].

The stations of the AGR series resemble each other more closely than do those of the Magnox series and so the general features can be described without special reference to any one station. Each reactor has a thermal power output of 1500 MW$_t$ obtained from 120 t of UO$_2$ enriched to about 2.1% in ^{235}U, and the nett electrical output is 620 MW$_e$ and thermal efficiency 41.6%. Cooling is by carbon dioxide pressurized to 4.2 kPa and with an outlet temperature of 650°C, leading to steam at 538°C and 16.3 MPa, with one 660 MW$_e$ turbine per reactor. The active core is only 9.3 m in diameter and 8.2 m high which compares with 17.4 m diameter and 9.1 m high for the Wylfa Magnox reactors which have a similar electrical output; consequently the power density is about three times as great at 2.7 kW$_t$/l. A prestressed-concrete pressure-vessel is used and this contains the core, gas circulators and heat exchangers. Improvements in temperature and efficiency are mainly due to the use of oxide fuel. Since UO$_2$ has a poor thermal conductivity compared with metal the diameter of a fuel rod is about half that of a Magnox rod, being made of 14.5 mm diameter pellets of UO$_2$ contained in a slightly ribbed can of stainless steel 0.38 mm thick. In order to obtain an adequate heat output from each coolant channel a fuel element is composed of a cluster of 36 rods arranged in rings of 6, 12 and 18 inside a graphite sleeve of 190 mm internal diameter.

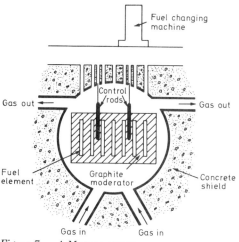

Figure 7   A Magnox reactor

Figure 8 shows a cross-section of an 18-rod cluster used in the Windscale AGR. Stainless-steel grids support the rods in the element which are about 1 m long, and eight elements are arranged on a stainless-steel tie-bar so that they can be inserted and withdrawn from a coolant channel as a single unit. Because of the large diameter of the compound fuel element, the pitch spacing in the graphite moderator is over twice that in a Magnox core at 460 mm square and the number of fuel channels is around 300 compared with about 6000 in Wylfa. On-load refuelling facilities have been retained, but since only one fuel channel per reactor needs to be changed in three days, the two reactors are closely linked and one charge—discharge machine is adequate for both reactors.

The need for enriched fuel is due to higher thermal-neutron capture in the stainless-steel cans, increased absorption in ^{238}U resonances owing to use of multi-rod oxide fuel, and decreased fast fission rate owing to use of small-diameter oxide fuel. The conversion ratio of ^{238}U to plutonium is about 0.44 compared with 0.8 in Magnox systems, so additional enrichment is needed to obtain adequate burnup, which can be about five times as great as Magnox in terms of MWD/t owing to the much greater resistance of oxide fuel to radiation damage. Maximum permitted fuel-centre temperature is 1500°C (fuel melting point 2800°C) and the maximum can temperature is 825°C. Two enrichment zones are used in the core, 2.1% for the inner zone and 2.6% for the outer zone in order to obtain more uniform heat generation across the core. Graphite moderator temperature is limited to around 350°C to reduce differential dimensional changes due to irradiation (dose is about five times that in a Magnox system), accumulation of Wigner energy, and the effects of radiation-induced C/CO_2 reactions. This is achieved by diverting part of the cool inlet gas at about 300°C through the graphite before mixing with the main stream before entering the fuel channels. Control of reactor power and shutdown is by boron-steel rods as for Magnox systems.

The risk of coolant loss is much less for a pre-stressed-concrete pressure-vessel with internal heat exchangers and gas circulators than for a steel pressure-vessel with external cooling circuits. Thus the risk of depressurization for an AGR, which would be accompanied by fuel overheating owing to the intense fission-product activity, and consequent release of activity is very low. This inherent safety in the system is demonstrated by the siting of two of the stations, Heysham and Hartlepool, close to large centres of population. The remaining three stations, at Hunterston, Hinkley Point and Dungeness, are sited adjacent to existing Magnox stations.

The AGR concept has raised the thermal efficiency of the gas-cooled graphite-moderated system to equal that of recent fossil-fuel stations, and it makes use of similar generating equipment. Uranium utilization is better than that of Magnox, being equal to about 0.8% of the natural uranium, but the spare plutonium production rate is only about 0.4 of that for a comparable Magnox station owing to lower conversion ratio and higher fuel burn-up. Unfortunately there have been considerable delays in the construction of the planned AGR stations in the UK, to some extent as a result of corrosion problems.

Any further large increases in gas temperature require the fuel to be in a more finely divided form to keep the fuel-centre temperature well below its melting point; and also higher-temperature cladding materials are needed. The development of coated-particle fuel has enabled the next step in temperature to be taken. Coated-particle fuel as developed in the UK consists of a small sphere of uranium dioxide surrounded by layers of pyrolytic carbon, silicon carbide and carbon and having a total diameter of about 800 μm. The carbon and silicon carbide layers form an impervious pressure vessel around the fissile kernel and thus take the place of the conventional canning. A small nominal 20 MW heat-output reactor, Dragon, owned by OECD, has been operational since 1966 at Winfrith to test the concept of a high-temperature gas-cooled reactor. This has a graphite core 1.6 m high of 1.07 m diameter made up of graphite blocks containing annular rings of coated particles bound by graphite and cooled by helium. The coated particles have a maximum surface temperature of 950°C with a maximum centre temperature of 1500°C, helium-gas outlet temperature 750°C, and coolant pressure 2 MPa. Because of the small core size, highly enriched fuel is necessary. This reactor serves as a test facility and from experience gained with it designs for large power reactors of this type have been formulated[6]. So far no orders have been placed in the UK for such systems but the general fuel and core design is now well established. Annular compacts of particles and graphite are placed in cans of graphite and cooled internally and externally. Graphite cans are used to prevent erosion of the coated-particle fuel compact by the gas stream and also partially to contain fission products released from

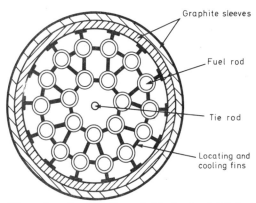

Figure 8   Cross-section of Windscale AGR fuel element

Graphite sleeves

Fuel rod

Tie rod

Locating and cooling fins

failed particles. Several of these elements are inserted in holes in a graphite-moderator block, and fuel changing is effected by removal of the complete moderator/fuel block. Helium coolant is used in preference to carbon dioxide to prevent chemical reaction, since the graphite moderator cannot be kept to low temperatures in this design. *Figure 9* shows schematically a possible HTR.

In the United States there is also interest in this type of reactor and one 42 MW$_e$ system has been operational at Peach Bottom since 1967. A much larger single reactor station at Fort St. Vrain is scheduled to become operational in 1974 and six more high-temperature reactors are planned. All these are designed to use conventional steam-generating plant and turbines and so the thermal efficiency remains similar to that of the AGR design. In comparison with the UK coated-particle fuel, the U.S. fuel has a core of ^{235}U carbide and thorium carbide. The Fort St. Vrain reactor is to use two types of particles: ^{235}U and Th carbides, overall diameter about 500 μm, and thorium carbide only, overall diameter about 800 μm. The thorium is converted into ^{233}U, another fissile isotope of uranium which does not exist naturally, by the following chain:

$$^{232}Th + n \rightarrow \,^{233}Th \xrightarrow{\beta} \,^{233}Pa \xrightarrow{\beta} \,^{233}U$$

^{233}U is in some ways a better fuel for thermal reactors than ^{235}U since the proportion of neutron reactions which do not lead to fission is much lower than for ^{235}U, 9% instead of 16%. Different particle sizes are used to enable the ^{233}U-rich particles to be separated from the depleted particles by screening. ^{233}U can then be separated and incorporated into fresh fuel, so that the ^{233}U is returned to the reactor, reducing the supply of ^{235}U needed. For the initial core there will be 19.5 t of thorium and 0.88 t of ^{235}U in a core 4.7 m high and 6 m in diameter with an average

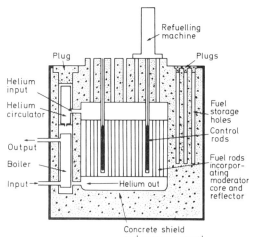

Plug

Refuelling machine

Plugs

Helium input

Helium circulator

Fuel storage holes

Control rods

Output

Boiler

Fuel rods incorporating moderator core and reflector

Input

Helium out

Concrete shield and pressure vessel

Figure 9   High-temperature gas-cooled reactor

rating of 6.3 kW$_t$/l (6.3 MW$_t$/m^3) and a thermal power of 842 MW$_t$.

The fuel pellets are compacted with graphite into rods of 12.7 mm diameter and 210 of these are inserted directly into a hexagonal graphite prism 360 mm across the flats. This graphite prism also contains 108 coolant holes parallel to the fuel rods, and the coolant stream does not directly contact the fuel compact. The complete core is built up from these graphite prisms. Helium is used as the coolant at a pressure of 4.93 MPa. It enters the core at 400°C and emerges at 785°C. Maximum fuel-centre temperature is 1260°C. Steam is raised at 538°C and 16.9 MPa to drive a single 342 MW$_e$ turbine. Nett electrical output is 330 MW and the thermal efficiency 39.2%. Off-load refuelling is planned to take place once yearly, changing about one-sixth of the core at a time. The system is contained in a prestressed-concrete pressure-vessel. These figures are design figures for the Fort St. Vrain reactor station which has not become operational at the time of writing. From information released it is difficult to estimate a value for fuel utilization.

Since the planned reactors of this type use conventional steam-generating plant and steam turbines they do not utilize the full potential of the system. Higher gas-outlet temperatures are possible and it is likely that the high-temperature reactor could be developed to produce electricity in a direct cycle by using a gas turbine. Very large gas turbines would be necessary compared with current practice, but significantly higher generation efficiencies would be obtained.

An interesting concept has resulted from a joint study by British Steel and the UKAEA which is to use a high-temperature reactor partially for process heat in reducing iron ore to metallic form and partially for electricity generation to supply an arc furnace for producing steel as an end-product[7]. The scheme is based on a reactor similar in general design to Dragon, having a thermal power of 2208 MW$_t$. Cooling is by pressurized helium and the power density has been downrated by 20% to give a helium outlet temperature of 950°C without raising the fuel-centre temperature. Process heat is used in the form of superheated steam at 900°C to produce hydrogen from methane in a steam reformer and to heat the hydrogen. 734 MW is used in this way. The remaining 1474 MW is also used to raise steam to drive a steam turbine and so produce 600 MW$_e$ electrical output for operating the arc furnace, thus utilizing a total of 1334 MW or 60.5% of the total heat output (2208 MW).

Conversion of iron ore to iron proceeds in four stages. First the crushed ore is preheated to 900°C by burning some of the hydrogen with air. Secondly the ore is reduced to FeO by reacting with hydrogen in a fluidized bed at 750°C. Stage three reheats the FeO to 900°C by burning hydrogen with air, and stage four reduces the FeO to iron sponge (92.5% iron) by reacting with hydrogen in a fluidized bed at 700°C. The iron sponge is compressed into billets and converted to steel in the arc furnace. Estimated annual output is 5.4 Mt of machined ingots, and estimated costs are slightly less than for conventional methods of steel production. Also the system does not consume coal or coke as a reducing agent, although it does require a methane supply (natural gas).

The comparatively small volume of fuel required by nuclear reactors in general for a given heat output also makes them of great potential interest for use in desalination plant, particularly in remote coastal regions.

Liquid-cooled Reactors

An alternative coolant to carbon dioxide or helium is water. Ordinary (light) water or heavy water (deuterium oxide, density about 10% greater than light water) may be used, and since both of these are good moderators the coolant and moderator may be combined. Enriched uranium must be used with light-water moderators since the hydrogen is a parasitic absorber of neutrons, but it is possible to use natural-uranium fuel with a heavy-water moderator. Heavy water is expensive since it is concentrated from ordinary water (in which it is present to 0.015%) by electrolysis, the rate of decomposition of H_2O being rather greater than of D_2O. Heavy water is mainly produced in Canada, USA and Norway, using their abundant hydro-electric power to operate the separation plant and both Canada and Norway have shown interest in heavy-water-moderated reactors.

Canada has based its nuclear power programme mainly on the CANDU type of reactor, *Figure 10*. This uses natural-uranium oxide fuel and is cooled and moderated by heavy water. The heavy-water coolant has to be under high pressure to prevent boiling, and steam is raised from ordinary water in heat exchangers. Seven reactors of the pressurized heavy-water type are in use in Canada and their total nett electrical output is 2300 MW_e. Four more are under construction and thirteen are planned. There is also one boiling-heavy-water reactor of 250 MW_e output but with this type there is a risk of considerable heavy-water loss through unavoidable steam leaks, which adds to the running costs of the system. A feature of all these reactors is that the pressurized coolant is contained in tubes surrounding the fuel elements and the bulk of the moderator is at low pressure and temperature in a calandria, which is a vessel pierced by tubes arranged on a regular lattice through which the pressure tubes pass. These tubes must be made of material which is a poor neutron absorber; zirconium alloys have proved most suitable for this purpose.

The Douglas Point CANDU reactor[8] has been operational since 1968. This is a single reactor station with a gross/nett electrical output of 220/208 MW_e and a thermal power of 701 MW_e, giving a thermal efficiency of 29.5%. It has a stainless steel calandria of 6 m diameter, 5 m long, with 306 horizontal zircalloy tubes 1.3 mm thick on a 229 mm square pitch. Each of these tubes contains a pressure-tube fuel channel 100 mm in diameter and 4 mm thick containing the fuel rod bundles and heavy-water coolant. The fuel clusters consist of nineteen rods of natural-uranium dioxide pellets of 15 mm diameter clad in 0.7 mm zirc-alloy. Each cluster is 495 mm long and ten of these fill a fuel channel. Total weight of uranium in the core is 49.5 t, core size is 4.5 m in diameter by 5 m long, and the average power density is 8.7 kW_t/l (MW_t/m^3). It is interesting to compare this with the Berkeley Magnox station described previously which has a similar thermal power. Although the lattice pitch is similar the core volume is much smaller and the power density much higher because (heavy) water is a better coolant than pressurized carbon dioxide. Heavy-water coolant enters the fuel channels at 249°C and leaves at 293°C. It is pressurized to 8.7 MPa to prevent boiling. Steam is raised via a heat exchanger at 250°C and 4.1 MPa and feeds a single 220 MW_e turbine. About 150 t of heavy water is contained in this system, and since the bulk of the moderator is separated from the coolant some novel control methods are possible.

For start-up or shut-down the heavy-water level in the calandria is varied to alter neutron leakage and so the multiplication factor. Boron oxide, which is a neutron absorber, or poison, is dissolved in the moderator and its concentration is varied to compensate for the effects of fuel burnup and fission-product build-up. Four conventional neutron-absorbing rods of stainless steel are used to control power variations and to suppress spatial oscillations in power. Finally this particular reactor uses eight booster rods containing ^{235}U to increase multiplication following a shut-down and so to overcome the effects of xenon fission-product poison build-up. Fuel utilization is about 1.1% of

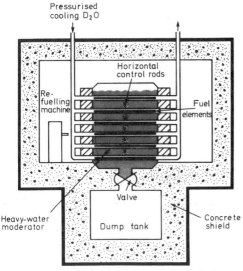

Pressurised cooling D_2O

Horizontal control rods

Re-fuelling machine

Fuel elements

Valve

Heavy-water moderator

Dump tank

Concrete shield

Figure 10   CANDU reactor (Douglas Point)

natural uranium and on-load refuelling is used. Heavy-water losses must be kept to a minimum owing to its high cost and it must not become contaminated with ordinary water, which would increase the parasitic neutron absorption.

Ordinary water is plentiful and cheap, it is a good coolant and a good moderator since it contains a high proportion of hydrogen. As already explained however, enriched uranium must be used to obtain a critical system. The slowing-down length for fission neutrons is small compared with heavy water or graphite and consequently a water-moderated reactor can be made very small without losing a high proportion of neutrons by leakage from the core. Also this means that the fuel rods must be closely packed, the volume of water typically being only twice the volume of uranium in the core. Therefore the pressurized-tube system of CANDU cannot be used and complete core must be contained in a pressure-vessel. Originally, compact water-moderated reactors were designed in the USA for submarine propulsion. Following from this the size was scaled to be suitable for land-based power stations, and the USA has mainly based its nuclear power programme on light-water reactors, both pressurized (indirect cycle) and boiling (direct cycle)[9]. These systems are generally similar in core design and vary principally in the operating pressure of the coolant and moderator. The pressurized-water reactor (PWR) (*Figure 11*) operates at a high enough pressure to prevent boiling and steam is raised indirectly via a heat exchanger. In the boiling-water reactor (BWR) (*Figure*

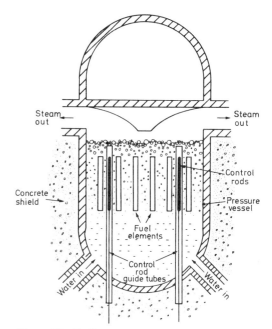

Figure 12   *Boiling-water reactor*

12) operating pressures are lower, boiling is permitted in the upper part of the core, and steam is raised directly. Moderator—coolant temperature must be kept well below the critical temperature for water ($374°C$) and consequently the thermal efficiency is limited to about 35%.

The characteristics of both PWR and BWR will be illustrated by considering one example of each type that has come into operation recently. For the PWR the Maine Yankee nuclear power station which reached full power operation in 1973 will be taken[10]. This is a single-reactor station with a nett electrical output of 793 MW_e (830 MW_e gross), a thermal power of 2440 MW_t, and a thermal efficiency of 32.5%. Core size is 3.6 m high and 3.5 m in diameter, giving an average power density of 75 kW_t/l (MW_t/m^3). This high power density is possible because of the water coolant and the close packing of the fuel. Fuel rods are made from pellets of uranium dioxide 9.64 mm in diameter and clad in zircalloy-4 cans of 0.66 mm wall thickness. A fuel element consists of 176 of these rods on a 14 mm lattice pitch and the elements are close packed to form the core. Maximum can temperature is $347°C$ and the maximum permitted fuel-centre temperature is $2005°C$, much in excess of the $1500°C$ limit for AGR. The total core loading is 87 t of uranium at three enrichments for more uniform power distribution, 2%, 2.4% and 2.9%. Light-water moderator—coolant enters the core at $282°C$ and leaves at $310°C$ and the system is pressurized to 15.8 MPa. A thick steel pressure-vessel is therefore needed and is a cylinder with spherical end caps of height 12.5 m and diameter 4.4 m, thickness 216 mm.

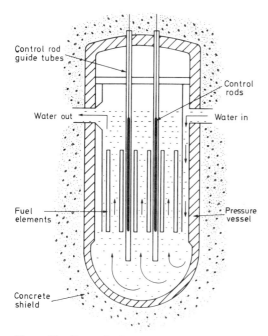

Figure 11   *Pressurized-water reactor*

Three external coolant loops containing heat exchangers and pumps are used, and steam is raised at 260°C/4.54 MPa to drive a single turbine. Reactor control is conventional by means of neutron-absorbing rods containing boron carbide. Fuel-element life in the core is about three years, and since on-load refuelling is difficult with a pressurised water system, one third of the fuel is changed off-load once a year. This necessitates a shutdown of about one month a year for the operation.

Owing to the close packing of the fuel, neutron capture in the ^{238}U resonances is higher than in AGR, which has a similar enrichment, and the enhanced plutonium production extends the fuel life. Uranium utilization amounts to about 1.1%, and spare-plutonium production rate from the PWR type of reactor is about 330 kg per annum from a 1000 MW$_e$ system.

The Pilgrim 1 BWR has been in operation since 1972[10] and has a nett output of 664 MW$_e$ (691 MW$_e$ gross), a thermal power of 1998 MW$_t$ and an efficiency of 33.2%. It might be expected that a BWR has slightly greater thermal efficiency than a PWR since there is no temperature drop across a heat exchanger. Core dimensions are 3.64 m high and 4.14 m diameter, which gives an average power density of 40 kW$_t$/l, about half that for an equivalent PWR. A lower power density is necessary since boiling is permitted and it is necessary to prevent burnout of the fuel owing to blanketing by steam bubbles. Fuel rods are again made from pellets of uranium dioxide, 12.7 mm in diameter, clad in 0.81 mm thick zircalloy-2 and arranged on a square pitch of 18.7 mm. 49 rods each 3.66 m long are contained in a fuel element and 508 fuel elements are close packed to form the core, which contains 112 t of uranium enriched to 2.12%. Maximum can temperature is 294°C and fuel-centre temperature maximum is 2300°C which is 800°C more than with the AGR. Light-water moderator and coolant enter the core at 276°C and the core is pressurized to 7.2 MPa. Steam supplied to the turbine is at 283°C and 6.8 MPa. Control of power and shut-down is by absorber rods containing boron carbide. As for the PWR, refuelling is carried out off-load, at 12-monthly intervals assuming full load factor, and one third of the core is replaced during the 30-day shut-down period. Uranium utilization is predicted to be within the range 0.8% to 1.1%, and spare-plutonium production rates are very similar to those in the PWR.

Future plans for light-water-moderated PWR and BWR systems include larger reactors up to 1300 MW$_e$ output. These are designed to use a steel pressure-vessel of similar size to those described for systems of about half this electrical output by increasing the power density. Further increases in power output would almost certainly necessitate a large core and larger-diameter and thicker pressure-vessels. The risk of failure of a thick steel pressure-vessel is much greater than for a steel-lined pre-stressed concrete system. For typical PWR and BWR vessels the critical crack size for sudden failure is less than the wall thickness so there would not be any warning leak before failure. This sudden failure would release large quantities of superheated water, resulting in high pressures and possible damage to the core. Unless a reliable and effective emergency cooling system is available the intense heat due to radioactive decay of fission products could melt the canning and fuel and so release large amounts of radioactive material from the core. With light-water reactors it is thus necessary to make provision for emergency cooling, such as a water spray to drench the core and to provide containment. Typical methods are a large containment building which will withstand several atmospheres overpressure (as for Maine Yankee), or a close-fitting auxiliary pressure-vessel with water-filled compartments to condense the steam. Off-load refuelling is general for the PWR/BWR types of reactor and necessitates a shut-down about one month long every twelve months. Therefore the maximum possible load factor cannot exceed 92%.

One method of improving the fuel utilization is to be tested in the Shippingport PWR at a future date. This is to use ^{233}U as fuel and ^{232}Th as the fertile material. Both materials would be in oxide form but the core would consist of blanket elements containing $^{232}ThO_2$ with several distinct regions of seed elements containing $^{232}ThO_2$ and a high proportion of $^{233}UO_2$. Most of the power will be produced in the seed regions, but since these are small with a large surface-to-volume ratio there will be substantial leakage of neutrons into the general blanket region. Leakage will be much greater than for a cylindrical core surrounded by a blanket and so the conversion ratio will be high. Control will be exercised by moving the seed elements partially in or out of the core. This will give a higher conversion ratio than by using absorbing control rods. The $^{232}Th/^{233}U$ breeding cycle has been chosen since it offers the best possibility in thermal reactors of a conversion ratio greater than unity. Estimates of conversion ratio for this type of reactor, sometimes called the Light-Water Breeder Reactor (LWBR), are slightly greater than one. Such a system should therefore give greatly improved fuel utilization compared with other existing thermal reactors, by making use of the large reserves of thorium.

Another type of water-moderated and cooled reactor is the Steam-Generating Heavy-Water Reactor (SGHWR) which is a combination of CANDU and BWR. Like CANDU the heavy-water moderator is contained in a calandria at low temperature and pressure. Light water, however, is used as the coolant and flows through pressure tubes inserted through channels in the calandria. In addition the water coolant is allowed to boil in the vertical coolant channels so that steam is produced directly from the core without the necessity for heat exchangers. Pressure-tube reactors are easily scaled to large power output simply by increasing the core size and hence the number of coolant channels. Pressure is contained by the coolant-channel pressure-tubes rather than by a pressure-vessel, so increase in size does not mean an increase in thickness. Tube thickness is low enough to enable leaks to become apparent before serious rupture of the tube, but nevertheless emergency spray cooling of the core is provided in case there is loss of coolant.

Since 1968 a SGHWR[11] has been operating at Winfrith (*Figure 13*). It has a nett output of 94.5 MW$_e$ (103 MW$_e$ gross), thermal power is 309 MW$_t$ and the thermal efficiency is 35%. Core size is 3.12 m diameter and 3.66 m high giving an average power density of 10.9 kW$_t$/l (MW$_t$/m^3). Power densities in heavy-water reactors are of necessity lower than for light-water reactors (see CANDU, PWR and BWR) because of the longer distance travelled by neutrons in heavy water during slowing down. Sufficient distance must be left between the fuel channels to allow the neutrons to slow down without high probability of capture in ^{238}U resonances. In this reactor the fuel channels are arranged on a 260 mm square pitch and there are 104 channels altogether. The heavy-water moderator is contained in a calandria which is pierced by zirconium tubes, 184 mm outside diameter with a 3.3 mm wall, and these contain the zircalloy-2 pressure-tubes of 130 mm inside diameter and 5.1 mm wall thickness. Diameter of the calandria is 3.7 m and its height is 3.96 m. As light water is used as coolant and it is a neutron absorber, enriched uranium has to be used as the fuel.

The Winfrith SGHWR comprises fuel rods 3.66 m long made from 14.5 mm diameter UO$_2$ pellets (similar in diameter to those for the AGR) clad in 0.66 mm zircalloy-2. A fuel element consists of 36 fuel rods arranged in rings of 6, 12 and 18. In place of a central fuel rod in the element there is a zircalloy-2 tube pierced by holes. This is the emergency spray cooling supply in case of loss of pressure and coolant. Enrichment is 2.3% or just over three times the content of natural uranium. Maximum can temperature is 307°C and fuel-centre temperature is limited to 1905°C. The light-water coolant is pressurized to 6.7 MPa,

enters the core at 275°C and leaves at 282°C. Steam is supplied to the single 100 MW$_e$ turbine through two coolant loops at 280°C and 6.3 MPa.

Several methods of control are used. Moderator height is varied for normal start-up and shutdown, boric acid is dissolved in the moderator, and its concentration is varied as burnup proceeds. Emergency shut-down can be carried out by dumping the moderator and injecting boric acid solution into tubes in the core. Core design is such that it has a positive void coefficient, which means that increase in the steam voidage in the coolant channels increases reactivity. This enables the reactor to have load-following characteristics. An increase in steam demand at the turbine stop valve lowers the pressure in the coolant tubes. More water is consequently turned into steam and the increased voidage causes the reactor power to rise. The power continues to rise until the power is high enough to restore the pressure to equilibrium value and so decrease the voidage to its equilibrium concentration. Reductions in power demand are followed in an equivalent manner by reduction in the voidage. This reactor therefore will respond to small power changes rapidly without the need for control-mechanism adjustment. Uranium utilization for the Winfrith SGHWR is about 0.9%.

It will be seen from the foregoing descriptions that water-cooled and moderated reactors appear to have an ultimate thermal efficiency less than 35%. This is similar to Magnox systems but below possible values for AGR, HTR and, as described later, fast reactors. Uranium utilization on a once-through basis is about double that for Magnox systems but similar to those for AGR and HTR. Pressure-tube reactors also appear to have a lower probability of serious accidents from loss of coolant than pressurized reactors with thick steel pressure-vessels. If prestressed concrete pressure-vessels can be developed for PWR and BWR, then their safety will be subject to much less doubt than is common at present.

Fluid-fuelled Reactors

This Section concerns a class of reactors which has been the subject of much study and of limited low-power experimental work, but which so far has not been developed to large-scale power generating systems. They differ from all the other reactor types in that the fuel is dissolved in the coolant and therefore the power-density distribution, localized burnup and reactivity lifetime of the fuel are unimportant. In general such reactors are thermal reactors, since fast reactor systems would require a large inventory of fuel in the coolant circuit owing to the high enrichment needed. Thermal reactors work at lower enrichments and require larger cores, so that the proportion of fuel outside the reactor core is much smaller.

The first system with fluid fuel was the Homogeneous Aqueous Reactor (HAR). This consisted in its breeder form of a spherical core containing ^{233}U nitrate in solution in the heavy-water

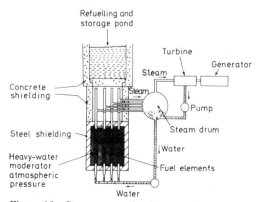

Figure 13   Steam-generating heavy-water reactor

moderator, which was separated by a thin zircalloy vessel from the breeder, a spherical shell of ThO_2 (thoria)—heavy-water slurry. Such a system is a pressurized-water reactor and so must have a thick pressure-vessel. At the necessary concentration for efficient breeding the thoria slurry has a thick consistency, is difficult consequently to pump and is also rather abrasive. The core solution also needs to be strongly acid to reduce radiolytic gas formation and consequently erosion, and corrosion of the zircalloy core vessel is a serious problem. Such a system offers the possibility of a breeding ratio greater than unity, especially as the breeding region is in a low neutron flux. This reduces the loss caused by burnup of the intermediate nucleus ^{233}Pa which has a high neutron absorption cross-section.

Another system which has been given consideration is the Liquid-Metal-Fuelled Reactor (LMFR) using liquid bismuth — a very weak neutron absorber — as the coolant with uranium dissolved in it in the core and a thoria/bismuth slurry in the breeder. A separate moderator is required so that the fuel-coolant runs through channels in a graphite moderator. No pressurization is needed with this coolant.

As early as 1959 the USAEC, which had a large interest in liquid-fuelled reactors, decided to stop development of the HAR and LMFR and concentrate on a third type, the Molten Salt Reactor (MSR)[12]. The original impetus had been for a high-temperature reactor for aircraft propulsion, and in 1954 the Aircraft Reactor Experiment (ARE), which had a core consisting of molten $NaF—ZrF_4—UF_4$ flowing through inconel tubes in a BeO moderator, operated successfully for nine days at outlet temperatures up to $860^\circ C$ and thermal powers up to $2.5 MW_t$. This project was abandoned in 1958 but left considerable information on the processing and use of fused salts.

Consequently a study of the possibility of using fused-salt reactors as large-scale power producers

was begun, using fluoride salts. These have viscosities at working temperatures similar to paraffin (kerosene), a volumetric heat capacity approaching that of water, and a very low vapour pressure. Melting points are high, however, in the region 400 to $500^\circ C$. In 1960 the Molten Salt Reactor Experiment (MSRE) was instituted at Oak Ridge. This was fuelled with ^{235}U contained in a $^7LiF—BeF_2—ZrF_4—UF_4$ fused salt having molar percentages 66—29—5—0.2. ^7Li was used instead of natural lithium since the other isotope, ^6Li, is a strong neutron absorber. It was circulated through a 1.4 m diameter core built from bare graphite bars and produced $7.3 MW$ of heat with a fuel outlet temperature of $654^\circ C$ (salt melting point $434^\circ C$). Heat produced was rejected to waste. At these temperatures the fuel vapour pressure is very low, less than 13 Pa, and the only pressure in the system was the helium blanket at 130 kPa. Xenon and krypton fission products were purged from the fuel with helium. This reactor reached full power in May 1966, was refuelled with ^{233}U in October 1968, thus becoming the world's first ^{233}U fuelled reactor, and was closed down in 1969.

From experience gained with this system it has been possible to specify the general features of a large-scale Molten Salt Breeder Reactor (MSBR). Power densities in these systems are high compared with Magnox or AGR, where the graphite moderator is intended to last the life of the reactor and consequently radiation damage to the graphite may become serious after about four years. Above a certain irradiation level graphite shows a very rapid change in dimensions with increasing dose. For this reason a two-fluid system, one fissile and one fertile, with graphite tubes separating the fluids was not considered feasible. Development of techniques for ^{233}Pa removal, which reduces neutron loss and burnup of ^{233}Pa, has made a single-region system feasible. A conceptual design for a $1000 MW_e$, $2250 MW_t$, 44% efficiency MSBR has the following features (*Figure 14*). The core is 5 m in diameter and has bare graphite bars with circulating channels between for the fuel. In the outer part of the core the proportion of graphite is reduced so that this region acts as a blanket. If necessary, to counteract radiation damage, the graphite core elements can be replaced. 1500 kg of ^{233}U are contained in the salt, composed of $^7LiF—BeF_2—ThF_4—UF_4$ with molar percentages 72—16—12—0.3 which melts at $500^\circ C$. At a coolant flow of $4 m^3/s$ the salt leaves the core at $704^\circ C$ and transfers its heat to a sodium fluoroborate secondary coolant, which finally raises steam at $538^\circ C$. A helium purge removes ^{135}Xe and other noble fission-product gases. In addition some fuel is by-passed to a chemical plant to isolate the ^{233}Pa and remove the salt-soluble fission products. The entire core contents are processed once in 10 days and poisons are kept to such a low level that only about 1% excess reactivity is needed for control purposes. Uranium can be removed as gaseous UF_6 by fluorinating the salt, liquid bismuth and lithium metals

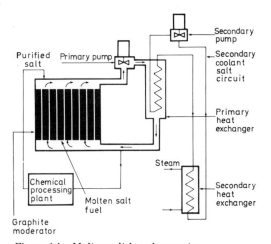

Figure 14   Molten-salt breeder reactor

extract the ^{233}Pa, then the fission products are extracted into liquid bismuth and from there into lithium chloride. Bismuth is used since it does not take up thorium. The breeding ratio is estimated to be 1.07.

Apart from the USA, India shows interest in molten-salt reactors and in the UK some work has been done to develop chloride rather than fluoride salts for possible use in a fast molten-salt reactor.

Fast Reactors

It is not essential to slow fission neutrons down in order to create a sustained chain reaction, provided that the uranium is enriched sufficiently. One example of a fast reactor is the fission bomb, which is constructed to have sufficient multiplication to ensure that its reactivity is much greater than that required for the prompt-critical condition. For a sphere of pure ^{235}U the critical mass is only about 50 kg (volume about 2.3 l). Plutonium is also a suitable fuel for fast reactors and the minimum critical size for a sphere of plutonium is 16.5 kg (volume about 1 l). For high-temperature operation of a fast power reactor metallic fuel is unsuitable and, as for thermal reactors, uranium or plutonium oxide (and possibly carbides or nitrides) must be used. In addition a coolant which is a poor moderator must be used. Water, with its high hydrogen content, is totally unsuitable and liquid metal coolants (e.g. sodium) are mainly favoured, high-pressure-gas cooling being a less likely alternative. Fast reactors therefore must be greater in critical mass than the minimum sizes quoted above but nevertheless they are small in volume, with a very high power density compared with thermal reactors.

The real value of the fast reactor lies in its ability to breed fissile material from fertile material more effectively than thermal reactors. Only thermal reactors operating on the ^{233}U/^{232}Th cycle, as mentioned earlier, offer any hope of creating more fissile material than they consume. This is because of the low number of neutrons per fission (2.44 for thermal systems) and the large wastage: absorption in structural materials, canning, coolant, moderator, non-fissile capture in the fuel and fertile material, and leakage. For fast fission of ^{235}U or ^{233}U the number of neutrons per fission is similar to that in thermal fission, but for plutonium it increases with the energy of the neutron causing the fission, rising to over 3 in the range of unmoderated fission-neutron energies. But at these high neutron energies cross-sections generally are low, in the region of 1 barn or less. Non-fissile capture in plutonium is small, the cross-section of stainless steel, used for canning and structural material, is also low and the fission products have such low cross-sections at these energies that their poisoning effect is small. ^{135}Xe for example is no longer a serious problem. Cross-sections in the MeV region are indeed so low that it is difficult to find a suitable material to act as control absorber. Boron is commonly used,

sometimes tantalum, although their cross-sections are only a few barns. Taking all these facts into account (e.g. high number of neutrons per fission and low parasitic cross-sections) calculations show that the plutonium/^{238}U cycle offers the best possibility of converting fertile material to fissile and providing excess fuel, so that the number of fast reactors can be gradually increased without needing a supplementary enriched-fuel supply.

The general form of fast reactors consists of a small highly rated core (typically 400 kW$_t$/l average) fuelled with plutonium (15–30% depending on power output) and ^{238}U (natural or depleted uranium). The core contains as little material of low atomic weight as possible to minimize degradation of the neutron energy by elastic collisions. Since the core is small, neutron leakage is high and the core is surrounded by a region called the breeder containing a high proportion of fertile material. Most of the excess fissile material is created in this blanket, and the fertile material in the core serves to prolong fuel life. Because of burnup of fuel in the core and also radiation damage to fuel and canning, both core and blanket fuel elements must be periodically removed from the reactor so that the plutonium created in the blanket can be fabricated into fresh fuel elements. With the very high fuel rating, radioactive heating from fission-product decay is very intense on removing the fuel from the core. It is necessary to allow a long decay time before the fuel is handled for processing, and consequently about as much fuel is held up outside the core awaiting processing as is used in the core. Therefore the fuel inventory (I) for a fast reactor is about twice the core loading (M). Breeding gain (BG) is defined as the excess fissile atoms (over and above those needed to keep the reactor working) created per fission, and so the excess-material production rate is also proportional to thermal power (M times fuel rating R) and load factor (LF). Therefore excess-plutonium production is proportional to

$$BG \times M \times R \times LF$$

Fuel doubling time (t_D) is the time needed to create enough excess fuel to start up another similar reactor. t_D in years is the plutonium needed for one more reactor $I \times$ enrichment (e) divided by the annual production rate:

$$t_D = \frac{I}{M} \frac{e \times 2.5}{BG \times R \times LF}$$

A third reactor can then be commissioned after a further $t_D/2$ years and a fourth reactor after yet another interval of $t_D/3$. Therefore the average doubling time in reaching four reactors from two is $1/2 + 1/3 = 0.833$ times the single-reactor doubling time. Eventually when many reactors are working the average doubling time approaches $0.693\ t_D$.

The neutron lifetime in a fast reactor is about 10^{-7}s, and so the rate rise at prompt-critical is much higher than for thermal reactors (see Section on Reactor Control, p 317). Plutonium has a delayed-neutron fraction of 0.26% compared with 0.65% for ^{235}U. For delayed neutrons however the average energy is about 1/5 that of prompt neutrons. In a small reactor such as a fast reactor the neutron leakage rate is high from the core, but the lower-energy delayed neutrons have a lower probability of leakage than the prompt neutrons so that the effective delayed-neutron fraction is greater than 0.26%, depending on details of the reactor design. Below prompt-critical the dynamic behaviour of a fast reactor is similar to that of thermal reactors and hence control is no more difficult.

For rapid power increases, such as occur near prompt-critical, only the fuel heats up since there is no time for the extra heat to be transferred to the coolant. In thermal reactors heating of the fuel alone increases the absorption of neutrons in ^{238}U resonances, owing to increased thermal agitation of the fuel (Döppler broadening of resonances) and so the reactivity is reduced and the power 'excursion' is damped. In fast power reactors there is unavoidably a proportion of low-atomic-weight nuclei such as oxygen in oxide fuel and sodium coolant. Consequently the fission neutrons are partially slowed down and there is a substantial neutron flux in the energy region of the ^{238}U resonances. Therefore, just as in thermal reactors, an increase in fuel temperature leads to increased neutron absorption in ^{238}U resonances and reactivity is reduced.

Since 1963 a fast reactor called DFR has been in operation at Dounreay[13]. This has a total heat output of 60 MW$_t$ and a nett electrical output of 13 MW$_e$(15 MW$_e$ gross). Core volume is only about 140 l and average power density 400 kW/l (MW/m^3). The DFR also serves as a test rig for development of fast-reactor fuel and components and so the core composition varies with time. Its fuel and coolant are not typical of later fast reactor designs as it uses metallic fuel (75% ^{235}U plus 25% ^{238}U) containing 10% of molybdenum; a sodium/potassium alloy, 70% sodium and 30% potassium, which is liquid down to almost room temperature is used as coolant. Also the coolant is pumped by 24 electromagnetic pumps which at present have only been developed successfully for low pumping rates. However, in common with all liquid-metal-cooled fast reactors, it works at low coolant pressures, around atmospheric, and so does not need a pressure vessel. The whole reactor is contained in a spherical containment shell which is intended to withstand the overpressure due to a coolant fire. Provision of this containment was probably unnecessarily pessimistic and the reactor has worked very reliably since commissioning.

Sodium-cooled fast power reactors are being most actively studied in the UK, France and USSR where the following prototypes have been constructed: Dounreay PFR (250 MW$_e$), Marcoule Phenix (250 MW$_e$), and Shevchenko BN-350

(350 MW$_e$). The Dounreay Prototype Fast Reactor (PFR), which achieved low power criticality in March 1974, is typical of these and will be described in more detail[14]. Its designed thermal power output is 600 MW$_t$ with a nett electrical output of 250 MW$_e$ (260 MW$_e$ gross) and a thermal efficiency of 40%. Its main aim is to serve as a demonstration and test facility for a future 1300 MW$_e$ output Commercial Fast Reactor (CFR) which could be the basis of a series of fast-reactor power stations. The general design shows many changes from DFR: plutonium/^{238}U fuel in oxide form instead of ^{235}U/^{238}U fuel in metallic form, sodium coolant instead of Na/K, mechanical pumps instead of electromagnetic, and no special containment. Mean enrichment is 26% where here enrichment means the fraction of plutonium in the plutonium-plus-uranium fuel. Two enrichments are used for flux flattening and more uniform power generation, 22% in the inner radial zone (about 40% of the core) and 29% in the outer radial zone. Each oxide fuel pin (the diameter is too small for it to be called a rod) is 5 mm in diameter and clad in 0.38 mm stainless steel. 325 of these pins are arranged on a triangular pitch in hexagonal stainless-steel sub-assemblies 140 mm across flats, which are themselves arranged on a 145 mm triangular pitch. 78 of these sub-assemblies make up the core and these are arranged in groups of six around a similar-sized central support (termed a leaning post) which contains either a control rod or an experimental irradiation facility. The core, which produces most of the 600 MW$_t$, is roughly cylindrical with a diameter of 1.45 m and height of 0.9 m, the volume therefore being 1400 l and the average power density 400 kW/l. Above and below the active region of the fuel assemblies are axial breeder-regions containing ^{238}U oxide and similar-size ^{238}U oxide breeder assemblies form the radial breeder. This is only one-assembly thick for 2/3 of the core perimeter and three-assemblies thick for the remaining 1/3 of the perimeter. In order to improve neutron economy the radial breeder is surrounded by a reflector made of stainless steel. Maximum permitted can temperature is 692°C and the maximum fuel-centre temperature is 1500°C.

Molten sodium is the coolant and no pressurization is necessary even at operating temperatures. The complete core is immersed in a large tank of stainless steel 12 m in diameter and 15 m deep. This vessel is surrounded by a mild-steel vessel to give double-walled construction and the whole is sunk into a concrete pit. Sodium leakage from the vessel would therefore be very well contained and the core could never become uncovered. A total of 920 t of sodium fills the vessel, which is covered by a shielding lid that supports the core structure and all other components in the vessel. This lid also seals the tank and contains the inert gas blanket over the sodium which prevents oxidation. Also contained in the core vessel are three 1 MW centrifugal pumps with a lower liquid-sodium-

References p 339

lubricated bearing and six sodium-to-sodium primary heat exchangers. Coolant inlet temperature to the bottom of the core is 400°C, the inlet pressure 700 kPa and the outlet temperature 585°C. *Figure 15* shows the schematic arrangement of the PFR core and cooling circuits. Three secondary coolant circuits, each containing 75 t of sodium (one per two primary heat exchangers), circulate hot sodium to three sodium/water heat exchangers which feed steam to a single turbine at 530°C and 16.2 MPa pressure. Use of a secondary coolant circuit ensures that the core vessel cannot be emptied by a leak in the external sodium circuits. In case of total power failure the intense fission-product decay heat from the core can be dissipated at first by the slow heating of the large mass of sodium in the core vessel. If prolonged, it is possible to bring into use an emergency natural-convection cooling circuit which discharges its heat to the atmosphere.

Refuelling will take place off-load in the PFR. Since 6 elements are removed together on one leaning post, the reactivity changes would be too great for on-load refuelling to be possible. At full power 1/6 of the core would be changed over a weekend once every 6 to 7 weeks. Initially, spent fuel will be stored in the core vessel so that the fission-product decay heating can be removed. When this has decayed sufficiently the spent fuel is then removed for processing. Burnup will correspond to consumption of about 8% of the atoms in the fuel and the breeding gain in PFR is estimated to be −0.52 in the core (core consumes more fuel than it breeds) and +0.51 in the breeder, giving a nett breeding gain of −0.01. This system will therefore not produce excess fuel, mainly because the breeder zone is of insufficient thickness over 2/3 of the core perimeter. Since this is a demonstration and test reactor, a positive breeding gain is not essential provided that it shows that breeding can be achieved with a larger system. For the 1300 MW$_e$ CFR design the calculated breeding gains are −0.33

(core) and +0.49 (breeder) giving a nett value of +0.16. This is intended to use similar fuel to the PFR but with lower enrichment (17.5%), since the system is larger and neutron loss by leakage is thereby reduced. With a fuel rating of 0.26 MW$_t$/kg average and an inventory 1.8 times the core loading the fuel doubling time is about 20 years. Oxide fuel is chosen because of its good resistance to radiation damage (fast-neutron irradiation in the core is so great that every atom in the stainless-steel cladding suffers one displacement per week on average), and is assembled at low density (8.8 t/m^3) to reduce swelling. A density increase to 9.9 t/m^3 is estimated to increase the breeding gain from 0.16 to 0.20 and so reduce the fuel doubling time from 20 to 16 years. Further changes to carbide fuels, assuming that they could be developed to have as good radiation-damage resistance as oxide fuels, would (it is estimated) about halve the fuel doubling time.

Other types of fast breeder reactor as well as the sodium-cooled type have been considered. These include steam-cooled and molten-salt-fuelled systems, but the next most favoured type to the sodium-cooled system is the gas-cooled fast reactor. So far only design studies have been made, for example by the Gas-Cooled Breeder Association with members drawn from various European countries[15]. Oxide fuel is again favoured but very high pressures of gas coolant are necessary, similar to coolant pressures in PWR reactors. Studies on an oxide-pin-fuelled system to produce an output of 1000 MW$_e$ have shown that a helium coolant pressure of 12 MPa would be necessary. Outlet temperature for the gas would be 587°C and the thermal efficiency 35.4% (pumping powers are high for such pressures). Core size is slightly greater than a comparable sodium-cooled system and the fuel doubling time is estimated to be 13 years. This is lower than for an oxide-fuelled sodium-cooled fast breeder because the lower density of the coolant gives less moderation of the neutrons. Other variants of this concept use coated particle fuel. The particles are not bonded as in HTR but the coolant gas flows through fuel elements containing beds of these particles. Again for an electrical output of 1000 MW helium coolant at 12 MPa or carbon dioxide coolant at 6 MPa is estimated to have a thermal efficiency of 36% and a fuel doubling time of 16 years. Some of the same criticisms levelled at the PWR apply to the gas-cooled fast breeder if steel pressure-vessels have to be used. Loss of coolant could create very severe emergency cooling problems since emergency water cooling might easily act as a moderator and produce an uncontrolled thermal reactor.

Fuel Resources and Utilization

Fission-fuel Resources

The non-communist world reserve of uranium in rich ore, defined (in 1974) as that from which uranium can be extracted for less than £6/lb U$_3$O$_8$

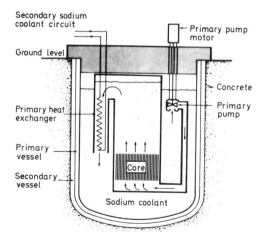

Figure 15   Prototype Fast Reactor (Dounreay)

(£13.3/kg) are probably not more than 3 million long tons, the distribution being as shown in *Table 4*. Dilute sources of uranium with consequently higher extraction costs are more plentiful and widespread. As shown in *Figure 16*, the uranium reserves thus increase rapidly with the acceptable extraction costs. Uranium is present in most regions of the earth's crust with a concentration of about one part per million and also in sea water at a lesser concentration.

It would probably not be feasible, on environmental as well as economic grounds, to recover a large fraction of the dilute sources of uranium in the earth's crust, but recovery from sea water appears more feasible in the long term, particularly if combined with desalination and recovery of other minerals.

Under present conditions only the richer uranium ores are being exploited. These contain between 2 and 6 lb U_3O_8 per ton. The uranium compounds are first extracted from the crushed ore by various chemical processes and dried and calcined to form a concentrated solid (so-called yellow cake) containing about 80% of uranium. Uranium is imported into the UK in this form and further refined at Springfields, where it is converted into uranium tetrafluoride, and then reduced to metallic uranium for the natural-uranium reactors by magnesium.

Enrichment by Separation of ^{235}U

As shown in previous Sections, much greater freedom of reactor design is possible if enriched fuel is used, and one method of doing this is to increase the ^{235}U content above the normal 0.72%. This must be done by physical methods since ^{235}U and ^{238}U are chemically identical. The uranium tetrafluoride gas is first converted to the hexafluoride by reacting with fluorine. Uranium hexafluoride is gaseous at temperatures above $56.4°C$ and the separation process depends on the fact that $^{235}UF_6$, being less dense, diffuses through a suitable membrane slightly more quickly than does $^{238}UF_6$.

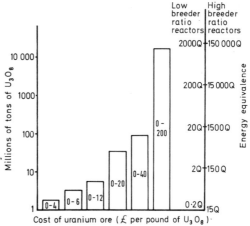

Figure 16   Uranium reserves as a function of refining costs of U_3O_8 [25]

The low ratio of diffusion rates (1.0043:1) means that the degree of separation in any single process is very small, but by operating a large number of diffusion chambers in cascade it is possible to attain the 1.5—3.5% ^{235}U enrichment necessary for the enriched thermal reactors. (Even higher enrichments are possible with larger numbers of chambers, and indeed weapons-grade ^{235}U was first produced by this process at Oak Ridge in Tennessee in a plant that covered over a hundred acres). Even at the low-enrichment level separation is a costly process and accounts for about 40% of the total fuel-element cost. It is hoped to replace diffusion by a centrifugal method which again uses uranium hexafluoride gas. Again the degree of separation in a single chamber is small and the chambers must be arranged in a cascade sequence.

Centrifugal separators are under construction at Capenhurst and Almelo as part of a collaborative programme between the Netherlands, W. Germany and UK[17,18].

Enrichment by Breeding

The fissile content of natural uranium may also be increased by adding fissile ^{239}Pu, which is produced to a greater or lesser extent in all reactors by the breeder reaction:

$$^{238}U + n \rightarrow {}^{239}U \xrightarrow{\beta} {}^{239}Np \xrightarrow{\beta} {}^{239}Pu$$

Being chemically different from the uranium in which it is produced, plutonium can be extracted from the spent fuel elements by remotely controlled chemical methods. The plutonium produced in reactors is mainly ^{239}Pu but contains an appreciable fraction of ^{240}Pu and lesser amounts of the higher isotopes. The actual isotopic composition of the plutonium is a function of the reactor type and the degree of burnup in the fuel element. Typical

Table 4 Reserves of rich uranium ore (in thousands of long tons). (Refining cost $\leqslant$ £6/lb U_3O_8)[16]

	Reasonably assured	Possible additions
USA	445	595
Canada	278	307
South Africa	280	38
Australia	107	10
France	42	31
Nigeria	30	39
Others	345	112
	1527	1132

Total 2 659 000 tons

annual plutonium production figures for 1000 MW$_e$ installations are as follows[19].

Magnox	550 kg/year (conversion ratios 0.8)
AGR	220 kg/year (conversion ratios 0.44)
PWR	270 kg/year (conversion ratios 0.5)

Fast breeder reactors have a conversion ratio approaching 1.2 and therefore their production of fissile material is in excess of their consumption; as shown in the previous Section, fuel doubling times of between 10 and 20 years should be possible.

By recycling the plutonium produced in similar reactors it is possible for some 70—80% of the total uranium (^{235}U + ^{238}U) eventually to undergo fission, whereas if the plutonium from a low-breeder-ratio reactor is recycled the total uranium utilization is limited to about 1%. The total uranium utilization (with recycling) is plotted as a function of conversion ratio in *Figure 17*, and the effect of this on the fuel value of the uranium reserves will be discussed in a later Section.

It is also possible to use ^{232}Th as the breeder blanket in a fast breeder to produce the fissile isotope of uranium ^{233}U

$$^{232}\text{Th} + \text{n} \rightarrow {}^{233}\text{Th} \xrightarrow{\beta} {}^{233}\text{Pa} \xrightarrow{\beta} {}^{233}\text{U}$$

and again ^{233}U is separable from ^{232}Th by chemical means.

Thorium is found in large quantities in the monazite sands in India where the reserves are estimated at 378 000 long tons, some ten times the uranium reserves of similar extraction cost[20]. In Canada, thorium is found to be closely associated mineralogically with uranium and economic reserves of 100 000 long tons are estimated[21].

It seems probable that the effective world thorium reserves may be of about the same order as those of uranium.

The Short-term Economics of Nuclear Power

It is difficult to discuss the short-term economic position of nuclear power in terms of competing generating costs from nuclear and non-nuclear stations in absolute terms in the present world inflationary situation, which affects both nuclear and conventional generating costs. An attempt will nevertheless be made to outline the main factors involved in the cost comparison between nuclear and non-nuclear stations and between the several reactor types.

The main difference between the nuclear and non-nuclear stations is that the nuclear capital costs are much higher but the running costs, including fuel, are very appreciably lower. The difference is most marked in the Magnox stations, where capital charges, including capital costs repayments and interest payments, make up over three-quarters of the total generating costs as opposed to one-quarter in the case of coal- and oil-fired stations, where fuel costs are the dominant factor.

Capital charges must be met whether the power station is operating at power or not, so the financial penalty for non-operation or operating at reduced load is much higher for nuclear stations. The main variables which affect the cost comparison are:
(1)   Interest rates (which affect capital charges).
(2)   Assumed economic life of station (also affects capital charges).
(3)   Load factor.
(4)   Value of by-products (e.g. ^{239}Pu in the case of reactors).

These variables are not primarily dependent on the technical merits of the power stations themselves. The interest rates are a function of the general economic situation; the assumed economic life of a station is not usually determined by the station actually wearing out but by its becoming uneconomic compared with later and more advanced stations; the load factor is governed mainly by the daily and seasonal variation in electricity demand rather than the serviceability of the stations. In a mixed system it is therefore more economic to run the high-capital-cost nuclear stations as base-load stations.

Clearly low interest rates, the assumption of long economic lives and high-duty cycles favour the competitive position of the nuclear stations.

In 1970, on the assumption of a 75% load factor, 7.5% interest rates, 20 years economic life for the nuclear station and 30 years for the conventional stations, Davis[22] estimated that the cost of electricity from the later Magnox stations would

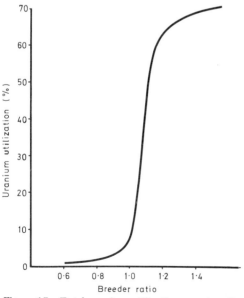

Figure 17   Total uranium utilization as a function of conversion ratio

be about equal to that from the best coal and oil stations, at about 0.3 p/kWh. An encouraging feature was that over the eight years during which the Magnox series was installed their generating costs in real terms fell by about 60% owing to detailed technological improvements. 1969/70 could be considered as the period in which nuclear power became commercially competitive in the UK. Since then the increase in interest rates has militated against the nuclear stations but the recent dramatic rise in fossil-fuel prices has more than balanced this.

For costing purposes the slightly enriched reactors occupy an intermediate position between the natural-uranium reactors and conventional stations in that compared to Magnox their capital costs are smaller, because of their smaller size for a given generating capacity, but their fuel costs are higher owing to cost of enrichment and a reduced plutonium credit.

The CEGB have recently issued the following forward estimate of capital and generating cost estimations for reactors commissioned in 1980[23] (*Table 5*).*

The basis of calculation was 10% interest charge, 64% duty cycle and 25 years economic life, assumptions which may seem to favour the system of lower capital cost.

The projected low capital cost of fast breeder systems together with high plutonium credit are estimated to lead to economically competitive generating costs from this type of reactor.

The Long-term Utilization of Fissile Fuels

Fuel reserves are often expressed in units of Q, this being the energy obtained by burning 46 500

* *Note added in proof* In a recent paper (*Atom* 1975 (Jan.), No.219, p 2) Sir John Hill, chairman of UKAEA, quotes recent actual generating costs from oil-fired, coal-fired and nuclear (Magnox) stations. Average costs for three stations of each type changed from near equality in 1972/73 to oil-fired 0.89 p/kW h, coal-fired 0.66 p/kW h, nuclear 0.40 p/kW h, reflecting the relatively increased fuel price for the first two and the low proportion of fuel cost for nuclear plant.

Table 5   Capital and generating costs for some nuclear power stations (CEGB figures)[23]. Projected to 1980 commissioning

	Capital cost (£/kW$_e$ installed)	Generating cost (p/kW h)
PWR	233	0.46
SGHWR	262	0.51
HTR	260	0.51
AGR	293	0.57
Magnox	366	0.72

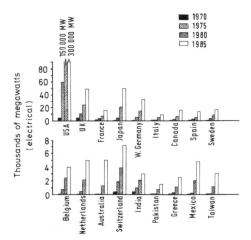

Figure 18   Present and projected nuclear-generating capacities in various countries

million tons (47 200 Mt) of coal†. The reserves of conventional fossil fuels are about 100 Q. The world consumption of fuel in 1970 was about 0.2 Q, but in recent years our fuel requirement has had a doubling time of only 10 years. Forward extrapolations, with admittedly large regions of uncertainty, then appear to indicate that the fossil reserves could be exhausted in 50–100 years.

The fuel value of 3 million long tons of uranium recoverable from rich ores would, if used in slightly enriched low-conversion-ratio reactors, be about 1 Q and contribute little to our overall fuel reserves. Indeed the present rapid expansion of the world nuclear programme based mainly on those reactor types could exhaust our rich uranium reserves by 1985[16].

The present and projected nuclear generating capacities of some of the leading nuclear nations[24] are shown in *Figure 18*.

Uranium costs represent a fairly small fraction of the total generating costs, even in enriched reactors, so that these might remain competitive even if uranium prices rose to perhaps £40/lb (£90/kg) as it became necessary to exploit less rich reserves. Even at this our uranium reserves would only produce about 20 Q of energy, one-fifth of our fossil reserves, if used in low-breeder-ratio reactors (*Figure 16*).

It seems obvious that fissile reserves can only make a significant contribution to the overall energy position if used in reactors with a conversion ratio in excess of unity, when the uranium utilization rises steeply to about 70% (*Figure 17*)[25]. Reserves recoverable at about £90/kg would then be equal to about 1000 Q in energy terms. With the higher utilization of uranium the reactor economics become even less dependent on uranium

† 1 Q = 10^{18} Btu = 1.055×10^9 TJ = 2.93×10^{14} kW h (N.B. See p 590, footnote to Table 2)

prices, and exploitation of uranium from such a dilute source as sea water at a cost of up to about £450/kg might become economic, representing a total fuel reserve of about 100 000 Q.

The main problem associated with a substantial future programme based on breeders is the accumulation of the necessary enriched fuel to build the cores in sufficient quantity.

The core of a 1000 MW$_e$ Fast Breeder of the Dounreay type requires an inventory of about 3 t of fissile material, so that if one-fifth of the 5 million megawatts installed capacity estimated for the year 2000 is to be met by fast breeders it will be necessary to accumulate some 3000 tons of fissile material by that date[24].

Even the most ambitious estimate for the expansion of the thermal reactor programme would make it difficult to reach the target in ^{239}Pu alone, and supplementation by ^{235}U from separation plants will be necessary. In order to have any reasonable hope of achieving the target thermal reactors with as high a conversion ratio as possible should be selected, which do not themselves require the use of separation effort, and this would limit the choice to CANDU or Magnox types. There is thus a conflict between exploiting the short-term competitive position of slightly enriched reactors and the desirability of accumulating a substantial fissile fuel reserve for the future programme.

When a substantial fraction of our generating capacity comes from fast breeders some form of supplementation of fissile material will continue to be necessary if one assumes that the present doubling time of 10 years for electricity demand will continue, since this is shorter than the fuel doubling time for the present type of fast breeder. Additional fissile material could be fed to the system either from thermal stations or from separation plant. A schematic diagram of a possible system

aimed at the long term is shown in *Figure 19*. Alternatively a fast breeder with a shorter doubling time is required and the gas-cooled fast breeder could be important in this context.

Environmental Considerations

Normal Reactor Operations

High fluxes of neutrons and γ rays are produced in all nuclear reactors and the accumulation of fission-product nuclei within the reactor is also an inevitable consequence of the fission process. In order to estimate the possible detrimental effects of the nuclear power programme on the general population and environment we must first discuss the biological effects of radiation and radioactive materials. α particles, β particles and γ rays ionize body tissue producing damage which depends on the intensity and type of radiation. α particles produce intense ionization over a range of about one-tenth of a millimetre, whilst γ rays have a range of tens of centimetres but produce less intense ionization. Fast neutrons are scattered elastically and inelastically by nuclei, displacing them from their normal positions and producing γ emission, whilst slow neutrons are more easily captured, often to produce radioactive isotopes.

For simplicity the biological damage produced by all types of radiation is often expressed in terms of a single unit, the rem, which is the product of the energy absorbed per unit mass (1 rad = 10^{-2} J/kg) multiplied by the 'quality factor' or Relative Biological Effectiveness (R.B.E.) of the radiation in question. β rays typically have a quality factor of 1.7, α particles 10 and neutrons from 3 to 10 depending on energy. The quality factor of X and γ rays is one by definition.

The I.C.R.P. (International Commission on Radiological Protection) are responsible for recommending acceptable levels of radiation dose for workers in the nuclear industry and for the population as a whole[27]. In order not to increase the average exposure of the general population significantly above the natural background (80—150 millirem a year) I.C.R.P. recommend that no member of the general population should be exposed to dose rates of greater than 0.25 millirem an hour (about 2 rem a year).

Personnel actually working in the nuclear industry, in power stations, fuel processing plant etc., like other people similarly employed with radiation or radioactive materials, are considered to form a separate group of so called 'designated' or 'classified' workers. The radiation doses to which they are exposed are carefully monitored using film badges and they are subjected to medical examinations and blood tests at least once a year. For workers in this special class the total whole-body cumulative dose must not exceed $5(N - 18)$ rems where N is the age of the person in years. This effectively means that the annual dose must be less than 5 rem.

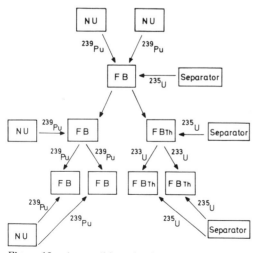

Figure 19   A possible mixed-reactor system for the long term

All reactors are shielded by thick concrete shields to reduce the radiation level outside to below that corresponding to the Maximum Permissible Level (MPL) for designated workers or members of the general public in regions to which these groups have access. The danger to the general public or reactor staff due to direct radiation during normal operation is therefore small. Designated workers may occasionally be exposed to dose rates higher than their permitted average value during fuel changes etc., but this must be compensated by reduced doses during other periods. This also applies to designated workers in the reactor fuel-fabrication and processing works. The absence of any statistically significant increase in the types of disease which could be radiation induced (e.g. cataracts, leukaemia and other forms of cancer) amongst the substantial numbers of designated workers during the past twenty years indicates

Table 6 Maximum body burdens of some fission-products and reactor fuels

Isotope	Critical organ	Maximum body burden (μCi)*
^{90}Sr strontium	Bone	1
^{91}Y yttrium	Bone	3
	G.I.	0.02
^{95}Nb niobium	Bone	44
^{99}Mo molybdenum	G.I.	0.01
^{96}Tc technetium	G.I.	0.3
^{103}Pd palladium	Kidneys	7
	G.I.	4
^{109}Cd cadmium	Liver	45
^{131}I iodine	Thyroid	0.6
^{133}Xe xenon	Total body	320
^{135}Xe xenon	Total body	100
^{137}Cs caesium	Muscle	98
^{140}Ba barium	Bone	1
	G.I.	0.7
^{140}La lanthanum	G.I.	0.9
^{144}Ce cerium	Bone	1
^{143}Pr praseodymium	G.I.	3
^{147}Pm promethium	Bone	25
	G.I.	0.3
^{151}Sm samarium	Bone	90
	G.I.	14
Natural uranium (soluble)	Kidney	0.04
	G.I.	10^{-3}
Natural uranium (insoluble)	Lungs	0.01
^{239}Pu plutonium (soluble)	Bone	0.04
	G.I.	0.02
^{239}Pu plutonium (insoluble)	Lungs	0.02

* 1 μCi = 3.7 × 10^4 disintegrations s^{-1}
G.I.: Gastro-intestinal tract

References p 339

that the dose-level restrictions and operating procedures are adequate.

In addition to the dangers associated with radiation from sources outside the body, there is an additional danger of irradiation from within due to swallowing or inhaling radioactive material, which may be retained in the body and often concentrated in specific organs. For example, radioactive iodine ^{131}I is concentrated in the thyroid, radioactive caesium ^{137}Cs in the muscle, radioactive strontium ^{90}Sr in the bone, and so on. The time for which ingested radioactive isotopes are harmful depends on their half-lives and the metabolism for the element. Short-lived isotopes represent a reduced hazard because their activity dies away quickly, whereas the hazard due to long-lived isotopes will be reduced if they are eliminated rapidly from the body. Clearly long-lived isotopes which are also retained in the body represent the worst hazard. After considering half-lives, metabolic rates and chemical toxicity I.C.R.P. have determined the Maximum Permissible Body Burden for each radioactive isotope. This is the maximum weight of the isotope which the body may contain without discernible clinical ill effects. The Maximum Permissible Body Burdens for the significant isotopes produced in the nuclear power industry are shown in *Table 6*. In the case of accidental over-exposure it is often possible to reduce the body content of a specific isotope by taking large doses of the non-radioactive element, which dilutes and displaces the radioactive isotope from the body tissue.

The possible dangers due to ingestion or inhalation of radioactive materials produced in nuclear power stations could be substantial in the absence of careful monitoring. Gaseous effluents from existing nuclear power stations are monitored with great care, both by the CEGB and by the Ministry of Agriculture. The results are made freely available to liaison committees consisting of CEGB staff, local residents and their expert advisers. Local milk is also sampled regularly since any ^{131}I or ^{90}Sr released from reactors would be concentrated in the milk of animals grazing in affected ground. No activity above normal background has yet been observed in the vicinity of an operating reactor in this country. The discharge of liquid effluent to the sea and neighbouring waterways is also strictly controlled and the activity concentrated in seaweed and fish life carefully monitored. Any increase in levels of radioactivity in the vicinity of operating stations has been shown to be negligible compared to the level of natural radioactivity in these areas. The main potential environmental danger comes from the treatment of the highly radioactive spent fuel elements. Active fission-product nuclei and plutonium build up in the fuel elements of operating reactors, and after 40% or so burnup of the original fissile material these must be removed from the reactor.

After initial underwater storage to allow the very short half-life activities to die away, the spent fuel elements from Magnox stations are transported by road to Windscale in Cumberland. They are contained in specially constructed flasks which are designed both to provide radiation shielding and to prevent the escape of radioactive material in the event of a road accident.

At Windscale the spent fuel elements are subjected to remotely controlled chemical processing to extract the plutonium and residual uranium. 99.98% of the fission-product nuclei are also extracted in this way and concentrated into a liquid of initial activity about 10^4 curies per gallon. The liquid is stored in robust water-cooled stainless-steel tanks housed in stainless-steel-lined concrete vaults and is under constant supervision. Some 700 cubic metres of this waste has been accumulated from the past 20 years of reactor operation in this country. It is recognised that storage in liquid form is not a completely satisfactory procedure and work aimed at solidifying the waste by a vitrification process is in progress. The long-term plan is to store the vitrified waste in specially constructed ponds near Windscale. The present ground area occupied by fission-product storage is 900 m^2 and it is estimated that this could increase to 7000 m^2 by the year 2000[26].

It is at present not possible to concentrate the other 0.02% of fission product activity and this is diluted and discharged to the sea off Windscale through a two-mile-long pipeline. Very extensive monitoring of the sea water, plants and fishes in the region show that the increase in the level of radioactivity to which the local population is exposed is less than 1% of the natural level. Whilst this situation appears to be satisfactory at present, it is generally acknowledged that as the scale of the nuclear power programme grows the percentage of fission-product waste discharged to the sea must be reduced even further, and underground storage of the vitrified concentrated waste may be necessary. The safe management of the plutonium extracted from spare fuel elements is vitally important because of its high toxicity.

Possible Accident Conditions

It seems clear that the environmental problems presented by the normal operation of nuclear reactors even in much greater numbers are manageable, and may well be considerably less than those accepted in the case of conventional power stations. We must therefore examine the possible hazards which may result from malfunction of nuclear stations.

The reactivity and consequently the power level of nuclear reactors are controlled by neutron-absorbing rods, and the intuitively obvious danger is that, owing to malfunction of rods or the servo mechanisms controlling them, the reactor power could increase in an uncontrolled manner. In order to prevent this all reactors are equipped with an auxiliary system for shutting the reactor down. This often consists of additional rods of neutron-absorber material normally held completely outside the reactor core and activated by a separate

control system which would drive them into the core and shut the reactor down if temperature or neutron-flux measurements indicated an unplanned increase in power. In some reactors, absorbing spheres rather than rods are used since these could enter a core even if it were deformed through excessive temperature rise. In most liquid-moderated reactors the safety or shut-down rods are augmented or replaced by systems injecting neutron-absorbing solutions (e.g. boron salts) into the moderator, whilst in CANDU the moderator may be emptied into a 'dump tank' thus shutting the reactor down.

There are also inherent factors in most thermal reactors which would limit the extent of any reactor 'runaway' or power excursion. As the temperature increases the moderator becomes less dense and therefore less efficient, and also because of thermal agitation of the ^{238}U nuclei the energy region over which their resonance trap operates is broadened, the so-called Döppler effect referred to in a previous Section. Both these factors reduce the fraction of neutrons captured by ^{235}U and reduce reactivity and consequently most reactors have a negative temperature coefficient which would limit the extent of any power excursion even in the unlikely event of the safety shut-down mechanisms failing to operate.

Loss of coolant is another possible hazard which could lead to overheating and damage to the reactor core. This could occur through blockage or rupture of the heat-transfer circuit or by failure of the main reactor pressure-vessel.

The resultant overheating of the core would activate the shut-down mechanism, but in some reactor types this would not represent a safe situation. Approximately 7% of the total heat produced in a reactor core is due to the fission-product activity and this is not immediately affected by the shut-down of the reactor. In highly concentrated cores such as the fast reactors, PWR, and BWR, the fission-product heating would be sufficient to cause rapid core-melting in the absence of effective emergency core-cooling. The adequacy of the core-cooling circuit in PWR particularly is currently the subject of concern particularly since this type of reactor has a steel pressure-vessel with possible danger of brittle fracture owing to temperature cycling and radiation. In the sodium-cooled fast breeder the high thermal capacity of the sodium pool surrounding the core provides an added safety factor even if sodium is lost from the circulating system.

Fission-product heating is less serious in other reactor types because the heat produced per unit volume is considerably less, but emergency core-cooling is provided in most types.

Reactor runaway or coolant failure may result in the melting or vaporizing of part of the reactor core, after which the reactor would become subcritical. The first priority in the event of such an incident is to limit the spread of radioactive fission products from the damaged core over the surrounding area. All reactors are contained in either a steel pressure-vessel surrounded by a concrete biological shield or a prestressed concrete vessel which serves both as a pressure-vessel and biological shield. In most enriched reactors a secondary containing vessel surrounds the whole reactor system including the primary cooling system. Filters prevent the discharge of particulate fission-product material to the atmosphere, so that the main dangers would be from gaseous fission products such as ^{131}I.

On only one occasion has there been such a release of ^{131}I in this country. This was from an early air-cooled reactor designed specifically for plutonium production and which did not incorporate the full safety measures now obligatory in power reactors. The release occurred at Windscale in 1958. Although it represented no real hazard to life, land over an area of 10 miles by 20 miles was contaminated with ^{131}I so that the milk from cattle in this area was considered to be unusable over a period of a few weeks (half-life of ^{131}I is 8 days).

The incident led to an increased awareness of possible reactor hazards. The safety of reactor design and operating procedures in the UK is the responsibility of the Nuclear Installations Inspectorate who are independent of both the United Kingdom Atomic Energy Authority and the Electricity Generating Boards. Reactor design must be such that no single failure could produce a release of radioactive material. A release is only possible if at least two independent components or systems fail simultaneously, and efforts are made to calculate the very small probability of this occurring, and also to calculate the maximum possible release of radioactive material which could result from this combination of faults. The safeguards against failures which could produce large releases must be correspondingly high[28].

It is therefore possible to plot a curve of the 'acceptable' frequency of accidental release as a function of the magnitude of such a release. The so called 'Farmer acceptable risk' curve is shown in *Figure 20*. All reactors operating in this country must be at least as safe as this curve implies. From the Farmer curve and consideration of the possible hazard represented by a release of a given magnitude it is deduced that one could live for a normal life span within 1 km of our existing power reactor types with only one chance in ten thousand of suffering serious damage to health from a release of activity[29]. This risk is small compared to the 6% risk of death or serious injury on the roads, in the factory or at home during a normal life span. The policy nevertheless has been to site nuclear power stations only in sparsely populated areas although this policy is now being relaxed to some extent as shown by the building of AGR stations at Heysham and Hartlepool.

Nuclear Fusion

As shown in *Figure 1*, the fusion of light nuclei is an exothermic process[30]. One of the most promising reactions as a source of energy is the fusion of two deuterium nuclei:

$$^{2}H + {^{2}H} \quad {\begin{array}{l} \nearrow \; ^{3}H + p + 4.0 \text{ MeV} \\ \searrow \; ^{3}He + n + 3.2 \text{ MeV} \end{array}}$$

This produces either a triton plus a proton or ^{3}He plus a neutron, and in each case about one-thousandth of the total mass involved in the reaction is converted into energy, about the same mass-to-energy conversion efficiency as that of the fission process.

The fusion of deuterium with tritium

$$^{3}H + {^{2}H} \rightarrow {^{4}He} + n + 17.6 \text{ MeV}$$

is even more efficient in that about four-thousandth of the total mass is converted into energy.

The fusion fuel reserves in the form of deuterium, the heavy isotope of hydrogen with an isotopic abundance of 0.015%, are estimated to be about $10^{10} Q$, many orders of magnitude larger than those of either fossil or fission fuels, and since the only radioactive product is tritium which is itself a fusion fuel, a fusion reactor programme would not produce large quantities of radioactive waste[31].

The rapid and uncontrolled fusion of deuterium nuclei is the basis of the 'hydrogen bomb' and nuclear fusion processes are responsible for the evolution of energy from the sun, but so far it has not been possible to produce a controlled fusion process on earth with an overall energy release. It was suggested by Sir George Thomson (1946) that a controlled fusion system with an overall energy release could be produced by heating deuterium gas or a mixture of deuterium and tritium[32,33], so that the thermal motion of the nuclei would be sufficient to overcome the Coulomb repulsive forces which prevent the near approach of the nuclei for fusion at normal temperatures. At high temperatures the electrons and nuclei of the gas become dissociated to form a so called 'plasma' made up of equal numbers of free nuclei and electrons, and there is an energy loss to the system due mainly to the emission of electromagnetic radiation, or 'bremsstrahlung', given off by the rapidly

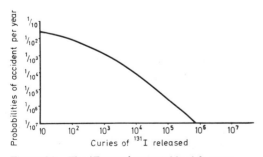

Figure 20   The 'Farmer' acceptable-risk curve

References p 339

vibrating electrons. This loss increases with temperature, but less rapidly than the calculated energy output due to the fusion process. Unfortunately the temperature at which the latter exceeds the former to give an overall energy output is extremely high, about 2×10^8 K for deuterium and 3×10^7 K for deuterium—tritium mixtures. The plasma must also satisfy the so-called Lawson Criterion $n\tau \geqslant 10^{14}$ s cm^{-3} where n is the plasma density and τ the time for which it can be held together, the so-called 'containment time'.

The required temperature far exceeds the melting point of any known material, so that the fusing plasma must be contained by magnetic fields rather than material vessels; and fusion research over the past twenty five years could be summarized as the search for magnetic field configurations which will contain a plasma at the required density in a stable manner for the time necessary to allow the proposed heating mechanism to achieve the fusion conditions.

Some heating can be achieved merely by passing an electric current through an ionized gas at low pressures, the ohmic heating process, and under these conditions the interaction between the current and the self-generated magnetic field around the plasma axis, the B_θ field, produces a compressive force towards the plasma axis and some degree of magnetic field containment. The resultant adiabatic compression of the plasma also provides a second heating mechanism, pinch force heating.

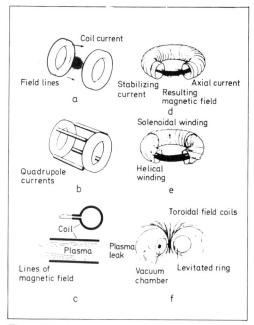

Figure 21   Some magnetic field confinement configurations [Reproduced from Pease, R. S. Phys. Bull. 1969 (Dec.), p 515 by permission of the Institute of Physics, Bristol ©]

In the linear pinch experiments of the above type, contact between the hot plasma and the metallic electrodes leads to the injection of impurity ions and low-energy electrons into the plasma and this represents an additional energy-loss mechanism. It can be avoided by using a toroidal discharge vessel so that the conducting plasma effectively forms the secondary of a high-current transformer. The stability of the discharge is improved by adding an axial magnetic field. This configuration was the basis of the British pinched toroidal discharge experiments Zeta and Sceptre in the late 50s[34-36] (*Figure 21d*).

Work at Princeton, USA, on their Stellerator experiments established that greater plasma stability could be obtained if the applied magnetic field were made to spiral about the axis of the discharge rather than be truly axial and this idea was incorporated in the Russian experiment Tokamak[37] (*Figure 21e*). Even greater plasma stability is theoretically possible if a conducting metallic ring is levitated magnetically to the axis of the discharge (*Figure 21f*).

Although the conditions of temperature and plasma containment have as yet fallen short of those required for a positive energy balance (*Table 7*), work has been directed towards the tentative design of a fusion reactor. The cross-section of a possible fusion reactor based on Tokamak is shown in *Figure 22*. It will be noted that molten lithium is proposed as coolant, and this also serves as a 'breeder blanket' in that the fusion neutrons interact with lithium to form tritium for use in future reactors.

$$^6\text{Li} + \text{n} \rightarrow {}^4\text{He} + {}^3\text{H}$$

$$^7\text{Li} + \text{n} \rightarrow {}^4\text{He} + {}^3\text{H} + \text{n}$$

The necessary neutron and thermal shielding between the plasma and the magnetic field coils necessitates the production of high magnetic fields over a large volume, and this can only be achieved economically by using cryogenic coils.

An alternative method of preventing contact between the plasma and the ends of the containing vessel is to increase the axial magnetic field strength near the ends. Electrons and ions are unable to penetrate the increasing magnetic field and are reflected back into the discharge. The region of higher magnetic field therefore acts as a 'magnetic mirror' and a configuration with such a mirror at both ends is known as a 'magnetic bottle' (*Figure 21a*).

A plasma so confined may be heated by adiabatic compression produced by the sudden increase in the average magnetic field (θ-pinch, *Figure 21c*,) or alternatively a reacting system of deuterons may in principle be built up by injecting high-energy deuterium ions into the system from accelerators. This is the principle of the Direct Current Experiment (DCX) at Oak Ridge, Tennessee, and of the Neutral Injector 'Phoenix' at Culham. Imperfect containment of the plasma at the ends is one

Table 7   Performance of some fusion experiments

Required conditions
$n\tau \geqslant 10^{14}$ s cm^{-3} $T \geqslant 2 \times 10^8$ K for D(dn) ^{3}He
$T \geqslant 3 \times 10^7$ K for T(dn) ^{4}He

Plasma compression experiments	n(cm^{-3})	τ(s)	$n\tau$	T(K)
Zeta	5×10^{13}	10^{-2}	5×10^{11}	10^6
Stellerator	10^{12}	10^{-2}	10^{10}	10^6
Tokamak	5×10^{13}	2×10^{-2}	10^{12}	4×10^6
θ—pinch	10^{13}	10^{-5}	10^8	2×10^7
Particle injectors DCX	10^9	10^{-4}	10^5*	300 keV (equivalent to 3×10^9)
Neutral injector	1.5×10^9	0.3	0.5×10^9*	20 keV (equivalent to 2×10^8)

*   Lawson's criterion not strictly applicable

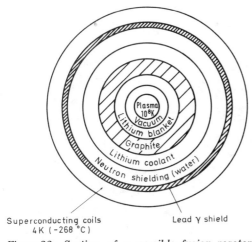

Figure 22   Section of a possible fusion reactor based on Tokamak

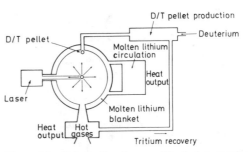

Figure 23   Schematic of a possible fusion reactor based on laser-induced fusion

of the main limitations of 'magnetic bottle' experiments and efforts are being made to improve this by superimposing a further magnetic field produced by four conductors parallel to the axis of the system (*Figure 21b*).

The performance of the various approaches to controlled fusion to date summarized in *Table 7* fall short of the required conditions for an overall energy release, but there has been a steady improvement in performance and in the understanding of the behavour of high-temperature plasmas as a result of these experiments[38].

A completely different approach is to use a laser beam to heat pellets of frozen deuterium and tritium. Theoretically the intensity of the laser beam can be programmed to produce a compressive shock wave in the vaporized tritium—

deuterium mixture to achieve the necessary fusion conditions. This work is in a fairly early stage of development, but a schematic of a possible fusion reactor based on this principle is shown in *Figure 23*[39].

Recently, with some justification, 30 000 MW fusion reactors have been predicted by the end of this century[38]. But a more widely accepted view is that a prototype machine demonstrating the possibility of controlled exothermic fusion by then would represent acceptable progress.

References

1   *Nuclear Engng* 1956, Vol. 1, 266
2   *Nuclear Engng* 1963, Vol. 8, 153
3   *Nuclear Engng* 1965, Vol. 10, 139
4   Fawcett, S. and Brown, G. *J. British Nuclear Energy Soc.* 1963, Vol. 2, 112
5   *Nuclear Engng* 1968, Vol. 13, 652
6   *HTGR Symposium, London, 1966, J. British Nuclear Energy Soc.* 1966, Vol. 5, 235
7   Valéry, N. *New Scientist* 1973, 59, 610

8   *Nuclear Engng* 1964, Vol. 9, 289
9   Walchli, H. and West, J. M. *Reactor Handbook*, 2nd edition, Vol. 4, Interscience, 1964, Chapter 16
10   *Nuclear Engineering International* 1974, Vol. 19, 303
11   *Nuclear Engng* 1968, Vol. 13, 416
12   Haubenreich, P. N. *J. British Nuclear Energy Soc.* 1973, Vol. 12, 145
13   Phillips, J. L. *Nuclear Engng* 1964, Vol. 9, 10
14   *Proc. Fast Breeder Reactor Conf.*, London, May 1966, Pergamon, 1967
15   Gratton, C. P. *J. British Nuclear Energy Soc.* 1972, Vol. 11, 331
16   Boxer, L. W., Hauserman, W., Cameron, I. and Roberts, J. T. *Proc. 4th Gen. Conf. on Peaceful Uses of Atomic Energy*, Vol. 8, I.A.E.A., Vienna, 1972, p 3
17   Avery, D. G., Bogaardt, M., Jelinck-Fink, P. and Parry, J. V. L. ibid., Vol. 9, p 53
18   Mohrauer, H. *New Scientist* 1972, 509
19   Stevenson, E. P. *Professional Engineer* 1974 (Jan.)
20   Dar, K. K., Yakaraman, K. M. V., Bhatnager, D. V., Garg, R. K. and Murthy, T. K. S. *Proc. 4th Gen. Conf. on Peaceful Uses of Atomic Energy*, Vol. 8, I.A.E.A., Vienna, 1972, p99
21   Davis, M., Bowie, S. H. U., Loosemore, W.R., Smith, S. E. and White, P. A. ibid., Vol. 8, p 59
22   Davis, M. P. 'Nuclear Energy Cost and Development' *Proc. Istanbul Symposium* published by I.A.E.A., Vienna, 1970
23   *Nuclear Engineering International* 1974, Vol. 19, 3
24   Ref.(A)
25   Seaborg, G. T. *Nuclear Energy* 1967 (Jan./Feb.), 16
26   Ref.(B)
27   *Recommendations of the International Commission on Radiological Protection*, Pergamon, 1958
28   Farmer, F. R. *I.A.E.A. Symposium on the Containment and Siting of Nuclear Power Reactors*, Vienna, 1967
29   Beattie, I. R., ibid.
30   Atkinson, R. E. and Houtemans, F. G. *Zeitschrift für Physik* 1929, Vol. 54, 656
31   Steiner, D. *New Scientist* 1971, 168
32   Thonemann, P. C. *Nuclear Power* 1956, Vol. 1, 169
33   Ware, A. A. *Engineering* 1957, Vol. 184, 610
34   Thonemann, P. C. *et al Nature, Lond.* 1958, Vol. 181, 217
35   Allen, N. L. *et al Nature, Lond.* 1958, Vol. 181, 22
36   Jones, W. M., Barnard, A. C. L., Hunt, S. E. and Chick, D. R. *Nature, Lond.* 1958, Vol. 182, 216
37   Golovin, I. N., Dhestrovsky, Yu. N. and Kostomarov, D. P. *Nuclear Fusion Reaction Conference*, Culham, UK, 1969
38   Pease, R. S. *Int. Conf. on Peaceful Uses of Atomic Energy*, Geneva, 1971
39   Nuckolls, J., Emmett, J. and Wood, L. *Physics Today* 1973 (Aug.), 46

General Reading

(A) Hunt, S. E., *Fission, Fusion and the Energy Crisis*, Pergamon, 1974
(B) Hill, Sir John, 'The Nuclear Fuel Industry', *Atom* 1974, No.207, p 2
(C) Bennet, D. J., *The Elements of Nuclear Power*, Longmans, 1972
(D) Glasstone, S., *Principles of Nuclear Reactor Engineering*, 1st edn, Macmillan, 1956
(E) Semat, H. and Albright, I. R., *Introduction to Atomic and Nuclear Physics*, 5th edn, Chapman & Hall, 1971
(F) Burcham, W. E., *Nuclear Physics, an Introduction*, Longmans

Total energy systems and heat salvage

Chapter 15

Combined cycles and total energy

15.1 General solutions

Transmission of energy
Combined cycles
Total energy

15.2 District heating

Factors relevant to the economics of district heating
District heating without electricity supply
District heating with electricity supply
District cooling
Consumer connections for district heating
Hot water versus steam for heat distribution
Development of district heating in Western Europe

15.3 Storing and pumping heat

Single, double, triple and quadruple heating mains
Types and design of heating mains
Calculation of heat and pumping losses
Expansion units and other auxiliaries
General economics of operating a district-heating system

Combined cycles and total energy systems are both forms of combined heat and power generation. They differ from power-station condensing steam turbines in that their purpose is to produce maximum *energy* efficiency. The primary purpose of central power generation is to produce maximum electrical efficiency. The maximum possible electrical efficiency is 40% based on today's technology and available materials but the average for Electricity Utilities is 28—30%.

The reason for this low efficiency is the thermodynamic limit on conversion of random (heat) energy to organized (mechanical or electrical) energy. The maximum efficiency is $(T_2 - T_1)T_2$, where T_2 and T_1 are the highest and lowest temperatures of the system in degrees absolute respectively. Taking T_2 at 843 K and T_1 at 303 K, the maximum efficiency approximates to 64%, the theoretical possible efficiency with a perfect gas, which steam is not. It is necessary to superheat the steam to avoid water droplets, both for power output and reduced turbine-blade wear.

In practice the average efficiency of a condensing steam turbine for central power generation would be fortunate to average 35% over its lifetime. As a world-wide average another 8% is lost in transformers, transmission lines and in distribution, giving the end user 27% of the input energy.

At today's fossil fuel prices and with a longer-term prospect of scarcity, this efficiency is broadly unacceptable: a critical factor in any national energy policy. Mainly for this reason, engineers involved in the energy field began to look elsewhere for more efficient solutions.

15.1 General solutions

Transmission of Energy

In today's Utilities, electricity is transmitted in bulk at from 6.6 kV to 735 kV, and transmission at 1000 kV or more is a future possibility. Even allowing for great advances in high-voltage technology, *Table 1* gives an indication of relative energy transmission costs.

It is clear from this Table that, relative to oil and gas, electrical transmission costs are much higher so casting doubt on the advantages of central generation.

Table 1   Relative transmission costs

Cheapest	Oil	1
	Gas	3
	Electricity	9
	Steam ⎫	
	Hot water ⎬	100
Most expensive	High-temperature air	—

Combined Cycles

As already stated combined cycles normally produce combined heat and power, but can produce power alone. The associated prime movers are gas and steam turbines.

The basis of the cycle is gas-oil or natural gas fed to a gas turbine; the exhaust heat of the gas turbine is recovered in a waste-heat boiler and converted into high-pressure steam; the steam is then expanded through a turbine which can pass the steam either to a condenser or to a process load. A condensing steam turbine would be used by an Electricity Utility and a back-pressure or pass-out turbine would be used in industries where there is a requirement for large quantities of steam.

Many such stations have been built and their efficiencies are known. In their condensing-steam-turbine form the overall efficiency has been slightly higher than with the best conventional steam turbines. The capital cost is lower, and the staffing cost considerably lower, than with conventional plant. While the overall thermal efficiency is slightly higher, the overall economic fuel efficiency would be lower on present (Jan. 1975) fuel costs. The fuels suitable in principle for gas turbines are gas-oil, natural gas, propane, butane or crude oil.

Suitability of Fuels

In a large (>20 MW$_e$) gas-turbine installation, the storage problems associated with LPG (i.e. propane and butane) would make them unsuitable as fuels. Crude oils would only be available in certain locations and so can be discounted. This leaves gas-oil and natural gas. At market prices (Jan. 1975) in the UK gas-oil is approximately 25% more expensive than residual oil. A possible increase in power generating efficiency of 2 or 3% could not justify moving from stations burning residual oil. If natural gas is available at prices equivalent thermally to residual oil (on a net CV basis) then it is well worthy of consideration.

Industrial Application

The industrial application of these cycles could be applicable in large process industries when natural gas is available at competitive prices. In an industry where there is a need for large quantities of low-pressure steam and an electrical maximum demand in excess of 20 MW$_e$ a combined cycle merits study.

The limiting factors in these cycles are the back-end temperature of the waste-heat boiler and the back-pressure on the steam turbine. In the waste-heat boiler the back-end temperature would be determined by the steam temperature: at least a 30°C differential between steam and back-end temperature should be maintained. This temperature differential is not too critical on low-sulphur fuels. The steam turbine should have a minimum absolute pressure range of 4:1 between inlet and outlet, i.e. 1600 kPa : 400 kPa, to make it an attractive investment.

Table 2   Comparison of combined with separate operation

Index	Item	Dimension	Combined operation	Separate operation
1	Inlet steam flow	(t/h)	395	395
	Inlet steam pressure	(kPa)	17 750	17 750
	Inlet steam temperature	(°C)	540	540
2	Condenser pressure	(kPa)	3.53	2.76
	Condenser flow	(t/h)	345	252
3	Feedwater temperature	(°C)	266	260
4	Steam-turbine output	(MW)	160	140
5	Gas flow	(t/s)	285	285
	Gas temperature	(°C)	307	307
	Gas pressure	(kPa)	105	101
6	Gas-turbine output	(MW)	49.5	50

Example

The example shown in *Table 2* gives a comparison between a combined cycle and separate operation. The Table illustrates that the typical type of use for combined cycles is more appropriate to central power generation than industrial use.

Total Energy

Total energy is a new name for an old concept. In the early years of this century many industries generated their own power by means of steam engines coupled to electrical generators. Many produced direct current but later produced alternating current. Because of the increased price of primary energy and the potential shortage it has become important to study methods of producing secondary energy in a more efficient manner.

Since central power generation is only averaging 28–30% overall efficiency a more efficient alternative must be found. One alternative is private power generation with waste-heat recovery, based on reciprocating internal-combustion engines, gas turbines, and back-pressure steam turbines or engines with their associated heat recovery or primary-heat-raising units.

At the present time *total energy* has come to mean a system which supplies all the electricity

needs of a site either in isolation or in parallel with a public Utility and all or part of the heating and cooling load of the site, using any one of the prime movers mentioned.

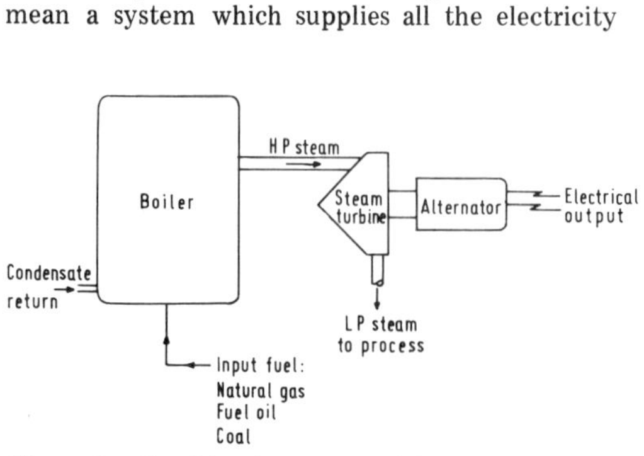

Back-pressure   Steam   Turbines   and   Engines (Figure 1)

These are well known in industry, and approximately 20% of the industrial electrical load of the UK is thus generated. The back-pressure steam engine is seldom found today but is by no means out-dated.

In the back-pressure steam turbine (Ch.10.7) system, the primary fuel is converted into steam at >1 MPa in a shell boiler (up to a pressure of some 2.7 MPa) or a watertube boiler (to a higher pressure) with some superheat at an efficiency of between 80 and 90% and then expanded through a multi-stage or single-stage turbine to a lower pressure, at which stage it is passed to process. The turbine shaft is coupled to an alternator to produce electric power.

The overall efficiency of such a cycle can be as high as 80%. In industries which have a heat demand in excess of 5 kg/h of low-pressure steam continuously to every kW and have a maximum electrical demand above 200 kW for more than 4000 h/yr, such a system is worth considering.

In order to calculate the electrical load associated with the inlet heat flow it can be assumed that an average of 60% of the heat flow through the turbine will be converted into electricity. A single-stage turbine will give slightly less than this figure.

Reciprocating Engines in Total Energy (Figure 2)

Reciprocating engines (Ch.12.1) can be divided into three groups:
1. Diesel engines (light and heavy fuel).
2. Dual-fuel engines.
3. Gas engines.
The total-energy cycle is the same for all three. It comprises a fuel input to the engine which provides three forms of energy recovery:

Figure 1   Combined power and heat: h.p. boiler steam turbine.   Courtesy Integrated Energy Systems Ltd

(a) Shaft-power which is converted into electricity.

(b) Exhaust-gas heat which is converted into steam or hot water.

(c) Jacket cooling-water which is recovered as hot water.

The overall cycle efficiency can be as high as 75%. The diesel engine will have the highest conversion of fuel to shaft-power, with the dual-fuel engine the next best. This is a very important factor when the net costs of the input fuels are approximately equal, as electricity has a financial value at least three times (UK, Jan. 1975) that of an equivalent amount of heat.

But the diesel engine is not necessarily an automatic choice over dual-fuel and gas engines, as natural gas is available in some countries at prices less than fuel oil.

Dual-fuel Engines. The advantage of a dual-fuel engine is that it can use both gas-oil and natural gas, so adding to the security of energy supply. Also it can take advantage of interruptible natural-gas contracts, that is to say one that can be interrupted for a period of up to 90 days in one year. Interruptible gas contracts are offered at a lower thermal cost than firm gas supplies, so that a prime mover that can use a second easily-stored fuel has real economic advantage.

The thermal efficiency of the dual-fuel engine is 36%, so that the availability of gas at a price 10% less than that of fuel oil makes it worthy of consideration.

Gas Engines. The spark-ignition gas engine has a thermal efficiency of 30% or less and can only use propane as a standby fuel. However with waste-heat recovery from the exhaust and recovery of the jacket cooling-water an operating fuel efficiency of 70% or greater is possible.

Natural gas at 140 kPa at the factory gate is necessary. It would be better if this pressure were available from the pipeline as the additional cost of compressors and continuous loss of power to the compressors is a distinct disadvantage.

If propane is used it would be wise to limit its usage to not more than 3 kt/yr or as an ideal

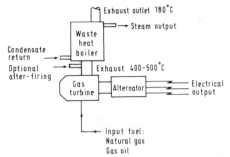

Figure 3   Combined power and heat: gas turbine. Courtesy Integrated Energy Systems Ltd

Energy distribution (%)

	Diesel	Dual-fuel	Natural gas
Shaft efficiency	41	39	29
Exhaust	32	34	41
Jacket water	14	13	21
Lube oil	4	4	5
Intercooler	6	7	—
Radiation	3	3	4
	100	100	100

average 2 kt/yr. The reason for this figure is that it approximately meets the economic level of storage and ease of delivery. With 2 kt/yr of propane and a load factor of about 50% it would be possible to produce at least 2 MW_e electrical power and 9000 MJ/h (2.5 MW_t) thermal flow.

Gas Turbines (Figure 3)

Gas turbines (Ch.12.2) are most suitable for total energy schemes where there is a high heat/electrical power ratio, i.e. 6 MJ/kW h or greater. These are now capable of meeting maximum demands in the range from hundreds of kW up to hundreds of MW in parallel units.

The very small units, i.e. 300 kW, have a thermal efficiency of about 13% while the larger units, i.e. 10 MW and greater, have thermal efficiencies of twice this value.

Types of Turbine. There are essentially two types of gas turbine — the radial and the axial gas turbine. In the context of total energy overall fuel efficiency is the criterion, not just high thermal efficiency. The difference between the two turbines respectively is that the incoming air is radially compressed and the turbine is of radial design, or axially compressed and the turbine of axial design.

We shall try to outline a fuel-efficiency comparison between the two. Intake and exhaust losses are neglected, and we take ambient temperature $15°C$, ambient pressure 100 kPa, a mechanical efficiency of 98% and an electrical efficiency of

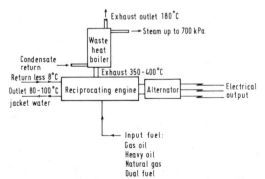

Figure 2   Combined power and heat: reciprocating engine. Courtesy Integrated Energy Systems Ltd

95%, and gas at a net calorific value of 37.26 kJ/l (37.26 MJ/m^3).

Fuel input (experimental):

Radial	Axial
22.2 MJ/kW h	21.1 MJ/kW h

Since 1 kW h = 3.6 MJ

Thermal efficiency:

	Radial	Axial
	16.2%	17%

Heat can be recovered from the exhaust gases. We shall assume heat in the form of dry saturated steam at a pressure of 700 kPa (heat content 2559 kJ/kg) and feedwater at a temperature of 50°C.

Exhaust mass flow (experimental) (kg/kW h):

Radial	Axial
33	30

Exhaust temperature (experimental) (°C):

Radial	Axial
550	525

Taking a value of 1.088 kJ/kg °C for the specific heat, an efficiency of 95% for heat transfer from gas to steam and an outlet (gas) heat-exchanger temperature of 200°C, then the heat recovery is:

Radial	Axial
11.93 MJ/kW h	10.75 MJ/kW h
Steam 4.66 kg/kW h	4.20 kg/kW h

Total fuel efficiency (electrical + thermal output):

Radial
= (11.93 + 3.6)/22.2
69.8%

Axial
= (10.75 + 3.6)/21.0
66.6%

These are rough calculations. Ideally both should be between 66 and 70% efficiency, and total efficiencies of up to 70% are achievable using gas turbines.

The figures shown can be improved by lowering the outlet temperature of the heat exchanger, and this can certainly be achieved on natural gas as it has little or no sulphur. When using gas-oil it should not fall below 200°C if there is 1% sulphur or more in the fuel, to avoid corrosion (cf. Ch.20.2).

Gas Pressure. The axial gas turbine requires a gas pressure of at least 1000 kPa whereas the radial gas turbines can operate at pressures of 650 kPa. The lower gas inlet pressure can be an advantage when compressors are needed. In many cases gas is supplied at between 50 and 100 kPa.

Worked Examples (for UK, Jan.1975)

It is difficult to illustrate the detailed advantages without taking a typical factory load and analysing

its suitability for a total energy scheme. This can only be done by considering one country at a particular point in time; but the reader can apply the principles and methods to his own situation.

Factory Load. The electrical maximum demand is 5 MW and the electrical consumption is 18 000 MW h/yr, the steam capacity is 15 t/h at 700 kPa, with a consumption of 60 000 t/yr.

Choice of Prime Mover. From the information given we can see that there is a ratio of just less than 1 kW h/3 kg of steam. At this type of balance all prime movers would be investigated.

To cover a load of 5 MW$_e$ we would consider 4 × 1300 kW engines or turbines for base load and one on standby, giving a total of five engines. The fuels suitable for engines of this size are natural gas, oil at a viscosity 4.2 mm^2/s (cSt), 247 mm^2/s and 864 mm^2/s.

Cost of Fuels. We shall assume (UK, Jan. 1975): 5.0 p/l for 4.2 mm^2/s oil, 4.4 p/l for 247 mm^2/s oil, and 4.2 p/l for 864 mm^2/s oil, and natural gas 11.4 p/100 MJ net CV at pressures of 700 and 280 kPa. Electricity at 2.00 p/kW h.

Worked Examples for Individual Prime Movers

Dual-fuel Engine. We shall take a 600 rpm engine. The fuel consumption will be 250 l/kW h with natural gas at a net CV of 37.26 kJ/l (37.26 MJ/m^3). Therefore natural-gas consumption is 9.31 MJ/kW h. The other fuel used is 4.2 mm^2/s oil for pilot fuel injection. This consumes 6% of full-load consumption, i.e. 0.013 kg/kW h. The oil has a CV of 45.6 MJ/kg and a cost of 6.00 p/kg.

Energy Input and Cost. The primary energy input is 9.31 + (0.013 × 45.6) = 9.91 MJ/kW h.

Input fuel cost:

Natural gas	9.31 × 11.4/100
	= 1.06 p/kW h
Gas oil (4.2 mm^2/s) 0.013 × 6	
	= 0.078 p/kW h

Total   = 1.138 p/kW h

Exhaust Heat Recovery and Value. The exhaust mass flow is 7.32 kg/kW h at a temperature of 440°C.

Heat recovery = mass flow × specific heat × heat-exchanger efficiency × (absolute temperature difference of gas)

= 7.32 × 1.088 × 0.95 × (440 − 200)
= 1815 kJ/kW h

We require the heat in a dry-saturated-steam form at a pressure of 700 kPa with feedwater at 50°C. The heat required to raise this steam is 2558.6 kJ/kg. Therefore steam raised = 0.71 kg/kW h.

In a normal shell boiler we would expect an efficiency of about 80% over the year. The net cost per 100 MJ is taken at 11.4 p and the gross price at 10.25 p/100 MJ. Heat value in steam is 10.25/0.8 = 12.8 p/100 MJ. Therefore exhaust heat value is $1.815 \times 12.8/100$ = 0.232 p/kW h.

Fresh-water and Lube Oil Systems. The cooling water is available at about 82°C and should be returned to the engine at 72°C. The quantity of heat available is 1.41 MJ/kW h; the heat value based on 10.25 p/100 MJ is 0.144 p/kW h.

The lubricating oil is available at 73°C and returns to the engine at 60°C. The quantity of heat available is 0.365 MJ/kW h and the heat value is 0.0375 p/kW h.

The total cooling-system recovery is 1.41 + 0.365 = 1.775 MJ/kW h with a value of 0.182 p/kW h.

Economic and Energy Balance.
1 kW h costs 2.00 p.
Input fuel costs 1.138 p/kW h.
Exhaust heat recovery has a value of 0.232 p/kW h.
Cooling system recovery is 0.182 p/kW h.
Total recovery is 0.414 p/kW h.
Savings are $(2.00 - 1.138 + 0.414)$ = 1.276 p/kW h.

These are only fuel savings. We must also establish maintenance, operation, insurance and lubricating oil costs.

Energy output (MJ/kW h) = 3.600 (elec.) + 1.775 (heat) + 1.815 (heat) = 7.19
Energy input (MJ/kW h) = 9.31 + (0.013×45.6) = 9.90
Efficiency = 72.5%

Maintenance. Maintenance is taken at 1½% of capital cost. We shall take a figure of £150/kW installed including standby, waste-heat-recovery plant and switchgear, with operation for 3600 h/yr. Therefore (15 000/3600) $\times$ (1.5/100) p/kW h = 0.0625 p/kW h.

Operation. This is a two-shift, five days/week, 48 weeks/year system. We would assume that one additional engineer is required over the existing staff, but shall cost two. Additional cost would be £8000/yr.
This gives 800 000/18 000 000 (factory annual load) = 0.035 p/kW h. Therefore operation and maintenance cost is (0.0625 + 0.035) = 0.0975 p/kW h.

Insurance. We shall assume a nominal insurance cost of 1% of product output. This gives 2.41/100 = 0.024 p/kW h.

Lubricating Oil. Based on lubricating oil at 13.20 p/l, oil consumption of 1.35 g/kW h, specific gravity 0.9, cost is $(1.35 \times 13.2)/(0.9 \times 1000)$ = 0.020 p/kW h.

Total cost of operation, maintenance, insurance and lube oil is 0.035 + 0.0625 + 0.024 + 0.020 p/kW h = 0.142 p/kW h.

Eventual Unit Cost and Savings. The real cost in p/kW h is: Fuel − heat recovery + (maintenance, operation, insurance and lube oil) = (1.138 − 0.414 + 0.142) = 0.866 p/kW h.

Based on 18 000 000 kW h and electricity at 2.00 p/kW h, the annual savings are $18 \times 10^6 \times (2.000 - 0.866)/100$ = £204 300. The capital investment to install 5000 kW with adequate standby would be about £750 000.

Possible Improvements. If there is a reasonable percentage of raw-water make-up, then if the heat from the air-coolers can be used an additional quantity of heat, 720 kJ/kW h, is available at a temperature of about 40°C and of value 0.074 p/kW h. The increase in revenue is £13 360/yr. The overall efficiency is now: 100(7.19 + 0.72)/9.90 = 80%. It is also possible to decrease the exhaust temperature from the heat exchanger, giving an overall efficiency of 81%. This is about the maximum theoretical efficiency, and if all the heat recovered can be used the annual savings are about £220 000/yr.

Notes. It is clear that only a percentage of the total heat required in a steam form at 700 kPa is available for use. The remaining low-grade heat can be used for preheating boiler feedwater (every 6°C rise in feedwater constitutes a 1% saving in fuel). In many cases it can be used for environmental heating. The remaining steam required can be raised by composite-firing the exhaust-boilers (giving a heat transfer of 95%) or by conventional shell boilers run in parallel with the exhaust-boilers.

We have assumed a 415 V three-phase supply at 50 Hz with an earthed neutral on one generator. By using electronic governing we would expect a quality of supply as follows: ±3% on voltage and ±0.4% on frequency. Additionally we would expect the 3rd harmonic to be less than 5% of the fundamental wave.

If this supply feeds an 11 kV factory bus-bar through an existing distribution transformer which would be star-wound on the 415 V side and delta-wound on the 11 kV side, then it is essential that the 11 kV delta is fed through an earthing transformer. The ideal solution and often the least expensive in the long term is to replace the transformers and install a 415 V delta-wound primary and 11 kV star-wound secondary.

Electrical Control and Governing. The electrical control and governing cannot be covered in this brief summary. It is sufficient to say that if there are two or more sets operating in parallel, then electronic governing is usually desirable, and also where possible to use alternators of similar make and characteristics; otherwise the complicated addition of chokes is necessary. In most applications of this size of unit, brushless alternators are suitable.

Heavy-fuel Engine (see note on costs, p 346). In using heavy-fuel engines for total energy schemes the first thing necessary is to define the oil. We shall assume 600 rpm engines and the following would be a limiting fuel specification at the injectors.

Water (volume %)	0.1
Ash (wt %)	0.1
Sediment (wt %)	0.01
Vanadium (wt %)	0.01
Sodium (wt %)	0.003
Sulphur (wt %)	3.5
Conradson carbon residue (wt %)	10.0
Viscosity at injectors	12.6 mm^2/s

This is only a guide-line as a fuel specification should always be presented to the engine manufacturer for agreement.

Fuel. In this case we shall take 247 mm^2/s oil at a cost of 4.40 p/l. When oil is used above a viscosity of 4.2 mm^2/s it is usually treated before passing to the injector. As shown in the specification the oil should be at 12.6 mm^2/s. To reduce the viscosity it is usual to preheat the oil and additionally to pass it through centrifuges to purify and clarify it.

Fuel Consumption and Energy Input. The fuel consumption is 0.216 kg/kW h at gross CV 43.03 MJ/kg, i.e. 9.285 MJ/kW h. The cost is 1.013 p/kW h based on 11.2 l/kW h.

Exhaust-heat Recovery and Value. The exhaust mass flow is 7.725 kg/kW h at a temperature of 400°C. The specific heat of the exhaust gases is 1.088 kJ/kg °C. Proceeding as before, if 200°C is the outlet gas temperature from the heat exchanger, heat recovery = 1.597 MJ/kW h. As dry saturated steam at a pressure of 700 kPa with a feedwater temperature of 50°C, i.e. 2.559 MJ/kg, this is equivalent to 0.624 kg/kW h of steam. With an efficiency of 80% in a shell boiler using the same oil, the heat value in steam is 13.64 p/100 MJ. Therefore exhaust heat value is 1.597 × 13.64/100 = 0.218 p/kW h.

Fresh-water and Lube Oil Systems. The cooling water is available at a temperature of 82°C and should be returned to the engine at between 71°C and 77°C. The heat available is 1.132 MJ/kW h. Using the above value, heat value = 0.154 p/kW h.

The lubricating oil is available at a temperature of 79°C and returns to the engine at between 63°C and 68.5°C. The quantity of heat available is 0.495 MJ/kW h, and the value 0.0675 p/kW h.

The total cooling system recovery value is (1.132 + 0.495) = 1.627 MJ/kW h, with a value of (0.154 + 0.0675) = 0.2215 p/kW h.

Economic Balance and Efficiency. 1 kW h costs 2.00 p, input fuel costs 1.013 p/kW h. Exhaust-heat recovery has a value of 0.218 p/kW h, cooling system recovery is 0.2215 p/kW h, total recovery is 0.4395 p/kW h. Savings are (2.000 − 1.013 + 0.440) = 1.427 p/kW h. These are only the *fuel* savings.

Energy output = (3.600 (elec.) + 1.597 (heat) + 1.627 (heat)) = 6.824 MJ/kW h
Energy input = 9.285 MJ/kW h
Efficiency   = 73.5%.

Maintenance and Heating. We shall take a figure of 3% of capital cost for maintenance and pre-heating of oil; £150/kW installed including standby, waste-heat-recovery plant and switchgear. Hours of operation 3600 h/yr. Therefore the associated cost is (15 000/3600) × 3.00/100 = 0.125 p/kW h.

Operation. We take the same cost as for the dual-fuel engine, giving 0.035 p/kW h.

Insurance. We shall assume a nominal insurance cost of 1% of product output. This gives 1×10^{-2} × 2.40/100 = 0.024 p/kW h.

Lubricating Oil. Based on lubricating oil at 13.20 p/l, sp. gr. 0.9, and oil consumption of 1.35 g/kW h, cost is 0.020 p/kW h. Total cost of operation, maintenance, and heating, insurance and lube oil is (0.035 + 0.125 + 0.024 + 0.020) = 0.204 p/kW h.

Eventual Unit Cost and Savings. The real cost is fuel − heat recovery + (maintenance, heating, operation, insurance and lube oil) = (1.013 − 0.440 + 0.204) = 0.777 p/kW h.

Savings. Based on 18 000 000 kW h/yr and electricity at 2.00 p/kW h, the annual savings are £220 000. The capital investment to install 5000 kW with adequate standby would be about £750 000.

Possible Improvements. As already stated there is a considerable quantity of low-grade heat available (1170 kJ/kW h) from the air-coolers. The value of this heat at 13.64 p/100 MJ is 0.1596 p/kW h. The increase in revenue would be about £31 000/yr. Therefore overall savings could be as high as £251 000.

The overall efficiency is now 84.8%, the maximum possible theoretical efficiency.

Electrical. The same conditions electrically apply as to the dual-fuel engine.

Gas Turbine (see note on costs, p 346). We shall assume 2700 kW$_e$ gas turbines as the balance of electricity to heat requirements is such that unit sizes of 1000 kW would produce too much heat and so discharge it to the atmosphere wastefully.

Fuel Consumption. We shall assume natural gas as the fuel with a net CV of 37.26 MJ/m^3, and ambient temperature at 15°C. Then the fuel consumption is 16.80 MJ/kW h at 11.4 p/100 MJ, costing 1.93 p/kW h.

Exhaust heat recovery, cooling 25.3 kg gas per kW h from 490 to 200°C, is as calculated previously 7.584 MJ, valued at 12.8 p/100 MJ, i.e. 0.971 p/kW h.

Fuel cost per kW h = (initial fuel cost − waste-heat recovery cost) = (1.93 − 0.971) =

0.96 p/kW h. Steam at a pressure of 700 kPa from feedwater at 50°C has a heat value of 2559 kJ/kg; therefore output = 2.96 kg/kW h.

For electrical/heat balance this turbine would seem ideal, but the eventual answer is usually decided on economic grounds.

Efficiency = 100 (electricity + waste heat)/(fuel input) = 100 (3.600 + 7.584)/16.8 = 66.6%.

Maintenance and Lube Oil Systems. Maintenance is relatively low on these turbines as it would take seven years for the primary inspection. We shall take a figure of 0.15 p/kW h for maintenance and lube oil.

Insurance and Operation. We shall take a nominal figure of 1% of output for insurance, giving 2.75 × 1/100 = 0.027 p/kW h. We shall retain the previous cost of 0.035 p/kW h for operation.

Cost of maintenance, lube oil, insurance and operation
= (0.15 + 0.027 + 0.035) p/kW h
= 0.212 p/kW h

Eventual Unit Cost and Savings. The eventual cost is: Fuel input — waste-heat recovery + maintenance, lube oil, insurance and operation = (1.93 — 0.971 + 0.212) = 1.17 p/kW h approximately.

The savings are £149 000 per year for an approximate investment of £800 000.

Electrical Output. The electrical output would be generated through a brushless alternator at a speed of 1500 rpm.

Analysis of Results

From *Table 3* it appears that the heavy-fuel engine is the most suitable choice of prime mover. This is only so if both the steam and hot water recovered can be used. If only the steam can be used then the gas turbine is the most suitable.

The case studied is not a particularly favourable one as it has a load factor of only 39.5%. If the price of electricity remained at 2.00 p/kW h and the load factor increased to 60%, the change to total energy would be a very good investment.

Total Energy using Waste Fuels

As defined total energy is combined power and heat, and so does not preclude using otherwise on-site waste for conversion to a useful form of combined power and heat.

The two most readily available waste forms of primary energy are waste gases at temperatures greater than 350°C, and wood-waste.

Waste Gases and Heat Recovery. Typical application of waste gases are distilleries that normally use coal or butane to heat their stills. The exhaust gases are available with a CO_2 content of 10% and a temperature of 482°C when using coal. A typical still might burn 150 kg/h. Then with eight stills there is an input of 1200 kg/h. Usually there are 18 kg of waste-gas per kg of coal, with 85% excess air.

If steam is required at 1050 kPa dry-saturated from a feedwater temperature of 77°C, its heat value is 2455 kJ/kg.

The gases will have a mass flow of 18 × 1200 = 21 600 kg/h. Taking the specific heat of the gases as 1.088 kJ/kg °C, the steam equivalent of the waste-heat recovery calculated as previously is 2570 kg/h.

If this steam is expanded through a single-stage turbine down to a pressure of 175 kPa it can produce 80 kW.

This is an example of total energy in its most financially attractive form.

Other Industries

Other industries that could benefit from such schemes are those making steel, glass and cement. A number of these schemes are applied in these industries.

Wood-waste. As an average, 1000 kg/h of wood-waste in a timber yard is reasonable. The net CV is about 19 MJ/kg. If steam (low-grade) and electricity are required on the site, using a shell boiler with a maximum steam pressure of 1.61 MPa and 55°C of superheat it can produce 5500 kg/h of steam, and if this is expanded through a single-stage turbine down to atmospheric pressure it can produce 400 kW of electricity.

The installed cost of such a scheme would be £100 000 including silos and wood-hogging equipment. On a 2000 h/yr basis, if electricity would otherwise cost at least 2.50 p/kW h (probably more because load factor is low) and steam 0.30 p/kg (based on residual oil at 4.1 p/l), there is an annual recovery value approaching £60 000. Allowing for

Table 3 Total-energy comparison of prime movers

	Fuel input (MJ/kW h)	Heat recovery		Cost (p/kW h)	Energy efficiency (%)
		Steam (MJ/kW h)	Hot water (MJ/kW h)		
Dual-fuel engine	9.91	1.815	1.775	0.866	72.5
Heavy-fuel engine	9.285	1.597	1.627	0.777	73.5
Gas turbine	16.80	7.584	—	1.17	66.6

References p 364

operation, maintenance, insurance and electricity standby connection charges, this form of combined power and heat could save £50 000/yr.

There are of course cases where there is a continuous electrical need and only a minor use for heat. Such cases should be considered but will not have as high a return on capital as a continuous total energy scheme.

Smaller Total Energy Schemes

In schemes with electrical maximum demands between 60 kW and 2000 kW the use of high-speed diesels (i.e. 1000 and 1500 rpm) and gas engines is common. For the diesel engine the only suitable fuel is gas-oil, and for the gas engine natural gas and propane. Many total energy schemes of these sizes are in operation throughout the world.

The system of analysis can follow the lines already demonstrated for the heavy-fuel and dual-fuel engines.

The Future

The future forms of total energy will hopefully be based on the Stirling engine[1,2] and especially on the fuel cell (Ch.13.4). The particular fuel cell suggested can be described as a continuous battery fed by hydrogen and oxygen; the ensuing products are electricity and hot water. There is an electrochemical efficiency of 60%. The present impediments to growth are the cost and life-span of catalysts. At present these incorporate platinum and are too expensive for commercial use.

Conclusions

Total energy schemes should be considered on any site where there is a use for electricity and steam (or hot water) for more than 2000 h/yr. Any site which has a heat to electricity ratio in excess of 3 kJ/kW h merits study.

15.2 District heating

District heating is the term given to the public supply of heat for space heating and domestic water heating, under arrangements where the consumers are treated as individual customers and are charged either at a flat rate or on the basis of metered consumption.

The incentive for the adoption of district heating exists in so far as, in a specific project, it is assessed to have definite economic advantages over alternative and potentially competitive forms of heating, such as electricity, individual central heating, and room-heating appliances. Since the establishment and maintenance of district heating systems *per se* is costly, district heating is most advantageous economically if it can make use of very cheap sources of heat, such as reject heat

from power stations and industrial processes, heat from the incineration of refuse, and geothermal heat. Most of the schemes which have been successfully established are of this kind, though in assessing projects for particular localities the use of conventional fuels may occasionally come up for consideration. It is important, in this connection, to distinguish between true district heating and 'group heating' schemes. The former is 'open-ended', the demand known only in general terms; in the latter a relatively small number of dwellings, e.g. a group of blocks of flats, or a definitely known load linked with a local authority, is supplied with heat from a central boiler house (virtually a captive market). In fact, there has up to the present time been very little experience with district heating in the UK, though numerous group-heating systems have been established. The scope of future development of district heating in this country is as yet unclear; its adoption in a particular situation will depend on careful assessment of a complicated set of economic and social factors.

Apart from economic advantage, there are social factors that are certain to increase in importance. District heating gives greater freedom to the architect, a factor less important than hitherto when fuel storage and ash disposal were common needs; and it gives the consumers greater flexibility in space allocation. There may also be some reduction in air pollution. Above all, district heating combined with electricity generation can (at a price) save a considerable part of the heat necessarily rejected owing to the low thermodynamic efficiency of electricity production; currently in the UK this loss is equivalent annually to the total heat in 55—60 million tons of coal and it is economically important since some £550 million p.a. worth of (mostly imported) oil is (Jan. 1975 prices) used in power stations. This is a notable example of the difference in point of view between energy utilization efficiency (a national problem) and efficiency in terms of cost effectiveness — a difference always present in defining the efficient use of energy but seldom so clear-cut.

In Western Europe, there has been considerable development of district heating in Sweden, Denmark and West Germany. In Eastern Europe, comprehensive networks exist in the Soviet Union, Poland, Czechoslovakia and Hungary, and most American cities have systems that supply steam for space heating in winter and for driving refrigeration systems in summer.

This Section of the Chapter will be devoted principally to a consideration of the main features of the several possible variants of the district-heating theme, progressing from the simplest concept based on the supply of heat alone from heat stations fired by conventional fuels, through the use in such systems of reject fuels, industrial waste heat and geothermal heat, to combined heat and electric power stations fired by fossil fuels, and finally to combined stations operating on nuclear fuels.

Some information about the present status of district-heating developments in Sweden, Denmark and West Germany is given at the end of this Section.

Factors relevant to the Economics of District Heating

Each district-heating project must be subjected to assessment to establish whether or not the overall cost of heat provided by it would be substantially lower than that of heating by individual appliances. Important considerations include the following:

(a) Cost of prime fuel. Obviously, the inducement to adopt district heating using reject heat or fuel becomes greater as commercial fuels become more expensive compared with recovered waste or energy. Equally obviously, for a commercially viable system savings in fuel costs must more than outweigh the real costs of providing and maintaining the system. Each case needs separate detailed assessment, taking into account, for example, the value of any relevant capital items (gas mains, electricity supply equipment, domestic-heating appliances) which may already exist in the district under consideration.

(b) Cost of boiler plant. Even in systems which depend mainly on waste heat, it is still necessary to install some conventionally fired boilers, to supply peak heat as required and also to serve as standby in case of partial or complete breakdown of any of the units normally providing heat to the system. *Figure 4* indicates that for a system designed to serve a population of 10 000 people, the cost of boiler plant and ancillaries was about £170 000 in the UK in 1972.

(c) Cost of pipe network. *Figure 5* shows the relation between the cost of mains and the capacity of the system for a number of different

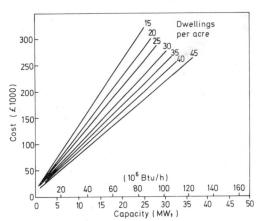

Figure 5   Cost[3] of mains (UK, 1972)

housing densities. It will be seen that the cost, which is of a similar order to that of the boiler plant, is highly sensitive to the housing density; serving 10 000 persons, it ranges from about £130 000 for a density of 45 dwellings/acre (4050 m^2) to about £225 000 for a density of 15 dwellings/acre.

In the past it used to be assumed that the economic lower limit for installation of district heating in Germany was about 160 GJ/h km^2 (44 MW$_t$/km^2) for existing urban areas and 100 GJ/h km^2 for new areas. With increased fuel costs, the economic limit is however coming down steadily, so that today even areas with less than 50 GJ/h km^2 (14 MW/km^2) can be provided economically with district heating.

In addition to the actual installed cost of the pipe network, a number of subsidiary related factors are of great importance. The network as installed must match the final designed capacity of the system. If the build-up of the heat load to the full capacity is slow, as in a new urban development, this means that the capital charges per unit of connected load will be extremely high. If, on the other hand, the scheme is to serve an established community, the cost of the network will be much higher than in a new development (see e.g. ref.5), so that although the build-up will be more rapid, unit costs will still be high. However, if the consumer network is under the jurisdiction of the operating authority these disadvantages may be partly offset by the captive market. And on the other hand if it were proposed to use the heat from a power station sited near the fuel source, long runs to an adequate population centre might be necessary in some countries. Ideally the build-up of heat and power loads in new developments may be kept in step by using modular units, so avoiding the worst penalties.

(d) Method of charging consumers. The common alternatives are to meter the heat supplied to each individual dwelling or to charge consumers on a flat-rate basis. Metering of individual consumers

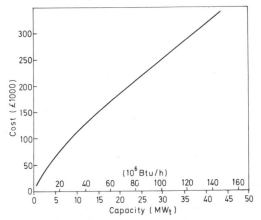

Figure 4   Cost[3] of boiler plant, chimneys, building etc. (UK, 1972)

References p 364

adds to the initial cost but is widely practised in Europe, particularly in Scandinavia. The meters used are frequently based on the evaporative principle and represent a design compromise between first cost and sophistication; an assessment and review of the characteristics of meters of this type has been made in the UK[4]. They are less widely used in the UK and the more usual practice so far has been to charge consumers on a flat-rate basis. However, the supply of unmetered heat encourages waste, particularly of domestic hot water, and this waste can rise to levels high enough to nullify to an appreciable extent the potential thermal (and economic) advantages of district heating. So far as heat supplied for space heating is concerned, waste can be controlled, at a cost, by the provision of thermostatic valves, but this solution cannot, of course, be applied to limit the waste of domestic hot water.

(e)   Concentration of assured consumers. It is important that the development should be able to rely on a sufficient body of consumers at adequate population density. Municipal housing developments present the most favourable conditions from this point of view. The standards of domestic heating offered to the consumers would be higher than those generally available to them in the past.

(f)   Equability of climate. The scheme must be designed to meet the most severe weather conditions likely to be experienced, and this obviously has its implications for cost. The absence, as in the UK, of periods of extremely cold weather is a favourable indication for district heating.

(g)   Interest charges. The economics of district heating, like those of other capital projects, must depend greatly on the ruling level of interest rates. But, at least in the case of new housing developments, other competing forms of heating will be similarly affected.

(h)   Matching of heating and electrical loads. This of course applies only to those cases where the district-heating scheme depends to a major extent on the use of rejected heat recovered from power stations. The modifications and complications which such use involves at the power stations are discussed later in this Chapter. The extent to which heat and electricity loads can be matched depends on a variety of factors on which it is impossible to generalize, e.g. the ratio of domestic to industrial heating load, and whether there is scope for combining the system for supplying heat with air conditioning and/or refrigeration. As would be expected, the ideal situation of a well matched heat and electricity load is rarely met with in practice.

Of the purely technical features pertaining to the district-heating system itself, the most important of the factors mentioned above in its effect on the cost of heat is the pipe network.

In the following paragraphs, the several possibilities for the supply of energy for district heating are discussed in more detail.

District Heating without Electricity Supply

Using Conventional Fuels Only

From the technical standpoint, a great deal of experience exists in the use of coal, oil and gas in furnaces of the capacities required to provide heat in district-heating schemes, so that the most difficult design problem is that of optimizing the boiler unit size and the standby provision for fluctuating and seasonally variable loads. A great deal will of course depend on whether the system is to serve both domestic and commercial consumers or domestic consumers only.

The potential economic advantages of a district-heating system fired by conventional fuels over the provision of heat by individual appliances depends mainly on the larger furnaces employed. First, it is possible to use the cheaper industrial grades of fuel bought in bulk, rather than the expensive premium fuels distributed to individual consumers, and secondly there will be some gain — perhaps 5—10% — in combustion efficiency. However, against this improvement in efficiency have to be set the heat losses in the distribution mains. The magnitude of these losses will obviously vary from one situation to another, but values quoted in the literature amount to at least 5%, so that gains in efficiency achieved at the boilerhouse will be largely cancelled out. It seems unlikely that savings in the cost of fuel will provide economic justification for district heating schemes in this category in any but the most exceptional cases. It may be pertinent to mention here that improvement in the thermal insulation of dwellings is likely to give rise to savings in fuel at least as great as that from the adoption of district heating *per se*, for much smaller capital expenditure[3].

Unconventional Sources of Heat

The sources of heat discussed under this heading, viz. municipal refuse, sewage, industrial waste heat and geothermal heat, are widely diverse in character, but they have in common the fact that they are at present largely wasted. To the extent that they could be put to use to replace conventional fuels they would augment indigenous fuel resources and thus reduce the reliance of a nation on imported fuel. In particular, there exists the possibility that heat from some of these sources could be made available at sufficiently low cost to make a decisive improvement in the economics of district heating.

Refuse Incineration[6]. In most urban societies, the amount of refuse thrown away per person each year varies between 200 and 300 kg. This refuse has a high content of moisture (*c.* 40%) and also a high percentage of inorganic non-combustibles, such as stones and bottles. The calorific value varies between about 6 and 9 MJ/kg, compared with about 30 MJ/kg for coal and 43—45 MJ/kg for oil. The material is low in density (*c.* 150—250 kg/m^3), and extremely variable in size. A trouble-some constituent is PVC plastic, because during combustion it yields extremely corrosive hydrogen chloride.

For district-heating purposes, it is necessary on economic grounds to use very large incinerators, with a minimum capacity of some 5 t/h, and preferably larger. The upper limit of size is normally governed by the distances over which refuse can be transported economically to the plant.

If it be assumed that the heat value of the refuse produced by an average family of four people is about 10 GJ per annum, while the heating demands for such a family are between 40 and 60 GJ per annum, then it seems that, taking into account heat losses during combustion etc, the refuse thrown away by between five and eight families is required to provide the heat needs of one. In theory, therefore, and in the most favourable circumstances, it seems that, until the district-heating network is built out to cover more than 15% of the city's dwellings, the whole needs of the system could be met by the combustion of refuse.

The pattern of operation of a refuse incinerator is largely governed by the presence of large quantities of hydrogen chloride and oxides of sulphur in the flue gases. Because hydrogen chloride and oxides of sulphur in flue-gas condensates attack steel and many of its alloys very rapidly at low temperatures, this means that the chimney exhaust temperatures have to be kept at above 220$°$C, which is appreciably higher than for other types of furnace. However, the rate of corrosion also increases drastically at temperatures above about 480$°$C, so that in practice the conditions of steam generation are limited to 4.3 MPa and 450$°$C. It is therefore clear that the efficiency of a boiler system fired by a refuse incinerator must be substantially lower than that of one fired by a conventional furnace. In addition, the specialized construction necessary for such furnaces, and the auxiliary equipment needed, add greatly to the capital cost. It has been found best to use the comparatively low-pressure steam produced in back-pressure or pass-out turbines, and to use hot water at about 110$°$C for covering the base load of the district-heating network. The electricity produced is a valuable by-product, even though the turbine efficiencies are of necessity rather low.

In many European schemes, refuse incinerators are used to provide the bulk of the heat required during the summer and other off-peak periods, while the main heating/power systems are being overhauled.

Summarizing, it may be said that the economics of district heating based on refuse incineration are not outstandingly attractive in their own right, but amenity considerations could probably, in certain cases, lead such schemes increasingly to be regarded with favour in the future.

Sewage. If sewage sludge is dewatered to a moisture content below, say, 70% and a calorific value of about 6 MJ/kg, it may be burned autothermally in a fluidized bed or a multiple-hearth furnace. The possibility of disposing of sewage sludge by incineration is attracting increasingly favourable attention from one amenity angle. It is then logical to consider using the heat released in the combustion of the sludge for raising steam at the relatively low pressure adequate for operating a district-heating system. There is also the possibility of using in this way not only fuel in the form of raw or fermented sludge, but also methane produced during its anaerobic fermentation. This latter proposal would have less force if some other profitable use (e.g. power for the sewage works itself as at Mogden, West London) existed for the methane gas.

Industrial Waste (or Rejected) Heat. It is common practice for large industrial undertakings to utilize their own waste heat for works purposes. Often, however, the amount of waste heat produced is far in excess of the amount which can be effectively utilized within the works. Disposal of this excess by river-water coolers entails expense, so that it is potentially an economically favourable source of heat for district-heating purposes. Each case must, of course, be considered on its merits. There will be many factors to be taken into account in matching the availability of the waste heat to the district heating needs, foremost among which is likely to be the cost of constructing the distribution network.

Geothermal Heat[7]. In a few localities in the world, the best known of which is Iceland, geothermal heat is sufficiently readily available to enable it to supply the district-heating needs of a whole urban population. The possibilities of applying geothermal heat in district heating are, however, not restricted to these relatively simple cases. At Carrierres-sur-Seine in France, for example, it has been found that the temperature of the ground water increases 1$°$C for each 30 m depth, and at a depth of 1580 m the temperature of the water is about 60$°$C. This water is pumped to the surface and utilized for district heating. In cases such as this, the hot water from the underground source may be used for supplying base load heat, the peak load being supplied by other sources, e.g. the waste heat from power generation. In schemes of this kind, the initial investment cost of drilling is high, though fuel costs are limited to pumping costs and interest on capital.

District Heating with Electricity Supply

District heating may be combined with electricity generation in either an independent scheme or within the public utility operation, depending on commercial factors and legislative controls. Apart from the effects of scale, the economics of an independent scheme where legally permitted will be greatly affected according to whether it is possible for the community served by the scheme to draw current from outside, or whether it has to be entirely self-contained.

In the generation of electricity, whether by steam turbines, gas turbines or diesel engines, more than one-half of the potential energy of the original fuel is normally wasted in the form of low-grade heat. In view of the immense scale of fuel usage in electricity generation, the prospect of recovering a substantial part of this heat and using it for district heating purposes with a positive cost benefit is obviously highly attractive.

The problem of recovering waste heat from generating stations based on steam turbines differs in one important feature from that of recovery from gas-turbine and diesel powered systems. While in the latter cases recovery of waste heat is straightforward and can be effected without significant modification in the operation of the power unit, with steam-turbine plant a change in basic design is required, viz. from condensing sets to back-pressure or pass-out turbines, and possibly additional boiler plant. These matters are discussed more fully below.

When the power station is operated in this way, there is a difficult problem in achieving the optimum balance between heat and electricity production. With back-pressure turbines, the ratio of heat to electricity production is higher than would be accommodated in a unit of the normal present-day capacity, while in units much below $100 MW_e$ the economics of electricity production are very unfavourable. Pass-out turbines, however, offer a more flexible solution to this problem; and the existence of a demand for steam for driving air-conditioning and refrigeration plant is, as was mentioned earlier, a favourable feature for the overall economics.

The technical and economic problems associated with the various types of combined systems are discussed in greater detail below.

Gas Turbines[9] (Ch.12.2)

Gas turbines, both open cycle and closed cycle, are used widely by the utilities in the production of electric power. Their advantages are that they can be started up rapidly, and so are suitable for the generation of power to satisfy peak requirements, and that they do not need large quantities of cooling water. On the other hand, the thermal efficiency of gas turbines is lower than that of steam generating sets because of the higher exhaust temperatures used. The efficiency of gas turbines can be improved by using them in conjunction with steam turbines. In the present context, however, this type of arrangement is rather costly in terms of capital, and because the feed temperature of the steam is then low, the thermal efficiency of the steam turbine is not high. A better proposition is to use directly the waste

heat in the exhaust gases for district-heating purposes. The heat available and the electric power generated are then roughly equal in quantity, and the economics are much improved if there is a demand for energy for air-conditioning. Open-cycle gas turbines are in use in the United States to provide both district heating and district cooling. The turbine drives a compressor which serves for the provision of chilled water, while the waste heat in the flue gases is used to heat the water in the district-heating circuit.

The capital cost of the gas turbine per installed kilowatt is relatively low but, because premium fuels are required, fuel costs are high.

It has been proposed to use closed-cycle gas turbines for 'total energy' (Ch.15.1) operations employing high-temperature helium-cooled nuclear reactors as power sources. According to one scheme, worked out by the Escher Wyss organization in conjunction with Westinghouse, the heated and compressed helium first gives up the bulk of its stored energy in driving the turbine; the residual heat is abstracted by the water of the district-heating system. It is said that overall efficiencies of more than 85% are attainable in such a system, with considerable flexibility as between electricity and heat outputs.

Diesel Engines[8] (Ch.12.1)

Diesel engines are used in a number of places as the basis for total-energy systems (Ch.15.1). The best known example in the UK of the use of diesel engines in a district-heating scheme is at Aldershot, where 18 MW (max.) of power and 110 GJ/h of heat are being supplied by Mirlees diesel engines at >60% efficiency.

Although the diesel engine operates at a high thermal efficiency (approaching 40%), at least one-half of the potential energy of the diesel fuel appears as heat in the cooling jacket and exhaust gases. There is little technical difficulty in reclaiming this waste heat and using it for district-heating purposes. The heat from the cooling jackets is taken off by means of so-called latent-heat or evaporative (Ch.16.3) cooling. The water is pressurized to about 200 kPa absolute and steam is flashed off, giving up its latent heat in a condenser to produce hot water at about $115°C$. Additional capital and operating costs are introduced in the process. The exhaust heat is employed for heating up water in a standard waste-heat boiler (Ch.16.1).

It is to be noted that with diesel power units, unlike steam power stations, it is impossible to vary the ratio of output of electricity and total heat. However, as in diesel power plant both jacket heat and exhaust heat are commonly wasted, the intrinsic cost of the recovered heat, as distinct from the cost of distribution, is very low since recovery despite some increase in costs does not entail any sacrifice in electricity production. Diesel-powered schemes, like those based on gas turbines, suffer from the drawback that fuel costs are high because of the need to use refined fuels.

Steam Turbines[8].

The condensing steam turbine in general use for power production employs a fuel which may be oil, coal, gas or nuclear power, to heat water to a temperature close to the critical point. Most modern power stations operate at a steam feed temperature of between 500 and 540°C and a steam pressure of 11—14 MPa. On the exhaust side of the turbine the steam temperature may be as low as 40—50°C and its pressure will be down to approximately 5 kPa absolute. Even under these conditions however the steam contains an appreciable amount of latent heat, which must be removed by coolers. Such coolers are large and costly.

In even the most up-to-date condensing turbines about 50% of the heat of the original fuel is lost in the coolers. If account is taken of the unavoidable stack loss of 10% and the loss of heat in ash, it becomes clear that the efficiency of power generation at a coal-fired plant is at best around 34%. In practice, owing to the fact that many power stations on the grid system are obsolescent, the overall generation efficiency of electricity production (UK) is little more than 30% (1973); 28% delivered taking into account transformer and transmission losses.

The operation of the total energy principle, i.e. the utilization of heat as well as electricity, would enable steam power stations to be operated more efficiently and more economically. However, in normal condensing stations, the waste heat at the condensing end is at too low a temperature for practical use, and if the total energy concept is to be applied the turbine design must be suitably modified to make utilization of this waste heat possible. The main possibilities are back-pressure turbines and pass-out turbines or intermediate take-off condensing turbines (ITOC sets).

Back-pressure Turbines. In order to increase the condenser water temperature to a useful level, turbines can be employed in which the low-pressure side is at approximately 110°C or even higher, instead of the very low levels employed with normal steam turbines.

The main established uses for back-pressure turbines are for the supply of industrial process heat, and in conjunction with refuse incinerators, where owing to corrosion considerations it is impracticable to use high steam feed temperatures.

The main disadvantages of back-pressure turbines arise from the inflexible ratio of heat to electricity produced and the very large quantities of heat produced by even the smallest and thus least economic turbine. For example at Battersea power station in London back-pressure turbines of only 1.5 MW$_e$ capacity are used in connection with the local heating scheme, and these are able to produce about 5 MJ of heat for each MJ

(3.6 kW h) of electrical energy. Further, these turbines have an internal efficiency of only 64% and a current-generating efficiency of no more than 11.5%. It would probably be uneconomic to use back-pressure turbines when the heating load fluctuates widely, seasonally and diurnally, as it does in a typical community. For the reasons advanced above, the operation of modern, large turbine units, which vary in capacity between 200 and 650 MW$_e$, is obviously ruled out.

Pass-out Turbines[10]. The basic design of pass-out turbines is similar to that of normal condensing turbines, and the units can be made in any required size, up to that common for large condensing sets. They have therefore become the heating/power station turbines in most general use throughout the world.

In the normal condensing turbine, provision is made to bleed off some steam to preheat the boiler feedwater. In pass-out turbines, however, extra steam can be bled off at different points in the plant and used to heat up water in the district-heating circuit. It is possible to control the degree of steam bleed-off to accord with the heat demand. When no heat is required, the extra steam bleeds can be shut off altogether, and the plant operated as a normal condensing turbine. The power-generating efficiency may then approach that of the normal turbine, the actual value depending on the rated bled-steam flow for which the turbine design is optimized.

The layout of a power/heating installation based on a pass-out (ITOC) turbine is shown schematically in *Figure 6*. Boiler feedwater is heated by means of bleed heat exchangers. The high-pressure steam is first passed into the high-pressure turbine section and then into the medium-pressure turbine. The steam bleeds for the feedwater preheat are taken off at the high- and medium-pressure sections. At the junction between the medium- and low-pressure sections, steam is bled off through automatically controlled by-pass valves at about 110—140°C, and passes through a valve into the second district-heating condenser/heat exchanger to heat the water to the correct circulating temperature. The rest of the steam passes to the condensing section and then to the first district-heating condenser, to raise the temperature of the return district-heating water. There is also a normal type of cooling system for operation without heat abstraction. In order to save the cost of installing cooling towers, the Stal-Laval system[11] uses precooling of the return water, so that this alone can carry the cooling load. The return water is cooled by being used for such purposes as the heating of swimming pools, snow melting in winter, pavement heating, etc.

There are a number of variants of this basic design of pass-out turbine. For example, Russian types, which provide circulating water at temperatures of up to 200°C, have a large number of step bleeds in order to obtain better steam economy.

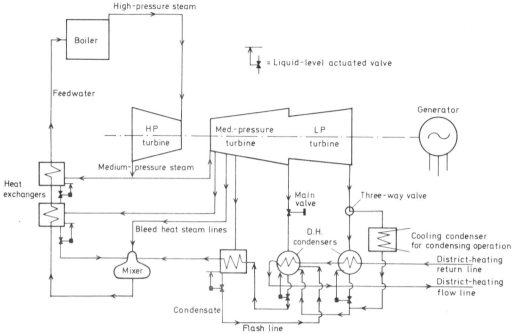

Figure 6   Typical modern ITOC turbine of 200 MW$_e$ *capacity*

During operation at maximum heat take-off, the pass-out turbine can attain a heat-plus-power efficiency as high as 85%, compared with an efficiency of only 36% when it is used for power generation only. But it cannot be assumed that because of this gain in overall efficiency the fuel cost for heating is zero. The means adopted for making heat recovery feasible leads at the same time to a reduction in the efficiency of electricity generation and hence to an increase in its cost. Moreover, a larger boiler plant will be needed to maintain the same electrical load, involving additional cost. The element of fuel cost which is to be assigned to heat production in a particular case will, of course, depend on the characteristics of the system.

There are other advantages of a combined system based on pass-out turbines. While electricity has, of course, to be used immediately on production, heat can be stored in the form of hot water, so that a pass-out station can be operated in a more economical way than a normal condensing power station. At night, when both heat and power consumption are low, the pass-out station is operated with by-pass at a maximum. The heated water is stored partly within the insulated pipe network, and partly in insulated storage vessels (accumulators). During periods of peak power usage, the by-pass operation is suspended and the plant produces only electricity and no heat; the heating demand is then satisfied from the energy stored in the hot water.

Advantages and Drawbacks. The potential advantages of district heating combined with power generation using pass-out turbines may be summarized as follows:

(*a*)   There is a substantial saving in overall fuel cost and demand for fuel.

(*b*)   A given plant can be run on more even load, leading to savings in plant capacity, with corresponding capital savings.

(*c*)   That part of the peak load problem at power stations which is caused by increased space-heating demands during very cold spells is ameliorated; the amount of standby plant which has to be provided and maintained is reduced, since additional heating load can be met by increasing the period during which the plant is on full by-pass operation.

(*d*)   Heating/power stations will produce less off-peak electricity than do conventional stations; where a problem exists in the profitable disposal of off-peak electricity, this will therefore be eased.

These potential advantages of heating/power stations have been discussed above in a generalized way; they can only be quantified in relation to a specific project. To present a balanced picture, the disadvantages must also be mentioned.

(i)   Additional capital cost is incurred, in the turbine and sometimes the boiler plant as well as in the heating network.

(ii)   The additional plant components mean lower reliability (cf. Ch.13).

(iii)   Operation of the plant as a whole is more complicated.

Apart from these disadvantages of the system itself, mention may also be made of the fact that interruption of primary fuel supplies, from whatever cause, will affect all the consumers of heat served by a combined scheme, with possibly serious social consequences.

The necessity to ensure maximum security in electricity supply may also react significantly on the economics of power/heating schemes. If the scheme is completely independent of the public supply, account must be taken of the possibility that to provide an acceptable level of security it will be necessary to install additional standby plant. If on the other hand district heating is combined with the public supply of electricity, account must be taken of the effects on power cost that may arise from the need to run plant on a load pattern different from that which would optimize the total system cost of the electricity grid.

Nuclear Heat Sources

The use of nuclear reactors in combined power/heating schemes poses particular problems, on which little experience has as yet been gained. At the present time only one nuclear reactor has been used for municipal heating, the Ågesta reactor in Sweden, which supplied the base load of the district-heating network of Farsta, a dormitory suburb of Stockholm. This is a pilot scheme, regarded as experimental and exploratory. After ten years of satisfactory operation, it has been closed; on the information gained, several much larger but basically similar stations are being built: Haninge (south of Stockholm), Barsebeck (3 units, for south Skane), and a third close to Gothenburg.

When nuclear reactors are used for district-heating purposes, it is economically desirable that (as with other energy sources) they should be sited as close as possible to the centres of population they are to serve. But safety is the paramount consideration, and if the reactor can be accommodated underground at a realistic cost this is advantageous. Underground siting is facilitated by compact design.

The Ågesta reactor was of the pressurized heavy-water type. It was sited inside a rock chamber 56 000 m^3 in volume, so constructed that each part had a rock-covering of at least 60 m. Because of this, and the advanced safety precautions at the plant itself, the installation was located at a distance of only 9 miles from Stockholm, and very close to populated areas. The installation is shown schematically in *Figure 7*.

The fuel rods in this experimental plant consisted of enriched UO_2 and were canned in Zircalloy. The operating temperature inside the stainless-steel pressure vessel was 220°C, and the heavy water gave up its heat to circulating normal water in the system of primary heat exchangers. The steam produced, which had a pressure of 1.4 MPa gauge, was used to drive either a back-pressure turbine in winter or a condensing turbine in summer. During winter operation the plant delivered 18 MW_e of electricity and about 100 MW_t of heat, and in the summer 31 MW_e of power but no usable heat.

District Cooling

As air conditioning becomes more widely adopted, there is likely to be a move away from individual cooling plants to centralized installations where chilled water is produced for distribution by means of underground supply and return lines. This is an already established trend; several plants serve networks for both district heating and district cooling, notably in the USA. One of the largest systems of this type is at Hartford, Connecticut. This has a refrigeration capacity of 53 MW, and is also able to deliver around 100 t/h of steam for supplying various heating networks. As has been mentioned earlier, the combination of district cooling with district heating can make an important contribution to the economy of power/heating systems.

Consumer Connections for District Heating[12]

The normal layout of a town's district-heating network is in the form of a ring circuit, with flow and return mains. This system is thus able to continue in operation even if a section of the pipeline network has to be shut down to enable repairs to be carried out.

The types of auxiliary equipment required depend upon whether the consumer is an industrial organization, a large block of flats, or a single small house.

For tall buildings it is normally necessary to heat up a secondary circuit of water by means of a calorifier installed in the basement, and to use a

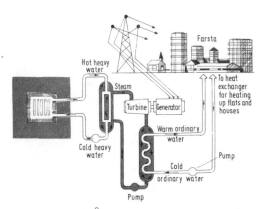

Figure 7 The Ågesta experimental pressurized heavy-water nuclear reactor provided heating and electric power to the town of Farsta near Stockholm

closed pumping system to distribute this secondary heated water around the building. For low blocks, direct heating can be employed. Accurate mechanical heat meters may be positioned in the basement, to register the total amount of heat abstracted by the block. The consumption of each individual tenant is either estimated by apportionment of this total or measured by means of small secondary heat meters. The problems of metering have been briefly discussed earlier in the Chapter (see pp 351—352). For smaller houses, small prefabricated connection units are available.

Hot Water versus Steam for Heat Distribution

Advantages of hot water: obtained more economically than steam, no wastage of purified water, safer and easier to handle, and the capacity of the network can be varied merely by altering the temperature difference between flow and return.

Advantages of steam: when supplying high-lying districts, operating refrigeration plant, and supplying industrial consumers.

In Europe the bulk of the heat comes from combined stations, which are disadvantageous for steam production, so that hot water is necessarily favoured. In the USA steam systems have predominated because of relatively low fuel prices, which reduce the emphasis on total energy, and because of the suitability of steam for air-conditioning (cooling) systems. There are indications that even in the USA steam may be on the way out, flashed-off steam from the hot-water mains being used for the 'chillers'.

Development of District Heating in Western Europe

Comprehensive district-heating networks exist throughout the whole of Eastern Europe, where very severe winter conditions are experienced. In Western Europe, the countries in which district heating has been most widely developed are Sweden, Denmark and West Germany.

Sweden[13]

In 1973, about 600 000 housing units were connected to district-heating systems, and it is estimated that by 1980 70% of all multi-dwelling blocks and 20% of single family houses will be served.

The bulk of the heat used for the various district-heating networks is the associated heat from power stations, which in some cases (e.g. Västerås) covers up to 98% of the total heat required for the operation of the system. Swedish district-heating/power stations operate at efficiencies of roughly 85%, i.e. 30% electrical and 55% thermal energy.

The total district-heating energy production during 1973 amounted to about 12 000 GW h (43 000 TJ), corresponding to an estimated saving of about £12 million per annum (at 1973 prices for fuel oil).

Figure 8 shows the layout of the Malmö district-heating network.

Denmark[14]

In 1973, about 30% of all dwellings in Denmark were connected to district-heating networks. This

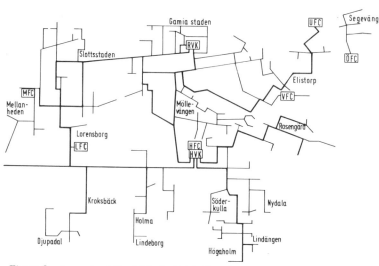

Figure 8   Layout of the Malmö district-heating network (Sweden). HVK and RVK: heating/power stations. MFC, LFC, HFC, VFC, UFC and OFC: boiler plants. Length of main pipeline network: 150 km in 1972

is a high proportion, considering that Denmark is a country with mainly small family houses. The adoption of efficient construction methods has made it possible to install district heating in areas with very low thermal-load densities. Oil-fired heating/power stations and refuse incinerators provide the bulk of the heat. The present connected capacity is about 25 TJ/h and the annual amount of heat supplied is running at the rate of about 50 000 TJ.

West Germany[15]

In 1972 there were 91 district-heating undertakings in West Germany with a combined connected capacity of 67.5 TJ/h. A total of 82 849 main consumers were connected, and the network length amounted to 4501 km.

The biggest scheme of all is in Hamburg, with a connected capacity of 11.1 TJ/h.

The average utilization factor for German district-heating systems is about 0.57, with industrial

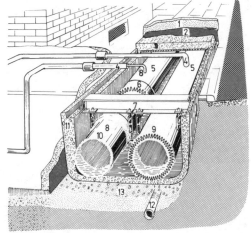

Figure 9   Twin district-heating main laid in concrete conduit, together with house connection. Courtesy Malmö Technical Authority and Leif M. Bell (adapted from a drawing by Leif M. Bell).

  1   Top layer of gravel
  2   Asphalt cover. Bitumenmat
  3   Concrete slab
  4   Feeder line. Plastic pipe culvert with cover of PEH or PVC and insulation of mineral wool, polyurethane foam or similar
  5   Main- or distribution pipeline
  6   Mineral wood insulation covered with plastic sheet
  7   Pipe support including steel beam, movable pendulum and support ring
  8   Steel tube
  9   Outgoing flow pipe. 120–80°C
10   Return flow pipe. 70–50°C
11   Prefabricated U-section from reinforced concrete
12   Drainage pipe of plastic
13   Crushed stone

· *References p 364*

use around 6400 h/yr, while domestic utilization is about 1600 h/yr at maximum capacity.

The total heat supplied by the networks annually is about 134 000 TJ. The average annual increase in capacity is 9.4%, and it is planned that by the turn of the century 25–30% of all dwellings in the Federal Republic will be connected to district-heating systems.

Heating/power stations provide 70.4% of the heat used for district heating, reject heat from industry 8.4%, and refuse incinerators and direct heating 21.2%.

15.3 Storing and pumping heat

Single, Double, Triple and Quadruple Heating Mains[16]

Heat is transported from district-heating supply plant to the consumers by means of insulated pipelines. The nature of these pipelines varies considerably, depending upon the type of terrain through which they pass, whether the area is already built-up, whether the water table is high or low, and so on.

The most common pipeline system is the double-ring main system (*Figure 9*), with an outflow temperature of around 110–140°C and a return temperature of about 60–70°C. The normal district-heating system uses water in a closed system, which is made alkaline by the addition of sodium hydroxide and is kept at an approximate pH of around 9.5. In addition, the water is virtually free from oxygen and is kept so by the addition of hydrazine and similar antioxidants. Pipelines are always constructed from mild steel, with the exception of steam-condensate returns (where fitted), which are made from stainless steel. Pumping is carried out at the works, using booster pumps where necessary in the circulating system. It is essential that the water inside the system is pressurized to such an extent that even at the maximum altitude there is always a positive pressure, sufficient to prevent the hot water from boiling within the pipes. Booster pumps and pressure vessels may be necessary, employing nitrogen gas inside the pressure vessel to maintain the necessary liquid head.

The temperature of the water in the flow line varies during the year, reflecting the different heating capacities needed in summer and in winter. If a network is being built out, one can carry the increased load temporarily by increasing the flow temperature. Obviously there is a limit to the flexibility of supply which can be obtained in this way, but it does enable the distribution network to be built out temporarily without immediate need for increased-diameter mains. Steam systems have to be preplanned much more accurately, as one mainly uses the latent heat in the steam which does not vary appreciably with small changes of temperature and pressure.

Heat can be taken from the pipeline either directly or indirectly. For small houses or low-rise blocks of flats, the hot water from the district-heating network is bled directly into the central-heating system, using a thermostatic valve, and the cooled water is then returned via the return main. Town water is heated via a calorifier to about 80°C for baths and for cooking purposes.

If one has to supply premises which are elevated beyond the pressure range of the district-heating supply system, i.e. houses on hills or tall blocks of flats, one needs to install substations. In these, the circulating district-heating water heats a separate flow system via heat exchangers, which has its own pumping system, powerful enough to reach even the most elevated consumer. Obviously, the construction of a secondary water circuit is expensive and also loses some heat, and such a system is therefore avoided wherever possible.

Single-pipeline System

A common trend of electric power production is the location of very large stations far from the consumer network, close to coalfields or near oil installations. Although in the UK it is rare to find an electric power station further than 60—70 km from a city, this is not so in the USSR where distances can be immense. Under such circumstances it becomes economic to heat water to temperatures of 200°C in stations equipped with a series of bleed heaters, and to transport this water by means of a single mild-steel insulated pipeline towards the nearest city or cities. The very hot water is then bled into a normal 120°C/60°C urban twin-pipe district-heating system.

As there is no return main the volume of water within the system increases, necessitating the disposal of some of the cooled circulation water. This is used for swimming-pool heating, for laundries, and for various process industries. In practice it is found that there is little need to run off excess water to the sewers.

The main problems involved are in water purification. The water sent out from the power stations is purified river water, treated previously by coagulation, filtration and ion-exchange techniques. Removal of oxygen and carbon dioxide is carried out by mechanical means. As the water is transported at about 200°C it is completely sterile. Heat losses are very low as the pipelines are well insulated. Costs of pumping are less than with cold water owing to the reduced viscosity.

In the USSR it has been found that the use of a single-pipeline system is still feasible even if the distance between the consumer city and the power station exceeds 225 km. On the other hand, the system has been found to be less effective when unduly low temperatures are experienced, because under such circumstances the ratio of heating to hot water required is too high and this means that

one needs to discard too much water to the sewers. Conditions in the UK, where the bulk of power production is carried out in very large power stations which are some distance but not really far from urban areas, and where climatic conditions are never extreme, would seem to be ideal for the application of the single-pipeline system of heat supply. Unfortunately no study has been carried out, even though the cost of transporting heat in the form of hot water can be shown to be less than half that of transporting the same amount of energy in the form of electricity. Furthermore, supplementing overhead high-voltage cables by buried pipelines would reduce the threat to the visual environment.

Three-pipe System

This system of supply is relatively rare, but virtually all the pipeline network in the Berlin district-heating system is of this nature. The three-pipeline system consists of two normal-sized mild-steel mains and a very much smaller one, all running parallel with each other. The intention is that in winter two flow mains and one return main are used for the supply of the hot water, while during the summer only the narrower pipe is used. It is possible to vary the supply temperatures in the two flow pipelines, so that the smaller one can run at a higher temperature than the larger one, enabling industrial loads to be carried more economically. The system is more flexible in cases of breakdown and repair, and in summer pumping costs and heat losses can be reduced considerably by using only the narrow-gauge pipe. Obviously the capital costs are higher than with the more common twin-pipe system, but under certain circumstances the system is economically extremely viable.

Four-pipe System

It is generally inadvisable to produce hot water for consumption at a central station and to transport it to the consumer, as the pipelines for oxygenated water must be made of more corrosion-resistant (and therefore more cathodic) materials than pipelines carrying deoxygenated and alkaline district-heating water. Four-pipeline systems are commonly used today when the district-heating supply of a community is combined with district cooling. There are then three insulated pipelines, carrying respectively the heated water supply, the return water, and the chilled water, in addition to an uninsulated pipeline for the cooling return water. With the present-day growth of complete air conditioning, where for appreciable periods of the year both heating and cooling capacity may be required, such systems are likely to increase in number. Even in very hot areas where heating needs during the day are zero, some space heating may be needed in the evenings. There is always the need to provide hot water via calorifiers. In temperate areas, there is a need for cooling even in winter in office blocks where large areas of glass face south, or where there is need to dispose of waste heat produced by electrical appliances, as in hospitals and television studios.

Types and Design of Heating Mains

As the basic mains used for district-heating purposes are made from mild steel, which is a rather corrodible material, the problems of insulating such mains to avoid both heat losses and corrosion damage are quite formidable. Approximately, the rate of corrosion doubles with every rise of $10°C$. Therefore a district-heating main with a surface temperature of say $90°C$ corrodes about 250 times as fast in the same circumstances as a cold-water, oil or gas main at around $10°C$.

As the temperature of the heating mains does not remain constant, expansion and contraction movements take place, putting strain on the insulation cover. Costs of digging out and repairing district-heating mains are very high indeed, and in consequence even the slightest corrosion damage can ruin a district-heating system financially. Because until fairly recently little was known about the corrosion aspects of buried insulated pipelines in the UK, several schemes failed completely for this reason. In Europe, because more technical expertise was available, this did not occur and the average annual maintenance charge for district-heating distribution systems is about 0.5% of the capital investment in pipelines.

Insulation Systems

The following are the most common:
1. Loose-fill insulation.
2. Mineral-wool insulation in trenches.
3. Gas-concrete in concrete conduits.
4. Foamglass insulation.
5. Prefabricated foam plastic-insulated pipelines.
6. Bitumen insulation.
7. Steel pipeline-within-a-pipeline systems.

Loose-fill Insulation[17]. The basis of this technique is to dig a trench in the soil, and to lay a bed of powder, which may be either of plastic, bitumen or pitch residue in origin. After compaction, the pipeline is placed on top and then covered in with some more of the same material. The substance close to the pipeline sinters and coats it, while the rest of the material, which has a hydrophobic character, protects the pipeline even when the groundwater rises to a level above the pipeline.

Methods of this sort have given good results under circumstances where straight pipeline runs only are envisaged, where the subsoil is stable, and where such factors as sidestream take-offs, junctions etc. are minimal. Unfortunately many other pipeline networks constructed in this way have given trouble. Subsoil movements have disturbed the continuity of the insulation layer, enabling ground water to get into contact with the mild-steel pipelines. Some of the materials are unable to withstand long-term expansion and contraction stresses, and many give troubles at expansion units and junctions. The method is very cheap, but care must be taken that it is not used in unfavourable circumstances.

Mineral-wool Insulation in Trenches[8]. Mineral wool is an excellent insulating material, but this method is not advisable for use where there is even a chance that the outer cover of the insulation might become wet. The system is best used in conjunction with pipelines laid in the cellars of buildings, inside tunnels, conduits and the like, where there is no question of water rising. To eliminate the chance of accidental wetting of the mineral-wool insulation, this is normally covered with polythene foil.

Gas-concrete Insulation[8,18]. Probably the most popular form of thermal insulation for district-heating pipelines in the past was the use of concrete conduits, filled with gas or aerated concrete. Under normal circumstances, when the gas-concrete remains dry, pipelines are preserved extremely efficiently as the alkaline nature of the insulation material protects the surface of the mild steel against corrosion. The system cannot be used if there is the slightest chance that the water table may rise to engulf the insulation material.

Foamglass Insulation[19]. Foamglass is a material made by mixing broken glass with carbon and heating inside moulds. The material is available in half sections, which can be glued with a special bitumen binder round pipelines, junctions etc. Although the material is itself virtually impenetrable by water, the joint is made completely safe by wrapping round with a bitumenized glass-fibre tape. Disadvantages are its softness and brittleness, but on the other hand there is no deterioration with age. If the material is suitably protected against mechanical and thermal shocks, foamglass insulation keeps water away from pipelines, even if these should be permanently submerged.

Prefabricated Foam Plastic-insulated Pipelines[20—23]. These have recently become much more popular (*Figure 10*). The plastic foam usually employed is rigid polyurethane, which adheres exceedingly well to mild steel, is elastic enough to withstand pipeline movement, and is impervious to water. These coated pipelines are then surrounded by an extremely rigid layer of either polyethylene or PVC sheeting. This protects the rather soft polyurethane foam against damage during handling and eliminates the chance of water penetration into cracks in the foam. The ends of the pipelines are always left uninsulated to enable butt jointing by welding or other techniques to be carried out. Joints are also left uninsulated. There are a number of different techniques practised by different firms active in the field to insulate these rather more tricky parts of the pipeline network *in situ*. If such pipeline systems are well constructed, they have considerable advantages over other methods in water resistance, durability and cost.

Bitumen Insulation[24]. In this system (*Figure 11*) a shallow trench is dug and a lightweight steel/aluminium shuttering system is placed in it. A layer of plastic foam is then placed at the bottom of this shuttering, the pipelines are put in place, and the bitumen is spread. This is elastic enough to withstand slight movements in the pipeline without cracking. The material withstands district-heating pipeline temperatures of up to 150°C and is totally waterproof. The system is considered to be particularly useful in places where there is a chance of slight ground subsidence. Costs are moderate.

Steel Pipeline-within-a-pipeline System[21]. In this system one, two or more mild-steel pipelines are individually insulated and are housed inside yet another mild-steel pipeline, which in its turn is thoroughly coated by means of a bitumen-impregnated fibreglass fabric. The various prefabricated units, which include expansion units, T-joints, branch joints, bends etc., are supplied in sections for the inner pipes to be butt-welded together and a steel cover to be installed, again by welding. The system is therefore completely watertight under any conditions of ground-water level, and is even suitable for leading district-heating mains underneath lakes or rivers. Prefabricated manhole covers are made to be incorporated with the other sections of the system. Great care is taken to ensure that no water can possibly penetrate into the gap between the inner and outer pipes. Welds are inspected using X-ray equipment and drain plugs are installed at intervals.

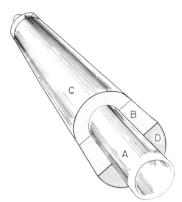

Figure 10   Prefabricated insulated pipeline section. Courtesy Ric-Wil
A: *Fibreglass reinforced plastic pressure carrier pipe (up to 1 MPa pressure)*
B: *Polyurethane foam insulation*
C: *PVC plastic outer jacket*
D: *Waterproof end barrier*

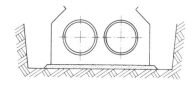

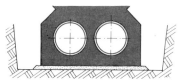

Figure 11   Insulation of district-heating pipelines with bitumen in sheet aluminium formers. Courtesy Lebit GmbH

Of all the different systems of pipeline insulation, this technique is the one least likely to fail. However it should be pointed out that costs are very high, especially when larger-diameter pipes are to be laid. Where severe conditions are unlikely to occur, it may be better to use a cheaper system.

Apart from the systems described above, prefabricated units are on the market, in which district-heating mains are co-ordinated with cold-water pipes, gas pipes, sewage runs and electric-power conduits. These are commonly used in new building complexes and save considerably on installation costs.

Calculation of Heat and Pumping Losses[7]

Heat Losses

The heat loss from a pipeline, q in W/m, is given by the following equation:

$$q = (\theta_p - \theta_o)/(R_i + R_s) \qquad (1)$$

where θ_p is the temperature of the pipeline and θ_o that of the external air (°C), R_i is the thermal resistance of the pipeline plus insulation and R_s of the soil cover (m °C/W). R_i is commonly evaluated from the following equation:

$$R_i = \ln (r_o/r_i)/2\pi k_i \qquad (2)$$

where r_i and r_o are the inner and outer radii of the insulation, and k_i is the thermal conductivity of the insulating material (W/m °C).

Values of k for materials involved in district pipeline insulation, at 50°C, are:

Material	k (W/m °C)
Loose fill (Gilsulate)	0.099
Loose fill (Protexulate)	0.115
Mineral wool	0.041
Gas-concrete (density 300 kg/m³)	0.067
(density 650 kg/m³)	0.131
Foamed polyurethane	0.042

Bitumen	0.700
Heavy concrete	0.500
Dry sand	0.35
Wet clay	2.3

R_S can be calculated from the Louden equation:

$$R_S = (2\ k_s)^{-1}\ (\ln d/r_O)\ (1 + (1 - (r/d)^2)^{1/2})$$

where k_s is the thermal conductivity of the soil (W/m °C), r_O is the outer radius of pipeline + insulation, and d is the depth of the centre of the pipeline beneath the surface of the soil.

Normal heat losses for pipelines average between 150 W/m pipe-run for a pipeline of approximately 50 mm diameter serving 50 consumers, and about 300 W/m for a 100 mm diameter pipe which serves about 600 consumers. It can therefore be seen that heat losses in properly insulated district-heating networks do not loom particularly large.

Pumping Losses

These are extremely complicated to evaluate as they contain head losses, friction losses and energy losses at enlargements, valves, joints, weirs, etc. The temperature of the water also affects the pumping costs, as high-temperature water has a lower viscosity and thus (though the flow is turbulent) needs slightly less pumping energy. The evaluation of pumping costs for steam is complicated by the fact that we deal in such a case with a compressible fluid with varied pumping energy along each given length. Detailed pumping costs are evaluated normally by the use of specialized computer programmes, which can be purchased directly as software from any of the major computer manufacturers. As an approximation, the head loss for water flowing at about 2.5 m/s is 15 m of water per 100 m of pipeline.

On the whole, pumping costs are a very minor item in the overall charges on a district-heating system, seldom amounting to more than 5% of the

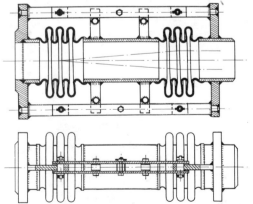

Figure 12 Articulated compensator for district-heating main. Courtesy Engineering Appliances Ltd

References p 364

annual charges on the pipelines and auxiliary equipment.

Expansion Units and Other Auxiliaries[25]

As the temperature of the water inside the pipeline system can vary appreciably throughout the year, and particularly during shut-down, it is essential to incorporate devices to permit expansion and contraction to occur without damage to the system.

Expansion problems must be considered in each particular part of the pipe and can often be solved by natural flexibility, offsetting pipe runs at intervals, changes in direction, or by the use of pipe loops. But bellows expansion joints are most commonly used for a compact and efficient solution. There are two different applications of bellows:

(a) The axial compensator, where the bellows is used in compression and extension along its axis, is useful for small amounts of expansion in crowded runs; good guiding and strong anchoring of the pipe are essential.

(b) The articulated and angular compensators, where the bellows is used in bending within hinge and tie-bar systems, are used for large expansion movements and at high level where only very light guiding and pipe anchors can be provided; these units are pressure-balanced and only impose very small forces on pipe systems (*Figure 12*).

Expansion units are usually located in inspection chambers fitted with manholes. To avoid stray-current corrosion of pipelines one commonly uses either electric drainage or sacrificial anode protection. In the latter case, the pipelines are connected to large lumps of zinc which are buried deep within ground water.

Particularly when the heat source is waste heat from electric power generation, it is necessary to fit heat accumulators in the system. These are large steel tanks up to 30 m high and 10 m in diameter, well insulated with mineral wool and polished-aluminium sheeting.

General Economics of Operating a District-heating System

By far the largest item in the operation of a district-heating system is the capital investment involved in building the pipeline network. For most of the network one assumes a life of 50 years, so that the annual depreciation rate is 2%, with an additional 0.5% for maintenance and repair. The heaviest charge is thus interest on capital. Heat losses and pumping costs usually account together for about 10% of the network charges.

The savings made are in the lower cost of fuel in the form of waste heat. Even when the district-heating system has to bear part of the plant costs of combined power/heat generation, the fuel cost is usually less than 30% of that which it would be if produced by the combustion of primary fuel.

Roughly the same cost savings occur when heat is produced by other waste-heat-utilization methods such as refuse incineration or geothermal sources. Savings in fuel cost when burning primary fuels in large boiler installations are much more marginal, seldom amounting to more than about 20% of the cost in individual appliances. However, in most continental systems, waste heat accounts for the bulk of the heat used and therefore one can expect to save between 60 and 65% of the total fuel cost. These savings are normally sufficient to pay for the large annual costs of the complex district-heating networks and may lead to a profit.

For example in the case of the Copenhagen district-heating system the budget works out as follows ($\approx$1972):

Income

Sale of heat	99.0%
Connection charges	1.0%
	100.0%

Expenditure

Cost of waste heat and distribution costs (heat losses, pumping, labour)	50.0%
Interest charges on network	8.4%
Depreciation of plant and network	9.4%
Pensions	1.4%
Plant development account	12.9%
	82.1%
Excess of income over expenditure	17.9%

The cost of the district heat to the consumer is evaluated on commercial principles so as to compete effectively with alternative forms of heating such as gas or electric fires, or small domestic central-heating systems.

References

1   Walker, G., *Stirling-Cycle Machines*, Clarendon Press, Oxford, 1973

2   Wootton, A. G. *Fuel, Lond.* 1974, Vol.53, 71

3   Denbigh, J. C., Personal communication (1972 prices, UK)

4   Sparham, G. A. *BCURA Mon. Bull.* 1966, Vol.30, 369

5   Winkens, H. *Heizkraftwerke Fernwärme International* 1974, 3.1, pp 1--2

6   Schwer, M. *Surveyor* 1973 (June 8), pp 29—33

7   Ref.(B)

8   Ref.(D)

9   Hohl, R., 'District Heating in Switzerland', *Brown Boveri Rev.* 1973 (June), Vol.60, 253—264

10 Muir, N., 'District Heating in Sweden', Paper read at *Inst. of Engineers of Ireland Conf.*, Dublin, Oct. 1973

11 Stal-Laval. Technical brochures

12 Hausanschlusse an Fernwärmenetze, VDEW, Frankfurt, 1966

13 HKW Västerås, 'Thermal Power Heat Station in Sweden', *Fernwärme International* 1973, 2.3, pp 57—67

14 Molter, F. J. *Fernwärme International* 1974, 3.1, pp 3—12

15 Anon. *Fernwärme International* 1973, 2.5, pp 125—132

16 Bau von Fernwärmenetzen, VDEW, Frankfurt, 1966

17 Protexulate Ltd. Technical brochures

18 Thermo-Crete GmbH. Technical brochures

19 Pittsburgh Corning Ltd. Technical brochures

20 Pipe Conduits Ltd. Technical brochures

21 Princes Development Ltd. Technical brochures.

22 Wanit GmbH. Technical brochures

23 Pan-Isovit GmbH. Technical brochures

24 Lebit GmbH. Technical brochures

25 Engineering Appliances Ltd. Technical brochures

General Reading

(A) Total Energy Conference, Brighton, Inst. Fuel, 1971. Vol.1 (Papers), 1971. Vol.2 (Discussion), 1972

(B) Diamant, R. M. E. and McGarry, J., *Space and District Heating*, Butterworths, London, 1968

(C) *Proc. 2nd Int. Conf. on District Heating*, Budapest, May 1973

(D) Diamant, R. M. E., *Total Energy*, Pergamon, Oxford, 1970

Chapter 16

Heat salvage

16.1 Heat recovery in waste-heat boilers

Flue-gas enthalpy
Boiler efficiency
Heat transmission
Flue-gas draught and mass velocity
Economic stack sizing
Use of extended surfaces
Heat-flux limits
Steam pressure level
Material selection
Tube cleaning
Abnormal operating conditions
Selection of boiler type
Examples of applications in industry

16.2 Recuperators and regenerators

Regenerators
Recuperators

16.3 Some other forms of heat and energy salvage

Recovery of raw materials
Use of oxygen for combustion
Dry-quenching of coke; waste-heat boilers; gas turbines
Recovery of potential heat
Blast-furnace stoves
Hot-blast cupolas
Charge preheating
Using low-grade waste heat
Fuel saving in industry

This is a Chapter dealing mainly with heat recovery using waste-heat boilers, recuperators, and regenerators. Economizers are regarded as an integral part of a boiler plant (see e.g. Ch.10.3). Heat salvage in connection with reciprocating engines and gas turbines is discussed in Chapter 12.3. This title in principle could be extended to a large part of the contents of this book, since savings from good operation, good maintenance and good housekeeping in general are discussed in most Chapters and account for a large part of the savings possible once the equipment has been designed and installed. A few more general examples of heat recovery at low temperatures have been added here in Section 16.3.

16.1 Heat recovery in waste-heat boilers

This Section is concerned mainly with heat salvage from flue gases at atmospheric pressure in water-tube boilers. However, most of the principles discussed apply equally to heat salvage in firetube boilers and from other fluids, and particular mention is made later on to heat salvage from high-pressure process-gas streams.

Flue-gas Enthalpy

The flue gases leaving an industrial furnace must be at a temperature greater than the lowest process temperature encountered in the furnace, and are normally considerably higher to ensure that high rates of heat transfer are used in the furnace. The sensible heat available in the flue gas is called the flue-gas enthalpy and is normally referred to a datum of $15°C$, i.e. flue gases at $15°C$ have zero enthalpy. More detail is given in Ch.8.1, and Ch.8 Figure 3.

Boiler Efficiency

The maximum possible amount of heat which can be salvaged in a boiler is normally considered to be the flue-gas enthalpy. The actual amount of useful heat salvaged is always less than this for two reasons: (1) heat is lost from the boiler to the surroundings; and (2) the flue gases cannot be cooled below the local water temperature, or in practice below the gas dewpoint.

The waste-heat boiler efficiency is defined as the useful heat salvaged divided by the maximum possible amount of heat which can be salvaged, and is given by:

$$\eta_B = \frac{H_1 - H_2 - Q_L}{H_1} \tag{1}$$

where η_B = boiler efficiency, H_1= inlet flue-gas enthalpy, H_2 = outlet flue-gas enthalpy, and Q_L = heat loss.

It is often adequate to assume that the heat loss Q_L is 2% of the flue-gas enthalpy change and that the mass specific heat of the gas varies little over these lower temperatures. Equation (1) then simplifies to:

$$\eta_B = \frac{0.98\,(T_1 - T_2)}{(T_1 - 15)} \tag{2}$$

where T_1 = inlet flue-gas temperature ($°C$), and T_2 = outlet flue-gas temperature ($°C$).

When recovering heat from the flue gases of a fired furnace, it is often more convenient to consider the effect on the overall efficiency of the inclusion of a waste-heat boiler. The overall efficiency is given by:

$$\eta_O = \frac{H_F - H_2 - Q_L}{H_F} \tag{3}$$

where η_O = overall efficiency, and H_F = abiabatic flame enthalpy (or heat liberated).

In this case, it is often assumed that Q_L is 3% of the heat liberated and equation (3) then simplifies to:

$$\eta_O = 0.97 - \frac{H_2}{H_F} \tag{4}$$

Since the variation of the ratio H_2/H_F with flue-gas outlet temperature depends only on the type of fuel fired, the excess air, the combustion air preheat, and the atomizing medium, it is possible to present equation (4) graphically for a particular fuel.

Flue-gas Outlet Temperature

To obtain maximum efficiency from a waste-heat boiler installation, it is necessary to keep the flue-gas outlet temperature as low as possible. The following limitations must be considered when setting the flue-gas outlet temperature.

At every point in the installation, the flue-gas temperature must be greater than the water temperature. Setting the flue-gas outlet temperature greater than the water inlet temperature does not automatically guarantee this, as is illustrated in *Figure 1*. The smallest temperature difference in the installation is known as the temperature approach.

The selection of the temperature approach is often an economic one based on a balance between the installation cost (which normally increases as the temperature approach decreases) and the operating cost (which increases as the temperature approach increases).

To ensure flexibility for control purposes, it is common practice to fix a lower limit of $30°C$ on the temperature approach.

Sulphur-containing fuels produce flue gases containing SO_2 and SO_3. Since all flue gases contain steam, it is important that, in such cases, all parts in contact with the flue gases should at all times be maintained above the dewpoint of the flue gas. This limit, further details of which are given in Chapter 20.2, may limit flue-gas outlet temperature and/or water temperature.

To obtain maximum overall efficiency from an installation, it is common practice to install a combustion-air preheater after the boiler. In such cases the flue-gas outlet temperature from the boiler can often be higher, compatible with the combustion air preheat temperature and the combustion air-preheater temperature approach (typically 350°C maximum and 30°C respectively, giving a maximum flue-gas outlet temperature from the boiler of 380°C).

Excess Air and Air Inleakage

The overall efficiency which can be obtained from an installation for a specified flue-gas outlet temperature is significantly affected by the amount of excess air. Increasing the excess air has the effect of significantly increasing the flue-gas mass flow without significantly increasing its enthalpy, so that the flue-gas temperature will be lower for the same heat content.

Air inleakage before and during passage through the waste-heat boiler installation has the same dilution effect as increasing the excess air. Since waste-heat boilers invariably operate under draught conditions, it is important that attention is paid to the detail of air exclusion.

As well as its effect on boiler efficiency, air inleakage can also have the following ill effects. Since the flue-gas mass flow is increased, the ducting, fan and stack must be sized for this increased value. Air inleakage may also result in local cooling of heat-transfer surfaces, resulting in thermal stress, tube leakage, or flue-gas condensation.

Heat Transmission

The modes of heat transfer which are important in waste-heat boiler design are summarized here. Appendix C 11 gives brief coverage of heat transmission in general and ref.1 fuller coverage of heat transmission in waste-heat boilers.

Conduction

In the design of waste-heat boilers conduction is normally only of importance in the transfer of heat along the external fin (if any) and through the tube wall.

A flow of heat can only be maintained along a fin if there is a temperature gradient in the fin. Consequently, the fluid transferring heat to the fin is doing so at a temperature difference which varies along the fin. It is normal to express the heat flow from the fluid relative to the temperature difference at the base of the fin, and to define the fin efficiency as a factor by which the fin surface area must be multiplied to allow for the reduced temperature difference towards the tip of the fin. *Figure 2* gives fin efficiencies for transverse helically wound fins of constant thickness.

The tube-wall resistance for boilers and economizers involved in heat recovery from flue gases is low and is often ignored; for superheaters it is often significant.

Convection

In waste-heat boiler installations convection is the chief resistance to heat transfer. Convective heat-transfer coefficients (App.C11) for the present purpose are often taken as proportional to (fluid mass velocity)$^{0.65}$, for flow outside tubes; inside tubes, the power is 0.8.

In the transfer of heat to the water or steam from the heat-transfer surface, convection is the mode of heat transfer in economizers and superheaters. With the exception of superheaters, the water-side resistance is relatively low; but, although it has little effect on the overall heat transferred,

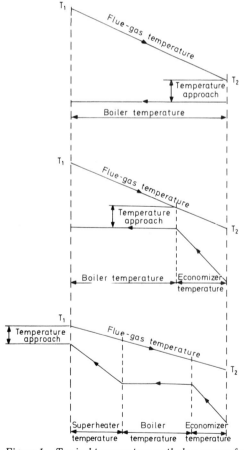

Figure 1   Typical temperature-enthalpy curves for waste-heat boilers

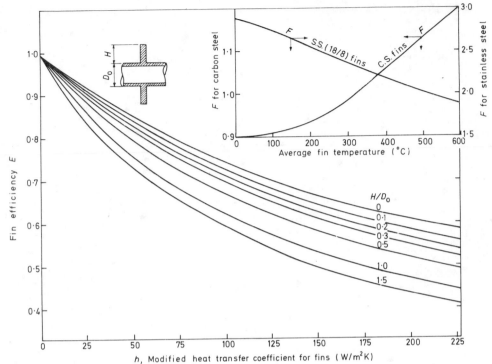

Figure 2   Fin efficiency for transverse fins of constant thickness. Graph based on $h = h_0$, carbon-steel fins, height 19 mm, thickness 1.58 mm, average fin temperature $315°C$.
For other conditions use:

$$h = h_0 \times F \times \left(\frac{1.58}{\text{thickness}}\right) \times \left(\frac{\text{height}}{19}\right)^2$$

where h_0 = calculated heat transfer coefficient for fins, and F = constant depending on fin temperature and material — see inset.

Value of H/D_0 for common arrangements:

Pipe nominal bore (mm)	50.8 (2 in)	76.2 (3 in)
H (mm) = 12.7	0.21	0.143
H (mm) = 19.05	0.316	0.214
H (mm) = 25.4	0.421	0.286

Pipe nominal bore (mm)	101.6 (4 in)	152.4 (6 in)
H (mm) = 12.7	0.111	0.075
H (mm) = 19.05	0.167	0.113
H (mm) = 25.4	0.222	0.151

(Courtesy Extended Surface Tube Company)

it can have a significant effect on the tube wall temperature, particularly if there is a flue-gas dew-point problem.

Radiation

Radiation is only of importance in the transfer of heat from the flue gases to the metal surface at flue-gas temperatures above $650°C$. The heat transferred by radiation is normally calculated in three stages:
1. The radiation from large cavities upstream and/or downstream of a tube bank. Since this results in a large gas-path length, it is usually very significant.
2. The radiation from the spaces around the tubes within the tube bank. Since this results in small beam lengths, it is not usually so significant, particularly with finned tubes.

3. The radiation from the walls enclosing the tube bank. This usually has a significant local effect.

Flue-gas Draught and Mass Velocity

The draught maintained in a waste-heat boiler is normally determined by the draught requirements of the equipment upstream and/or downstream of it. Typical examples are as follows:
(a)   A waste-heat boiler in the convection section of a fired process heater, operating under natural-draught conditions, could have a draught of approximately −25 Pa (below atmosphere) at inlet and −300 Pa at outlet, fixed by the requirement of −12 Pa draught at the bridgewall.
(b) A waste-heat boiler in the convection section of a downfired reformer furnace, operating with an

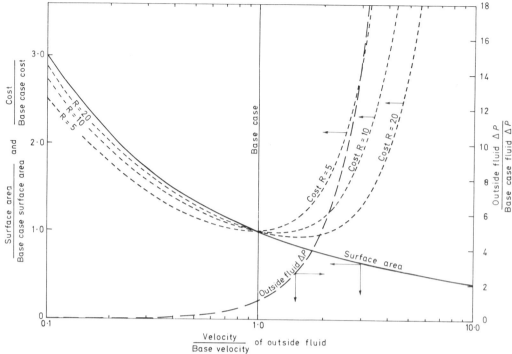

Figure 3   Variation of waste-heat boiler cost with outside fluid velocity.

Basis: No fouling, no radiation
　　Finned surface: bare surface = 8:1
　　Fins 25.4 mm high × 1.27 mm thick carbon steel
　　Base case outside-fluid coefficient = 56.75 W/m² K, inside-fluid coefficient = 5.675 kW/m² K

$$R = \frac{\text{Cost of base-case surface area}}{\text{Cost of base-case outside-fluid } \Delta P}$$

Costs are assumed $= C_1 \times$ area $+ C_2 \times \Delta P$ with no material specification changes, and neglecting any change in inside-fluid ΔP.

(Courtesy Extended Surface Tube Company)

induced-draught fan, could have a draught of approximately −1000 Pa at inlet and −2000 Pa at outlet, fixed by the requirement of −12 Pa draught at the top of the furnace tubes. (10 Pa ≈ 1 mm w.g.)

If the designer of a waste-heat boiler installation has any freedom in selecting the draught, it should be as close to zero as possible to minimize air inleakage and fan/stack requirements, but should never under any operating condition become less than zero (gauge pressure positive) since this would result in the leakage of hot flue gases around the structural steelwork.

The flue-gas heat-transfer coefficient increases with flue-gas mass velocity, but the flue-gas pressure-drop increases faster, approximately in proportion to the 1.8 power.

The selection of flue-gas mass velocity therefore requires an economic balance between the cost of the heat-transfer surface (which usually decreases as the mass velocity increases), and the cost of generating flue-gas draught (which increases as the mass velocity increases). A typical variation in total cost with mass velocity is shown in *Figure 3*.

Although the required surface area decreases as the flue-gas mass velocity increases, the increased tubewall temperature can, in some cases, result in a material-specification change that would increase the specific cost of the surface. This is unlikely for economizers and boilers, but quite possible with superheaters.

Typical optimum mass velocities in process-plant applications are 1.5–2.5 kg/s m² for natural-draught conditions, and 4.0–8.0 kg/s m² for forced-draught conditions.

Economic Stack Sizing (Ch.19.4, App.C7)

For forced-draught applications, stacks are normally sized on velocity limits and minimum height consistent with stack-emission and plant-noise limits.

For natural-draught applications, stack sizing is based on the calculation of the cheapest stack geometry to give the required draught at the base of the stack, subject to height and velocity limits as in forced-draught applications and, in the case of free-standing stacks, aspect ratio limits (commonly 15:1).

If certain cost assumptions are made the equations involved can be solved graphically to provide the cheapest stack geometry. It is important that stacks should be designed for the worst possible operating conditions; in particular that allowance is made for increased excess air due to mal-operation of the furnace.

Use of Extended Surfaces

As has been mentioned previously, the flue-gas heat-transfer coefficient is normally very low compared with the water heat-transfer coefficient, typical values being 60 and 6000 W/m^2 °C respectively. In these circumstances, it is beneficial to provide extended surface area in contact with the flue gas relative to the surface area in contact with the water.

In view of the above, extended surfaces are normally used as much as possible in waste-heat boilers, but subject usually to the following restrictions:
1. A minimum of two bare rows of tubes is normally incorporated in any tube bank facing significant cavity radiation (usually those facing a firing chamber).

2. The fin-tip temperature, with appropriate temperature margins, must be within the temperature limit for the fin material. The tube-wall temperature, with appropriate temperature margins, must be within the tube-wall temperature and stress limit.

Extended surface areas used in waste-heat boiler applications normally consist of transverse helically wound fins (cut or uncut) or studs, welded to the base tube. Helically wound fins provide the greatest surface area extension for a given weight of finning material; studs are more robust and easier to clean.

It is important when using extended surfaces that the tubes are suitably supported, to avoid damage to the fins or studs. It is normal, in the case of horizontal finned tubes, to ensure that the support ledge is at least four fins wide.

Heat-flux Limits

Since waste-heat boilers normally recover heat from relatively low-temperature flue gases, heat flux levels very rarely approach the critical value. This is particularly true for vertical tubes. But water-tube boilers employing horizontal tubes can give rise to problems since critical heat fluxes in horizontal tubes can be as low as 20% of the corresponding values in vertical tubes. Accurate correlations are not yet generally available for predicting critical heat fluxes in horizontal tubes, but the following provides a qualitative guide:
(*a*) Water inlet velocities greater than 1.5 m/s are unlikely to give problems.
(*b*) The smaller the tube diameter the higher the critical heat flux.
(*c*) The higher the water pressure, the lower the critical flux.
(*d*) A tube inclination as low as 10 degrees can result in a significant increase in the critical heat flux.
(*e*) Peak heat fluxes below 240 kW/m^2 are unlikely to give problems.
It is important when calculating the peak flux to which a tube is subjected that allowance is made for possible coincidence of the peak radiant flux and the peak convective coefficient on the tube.

Steam Pressure Level

The steam pressure level to be used in a waste-heat boiler is often pre-determined by the equipment the steam will serve. When it can be varied to suit the waste-heat boiler, the following criteria should be considered.

Raising the pressure increases the saturation temperature and the amount of feedwater heating required. The effect of this on the maximum possible boiler efficiency depends on the relative magnitude of the heat capacities of the feedwater and the flue gas. For the production of saturated steam from boiler feedwater at 100°C, using a minimum temperature approach of 50°C, *Figure 4*

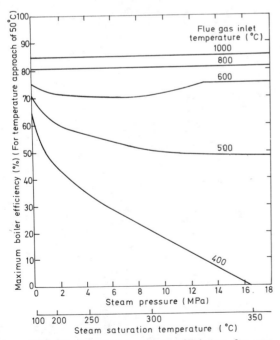

Figure 4 Maximum boiler efficiency for production of saturated steam from boiler feedwater at 100°C

gives the maximum possible efficiency which can be obtained. It will be seen that at flue-gas inlet temperatures above 600°C the maximum efficiency is not affected by the steam pressure level, since the approach at the water inlet temperature of 100°C controls.

As well as affecting the maximum efficiency which can be obtained, raising the steam pressure level also affects the quantity of surface area required, owing to the change in temperature difference. *Figure 5* shows the relative change in surface area when designing to the maximum efficiency from *Figure 4*, assuming constant heat-transfer coefficients.

The steam pressure level does of course have a significant effect on the thickness of the pressure-containing parts, both internal and external to the boiler. In particular, it can have a significant effect on the thickness and material selection of superheaters and may limit the superheat level and/or superheater flow arrangement owing to the limitations on superheater materials.

Increasing the steam pressure level decreases the density difference between steam and water and, for boilers operating under natural circulation, will necessitate a higher steam-drum elevation. In general, it also decreases the critical heat flux.

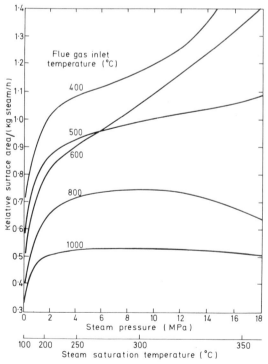

Figure 5   Relative surface areas for maximum efficiencies given in Figure 4

Material Selection

Waste-heat boilers are always placed downstream of other items of process equipment, and consequently their operating conditions will always be affected by the conditions in the upstream equipment. It is important that the worst possible combination of operating conditions is determined when fixing the design conditions for material selection. In particular the following factors should be considered:
1. Undersurfacing of upstream equipment so that increased inlet temperatures and mass flow rates to the waste-heat boiler are obtained.
2. Increased duty requirement from upstream equipment so that heat release is increased and inlet temperatures to the waste-heat boiler are increased.
3. Oversurfacing or turn-down requirement on upstream equipment so that hot flue gases are by-passed into waste-heat boilers.
4. Turn-down requirement and type of control for waste-heat boiler, particularly if a superheater is included.
5. Increase of steam-drum pressure up to relief-valve setting.

Tubes

The design temperature for tube-material selection should be calculated from the worst combination of: water/steam temperature, water/steam heat-transfer coefficient, fouling resistance, flue-gas temperature, peak flue-gas convection coefficient, and peak flue-gas radiative flux.

For waste-heat boilers and economizers, the tube material is normally carbon steel. If the heat is recovered from process gases containing hydrogen, it is important, particularly at high steam-pressure levels, to check the tube material for hydrogen embrittlement using ref.2, in which case a molybdenum or low-chrome steel may be necessary.

For superheaters, particular care must be paid to tube material selection. Superheaters operate with much lower steam-side heat-transfer coefficients, and higher steam temperatures and consequently higher tubewall temperatures are obtained. In addition, superheater metal temperatures are far more susceptible to fluctuations in operating conditions.

In particular, care must be taken in the choice of the method of control of the superheated-steam outlet temperature (attemperation). The easiest method of control is attemperation (direct or indirect) at the exit of the superheater. This has the disadvantage of allowing the superheat temperature at the exit to rise above the required value, and this rise must be designed for (including allowance for any boiler oversurface). Alternatively attemperation can be carried out part way through the superheater, but care must be taken that sufficient control can be achieved in this way since it is detrimental to superheater tubes to attemporate below saturation. A third method of control is to provide a steam bypass around the superheater, but this results in an even higher rise in the metal temperature owing to the decreased steam heat capacity and the decreased steam heat-transfer coefficient.

It is often necessary, owing to the temperature and/or the thickness, to use more exotic superheater tube materials than carbon steel and low-chrome steel. In such situations, the most economic material to use is austenitic stainless steel. Considerable problems have been encountered in the past over the use of this material owing to stress-corrosion cracking of the tubes, particularly if the feedwater treatment is unreliable. Alternative materials which can be used are ferritic stainless steels or Incolloy alloys.

Extended Surfaces

The design temperature for the selection of extended-surface material should be based on the maximum tubewall temperature and the corresponding temperature rise from the base to the tip. Since extended surfaces are not pressure-containing parts, it is common practice to permit higher metal temperatures than for the corresponding tube materials.

Manufacturers of extended surfaces often restrict the range of extended-surface materials to carbon steel, 13% chrome steel, and austenitic stainless steel. There are usually no restrictions on the combination of base tube and extended-surface materials for extended surfaces which are welded on.

Supports

The design temperature for tube supports must allow for any mal-distribution of flue-gas temperature across the duct.

Tube supports are normally castings and, for flue gases from clean fuels, are fabricated in heat-resisting cast iron or 25% Cr/12% Ni material. More exotic materials such as 50% Cr/50% Ni may be necessary if the fuel fired contains salts with a low ash-fusion point such as alkali sulphates or vanadates, which can cause corrosion. Further details of this are given in Chapter 20.1.

Tube Cleaning

It is important at the design stage of a watertube waste-heat boiler installation that allowance is made for cleaning the heat-transfer surfaces in contact with the flue gas.

When the fuel fired is relatively clean so that little soot or ash is normally formed, it is usually adequate to pitch the tubes so that the tube bank can be mechanically cleaned during plant maintenance periods or in the event of mal-operation causing blockages.

For fuels which produce significant quantities of soot or ash, it is common practice to install air- or steam-operated soot blowers within the tube bundle. These are normally positioned every four tube rows and, for maximum efficiency, require an increased row pitch at these points. Sufficient soot blowers should be installed to provide coverage over the entire flue-gas duct. Sufficient access platforms must be provided external to the flue-gas duct, particularly in the case of retractable soot blowers. Further details of soot blowers are given in Chapter 10.5.

Abnormal Operating Conditions

It has already been emphasized that, in calculating metal temperatures for material selection and pressure design, the worst possible combination of operating conditions should be considered. In addition other effects of abnormal operating conditions should be considered, including the following:

1. If the waste-heat boiler installation comprises several heat-recovery items among which the economizer duty may be split, it is sometimes possible that abnormal conditions in one item can result in boiling conditions in the economizer of another item. If high steam purity is required, this condition should be avoided since the steam produced will not normally pass through the steam-drum separation devices.

2. Failure of the boiler feedwater supply and the resultant loss of level in the steam drum will normally result in action being taken to reduce the heat input to the boiler. Sufficient water hold-up must however be provided in the steam drum to ensure that the boiler does not run dry while the heat source is reduced, although as the hold-up time is increased the steam drum size and cost increase considerably. Steam drum hold-up times of 2 to 5 min at design capacity are typical on process-plant applications.

3. Increased flue-gas temperatures during abnormal operating conditions will increase the heat loss and may affect the refractory selection.

4. Increased flue-gas and metal temperatures will result in increased thermal expansion. Sufficient clearance must always be maintained to permit free longitudinal expansion of the tubes. Tube-support plates should be designed in sections to keep the thermal gradient across the plate less than 250°C.

Selection of Boiler Type

Waste-heat boilers are in general associated with heat salvage from relatively low-temperature, low-pressure flue gases. There are, however, often applications for heat salvage from process-gas streams which are often at high temperatures and/or high pressures. The following Section gives general guidelines for the selection of boiler type for both cases.

Watertube

Watertube boiler types have two main advantages; they can utilize extended surfaces on the gas side, and they can more economically withstand the water-side pressure. Consequently, their main applications are as follows:

1. For heat recovery from low-pressure flue gases where the flue-gas mass velocity must be limited to maintain reasonable draught conditions and consequently the gas-side heat-transfer resistance is considerably higher than the other resistances. In such circumstances extended-surface tubing can be used to economic advantage.

2. In waste-heat boilers operating at high steam-pressure levels where the thickness and cost of the shell of a firetube boiler would be prohibitive. This is less of a restriction if high-strength steel shells and detailed stress analyses are carried out as discussed later.

3. Watertube designs are advantageous if provision is to be made for auxiliary firing to augment the heat available from the flue gases. In such cases the auxiliary firing chamber can more conveniently be accommodated with a watertube design.

Firetube

Firetube boilers have three main advantages: they can more easily be designed to withstand high pressures on both sides, in some circumstances the boiler shell can be utilized as the steam drum, and they are less susceptible to water-side blockage due to debris and mal-operation. Their main applications are as follows:

1. For heat recovery from high-pressure gases where all gas leakage must be eliminated. Typical of such applications is heat recovery from process-gas streams.

2. Heat-recovery installations where the purity of the steam is not critical and gravity separation in the shell of the boiler is acceptable. In such cases, the steam drum can be eliminated, provided adequate hold-up time can be maintained in the boiler shell. Designs such as these are usually limited to steam pressure levels of approximately 3 MPa, since they usually result in large untubed areas of the tubesheet with consequent stressing problems.

3. Heat-recovery installations operating with high steam pressures and/or high heat fluxes where it is imperative to avoid tube blockage.

Firetube boilers have been designed and built for steam pressure levels of 12 MPa recovering heat from process-gas streams at 3 MPa and 1200°C. In such conditions the most critical section of the design is the tubesheet, and it is essential that a fully detailed fundamental stress analysis should be carried out covering both pressure and thermal effects, to ensure that the design complies with criteria such as those laid down in ASME, Section VIII, Division 2, Appendix 4. It is also important

in such conditions to ensure that the peak heat flux at the tube inlet is maintained below the critical value. Allowance should be made for the vapour blanketing effect of vertical tube rows on the critical value, and the effect on the peak value of enhanced convective heat-transfer coefficients at the tube inlet.

Examples of Applications in Industry

An idea of the possible savings made by a waste-heat boiler can be obtained from data given on page 558 of *The Efficient Use of Fuel* (1958). The following efficiencies in terms of steam output/basic fuel to process can be deduced (assuming that net CV of coal used was 28 MJ/kg):

Cement kiln $\approx$20%

Glass furnace   <10%

Malleable melting furnace and puddling furnace $\approx$50%

O.H. steel furnace, and steel reheating furnace $\approx$ 2%

This wide range illustrates the different opportunities for savings presented by different industries; the smaller gains in overall process efficiency would in general be matched by smaller waste-heat boiler sizes.

16.2 Recuperators and regenerators

The recovery of heat from the waste gases leaving a furnace, by preheating the air and sometimes, with fuels of low calorific value, the gas, is known as recuperation. In a true recuperator the medium to be heated and the waste gases are separated by a partition, which may be made from metal or refractory, and the unit operates continuously, the heat passing through the partition wall which should be as thin as possible.

A regenerator operates discontinuously, and consists of at least two large brick-filled chambers. The hot waste gases pass through one of them, heating up the refractory, whilst the air (or gas) passes through the other in reverse direction, being heated up by abstracting heat from the refractory. After a predetermined time (the reversal period) reversing valves are operated and the roles of the two chambers are interchanged.

Regenerators

Regenerators are used where very high preheat temperatures are required, as with steel or glass making. Among the advantages as compared with a recuperator, the regenerator:

1. is simpler in construction than a recuperator,
2. accepts extremely high waste-gas temperatures,
3. yields very high preheats, and

4. is more robust than a recuperator.

Some of the disadvantages are:

1. it needs reversal gear,
2. preheat temperature varies during the cycle, and
3. it limits burner design, since it can withstand only low pressures.

The limit to burner design arises from the fact that the waste-gas offtake of one cycle must become the burner port during the next cycle. Modern burners require at least 1—1.5 kPa to operate correctly.

The driving force of a regenerator during both halves of the reversal period is the temperature difference between the refractory and the circulating gases. The heat stored during the heating period is the only source for heating the air during the cooling period, and many factors are involved. In a very thorough investigation reported at the first Waste Heat Recovery Conference, in 1948, Finlayson and Taylor[3] evaluated a large number of these factors and some of their findings are given below.

Effect of Regenerator Size

As could be expected, increase in size gives increased performance, but at a diminishing rate. As with recuperators, the higher the preheat, the greater the increase in heating surface required for a given increase in preheat temperature.

Total Mass of the Filling

A regenerator, being a form of heat accumulator, is sensitive to the mass of refractory. If this is too small, the temperature rise and fall is too rapid, and the reversal period becomes very short. Bearing in mind that there is a disruption to the air and the waste-gas flow during the actual operation of the reversing procedure, it will be appreciated that with very short cycles this can lower the efficiency considerably. The optimum regenerator size will depend upon the thickness of the checker brickwork, and its physical properties.

Physical Properties of the Filling

From a theoretical point of view the best filling would be a refractory with a good thermal conductivity and a high heat capacity. It must also be strong enough at working temperature to support its own weight.

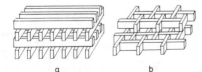

Figure 6   Regenerator brickwork: (a) checker packing (straight); (b) chimney packing

References p 390

The chemical nature of the refractory is of great importance since material is always carried from the furnace as a suspension in the waste gases and may react with the checkerbricks. It is usual, for example, to use silica refractories with the open-hearth furnace, but for glass melting basic refractories are recommended. For coke ovens, fireclay bricks are used.

Varying the Load

It was shown that the heat transfer improves with higher rates of flow, indicating that if higher velocities are used a smaller unit will be required.

Varying the Velocity

This was considered separately for in-line and staggered filling (*Figure 6*). It was found that the same law, namely h_c varies as $u^{0.5}$, holds for both arrangements, but that the straight-through filling gave less pressure drop for the same heat transfer than the zig-zag filling. In general the heat transfer appeared to be proportional to the pressure loss raised to a power between 0.25 and 0.33.

Effect of Brick Thickness

According to the tests carried out by Finlayson and Taylor brick thickness had very little effect on the heat transfer. Etherington[4] favours a thicker brick only on the grounds that it will be mechanically stronger, will not slag away so quickly, and will not overheat so readily during the highest temperature phases of the furnace operation.

Effect of varying the Length of Reversing Period

Reducing the length of the period from 30 min to 10 min gave an improvement in heat transfer equivalent to approximately 30% more regenerator surface when the bricks were ½ in (12.5 mm) thick but equivalent to only 7% when the bricks were 2 in (51 mm) thick. This agrees with the practical observation that it is worthwhile to change to short cycles when heavily loading regenerators constructed of thin bricks.

Effect of Operating Conditions

Hulse[5], at the second Waste Heat Recovery Conference in 1961, discussed the problem of the effect on overall heat-transfer rates of the mal-distribution of flow in many types of heat exchangers. With an open-hearth furnace regenerator, he found that bad distribution of the gases reduced the preheat by more than 200°C. The use of suitably placed baffles can easily give 100—150°C more, and as a consequence save up to ≈ 415 MJ/t of fuel. His views were confirmed at the same conference by Majorcas and co-workers[5], who discussed a horizontal regenerator on an open-hearth plant, and concluded that the aerodynamic design of a regenerator is as important as its thermodynamic design.

In designing a regenerator, the optimum conditions cannot be met since provision must be made to be able to deal with all the waste gases and air used until the end of the furnace campaign (period of operation), which with a glass tank may last three years and show an increase in fuel input at the end of perhaps 50%: by then the regenerator may be partially blocked by dust, slag, or spalled brick.

Moorshead and Whiteheart[5] discussed this problem at the same conference and underlined the difficulties of assessing regenerator performance. Up to 25% of the heat recovered may be lost from the regenerator structure, and in addition cold-air infiltration may occur, so that a good case may be made for the insulation of the regenerator structure and for boxing it in by steel plates to avoid infiltration. Attention should always be paid to accessibility, since the efficiency may be maintained by the removal from time to time of dust from the checkers by steam-lancing; but the correct choice of brick quality can reduce the blockage problem in some cases, by avoiding chemical reaction between the dust and the bricks.

Methods of obtaining higher preheats might be summarized as: (1) close attention to brickwork and structural details to prevent air infiltration; (2) getting maximum uniformity of flow of waste gases and air; (3) regular cleaning of regenerator passages; (4) insulation.

Other Forms of Regenerators

There are forms of regenerator which operate continuously, making the differentiation between regenerators and recuperators even more tenuous. Two types are at present in commercial use. One is the pebble air-heater in which the waste gases pass through a cylindrical vessel and heat a stream of small pebbles which are flowing downwards through a restricted throat and thence through a stream of air to which they give up their heat. At the bottom the pebbles are picked up and transferred to the top for recycling. The waste gases must not contain solid or liquid particles, which would clog up the system. Very high rates of heat transfer can be obtained, and the system can deal with very high temperatures; but sealing between the two media is rather difficult.

Rotary Regenerators

In a rotary regenerator the waste gases and the combustion air travel in opposite directions whilst a continuously rotating element cuts through both streams. The rotating element is a porous structure consisting of a mass of thin plates of metallic or ceramic material combined in most cases in the 'flame-trap' system of alternate straight and corrugated layers. The best performance is given by a fine pore size with a high surface-area/volume ratio. This, however, and the length, are limited by the fouling from solids carried in the gases.

The heat transfer is obtained by a wiping action. Both sides of the surface are alternately swept by hot gases and air. Deposit on the surfaces does not reduce the preheat since, in contrast to the recuperator, thermal conductivity does not play much part; but of course the pressure loss increases with fouling.

Two distinct groups of rotary regenerators have been or are being developed. One group, such as the Ljungström of which the Howden (*Figure 7*) is an example, operates at *low temperatures* with the matrix rotating at about 2 rpm, and is used for preheating the combustion air in power stations. The problems[6] with this type of unit are essentially connected with leakage or corrosion at low temperatures. The question of leakage has largely been solved, but the problems of corrosion and fouling are still being actively investigated. J. Howden and Co. claim to have carried out a comprehensive investigation into the effectiveness of metals and metallic coatings in combatting corrosion, and offer a vitreous enamel coating that reduces corrosion considerably and at the same time protects the heating surface from damage by burning soot deposits.

A very good assessment of rotary air heaters was given by Chojnowski and Chew at the Heat Exchanger Conference in Paris, June 1971[7]. They stated that on units installed with modern coal-fired boilers the chief remaining problem was severe fouling of the elements, with resulting high draught losses. In extreme cases the additional draught losses could exceed the fan capacity available, and might lead to reduction in plant load below the nominal maximum level.

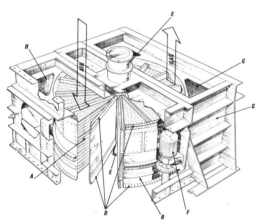

Figure 7   Rotary regenerator for low temperatures (Courtesy James Howden & Co. Ltd)

A *Heating surface elements*
B *Rotor in which the elements are packed*
C *Housing in which the rotor rotates*
D *Seals and sealing surfaces*
E *Support and guide bearing assemblies*
F *Drive mechanism*
G *Gas by-pass*
H *Air by-pass*

High-temperature Rotary Regenerators

Attempts have been made over a number of years to develop a rotary regenerator to operate at temperatures similar to those encountered with conventional regenerators. The Air Preheater Company of New York have installed rotary regenerators on glass-melting furnaces to operate in the range 600—700°C inlet temperature, and have obtained consistent air preheats of about 550°C. In these cases the rotary regenerator has been installed between the reversing valve of the conventional regenerator and the stack. One installation is on a fibre-glass tank where there is no conventional regenerator; half of the gases at about 1315°C then go directly to the stack, the other half being diluted with cold air to about 650°C before entering the rotary regenerator. The use of air at 450°C in place of cold air increased the heat-to-glass transfer from 2.0 MW to 2.9 MW. With those furnaces that have fixed regenerators, the effect of introducing hot air into the checkers in place of ambient air is a reduction in the temperature cycling within the mass of intricate and expensive refractory material making up the checker pattern. Increased service life of the refractories results, owing to a reduction in thermal shock, spalling, cracking etc. The need for cleaning the brick regenerator also diminishes with the introduction of high-temperature air.

A ceramic rotary regenerator is currently being studied by the BSC Corporate Engineering Laboratory[10], for use on soaking pits. A computer model of the proposed system has been studied which shows that preheats of at least 1000°C will be possible. They are basing their design on a silicon nitride matrix, which has been proved to withstand attack from a variety of corrosive environments up to a temperature as high as 1400°C and is capable of being fabricated into the complex shapes necessary for a compact regenerator matrix.

The Rover Gas Turbine Company has been developing a small rotary regenerator for use with their gas-turbine engine, and at the Paris Heat Exchanger Conference mentioned above R. N. Penny concluded that the small gas turbine, which is vitally dependent on the periodic-flow heat exchanger, will in the future make use of dramatic temperature increases as ceramic materials with outstanding thermal properties are developed. The ceramic regenerator, he said, has now developed to an advanced stage, and has assured the technical success of the small gas turbine in vehicle applications. This low-cost compact device, capable of recovering over 90% of the waste exhaust heat for further useful application, will have an important part to play in other as yet unexplored applications where efficient heat transfer is a requirement.

Fluidized Sand-bed Regenerator

In a process for making smokeless fuel from briquetted coal dust, the 'green' briquettes are heated by floating on a very slightly inclined fluidized sand bed. The sand travels downhill, forming a fluidized stream in which the briquettes travel (having almost the same density as the sand). At the end of this heating chamber, the briquettes are separated from the sand by a sieve and the sand is recycled. The fluidizing air leaves the furnace containing a small amount of combustible gases. This is burnt, using a small amount of new fuel, in a pressurized combustion chamber, and the hot products are then used to fluidize and heat the recycled sand. This is a true salvage process, since the gases would have to be burnt in any case.

It is interesting to consider that with a sand particle of 0.5 mm diameter, 1 m³ of sand would expose over 12 500 m² of surface. This explains why the exchange of heat between the gases and the sand is completed in 20—30 mm of height of air lift.

Recuperators

The heat transfer in a recuperator between the waste gases and the heating surface is either by convection, which depends on the velocity of the waste gases across the heating surface, or by radiation from the water vapour and carbon dioxide in the waste gases, which depends on the thickness of the layer, the concentration of these components in the gases, and on the temperature. A small but very important contribution, which can be more of an embarrassment than a help, is derived from the solid radiation from the refractory surfaces adjacent to the recuperator.

Because radiation increases very rapidly as the temperature rises whereas with convection temperature has only a relatively small effect, there are two main groups of recuperators: the radiation type, *Figure 8*, which is used at temperatures up to 1500°C, and the convection type, operating at temperatures below 1050°C.

Originally all recuperators were made from ceramic material, fireclay or silicon carbide in general, but in spite of ingenious designs, these all suffered from leakage from the air (pressure) side to the waste gas (suction) side. Even with very low air pressures, leakages up to 50% or more were very common.

These recuperators were gradually replaced by cast iron and cast chrome alloy units which can withstand air pressures of 7.5 kPa or more. During the past 30 years, as drawn alloy tube and rolled plate became more readily available, still more advanced fabricated recuperators have been developed.

The problems associated with the selection of recuperators and of the materials from which they are constructed are many: they are due largely to our inability to predict the entire conditions under which the heat exchanger will operate, or to lack of knowledge of the precise characteristics of the available materials.

Radiation Recuperators

The Escher recuperator was introduced into the UK in 1949, having been developed in Australia during the war[8]. The heating surface takes the form of two concentric cylinders, the waste gases passing through the centre and the air through the annulus. Since then, over 600 have been installed varying in size from 2.4 m to 41 m long and from 380 mm to over 2 m diameter. It has been developed into a combined unit including a convection section.

The simple smooth surface of the inner shell makes the radiation recuperator especially suitable for use with very dirty gases, since no cleaning is required. It presents a negligible resistance to the flow of waste gases and also takes the place of the chimney; thus space requirements are reduced to an absolute minimum (*Figure 9*).

Radiation recuperators have also been made with a ring of tubes taking the place of the inner and outer shells of the concentric-tube type. There are certain advantages in this design, but it is more difficult to install and to repair. In a survey of more than 60 recuperators installed in steelworks, it was concluded that the radiation recuperator gave the least trouble, and had the longest life[9].

Figure 9   Escher air and gas recuperators installed on soaking pits (Courtesy Stein Atkinson Stordy Ltd)

Convection Recuperators

These are of two types, those using cast tubes and those using drawn tubes in bundles of one form or another. The cast tubes, with fins on the inside as well as on the outside, may be bolted together in a header box when the temperature and pressures are low (needle-type recuperator), or may be

Figure 10 Composite tube recuperator (Courtesy Thermal Efficiency Ltd)

cast with integral expansion joint and flange at each end to enable them to be welded together to form banks of any size (known as the 'composite' type), *Figure 10*.

The steel-tube recuperators, *Figures 11* and *12*, are made in many forms and can be adapted to suit the location. The tubes are grouped in

Figure 8   Section through Escher type radiation recuperator (Courtesy Stein Atkinson Stordy Ltd)

bundles, which are attached at either end to header boxes. The shape of the tubes and end plates is designed to allow the individual tube bundles to expand without restriction, since the various groups may operate at different temperatures. Whereas the thickness of the cast tubes is usually 8 mm, that of the drawn tube is 3 mm.

Convection recuperators may be installed in the flue, or may be freestanding above the ground. In most cases, waste gases flow outside the tubes in a single pass. However, with dust-laden gases it is easier to keep the tubes clean if the gases pass inside them. In these, the 'flue-tube' recuperators (*Figure 13*), the tubes have 50–100 mm inside diameter. The air flows on the outside over a number of passes guided by a series of baffle plates.

Another type of convection recuperator has the inlet and outlet boxes arranged one above the

Figure 13   Flue-tube recuperator (Courtesy Metallurgical Engineers Ltd)

other in such a way that the tubes from the top box pass downwards concentrically with tubes passing downwards from the bottom box. These lower tubes are sealed at the lower end, which extends a little below the end of the upper tube. The air passes through the inner tube and upwards along the annulus between the tubes, to the hot box. Another simple form is the 'hairpin' type, in which the tubes are shaped as an inverted-U one leg of which connects to one box and the other to the other.

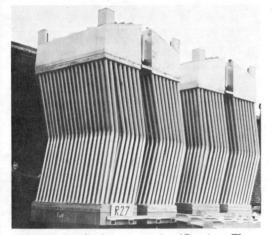

Figure 11   Steel-tube recuperator (Courtesy Thermal Efficiency Ltd)

Figure 12   Channel-type recuperator (Courtesy Metallurgical Engineers Ltd)

Ceramic Recuperators

As stated above, in the UK refractory recuperators have been replaced by metallic ones. In the USA, refractory recuperators are still used on soaking pits, followed by a metallic one. The hot air from the latter is used at the burner to induce the air through the refractory recuperator. This 'jet-pump' system is said to enable sufficient pressure to be obtained at the burner to give reasonable flame direction without the air leakage of the earlier installations.

In addition, at least two separate high-pressure refractory recuperator developments are in progress at the present time, intended for use with a soaking pit[10].

Selection of Recuperators

Each type of recuperator has advantages over the other types under some circumstances. Below are some of the considerations which affect the choice:
1. Initial cost.
2. Operating temperature.
3. Required preheat and turn-down.
4. Pressure level and pressure loss.
5. Campaign life; reliability.
6. Ease of repair and site considerations.
7. Resistance to corrosion.
8. Resistance to fracture.

Cost of Recuperators

Cost is not necessarily the most important criterion in selecting a recuperator, but in making comparison of costs care should be taken that all the incidental costs and savings are included. For example, a radiation recuperator does not normally need a chimney, but the cost of a chimney should be included in the price of a comparable convection type, as should the cost of additional flues and foundations. On this basis the cost of a radiation recuperator, including its supporting structure, lies between the multi-tubular type in a flue setting and the high-pressure cast-tube type. This relation does not always hold true for the very largest radiation type, where the cost of the supporting structure becomes very high.

Cost, however, should not be the main criterion in those cases where the recuperator forms an essential part of a furnace which itself is the hub of the works, such as a hot-blast cupola or a pusher furnace. In this case continuity of operation and speed of repair are of much greater importance than initial cost, since advantage in cost can quickly be lost if production comes to a standstill, even for a short time. In contrast, loss of one soaking pit out of a battery of say 30 would not necessarily be a disaster.

Operating Temperature

Every recuperator has a temperature limitation, governed by the maximum metal temperature the recuperator can withstand. The metal temperature can be represented by:

$$t_p = (h_g t_g + h_a t_a)/(h_g + h_a)$$

where t_g and t_a are the waste gas and air temperatures and h_g and h_a are the respective rates of heat transfer.

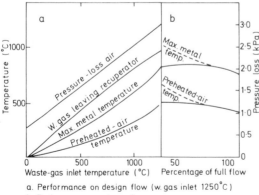

a. Performance on design flow (w. gas inlet 1250°C)
b. Performance on turn-down

——— Dilution
- - - - No dilution

Figure 14 Performance data of a recuperator (Courtesy ISI June 1973)

If h_g and h_a are equal (as is often approximately correct with convection recuperators) the metal temperature would be halfway between the waste gas and air temperatures. This is why it is necessary to limit the waste gas inlet temperature to 1050°C with this type of unit.

Radiation recuperators can accept waste gases at temperatures up to 1500°C without dilution, by using high velocities on the air side and hence high values of h_a. The pressure loss on the air side is higher than is usual with convection recuperators.

During turn-down the preheat temperature tends to rise, since the area of heating surface per unit volume of air increases. With radiation recuperators the rate of heat transfer on the waste-gas side tends to decrease less rapidly than on the air side, since it is not so dependent on velocity. Hence the metal temperature tends to rise considerably on low flow.

With the heat-resisting metals available a working temperature of 1000°C or more is often quoted, but it should be remembered that most of the troubles associated with recuperators are worse at high temperatures, so that it is wiser to limit the maximum metal temperature to about 850°C, even with the best materials, to allow for errors and accidents.

The ideal recuperator should give the highest preheat for the lowest cost, and should be safe under all conditions. This is of course impossible, and one of the big problems is to assess the true range of operating conditions. If the waste-gas entry temperature is overestimated, the specified preheat will not be obtained. If it is underestimated, the recuperator may be overheated. Solid material may be carried over with the gases and, if it does not cause corrosion, may deposit on the heating surfaces and so lower the preheat. The flow of gases may not be uniform, so that one part of the recuperator may be heated more than another. Radiation from surrounding brickwork may cause local overheating. It is also possible for furnace control instruments to fail and so cause combustion to take place in the recuperator.

Performance Data

In order to assess the margin of safety in any given recuperator proposal, the performance under a variety of circumstances needs to be known. The use of a computer enables this to be done quickly. *Figure 14* is typical of the information now offered to clients when tendering for recuperators. The range of metal temperature is clearly seen, as is the point at which waste-gas cooling becomes necessary.

Maximum Preheat

In deciding what preheat to ask for in a particular case, two important questions should be posed.

First, the cost, and secondly the margin of safety. In a rough-and-ready way it can be said that each additional $100°C$ of preheat doubles the amount of heating surface required up to that point. In addition, as the preheat increases it becomes necessary to use more expensive alloys.

If the recuperator is an essential part of an integrated plant it is wiser to decide on a moderate preheat, say $550°C$ rather than $750°C$. The greater fuel saving from the higher preheat would be more than lost if there were a breakdown.

On furnaces where conditions are stable, such as on a glass tank, the higher preheats are quite practicable. Some degree of safety can be gained by constructing the recuperator in two stages, operating in parallel flow or counterflow. *Figure 15* shows the result of a series of calculations using radiation recuperators. The best choice is 4, since the metal temperature at the bottom is lowest with this arrangement, and the hot-air outlet is at the lower end of the top recuperator where no weight is carried. Similar arrangements are possible with convection recuperators, and 'shock elements' are frequently installed on the waste-gas entry side.

Causes of Recuperator Failure

Recuperators fail from many causes most of which can be grouped into the following:
1. Incorrect data for design.
2. Design errors.
3. Fabrication faults.
4. Material faults.
5. Operational faults.

Incorrect Data for Design. The difficulty of assessing in advance the conditions under which the recuperator will have to operate has already

been mentioned. Because of this problem it is important to have some margin of safety. Frequently it is necessary to operate with higher flows than design values, and unless sufficient margin in pressure is included on both air and waste-gas sides, difficulties will arise. For this reason it is usual to design for about 80% of the fan capacity, but to ensure that the fan pressure will be high enough to pass all the available air through the system.

Design Errors. Fitzgerald, Howarth and Roberts[7] discussed some of the faults from a user's point of view. The faults enumerated were:
1. Incorrect choice of materials.
2. Uneven temperature distribution.
3. Insufficient allowance for thermal expansion.
4. Inadequate spacing between tubes.

Different types of recuperators exhibit different faults, and some of those attributed to design errors can equally be classed as due to user abuse. It is very often difficult to pinpoint the cause of trouble because of the damage which occurs after the original incident, particularly if, because of production requirements, it is expedient to continue operation after a recognized fault has occurred.

Fabrication Faults. These may include faulty castings, inadequate preparation for welding, and incorrect welding procedure. Some years ago a whole series of faulty welds were encountered from a fabricator who for years had produced flawless work. The ownership, management, and labour force had changed over quite a short period, and the old methods were forgotten or were not known. Since then ultrasonic or X-ray testing of welds has been carried out with completely satisfactory results.

Material Faults. In discussing the properties of heat-resisting steels with a metallurgist at the National Engineering Laboratory, with particular reference to creep stress, it was stated that the curves showing this property against time were a little misleading, and should be taken only as indicative, since the spread of results for a particular steel is so wide that the best result for a lower quality steel may be higher than the worst figure for the same test on a more expensive alloy.

Rolling faults, such as lamination, and the supply of the wrong material for the specification, are fortunately rather rare but should not be completely ignored.

Operational Faults. These include such things as forgetting to leave switched on a dilution-air fan or a control instrument, or misplacing a thermocouple so that the real temperature is much higher than that indicated. Running a furnace at a higher fuel-input rate than specified can be considered a fault if the additional fuel is not completely burnt in the furnace, since this can endanger the recuperator, particularly with liquid fuels.

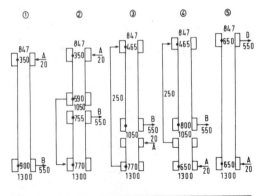

Length (%of ①)	100	101	105	106	113
Alloy (%of ①)	100	102	109	118	144

Figure 15   Selection of a recuperator system. A series of temperature ($°C$) calculations using radiation recuperators. The best choice is 4, since the metal temperature at the bottom is lowest with this arrangement (Courtesy ISI June 1973)

References p 390

It is not the part of the recuperator supplier to specify the method of furnace operation. Production must always be the first consideration. For example, on a five-zone pusher furnace one client found that production was best obtained by firing the preheat zone at a much higher rate than originally envisaged from experience at other works. This resulted in corrosion from the fuel-oil ash, as described below. Efforts should be made to accept this method of firing, if it is agreed that it gives better production, and to find means of minimizing the effect on the recuperator.

Sources of Failure and the Remedy

The following list of possible sources of trouble includes those associated with many different types of recuperators, and are not necessarily applicable to a particular case.

Leakage of Air. All recuperators must have a critical portion where the heating surface has to isolate the air from the waste gas.

Ceramic recuperators depend for the seal on the cement joints over which the refractory slides as it expands and contracts. This rubbing action can loosen a little cement as a powder. If the air is under appreciable pressure this powder is blown away, resulting in a fine gap. Because of this, the pressure on the air side is limited to about 250 Pa gauge or less. Even then there is usually an initial air leakage of about 15%.

Cast-metal recuperators of the needle type have the tubes bolted together with suitable jointing material, to form a unit. Under the influence of repeated heating and cooling these bolts tend to loosen, and leakage may occur.

Radiation recuperators have 'floating' hot-air boxes which are connected to the fixed air main by a form of joint that allows for the relatively large expansion. This joint may be of the packed-gland type. Under the effect of temperature and lateral pressure from the expansion or thrust from the recuperator, the packing may become hard and unyielding, giving rise to leakage. A metallic expansion bellows overcomes this difficulty.

Cracks. Cracking occurs with ceramic recuperators if sufficient allowance is not provided for expansion. One of the difficulties with this type of recuperator is that leakage may occur if such allowance is made.

With metallic recuperators, cracking can occur for the following reasons:

1. Jamming. Free expansion is prevented somewhere in the system by jamming. This may be due to severe overheating causing the maximum allowable expansion to be exceeded, even locally.

Dislodged brickwork in the flues may cause jamming in some circumstances.

Creep, the permanent extension which occurs if a length of steel is held under load for long periods, can cause trouble by taking up some of the available clearance. It can also initiate 'elastic instability' especially with radiation recuperators that are operating at high temperatures and pressures. The instability causes the inner shell to collapse toward the centre, forming 'lobes', usually four, during which cracks may be formed. Elastic instability depends on the asymmetry of the cylinder, the unsupported length, the metal thickness, the external pressure, the tube diameter, and the temperature.

In one case the recuperator had operated quite successfully for about 18 months, and according to the experts consulted was initially quite stable, in spite of the inevitable out-of-roundness as fabricated.

2. Thermal Stress. This occurs because of the difference in temperature across the thickness of the metal, and is equal to:

$$\beta E \Delta t / 2(1 - \mu)$$

where $\Delta t = QL/k$, E is the modulus of elasticity, β is the coefficient of thermal expansion, μ is Poisson's ratio, Δt the temperature difference, L the metal thickness, and k is the coefficient of thermal conductivity. This is the normal stress at equilibrium. Any sudden change in heat transfer to the surface results in a rise in skin temperature of the metal, which gives rise to an additional stress centre at the same position.

3. Sigma Embrittlement[11]. Sigma is a brittle phase which is formed in certain austenitic steels by a slow process of change, possibly over several thousands of hours. The temperature range in which this change occurs in 650 to 900°C. Above 900°C the brittle phase is reabsorbed. This brittleness is latent at temperatures above 400°C. Too rapid heating up from cold or too rapid cooling down might cause fractures in an old recuperator, where the change to sigma-phase had already occurred. Certain elements, especially nickel, reduce the tendency to form sigma-phase, whilst others, particularly silicon, increase the tendency. For this reason 25% chromium 25% nickel rather than 25:16, and less than 0.8% Si rather than the more usual 1.5% Si, are recommended.

Corrosion. There are four main forms of corrosion encountered in recuperator practice: oxidation, sulphur attack, attack from molten salts, especially from fuel-oil ash, and attack by carbon monoxide in the gases.

Oxidation. Those parts of a recuperator which can reach a temperature of 500°C or more will be subject to oxidation by oxygen, water vapour, or carbon dioxide in the waste gases or in the gas being heated. Alloys are readily available to minimize the oxidation up to about 1100°C, but ultimately every recuperator will suffer scaling and will need replacement of the hottest parts.

The intensity of the attack will depend on the temperature of the surface as well as on the concentration of the oxidants in the gases.

Sulphur Attack. The action of sulphur in the waste gases depends on whether the gases are oxidizing or reducing and on the nature of the metal as well as on the temperature of the gases and the metal surface.

At the hot end of the recuperator, if the gases contain more than about 2% of oxygen, and in the absence of vanadium and alkali oxides, the sulphur in the waste gases (in the form of SO_2 or SO_3) is little more corrosive than CO_2 or H_2.

At the cold end of the recuperator, however, acid corrosion (Ch.20.2) may occur if the metal temperature drops below about 130°C, corresponding to the 'acid dewpoint' of the gases[12]. The sulphur in the fuel when burnt with excess air first forms SO_2, which at lower temperatures is further oxidized to SO_3. This is a slow reaction, but the conversion of 1–5% is sufficient to account for severe low-temperature corrosion. This can be quite serious on counterflow recuperators, especially of the radiation type which are usually more exposed than the other types, and which generally have the colder surfaces constructed from mild steel. The solution is to metal-spray the mild steel with aluminium, or to use low-alloy steel.

If the sulphur-containing gases are reducing, i.e. if insufficient air is provided for complete combustion, the sulphur is converted to hydrogen sulphide. This constituent does not seriously attack mild steel, but can cause serious corrosion at higher temperatures with nickel–chrome alloys. For this reason it is usual to recommend that an oxygen content of about 2% be maintained in the waste gases.

Fuel-oil Ash. All except the lightest of oil fuels contain a small amount of inorganic matter which, on combustion, gives rise to a series of compounds, essentially mixtures of alkaline sulphates with vanadium pentoxide. These form low-melting-point eutectics which are carried in the molten state in the gases. They deposit on the heating surface and exert a fluxing action on the protective oxide film, thus exposing fresh metal to further oxidation. The fluxed oxide is oxidized further to a compound insoluble in the molten salt which is then freed to renew the fluxing action. This process results in a very characteristic effect, the action being very localized and leading to a hole in sheet material while the metal adjacent to it may retain its full thickness.

This form of corrosion, which is very troublesome in boiler practice, has been studied quite extensively, and a number of solutions have been suggested, including: removal of the vanadium from the oil, protective coating of the metal, additives to the oil, and change of material.

Only the last of these proposals seems likely to be generally effective, although the injection of powdered dolomite into the oil flame on the boilers at several powerhouses is said to be very effective. A fairly new alloy containing 50% chromium and 50% nickel has given many times the life of 25/20 alloy in a vanadium pentoxide

climate. Another alloy containing 60% Cr and 40% Ni is also recommended. These alloys are, of course, expensive, but if their use can overcome a difficult corrosion problem the extra cost can be justified. This form of corrosion is only serious at temperatures over about 580°C, so that the expensive alloy would only be needed in a relatively small part of the recuperator.

Carbon Monoxide. Where the metal temperature is high (over 900°C) and the gases contain large amounts of carbon monoxide, there is danger of metal erosion. The mechanism of this form of attack seems to be that the carbon monoxide reduces the protective oxide film to spongy metal which allows further oxidation of the underlying metal as soon as circumstances permit. This trouble was encountered on a soaking pit, where metal at the hot end of the recuperator was eroded over a small area to paper thickness. Investigation showed that the damage exactly coincided with an area of the inner shell where direct radiation from the brickwork of the flue fell on the metal surface. This pit was operating with coke breeze bottoms, and it is known that for several hours after a new bottom has been made the waste gases contain a high proportion of carbon monoxide. This gradually diminishes and the gases become oxidizing again, but if during the early stages it coincides with a high temperature, then erosion must occur.

Blockage. Very dirty gases can cause blockage in a recuperator, especially where the heating surfaces are horizontal. Convection recuperators of the needle or composite-tube type are more prone to this trouble, since the spaces must be kept small to yield high convection coefficients. Leaving off some fins, or at least spacing them more widely apart, will reduce but not eliminate the trouble. Continuous shot cleaning (Ch.10.5), which is frequently resorted to, is not a complete answer, since the cleaning is never complete and even slight overheating can cause the shot to fuse together. Radiation recuperators which are mounted vertically do not suffer from this trouble, and are in fact used with gases containing $21 g/m^3$ (9 gr/ft^3) of dirt without needing cleaning for years. Similarly, the Escher type of combined radiation/convection recuperator, using hollow fins, can be used with dirty gases[13].

A case of blockage which occurred some years ago was identified as being due to elastic instability. A section of the inner shell collapsed inwards with the lobes in the form of a four-leaved clover. This is discussed above (p 382); the recuperator was operating under high pressure.

When gas is preheated in the cast type of recuperator, there is a danger of decomposition with the formation of carbon particles which can block the recuperator. This has been observed even with blast-furnace gas, with which there is a critical temperature, about 300°C, above which it can be catalytically cracked to form carbon dioxide and carbon. This decomposition does not occur with steel recuperators, and blast-furnace gas is regularly preheated to 600°C without trouble.

Safeguarding of Recuperators

The aim of safeguarding is to ensure that the eventual need to replace the unit because of normal wear and tear is delayed as long as possible. The various methods employed are described below, with comments.

The usual method of protection is to cool the ingoing waste gases to a temperature low enough to limit the metal temperature. This is done either by the introduction of cold air into the flue before the recuperator, or by the use of water sprays. Whichever method is used, it is essential that the cooling be progressive, and not 'on—off', since cooling too quickly can cause cracking of the unit by the very sudden contraction, which will not be uniformly distributed. At the very least, the protective oxide film on the surface of the heat-resisting steel will be cracked off, exposing the metal to further oxidation.

Gradual introduction of the cooling medium can also cause trouble in some cases. If, for example, the waste gas contains combustibles owing to shortage of air, the first effect of cold-air dilution is to increase the temperature by combustion. Similarly, if water sprays are used the first effect would be to increase the heat transfer by gaseous radiation, owing to the increase in water content, which may outweigh the drop in temperature. Both these effects are overridden as the cooling of the waste gases progresses, but are nevertheless undesirable.

Cooling by air dilution has a further disadvantage. If this has to be done on full flow, the increase in total volume of waste gases can sometimes cause excessive pressure loss. Water-spray cooling, on the other hand, although it avoids this difficulty, can lead to trouble since any dissolved solids in the water may deposit on the heating surfaces, and can cause corrosion if not blockage. There is also the danger of the sprays becoming blocked if the gases are dirty, or filling the flue with water if the evaporation is not complete.

Hot-air Blow-off.

In most cases the best way to maintain the required metal temperature is to pass more air through the recuperator than that corresponding to the amount of fuel being burnt. The excess air is then blown off before the main flow is distributed to the burners.

If, as on a multi-zone furnace, the air is metered on the hot side there is no problem, and within the limits of the combustion-air fan this method of protection is very successful.

If the air is metered cold, as on a soaking pit, another method may be used. The combustion air supply is split into two streams, both of which are separately metered and provided with a control valve. The hot air which is blown off is also metered, and a control unit matches the air through the by-pass main to this, so that the air supply to the burner remains correct. If the temperature of the hot air varies very much some form of temperature compensation must be provided. In the classical case of a glass tank, conditions in the furnace must be maintained steady for technical reasons, and the blow-off facilitates this by giving a constant preheat. The recuperator would be made larger than necessary so that at the end of the campaign, when the fuel input may be up to 50% more than at the outset, due to deterioration of the brick work, the preheat temperature may be maintained. This excess heating surface would give a much higher preheat at the outset, were it not for the blow-off.

Selection of Control Factor and Position.

Control of the waste-gas temperature is no guarantee that overheating will not occur, since, on low flow, the preheat and metal temperatures increase rapidly, even with constant waste-gas temperature.

Preheat temperature is a more reliable guide for recuperator protection, but even here anomalies may occur, particularly with some arrangements of two-stage operation, where one stage may be overheated although the preheat is normal.

Direct measurement of the metal temperature is obviously the best method provided that the position chosen represents the maximum metal temperature and the thermocouple (source of measurement) may be replaced without interrupting the operation of the furnace. A combination of metal temperature and preheat temperature seems to be the best practical control factor, either temperature overriding the control setting of the other if circumstances so require.

A further method of protection has been used on almost every Escher radiation recuperator. This makes use of the overall expansion of the recuperator, which is related to the metal temperature. In general, this is only used as a final safeguard, operating a limit-switch to cut off the fuel supply. Of course the operating temperature is then set within the normal safeguard temperature.

Other Forms of Safeguard.

Air cooling of strategic parts of the recuperator, such as the first pass, or of flue connections, may be provided in addition to the above.

Maintenance

From many years of experience there is no doubt that a high proportion of recuperator failures can be attributed to bad maintenance, either of the recuperators themselves or of the control equipment. It could be argued that these control systems are too complicated, particularly where an installation consists of more than one recuperator, and distribution of air and waste gases has to be provided. However, the need for protection increases with increasing preheat temperature since peak metal temperatures may approach or exceed

Table 1   Heat-resisting ferrous alloys.   Courtesy The Macmillan Press Ltd, Basingstoke (*Macmillan Engineering Evaluations*, 1971, p 166)

Material	Composition					Max. work temp.	Cracks		Atmosphere slightly				Fuel ash
	C	Si	Cr	Ni	Other elements		Thermal shock	Sigma	H_2	CO	H_2S	SO_2	
Castings													
Irons:													
Grey	3.2	1.5			0.7P	600	B				G	G	F
Low alloy	3.2	1.8			0.2 P max	650	B					G	
Silicon	2.0	6				850	B						
Austenitic	2.0	5	5	30		900	G						FB
30% Chrome	1.5	1.0	30			1050	B	B					
Steels:													
Ferritic	0.25	1.0	28	2		1100	B	B			G		
Austenitic	0.4	1.5	25	20		1150	F	F			FG		FB
Fabrications													
Mild steel	0.23	0.11				450	G	G		G	G	FP	
Al-coated MS	0.23	0.11			Al-sprayed	650	G	G		F	G	G	G
Al-Si steel	0.03	2.5			1.0 Al	750	G	G		F	G	G	
PKI*	0.13	0.5	14–18			750	G	B	G		G	G	G
18/8 (320)	0.1	0.8	18	8.5	0.6 Ti	800	G	FG			B	F	FG
25/20 (310)	0.12	0.8	25	20		1150	G	FP			VP	F	FB
20/30 (330)	0.15	2.0	20	35	0.25 Cu, 1 Mn	1100		VG			VG		
Inconel*	0.1	0.7	16	75+	3.75 Al	1200	VG	VG			B		
Incoloy 800*	0.1	1.0	19/22	30/34	0.5 Cu max	1000	G	G	G	G	B	F	FB
Nimonic-75*	0.15	1.0	19.5	bal	0.2–0.6 Ti, 0.5 Cu	1200	G	G	G	G	B	F	FB
Nichrome*		1.0	20	80		1150	G	G			B		VB
Kanthal*			23	—	4.5 Al, 1 Co	1300	F	B			G	G	G

* Trade name.   VG = very good, FB = fairly bad, VP = very poor

the limit set by available alloys. This is discussed above, under *Maximum Preheat* (p 380).

Heat-resisting Alloys

Heat-resisting alloys fall into four main classes.
1. The ferritic alloys. These are straight chrome steels with 11—27% chromium and low carbon contents. They are used widely in Germany and cover a range up to 1170°C. They are difficult to weld and have rather low strengths.
2. The martensitic alloys. These are similar to the ferritic alloys in generally containing no nickel, but have higher carbon contents and also present some difficulty in welding.
3. The austenitic alloys. This group covers a range from 10—26% chromium and 6—36% nickel, with other alloying elements such as titanium, molybdenum and columbium. They often have more than twice the strength of the corresponding alloys of groups 1 and 2.
4. Special alloys. This group has been included to cover the special alloys used for particular purposes. The Nimonics, the Correnels, the Incoloys, and the relatively new 50/50 alloys are of this group. In general, their characteristics are less well known, but their value has been well proved for special purposes.
Table 1 shows a selection of typical alloys in these groups. With this large range of alloys it can be understood how difficult it is to select the most appropriate alloy, particularly when considered from the point of view of permissible stress. The alloys of groups 1 and 2 have much lower coefficients of thermal expansion than the corresponding alloys of group 3, so that lower thermal stresses are set up, which partly offsets the lower

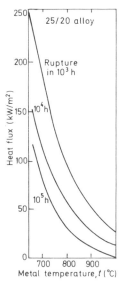

Figure 17 Limiting rate of heat transfer with 25/20 alloy (Courtesy ISI June 1973)

strength. This lower expansion also assists in maintaining the protective oxide film.

Some of the properties quoted for one quality of 25/20 alloy are given in *Figures 16* and *17*. This information is used to calculate the limiting rate of heat transfer through a plate 6.4 mm thick. Within the reliability of these data such charts could be extremely useful in designing heat exchangers in respect of thermal stress.

16.3 Some other forms of heat and energy salvage

Recovery of Raw Materials

Energy, in one form or another, is consumed in every industrial process. Any economy in the consumption of raw materials represents a saving of energy. Stora Kopperberg, the Swedish steelworks, have, for example, developed a method of recovering acids and metal from pickling baths. An environmental problem has thus been completely solved, and valuable raw material has been retained in the manufacturing process, so that the result can be considered as very satisfactory.

Use of Oxygen for Combustion

M. W. Thring has pointed out[5] that combustion should always be carried out with as high a temperature as possible by preheating the air and by minimizing excess-air dilution because this reduces the irreversibility of the combustion process. The use of oxygen to replace all or part of the air is a means of energy conservation for two reasons:

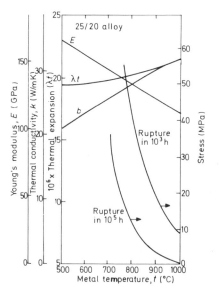

Figure 16 Properties of 25/20 alloy (Courtesy ISI June 1973)

first, the combustion takes place at a higher temperature, and secondly there are less waste gases (less nitrogen) to carry away valuable heat in the products of combustion. In a cupola, oxygen boost is used to increase the metal temperature especially after a stoppage, and allows a greater output from a given cupola. Oxygen may also be used to upgrade fuels like blast-furnace gas, by raising the flame temperature.

Dry-quenching of Coke (App.B6); Waste-heat Boilers (Ch.16.1); Gas Turbines (Ch. 12.2, 12.3)

These forms of heat recovery are dealt with elsewhere in this volume.

Recovery of Potential Heat

Many processes give rise to combustible gases, often of low calorific value, yet capable of useful work if correctly applied. In addition to blast furnaces and coke ovens, iron cupolas, zinc smelting furnaces and electric-arc melting furnaces all yield combustible gases. Quite an appreciable amount of methane is recovered from upcast air from coal mines.

Cupola gas, in particular, has a very variable calorific value, since it is sensitive to change in the make-up of the charge passing down the furnace. The carbon monoxide content can vary during the day from less than 10% to more than 29% in some cases. To achieve stable combustion with poor-quality fuel, special care must be taken in the combustion-chamber design to ensure that the hot gases are entrained by recirculation into the incoming air/gas mixture. Preheating the air or gas is also used frequently to stabilize combustion. The hot gases from the combustion chamber are usually led through a heat exchanger to preheat the blast.

Blast-furnace Stoves

The waste gases from blast furnaces were first used as long ago as 1814 in France for making steel, but it was not until the mid-century that they were used for preheating the air blast.

For over 100 years blast-furnace gas has had a CV of about 3.8 MJ/normal m^3, but during the past 15 years practice has been much improved by better use of the heat in the gases ascending the shaft to preheat the charge, and because of better ore preparation; currently therefore the CV is as low as 3.2 MJ/m^3. At the same time improvements have been made to the blast heaters.

The Cowper stove is a special form of regenerator which is direct-fired, and which can operate under high pressure and high temperature. In recent years much work has gone into the up rating of Cowper stoves by the use of better refractories and high-intensity burners, with increased size of combustion chamber. With the conventional system four stoves are required for each blast furnace; with the new system, such as that installed by Surface Combustion at the Jones and Laughlin Cleveland works after careful study by analog computer, it has been possible to raise the blast temperature by 200°C with a reduction in the stove size. It is claimed that many existing systems have sufficient capacity when modified to enable high blast temperatures to be attained without replacement by new and larger units. At least 1050 normal m^3/min can be burned in a stove only 6.7 m in dia. with a checker mass of 800 t and a surface area of 15 700 m^2.

Hot-blast Cupolas

Big savings in the coke requirement for iron-foundry cupolas have been achieved by the use of blast heated to 450—500°C in heat exchangers fired by the hot dirty gases evolved of necessity during iron melting. The Clean Air Act requires that the gases be burned to remove carbon monoxide and cleaned before being released into the atmosphere, and it is obviously good economics to use the heat recovered to preheat the blast. The heat exchanger must be of a type to withstand the very dirty gases without fouling, and the Escher combined radiation and hollow-fin recuperator has proved most suitable, operating for very long periods, even years, without the need for cleaning[13].

Another form of saving in cupola practice has been found to result from the extension of the melting area by the use of two sets of tuyeres. These have been shown to bring nearly all the benefits obtained from the use of hot blast, but of course leaves the potential heat in the top gases unused. It should be practicable to combine the two systems, and so obtain at least some of the benefits of each.

Hot Cooling

Many furnace parts need to be force-cooled. Furnace door frames and particularly skids in reheating furnaces are typical examples.

In cold-water systems the thermal energy extracted through the cooling elements attains a low temperature level and not only is it useless, but it necessitates the installation of some further water-cooling system. In the evaporative system[14] on the other hand the thermal energy extracted remains available at the higher temperature level and can be used as saturated steam. The operating conditions are also improved, and down-time is much reduced because of reduction of corrosion of the cooled member which remains above the acid dewpoint.

Hot-cooling systems can have either natural (thermal syphon) circulation or forced (pump) circulation. With natural circulation the steam drum must always be positioned above the cooling elements. In this case the water in the down pipes (between the steam drum and the cooling elements) is of considerably higher density than the steam/water mixture in the risers (between the cooling elements and the steam drum). This creates, if the plant is adequately dimensioned, sufficient driving force to generate the required circulation.

With forced circulation, pumps are required to maintain the necessary flowrate. The only stipulation is that the pumps must be positioned below the steam drum.

Charge Preheating

The direct recovery of energy from waste gases can be achieved in metallurgical furnaces by preheating the incoming raw materials. The majority of metallurgical furnaces have developed over the years from low-powered units having a large degree of charge preheating to high-powered, high-output units having higher waste-gas temperatures but with correspondingly lower efficiencies. The development of high-powered units has necessitated the use of external energy recovery devices of the types mentioned previously, in order to maintain good fuel economy. The trade-off between charge preheating and external heat recovery depends mainly on the relative capital costs of providing extra furnace capacity for the charge preheating and of extra equipment.

In the main, the effectiveness of charge preheating depends upon the physical and dimensional properties of the material being heated together with the handling methods used. For example the

reheating of steel slabs[15] before rolling is not conducive to in-furnace charge preheating, because of the relatively small surface area of the slabs available for heat transfer. *Figure 18* illustrates how the furnace design and heat-release pattern can alter the output of continuous furnaces to provide gains in productivity at the expense of increased waste-gas loss and lower efficiency. It will be noted that whereas the output of the highest-rated furnace is 2.5 times that of the lowest-rated design, the waste-gas (Celsius) temperatures are only 50% greater. Thus in the metalworking industries the trend has been towards compact high-output slab-reheating furnaces having high waste-gas temperatures, and external recuperators to provide combustion air preheated to high temperatures.

In some instances it is economic to duct waste gases from high-temperature high-cost furnaces into low-temperature low-cost preheaters containing the material to be heated. An example of this is to use the waste gases from steelmaking processes to preheat scrap charges[16].

However in one specific type of furnace, namely the vertical shaft type, it is possible to increase size and output while at the same time reducing waste-gas losses. Shaft furnaces are generally used for the continuous heating and melting of scrap metals (cupolas) and for the heating, reduction and melting of metallic ores (blast furnaces). In the latter case the height of the shaft determines the degree of charge preheating; to facilitate heat transfer and chemical reactions, all the materials

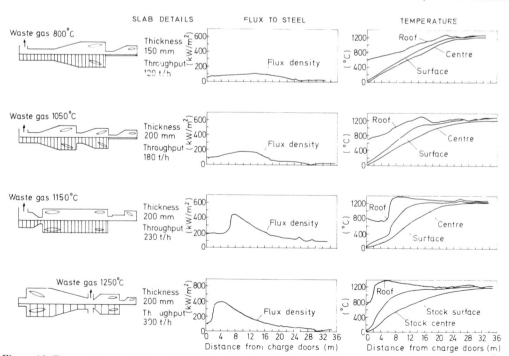

Figure 18 Temperature and flux distribution in 7 m wide continuous steel-slab reheating furnaces[15] *(Courtesy ISI)*

for this type of furnace are ideally sized within the range 20—60 mm.

Using Low-grade Waste Heat

High-temperature waste gases are required to generate economically medium- and high-pressure steam for electricity production. In many industries waste gases arise at temperatures lower than that desirable for power generation, and large quantities of energy are frequently wasted. In recent years considerable attention has been paid to methods of recovering such low-grade energy, particularly as part of total-energy systems (Ch.15.1), and more recently by combining industries using low-temperature energy on sites adjacent to large-scale producers of low-grade waste heat.

Absorption-refrigeration Cycles

Absorption-refrigeration cycles utilizing ammonia and water as the refrigerant and secondary fluid are in operation in several industries where process temperatures down to $-55°C$ are required together with large thermal capacities[17]. Large absorption-refrigeration systems are found for example in petrochemical plant, where on the one hand product-gas streams laden with water vapour at temperatures in excess of $100°C$ release large amounts of heat while on the other hand substantial refrigeration capacities at $-45°C$ are required for scrubbing processes. Preliminary cooling, prior to liquefaction and separation of gases, can also be achieved by this technique and in future it might be used to reduce the production costs of high-purity oxygen on steelworks sites.

Thermoelectric Power Generation

In recent years the development of new semiconductor thermoelectric materials has created renewed interest in the possible application of thermoelectricity for generating power (Ch.13.5). The efficiency of thermoelectric generators is determined by the Carnot efficiency and the efficiency of the thermoelectric materials used[18]. To date practical devices have been developed only for use at low power levels and temperatures up to $1000°C$, with efficiencies of around 10%, and application to the recovery of industrial waste heat is not likely in the foreseeable future.

Enhancing Chemical Processes

The reaction rates of many biochemical and chemical processes can be increased by the addition of thermal energy at low temperatures. Generally this can be readily achieved within integrated chemical complexes, but in some instances chemical processes require to be sited away from heat-producing equipment. The use of waste heat in sewage-treatment plants is now being given serious consideration. Studies have shown that the biological processing of sewage can be increased by a factor of ten by raising temperatures using low-grade heat[19]. The biological treatment of industrial effluents is widely used by industry, and the use of low-grade heat to enhance the rates of these processes is an area for further development.

District Heating (Ch.15.2), Agriculture and Aquaculture

Recent developments in Sweden have established the viability of pumping hot water and low-pressure steam for distances up to 80 km from nuclear power stations[20]. Over 40 Swedish towns already use local district-heating systems to provide combined electricity production and hot-water distribution at an overall efficiency of about 85% compared with conventional maximum electricity production efficiencies of around 35%. Work in the UK, where over 280 district-heating schemes have been reported[21], and the USA has progressed along similar lines but the major advance has been in Sweden where special piping and district-heating systems have been developed to provide low heat loss at low capital cost. The cost of producing and pumping hot water and low-pressure steam is considered in Chapter 15.3 and by Margen[20] in relation to nuclear-heat/electric stations. The production and transport costs from industrial complexes using low-grade waste heat should fall approximately within the range of costs indicated in these references.

Fish farming and crop raising have been investigated for many years by power generation companies in many countries around the world. Saltwater fish farming is at an early stage of development, and presents many problems which will require solution prior to commercial application. Freshwater fish farming using warm water from power generating plant is practised[22].

Crop raising supplemented by low-grade energy is more advanced, and provides a means of obtaining high-free-energy and high-molecular-weight material from low-enthalpy low-free-energy sources. New techniques in hydroponic feeding, soil-less cultivation and vertical, sealed, greenhouses using conveyors and industrial waste heat offer the possibility of raising large amounts of vegetables at low capital cost[23]. An air-conditioned tower greenhouse heated by low-grade waste heat and suitable for siting near industrial complexes has been projected (*Figure 19*).

Fuel Saving in Industry

Paul Tate[24], in a paper with the above title, summarized his remarks in the following rules:
(i)   Wherever possible replace wasteful apparatus.
(ii)   Replace manual control by automatic control.
(iii) Reduce the waste energy at each stage of production to a minimum.

References p 390

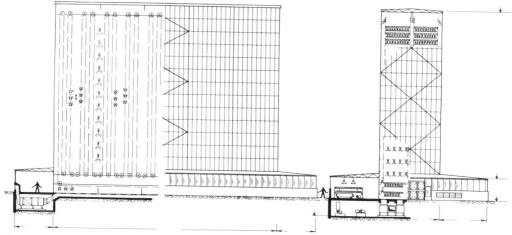

Figure 19 The Ruthner continuous tower greenhouse

(iv)   If possible cut out stages in production; e.g. three stages each 60% efficient result in an overall efficiency of about 20%.

(v)   If some waste energy is unavoidably produced in one part, try to use it as a supply of energy for another.

(vi)   Where good results are obtained, publicize them and emphasize attainments.

(vii)   Once levels of economy have been attained, maintain them by maintaining interest.

To these could be added:

(viii)   Never operate a process at a higher temperature than is really necessary, since the heating costs increase disproportionately with increasing temperature. A slightly longer soak or lower output, for example, might result in quite a large fuel saving.

Olson[25], in *Industrial Heating*, January 1974, said that the following four points should be remembered:

1. A saving of 10—15% can be made by close control of fuel/air ratio.

2. A recuperator can save 10—30% of the fuel.

3. Water-cooling losses can be kept low by maintaining the insulation on the water-cooled members.

4. The Fuel Department and the Fuel Engineer are important factors in the fight to save fuel, and their purpose should be re-emphasized.

References

1   Extended Surface Tube Company, *ESTUCO Welded Fin Tubes — Heat Transfer Design Data*

2   Nelson, G. A. *Hydrocarbon Processing* 1970 (Dec.), p 88

3   Ref.(A)

4   Ref.(B)

5   Ref.(C)

6   Thurlow, G. G., 'The Rotary Air Preheater', *J. Inst. Fuel* 1949, Vol.22, 219

7   Institute of Fuel and Institut Français des Combustibles et de l'Energie (Organizers), Heat Exchanger Conference, Paris, 1971

8   Escher, H., 'Metallic Recuperators', *Iron and Steel*, 1947 (Oct.)

9   Flux, J. H., Edwards, A. M. and Howarth, H. *J. Iron Steel Inst.* 1971 (Jan.)

10   Laws, W. R. *Iron & Steel International* 1974 (April)

11   Foley, F. B. and Krivabok, V. N. *Metal Progress* 1957 (May)

12   Blum, H. A., Lees, B. and Rendic, L. K. *J. Inst. Fuel* 1959 (April)

13   Kay, H., 'Recent Improvements in the Design of Hot Blast Cupolas', Paper No.6 of the Brussels Conf. on Heat Utilisation, 1964; *Foundry Trades J.* 1966 (April 14 and 28)

14   Westerhoff, F. *Iron and Steel* 1972 (Dec.), Vol.45, No.6

15   Laws, W. R., 'Trends in Slab Reheating Furnace Requirements and Design', *ISI Slab Reheating Conf. Proc.*, 1972 (June), pp 1—12

16   Laws, W. R. *Steel Times* 1971 (June), p 505

17   Richter, K. and Schumacker, G., 'Design, Operation and Comparative Evaluation of Industrial Ammonia—Water Absorption Plant', *Inst. of Refrigeration Rep.*, 1971 (Feb.4)

18   Cadoff, I. B. and Miller, E., *Thermoelectric Materials and Devices*, Reinhold, London, 1960

19   Ref.(D)

20   Margen, P. H., 'The Future for District Heating', *DHA District Heating Conf. Paper*, London, 1974 (May 21)

21   Haseler, A. E., 'New Techniques for Telethermics', *63rd District Heating Conf. Paper*, Chicago, 1972 (June 12—15)

22 Yee, W. C., 'Thermal Aquaculture: Engineering and Economics', *Environmental Science & Technology* 1972, Vol.6, No.3, 232

23 Ruthner, O. 'Industrial Production of Plants', *First European Bio-Physics Congress*, Vienna, 1972 (Sept.)

24 Tate, P., 'Efficiency and Conservation', *Energy World* 1974 (July)

25 Olson, *Industrial Heating* 1974 (Jan.)

General Reading

(A) *Waste Heat Recovery from Industrial Furnaces*, Chapman & Hall, London, 1948

(B) Etherington, H. & G., *Modern Furnace Technology*, Charles Griffin, London, 1961

(C) *Waste Heat Recovery*, Chapman & Hall, London, 1963

(D) *Air and Water Pollution*, Adam Hilger, London, 1972

Materials and control

Chapter 17

Refractory and insulating materials

17.1 Refractories

Testing
Silica
Magnesite
Dolomite
Chrome and chrome—magnesite
Aluminosilicates
Carbon
Special refractories
Glossary of terms commonly used in connection with
refractories and furnaces

17.2 Insulating materials

Raw materials
Brick manufacture
Properties of insulating bricks

17.1 Refractories

The connection between refractories and energy might at first seem tenuous. In fact it is strong and direct. High-temperature reactions are rarely carried out in the open air. True metal can be smelted and refined in the laboratory or field on a cupole, but in industry virtually all such reactions, whether they involve smelting and refining of metals or the production of glass, cement or steam, are carried out in refractory containers — often protected by water-cooling. The vital role of refractories in conserving energy was well illustrated by the development of the all basic open-hearth furnace used for steelmaking. Here it was soon realized that the enthusiasm of the refractories technologist in trying to achieve record lengths of life must be tempered by the fact that thin brickwork and leakage through cracks mean high fuel costs, together with a drop in output rate that could easily offset any savings in refractories cost per ton of steel. Nevertheless, improved refractories mean longer time between shut-downs and therefore not only output gains but fuel savings owing to less frequent heating-up with its non-productive fuel usage.

The container — usually referred to as a furnace or combustion chamber — is normally built of refractories, though, for low temperatures, insulating bricks or even ceramic fibre blankets can be used. The shape of the container varies greatly with the process and method of operation. With the traditional open-hearth type used for metal melting, e.g. steel or copper, it resembles a house, often with cellars filled with regenerator bricks. The combustion chamber of a boiler is also a refractory box, but here most of the inner surface consists of boiler tubes containing flue gas or water. For cement and refractory-clinker production it takes the form of a long rotating tube — presenting a particularly difficult lining problem, partly because expansion allowances are critical but also because rotation reduces stability. Electric furnaces, such as are used for smelting aluminium or making steel, are big consumers, and have an energy cost similar to that of fuel-fired furnaces.

Testing

The selection of refractory brick for a specific application has much in common with the selection of a man for a job. The first task is to write a job specification, the second to interview the candidates, and the third to choose the 'best-buy', not on some specific qualification but on a hunch based on all the information available. The test data on refractories may be one's own or have been supplied by the manufacturer. It will be compared with that for previously used or rival materials, and a choice made on an overall assessment. The insistence on some particular property, say maximum permissible specific gravity

in a coke-oven brick, may be essential, but can in the absence of full knowledge prove a limitation. Thus insistence on a thermal-shock resistance of 30 reversals (see later) on a basic roof brick could have held up the improvement that undoubtedly happened when magnesite—chrome bricks of rather lower shock resistance replaced chrome—magnesite bricks of lower slag resistance.

The great advantage of testing, particularly of new products, is that it enables the skilled operator to say, with 8 or 9 chances out of 10 of being right, that brick A is better than brick B for a particular application. This is a great deal better than guessing, and can lead not only to the rapid trial of improved materials but to the elimination of refractories whose use would have been dangerous. Furthermore, testing of used products often indicates both causes of failure and ways in which a particular material can be improved.

Standard Test Methods

In spite of many years of collaborative work there are still no internationally agreed standard methods of testing refractories. There are however British Standards (notably BS 1902:1966), American (ASTM), German (DIN) Standards, and a host of others, mostly at an earlier stage of development.

A selection of the more important properties evaluated in BS 1902:1966 are listed below:

Section 2. Density and porosity
  Porosity is defined as the ratio of the pore volume to the entire volume. The test is easy and cheap to do and among the most useful.
Section 3. True specific gravity
  Defined as the ratio of the weight of unit volume of a substance to that of the same volume of water. Particularly useful in determining the adequacy of firing of silica bricks.
Section 4. Powder density
  Similar to true specific gravity but includes closed pores.
Section 5. Permeability to air
  A measure of the rate at which fluids pass through a porous body. Values are frequently determined for refractories, but rarely used.
Section 6. Refractoriness
  An index of heat resistance, often determined by comparing the collapse of a cone made from the test brick with that of standard Seger cones.
Section 7. Refractoriness under load
  Resistance to the combined effects of heating and loading. The standard method is in compression but this is being rapidly replaced by a hot-modulus-of-rupture test made on a $6 \times 1 \times 1$ in bar* (see *Figure 1*).
Section 8. Permanent dimensional change on reheating — linear or volume

---

* 1 in = 25.4 mm

Section 9. Cold crushing strength
Mechanical strength in compression, e.g. of 3-in cubes.

Section 10. Thermal expansion
That part of the expansion of a material on heating, which reverses on cooling. An essential figure in considering expansion allowances — though these are rarely more than half the theoretical value.

Appendix B. Thermal shock resistance
Small testpieces (3 × 2 × 2 in) are heated to 1000°C and then shock-cooled, the procedure being repeated until failure occurs.

Other vital properties are slag resistance (difficult to quantify in a useful manner), thermal conductivity and specific heat.

Interpretation of Test Data

Table 1 summarizes the properties of the principal refractories. Their usefulness in considering the best material for the job can be illustrated by reference to specific materials. Thus high alumina (85%) may have been suggested as a probable improvement over medium alumina for a particular application — say the combustion chamber of a blast-furnace stove operating at 1500°C under a load of 0.173 MN/m² (25 psi). Comparison of the two sets of data shows that:

The porosities are of the same order. The refractoriness of the bricks are both above the stated operating temperature, though that of the high alumina (1850°C+) is well above that of the medium alumina (1690°C). The latter shows 5% subsidence in the maintained refractoriness under

load temperature. Unfortunately a rising-temperature test was used on the 85%-alumina material, but the result is more encouraging — 10% but at a much higher temperature (1730°C). Other data not included in this Table show that the 85% alumina has an exceptionally good hot modulus at 1400°C, namely 10.4 MN/m² (1500 psi). Both bricks show excellent thermal-shock resistance but background experience suggests that this is likely to deteriorate more rapidly with medium than 85% alumina owing to glass formation. The conductivity of the high-alumina brick is appreciably higher.

Other things being equal the choice would clearly go in favour of the high-alumina material — at least for a trial. This should show whether its properties are sufficiently better to warrant the much higher price.

Silica

Raw Materials

Although silica, either free or combined, constitutes a large part of the earth's crust, relatively little of it is suitable for brick manufacture. Thus sand even of high purity does not make good silica brick, but quartzite often does. The United Kingdom is fortunate in having large amounts of quartzite, though only part of this is suitable for brickmaking. Most of the early bricks were made of ganister, but Welsh quartzite was among the first materials used to make such bricks for the Siemens open-hearth furnace. These are still known on the Continent by the Welsh word Dinas. For so-called super-duty silica bricks the alumina content must be low (less than 0.5%) and the raw material capable of yielding a brick of less than 20% porosity.

Manufacture of Silica Bricks

This is simple in theory though tricky in practice. It consists of grinding (with or without sieving and remixing) to get a dense packing (*Figure 2*), hydraulic pressing, drying, and finally firing to 1450–1500°C in a batch or tunnel-type kiln. During this firing, the quartz goes through the so-called alpha to beta change at about 575°C (see *Figure 3*). This is reversible, but continued firing changes most if not all the quartz to two other forms of silica (cristobalite and tridymite) that have their own reversible expansions. The formation of these new forms occurs with an expansion, and the bricks are therefore somewhat larger than the mould. Because cristobalite has a most unfortunate and substantial alpha-beta change that occurs at 200/300°C, attempts have been made to produce volume-stable bricks consisting essentially of tridymite whose reversible expansion characteristics are more modest. Incidentally iron oxide has been added, particularly in the USSR, to catalyse tridymite formation at the expense of that of cristobalite.

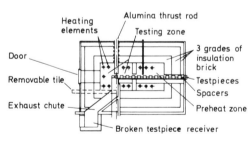

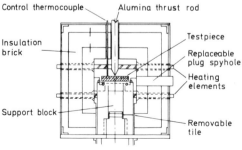

Figure 1   Hot-modulus-of-rupture apparatus (J. H. Chesters, Refractories: Production and Properties, Iron and Steel Institute, London, 1973, Fig.18)

Properties of Silica Bricks

Typical properties of low-alumina silica bricks are given in *Table 1*. It will be seen that these have a silica content of 96.3% and an alumina content of 0.3%. The lime (1.8%) has little effect on the refractoriness, and adds greatly both to the ease of brickmaking and the strength of the product. The porosity is relatively low — 20% — and the bulk density high (1.86). Using the simple relation:

$$\text{Bulk density} = \frac{(100 - \text{apparent porosity})}{100} \times$$

$$\times \text{ apparent specific gravity}$$

it can be shown that the apparent specific gravity is 2.33 (cf. *Figure 4*). This compares with the true specific gravity of quartz 2.65, cristobalite 2.33, tridymite 2.27 and quartz glass 2.21, and suggests that there is little residual quartz to cause growth in service.

The refractoriness (1730°C) is typical of the super-duty or first quality silica brick, though it is above the melting point of pure silica — 1723°C! This is explained by the fact that the cone test involves a rising temperature, and although the surface may have glazed at 1723°C or lower the full collapse comes at an appreciably higher temperature. Very important is the permanent linear change on reheating for 2 hours at 1500°C, namely 0.4% — appreciable but low compared with the sort of growth obtained with softer-fired products containing a lot of residual quartz.

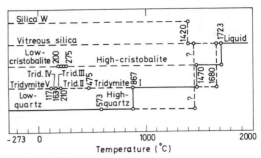

Figure 3   Classification of phases of silica at 1 atm pressure, as revised by Sosman (J. H. Chesters, Refractories: Production and Properties, *ISI, London, 1973, Fig.33)*

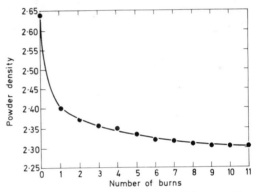

Figure 4   Influence of repeated firing on the powder density of silica bricks (after Hugill and Rees)
(J. H. Chesters, Refractories: Production and Properties, *ISI, London, 1973, Fig.49)*

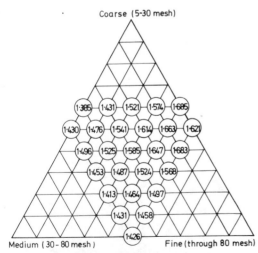

Figure 2   Influence of grading on bulk density (g/ml) of silica bricks (after Hugill and Rees) (J. H. Chesters, Refractories: Production and Properties, *ISI, London, 1973, Fig.48)*

The thermal shock resistance is very low, only 1—2 reversals in the standard test. This compares with 30+ for many materials, e.g. fireclay. For this reason large silica structures, such as coke ovens, must be heated up very slowly, often over many weeks. For open-hearth furnaces times of 24 hours or less have been successfully employed by slowing down during the most critical (200—300°C) stage. The importance of this is shown by the thermal expansion curves given in *Figure 5*, from which it will be seen that the bricks having specific gravities in the 2.32—2.35 range expand by about 0.5% between 200 and 300°C. It should be added that above 1000°C a hard-fired brick is virtually volume-constant — a great advantage in a structure such as an open-hearth roof or the dome of a blast-furnace stove where the temperature rarely falls below this level. It should also be noted that both the thermal conductivity and the specific heat of silica brick are similar to that of fireclay brick.

Table 1   Properties of selected refractory bricks (based on J. H. Chesters, *Refractories: Production and Properties*, Iron and Steel Institute, London, 1973, Appendix 11)

Brick type	Chemical analysis (%)		Apparent porosity (%)	Bulk density† (g/cm³)	Refrac-toriness (°C)	Cold-crushing strength‡ (MN/m² or N/mm²)	Refractoriness-under-load	
							Rising temp., 50 lbf/in² (0.345 MN/m²)	Maintained temp., 25 lbf/in² (0.173 MN/m²)
Siliceous fireclay	SiO₂ TiO₂ Fe₂O₃ Al₂O₃ CaO + MgO K₂O + Na₂O	59.0 0.6 5.9 25.0 0.6 2.3	20.6	2.12	1595– 1605	30.9	—	—
Medium-alumina fireclay	SiO₂ TiO₂ Fe₂O₃ Al₂O₃ CaO + MgO K₂O + Na₂O	56.8 1.4 3.4 36.1 1.1 1.2	22.1	1.97	1690	22.2	—	5% subsidence at 1490°C
Low-alumina silica	SiO₂ Al₂O₃ Fe₂O₃ CaO K₂O + Na₂O	96.3 0.3 0.7 1.8 0.09	20.0	1.86	1730*	36.0	—	—
Semi-silica (sand)	SiO₂ Al₂O₃ Fe₂O₃ TiO₂ CaO MgO K₂O + Na₂O	88.9 9.3 1.0 0.4 0.2 0.3 0.4	23.6	1.93	1630*	10.2*	25 lbf/in²* (0.173 MN/m²) Initial de-formation 1450°C, fail 1520°C	
Magnesite (low iron)	SiO₂ Fe₂O₃ Al₂O₃ CaO MgO	1.0 0.5 0.2 2.5 96	17–21	2.80– 2.95	—	19.3– 48.3	DIN 2 kgf/ cm² (0.193 MN/m²) 1750+	—

* Properties obtained on bricks of similar type, but not from same batch
† 1 g/cm³ = 62.4 lb/ft³ = 10³ kg/m³
‡ 1 MN/m² = 1 N/mm² = 145 lbf/in² (psi)
§ 1 W/m K = 6.9 Btu in/ft² h °F

Applications

As recently as 1960 the steel industry consumed about three-quarters of all the silica bricks made in the UK, the other main users being the carbonizing (gas and coke-ovens) and glass industries. By 1969 the steel industry intake had dropped to a third and the carbonizing industry to a quarter of the 1960 figures. The cause was, of course, the partial replacement of the open-hearth furnace by LD converters, which have basic linings, and of coal-gas plants by natural gas. The

glass industry increased its usage, and a new user (the blast-furnace stove) appeared, where alumino-silicate refractories were partially replaced by silica to meet the higher operating temperatures. Even arc-furnace roofs, once built almost exclusively of silica, have largely changed to 85% alumina or even basic brick — again to meet the conditions experienced in hard-driven furnaces.

Note should however be taken here of semi-silica bricks (see *Table 1*) containing about 88–93% silica, made from natural or artificial

Brick type	Permanent linear change on reheating (%)	Thermal-shock resistance	Slag resistance	Thermal expansion 20—1000°C (%)	Thermal conductivity § 500°C mean (W/m K)	Specific heat	Comments
Siliceous fireclay	—	Moderate 10+	Moderate for coal ash, basic slag, and iron oxide	0.5*	≈1.3*	0.25*	Good general-purpose brick for reheating furnaces, ladles, etc.
Medium-alumina fireclay	2 h at 1410°C −0.35	Excellent 30+*	Good for basic slag and iron oxide	0.75*	≈1.3*	0.25*	Excellent general-purpose brick for reheating furnaces, doors, checkers, boilers, ladles, etc.
Low-alumina silica	2 h at 1500°C +0.4	Low 1—2*	Good for acid slags and iron oxide	1.25* irregular	≈1.4*	0.26*	Hard-fired low-alumina silica for open-hearth and arc roofs, coke ovens, etc.
Semisilica (sand)	Approximately volume-stable	Moderate	Moderate resistance to slags high in iron oxide and lime or coal ash	0.66* irregular	≈0.9	0.26	For lower courses of checkers, reheating furnaces and coal fire-boxes
Magnesite (low iron)	—	—	Excellent for iron oxide, cement, etc.	1.3	(900°C mean) ≈4.3	0.30*	For steel-furnace hearths, oxygen converters

* Properties obtained on bricks of similar type, but not from same batch
† 1 g/cm^3 = 62.4 lb/ft^3 = 10^3 kg/m^3
‡ 1 MN/m^2 = 1 N/mm^2 = 145 lbf/in^2 (psi)
§ 1 W/m K = 6.9 Btu in/ft^2 h °F

(continued)

sand—clay mixtures. These give good service in the lower courses of regenerators and in reheating furnaces — including roofs. They are more volume-stable than fireclay brick and can be insulated without the risk of the collapse experienced even with 42% alumina fireclay brick. They show in addition a remarkable resistance to coal-ash slags, and checker dust, attack being largely limited to the surface layers.

Magnesite

Raw Materials

Natural magnesite ($MgCO_3$) was the original starting point for brick manufacture. For over a century breunnerite (a solid solution of magnesium and iron carbonates) has been used in Austria. This is frequently associated with dolomite, but can be separated from it by magnetic

Table 1 (continued)

Brick type	Chemical analysis (%)		Apparent porosity (%)	Bulk density† (g/cm³)	Refrac- toriness (°C)	Cold- crushing strength‡ (MN/m² or N/mm²)	Refractoriness-under-load	
							Rising temp., 50 lbf/in² (0.345 MN/m²)	Maintained temp., 25 lbf/in² (0.173 MN/m²)
Chrome– magnesite	SiO₂ Fe₂O₃ Al₂O₃ Cr₂O₃ CaO MgO	2.1 9.5 15.7 18.7 0.6 53.1	17–21	3.03–3.11	–	–	–	1 h at 1600°C (28 lb), less than 0.5% subsidence
Dolomite (semi- stable)	SiO₂ Fe₂O₃ Al₂O₃ CaO MgO	2–3 1–2 1–2 48–50 36–40	18–20	2.73	–	58.6	–	1 h at 1600°C 2.5% subsidence
High- alumina (85%)	SiO₂ TiO₂ Fe₂O₃ Al₂O₃	6.5–9.5 2.4–3.0 1.5–2.0 85.8–88.2	17–21	2.82–2.97	1850+	55.2	28 lbf/in² (0.193 MN/m²), 10% subsi- dence at 1730°C	–
Forsterite	SiO₂ Al₂O₃ Fe₂O₃ MgO Cr₂O₃	38.2 1.9 7.3 51.9 1.0	20.3	2.76	1730+	21.9	–	1 h at 1600°C 6.2% subsidence

* Properties obtained on bricks of similar type, but not from same batch
† 1 g/cm³ = 62.4 lb/ft³ = 10³ kg/m³
‡ 1 MN/m² = 1 N/mm² = 145 lbf/in²(psi)
§ 1 W/m K = 6.9 Btu in/ft² h °F

methods after firing during which the iron converts to magnesio-ferrite. The other main natural form is the compact crypto-crystalline variety found for example in Greece and Turkey. This is low in iron and consequently difficult to dead-burn, i.e. shrink to a form showing little further change in specific gravity on refiring and good resistance to hydration.

In 1936, H. H. Chesny suggested to the Steetley Company that synthetic magnesia could be made at a price competitive with the natural material by reacting slaked doloma with sea-water to produce magnesium hydroxide, which, having an extremely low solubility, is precipitated:

$$2(CaO + MgO) \text{ (dolime)} + MgSO_4MgCl_2 +$$

$$4H_2O \text{ (sea water)} \rightarrow$$

$$4Mg(OH)_2 + (CaSO_4 + CaCl_2) \text{ (spent sea water)}$$

Within a few weeks a trial was made at Steetley, using a barrel of sea-water from Hartlepool, and dolomite from Coxhoe, County Durham. The magnesium hydroxide was rushed to the Central Research Department of United Steel, where, in three days, the material (a few pounds) was dead-burned and bricks (one-inch-diameter cylinders) made and tested. Natural magnesite from Austria was included in parallel experiments, and the conclusion reached that the sea-water material should be quite satisfactory. The first plant was built by Steetley in 1938. There were many teething problems, notably high impurity content, but these were overcome, and today the process in its various forms provides about half the world's requirements, including some of the purest and most fully shrunk material.

Production of Dead-burned Magnesite

This can be carried out in shaft or rotary kilns, the latter being standard for the sea-water

Brick type	Permanent linear change on reheating (%)	Thermal-shock resistance	Slag resistance	Thermal expansion, 20–1000°C (%)	Thermal conductivity § (W/m K)	Specific heat	Comments
Chrome–magnesite	2 h at 1700°C 0 to +1.0 2 h at 1800°C −0.1 to +0.6	Excellent 22–30+	Excellent for iron oxide and basic slags	0.8	≈1.4* (1000°C mean)	0.25*	For open-hearth roofs and sidewalls, arc sidewalls, soaking pits
Dolomite (semi-stable)	2 h at 1500°C, 0 to 0.2	Excellent 30+	Similar to, but less than, magnesite	1.1	—	—	For arc-furnace side-walls, cement kilns, etc.
High-alumina (85%)	2 h at 1500°C 0.0 to −0.7 2 h at 1600°C −0.1 to −1.2	Excellent 30+	Good resistance to iron oxide and basic slags	0.65	≈1.9 (1000°C mean)	0.25*	For blast-furnace stove-combustion chambers and checkers, arc-furnace roofs, reheating-furnace hearths
Forsterite	2 h at 1500°C +0.4	Moderate	Intermediate between silica and chrome–magnesite	1.0*	≈1.0 (400°C mean)	0.27	For glass-tank checkers, cement kilns, copper furnaces, and storage heaters

* Properties obtained on bricks of similar type, but not from same batch
† 1 g/cm³ = 62.4 lb/ft³ = 10³ kg/m³
‡ 1 MN/m² = 1 N/mm² = 145 lbf/in² (psi)
§ 1 W/m K = 6.9 Btu in/ft² h °F

product. Although magnesium hydroxide filter cake can be charged direct to a rotary kiln, prior briquetting to almond size is favoured, particularly for the very pure material. The maximum firing temperature varies with the magnesite but is generally in the range 1650–1750°C.

Manufacture of Magnesite Bricks

At one time magnesite clinker or the ground batch was 'soured', i.e. mixed with water and allowed to stand for several days before pressing. This was made necessary by the presence in the clinker of soft-fired ('caustic') material and of free lime from calcined dolomite inclusions. Without souring these lead to cracking in the kiln owing to the substantial expansion associated with hydration in drying. With well burned material of good quality this step can be omitted,

the clinker being ground to give a dense packing, mixed with a small amount of water and dry-pressed at 69.0–138 MN/m² (10 000–20 000 psi). Drying should however be carried out at low temperatures to minimize hydration, which can be substantial even with so-called dead-burned magnesite if the moisture contents and drying temperature are too high. Until about 1932, cracking losses in the kilns were about 10%, but with improvements in making and drying based on research these were reduced in most plants to nearer 2%.

The firing is now mostly done in gas- or oil-fired tunnel kilns, a maximum temperature of 1550–1750°C being typical, with an overall schedule of 3–5 days. For certain purposes, e.g. converter linings, the fired bricks are pitch-impregnated — often *in vacuo*.

Properties of Magnesite Bricks

The properties of a high-quality low-iron type of magnesite are summarized in *Table 1*. It will be seen that the MgO content is 96%, the CaO/SiO_2 ratio about 2:1 and the porosity in the range 17—21%. It has a high fail point (1750°C+) in the refractoriness-under-load test, and excellent resistance to slags, particularly those high in iron oxide (cf. *Figure 6*). No data are given for modulus of rupture but other bricks of the same general type show values as high as 6.9 MN/m^2 at 1500°C. They also show low shrinkage on refiring (0—0.5% linear) after 2 hours at 1700°C.

Most manufacturers of magnesite brick also make an unfired chemically bonded type. The testing of such brick is complicated by the fact that they change on heating, owing for example to the breakdown of a chemical bond, and the subsequent formation of a ceramic one. A brick heated from one face, as in an arc furnace sidewall, may therefore show a weak zone at intermediate temperatures. Some users quote data obtained on a sample fired at, say, 1100°C. The use of unfired bricks has proved economic, particularly in positions of high thermal shock, in spite of the fact that they are rarely much cheaper.

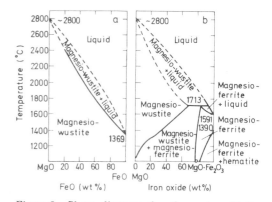

Figure 6   Phase diagrams for the system MgO—FeO (ex Muan)
(J. H. Chesters, Refractories: Production and Properties, ISI, London, 1973, Fig.60)

Energy savings suggest they should always be used if the cost per ton of product (e.g. steel) is no greater than that of fired brick.

Mention should also be made of forsterite bricks, which can be thought of as magnesite bricks of very high silica content. One expert defines them as containing 35—55% MgO and having a MgO/SiO_2 ratio of 0.94—1.33. With British bricks this ratio is generally much higher. Typical properties for a British product are given in *Table 1*.

Applications

These are very numerous. Among the most important for magnesite are:
Hot metal (iron) mixer linings
Basic oxygen (steel) converters
Arc furnaces
Open-hearth-furnace sidewalls and hearths
Induction furnace linings (both coreless and channel type)
Vacuum degassing, vessels, ladles and extension tubes
Copper converters and reverberatory furnaces
Lead dross furnaces
Nickel converters
Glass-tank melters
Rotary cement kilns

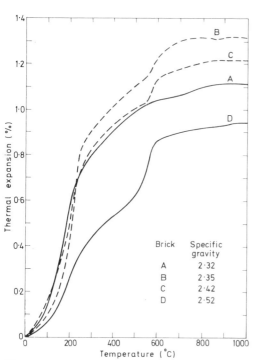

Figure 5   Thermal-expansion curves for silica bricks of varying degrees of conversion (cf. specific gravities)
(J. H. Chesters, Refractories: Production and Properties, ISI, London, 1973, Fig.54)

Dolomite

This refractory, made by calcining the natural mineral $CaMg(CO_3)_2$, consists essentially of a mixture of lime and magnesia. As lime has much less resistance to attack by iron oxide the durability of dolomite in, say, the bottom of an open-hearth furnace is appreciably less than that of magnesia but not by so much as simple theory would suggest. The explanation of this anomaly, well illustrated by the lining of oxygen converters, would seem to be the protection afforded by carbon — a breakdown product from tar-bonding or impregnation. A great deal of work has been done on the mechanism. As a result it is suggested that the main effect of carbon is the reduction of iron-rich slags to metallic iron which does not attack either magnesia or lime. Other factors are the blockage of pores by carbon and its non-wetting character which may discourage the entry of slag by surface tension.

Raw Materials

Dolomite is fortunately widely available both in the UK and abroad. It varies greatly however both in chemical and physical condition. The lime content of British dolomite falls in the range 29—31% and the magnesia content 19—22%. It is generally important that the silica be low — say less than 2%. The mixed oxides (R_2O_3 — including Fe_2O_3) vary considerably, but are generally less than 2%. Relatively high iron oxide can be advantageous as it promotes densification on firing. Indeed it is often added to dolomite to improve both sintering and hydration resistance of the resulting 'doloma', the name given in the UK to the mixed oxides resulting from calcination.

Production of Doloma

If calcination is stopped after the carbon dioxide has been driven off the product is too porous and reactive for refractory use. Firing is therefore continued to about $1700°C$, at which temperature the material is well shrunk and has a bulk density of at least 2500 kg/m³ (2.50 g/cm³). For severe duty much higher bulk densities, over 2800 kg/m³, are sought. The MgO + CaO content may be as high as 99%, though 96—98% is more usual. The calcination is increasingly carried out in rotary kilns though excellent material can be produced in cupolas. With certain dolomites a two-stage process is employed. Light calcination to produce an active oxide is followed by pelletization and dead-burning, often with iron oxide addition, to yield bulk densities as high as 3200/3300 kg/m³. A relatively recent development in Japan has been the production of clinker made from a mixture of doloma and magnesium hydroxide produced by the sea-water process. This enables micro-mixing to be achieved over a wide range of MgO/CaO ratios. Thus a high ratio may be employed for an area in an LD vessel where attack is particularly severe and a low ratio in the remaining regions. This leads to balanced wear, hopefully at minimum cost.

Manufacture of Dolomite Bricks

During the period 1934—45, increasing quantities of stabilized dolomite brick were used particularly in the UK. They provided a substitute for magnesite bricks in basic hearths. Before 1934 most efforts to produce a stable brick failed

References p 410

either because the product contained free lime and hydrated on standing, or because dicalcium silicate, with its unfortunate beta to gamma inversion, was present or formed in service. The solution arose when it was noted that bricks used in an open-hearth furnace backwall dusted on the outside of the furnace and others on the inside. X-ray examination showed that the two effects were quite distinct — the former being due to CaO changing to $Ca(OH)_2$ with the well-known slaking growth of quicklime and the latter to inversion of beta $2CaO.SiO_2$ to gamma $2CaO.SiO_2$ with a 10% volume expansion. These difficulties were overcome by adding sufficient serpentine ($3MgO.2SiO_2.2H_2O$) to yield a brick consisting essentially of magnesia + tricalcium silicate, and ensuring that any small amount of $2CaO.SiO_2$ present in the brick or formed in service was prevented from inverting by the addition of small amounts of boric and/or phosphoric acid. For complex reasons such bricks are no longer made in any quantity in the UK, but substantial amounts of semi-stable brick are still employed. These are made by pressing and firing a doloma—tar batch followed by tar impregnation to give reasonable shelf life, say 3—6 months.

Properties of Dolomite Bricks

These are best considered in comparison with magnesia — see *Table 1*. It will be seen that the semi-stable type has a porosity of 18—20%, a high cold-crushing strength, and quite a good refractoriness under load. It is almost volume-stable at $1500°C$, and has good thermal-shock resistance. Its resistance to basic and in particular iron-oxide-rich slags is high but not as good as that of magnesia.

Applications

The major user of dolomite products is the steel industry. Open-hearth furnaces used as much as 45 kg of doloma per ton of steel — mainly for fettling hearths. The oxygen converters in the UK also make use of dolomite linings — usually in the form of doloma—tar tempered blocks. The total consumption of dolomite for this purpose is however nearer 4.5—9.0 kg/t than the 45 kg/t mentioned for the open-hearth. The consumption in basic Thomas converters is nearer 22.5 kg/t, but these are rapidly being replaced by oxygen converters of one type or another. Semi-stable bricks have been used in numerous applications, e.g. converters, and arc-furnace sidewalls.

Chrome and Chrome—Magnesite

The term chromite is often used to include all chrome ores and concentrates. Strictly speaking it is the name for the spinel-type mineral $FeO.Cr_2O_3$. Chrome ores consist in the main of a solid solution of this mineral with other spinels giving a more general formula $(Mg, Fe^{2+})O.(Cr, Al,$

$Fe^{3+})_2O_3$. The gemstone spinel $(MgO.Al_2O_3)$ may be considered as the prototype, with the magnesium partly replaced by divalent iron and the aluminium partly replaced by trivalent Cr or Fe.

Raw Materials

It is a remarkable fact that few large steel-making countries, with the exception of the USSR, have their own major chrome-ore deposits. Among the biggest chrome-ore producers are Cuba, Greece, India, the Philippine Republic, Rhodesia, S. Africa, Turkey, USSR, and Yugoslavia. The chemical analysis and physical texture varies greatly not only from country to country but from lump to lump. Thus certain ores from the Philippines and Turkey are high in alumina while other ores are high in chromic oxide. The latter varies greatly — from about 30 to 60% of the total even in refractory-type ores. The gangue content is also important and very variable. It consists in general of minerals of the serpentine type — present in amounts from 2 to 40%. Another, though relatively rare, constituent of the gangue is calcite $(CaCO_3)$, which can have a disastrous effect on the hot strength of the chrome plastics used in reheating furnace hearths. Nor is it just the amount of gangue that matters. Distribution can be equally important. Thus certain apparently large grains of chrome are in fact conglomerates, the grains being separated by a fine coating of serpentine. Massive or coarsely-crystalline-type ores such as the Turkish, Grecian, Philippine and Cuban are often preferred for brickmaking.

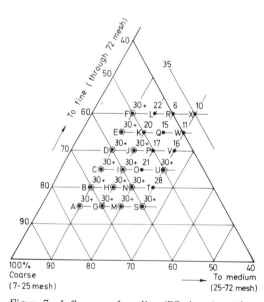

Figure 7 Influence of grading (BS sieves) on the thermal-shock resistance of 75/25 chrome—magnesite testpieces (after Chesters)
(J. H. Chesters, Refractories: Production and Properties, ISI, London, 1973, Fig. 120)

Manufacture of Chrome, Chrome—Magnesite and Magnesite—Chrome bricks

Although chrome bricks played a useful part in the early development of refractories they have largely disappeared, their task, e.g. as a neutral course, having been largely taken over by chrome—magnesite. They had two serious weaknesses, namely low hot strength and a tendency to burst when heated in contact with iron oxide. The early chrome—magnesite bricks were of the 70-30 type but the demand for greater slag resistance has led to a greater use of the 30-70 or so-called magnesite—chrome brick which, incidentally, shows a much lower bursting tendency.

The manufacturing techniques are closely parallel to those described for magnesite, i.e. controlled grading to give good thermal-shock resistance (see *Figure 7*), hydraulic pressing, drying, and firing at a very high temperature (1700°C+) for some direct-bonded products.

Unfired bricks are also available — generally with a chemical bond such as can be obtained by adding sulphuric acid. As with magnesite there is the risk of intermediate-zone weakness, but this does not prevent their use even in roofs where steelplates, either separate or integral with the brick, are used to weld the assembly into a monolith.

Properties of Chrome, Chrome—Magnesite and Magnesite—Chrome Bricks

The virtues and limitations of straight chrome bricks were listed above. Their neutral character is well illustrated by *Figure 8*, as are the limitations of other pairs of refractories when heated in contact with one another. *Table 1* lists the properties of a good-quality British chrome—magnesite brick. It will be seen that its porosity is given as 17—21% compared with the 25—30% typical of earlier products. It has excellent hot strength at 1600°C, good thermal-shock resistance, and is virtually volume-stable on reheating. Taken with its high slag resistance and relatively low thermal conductivity (approximately equal to that of silica), these properties help to explain its success when used for roofs of all basic steelmaking furnaces. Hot-modulus-of-rupture tests carried out on other samples of the same brand show figures as high as 9.7 MN/m² at 1400°C.

Applications

Among the principal applications of chrome—magnesite and magnesite—chrome brick are:
   Basic oxygen converter (steel) linings
   Arc furnace linings
   Open-hearth-furnace roofs and sidewalls, down-takes and end walls
   Copper converter linings
   Copper reverberatory-furnace linings
   Glass tanks — burner blocks and regenerators
   Cement rotary kilns

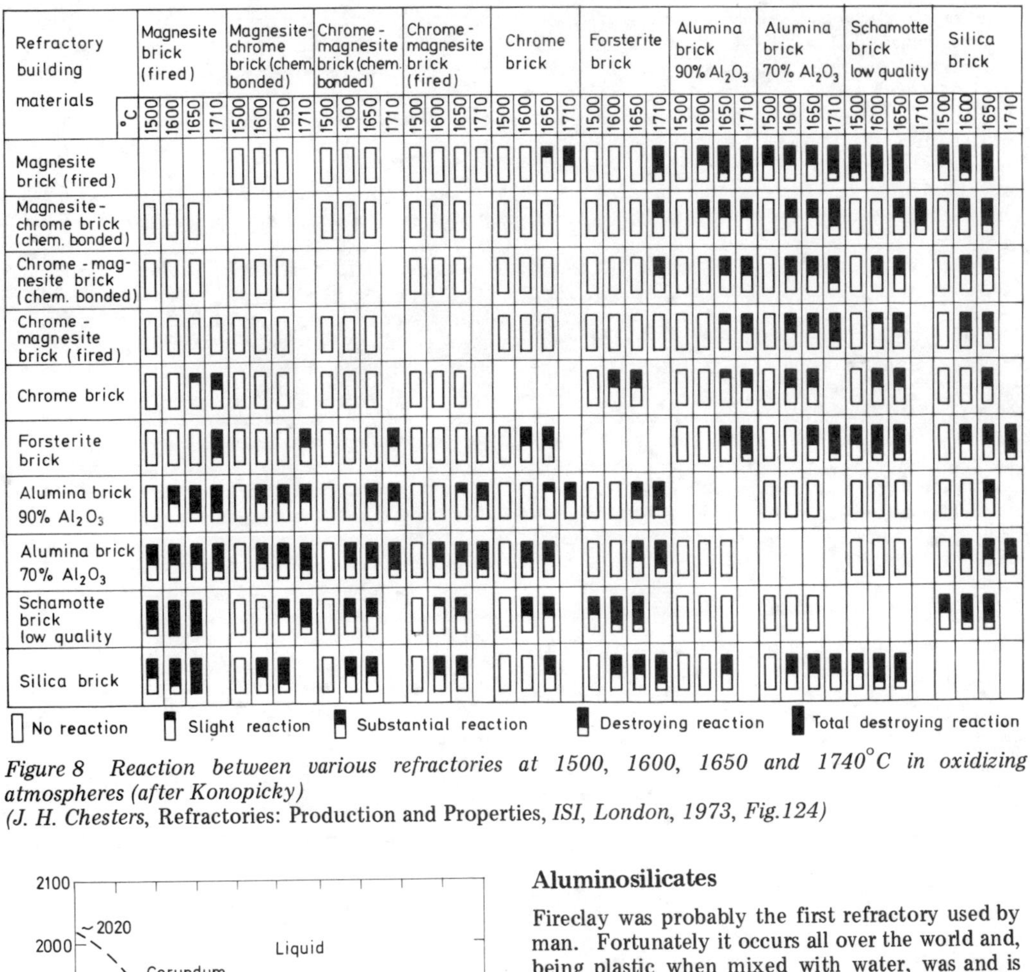

Refractory building materials	Magnesite brick (fired)				Magnesite-chrome brick (chem. bonded)				Chrome-magnesite brick (chem. bonded)				Chrome-magnesite brick (fired)				Chrome brick				Forsterite brick				Alumina brick 90% Al_2O_3				Alumina brick 70% Al_2O_3				Schamotte brick low quality				Silica brick			
°C	1500	1600	1650	1710	1500	1600	1650	1710	1500	1600	1650	1710	1500	1600	1650	1710	1500	1600	1650	1710	1500	1600	1650	1710	1500	1600	1650	1710	1500	1600	1650	1710	1500	1600	1650	1710	1500	1600	1650	1710
Magnesite brick (fired)																																								
Magnesite-chrome brick (chem. bonded)																																								
Chrome-magnesite brick (chem. bonded)																																								
Chrome-magnesite brick (fired)																																								
Chrome brick																																								
Forsterite brick																																								
Alumina brick 90% Al_2O_3																																								
Alumina brick 70% Al_2O_3																																								
Schamotte brick low quality																																								
Silica brick																																								

☐ No reaction ◱ Slight reaction ◪ Substantial reaction ◼ Destroying reaction ◼ Total destroying reaction

Figure 8   Reaction between various refractories at 1500, 1600, 1650 and 1740°C in oxidizing atmospheres (after Konopicky)
(J. H. Chesters, Refractories: Production and Properties, ISI, London, 1973, Fig.124)

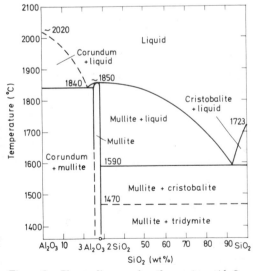

Figure 9   Phase diagram for the system Al_2O_3—SiO_2, based mainly on original work of Bowen and Greig with modifications on more recent data (ex Muan)
(J. H. Chesters, Refractories: Production and Properties, ISI, London, 1973, Fig.135)

Aluminosilicates

Fireclay was probably the first refractory used by man. Fortunately it occurs all over the world and, being plastic when mixed with water, was and is singularly useful for making refractory containers, such as crucibles, for metal or glass melting. It is still the most generally used refractory, but must frequently be replaced by more refractory aluminosilicates. A convenient breakdown (cf. *Figure 9*) of the whole range from semi-silica to pure alumina is as follows:

		Alumina (%)
(1)	Fireclay	25—45
(2)	Sillimanite and other Al_2O_3-SiO_2 type minerals	45—65
(3)	Mullite	65—75
(4)	Bauxite-based	75—90
(5)	Corundum	90—100

Raw Materials

Clay is a general term applied to 'fine particle size hydrous aluminosilicates that develop plasticity when mixed with water'. Refractory, i.e. fire-clays, could be defined as those clays having a melting point of at least 1500°C. This would include most fireclay brick, though a few, that nevertheless prove useful in low-temperature applications, might melt at a lower temperature owing to their high flux content — in particular alkalis, and iron oxide.

The UK is fortunate in having numerous and well distributed sources of fireclay each with their own characteristic, though by no means constant, analysis. In general these contain, in addition to clay minerals of the general type $Al_2O_3.2SiO_2.2H_2O$, about 1—4% Fe_2O_3 and 1—2% of alkalis.

None of the other minerals listed in (2) to (5) above occur in useful deposits in the UK. They are therefore imported, e.g. sillimanite from Assam and kyanite and andalusite from Africa or India. Mullite, so named because of its discovery in the Isle of Mull, Scotland, and having the formula $3Al_2O_3.2SiO_2$, is not naturally available in bulk but can be manufactured by reaction at high temperatures of alumina and silica or other suitable combinations to give a 3:2 ratio in the product. Bauxite, like chrome, is relatively rare, and must be imported by most of the large industrial countries, including the UK.

Manufacture of Aluminosilicate Bricks

Most present-day fireclay bricks are made by dry-pressing or extruding a column of clay followed by cutting and pressing to give the desired shape. In the dry-pressing process the clay batch generally contains at least 20% of grog (often referred to as schamotte or chamotte), and a relatively small amount of water. The grog which is precalcined fireclay comes in part from waste material (either new or used) and part from material specially fired as rough lumps in a periodic or a rotary kiln. The drying and firing is increasingly carried out in tunnel kilns where substantial use is made of heat from the cooling bricks. The methods used in making the other aluminosilicate bricks have more in common with those for basic bricks, e.g. magnesite. The materials, being hard, are crushed, graded, mixed with water, hydraulic-pressed, dried and fired to temperatures that tend to increase with the refractoriness of the product. Many fireclay bricks are fired at temperatures around 1000—1200°C, but sillimanite, mullite, bauxite-based and corundum bricks are generally fired in the temperature range 1500—1800°C.

Properties of Aluminosilicate Bricks

Table 1 includes data for two types, namely medium alumina fireclay and high alumina (85%). The former are typical for the most commonly used (35% alumina) material, the latter are the most used of the high alumina types. The much higher bulk density of the high alumina material (2820—2970 kg/cm^3 compared with the medium alumina 1970 kg/cm^3) is due to the high specific gravity of the constituent minerals, and not to lower porosity. The fireclay brick shows a typical refractoriness of 1690°C compared

with over 1850°C for the high alumina brick. More important however is the far greater strength at high temperature of the 85% material. This is illustrated by the not-strictly-comparable data in the Table but much more by hot-modulus-of-rupture tests which for the 85% can be as high as 10.3 MN/m^2 at 1400°C.

Up to 1400°C most well-fired medium alumina bricks show little shrinkage and may even show permanent expansion. At higher temperatures they are liable to shrink (or slump) whereas the 85% type rarely shows a linear change of more than 1% after 2 hours at 1600°C. Both types can show excellent resistance to thermal shock, owing in part to their relatively low thermal expansion. They both resist attack by iron oxide and basic slags, but in this respect the high-alumina is much superior though not necessarily cheaper per tonne of product. Their specific heats are very similar, but the high-alumina brick has a markedly higher thermal conductivity.

Mullite bricks vary greatly in properties according to the raw materials and manufacturing techniques employed. They are not dissimilar in properties to the bauxite type. Corundum bricks might well be classified as 'special' refractories — see later. They are sometimes electrocast and then have virtually zero porosity and extraordinary hot strength. They tend to be volume-stable even at 1800°C and are therefore extremely valuable for such jobs as burner quarls on really-high-temperature furnaces. Their resistance to basic slags, particularly iron oxide, is high (see *Figure 10*), but not so great as that of magnesia. Like most of the aluminosilicate refractories they are available not only as bricks and blocks but also as castables, mouldables, plastics and cements. This can be of great service where complex shapes or excessive cutting of brick would otherwise be necessary. A good example is the central portion of an arc furnace roof where 85%-alumina bricks are often replaced by phosphate-bonded 85%-alumina ramming.

Applications

These are so numerous as to be difficult even to summarize. The following are merely some of the major applications:

Fireclay
  Iron: Cupolas, blast furnaces and stoves, torpedo ladles, hot metal mixers
  Steel: Open-hearth and arc furnaces, ladles, casting pits, soaking pits, reheating furnaces
  Non-ferrous: Copper reverberatory and lead dross furnaces
  Glass: Pot and tank furnaces
  Carbonization: Coke ovens and gas retorts
  Power: Boilers
  Cement: Rotary kilns
  Chemicals and petrochemicals: Reactors and stills
  Waste disposal: Incinerators

Sillimanite type

  Iron: Blast-furnace linings and stoves
  Steel: Arc furnaces, ladles, soaking pits, reheat-
    ing furnaces
  Glass: Tank furnaces
  Cement: Rotary kilns
  Power: Boilers

Mullite

  Iron: Blast-furnace stacks and stoves
  Steel: Arc furnaces, ladles, reheating furnaces
    and electric soaking pits
  Glass: Tank furnaces
  Cement: Rotary kilns
  Ceramic: Kilns

Bauxite-based

  Iron: Blast-furnace stoves, torpedo ladles, hot
    metal mixers
  Steel: Arc and open-hearth furnaces, ladles,
    vacuum degassing, continuous casting, reheat-
    ing furnaces
  Non-ferrous: Aluminium melting

Corundum

  Iron: Blast furnaces and stoves, channel-type
    induction furnaces
  Steel: Arc furnaces, continuous casting and re-
    heating furnaces
  Cement: Rotary kilns
  Ceramic: Kilns
  Non-ferrous: Aluminium melting furnaces
  Glass: Tank furnaces

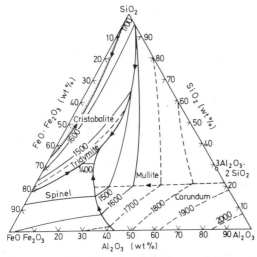

Figure 10 Diagram showing phase relations at liquidus temperatures in the system iron oxide— Al_2O_3–SiO_2 in air, based on work of Muan with minor changes based on recent work for the system Al_2O_3–SiO_2 (after Muan)
(J. H. Chesters, Refractories: Production and Properties, ISI, London, 1973, Fig.139)

References p 410

Carbon

It is strange that carbon, being alone or combined with hydrogen the most important of all fuels, should also be an important refractory. Its melting (or sublimation) point (about $4000°C$) is however higher even than that of the special refractories like zirconia, and were it not for the ease with which it burns it would be widely used as a refractory. Its main application is in blast furnaces, where it provides the best answer to breakout problems, and in aluminium smelters. It is also used in large quantities in arc furnaces but as graphite electrodes rather than as a re-fractory. Another important application is as a moderator or fuel container in nuclear reactors. Its properties vary greatly with the raw materials used and the manufacturing procedure. Blast-furnace blocks were originally made from coke and tar, but the coke is now often replaced, in whole or part, by electrically calcined anthracite or by natural or artificial graphite.

The porosity of all these products is typically 15—20%, and the cold-crushing strength 13.8—103.4 MN/m^3. Such bricks are virtually volume-stable at $1500°C$ or higher, and have thermal conductivities 4 to 40 times as high as that of fireclay brick. They are however readily oxidized even at dull red heat and must not there-fore be used in contact with air, water vapour or carbon dioxide.

Special Refractories

These are defined as those oxides, carbides, nitrides, silicides, borides, and their compounds which have specific properties such as high melting point, good erosion or spalling resistance, or superior strength that cannot be met from the range of tonnage refractories dealt with in the previous Sections. This includes common materials in unusual forms, e.g. fusion cast, as well as uncommon materials such as borides and silicides. Some of these materials, e.g. silicon carbide, have already played a major part in saving energy, e.g. as boiler refractories having high resistance to coal ash and at the same time a high conductivity. Others such as silicon nitride are as yet little applied, but are among the more promising for future developments, such as ceramic combustion chambers and blades for gas turbines. It is worth noting that of the two hundred or so compounds in the carbide, nitride and boride group only a dozen or so have been evaluated in furnace trials.

Manufacture of Special Refractories

It is not surprising that the need for such exotic materials has led to the use of equally exotic forming methods, e.g. slip casting, variable-frequency vibrating compactors, isostatic pressing and hot pressing. Further, that these procedures are also being increasingly used to get improved properties in more conventional refractories such as magnesia. It must however never be forgotten that the achievement of one property, say low porosity, can result in a disastrous deterioration in another, e.g. thermal-shock resistance.

Properties of Selected Special Refractories

The materials described below have been selected because they are being used in increasing quantities in important applications:

Zircon and Zirconia. Both these materials have already proved outstandingly useful in connection with continuous-casting nozzles. Zirconia (ZrO_2) is the most durable, but zircon is far cheaper and easier to obtain in shock-resistant form. Zirconia (like silica) has its inversions but these can be minimized by stabilization with a few percent of lime. Strangely enough a mixture of stabilized and untreated ZrO_2 is said to be better than the fully stabilized product. Zircon occurs, e.g. as a beach sand, in a number of countries, notably Australia, India and the USA. Bonded with a small amount of clay it yields a brick having excellent thermal-shock resistance but poor slag resistance. For superior slag resistance it must be ground to give a dense grading and only a minimum of bentonite or organic bond used in forming.

Silicon Carbide. The raw material used for brickmaking is prepared by electrically heating petroleum coke, pure silica sand, sawdust and a little common salt to a temperature of about $2500°C$. During this process the silica is reduced to free silicon which combines with the coke to yield silicon carbide. The sawdust burns out and keeps the mass porous, thus enabling gas to escape. The salt is stated to assist volatilization of impurities.

Silicon-carbide bricks and special shapes are made by adding a plastic fireclay to a dry-graded and mixed batch of silicon-carbide grains. After pressing (or ethyl silicate bonding) the shapes are fired to at least $1300°C$. More special forms are the nitrogen bonded (achieved by adding silica to the batch and firing in nitrogen), the self-bonded made for example by pressure sintering, and that made by pyrolytic deposition. Even material of 20% porosity shows extremely high cold-crushing strength and excellent hot strength whether in the conventional refractoriness-under-load, modulus-of-rupture, or creep tests. It has, in addition, unusually good thermal-shock resistance owing to its high thermal conductivity, low thermal expansion and high modulus of rupture.

Fusion Cast Refractories. These are increasingly used either as more or less solid blocks formed by casting directly from the liquid, or bricks made from crushed fused grains. The latter are of course more porous. Among the principal examples are electrocast mullite or alumina as used for glass tanks, and electrocast magnesite—chrome as used for high slag-attack areas in steel furnaces, e.g. arc-furnace sidewalls, and certain areas in open-hearth-furnace roofs.

Cermets. These deserve mention if only because they are used in the form of thermocouple sheaths for measuring high temperatures in certain operations, e.g. molten steel melting and in arc-furnace sidewalls. They consist of mixtures of metal and oxide, e.g. molybdenum—alumina and chromium—alumina, generally fired in special atmospheres, e.g. hydrogen, at temperatures around $1700°C$. In addition to high melting point and good slag resistance they generally possess excellent resistance to thermal shock. At the moment they are expensive, but their properties are certainly such as to warrant further study.

Glossary of Terms Commonly Used in Connection with Refractories and Furnaces

Bloating. The permanent expansion accompanied by the formation of a vesicular texture, which occurs when some types of clay or brick are heated.

Bond. A material added to (or already present in) a batch of refractory material, whose function is to promote strength either in the green, dry, or fired state. The same term is applied to the various methods of building bricks whereby adjacent courses are tied into one another.

Bursting. Disintegration following a permanent increase in volume, particularly of chrome—magnesite refractories which have absorbed iron oxide.

Calcination. Heat treatment applied to certain rocks and minerals to effect dissociation (and/or produce a change in physical structure); e.g. clay is calcined to drive off combined water.

Castable refractory. A hydraulic setting refractory suitable for casting into shapes and usually bonded with aluminous cement.

Ceramic. A general term applied to all material made from clayey and earthy substances by the application of heat.

Chamotte. Refractory clay that has been specially fired for use as a non-plastic material (cf. Grog).

Corrosion. Wearing away of a material, e.g. furnace brickwork, by chemical action of fluxes.

Crypto-crystalline. A crystalline structure in which the crystals are too small to be readily studied under the microscope.

De-aired. The term applied to a brick which has been moulded or formed from a batch subjected to a partial vacuum.

Erosion. The wearing away of a material; usually applied to wear caused by physical rather than chemical forces (cf. Corrosion, Abrasion).

Eutectic. The composition having the lowest melting point in a series of two or more components.

Extrusion. A shaping process in which a plastic body is forced through a die.

Flux. A material which lowers the fusion point of a refractory material.

Gangue.   Accessory minerals associated with relatively valuable minerals.

Gap-sized Grading.   A material from which the intermediate sizes have been screened, usually with the object of producing a brick of high bulk density.

Green-strength.   The strength of a ceramic body in the moulded but unfired state.

Grog.   Non-plastic material, usually prefired, added to a brick batch to reduce drying and firing shrinkage, or obtain special properties, e.g. high thermal-shock resistance.

Gunning.   The application of a refractory material by means of a cement gun.

Header.   A brick laid on the flat or on edge with its length perpendicular to the plane of a wall.

Isomorphous.   Minerals of analogous chemical composition and similar crystal structure. Such minerals normally form solid solutions, e.g. $2MgO.SiO_2$ with $2FeO.SiO_2$.

Jamb.   A vertical structural member forming the side of an opening in a furnace wall.

Key brick.   A tapered brick used for closing and also tightening up a curved arch.

Liquidus temperature.   The temperature at which crystallization commences when a fully melted material is slowly cooled, or melting is completed on heating up.

Mineralizer.   A small quantity of flux added to a brick batch or refractory aggregate to promote crystal growth or compound formation, e.g. the lime added to a silica-brick batch to accelerate the conversion of quartz to tridymite and cristobalite.

Modulus of rupture.   The transverse strength of a material.   It is given (British specification) by the formula:

$$M = \frac{3wl}{2bd^2}$$

where M is the modulus of rupture in psi, w is the total load (lbf) at which failure occurs, l is the distance between supports (in), b is the width of the testpiece (in), and d is the thickness of the testpiece (in).   (M' (kPa) = 6.895 M.)

Monolithic lining.   A lining containing no joints which is formed by ramming or sintering into position a granular material.

Orton cones.   Standard pyrometric cones as used in the USA.

Packing density.   The density of an aggregate packed in a container under controlled conditions (g/ml or lb/ft^3); e.g. the packing density of moulding sands as determined with an AFA rammer.

Palletizing.   A method of transporting bricks in which the latter are stacked on wooden or metal boards to facilitate handling.

Phase.   A physically homogeneous but mechanically separable portion of a system.

Plasticity.   The property of a material by virtue of which it can be moulded into any desired form, and which retains that form when the pressure of moulding has been removed.

Plumbago.   Clay—graphite refractories.

Port.   An opening in a furnace through which fuel or air enters or exhaust gases escape.

Powder density.   The density of a material in powder form.   This and the bulk density include both intra- and inter-particulate pores; in the second, some degree of compression and vibration may have been applied before determination.

Pyrometric cones.   Small pyramid-shaped pieces of mixtures of minerals which melt at definite temperatures under standardized conditions.   They are used as a basis for comparison in the determination of the Pyrometric Cone Equivalent or refractoriness of refractory materials.

Pyrometric Cone Equivalent (PCE).   In the determination of refractoriness, the test cone is heated in company with standard cones whose deformation temperature is known.   The PCE value is that of the cone or cones which deform at the nearest temperature to that at which the test cone deforms.   A cone is said to have deformed (or melted) when it has bent over until the tip is on a level with the base.

Recuperator.   A continuous heat-exchanger in which heat is extracted from the products of combustion and returned to incoming air through metal or refractory walls.

Reducing atmosphere.   One having a deficiency of oxygen, such as occurs for example in parts of furnaces where combustion is incomplete and iron exists either as the metal or as ferrous oxide.

Refractoriness.   A term used as an index of the heat-resisting properties of refractories.   It is usually determined on a sample in the form of a cone cut or prepared from the ground refractory (see Pyrometric Cone Equivalent).

Refractoriness under load.   A measure of the resistance of a refractory to the combined effects of heat and loading.   Often expressed as the temperature of shear or 10% deformation when heated under 25 or 50 lbf/in^2 (172.5 or 345 kPa).

Regenerator.   A cyclic heat-exchanger which alternately receives heat from combustion products and transfers heat to air or gas used in combustion.

Reverberatory furnace.   One in which fuel is burned at one end and the gases pass over the charge before leaving by the stack.

Schamotte see Chamotte.

Seger Cones.   Standard pyrometric cones as used in the UK and Germany.

Sill.   Horizontal structural member forming the bottom of a door in a furnace wall.

Sintering.   The bonding of powdered materials by solid-state reactions at a temperature lower than that required for the formation of a liquid phase.

Skewback.   A course of brickwork having an inclined face from which an arch is sprung.

Slag.   Material formed by fusion of oxides in metallurgical processes.   May also be applied to fused reaction product between a refractory and a flux.

Slip casting.   Process in which the material to be cast is ground, mixed with sufficient water to give a creamy liquid, and then poured into plaster moulds which rapidly absorb the added water, leaving a solid body having the inside shape of the mould.

Soldier course.   A course of bricks laid on end.

Solid solutions.   Certain groups of crystalline minerals having the property of dissolving in one another either in all proportions or over a limited range of composition.   Such groups are said to form solid solutions.

Spalling.   'Breaking or cracking of refractory brick in service, to such an extent that pieces are separated or fall away, leaving new surfaces of the brick exposed' (ASTM definition).

Specific heat.   The ratio between the amounts of heat required to raise the temperature of unit mass of a substance and of unit mass of water through one degree.

Stretcher.   A brick laid on the flat with its length parallel to the plane of the wall.

Vitrification.   A process of conversion of a substantial part of a refractory body into a glass. It is normally accompanied by a reduction in porosity and an increased spalling tendency.

Warpage.   The deviation from the intended surface of a refractory shape resulting from distortion during manufacture.

Wicket, temporary.   A door built in a furnace or kiln, e.g. the temporary closures on checker-chamber ends.

Wire-cut brick.   Brick cut by wire from a column of extruded clay and not re-pressed.

Bibliography on Refractories

General, including methods of testing

Budnikov, P. P., *The Technology of Ceramics—Refractories*, Edward Arnold, London, 1964

Ceramics — a Symposium, British Ceramic Society, Stoke on Trent, 1953

Chesters, J. H., *Steelplant Refractories*, United Steel, Sheffield, 1945, 1957, 1963

Chesters, J. H., *Refractories: Production and Properties*, Iron and Steel Institute, London, 1973

Chesters, J. H., *Refractories: Iron and Steelmaking*, Metals Society, London, 1975

Jourdain, A., *La Technologie des Produits Ceramiques Refractaires*, Gauthier-Villars, Paris, 1966

Kingery, W. D., *Introduction to Ceramics*, Wiley, New York, 1960

Konopicky, K., *Feuerfeste Baustoffe*, Verlag Stahleisen, Dusseldorf, 1957

'Methods of Testing Refractory Materials', British Standard 1902: Part 1A: 1966

Norton, F. H., *Refractories*, 4th edn, McGraw-Hill, New York, 1968

Silica

Sosman, R. B., *The Properties of Silica*, Chemical Catalogue Co. Inc., New York, 1927

Sosman, R. B., *Phases of Silica*, Rutgers University Press, New Brunswick, 1965

Magnesite

Rait, J. R., *Basic Refractories: Their Chemistry and Performance*, Iliffe, London, 1950

Dolomite

Basic Furnace Linings Committee, *The Development of Monolithic Dolomite Linings*, 1st and 2nd reports, The Iron and Steel Institute, London, 1946

Aluminosilicates

Grimshaw, R. W., *The Chemistry and Physics of Clays*, Ernest Benn, London, 1971

Carbon

Mantell, C. L., *Carbon and Graphite Handbook*, Interscience, New York, 1968

Walker, P. L., Jr, *Chemistry and Physics of Carbon*, Vol. 1—5, M. Dekker, Inc., New York, 1966—69

Special Refractories

Battelle Memorial Institute, *Engineering Properties of Selected Ceramic Materials*, American Ceramic Society, Ohio, 1968

17.2 Insulating materials

Insulation rarely adds to refractory life, and indeed may even reduce it.   It does however save energy — the heat loss through many refractory structures can be halved by judicious use of external insulation.   In some furnaces, e.g. the large structures operated at relatively low temperature for annealing boiler equipment, the walls are built completely of insulating material using either bricks, monolithic insulation or fibre blankets.   This not only greatly reduces heat loss through walls, but also heat storage and operating time.

Raw Materials

Insulating materials owe their low conductivity to their pores, while their heat capacity depends on the weight per unit volume of solids, and of course its specific heat.   One of the most widely used materials is diatomite, also known as kieselguhr, which is made up of a mass of skeletons of minute aquatic plants — deposited thousands of years ago on the beds of seas and lakes.   Chemically this consists essentially of silica contaminated by clay and organic matter.   Among the larger deposits are those of Denmark, Germany, N. Ireland, Portugal, USA, and the USSR.

Aluminosilicate insulating bricks for example are generally made from clay, though porous high-alumina bricks, made from sillimanite or bubble alumina, are available.   Silica insulating bricks are made from a batch similar to that of normal silica, but more finely ground.

Asbestos, whose use is said to have been discovered by Marco Polo, is one of the most widely used materials (but cf. p 488). This is a hydrated magnesium silicate occurring in many different forms. Having a fibrous texture, it is particularly suited to the production of coatings and sheet, though it is also available in brick form. A material with similar applications is vermiculite, a hydrated biotite (magnesia—iron mica). When suddenly heated it expands like popcorn to give a highly porous product that can be used for brickmaking, but is more often employed as a plaster. Basic insulating materials, made for example by foaming high-magnesia mixtures, have also been made, but are relatively little used. A new and quite serious competitor is calcium silicate made by reacting diatomite with lime. The blocks made from this generally contain additives, e.g. asbestos and calcite.

Brick Manufacture

The methods used in making insulating bricks vary greatly even for a given general type such as diatomite. The variation is due in part to the overall economics, but also to the demand for different products for different applications. Thus diatomite bricks are frequently subdivided into a solid grade having good strength and moderate insulating value and a porous grade that is weaker but has a lower conductivity. Certain diatomites, e.g. the Californian, can be sawn into slabs on site, either for immediate use or for light firing — say to 950°C. More often moulding, drying, firing and subsequent trimming to size is employed. Where additional porosity is required, carbonaceous material, e.g. charcoal or sawdust, can be added to the raw batch.

Whichever of these routes is employed, there are unique problems in drying and firing owing to the high moisture content (often 50%) of the raw batch, and the presence of combustibles that make difficult the following of a strict firing schedule. Another problem is the very high 'mould to fired' shrinkage which often causes distortion — a further reason for a final trim to size and shape.

Aluminosilicate insulating bricks are generally made by adding a combustible such as sawdust to the raw batch. The resulting pore-size distribution and therefore also the conductivity are controlled by sizing the material added. Other methods of achieving the desired porosity are: the addition of a material that can be volatilized, or the foaming of the batch prior to forming. None of these methods is completely satisfactory and as a result wastage in brick production can be very high.

Silica bricks of high porosity can also be made by using combustible additions. Some of the very light-weight type are however made by foaming, e.g. by hydrogen generation from additives such as aluminium and lime. Such bricks have porosities as high as 70% compared with nearer 50% for those made by addition of sawdust.

Properties of Insulating Bricks

Insulating bricks are classified in BS Specification 2973:1961 according to the temperature at which they show a permanent linear change of 2% or more. The steps suggested are 1100, 1250, 1350, 1450, 1500, 1550, and thereafter in 50°C steps. These might also be described, possibly a little optimistically, as 'maximum safe temperatures of operation'. Other data are however needed to assess the suitability of an insulating material for use in a particular position. These are illustrated by *Table 2*, which gives selected data for the principal brick types employed in furnace construction. It will be seen that these cover the range 800—1600°C — the lowest being the porous-type diatomite brick, which is also the best insulator, and the highest the high-alumina, and doubtless the silica — only tested to 1400° but probably safe at 1600°C.

Table 2   Properties of selected insulating bricks

Type	Thermal conductivity* at 400°C mean (W/m K)	Max. safe temp. (°C)	Cold crushing strength[†] (MN/m²)	Porosity (%)	Bulk density[‡] (g/cm³)	Permeability[§] (cm²/s cm w.g.)[‖]
Diatomite solid grade	0.25	1000	2.8	52	1.09	0.05
Diatomite porous grade	0.14	800	1.2	77	0.54	0.12
Clay	0.30	1500	2.7	68	0.56	0.64
High-alumina	0.28	1500—1600	3.1	66	0.91	1.20
Silica	0.40	1400+	4.1	65	0.83	0.09

* 1 W/m K = 6.9 Btu in/ft² h °F
† 1 MN/m² = 1 N/mm² = 145 lbf/in² (psi)
‡ 1 g/cm³ = 62.4 lb/ft³ = 10^3 kg/m³
§ For test method see BS 1902:Part 1A:1966, Section 5: through 1 skin
‖ 1 cm²/s cm w.g. ≡ 1.02 m²/s MPa

The cold-crushing strength gives a rough indication of probable resistance to wear in handling. The porosity is of general interest and the bulk density directly related to heat capacity. Permeability is rarely specified, but does indicate the probability of gas permeation.

The importance placed on different properties is best illustrated in relation to particular applications, e.g. the extensive walls of a checker chamber on an open-hearth furnace or glass tank, heat loss through which can be halved by proper insulation, with a consequent gain in air preheat and overall economy. For the top third the silica wall may be backed up by clay-type bricks whose conductivity (*Table 2*) is only about 20—25% of that of silica. These are unlikely to shrink since the interface temperature will not reach 1500°C even with substantial thinning of the inner brickwork. In the bottom two-thirds of the wall solid or porous-grade diatomite bricks should do the job, and have thermal conductivities only half of that shown for the clay type.

The use of porous high-alumina or silica brick of the type described would not offer any advantage and would be far more costly. Such bricks are more likely to be applied in the inner linings of high-temperature kilns. Thus porous silica has been used very successfully for the construction of the crowns of kilns used to fire silica bricks to about 1450°C, while porous alumina provides an ideal lining for the small periodic kilns used for batch firing of special bricks and shapes to temperatures around 1600°C.

The use of asbestos is now discouraged (see p411).

Chapter 18

Instrumentation and control

18.1 General techniques of measurement

Performance specifications
Measurement techniques
Temperature measurement
Pressure measurement
Flow-measuring systems
Density measurement
Smoke density measurement

18.2 Automatic control systems

Automatic valves
Analogue control systems
Digital system applications
Metering-pump-speed control system
Interfacing pneumatic controllers with digital computers

18.3 Special measurements used in fuel technology

Temperature
Total heat flow
Combustion control
Gas-borne particles in flue gases
Chimney emissions
Fireside fouling and corrosion
Water and steam purity

18.1 General techniques of measurement

Efficiency in the use of primary energy can be achieved and maintained only by the careful control of the conditions under which it is converted to produce heat or mechanical/electrical energy. To control any physical condition, a primary requirement is that its value be known; to obtain this information appropriate measurements are required. These include measurements of pressure, temperature, rate of flow, density and gas composition. Associated with these basic measurements are the transmission of their values over a distance, distant monitoring and recording instruments, and remote control of the process.

The rapid expansion of instrumentation in recent years has led to the appearance of a wide range of measuring equipment for sensing and registering the value of virtually all physical variables encountered in industry. Even when the range of measurements is restricted to the above list, a detailed description of all the types of instruments that can be used in various circumstances would be beyond the scope of one Chapter, which must, therefore, be limited to considerations of the general principles involved, along with brief notes on installation. Details of specific equipment can readily be ascertained from catalogues and other literature issued by the various manufacturers, but the range of choice is indicated here and the Chapter should assist in selection.

An installation for measuring and controlling the value of a physical variable can generally be divided between (*a*) the detecting element — which senses the value of the variable under consideration, (*b*) the measuring element — which registers this value in appropriate units, and (*c*) the control system — which permits the process to be manually or automatically operated and held to prescribed conditions.

The fundamental need to obtain good basic measurements cannot be over-emphasized and the importance of selecting the best transducer* for the job is only parallelled by the necessity to install the transducer to comply with the manufacturer's design criteria.

Although a very wide range of variables may be measured on process and production plant, these can be reduced to a small number for the great majority of installations. Among the most common are:

Temperature
Pressure
Density
Flow
Displacement

---

* A device for transferring power between systems

References p 480

These measurements are supplemented by gas-composition analysers of various types:

Carbon dioxide
Carbon monoxide
Oxygen
Sulphur oxides
Nitrogen oxides
Smoke
Grit and dust

These are considered in Ch.18.3.

The majority of transducers produce an analogue output. That is, the output signal from the transducer (whether it be pressure, current or voltage) is proportional to the measured variable (or some known function of it). A few produce a digital output, either a digit (pulse or step) for every successive increment of the input, or a coded discrete signal, representative of the numerical value of the input.

Transducers which produce a frequency output are strictly analogue in nature, but are often called quasi-digital because of the ease with which frequency can be converted to digital form by counting over a determinate period.

For each of the physical variables there is a choice of transducers. The means of measurement typically suited to the measurement of these parameters are described, with some of their limitations.

Notwithstanding some degree of standardization of transducer outputs, there exists a choice of signal levels from millivolts d.c. from thermocouples, 0—5 mA, 0—10 mA, 4—20 mA from process-industry transducers, demodulated single-phase a.c. and 400 Hz synchro outputs — all requiring individual matching and means of adjusting zero offset and span. Pneumatic signals may still be in the form 3—15 lbf/in^2 gauge (psig) or kgf/cm^2 gauge but increasingly read directly in pascal or bar derivatives.

Electrical transducers producing outputs varying from low level d.c. to fm (frequency modulated) are widely used, but pneumatic transducers and controllers are a popular choice, especially in the petrochemical industries. Transducers for data acquisition tend to be associated with electrical signal output to minimize transmission loss. The cost of pneumatic-to-electrical converters in a pneumatic system would be considerable if widespread data logging were required. When the best available transducer for each of the variables is selected, it will become evident that the chances are that a 'mixed bag' of d.c. low level and high level voltages, a.c. demodulated signals and variable-frequency pulse trains may result. Consequently, a generalized data-acquisition system based on standard inputs is seldom feasible.

In recent years there have been moves towards some degree of standardization of output levels from transducers, or more particularly the amplifiers or transmitters integral with them. This is particularly true where a comprehensive range of modules is offered by an instrument supplier specializing in complete systems.

Performance Specifications

Performance specifications for transducers, instruments and associated instrument systems require most careful study to understand their significance fully and may require reading several times before the effective performance can be interpreted. Much of the literature on instruments is presented in such a manner as to highlight the most impressive features. Much progress has been made in improving practice in this respect in recent years (British Standard 4462 sets down guidelines for manufacturers), but careful consideration and calculation may be required before the user can be certain of achieving the results he desires, and all too often claims prove to be almost impossible to substantiate in practice.

The primary process requirement is the ability to produce a consistent product continuously for twenty-four hours. This applies not only on process plant; similar reasoning applies for most other forms of instrument system from simple data logging to a direct digital-control system. What is meant by accuracy and repeatability and what factors contribute to them? Are they characteristic of the transducers used for taking primary measurements, controllers for making corrections, or processors for calculating corrections? In order to appreciate the import of these questions it will be useful to write down a list of all the factors that contribute to the efficient working of a simple instrument system comprising say, a transducer and associated indicator. These factors can then be broken down into their constituent sources of error to obtain the likely total error in a simple system. An example of a typical problem is fully described[1] which also includes the effects of such factors as:

Power-supply variations
Electromagnetic interference
Mechanical shock
Mechanical vibration
Humidity variation
Secular drift

Some instrument manufacturers go to considerable lengths to define the limiting environmental conditions for their instruments based on experience or controlled evaluation, but the final judgement as to the suitability of the instrument for any given application rests with the purchaser. The exception to this rule occurs when the system performance is specified and a supplier is selected to install the complete instrument system.

Measurement Techniques

The examples[1] illustrate the main problem areas encountered in defining transducer performance, where laboratory measurement techniques can be used to convert the transducer response into numerical or graphical analogue form relating input to output. Such ideal conditions seldom obtain in practice and we must also consider alternative forms of output to determine what criteria should be applied to them in order to ensure that the transducer output signal is not degraded by any subsequent processing of transmission media.

A wide variety of different forms of transducer output signal may exist, for example:

Pneumatic 3—15 lbf/in^2 (psig).
Direct voltage, several volts (potentiometric devices).
Direct current, 0—5 mA, 0—10 mA, 4—20 mA (process-control signals).
Low level d.c., mV (thermocouples).
Low level a.c., mV (magnetic flowmeters).
Demodulated a.c., several volts (displacement pick-offs).
Pulse trains (turbine flowmeters or pulse tachometers).
Frequency (density meters).

With the exception of pneumatic transmission, this is a formidable list at first sight, but from a measurement point of view a number of basic rules apply whether the signal is d.c., a.c. or a frequency.

D.C. Measurement

Typical d.c. measurement situations are illustrated in elementary form in *Figure 1*.

Diagram (a) shows a current source of impedance r, generating current i and feeding load R. Some of the current is diverted through the source impedance and the current reaching the load is reduced by the factor $r/(R + r)$. This represents an error which is determined by the ratio of the source impedance r to the load impedance R. For the error to be less than 1%, r/R must exceed 100; for it to be less than 0.1%, r/R must exceed 1000.

Diagram (b) represents a voltage source in which again the voltage appearing across the load is reduced by the factor $R/(R + r')$. The ratio R/r' this time determines the error, and as before ratios of 100 and 1000 must be attained to keep errors below 1% and 0.1% respectively.

These diagrams give a clue to the preference accorded to direct current as a transmission signal when data have to be transmitted over considerable distances. Where a voltage signal is used, the resistance of the transmission line (not insignificant

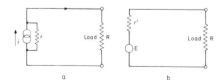

Figure 1   Generator load equivalent circuit

when distances of a mile or more are involved) must be included as part of the effective source resistance r'. Total source resistance (true source plus transmission line) may then reach upwards of a thousand ohms, requiring load impedances in the region of megohms. On the other hand, the use of feedback makes it possible to achieve current sources of very high internal impedance. Current leakage between cables caused by poor insulation could reduce the value of r, but this is a factor much more easily kept under control than line resistance. There are thus good reasons why direct current signals are widely used for long distance telemet.

Summing up these two points, in voltage-dependent systems the load impedance must be at least 1000 times the source impedance to obtain an accuracy of readout of 0.1% whereas in current-dependent systems the source impedance must be at least 1000 times the load impedance.

A.C. Transducer Signal Measurement

In certain applications a.c. powered transducers are popular because of the absence of parts such as brushes which are prone to wear and potentially add to unreliability; but appreciable errors may arise during measurement if certain fundamental precautions are not observed.

Alternating voltages encountered in a.c. transducers normally approximate to a pure sine wave. Any wave form that is not a pure sine wave may be expressed as the sum of a number of sine waves all bearing a harmonic relation to the fundamental (i.e. to the lowest frequency present). In practice, a.c. voltages suffer distortion and seldom remain pure sine waves for long, particularly when derived from industrial mains supply and used to feed transducers comprising electromagnetic elements. For this reason the supply may be derived from a separate oscillator to ensure effective control over both excitation waveform and voltage. The possible effects of waveform have been discussed[1] fully. When an a.c. transducer is calibrated on one source of a.c. power and subsequently used in service on another, there is a risk that the original calibration results will not be achieved in practice. It also follows that if for any reason the service power supply is shared between other loads, such as transducers or rotating power machines of intermittent duty, then variation in harmonics induced by such loads will also affect the transducer accuracy, and will most probably be represented by a shift of the zero.

It is highly desirable therefore that to minimize the possibility of errors from waveform distortion, transducers requiring a.c. excitation should be powered from an independent power supply and modulated and buffered to provide a high-level d.c. signal as close to the transducer as possible to minimize extraneous effects.

Signal Interference

Electrical interference is present in any instrumentation system which is either powered from a mains supply, or is operating in an environment in which electrical power is being generated or transmitted at power, audio or radio frequency. The level of interference depends on the extent of the precautions taken in linking elements of the system together, and on the design features of the elements themselves. This highly specialized subject has been described[1] in detail.

Temperature Measurement

Numerous types of temperature measuring and indicating instruments are available, but a detailed review of these is outside the scope of this book. The types of instrument most suited to data-acquisition applications are however confined to those most capable of remote electrical transmission, without appreciable translation loss with consequential inaccuracy. This tends to narrow the field for practical system applications where highest accuracy is required to thermocouples, resistance thermometers, and thermistors; but certain forms of radiation pyrometers, e.g. for furnace operation, are used where environmental conditions are severe.

Indirect temperature sensors, i.e. pressure thermometers, which convert temperature changes to pressure changes that are detected by means of a pressure transducer or direct-indicating pressure gauge such as a Bourdon tube assembly, are popular for some applications, because they are reliable and relatively inexpensive.

The now universally accepted standard unit of temperature measurement is the Celsius scale (one degree Celsius is quantitatively the same as the degree Centigrade) and based on the temperature of melting ice and the boiling point of pure water at a specified atmospheric pressure. The former is taken as zero while the latter is $100°$. It is extended below $0°C$ to $-273.16°C$ absolute zero.

Additional calibration points recognized for the international scale are defined in 'The Units and Standards of Measurement Employed by the National Physical Laboratory' (HMSO) and six such points for 'Temperatures of Equilibrium' are summarized below.

Liquid and gaseous oxygen at the pressure of 1 standard atmosphere (101.3 kPa) (oxygen point): $-182.97°C$.

Ice and air saturated water at the pressure of 1 standard atmosphere (ice point): $0.000°C$.

Liquid water and its vapour at the pressure of 1 standard atmosphere (steam point): $+100.000°C$.

Liquid sulphur and its vapour at the pressure of 1 standard atmosphere (sulphur point): $444.60°C$.

Solid silver and liquid silver at normal atmospheric pressure (silver point): $960.5°C$.

Solid gold and liquid gold at normal atmospheric pressure (gold point): $1063°C$.

In addition to the above six points the freezing points of various other metals may sometimes be conveniently used for calibration: mercury $-38.87°C$; tin $+231.8°C$; lead $+327.3°C$; zinc $+419.4°C$; antimony $+630.5°C$.

The temperature ranges for various types of instrument are summarized briefly as follows.

Expansion types:
   Absolute zero to $+600°C$
Resistance thermometers:
   Absolute zero to $+600°C$ (Standard)
   to upwards of $+1300°C$ (Special)
Thermocouples:
   Absolute zero to $+1300°C$ (Normal life)
                      $+1500°C$ (Medium life)
                      $+1700°C$ (Short life)
Radiation Pyrometry
   From about $+600°C$ to upwards of $+1800°C$.

Expansion Thermometers

All these instruments employ, in one form or another, the physical property, possessed by most substances, of expanding or contracting with rise or fall of temperature. In the case of a gas, if its volume is kept constant, its pressure changes with temperature.

Solid-expansion Thermometers. These instruments are operated by the change in length of a metal rod or the deflection of a bi-metallic strip. Their use is generally limited to simple types of thermostats and similar equipment, and they are rarely used in industry for the straightforward measurement of temperature.

Liquid-expansion Thermometers. The simplest of these instruments is the familiar mercury-in-glass thermometer. Special thermometers of this form are a sub-standard for temperature measurement and may be used with due precautions for checking

Figure 3   Liquid expansion thermometer

the accuracy of other types of temperature-measuring instruments. Mercury-in-glass thermometers are widely used in industry, over moderate temperature ranges, for routine indications of temperature, and various robust forms, e.g. sheathed in metal, are available as sub-standard or reference instruments for occasional measurements. Alcohol, toluene, and xylene are also used as alternatives to mercury for filling thermometers, especially in the case of alcohol, for temperatures below that at which mercury freezes (minus $38°C$).

The limitations of their use to indicate temperatures, and their inherent fragility, are important practical disadvantages of instruments of this type.

For recording purposes, there are other forms of fluid-filled systems. Basically they all consist of a temperature-sensitive element, which is formed by a metal 'bulb', joined to the measuring element by a length of capillary tube. *Figure 2* shows some types of bulb commonly used.

The liquid-sealed system very widely used in industry is the mercury-in-steel thermometer. The measuring element consists of a Bourdon-tube gauge, which acts as a volume measuring device (*Figure 3*).

The purpose of the measuring element is to register the change in volume of the mercury in the bulb, in terms of temperature, without being influenced by any change in volume of the mercury within the capillary tubing, due to changes in ambient temperature. To minimize this temperature effect, the capillary tubing is of extremely fine bore, so that the volume of the mercury it contains is very small compared with that contained in the bulb. Various methods are used to obtain automatic ambient-temperature compensation. For example, a wire of low expansion material may be run throughout the length of the capillary tube. The diameter of the wire and bore of the capillary are so proportioned that, with change in temperature, the change in volume of the mercury equals the change in the free space in the capillary. Errors due to the temperature variation of the measuring element itself can also be compensated for, e.g. by introducing a bi-metallic strip into the movement.

When the liquid used for filling is other than mercury, the coefficient of expansion of the liquid is generally greater. In such cases the accuracy of registration is influenced to a greater degree

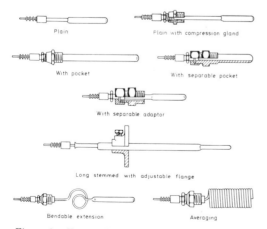

Plain   Plain with compression gland

With pocket   With separable pocket

With separable adaptor

Long stemmed with adjustable flange

Bendable extension   Averaging

Figure 2   Types of thermometer bulbs

by temperature changes of the capillary tube, especially when the 'bulb' is small and the capillary tube long.

Constant-volume Gas Thermometers. The general principle of construction of these thermometers is the same as for liquid-expansion thermometers. The system is filled with an inert gas and the measuring element operates as a true pressure-measuring device, since it registers the change in pressure of a constant volume of gas.

The effect of temperature changes of the capillary system increases as the temperature of the bulb rises, so that only partial compensation for this effect is possible, e.g. by a second capillary run adjacent to the main one and connected not to the bulb but to a second pressure-measuring element, e.g. Bourdon tube. The two pressure-measuring elements are linked by a differential lever operating the registering mechanism (*Figure 4*), which registers, in terms of temperature, the difference in the gas pressures within the two systems.

Constant-volume gas thermometers are not subject to any head error, but it may be necessary to correct for the altitude above sea level. Fluctuations in the day-to-day barometric reading are usually sufficiently small to be ignored but if desired can be corrected by adjustment of the zero of the gauge.

The advantage of gas thermometers over the liquid expansion types is that they can be used to measure the lowest temperatures encountered in industry.

Vapour-pressure Thermometers

In these, the same form of system is used as in the other fluid-filled systems already described, i.e. a bulb and a pressure-measuring element connected by a capillary tube. In this instance, however, the bulb is partly filled with a volatile liquid. Change in temperature of the bulb causes a corresponding change in pressure of the saturated

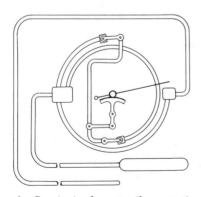

Figure 4   Constant-volume gas thermometer with automatic compensation for ambient-temperature variation

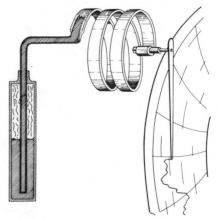

Figure 5   Vapour-pressure thermometer

vapour above the liquid, this pressure being registered by the measuring element in terms of bulb temperature (*Figure 5*).

Among the volatile liquids used are methyl chloride and toluene. The nature of the filling medium is determined by the maximum and minimum temperatures that are to be registered by the thermometer.

Vapour-pressure thermometers have certain characteristics resulting from the fact that the system is filled with a combination of liquid and vapour. For example, if the measuring element is mounted below the bulb and its temperature is lower than that of the bulb, the capillary and pressure-responsive element will be filled with liquid, and so a head error is introduced.

The main characteristics of the expansion-type thermometers are given in *Table 1*. The measuring element is usually provided with a circular scale or circular chart.

When installing fluid-filled temperature-measuring equipment, it is of paramount importance to give adequate protection to the capillary tube against accidental damage and mechanical shock. It should, therefore, be carefully supported throughout its length. The essential point is that there must be no possibility of relative movement between the bulb and the capillary tube — not always an easy matter to ensure in practice. Particular care should be given to the capillary at its junction with the bulb, which should not be strained in any way. Sufficient length of capillary should be provided to enable the bulb to be withdrawn from its pocket, or other form of mounting, for examination. Any damage to the capillary renders the equipment useless, and it must then be returned to the supplier for repair.

In view of the effect of ambient temperature variation, a path for the capillary should be selected that does not lie through high- or low-temperature zones or zones subjected to extremes of temperature. These remarks apply to the positions of exposed portions of the measuring element also.

Table 1   Characteristics of expansion thermometers

Type of equipment	Application	Overall industrial temperature range	Distant reading
Metal expansion	Gases	0 to +400°C	Not suitable
Liquid in glass	Solids and fluids	−200°C to +500°C	Not suitable
Liquid in steel	Solids and fluids	0 to +600°C	Up to 60 m
Gas expansion	Liquids	0 to +550°C	Up to 60 m
Vapour pressure	Solids and fluids	−20°C to +350°C	Up to 60 m

Figure 6   Basic principle of thermocouple measurement

Electrical Thermometers and Pyrometers

Thermoelectric Pyrometers. These instruments consist basically of a circuit of two dissimilar metals or alloys fused together at each end. When one end, the 'hot junction', is subjected to a high temperature, an electromotive force is generated proportional to the difference in temperature between the hot junction and the other, 'cold' junction. The current resulting from the emf is measured by a galvanometer or similar electrical instrument located at or near the cold junction. The two wires form a thermocouple, and the selection of the alloys or metals to be used is influenced by a number of factors, such as resistance to high temperature, corrosion and oxidation, and the temperature/emf relation. Relatively few metals and alloys are suitable for making thermocouples. There are two classes of thermocouples for industrial use — base metal for temperatures up to about 1200°C and rare metal for temperatures up to 1500°C.

A thermocouple is one of the simplest forms of transducer because it is analogous to an electro-voltaic cell and contains no moving parts. In *Figure 6* the temperature of junction 1 is T_1 and the temperature of junction 2 is T_2. A voltage E measured at points A and B will be approximately proportional to the temperature difference $T_1 - T_2$.

The effective emf of a thermocouple is the algebraic sum of the Peltier and Thomson emfs. The Peltier effect is attributed to the absorption of heat at the hotter junction and the evolution of heat at the cooler of them, the effect being dependent on the current flowing in the loop. The emfs developed in the process at both junctions are known as Peltier emfs. The Thomson effect is attributed to Lord Kelvin who stated that if a temperature difference existed between the ends of a homogeneous wire, an emf should be generated proportional to the temperature difference.

If E is the value of the emf, T_1 the temperature of the hot junction and T_2 the temperature of the cold junction, then from the combined Peltier and Thomson effects $E = a(T_1 - T_2) + b(T_1 - T_2)^2$ where a and b are constants depending on the metals used.

The second term is of minor significance (coefficient b is small) and for limited temperature ranges is ignored. An alternative relation which applies when one junction is kept at ice temperature is expressed:

$$\log_{10} E = c \log_{10} T_1 + d$$

In this case c and d are constants and $T_2 = 0°C$ (reference temperature).

For industrial work, the thermocouple is constructed on the lines shown in *Figure 7* and is contained in a protecting sheath, the outer end of which carries a terminal box. The connecting leads from the terminal box to the cold junction may be of the same material as the thermocouple, but to save expense, especially where a considerable length is involved, compensating cable having the same temperature/emf characteristic as the thermocouple over the limited temperature range to which it is subjected may be used for making the connection to the cold junction. The cold junction should of course be located away from the heated zone. For the connection from the cold junction to the measuring instrument, standard copper cable may be used. The cold junction may be maintained at a constant known temperature, or the instrument reading may be corrected for the actual cold junction temperature (*Figure 8*).

The most satisfactory arrangement is that in which the cold junction is located within the instrument with automatic compensation for cold junction temperature, e.g. bi-metallic compensation in a direct-deflection type instrument. Direct-deflection type indicating and recording instruments are available.

Circuit Applications of Thermocouples for Remote Temperature Measurement. In order to measure the temperature of a point relative to 0°C for example, a second thermocouple is required from which a reference voltage is obtained corresponding to a reference temperature which is known,

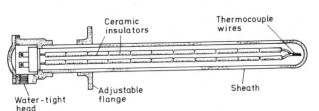

Figure 7 Construction of a thermocouple

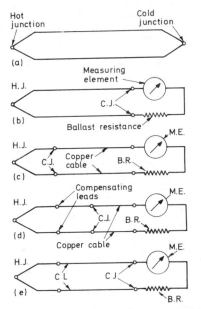

Figure 8 Development of the thermoelectric pyrometer circuit

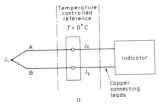

Figure 9 Thermocouple connections for 0°C reference temperature

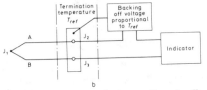

Figure 10 Thermocouple connections for floating reference temperature

such as an ice point at $0°C$ for a controlled temperature at some point higher than normal ambient temperature, e.g. $60°C$.

The thermocouple sensor emf is measured, and the temperature corresponding to this emf is derived from tables or prior calibration and added to the reference temperature to obtain the value of the unknown temperature. It is sometimes convenient to control the reference temperature at $0°C$ because this is relatively easy to maintain by means of a thermos flask filled with ice, and computation is straightforward. Where lack of maintenance or ice precludes the use of this facility an electronically controlled oven is sometimes used to control the temperature at some other value. A voltage equivalent to this value must be added to the thermocouple sensor output to obtain a voltage equal to that which would have been measured if the reference temperature had actually been $0°C$.

The terminals of the measuring instrument to which the thermocouple will eventually be attached are made of copper, which may be typically silver or nickel plated, but in any event the connecting leads must eventually consist of identical metals, ideally copper, to eliminate yet another thermoelectric junction that would occur if the thermocouple leads were connected directly to the instrument terminals. A schematic of the circuit required is shown in *Figure 9*.

Here the thermocouple metals A and B each make junctions J_2 and J_3 with copper connecting leads. Both J_2 and J_3 must be maintained at constant temperature. If this is not $0°C$ then a compensating voltage must be connected in series to back off the equivalent departure of the reference temperature from zero. Thus, the reference temperature may be allowed to fluctuate with ambient temperature changes, provided this temperature is accurately sensed at each junction and a precision voltage is derived from this measurement to provide a backing-off signal proportional to the reference temperature. Note that if one of the thermocouple materials is copper then only one reference junction is required. This backing-off technique is illustrated in *Figure 10*. Although a difficult process to instrument, it is favoured by some manufacturers for portable instruments or systems requiring minimum maintenance.

In order to obtain a higher output voltage it is possible to connect in series two or more thermocouples in which the junctions are insulated from each other and their lead terminations are maintained at the reference temperature. The total output will then be the sum of the outputs of the individual thermocouples.

When using thermocouples, it is good practice to twist the leads together to minimize common-mode interference.

Where an existing multi-point recorder is used and it is required to extend the system by the addition of, say, a data logger, the addition of a high-input-impedance data amplifier to the input of the recorder may be necessary because most recorders are potentiometric instruments whose input impedance becomes relatively low before balance is reached. If a thermocouple is being scanned by the data logger at the same time as it is being monitored by a recorder, these loading effects may be sufficient to produce large apparent errors.

When a long cable run separates the indicator or recorder from a multiple-thermocouple installation, and copper wiring must be used, remote switching as shown in *Figure 11* may be adopted in conjunction with a terminal block suitably thermally insulated to maintain a thermal gradient common to all connections. Since all terminals, including the reference junction, are kept at constant temperature the actual temperature is immaterial. The switch may be either electromechanical, such as a uni-selector, or consist of electronically controlled relays. This technique is sometimes referred to as remote-head scanning.

Operating Characteristics of Thermocouples. Thermocouple compensating-cable is expensive and the length is thus kept to a minimum before terminating in copper. A thermocouple characteristic is non-rectilinear except when used over a limited temperature range and must be corrected particularly if digital print out is required. Normal junctions are limited to about 1000°C and at this limit the accuracy is of the order of $\pm 5^{\circ}$C, i.e. $\pm 0.5\%$. Rare-metal thermocouples can be used up

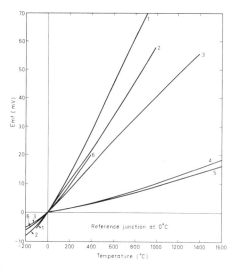

Figure 12   Typical thermocouple characteristics

to 1600°C but physical changes of the junction material eventually limit the life of the junction at these elevated temperatures. Even when used at lower temperatures of some one or two hundred degrees Centigrade, the best calibration accuracy obtainable is about $\pm 0.5\%$ over the working range. Typical thermocouple characteristics are shown approximately drawn in *Figure 12*.

Standard charts are available for various metal combinations[2,3]. These publications, called 'Reference Tables for Thermocouples', may be obtained from the British Standards Institution. These tables give both the emf generated for a specific temperature difference between the cold junction at 0°C and the hot junction, and the temperature difference required to generate a specific emf.

Thermocouple Installations. For temperatures not above about 500°C, a mild-steel sheath may be used, while for temperatures up to 900°C, specially treated steel or nickel—chromium alloy is required. Above 900°C, refractory metals and other components should be used. Rare-metal thermocouples are susceptible to contamination by metal vapours, and impermeable refractory sheaths should be used for temperatures above about 500°C. To give mechanical strength, the refractory sheath may be protected by an outer one of metal or refractory, but at the expense of increasing the time response of the hot junction of the thermocouple to change in temperature.

The length of service of thermocouples depends on the work they are used for — if working below the normal maximum temperature a continuous working life of 1000 h may be expected, increasing to say 2000 h or even more when working at appreciably lower temperatures. Therefore, the condition of thermocouples must be checked at regular intervals in accordance with experience.

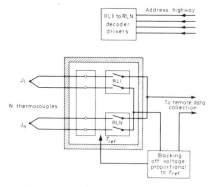

Figure 11   Remote-head scanning of multiple thermocouples

To prevent errors due to radiation from neighbouring surfaces, e.g. the walls of a flue, when measuring gas temperatures the suction pyrometer may be used. Essentially it is an open-ended tube containing a bare thermocouple surrounded by radiation shields. This assembly is inserted into the duct or flue and a sample of the gas is continuously drawn through it. The gas velocity is increased until no change is observed in the measured temperature. There are a number of designs of this particular form of equipment (*Figure 13*). A detailed discussion of suction pyrometry in theory and practice has been given by Land and Barber[4]. The method cannot be used by an inexperienced observer.

A summary of the characteristics of electrical temperature-measuring instruments is given in *Table 2*.

Because of the low emf and currents involved in equipment of this type, the electrical circuit should be handled and protected with special care. High resistance at joints in the circuit can cause faulty registration, and so the number of connections in the circuit should be kept to the minimum possible. Junction boxes and plug connectors should preferably not be used, and joints and connections should be soldered. The accuracy of thermoelectric pyrometers of the null-balance form is not affected by circuit resistance variation,

which affects only the sensitivity. The connecting cable must be protected from oxidation and corrosion, and it is of vital importance that it be protected also from stray electrical leakage from nearby power or lighting circuits. A path should be selected for the cable that will keep it away from high temperature or damp zones. If damp zones cannot be avoided, lead-covered cable suitably supported may be used. In hot dry locations a woven asbestos covering may be used. The insulation between the cores must be of a high order — that normally used for high voltage is satisfactory.

The above general remarks apply both to compensating cable (thermoelectric pyrometers) and copper cables.

When switches are used, e.g. when a number of thermocouples are operating with one measuring element, these should be of the double-pole type with, preferably, uncorrodible metal contacts.

The most important items of maintenance are regular inspection of the condition of all joints and contacts and regular accuracy checking, especially with thermoelectric instruments. Replacement thermocouples should always be kept in stock.

Resistance Thermometry

The fundamental difference between a thermocouple and a resistance thermometer is that, whereas the former generates a thermoelectric voltage, the latter is primarily a temperature-sensitive resistor that will only produce an equivalent voltage change if a constant current is passed through it as in *Figure 14*. This resistance may

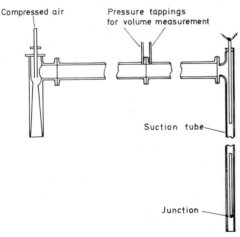

Figure 13   Suction pyrometer

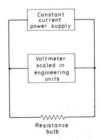

Figure 14   Application of voltmeter and constant-current power supply for direct reading of temperature

Table 2   Characteristics of electrical temperature measuring instruments

Type of equipment	Application	Overall industrial temperature range	Distant reading
Electrical resistance	Solids and fluids	$-240°C$ to $+600°C$	Suitable
Base-metal thermocouple	Solids and fluids	$-200°C$ to $+1100°C$	Suitable
Rare-metal thermocouple	Solids and fluids	$0°C$ to $+1450°C$	Suitable
Suction pyrometer	Gases	According to construction	Not generally suitable

References p 480

be typically 100 Ω at $0°C$ and 140 Ω at $100°C$ for platinum wire. Although nickel has been used, platinum wire is now used almost exclusively.

The most significant characteristic of a platinum resistance thermometer is the repeatability of the relation of resistance to temperature. Between the two reference points of $-182.97°C$ and $+630.5°C$, defined by the boiling point of oxygen and freezing point of antimony respectively, the relation is defined by the Callendar—Van Dusen equation:

$$\frac{R_t}{R_0} = 1 + a/t + \delta \left(\frac{t}{100} - 1\right) \left(\frac{t}{100}\right)$$

$$- \beta \left(\frac{t}{100} - 1\right) \left(\frac{t}{100}\right)^2 \quad (1)$$

where R_t is the resistance of the element at $t°C$, R_0 is the resistance of the element at $0°C$, and a, δ and β are characteristic constants for each sensor. Typical values are: $\delta = 1.49$, $\beta = 0.11$ for negative values of t, and $\beta = 0$ for positive values of t.

The International Practical Temperature Scale is not at present defined by a platinum resistance thermometer above the antimony point but may be redefined up to the gold point ($1063.5°C$). For industrial applications equation (1) applies up to about $800°C$; thereafter the use of a gold point reference, if used, introduces some doubt because the nominal gold point is suspected to be in error by as much as $1.5°C$.

Below 90 K ($-183°C$) the calibration of a resistance thermometer cannot be established from a simple equation, and interpolation techniques defined by R. J. Corrucini of the National Bureau of Standards and described in ref.5 have to be used. It is possible to calculate the resistance/temperature relation to an accuracy of ±0.05 K of the thermodynamic scale by measuring the resistance of the thermometer at three known temperatures.

A simplified form of resistance/temperature relation for a resistance thermometer becomes

$$R_t = R(1 + at + bt^2 + ct^3 + \ldots) \quad (2)$$

where R_t is the resistance at temperature t, R is the resistance at some lower temperature usually specified as $0°C$, and a, b, c are constants. The number of constants involved depends on the temperature to be measured but, in general, two are sufficient up to about $550°C$. For small ranges a rectilinear relation is sometimes assumed:

$$R_t = R(1 + at) \quad (3)$$

If the results of equations (1) and (2) are compared for $500°C$ based on platinum[6], the difference in calculated resistance is about 4%; consequently the straight-line approximation of equation (3) should only be used where limited excursions of temperature are known to occur.

The element is normally sealed in glass when used for temperatures up to $150°C$ and ceramic for use in temperatures up to $850°C$. This sealing has the two-fold advantage of providing structural strength and protection from chemical attack. Construction of a typical commercial resistance thermometer is shown in *Figure 15*. Resistance thermometers are sometimes used above $850°C$ with the penalty of reduced life.

In operation, a stabilized power supply with some form of measuring bridge is required. By choosing the appropriate energizing current, the temperatures may be indicated directly in any desired engineering unit. The length of interconnecting cable between the resistance element and measuring bridge is immaterial since the leads can be made to be relatively temperature-insensitive. But where the greatest accuracy is required, or where the power supply and bridge may be time-shared as with some data-logger systems, a three- or four-wire system of 'Kelvin connections' is used as shown in *Figures 16* and *17*. These ensure that the points of voltage measurement only relate to the temperature-sensitive element and not to extraneous voltage drops along the current supply cable from the energizing supply.

Thermistors

Thermistors are a comparatively recent development and provide a useful alternative to a resistance thermometer for certain applications. The fundamental differences in operating characteristics from a resistance thermometer are:

A negative temperature coefficient, i.e. the resistance decreases as the temperature increases.

The temperature coefficient, approximately 4% per degree Celsius, is nearly ten times as sensitive.

Thermistors consist of a compound of nickel, cobalt, manganese, uranium and other oxides, blended and fired under carefully controlled conditions to obtain a specific resistance/temperature characteristic within an ohmic range 100—450 000. This latter feature enables a variety of circuit applications to be considered, from direct electrical connection to a reliance on complex electronic

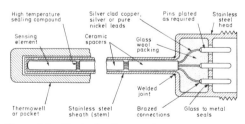

Figure 15   Typical resistance thermometer — physical construction

circuits. The characteristic may be adjusted by the addition of appropriate padding resistors in series or parallel to vary the slope of the resistance/temperature relation:

$$R_T = R_0 e^{b/T} \tag{4}$$

where R_T = resistance of the thermistor at absolute temperature T, b = constant over a small temperature range, and R_0 = resistance at some standard temperature.

Advantages and Disadvantages of Electrical Pyrometers. When it is required to measure the temperature of a liquid, gas or some substance in a process, the choice almost invariably falls between a thermocouple or a resistance thermometer. Frequently, size dictates the ultimate choice and here the thermocouple offers advantages because the thermojunction may be attached directly by point contact or used as a freely supported probe; for example, when measuring the temperature of air or non-corrosive gas.

Unfortunately the majority of chemical compounds and gases have some corrosive effect on both thermocouples and resistance thermometers and this, coupled with an environment subject to fluid pulsations or mechanical vibrations, rules out the use of an unprotected probe. Consequently some form of protective housing is required as indicated in *Figure 15*.

It will be useful to compare the principles of operation of thermocouples and resistance thermometers to assess their respective limitations for typical industrial applications.

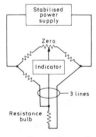

Figure 16   Three-wire system of compensation

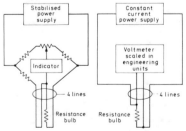

Figure 17   Alternative four-wire systems of compensation

References p 480

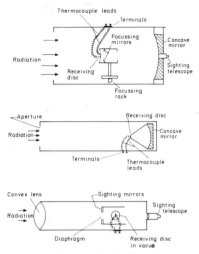

Figure 18   Types of radiation pyrometer

When comparing costs between resistance thermometers and thermocouples, it is important to consider also the cost of the mounting pocket because although the cost of a thermocouple may at first sight appear to be significantly cheaper, say 30% of the net cost for a resistance thermometer, the cost of the pocket and terminal housing may add a significant cost to either; the cost advantage thus almost disappears.

Radiation Pyrometers

Instruments of this type utilize the emission of radiant energy from the hot body for measuring its temperature. In general, the instrument makes no actual contact with the hot body, this method being used for temperatures that are higher than can be satisfactorily dealt with by a thermoelectric pyrometer, and for measuring the temperature of inaccessible bodies, e.g. the inside of a furnace or kiln.

Three kinds of pyrometer operating on this general principle are available, namely (*a*) total radiation, (*b*) optical, and (*c*) light-sensitive (photoelectric).

Total-radiation Pyrometers. These instruments measure the intensity of all the wavelengths of the radiation emitted from the hot body. These rays are focussed on to a temperature-detecting element by means of lenses or mirrors (*Figure 18*). The detecting element is usually a thermocouple, though a resistance element or even a bi-metallic strip may be used. The thermocouple e. ment is made of very fine wire and its cold junction may be contained in the radiation-receiving unit but protected from radiation from the hot body. In these circumstances, both hot and cold junctions are very nearly equally affected by ambient temperature variations. Alternative means are available for obtaining cold-junction temperature compensation. The remaining part of the equipment is identical with a thermoelectric pyrometer, i.e. the thermocouple is connected to the temperature-measuring element by means of an electric cable. The temperature scale is not uniform (as in resistance thermometers and thermoelectric pyrometers) but opens out rapidly towards the upper end, because the radiant energy emitted from a body is approximately directly proportional to the fourth power of its absolute temperature.

For accurate temperature measurement, the area of the hot body under observation must fill the field of 'vision' of the radiation receiver, which, therefore, must be correctly sighted. There is, therefore, a definite relation between the diameter of the hot body and its distance from the focussing lens or mirror. In addition, black-body conditions must be obtained as nearly as possible to ensure that the entire radiation emitted from the body is solely due to its temperature. The nearest approach to ideal black-body conditions is the measurement of temperature in an enclosed furnace, the only aperture being the peephole through which the interior is viewed. Whatever the colour of the furnace and its contents, when cold, its interior will behave as though quite black. Therefore, when it is hot, the radiation emitted from the walls and material it contains will have a spectral distribution solely determined by the temperature. When ideal black-body conditions are not obtained, a correction must be applied to the observed reading.

Optical Pyrometers. These are used for manual operation, i.e. for the spot readings of temperature. The operation consists of matching a standard source of light, e.g. from the filament of an electric lamp, with the light emitted from the hot body. The intensity of light from the standard source can be varied until the filament merges into the background formed by the hot body (disappearing-filament method). The current flow through the lamp for this condition is measured, in terms of temperature, by an electrical instrument.

Alternatively, a grey wedge is interposed between the standard source of light and the hot body (wedge pyrometer). The position of the wedge at the matching point is a measure of the hot body temperature.

There is also the polarizing pyrometer, which polarizes the light from the two sources in planes

at right angles, an adjustment being made to obtain a balance of brightness.

Generally a monochromatic screen is used in optical pyrometers so that matching is made on a particular wavelength, usually of red light, there being an accurately known relation between temperature and intensity of radiation for each wavelength. As in the instance of the total radiation pyrometer, an effort should be made to work under black-body conditions.

Light-sensitive Pyrometers. In photoelectric pyrometers the sensitivity of a photocell to red and adjacent regions of the spectrum is used. The electrical output is proportional to the colour appearance of an incandescent source:

Just glowing red (in daylight)	560°C
Blood red	620°C
Medium cherry red	760°C
Light cherry red	810°C
Medium orange	885°C
Dark yellow	975°C
Light yellow	1150°C
White	1250°C

The photo-electric cell is used for measuring the intensity of light emitted by the hot body, usually in the red and infrared bands of the spectrum. The magnitude of the current through the cell (which must be energized by the emf from an external source) is determined by the intensity of radiation. The current, which is extremely small, is amplified electronically and fed either into a direct-deflection indicator or recorder, or into a null-balance instrument of conventional form.

The rectifier or barrier-type cell generates its own emf which is sufficient to operate a sensitive measuring instrument.

The advantage of the light-sensitive method of measurement is that the response of the detecting element is practically instantaneous. It can, therefore, be used for measuring the temperature of rapidly moving objects. As with other radiation types of instrument, the Stefan–Boltzmann radiation law is assumed and the accuracy is therefore affected by the departure from black-body conditions. General characteristics of radiation pyrometers are given in *Table 3*.

One of the most important points to note when installing and using radiation pyrometers is to ensure that the field of vision of the pyrometer is

Table 3   Characteristics of radiation pyrometers

Type of equipment	Application	Overall industrial temperature range	Distant reading
Total radiation	Radiating surfaces	500°C upwards	Suitable
Optical	Radiating surfaces	700°C upwards	Not suitable
Light sensitive	Radiating surfaces	700°C upwards	Suitable

completely filled by the hot-body radiation. In some total-radiation pyrometers no adjustment is provided, so there is a maximum 'distance factor', i.e. relation between maximum permissible distance from the sighting tube to the hot body, and the smallest dimension of the hot body. Other forms use focussing lenses. To facilitate the setting, a sighting hole is frequently provided at the back of the tube.

The tube must be designed for the particular service under consideration. For example, when measuring the temperature in chambers under a pressure or suction, e.g. a steel furnace, the front end of the tube may be protected by a mica window, air being blown over it to keep it free from dust and dirt. Any protection calling for the interposing of a window or similar device between the sighting tube and the source of radiation calls for special calibration of the instrument by the manufacturer. It is necessary to select a focussing point that will not be influenced by direct or reflected glare from flames. Of course, it is quite impossible to measure the temperature of a body through intervening flames. When it is undesirable to have an open sighting hole in a furnace, kiln, etc., a refractory tube with a closed end can be fitted into the furnace wall and the radiation pyrometer focussed on to the inner end of the tube. This method generally results in a lower temperature reading. In certain cases, air-cooled or water-cooled sighting tubes are necessary. Attention is drawn to errors caused by intervening gases between the source of radiation and the pyrometer; these, notably carbon dioxide and water vapour, may be invisible. Mention has already been made of the effect of departing from the ideal black-body conditions; when such a departure is suspected, the advice of the instrument manufacturer should be sought.

The most important item of routine maintenance is to keep glass windows, lenses and mirrors scrupulously clean.

For permanently installed instruments, e.g. total-radiation pyrometers, cable runs and insulation requirements follow the same lines as for electrical temperature-measuring instruments, as does the installation and maintenance of the electrical measuring element itself.

Some Practical Aspects of Temperature Measurement

The art of accurate temperature measurement lies in the correct selection of the point of measurement and in ensuring that the temperature-sensitive element is actually subject to the temperature to be measured but not influenced by extraneous factors. The following notes describe the salient factors to be observed in temperature measurement.

When temperatures of liquids are being measured, it is essential that the liquid be in an agitated condition to attain homogeneity and eliminate temperature gradients. The sheath or pocket of the temperature-sensitive element should be long enough to prevent conduction of heat or cold from the surrounding surface — the length immersed should be 10 to 20 times the diameter. It is desirable to fill any air space between a temperature-sensitive bulb and its pocket with a good heat conductor, e.g. oil if circumstances permit.

When the temperature in a small pipe is to be measured, the temperature-sensitive element should be inserted into a right-angle bend so as to obtain full immersion.

Build-up of foreign matter on the temperature-sensitive element will cause a serious time lag with consequent errors in reading. Cleaning at regular intervals may, therefore, be necessary.

When measuring the temperature of a gas, in addition to ensuring that the gas is in an agitated state at the point of measurement it must be remembered that the temperature-sensitive element may be affected by radiation from hot or (more usually) cold surfaces within its area of 'vision'; reference has already been made to the suction pyrometer. For 'still' gas temperature measurement capillary-type bulbs are available long enough to smooth out temperature gradients and having a large heat-transfer area in relation to their mass.

Surface temperature measurement presents its own particular difficulties. On smooth machined surfaces good reproducibility is obtained; air spaces between the temperature-sensitive element and the surfaces of cast metals and refractories may result in unreliable readings. When using a mercury thermometer, one with a small narrow bulb is to be preferred, the bulb being pressed to the surface with a thin plate of metal and not a non-conductor of heat. Air spaces may be filled with a paste of copper or aluminium powder in oil. For surface-temperature measurement, various forms of thermocouple are available. These are of thin, flexible strip or similar construction and tend to conform to the shape of the surface on to which they are pressed. Thermocouples embedded in the surface may also be used.

A feature particular to all classes of electrical temperature-measuring equipment is that since the connecting link between the detecting and measuring elements is an electric cable, it can easily be disconnected and replaced by another. One measuring element can thus be used to register the temperature at a number of distant points, either by manual selection or, in the case of recording, by the automatic selection in a predetermined sequence. In manual selection for taking spot readings, the number of detecting elements that can be used with one instrument is virtually unlimited. In the case of recording, the automatic selection is limited to a dozen or so points. For industrial use, resistance thermometers are not recommended for temperatures above $600°C$; for higher temperatures, the thermoelectric principle is to be preferred.

In conclusion, the thermocouple is favoured for fast response, small physical size and high working temperature at lowest cost. For highest accuracy, linearity, stability, physical strength, resistance to corrosion, avoidance of temperature errors in connecting cables and ease of conversion of output to engineering units, the resistance thermometer or the thermistor is to be preferred. Only where fastest possible response is required, and a thermocouple can be used directly mounted by point contact, does the thermocouple offer any significant advantage since the use of a reference junction must be considered to be an operational impediment regardless of how it is achieved.

Where the environment is particularly hostile, e.g. in furnace operation, one of the various forms of pyrometers available may be the only practicable means of obtaining access to the temperature of the sample.

Pressure Measurement

Remote pressure measurements, like any other measurement problem, resolves itself into broad areas, each the subject of fundamental limitations. The problem areas may be specified as:

Calibration standards

Pressure/displacement or force conversion.

Displacement or force to electrical or pneumatic conversion.

The practical difficulties posed by these problem areas may be overcome individually by laboratory techniques. Singly they may approach a theoretical limit around 0.01% accuracy, but their combined errors may result in an accuracy that is seldom better than 0.1% and more frequently 1%. An accuracy of measurement of 0.1% (that is to say, the combined errors comprising repeatability, linearity, hysteresis, etc.) is very difficult to achieve in typical system environmental conditions of high and low temperature, shock, vibration, corrosive fluids or gas and random pressure overload conditions.

This Section is intended to highlight the fundamentals of accurate measurement and discuss forms of transducer that produce a pneumatic pressure or an electrical output voltage rectilinearly related to pressure in typical environments. Commercially available transducers each having particular merits in relation to systems applications will be briefly reviewed to illustrate the basic conversion techniques. The examples described are of similar performance. At best the salient characteristics are typically:

Linearity ±0.25% to ±1%.

Hysteresis 0.05% to 1%.

Repeatability ±0.25% to ±0.5%.

Thermal zero shift ±0.01% to ±0.02% full scale per °C.

Thermal sensitivity shift ±0.01% to ±0.02% per °C.

Definitions of the terminology used above follow later.

Pressure is one of the most commonly measured variables in the process industries. It is not only measured for its own sake, but instruments that measure pressure also inferentially measure temperature, liquid level, density, fluid flow, viscosity; and pressure-sensing devices form the basis of all analogue pneumatic computers, including controllers. Pressure (P) is defined as force per unit area, and has the dimensions of (mass)/(unit time)^2 (unit length).

The range of units is very great as will be discussed later. In general lbf/in^2 and kgf/cm^2 will often still be found. The usual symbol for pounds force per square inch is psi. The addition of 'a' to units implies above complete vacuum; the addition of 'g', gauge, means above atmospheric pressure. As we shall see later, all pressure-measuring devices measure pressure differences, and the reference pressure therefore has to be stated. For example, 30 psig is the same as 44.7 psia if the atmospheric pressure is 14.7 psia. Pressure may be regarded as a potential; pressure difference, commonly called dP or ΔP, is what causes fluids to flow against a resistance usually of a frictional nature due to roughness of the wall of the conduit containing the fluid and to internal friction (viscous forces) in the fluid itself. In industrial processes the phenomenon of pressure exists only in fluids, i.e. in liquids or gases.

The many equivalent units at present used, and their interrelation, are discussed more fully in Ch.10.7 with special reference to steam pressures, where it is suggested that the SI unit based on the pascal (or newton per metre^2) is the most logical, consistent and ultimately the most convenient to use in order to reconcile the welter of conflicting and sometimes vaguely defined alternatives.

Pressure Indicators

Direct reading instruments for measuring pressure may be classified as (a) 'wet'-type and (b) 'dry'-type instruments, whereas transducers which convert pressure to electrical or pneumatic signals for remote transmission and indication are almost exclusively 'dry' type.

'Wet'-type Instruments. These are restricted to the measurement of low gauge pressures and suctions (sub-atmospheric pressures) of a few tens of kPa.

A common application is for the measurement of draught and furnace air pressure. In its simplest form, the 'wet'-type pressure-measuring instrument consists of a glass U-tube partly filled with liquid (*Figure 19*). Pressure or suction applied to one limb of the U-tube produces a displacement of the manometer liquid. The vertical difference in level between the meniscuses in the two limbs is a measure of the pressure applied. In order to confine the displacement of the liquid to as far as possible one limb, the other is made of large cross-sectional area, thus forming a reservoir; the reservoir may be made of metal (*Figure 20*). The nature of the manometer liquid depends on the application. For low air and gas pressures oil may be used, and the glass tube may be inclined to increase the displacement of the liquid and thus give more accurate readings. For liquid pressures and higher ranges of air and gas pressures, mercury is used as the manometer liquid.

The simple manometer can give only local indication of pressures and it is inevitably rather fragile. For indication on a dial and also for recording pressure variations on a chart, float-operated mercury-filled U-tubes, sealed-bell and ring-balance type instruments are available, but have tended to be superseded by transducers. These earlier methods were described in detail in ref.7.

'*Dry-type Instruments.* These comprise instruments in which the applied pressure causes displacement of a diaphragm or similar element, the magnitude of this displacement being a measure of the pressure.

Through a very light magnifying movement, the deflection is transmitted to an indicating pointer (*Figure 21*). Pressure-measuring instruments of

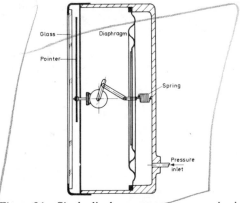

Figure 21   Single-diaphragm pressure-measuring instrument

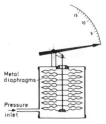

Figure 22   Multi-diaphragm pressure-measuring instrument

this type are limited to gases and the design with an edgewise scale and suitable for wall or panel mounting is widely used in steam-generating and similar plant.

Another form of gauge, also for measuring low pressures, uses corrugated metal diaphragms. When a number of diaphragms are used the assembly may be likened to a metal concertina. When pressure is applied to the diaphragm assembly, the resultant deflection is determined by the flexibility of the unit (*Figure 22*). Gauges of this type are available for recording the pressure (or suction) on a paper chart and are suitable for pressure ranges varying from 100 Pa—1 kPa above or below the prevailing atmospheric pressure ($\approx$100 kPa) to +350 kPa gauge, or for a hard vacuum.

A development of the capsule is the bellows unit, i.e. a corrugated metal cylinder of beryllium—copper or similar alloy. This is usually spring-loaded, the amount of deflection for a given pressure being determined by the combined resistances of the bellows and of the spring. The range of maximum pressures is approximately 1—700 kPa and the unit can be used for measuring both liquid and gas pressures.

For higher pressures, up to the maximum encountered in industrial practice, the Bourdon-tube element is used. This consists of a tube of elliptical cross-section bent into the shape of the

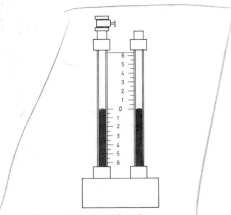

Figure 19   Double-tube manometer

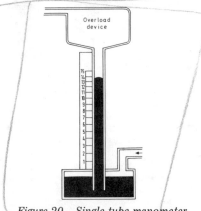

Figure 20   Single-tube manometer

References p 480

letter 'C'. For general purposes this is made of phosphor bronze or beryllium—copper. The free end of the tube is sealed; the other end is anchored and the pressure to be measured is applied to its interior at this end. Pressure applied to the interior of the tube tends to change its shape from elliptical to circular and, in so doing, to straighten the tube. The resultant movement of the free end of the tube, via a sector and pinion or similar mechanism, is calibrated to indicate or record in terms of pressure (*Figure 23*). In another design, the tube is wound into the form of a helix. This gives a larger displacement of the free end of the unit and the pen or pointer may be fixed directly on to the free end of the tube (*Figure 5*).

An instrument that could with advantage be more widely used on small steam-raising plant is the 'critical pressure' gauge. This is for measuring not the critical pressure of the steam but some preferred or desired pressure, with a false-zero gauge, so as to obtain a higher reading accuracy and an obvious indication when the actual pressure is not the desired pressure. In relation to its value as an aid to efficiency it is probably the least expensive of instruments. It consists of a loaded diaphragm, adjusted to give centre-of-scale reading at the selected preferred steam pressure. Full-scale deflection is given by a change in pressure of 10—50 kPa above or below the selected value, so that the gauge gives an immediate indication of changes in steam demand. The gauge is protected against possible damage from pressure variations beyond its scale range.

For certain applications, as for example in evaporators, it is important to maintain a predetermined absolute pressure. A common design of absolute-pressure gauge is one having two pressure elements, e.g. two bellows units, connected in opposition. One unit is evacuated so that its expansion and contraction are influenced by changes in atmospheric pressure; the other is subjected to the pressure being measured. The displacement of the element is then proportional to the atmospheric pressure plus the gauge pressure. This displacement is registered in the appropriate units.

The installation of a pressure-measuring device should present no difficulty provided that a few simple rules are observed. When measuring the pressure of a gas in motion, e.g. air flow through a duct, the plane of the pressure tapping must be at right angles to the direction of flow, otherwise a measurement will be obtained of the static pressure plus or minus a pressure which is a function of the velocity head. For a similar reason, a parallel portion of the duct should be selected for the point of measurement. This applies to liquid measurement also.

It is important to keep the impulse pipe clear so that the pressure can be accurately transmitted from the point of measurement to the measuring element. This pipe should, therefore, be laid in such a manner that it does not become choked with scale; accumulated moisture (in gas measurement) or air pockets (in liquid measurement), must also be avoided especially when low gauge pressures are being measured. Dust and moisture are liable to give trouble when dealing with hot flue gases.

When dust and moisture are expected near to the tapping point, an impulse pipe of generous proportions should be used, and facilities for cleaning the bore without undue trouble should be provided. The pipe should slope upwards from the tapping point to allow any moisture condensed from the cooled gas to drain back.

Where possible, gas pressure tappings should be made in the top of a horizontal conduit, and liquid pressure tappings in the side. For obvious reasons, pressure tappings should never be made in the invert of a pipe or conduit.

For liquid pressure measurement, the impulse pipe will be charged with the liquid. Therefore if the gauge is mounted at a level above the tapping point, it will register the pressure at the tapping point, minus the pressure due to a column of the liquid equal in height to the vertical distance between the tapping point and the gauge. The converse is true if the gauge is mounted below the tapping point. This discrepancy may be negligible when a high liquid pressure is being measured, but in other circumstances it becomes significant and must be taken into account.

No instrument should be subjected to unduly high temperatures. When dealing with hot fluids adequate cooling is, therefore, required between the tapping point and the gauge; the impulse pipe itself may be adequate for this purpose. In steam-pressure measurement the impulse pipe must be charged with condensate and a water-filled syphon or similar device must be mounted near to the gauge. A gauge should never be too hot to handle

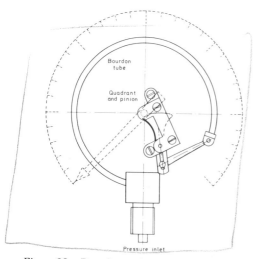

Figure 23   Bourdon tube pressure-measuring instrument

with comfort. In theory at least, a wet-type gauge of the mercury-in-glass type could be used for low-pressure steam measurement, but there are obvious risks. For steam-pressure measurement, a dry-type gauge of suitable form is, therefore, invariably used. A dry-type gauge should never be over stressed. The instrument selected should, therefore, be one with a maximum reading about 50% higher than the pressure normally to be measured. Dry-type draught gauges are particularly susceptible in this respect — they should not be 'tested' by blowing into them. 'Work within the middle of the range' is often a good precept, though some instrument dials are so calibrated as to be important exceptions to such a rule. In common with all other measuring instruments, pressure gauges should be mounted in a position free from vibration. If the gauge is of the indicating pattern, its function is to facilitate minute-to-minute plant operation; it should therefore be placed in a readily visible position. If of the recording pattern, it should be located at a point where the atmosphere is free from dust and dirt because the case has to be opened at regular intervals for changing the chart etc.

For ease of maintenance, there should be an isolation valve at the tapping point and a valve immediately adjacent to the gauge. The latter may be used to damp down pressure pulsations. For draught measurement and similar applications, it is useful to fit a three-way cock at the gauge to enable the gauge to be vented to the atmosphere for checking purposes.

Pressure Transducers

Pressure measurement may be made in terms of gauge, absolute or differential pressure. The simplest type of pressure transducer is based on a chamber containing a Bourdon tube or capsule. The capsule is sealed as illustrated in *Figure 24* except for the inlet pressure connection and an electrical pick-off to detect deflection under pressure. In this form the transducer measures 'gauge' pressure, which is the pressure relative to atmospheric pressure. As the name implies it is the same pressure as would be indicated on a simple

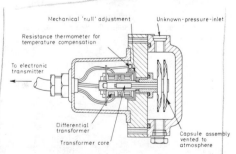

Figure 24   Gauge pressure transducer. Courtesy KDG Instruments Ltd

References p 480

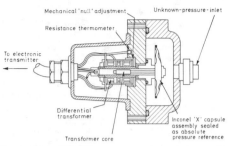

Figure 25   Absolute-pressure transducer. Courtesy KDG Instruments Ltd

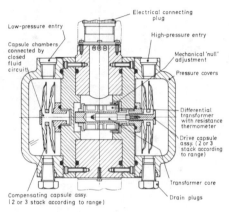

Figure 26   Differential-pressure transducer. Courtesy KDG Instruments Ltd

Bourdon-tube pressure gauge. If the capsule is now evacuated and sealed and the pressure under rest admitted as before to the chamber, the transducer will measure absolute pressure since the force acting to deflect the capsule is produced by the difference between the internal and external pressures acting on it. With the internal pressure 'zero' (vacuum), the effective force is proportional to the absolute pressure (*Figure 25*). The same form of construction but with connections permitting two unknown pressures to be admitted, one to each chamber as in *Figure 26*, enables a differential pressure to be measured; this is widely used with orifice-plate flowmeters. The pressure-element assembly must be carefully designed and manufactured from selected materials since it contributes one of the major sources of inaccuracy, i.e. non-linearity and hysteresis. Differential-pressure measurements at high static temperatures are sometimes made with liquid-filled nesting capsule assemblies which provide complete overload protection, a very necessary feature. The liquid filling suffers from temperature effects, whilst a further source of error is the variation in compressibility of the liquid with change in static pressure.

Errors from temperature variations occur from two main sources:

Zero errors. These arise from differing rates of linear expansion in the assembly owing to the necessity of using dissimilar materials. In liquid-filled assemblies this can be of the order of 0.2% per °C but may be reduced to 0.01% per °C for say 700 kPa pressure range, by the use of a twin assembly for differential-pressure measurement. Flush diaphragm transducers may commonly exhibit the smallest zone errors, of some 0.01% per °C for a given pressure range. Care should be taken in ascertaining at what pressure range zero errors are specified — the zero shift may be greater at the 1% pressure figure than at 10—100% of range, for a given temperature.

Sensitivity errors. These arise from variations in the modulus of elasticity of the Bourdon tube, capsule or diaphragm material. For materials normally used in pressure-element construction the temperature coefficient lies in the range +0.03% to 0.05% per °C. The force-balance type of transducer suffers less from this effect because it is fundamentally a null-seeking system in which spring restraints imposed by the pressure container are deliberately kept to a minimum and temperature effects are consequently several orders in magnitude less than in those designs relying on the spring-containing forces of the capsule or Bourdon tube.

Certain forms of displacement-type pressure transducer incorporate a temperature correction system based on measuring the actual temperature of the chamber by means of a resistance thermometer or thermistor and using this to compute an electrical correction applied to the transducer output signal. Errors in range and zero may each be reduced by such an electro correction system to better than 0.01% per °C.

After reviewing standards of measurement and the terminology used, more detailed examples of pressure/electrical conversion techniques will include the following:

Potentiometric.
Differential transformer.
Unbonded strain gauge.
Force balance (electrical or pneumatic).
Piezo crystal.

Calibration Standards. Pressure measurement, like any other form of measurement associated with physical variables, ultimately relies on some reference standard against which the elements comprising the pressure-measuring system must be calibrated. The accuracy of any such system therefore relies on the manner in which calibration of the working pressure-transducer is related to the reference standard from which fundamental units of pressure are derived under controlled conditions.

A choice of three basic calibration techniques may be adopted, depending on the potential accuracy of the pressure transducer and use to which it will be put in service.
1. Calibration of the working pressure-transducer by comparison with a laboratory 'secondary' or 'transfer' pressure standard that is repeatably accurate to an order better than the transducer under test.
2. Calibration of the transfer pressure standard by comparison with a laboratory 'primary' pressure standard. This will again generally be capable of repeatable accuracy to an order better than the transfer standard and may necessitate a temperature-controlled environment.
3. Calibration of the 'primary' pressure standard by comparison with national fundamental standards. This entails a series of precise measurements which may be undertaken by the National Physical Laboratory in the UK, the National Bureau of Standards in USA or other national authorities.

For calibration of transducers to an accuracy of reading of 1% a secondary pressure standard comprising a specially engineered Bourdon tube assembly may be used, enabling repeatability and rectilinearity of one part in 500 to be achieved for pressure ranges up to 66 MPa. Alternatively, secondary pressure standards may be based on force-balance techniques, illustrated schematically in *Figure 27*, which in this form possess the advantage of mobility for laboratory, workshop or field use. The response time of such a system may typically be 0.1 s while the short-term resolution and repeatability may be defined as 0.001% and 0.02% respectively of full scale. Primary pressure standards are more commonly known as 'dead-weight testers' because they rely on controlled conditions of mass and dimensions to derive a pressure reference illustrated schematically in *Figure 28*. A dead-weight test comprises a screwed ram as a means of developing pressure and a piston supporting a number of known weights by floating in a chamber filled with oil, the latter connected to the transducer under test. In operation, a weight corresponding to the pressure required is placed on the piston, then the screw is rotated until sufficient pressure is obtained to raise it. The piston is then spun to minimize friction which might introduce errors of the pressure reading measured at the output of the transducer.

Under carefully controlled conditions of temperature and cleanliness, it is possible to achieve a resolution of 0.002% of reading, an accuracy

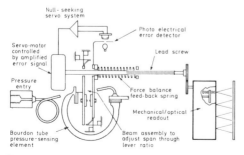

Figure 27   Force-balance pressure transducer. Courtesy Dresser Industries Inc.

of 0.015% of reading on low ranges e.g. 2–350 kPa and 0.025% of reading on higher pressure ranges, e.g. 100 kPa to 3.5 MPa. Models are available for use up to 70 MPa.

A liquid column or U-tube can be used as a standard within the limitations of physical space considerations, typically up to 2 m. A liquid column is suitable for use with water, oil or mercury, the latter giving a range equivalent to approximately 20 kg. The important point to remember is that the meniscus changes in contour according to whether the column is rising or falling, thus giving rise to a small error in visual readout. The use of a large-bore tube minimizes this effect and assists in obtaining readings free from parallax errors.

Further to the general definitions[1] applicable to instruments and systems, the following definitions of terms are more specifically applied to pressure transducers.

Potentiometric-type Transducers. The simplest form of transducer is undoubtedly a Bourdon tube or capsule driving the wiper of a potentiometer by means of a suitable linkage. The potentiometer may be linear or wound to some predetermined non-linear low and may be of wire-wound, metal-film or conductive-plastic type. Earlier potentiometric transducers were often unsuitable for industrial applications where the transducer output was required to be processed for subsequent control purposes and where a noise-free signal was essential and the inevitable hunting of the system about a pre-determined datum point often caused excessive wear, dead spots and eventual failure.

There have, however, been considerable advances recently in potentiometer technology which have resulted in greatly improved techniques of construction; useful life figures in excess of 5×10^6 operations are now quite normal. This could mean, in most cases, 5 years trouble-free operation. Examples of differential, gauge and absolute potentiometric transducers are obtainable from various specialist potentiometer manufacturers. Noteworthy mechanical features of the more sophisticated designs eliminate pivots and the potentiometer wiper assembly operates via a flexure spring linkage from the Bourdon tube or pressure capsule as shown in *Figure 29*.

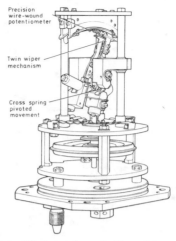

Precision wire-wound potentiometer

Twin wiper mechanism

Cross spring pivoted movement

Figure 29   Wirewound potentiometric pressure transducer. Courtesy Penny and Giles Transducers Ltd

A range of such transducers is available from UK manufacturers for working at pressures up to 80 MPa. Resolution is defined by the numbers of turns available, up to a maximum of 1000, and the repeatability is claimed as 0.2% to 1% depending on the particular variant.

Differential-transformer and Variable-reluctance type Transducers. These do not impose any kind of restraint on the prime mover, i.e. pressure element. They have no moving parts subject to frictional losses or wear and provide an output having infinitesimal resolution with a high order of linearity. This latter feature enables the signal to be amplified for subsequent transmission or signal conversion for control applications. The differential-transformer type of pick-off can therefore be incorporated in pressure transducers which operate in either laboratory or industrial environment, and provide an electrical pressure transmission system which has intrinsically a high-accuracy capability.

A practical design of a combined transformer and associated electronic transmitter is illustrated in block schematic form in *Figure 30* in which an oscillator (B) is amplitude-modulated by a bias voltage generated by the output of the d.c. comparator amplifier (A). This amplifier compares a feedback signal obtained from the transducer secondary circuit with a reference signal. The oscillator current feeds the primary of the differential transformer.

The transducer secondary-output signal voltages are summed and differenced in the circuit (D). The difference signal is proportional to the pressure applied to the transducer and is passed to the output amplifier (denoted by E, F, G). The mean signal is converted into d.c. at the phase detector (C) and is fed back to the amplifier (A) to control the transformer input current. The mean signal

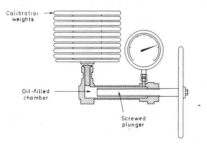

Calibration weights

Oil-filled chamber

Screwed plunger

Figure 28   Dead-weight tester

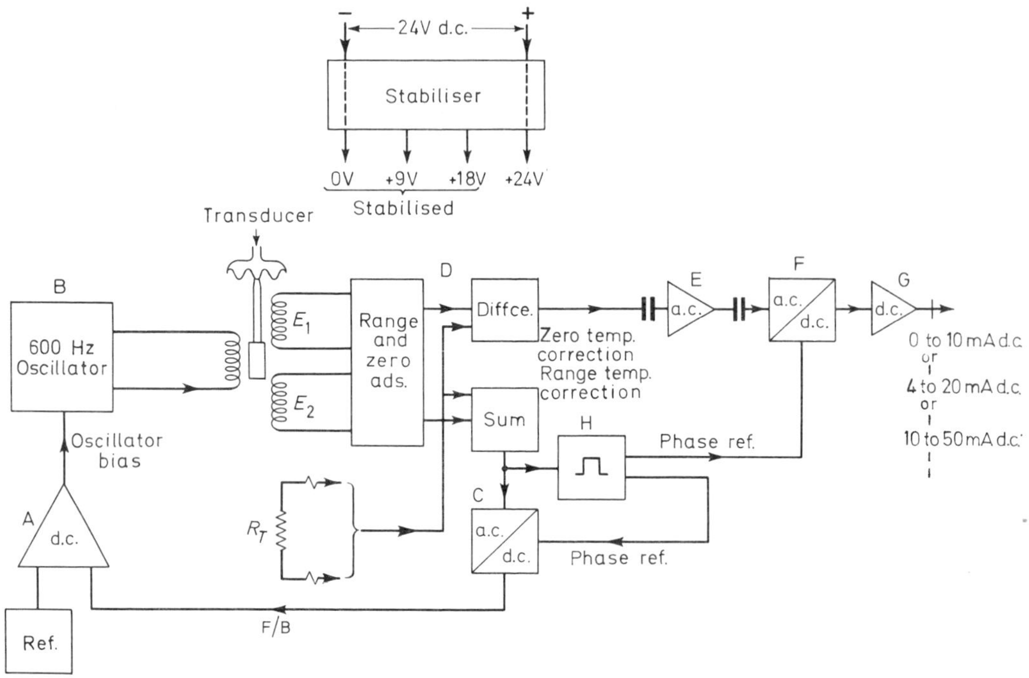

Figure 30   Application of differential transformer to pressure transducer-circuit.   Courtesy KDG Instruments Ltd

is also used as a phase reference to switch the phase detectors (C and F) via the pulse generator (H).

A resistance-thermometer element (RT) can be used to provide a signal to offset changes in the modulus of elasticity of the transducer pressure element. This signal is used in circuit (D) to change the effective value of $E1 + E2/2$ in the feed-back loop to the amplifier (A). Zero shifts in the transducer due to linear expansion are offset by using the signal from RT in circuit (D) to modify the difference signal $E1 - E2$.

The difference signal $E1 - E2$ is first amplified by means of the a.c. amplifier (E) before being applied to the phase detector (F). The rectified signal thus obtained is smoothed before being applied to the d.c. amplifier. A choice of output signals, current or voltage is obtainable from the d.c. amplifier (G).

Examples of absolute and differential-diaphragm transducers are illustrated in sectioned diagrams *Figures 25* and *26* respectively.

For a typical commercial specification, the output voltage can be 0 to 5 V or 1 to 5 V d.c., the span can be set to within ±0.3% output impedance less than 50 Ω. The output current can be set to 0 to 10 mA, 2 to 10 mA, 0 to 20 mA or 4 to 20 mA d.c., with load limits of zero to 1500 Ω for the 10 mA range and zero to 750 Ω for the 20 mA range. In the worst case the effect of load variations on the current output is ±0.1%. The ripple in the output circuit is less than 0.1%

(r.m.s) of the signal level, or 6 mV whichever is the greater. The linearity over the range 5% to 100% is better than ±0.1%.

Strain-gauge Pressure Transducers. Strain-gauge pressure transducers incorporate a number of desirable features. The form of construction enables a compact assembly to be engineered to perform accurately and reliably in extreme environmental conditions and with highly corrosive fluids.

The output takes the form of varying d.c. voltage that does not depend on brush contacts (as with potentiometric pick-offs), and the signal voltage is sufficiently high to be compatible with most data acquisition systems.

Various methods are used to construct strain-gauge pressure transducers but all are based on one of two principles, namely 'bonded' or 'unbonded'. The bonded form is the simplest form of construction because it involves the attachment of a fine wire or piezo-electric strain gauge by direct adhesion to the pressure diaphragm. The unbonded type gauge is possibly the more versatile form of construction. It comprises a fine tungsten—platinum resistive wire of some 5 μm diameter wound around sapphire posts which are mounted on a star-spring structure. The wire filaments are cemented with high-temperature epoxy resin to ensure maximum stability of the sensing element in harsh vibrational environments. A wide range of sensitivities is obtained with this type of pressure transducer by varying the diaphragm thickness and force-summing area, star-spring thickness and strain-gauge wire resistance.

One manufacturer has produced transducers covering pressure ranges from as low as 0—15 kPa to 0—66 MPa and temperatures between −160°C and +325°C. Numerous variants are available, specifically designed to be resistant to nuclear radiation or corrosive liquids, with accuracies (including combined linearity and hysteresis errors) of ±0.5% of full range output. Two sensing arrangements are used; one utilizes a rhombic structure described below, the other a 'flat' sensing device, both being designed with a view to minimizing the effects of vibration and acceleration by making the mass of the sensitive element as small as possible in keeping with the spring tension.

A schematic diagram of an unbonded strain-gauge pressure transducer element is shown in *Figures 31* and *32*.

The heart of the transducer comprises the strain gauge which is subjected to strain applied in the windings by means of forces that are in turn applied in a controlled manner by the pressure force being measured. A typical transducer assembly is shown sectioned in *Figure 33*. The pressure causes displacement of the diaphragm. A force rod connected to the centre of the diaphragm transmits the force (proportional to the applied pressure) to the sensing element. Transducers of this type have been developed to meet requirements for medical, aerospace, nuclear and industrial environments.

A transducer of particular interest for industrial applications is a bonded strain-gauge instrument designed to measure a low differential pressure in the presence of a high line pressure up to 20 MPa. The unit converts differential pressures over the range 0—250 cm to 0—750 cm of water into a proportional electrical output of 0 to 15 mV d.c.

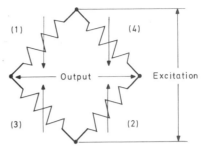

Figure 32   Bridge circuit configuration. Courtesy Bell and Howell Ltd, Electronics and Instruments Division

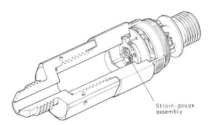

Figure 33   Unbonded strain-gauge pressure transducer. Courtesy Bell and Howell Ltd, Electronics and Instruments Division

Positive overpressure protection is provided by mechanical stops, enabling the transducer to withstand overpressure of 20 MPa applied to either side without deviation in specifications. One such design is claimed to withstand 24 MPa without diaphragm rupture. An electrical network mounted in a junction box provides compensation for changes in zero pressure output and full-scale pressure output due to changes in temperature. The unit has provision for external shunt calibration.

Force-balance Transducers. These may be electrically or pneumatically operated: the motion or force exerted by the differential pressure acting across a flexible diaphragm is matched or balanced by a motion or force produced by an electrical current in a coil.

Pneumatic Force-balance Transducer. The pneumatic transducer is the commonest, and will be dealt with first. The balancing part of the system is essentially a pneumatic pressure divider (or pneumatic potentiometer), which consists of a fixed restrictor (resistor) and a variable restrictor (flapper-nozzle) in series in a small-bore (1 or 2 mm) pipe, venting to atmosphere. The pressure in the space between the fixed and variable restrictors is related to the pneumatic resistance of the flapper-nozzle, i.e. to the clearance between the nozzle and the flapper (*Figure 34*). The relation is non-linear and the sensitivity is extremely high; in the usual working range of 140 kPa gauge supply

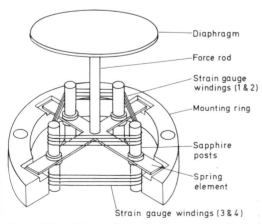

Diaphragm

Force rod

Strain gauge windings (1 & 2)

Mounting ring

Sapphire posts

Spring element

Strain gauge windings (3 & 4)

Figure 31   Unbonded strain-gauge pressure transducer

pressure, the intermediate pressure changes by about 85 kPa for a change in flapper-nozzle clearance of about 0.38 mm. A further defect is that the pressure indicated on the gauge is derived from air passing the upstream fixed restrictor, and is thus a restricted flow. These defects are overcome by two means:

1. The addition of a 'negative feed-back' bellows or diaphragm.
2. The use of a relay whose output is a pressure which is greater than the input by a constant factor, usually 4, 5 or 6.

The use of the relay permits a relatively small part of the non-linear flapper-nozzle clearance/pressure relation to be used, greatly improving the linearity. The feed-back reduces the sensitivity of the system by always 'backing-off' a large part of the input movement of the flapper.

Motion Balance. This type of transducer converts a motion input at the flapper into a pneumatic signal. The overall accuracy is about 0.5%, and a bellows, diaphragm, capsule, Bourdon tube etc., may drive the flapper, but this system is rarely used because of the mechanical linkages involved which wear and produce backlash.

Force Balance. This is much more common and more accurate. It may be instantly distinguished from the motion-balance type by the presence of a pivot on the main beam (*Figure 34*).

The force input is shown as a differential pressure, the difference between the force exerted by two pressures, one in each bellows. The bellows need not be of the same effective area, so that 'scaling' may be introduced — for example, if one bellows is twice the effective area of the other then a pressure in it need only be one half of the pressure in the other for the device to give zero output. Usually two bellows are not used, but instead a diaphragm with the pressures to be measured applied to either side. In all the above if one pressure is atmospheric the output will be 'gauge pressure' and if alternatively one side is exhausted and sealed the output will be 'absolute pressure'. For protection against overload the two diaphragms are commonly used, rigidly connected at their centres by a bar passing through a hole in a plate, the space between the diaphragms being filled with a liquid which is incompressible. On overload, one diaphragm will rest against the

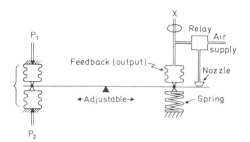

Figure 35   Pneumatic force-balance transmitter — sectioned. Courtesy Foxboro-Yoxall Ltd

central plate and the amount of liquid present is not sufficient in this condition to extend the other diaphragm beyond its elastic limit. The amount of liquid is very small, as the force-balance nature of the instrument ensures a virtually nil displacement of the diaphragms whatever pressure differential, within the desired range, is applied (*Figure 35*). Referring to *Figure 34* it will be seen that movement of the pivot along the beam alters the range of the instrument for a fixed change in output and this can also be seen in *Figure 35*. The output of all pneumatic transducers of the types described above is 20—100 kPa gauge and this is universal and worldwide. The chosen starting point of the measurement, which may be zero differential pressure or any differential pressure above or below zero differential pressure may be chosen by adjustment of springs, such as the zero adjustments spring of *Figure 35*, and the range may be chosen by shifting the pivot on the beam. All pneumatic transducers are standardized to have the same output range and are mutually compatible. For this reason all pneumatic receiver gauges are calibrated with their zero mark at a time pressure of 21 kPa gauge (3.0 psig), and their maximum mark at 10.5 kPa gauge (1.5 psig). The scale in the span 21—105 kPa gauge (3—15 psig) may be calibrated in any units which match the zero and range setting of the transducers described above. Such transducers are commonly called transmitters. On changing the range of such a system it is only necessary to change the dial or the dial calibration of the receiver gauge.

The safety feature of having only air at a pressure up to 105 kPa in transmission lines and in control rooms instead of perhaps very high pressures of possibly hazardous fluids is obvious.

Electrical Force-balance Transducers. In its simplest form the transducer comprises a pressure capsule, displacement transducer such as a differential transformer or capacitance pick-off, and an associated oscillator/amplifier driving an electromagnetic force element to constrain displacement of the capsule by means of a suitable mechanical linkage. The linkage may also be designed to introduce some measure of temperature compensation.

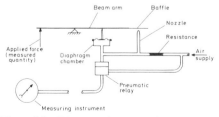

Figure 34   Diagram of pneumatic transmitter

In operation the force-balance transducer behaves like a null-seeking servo system in which the aim is to keep displacement of the capsule to a minimum by making the electronic gain as high as possible. This minimizes the spring effects of the capsule. A high-stability readout resistor connected in series with the current feed to the electromagnetic force element provides a voltage readout signal directly proportional to current. This voltage is capable of being calibrated in units of force and hence of pressure.

The electronic force-balance system is potentially capable of infinite resolution. Combined effects of non-repeatability, hysteresis and non-linearity can be kept to within 0.05% over a wide temperature range, and for laboratory applications, 0.01%.

Electrical force-balance transducers have been developed for gauge, absolute and differential pressure measurement. Transmission current may be 0—10 mA, 0—15 mA or 0—20 mA d.c. for line resistance of 2000 Ω or 3000 Ω by means of a repeater unit.

An example for obtaining high mechanical advantage in keeping with preserving a design that is inherently temperature-compensated may be studied with interest; it relies on the use of two bellows of slightly differing diameters and connected to work in opposition, the difference force being proportional to the pressure under test. An example of such a transducer is the absolute-pressure transmitter illustrated schematically in *Figure 36*.

Variations of vacuum applied to the two-ply bellows (A) are transmitted through a series of levers to the force-balance beam (B). The beam carried a pot coil (P) moving in the field of a permanent magnet (M) and in series with the transmission line. The beam also carried the centre vane of capacitor (C) connected to an oscillator/amplifier to provide a restoring force. Barometric compensation is achieved by an evacuated bellows (D) connected to the primary lever. Zero sup-

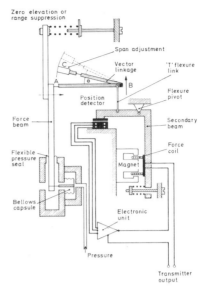

Figure 37   Kent 'Deltapi' force-balance pressure transducer. Courtesy Kent Instruments Ltd

pression is controlled by a compensation spring acting on the measuring bellows.

In the Kent 'Deltapi' series of industrial force-balance pressure transducers, an ingenious linkage system is used to obtain the necessary mechanical advantage between force transducer and pressure-sensing element, as shown schematically in *Figure 37*. By separating the measuring unit from the transmission unit, a choice of pressure capsules or diaphragms may be supplied to meet requirements for measuring gauge, absolute, or differential pressures from as low as 0—1.25 kPa (0—120 mm water gauge) to a maximum of nearly 20 MPa with maximum overrange pressure for the latter of 27 MPa. The materials used for the measuring unit are selected from a range of alternatives chosen for compatibility with the characteristics of the measuring fluid. The transducer assembly has been designed to conform with intrinsic safety requirements.

The total movement of the position-detector is less than 12 μm corresponding to a 4 to 20 mA change in output current, with the result that the beam remains in a substantially constant position and the output current varies in proportion to the applied forces and hence to the applied pressure. Zero adjustment is achieved by applying an adjustable spring force to the secondary beam. Zero levation or range suppression (when fitted) is achieved by applying an adjustable spring force to the force bar.

Range reversal (i.e. 20 mA output at minimum range and 4 mA at maximum) may be achieved by reversing the pressure connections to the capsule, and, via the range-suppression mechanism, moving the span completely below the zero

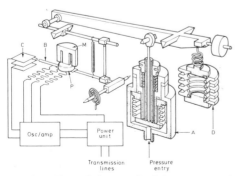

Figure 36   Force-balance absolute pressure transducer

datum. A typical specification of performance is given below.

Ambient temperature limits:

Body	$-40°$C to $+120°$C
Electronic Unit	$-40°$C to $+100°$C

(When electronic unit is mounted remotely (e.g. for environments where ambient temperature may exceed $+100°$C) the maximum permissible capacitance of connecting cable must not exceed $0.015\,\mu$F.)

Accuracy (including non-linearity and hysteresis)	$\pm0.5\%$ of span
Non-linearity	0.24% of span
Zero temperature coefficient	0.02% per $°$C
Output current	4 to 20 mA d.c. two-wire operation
Output impedance	2 MΩ
Min. volt. drop across transmitter	12 V
Power supply requirement	20—28 V d.c. (20—50 V d.c. depending on load requirements)
Max. load resistance:	400 Ω at 20 V 1900 Ω at 50 V

Another type of force-balance system relies on the use of a calibrated spring interposed between the pressure-sensing element and force-restoring mechanism such as motor driven lead-screw. This method, by virtue of the relatively large displacement of the force transducer (in this case the revolutions of the motor extend a low-rate spring), conveniently provides combined means for both analogue and digital readout with very fine resolution.

A mechanical or electronic digital position-indicator coupled to the motor shaft of this form of transducer is calibrated to read directly in units of pressure and an accuracy of 0.05% of full scale is obtained. Speed of response may be between 1 and 3 s for full-scale travel, depending on the span; and pressure ranges from 2.5 kPa to 66 MPa are obtainable.

Solid-state Pressure Transducer. The advances in solid-state technology in recent years have resulted in attention being given to the elimination of moving parts or frictional members in pressure transducers and reliance instead on the direct conversion of the mechanical strain developed in the pressure sampling chamber, into a resistive or voltage change.

Piezo-electric Transducers. Mention has already been made of the piezo-electric form of conversion in relation to certain forms of bonded strain gauge. The most widely known is probably the special type of ceramic made from lead zirconate titanate. It is obtainable in various grades and marketed by Brush Clevite Co. Ltd under the trade name PZT. The values of strain sensitivity for PZT gauges are about ten thousand times as great as for normal resistance-wire gauges. The gauge is cemented to the pressure sampling chamber, and when strains occur in the surface under pressure the stresses induced in the gauge give rise to voltages produced between the gauge electrodes owing to the piezo-electric effect. The voltage produced by the gauge is directly proportional to the unit strain and hence pressure.

In addition to high strain sensitivity, PZT gauges possess the major advantage that they are self-generating; consequently the associated instrumentation is considerably simplified. The electrical output impedance of PZT gauges is almost entirely capacitative however, and for low-frequency or d.c. operation very high impedance matching devices are necessary and a low-loss shunt capacitor or charge amplifier must be used. The effect of using a shunt capacitor reduces drift due to leakage but causes the output voltage to be reduced, typically from volts to millivolts. The voltmeter used for measuring this voltage must necessarily possess a very high input impedance such as that obtained with an electrometer or certain forms of high-grade digital voltmeter.

Semiconductor Transducers. A pressure-sensitive transistor known as a Pitran has been developed in the USA. The Pitran piezotransistor is basically a silicon *n-p-n* planar transistor that has its emitter base junction mechanically coupled to a diaphragm located in the top of a TO-46 can. When a pressure or point force is applied to the diaphragm a large reversible change is produced in the transistor characteristics. An unamplified linear output of at least 20% of the supply voltage can be achieved with supply voltages from less than 1 V to more than 50 V.

Transducer Applications. The alternative forms of construction and conversion techniques embodied in the pressure transducers described in this Section each possess individual features that need to be considered very carefully before the true specification of performance can be assessed. The industrial user, for example, requiring trouble-free 24-hour day continuous operation may opt for a transducer which will be totally unsuited for, say, aerospace applications where a finite life in a more predictable environment are primary requirements.

From the systems-engineering standpoint a number of application problems may predominate, not least that of cost. Undoubtedly, the simplest form of transducer may appear to be one that is self-energizing and produces a d.c. voltage proportional to pressure, but this advantage may be largely mitigated if an expensive high-impedance measuring circuit is required or if considerable distance exists between the transducer and nearest data-collection point.

For all-round versatility and accuracy of performance the differential transformer or variable reluctance transducer tends to be preferable, but where system-space and weight limitations predominate the strain gauge and potentiometric transducers are probably more attractive.

The force-balance transducer is capable of several orders of accuracy better than most other types but in its most accurate (0.01%) form is several times more expensive than, say, a straightforward 1% differential transformer or potentiometric transducer. Irrespective of the type of transducer selected, common problems such as overpressure, transient pressure changes or pressure pulsations, vibration, and thermal cycling, all require the necessary analysis and precautionary measures. It is usual to connect any pressure-measuring instrument (or transducer) to the container of the fluid whose pressure is to be measured by means of small-bore tubing, 6 mm to 12 mm. If liquid pressures are being measured the head effect of the liquid in this connecting tube must be considered. For example if the pressure of water in an overhead tank or pipe is to be indicated at ground level, different methods may be used. If an ordinary Bourdon-tube gauge is used, and the overhead vessel is 6 m above the gauge, then even with zero pressure in the vessel the gauge will read the equivalent of 6 m of water, that is, 60 kPa.

This may be dealt with (1) by attaching a label to the pressure gauge instructing that the above values be subtracted from any indication, (2) by offsetting the zero position of the pointer on the dial from the mechanical (no-load) zero position of the pointer, (3) by using a pressure gauge with a 'suppressed zero' by pre-loading the Bourdon tube with a spring so that it starts to indicate at the value of static pressure provided by the column of water, (4) by use of a transducer mounted at the vessel, pneumatic or electric transmission being used to an indicator at ground level, (5) by use of a transducer at ground level, offsetting its zero as described under transducers, or (6) by using a differential-pressure indicator or transducer with a dummy pipe full of water connected to the non-processed side of the instrument and carried up to the vessel level. In this last simple but very accurate compensation method it is necessary to ensure that the balancing pipe is always full of water to the correct level. A similar difficulty arises in measuring the pressure of a gas saturated with a liquid which may condense in connecting piping. In measurement of steam pressure it is always necessary to condense the steam entering the pipe to the instrument and to fit a syphon or pig-tail in the line to keep a plug of water blocking the entry of steam to the gauge, which would be damaged by the high temperature of the steam.

Pressure surges in the forms of transients or pulsations may require some measure of filtering by means of a suitable constriction at the transducer inlet, the exact diameter and length depending on the natural frequency characteristics of the pressure-transducer sampling chamber. The constrictions consist of small orifices or capillaries

(sometimes in the form of the helical channel produced by inserting a threaded rod or bolt in a closely fitting cylindrical case, which permits adjustment by screwing the bolt in or out), but they are prone to blockage and need careful maintenance. With force-balance or motion-balance transducers a pneumatic damper (light piston in a cylinder of air) or a hydraulic dash-pot may be applied to the beam.

With pneumatic transducers an 'inverse-derivative' may be inserted in the signal output line. This is really a low-pass filter which attenuates a pneumatic signal as a function of its frequency, also supplying power amplification.

For some duties on corrosive media the pressure gauge may be filled with a chemically inert liquid and its connecting tube closed with a flexible thimble of a material resistant to the measured fluid. Devices are obtainable which will isolate pressure gauges at pre-set values of rising pressure, so preventing overload damage, but the gauge indication in conditions of overload will be misleading.

Temperature extremes may sometimes be avoided by the use of adaptors to reduce heat transfer to the transducer; for pressure measurement at extreme temperatures a water-cooled adaptor may be necessary.

Flow-measuring Systems

Flow measurement is perhaps one of the most important parameters to challenge the resources of the instrument engineer. To many industries where transactions are computed on the transference of bulk liquids it is probably the most important.

A variety of flow-measurement techniques have been developed suited to gas, liquid or slurry media and the form of containment such as pipe or duct. The latter frequently dictates the form of measurement to the point of restricting choice to only one or two possible solutions. For example, the sheer size alone of a waterway may preclude any form of metering device that first requires a circular-cross-section feedpipe.

Flow measurement may be required either as flow rate or as a quantity, the latter being the time integral of the former. For control purposes the rate meter is most important, and in control applications the performance of any subsequent control system is critically dependent on the accuracy and repeatability of the basic flow measurement. For stock transfer measurement and costing, direct quantity measurement by means of a bulk meter is more applicable. These two distinctive measurement problems have resulted in ranges of flow meters aimed at providing the required information for these particular applications. Bulk metering may be further defined in one of two units of measurement, volume or weight (mass). Several techniques employed for mass metering rely on some form of computing such as temperature correction of volume, assuming that the temperature—volume—density relation of the liquid is accurately known. Very few mass flow meters exist that measure true mass flow directly.

A further distinction, necessary for technical reasons, may be made between various forms of flow-metering instrumentation. Flow meters may be classified generally as belonging to one of two categories, inferential or absolute (positive displacement). Examples of these two categories may be (*a*) when in the relation between flow and pressure measurement of the related pressure is expressed in units of flow (inferential), and (*b*) a vane or piston meter (absolute). For application in data-acquisition systems it is relevant to raise the question of analogue versus digital data-transmission of flow measurement, because the ultimate purpose to which the measurement information will be applied is a key factor in deciding on the optimum form of instrumentation. Generally speaking the inferential type of flow meter relying on pressure (or head) measurement, or on voltage generation in a magnetic field as in a magnetic flowmeter, are basically analogue types of instrument, whereas positive displacement meters such as vane or piston meters may be adapted for analogue or digital (pulse train) outputs. Semi-positive displacement meters, such as turbine meters, tend to be suited only for digital pulse train outputs.

The common factor that defines the ultimate performance of any flowmeter with either an analogue or digital output is the minimum detectable change of electrical signal for a given minimum detectable change of flow. Thus an analogue output is limited by electrical noise superimposed on a d.c. signal, together with d.c. drift; a digital pulse-train output is ideally required to contain ten times the minimum number of pulses to define the smallest desired increment of flow rate, so that the inevitable ambiguity of digital readout, of plus or minus one count, is immaterial to the final result.

This Section covers flow measuring systems and illustrates the techniques used for flow-rate and mass-flow measurement best suited for various media and conduits. Flow-measuring systems may be grouped as follows:
1. Inferential methods of measurement based on differential pressure obtained from pitot, weir, or orifice plate.
2. Inferential methods of measurement based on mechanical turbine meters or electromagnetic flow meters.
3. Positive-displacement methods of measurement based on vane- or piston-type meters.
4. Mass-flow meters.

Differential-pressure-related Systems

General Performance Characteristics. Pressure-related flow-measurement techniques constitute a substantial proportion of flow-measuring systems in a wide field of applications, from air-speed measurement to that of water flow in rivers. In general, the techniques derive from one of two broad principles, either inducing a pressure drop by means of a constriction, or transforming the kinetic energy of the flowing liquid or gas into potential energy in the form of static head.

The use of a constriction in the path of the liquid and measuring the resulting pressure drop has been used for centuries and practical application of this principle is based on the original theoretical work of Venturi in the 18th century and Bernoulli in the 19th century. It is the practical limitations of accuracy that such systems eventually suffer that has prompted the evolution of more direct methods of flow measurement such as magnetic flowmeters and positive-displacement meters.

The fundamental equation for any flow-measuring device which changes the local acceleration of a fluid in consequence of a change of area of the passage in which the fluid flows is based on Bernoulli's theorem and takes the form:

$$O = Cd^2 (2gh)^{0.5}$$

where O = volume flow rate, C = coefficient of discharge = actual flow rate/theoretical flow rate, d = diameter of primary device, g = gravitational constant, and h = resultant differential pressure as head of fluid. The coefficient of discharge C is a convenient way of quantifying a number of variables that each contribute to errors in theoretical performance of differential-pressure flow-metering devices after installation. A breakdown of such errors might include:

1. Variations in pipe size.
2. Physical location where long straight sections do not occur immediately before and after the device (refer to BS 1042 [8] which specifies tolerances).
3. Expansibility, an effect whereby the liquid undergoes a change in density as it passes through the primary device. The expansibility correction factor is denoted by ϵ when the density is measured upstream and by δ when measured downstream. It is possible to calculate the correction for Venturi tubes and nozzles but only empirical values are known for orifices, thus a tolerance must be allowed for. According to BS 1042 [8] a tolerance on flow measurement is one-third of the value of the correction for orifices and one tenth for Venturi tubes and nozzles. Alternatively, to avoid these additional tolerances the ratio of upstream to downstream pressure should be kept as close to one as possible.
4. Reynolds number: The dimensionless criterion which relates viscous forces and inertial properties of the fluid and is a convenient measure for comparing the performance of geometrically similar

flow-measuring devices under different fluid conditions and may be written

$$(Re) = \frac{du\rho}{\mu}$$

where d = throat bore, u = velocity, ρ = density, and μ = viscosity.

5. Experimental corrections for coefficient of discharge have been obtained to which again must be added a tolerance of one-third of the correction for Venturi and Dall tubes, square-edged orifices and nozzles. If the Reynolds number is sufficiently high a correction may be unnecessary, but if applied it is related to 70% of the maximum flow rate. Certain orifices are designed to provide a constant value of coefficient of discharge down to low Reynolds numbers, which may be encountered at low flow rate or with fluids of high viscosity.

6. Internal surface roughness of pipes: In BS 1042 [8] the conditions for specifying corrections are described; corrections are based on internal surface finish of commercially obtainable pipes, and/or use of small-bore piping.

There are various forms of differential-pressure flow-measuring devices. The subject is extremely specialized and the reader is recommended to study the references listed[9,10].

However, broadly speaking, there are three groups of primary device, and these are described in the following paragraphs:

Constant area—variable head.
Variable area — constant head.
Variable head— variable area.

Constant Area—Variable Head Devices. Orifice plates, venturis and nozzles all provide a simple means of inducing a pressure drop to the fundamental relation $u = k \, h^{0.5}$ where k is a constant

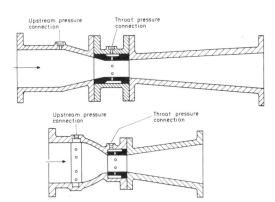

Figure 39   Typical venturi tubes

related to the ratio of pipe to orifice, the density and viscosity of the liquid, and h is the pressure drop in head of fluid caused by the restriction. Thus the velocity is proportional to the square root of the measured pressure drop. *Figure 38a* illustrates a simple square-edged orifice plate with two forms of connection. One, denoted d and d_2, locates the connections of one pipe diameter upstream, and half a diameter downstream at roughly the 'vena contracta' point, the point where the fluid jet issuing from the orifice has its smallest diameter. The second type has 'corner' connections, illustrated in *Figure 38b*; the pressure tappings are made to each face of the orifice plate. BS 1042 [8] sets out a very thorough treatment of the theory and operation, together with a guide to the installation of orifice plates.

The coefficient for an orifice plate is about 0.6, lower than for a venturi, while the permanent head loss is normally 60% to 70% of the differential pressure. This means that although cheap to install, an orifice plate may cause six or seven times the pumping loss of a venturi which has a coefficient of 0.95 to 0.9%. Typical venturis are shown schematically in *Figure 39*.

BS 1042 recommends that for minimum flow resistance there should be a parallel section of specified dimension carrying connections, an inlet core, a throat, again parallel, and carrying connections, an outlet core which is longer than the inlet core, and a final parallel core. The advantage of the venturi is self-evident; the streamline section tends to be self-cleaning and results in low permanent head loss, of about 10% of the differential pressure. Thus pumping losses are considerably less than when an orifice plate is used.

Variations on the concentric orifice plates are the eccentric and segmental orifice plates, useful for measurement of slurries and dirty liquids.

The flow nozzle (*Figure 40*) exhibits performance that lies somewhere between that of an orifice plate and venturi. It also is specified in BS 1042 [8] and is essentially a streamlined orifice plate. The downstream connection is marked by the flow nozzle itself. A flow nozzle has the advantage of

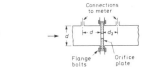

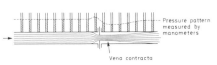

Figure 38a   d and d_2 orifice plate and subsequent pressure pattern

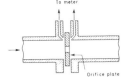

Figure 38b   'Corner'-type orifice plate

References p 480

reducing the permanent head loss associated with a simple orifice plate, and can be installed retrospectively between existing flange fittings.

Pitot tubes are one of the most common forms of flow-measuring device used industrially for transforming the kinetic energy of the liquid or gas into potential energy in the form of static head. In principle, a small sample of the fluid to be measured is brought to rest and the resulting impact head is subtracted from the static pressure. This is shown schematically in *Figure 41*. For liquid velocity u, mass of liquid brought to rest m, and head h of flowing liquid, by equating kinetic energy to static head (potential energy):

$$\tfrac{1}{2}\, mu^2 = mgh$$

or

$$h = \tfrac{1}{2}\, u^2/g \quad \text{or} \quad u = (2gh)^{0.5}$$

From this we can deduce that a pitot is only capable of point readings of velocity and does not represent the average flow in a uniform cross-section of pipe. It is further sensitive to yaw angles in excess of $10°$. The square root of pressure must be calculated to obtain a direct measure of the velocity. Measurement of low flow at the bottom end of the scale thus becomes difficult to make accurately. The overall accuracy is of the order of a few per cent of full scale. It is often used to determine velocity profiles in experimental studies.

An interesting adaptation of the pitot, the 'Annubar' developed by the Dietrich Standard Corporation, overcomes many of the disadvantages of the simple pitot and obtains a more representative sample of liquid flow than the point reading obtained from the former. The device is shown in *Figures 42* and *43*. It consists of a small pair of sensing probes mounted perpendicularly into the flow stream by means of a conventional

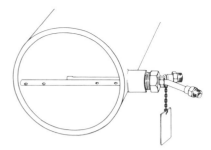

Figure 42   'Annubar' flow sensor

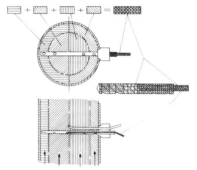

Figure 43   'Annubar' principle of operation

threaded fitting. The downstream pressure-sensing probe has one orifice which is placed at the centre of the flow stream to measure downstream pressure. The upstream sensing bar has multiple orifices. These orifices are critically located along the bar so that each orifice detects the total pressure in an annular ring. Each of these rings has a cross-sectional area exactly equal to the other annular areas that are detected by each of the other orifices. Inside the upstream bar is another sensing tube (on sizes 50 mm and larger) which is precisely located between orifices to sense the true average head of the entire cross-section of the flow stream, regardless of flow regime.

The accuracy of the system is claimed to be ±0.5 to $\pm1.5\%$ for most conditions. This accuracy is also permanent and repeatable throughout the normal life of the system. Its large sensing parts are unaffected by erosion and typical scale formation. Like some other primary elements, however, its use is not recommended for heavy slurries, without purging. A further advantage of this system allows the upstream bar to be independently rotated so that all sensing parts are pointing downstream in a totally protected position. This permits the element to be left in place for years without any ill-effect to it or significant restriction to the system.

It handles absolute pressures from approximately zero up to 17 MPa and is available in brass or stainless steel. The wide operating range

Figure 40   Flow nozzle

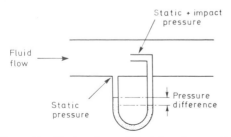

Figure 41   Pitot tube.   Principle of operation

is obtainable in standard models without modification. Another worthwhile advantage is that the entire element, on line sizes above 50 mm, can easily be pointed in either direction to detect reverse flows without shutting down the system.

Variable Area—Constant Head Devices. This technique relies on maintaining a constant differential pressure by varying the effective restriction as the flow rate changes. Possibly the most highly developed instrument of this type is the rotameter. It comprises a float reminiscent of a whipping top suspended in a tapered tube by the liquid and arranged to rotate by means of notches, to prevent sticking to the walls of the tapered tube. As the flow rate increases the float rises to permit a larger area of flow. The principle is shown schematically in *Figure 44*. The operation is almost independent of density changes in the liquid. A further advantage is that because the position of the float is directly proportional to flow rate it is unnecessary to obtain a square root as with differential-pressure systems.

The accuracy can be of the order of 0.5% for flow rates above 10% of maximum, i.e. the normal working range if 10:1. The range can be increased by connecting several meters in series. Each meter is arranged to have an expanded section at the top of the tube so that when the float reaches the top there is unrestricted area for the fluid and flow is measured on the next free float. The float-type variable area meter requires vertical mounting and this limitation, coupled with a rather long section of pipe, may limit its application in certain installations. Its undoubted advantage is simplicity of operation, rectilinear output, and good accuracy.

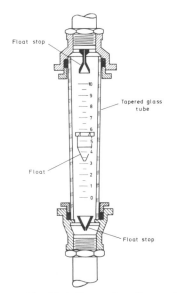

Figure 44   Variable-aperture flow meter

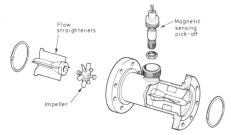

Figure 45   Turbine meter — exploded view

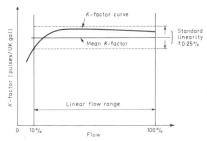

Figure 46   Typical linearity curve for turbine flowmeter

If the fluid is transparent, then direct reading of flow may be obtained by external calibration of a tapered glass tube. For liquids operating at high pressures a steel-walled tube may be supplied and remote indication of flow rate may be arranged by means of a magnet implanted in the float, detected by a servo driven magnet-follower mounted externally to the tapered flow-metering section.

Variable Head—Variable Area Devices. The metering techniques under this heading are used for very large flows of liquid in open ducts or channels such as rivers, effluent etc., and thus have somewhat limited and specialized application.

Turbine Meters. There have been extremely rapid developments in turbine meters in recent years, partly because of the advances in electronics technology and the ease with which the output may be used to indicate rate of flow or be integrated to provide total flow. An exploded view of a typical turbine flowmeter is shown in *Figure 45*.

A turbine flowmeter is known as an inferential meter because the operation depends on the relation between a rotating form of turbine and the stream velocity of the liquid in which it rotates. The turbine is consequently subject to variations in performance caused by the characteristics of the liquid, friction or bearings, etc., to the extent that, at flow rates approaching zero, the slippage may amount to 100%. That is, the turbine will eventually stall and register zero output.

Turbine meters range for 1—30 cm dia. and more and, subject to the limitations mentioned above, provide an output rectilinearly related to flow to an accuracy approaching 1/4% as shown in *Figure 46*. While turbine meters can be obtained

Table 4   Turbine-flowmeter specifications

Repeatability	Better than 0.05%	⎫	Under
Linearity over standard flow range	±0.5% (selected) better than ±0.25%	⎬	defined
Linearity over extended flow range	±1%	⎭	conditions
Maximum internal fluid pressure	21 MPa (3000 psi)		
Pressure drop at standard flow range	Approx. 28 kPa		
Pressure drop at extended flow range	Approx. 85 kPa. Line pressure up to 24.5 MPa		
Temperature range	-20 to $+200^\circ$C depending on choice of -50 to $+400^\circ$C sensing probe		
Sensing probe coil inductance (MIP.8Mk.2)	-20 to $+200^\circ$C		
Coil dc resistance (MIP.8Mk.2)	1.1 kΩ (approx.)		
Minimum voltage level at 10% flow range	20 mV r.m.s. open-circuit		
Output frequency at full scale standard flow range	Approx. 1000 Hz up to 50 mm size. Between 500 Hz and 250 Hz larger sizes.		
Material specification			
Flowmeter body	Stainless steel EN58J		
Bearing housing	Stainless steel EN58J		
Support tubes or spiders	Stainless steel EN58J		
Rotor	Stainless steel Firth Vickers 520		
Bearings	Hardened journal—depends on application.		

		Size		Standard flow range		Extended flow range		
	Type No.	(in)	(mm)	(gal/min)	(m^3/min)	(gal/min)	(m^3/min)	Pulse/gal
Six	⎰ M7/0500/3	½	12	0.5–3	2.3–14×10^3	0.35–4.5	1.6–20×10^{-6}	20 000
Sizes	⎱ M7/1500/150	1½	37	15–150	68–682×10^{-3}	4.0–225	18–1023×10^{-6}	400
Nine	⎰ M7/2/300	2	50	25–250	0.1–1.1	5–350	0.02–1.6	240
Sizes	⎱ M7/16/18 000	16	400	1800–18 000	8.1–81	1000–20 000	4.5–90.1	0.83

NOTES: When using flowmeter above standard flow range it is advisable to ensure that fluid pressure is above 210 kPa to prevent cavitation.

Flow ranges in the Table are quoted with magnetic pick-up. The use of inductive pick-up such as MIP.18 or MIP.119 permits the minimum flow rate to be halved in most cases.

Figures in the Table are based on water calibration specific gravity 1.0, viscosity 1.0 mm^2/s (cSt). Higher fluid viscosities tend to restrict operating range. For specific details refer to manufacturer.

for directly-driven counters, the most suitable form of output is an electrical pulse train in which the frequency is proportional to rate of flow. The pulses are generated by arranging a pick-up coil to be mounted in close proximity to the turbine rotor, which then behaves like the classic toothed-wheel rotor of a digital tachometer. Passage of the rotor past the coil induces an emf which can be transmitted some distance if required in frequency form, to be measured on a frequency-to-digital converter, or converted to a d.c. analogue voltage in a frequency-to-voltage circuit for indication on a conventional moving-coil meter. Very little loss of accuracy occurs during the frequency-to-voltage conversion process. Current or voltage outputs can be accurate to ±0.25%. Errors will usually depend on the accuracy of the indicating instrument, typically 1% on a moving coil meter or 0.25% on a potentiometric recorder. The pulse output may be summed on a simple digital totalizer giving totalized flow without loss of accuracy.

In digital form the output may be combined with local temperature measurement, which for many liquids of known composition, such as certain gasoline products, may result in a temperature-corrected volume reading proportional to mass flow.

A practical application of the turbine principle to metering is the Meter-flow range of meters for handling flows in pipe sizes 12–40 mm; corresponding characteristics are given in *Table 4.*

Electromagnetic Flowmeters

Electromagnetic flowmeters are particularly suitable for the flow measurement of slurries, sludge, and any electrically conducting liquid. The concept is based on an original suggestion by Faraday, as an application of his laws of electromagnetic induction, but it is only in recent years, when modern electronic circuits became available, that the principle could be applied. This was because the meter is basically an elementary voltage generator and the signal voltages are very small, a few millivolts at very low flow rates.

A schematic of an electromagnetic flowmeter is shown in *Figure 47*. It consists basically of a pair of insulated electrodes buried flush in opposite sides of a non-conducting, non-magnetic pipe carrying the liquid to be measured. Around the pipe are mounted a pair of coils which are energized to produce a magnetic field. The arrangement is now analogous to a conductor moving across a magnetic field, in which a disc of liquid bridging the electrodes represents the conductor.

The relation between induced voltage, magnetic field and velocity of conductor is $E = Blu$, where E is emf in volts, B is flux density in tesla (weber/m^2), l is length of conductor (pipe diameter) in metres, and u is velocity of conductor (flow) in m/s.

The meter relies on a high-gain amplifier of excellent stability to process the induced voltage into a useable form and when specially calibrated accuracies of the order of 0.5% are attainable, as shown in *Figure 48*. The concept is applicable to pipes of any size within the practical limitations of constructing a powerful magnetic field (*Figure 49*). From a fluid-handling point of view the main advantage obtainable from the use of a magnetic flowmeter is the absence of any obstruction that may cause pressure drop and hence increase operating costs, particularly if heavy slurries are being handled and pumping costs are high. A further advantage is that the output is rectilinearly related to flow. A practical development of the magnetic flowmeter is the Kent 'Veriflux' range of sensing heads and converters for handling flows in pipe sizes 3—1800 mm.

A Kent 'Veriflux' flowmeter system comprising a sensing unit (the detector head), a transmitting unit and receiving instruments as required is shown schematically in *Figure 50*. The duties of the

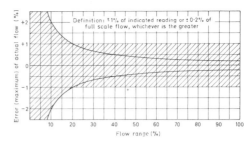

Figure 48   Typical accuracy/flow characteristic

transmitting unit may be broadly defined as:
1. To amplify the relatively small emf signal generated in the Veriflux detector head and to convert it into a standard current or voltage output suitable for the widest possible range of receiving instruments.
2. To reject unwanted signals generated in the detector head.
3. To compensate the effects of varying mains voltage and frequency of the flow signal.

The converter is precalibrated in velocity units before despatch and may be interchanged with any other converter on site without affecting the accuracy of the installation. A built-in test facility provides a complete operational check of the converter electronics. All plug-in, printed-circuit cards are pre-calibrated for ease of maintenance.

A very wide range of receiving instruments can be employed in conjunction with the sensing and transmitting unit. The converter will operate proprietary types of potentiometric recorders, deflection-type indicators and recorders, integrators, and electropneumatic transducers. *Table 5 (p 448)* shows the abridged specification for the Kent 'Veriflux' system.

Positive-displacement Meters

Positive-displacement meters, as their name implies, respond directly to movement of liquid and are capable of measuring flow rate to an accuracy better than ±0.1% and repeatability ±0.01%. They are commercially referred to as bulkmeters and are particularly suited to applications such as aviation refuelling, ship bunkering, tank calibration, engine test rigs, proving other forms of metering, and blending processes. Typical examples of meters for such applications are manufactured by Wayne Tank and Pump Co., and Avery Hardoll on whose basic range these notes have been compiled. A typical positive-displacement meter is shown sectioned in *Figure 51*.

It comprises a casing containing a rotor assembly fitted with four blades in opposing pairs, each pair being mounted on rigid tubular rods. Above the rotor casing is bolted an inlet and outlet manifold whilst a calibrating mechanism and direct-reading mechanical counter are bolted on the front cover. The only moving parts within the fluid

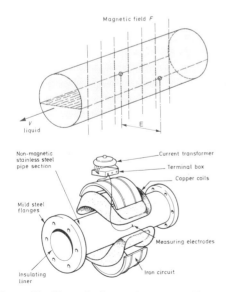

Figure 47   Magnetic flowmeter construction

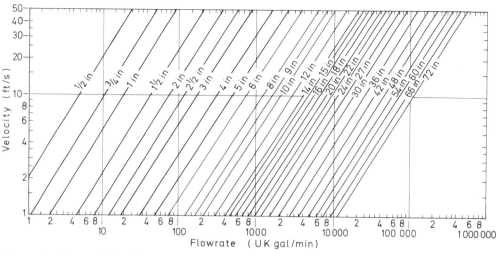

Figure 49   Flowrate/pipe-size relation — magnetic flowmeter

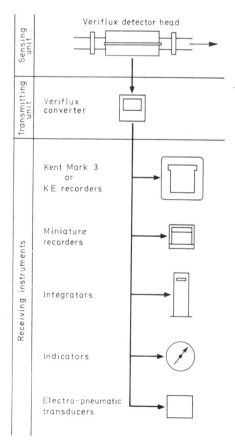

Figure 50   Kent 'Veriflux' flow-measurement system

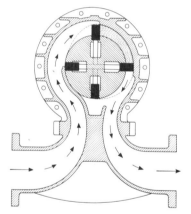

Figure 51   Positive displacement meter. Principle of operation

being metered are the rotor and rotor blades, which are constantly immersed.

In operation, fluid enters the meter through the inlet manifold and causes the rotor to revolve in a clockwise direction by pressure on the blades. The proximity of the rotor to the casing forms an efficient seal, whilst the profile of the casing ensures that the blades are guided on to the measuring crescent, where the combined effects of gravitational and centrifugal forces cause a very efficient seal to be formed. An extension shaft driving through a pressure-tight gland in the front cover of the meter transmits the rotor revolutions through calibrating gearing and thence to a counter or pulse generator for remote indication. In the range, two basic categories of meter are offered, one for general bulk metering in ranges 6.6—80,

12.3–15 and 17.7–230 m³, while for high-flow rate bunkering ranges 41.5–415 m³/h and 83–830 m³/h are obtainable. Typical calibration curves are shown in *Figure 52* for both types of meter.

Reference has been made to a calibrating mechanism interposed between the rotor shaft and readout counter. The unit comprises a stepless friction wheel and disc, the ratio being adjusted by means of a calibration screw. On BM series meters one complete turn of the screw alters the calibration by approximately 0.1% while similar rotation on LBM meters alters the calibration by 0.23%. Thus, calibration adjustment can be effected at any chosen flow rate and this feature, coupled with an inherent repeatability of ±0.01%, enables very accurate readings to be taken.

For temperature compensation, a compensator can be supplied which automatically adjusts the indicated volume on the counter to an equivalent volume at the recognized standard temperature (60°F or 15°C). If another 'standard' temperature is required in special circumstances, the temperature compensator can be calibrated accordingly. The unit is not affected by ambient temperature and the response time is some 15 s depending on fuel characteristics. This development enables the range of bulkmeters to be used for applications such as the loading of heated oils.

The unit is shown schematically in *Figure 53*, and basically consists of a temperature-sensitive fluid reservoir tube which is connected to a bellows compartment, the assembly being immersed in the liquid being metered. Any temperature variation from standard is immediately sensed by the reservoir tube which is transformed by the bellows

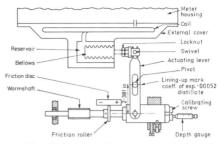

Figure 53   Diagrammatic arrangement of temperature compensating unit

into linear movements. This movement, through levers, adjusts the existing bulkmeter calibrating mechanism by repositioning a friction wheel on a disc. Because the temperature-sensitive components are fully immersed in the liquid being metered, they are not affected by atmospheric temperature. As the cubical expansion coefficient differs from liquid to liquid, a dial and knob are fitted to the front cover of the unit for adjustment as necessary. The dial is graduated in coefficients of expansion and a pointer on the dial is set to the appropriate position by the knob. Rotating the knob also adjusts a variable fulcrum lever so that the friction-wheel movement is in proportion to the coefficient of expansion of the liquid concerned for a given temperature rise. The knob is normally locked in position and can be sealed by Customs authorities.

In addition to local direct reading of volume, fully flameproof transmitters are available for remote digital read-out of volume or rate of flow, and remote analogue read-out of rate of flow. These transmitters enable positive displacement meters of the type described to be incorporated as the primary sensing elements in a variety of system applications, e.g. preset systems, servo blending systems, railcar loading and road-tanker loading.

Alternative forms of transmitters available are described in the following paragraphs.

Pulse Transmitters.

Micro-switch Transmitter. The switch is actuated by a rotating cam, which is geared to the calibrating mechanism drive so as to give the required number of pulses per unit of measured volume. It is used in conjunction with low-speed, heavy-duty electromechanical counters, e.g. for measurement of jet-engine fuel consumption.

Magnetic Reed-switch Transmitter. A single-pole, changeover reed-switch is positioned in the field of a fixed permanent magnet, and a chopping disc having ten blades is placed so that the blades rotate between the switch and the magnet, thereby shunting the flux and causing ten switching actions per disc revolution. This transmitter is used with medium-speed solid-state counters or high-speed electromechanical counters, for digital read-out, also for data acquisition and presetting in solid-state logic systems.

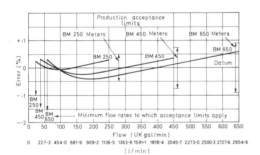

Figure 52   Typical accuracy curves for bulkmeters

References p 480

Table 5   Abridged specification — Kent 'Veriflux' system. (For detailed specifications refer to the appropriate data sheets)

Available sizes	All standard English and metric sizes from 1/8in—72 in (3 mm to 1.8 m)
Flow—velocity ranges	English version. From 0—2 ft/s to 0—120 ft/s Metric version. From 0—0.5 m/s to 0—40 m/s Refer to *Figure 49*
Accuracy	Typically ±1% of indicated reading or ±0.2% of full-scale deflection, whichever is the greater (see *Figure 48*)
Minimum conductivity of metered liquid	1 mS/m
Electrical outputs	Voltage and current outputs including: 0—10 mA; 10—0—10 mA; 0—20 mA; 20—0—20 mA; 0—15 mA; 4—20 mA.
Power consumption	Detector heads: (at 240 V) from 20 VA at 2 A to 2000 VA at 9 A dependent upon size. Power-factor correction can be incorporated if required. Converter: less than 10 VA
Pressure limitations	(Where suitable end-fittings are specified) continuous operation at pressures up to 150 psi (1 MPa) is permissible. Detector heads for higher pressures can be supplied.
Temperature limitations	Detector heads can be supplied suitable for temperatures of up to 200°C.

Photo-transistor Transmitter. A rotating slotted disc is interposed between a light source and a photo-transistor. With the appropriate power supplies connected, pulses are generated by the photo-transistor as the disc rotates, and are passed to an amplifier/trigger unit for shaping as required by the counter in use. Pulse output can be either square wave or differentiated with or without an emitter-follower stage. Uses include meter-proving in conjunction with solid-state high speed counters, and data acquisition and presetting in solid-state logic systems. A version of this transmitter has been used for a jet engine test-rig, for high-accuracy rate of flow indication.

Analogue Transmitter.
    D.C. Tachogenerator. Driven from the bulk-meter gearing. Voltage output is proportional to rpm and hence to flowrate indication; also used for providing a monitoring feedback where servo-control is required as part of a system.

The choice of remote transmitting generator depends largely on the resolution and repeatability required. By selecting an adequate number of pulses per revolution of the rotor, for example, the minimum increment of flow may be counted. Analogue forms of read-out are restricted in the limit, by the linearity of the generator over the flow range required, and the signal/noise ratio at low flow rates, noise being a function of commutator ripple.

Mass Flow Measurement

General Principles and Applications. Volumetric flow measurement is a direct means of measuring bulk transfer of liquids, but for many applications it does not provide sufficient quantitative data on the product. Whereas, in the limit, the total volume may be checked visibly in a tank or container suitably calibrated in units of volume, what is frequently required ultimately, particularly for costing purposes, is a measure of the total weight or, as an intermediate step, the mass flow. Typical requirements for mass flow measurement are for liquid fuels and liquid foodstuffs where regulations governing their sale require close tolerances of weight measurement. In the case of petroleum-based fuels, there is a direct relation between the specific gravity and calorific value; for applications such as aircraft tank gauging and tanker bunkering accurate knowledge of the total weight and distribution of weight is of vital importance to the stability and possible survival of the craft.

In the past, weight measurement tended to be obtained by analogue computing techniques, from a knowledge of pressure, temperature and an assumed density relation, but work in recent years has resulted in the development of true mass flow meters and sophisticated digital mass metering systems based on density-corrected volumetric displacement meters.

These techniques include:
Inferred mass flow measurement (volumetric displacement corrected for density inferred from temperature and pressure).

Direct mass flow measurement (measurement of momentum).

True mass flow measurement (density-corrected volumetric displacement).

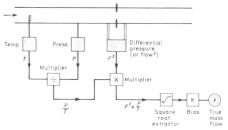

Figure 54   Elementary computation of mass flow

Inferential Mass Flow Measurement Systems. Referring to the fundamental relations

Volumetric flow $Q = k(2\Delta p/\rho)^{0.5}A$ and Mass flow $= Q\rho$ it is possible to substitute for density in a gas the relation:

$$\rho = PM/R\theta$$

where P = upstream pressure, M = molecular weight of the gas, A = pipe area, Δp = pressure loss, θ = absolute temperature, R = gas constant, and ρ = gas density. An elementary mass metering system based on the above relations is shown schematically in *Figure 54*, where measurements of temperature and pressure are made with traditional instruments and mass flow is computed in electrical or pneumatic analogue form.

The assumption made in the above calculations is that the gas obeys Charles' and Boyle's laws, but the assumed relation is inaccurate for a multi-product process such as the metering of coal gas and natural gas through the same pipe for example because M varies with each product. Other sources of error, such as non-rectilinear pressure/density relation, can be much more severe, as illustrated in *Figure 55*, where for carbon dioxide the deviation from Boyle's law is as much as a factor of five at 100 atm (10 MPa).

Although this example in gas metering is possibly severe, it serves to illustrate the potential limitations in obtaining accurate measurement of mass flow by such means, excluding of course additional errors arising from the transducers themselves.

For the precise measurement of flow of a gas at varying pressure and temperature it is thus necessary to determine the velocity of flow and the density of the flowing gas, and from these values calculate the actual quantity of flow. For on-line high-accuracy measurement this is impracticable and the use of a computer is essential.

An example of such a computer is shown in *Figure 56*, designed specifically for the measurement of gas flow, for use in conjunction with existing primary and secondary measuring devices.

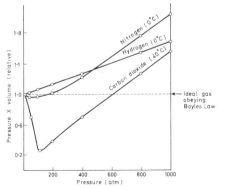

Figure 55   Non-linear pressure/volume relations of gases

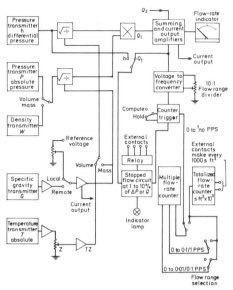

Figure 56   Practical gas-flow computer. Courtesy KDG Instruments Ltd

The instrument is designed to accept input signals from conventional differential pressure, density or pressure plus temperature transmitters and to provide an output signal directly proportional to the actual rate of flow specified in standardized units. The output signal can be related to the input signals by equations such as:

$$W \propto \left(\frac{\Delta p}{R\theta}\right)^{0.5} \quad \text{or} \quad \propto (\Delta p\rho)^{0.5}$$

The appropriate equation is selected by a single switch.

Normally, the first equation is more economical in terms of measuring instrument cost, but for some gases the variation in gas 'constant' requires that for high accuracy the second equation must be used, thus requiring the measurement of density.

The computer has an accuracy of better than 0.1% of span for any flow between 10% and 100% of span. This accuracy, together with long-term stability, is achieved by use of a pulse-width modulation variable-area multiplier technique for all multiplying, dividing and square-root functions. The technique involves the use of an integrating amplifier which causes its output to ramp up and down at different but rectilinear rates dependent upon the values of the input signals. The output of the amplifier is fed to a trigger circuit which provides pulses the length and spacing of which are determined by the ramp slopes of the integrating amplifier. The pulses are then averaged in a resistor-capacitor network to provide an output as a function of one, two or three separate input signals. The circuit is so designed that the output is dependent only upon the values of the input signals, the values of the resistors in the integrating circuit and the function to be performed. By eliminating the parameters of capacitors, etc., from the output equation the computing technique is made accurate and stable.

An additional feature available on the computer is an integrator using a similar integrating amplifier technique. This provides a digital output calibrated in units of totalized flow. The totalizer flow counter may be used for cost metering, particularly for large natural methane installations.

Preset Volume—Weight Conversion. For certain specialized flow metering applications such as luboil blending, aircraft fuelling and marine bunkering where the properties of the fuel are known within defined limits, simple volume-to-weight conversion may be obtained by what amounts to a scale factor correction based on foreknowledge of specific gravity. A practical application of this technique to a positive displacement is the addition of a 'volume—weight adaptor'. The use of this adaptor presupposes that the specific gravity of the fluid is known to an order of accuracy approaching the basic accuracy of the meter.

True Mass Flow Measurement Systems. True mass flow measurement by means of density-corrected volumetric displacement metering offers the most accurate solution for gases or liquids, because by taking due care to ensure that the temperatures of primary sensors for measuring volumetric flow and density are the same, the effect of temperature on either volume or density is nullified by the subsequent multiplication process. This point may be further illustrated from the fundamental relations governing volume and density where to a first-order approximation for given mass W:

Volume $V = V_0(1 + at)$

Density $= W/V_0(1 + at)$

W = volume $\times$ density = independent of t

From transducers available that are potentially capable of the required accuracy of individual measurement of density and volumetric displacement (0.01 to 0.1%) two methods of mass-flow computation emerge, one based on the use of a frequency-domain transducer for density measurement, the second based on force balance.

The first system of computation relies on binary rate multipliers for multiplying frequencies corresponding to density and rate of flow, while the second system (not widely adopted but included for technical interest) relies on traditional differential-digital-analyser techniques of successive addition of units of weight at a frequency corresponding to rate of flow.

Direct Mass-flow Measurement. The measurement of mass flow by direct means requires the momentum of the liquid to be measured. Work on this problem, notably by Elliott Bros., in recent years, has been prompted by requirements for accurate measurement of fuel consumption in high-performance jet aircraft. The additional limitations in space and weight imposed in such applications tend to preclude the use of mass-flow measuring systems based on individual measurement of related parameters such as volumetric flow, temperature and pressure, computing mass flow from these.

Two techniques for measurement of momentum of liquid are applied, one providing an analogue output, the second providing a digital output. The first is an elementary method and relies on the measurement of the deflection of a vane in a specially shaped chamber, the momentum forces on the vane being restrained by a calibrated spring. The vane is magnetically coupled to a potentiometer, thus providing an analogue signal.

The second method, capable of higher accuracy, relies on imparting an angular momentum to the fluid and measuring the corresponding deflection angle of the impeller relative to the input drive. A spring-restrained connection is interposed between the drive and impeller, and pick-off coils are used to detect pulses which are time displaced proportional to spring deflections, thus providing an output in digital form.

Spring-restrained, Variable-orifice Transmitter. An example of this form of flow transmitter is the Elliott flowmeter which forms part of a flow-metering system designed to provide a visible indication of the quantity and rate of fuel consumption. The information is conveyed in the following manner: a pointer traversing a dial gives continuous information on the rate at which fuel is being consumed; a counter drum records the weight of the fuel consumed over a given period of time, determined from the last operation of the reset button.

The transmitter is fitted in the fuel-feed pipeline and is used to sense the fuel flow-rate to produce an electrical signal which has the required flow-rate information contained within the output. The unit embodies design features which enable maximum accuracy to be obtained regardless of fuel type and fuel temperature in the widely varying ambient conditions to which the transmitter may be subjected during operational use.

In operation, the transmitter is fitted to the filter on the power plant so that fuel being consumed passes through the metering chamber and displaces a vane which attains a position where the pressure exerted by the flow of fuel is equal to that of the vane control-spring tension. The metering chamber is designed so that the vane displacement, which is magnetically coupled to the wiper of the potentiometer, is rectilinearly related to the flow rate. The flow-rate signal voltage is taken from the wiper of the potentiometer to an amplifier stage of the rate-of-flow indicator.

Elliott transmitters have been developed to measure flow rates up to 7500 lb/h (3500 kg/h) in two sizes designated 7801—12 000 and 7801—

13 000. The specification for Flow Transmitter type 7801—12 000 is given in *Table 6* and *Figure 57*.

Angular Momentum Mass Flow Rate Measurement Transmitter. Fuel entering this type of transmitter is first passed axially between radial straightening vanes to remove swirl and other disturbances. The fuel then passes through the rotating measurement assembly. This consists of a multi-vane impeller, connected coaxially to a surrounding drum by means of a spiral spring. The drum entirely shrouds the circumference of the impeller, eliminating viscous drag on the impeller. The drum is rotated at a constant speed of 100 rpm by a synchronous motor unit. This low speed enables an extremely long life to be achieved from the ball bearings which support the transmitter rotating mechanism. The motor housing is supported in the centre of the transmitter by webs, machined integrally with the transmitter body.

As a result of the fuel passing through the annular space in the measurement assembly, the impeller is deflected relative to the drum owing to the torque required to impart an angular momentum to the fuel. Since the spiral spring has a linear torque/deflection characteristic, the deflection angle is a measure of this angular momentum. This assumes a constant angular velocity which applies in this system since the motor speed is crystal controlled. It follows therefore that the greater the mass-flow rate, the larger the deflection of the impeller relative to the drum. The deflection is measured by attaching small powerful magnets, two each, to drum and impeller

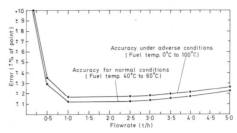

Figure 57   Calibration limits for spring-restrained variable-orifice transmitter

circumferences 180° apart, and placing two pick-off coils on the outside of the non-magnetic transmitter body. As each magnet is rotated past its respective pick-off coil, an electrical pulse is produced in the coil. The time displacement of the impeller pulses is thus directly proportional to mass-flow rate.

Solid-fuel Meters. There are many machines on the market for the continuous weighing of solid fuel, as distinct from weighing in batches. The general principle involved in the operation of these machines is to supply the fuel to a constant-speed belt conveyor, a length of which is 'live'. That is to say, it is supported by a counterbalanced weigh-bridge assembly. The change in weight of this section, caused by alterations in the amount of material passing on to it, is usually detected by a null-balance measuring system. The counterbalancing force is continuously adjusted to maintain the system in equilibrium, the amount of adjustment required being a measure of the weight of fuel. By mechanical means, e.g. by detecting the position of the balancing mechanism at short regular intervals, the weight passed can be integrated on a counter.

Although these machines (whose construction involves the use of knife edges, levers and tie rods) are highly accurate, their design and operation hardly fall within the bounds of the field referred to under the general term Instrumentation.

For the measurement of solid fuel, such a high degree of accuracy may not be required, since errors in net weight resulting from variations in the moisture content of the fuel will greatly exceed any weighing error. The gross weight of the fuel can be measured consistently with an accuracy within 1%, but the difference between the gross and net weights can be up to 15%, or more, depending on the amount of moisture in the fuel. The latter can vary rapidly with change in atmospheric conditions.

If the usual commercial tolerance of the order of ±2% can be accepted (as it is for the metering of fluid flows) there is no reason why similar principles should not be employed. For example, the dead weight of the live section of the conveyor belt could be counterbalanced and the weight due to the material passing over the belt applied to a pneumatic force-balance transmitter. Within the

Table 6   Specification for flow transmitter type 7801—12 000

Voltage requirements	13 V to 14.5 V dc from indicator fuel-flow rate
Fuel-flow calibration range	200 to 3500 kg/h
Electrical connector	Terminal block with 6-32 UNC terminal screws
Maximum working pressure	660 kPa (100 psi)
Pressure drop at 3500 kg/h calibrated flow not more than	7 kPa (1 psi)
By-pass maximum pressure drop at 3500 kg/h with vane at zero flow condition	25 kPa (4 psi)
Accuracy	Refer *Figure 57*
Weight	2.5 kg (5.5 lb)
Operating fuel temperature range	+5°C to +80°C
Abnormal temperature range (No derangement)	−40°C to +100°C

range of the equipment, the output air pressure of the transmitter would vary between 20 and 105 kPa, and this would be registered in terms of weight on a conventional form of measuring instrument.

For measuring the quantity of fuel fed to small boilers, an 'inferential' method has been used successfully for many years. The operation of these 'weighing' machines depends on the assumption that the density of any particular type of dry coal, e.g. bituminous, is practically constant in whatever district or country the coal is mined and is independent of the size of the lumps or particles.

Therefore, if the volume is continuously measured, once the meter has been rated for a particular type of coal, it can be calibrated directly in weight units. It should be noted, however, that assessing weight by volume versus direct weight measurement is a controversial matter. In ref.11 it is stated that 'The bulk density of loosely packed coal is not a quantity reproducible to within 5%, except under very closely controlled conditions of measurement. Compaction may increase the value by amounts up to 20%.' The publication gives tables of bulk density of loosely packed fuel which indicate that the density varies not only with the grading of the coal but the dimensions of the container used for the measurement. Nevertheless, many of these volumetric coal meters appear to give satisfactory service.

The principle of operation of a volumetric-type coal meter, applied to a conveyor, is illustrated in *Figure 58*. The deflection of the measuring door is directly proportional to the volume of the coal, and this deflection, together with the speed of the belt, is transmitted to a counter which automatically and continuously multiplies these two variables. The counter, therefore, registers the total volume passed, and can be rated to register directly in weight units.

The method is applicable to all types of mechanical stokers, e.g. chain-grate and ram stokers, and can also be used for coal supplied down pipes and chutes. In the latter cases, the measuring element is an endless chain which is carried downwards with the coal.

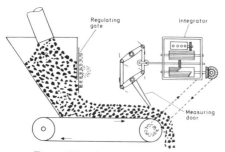

Figure 58   Volumetric coal meter

Applications of Flow Measurement

The advantage of inherent simplicity of differential-pressure flow-measurement devices may be offset by several operational limitations associated with their use. The first limitation is the creation of permanent pressure loss in the line. For example, the conventional orifice plate, flow nozzle and similar devices interfere with the flow being measured so as to cause a permanent pressure loss. This loss must be taken into account when calculating pump and associated motor sizes.

A second limitation is the restricted flow range over which accuracy can be sustained, typically 4:1. In the case of an orifice plate, venturi or flow nozzle for example, results may be accurate around the 'designed flow rate' and for the relevant section in which the system is installed, but if the flow deviates outside the calibrated range the accuracy falls off. A 'best' figure for calibration is 0.5% under known conditions of temperature, viscosity and density of the liquid that may seldom be repeated ideally during regular use.

A third drawback is the deterioration of accuracy with use, brought about by wear and tear of the orifice plate by the liquid; consequently provision must be made to replace the orifice plate and recalibrate. Last but not least is the performance (i.e. accuracy and repeatability) of the secondary measuring device, the differential-pressure transducer.

The next form of inferential flow-measuring system, based on a magnetic flow meter, suffers none of the operational limitations associated with orifice plates, nozzles and weirs and consequently may be a preferred solution where the liquid is electrically conducting and an accuracy of 1% is adequate. Advances in electronic technology have enabled stable, sensitive readings to be obtained at an analogue output level adequate for practical instrumentation systems.

Turbine flowmeters have gained increasing popularity in recent years, notably because of the relative simplicity of construction and pulse-train output that lends itself to remote indication of flow directly in digital form. Accuracy of the order of 0.25% is typical, while for selected flow conditions calibration to the order of 0.1% is claimed.

Positive-displacement meters offer the most accurate solution to flow measurement, and the vane-type described is possibly the most widely used. Other forms such as piston meters are used mainly as metering pumps for applications such as petrol filling stations. Positive-displacement (p.d.) meters tend to introduce larger frictional losses than say a turbine meter, but the repeatability is an order better. Certain fluids such as foodstuffs are seldom metered by p.d. meters because of problems of in-line cleaning, while fluids containing abrasive particles cause rapid wear. The choice of readout, be it direct indication, electrical pulse-train or analogue voltage, offers flexibility in the application of p.d. meters, particularly for on-line blending.

Mass metering systems would appear to offer the most scope for development because of the relatively few solutions presently obtainable. The fundamental problem of measuring true weight of liquid flowing to an accuracy no worse than 0.1% over a flow range of 10:1 can be appreciated from a study of the limitations of the various forms of volumetric flow meters described in this Chapter.

For ultimate accuracy of liquid or gas flow measurement the best solution would appear to be the density-corrected flow meter solution relying on digital counting techniques for computation, but such a system relies on individual measurement of density and flow. The 'momentum' type of mass flowmeter, based on measuring the energy transfer of the liquid under controlled sampling conditions, offers the most compact solution for such applications as aircraft fuel metering systems.

The flowmeters which have been used as examples in this Section are only a few of a very wide range of flow-measurement devices which are used in the process industries, although they cover the major usage. Other types which are worthy of mention include the ultrasonic meter, in most of which a beam of ultrasonic sound waves is propagated upstream against the flow and another downstream with the flow. One of these is retarded in its propagation rate by the flow while the other is accelerated. Pulses simultaneously transmitted in two channels therefore arrive at different times and this time lapse is a measure of flow. These two pulse channels can be arranged to be regenerative and the beat between the two channel frequencies is a measure of flow rate.

Many of the more recent developments in flowmetering use vortex shedding principles in which a drag body in the flow stream sheds a vortex whose periodic pressure fluctuations are a measure of flow rate. Mass flow meters have been developed which use the angular momentum imparted to the flow and the torque required to straighten the flow is related to the mass flow rate. Correlation of pressure or temperature disturbances in the flow stream at two points along its path have been used to determine flow rates, and a 'gyroscopic' flowmeter in which the flow is led around a loop which is vibrated transversely has been developed but has not seen significant application. These flowmeter types add to the range available to the systems engineer but in terms of usage they are likely to remain in a minority for some time to come.

Density Measurement

Measurement of the density of solids is comparatively straightforward but the accurate measurement of the density of a liquid or gas is one of the most difficult problems that faces the instrument engineer. Fundamentally the density of a substance is defined as the mass per unit volume, but an equally useful measure is the specific gravity, which is the ratio of mass or weight of the substance to that of the equivalent volume of distilled water at a defined temperature, usually $16°C$.

A choice of measurement techniques is open to the system designer, but the type of instrument selected does not automatically ensure that the readings obtained therefrom will necessarily match the calibration performance figures claimed by the manufacturer. There are many problems in obtaining a representative sample of the substance under test, coupled with uncertainties about the physical properties of the substance that make the measurement of density to an accuracy better than 1 part in 1000 extremely difficult. Under ideal laboratory conditions it is possible for accuracies of density measurement to be indicated to 1 part in 10 000 but the question of the ultimate density standard remains. An analysis of the potential sources of error that may occur in measuring a sample by laboratory means, and translating this to an automatic instrument reading, may invalidate claims for higher accuracy than 1 part in 10 000.

The various forms of instrument technique used for density measurement may be classified into one of two types, direct and inferred. The only truly direct method of density measurement is to weigh a sample in a container of known volume although certain inferred methods such as measuring the natural frequency of a sprung mass of sample of known dimensions yield an equivalent accuracy. Inferred methods such as those relying on absorption of nuclear radiation or buoyancy of a float are less accurate, particularly for liquids containing solids in suspension.

It is proposed to discuss briefly typical problems associated with liquid density measurement, compare the basic characteristics of two instruments potentially offering the most accurate means of measurement, then follow by problems of smoke measurement and gas analysis.

Application Problems

Successful continuous measurement of density depends on four factors: obtaining a representative sample, minimizing contamination of the instrument, and eliminating physical disturbances such as mechanical shock and vibration, fluid surges; and last but not least, obtaining quantitative data on pressure/temperature/density relations of the sample under test. These may be reviewed in turn.

Obtaining a representative sample is more difficult than it seems and frequently may involve the use of a bypass system which allows a proportion of the main stream to be routed to the instrument under controlled conditions of flow and pressure. On certain liquids however, such as slurries, the problem would be to ensure thorough mixing at the point of off take. Equally important is the prevention of air inclusions that are present in most liquids. Certain petroleum products contain impurities such as sand and water that either directly or indirectly invalidate instrument readings. Filtering is very desirable on most density-meter installations. The added cost is usually far less than the cost of an instrument overhaul.

Physical disturbances occur in most applications because of the presence of machinery associated with any process involved in liquid handling. Certain instruments such as direct-weighing types suffer more in this respect than flotation types, and vibrating mass types suffer least. Liquid surges, by virtue of their change in momentum, may cause temporary apparent density changes to be detected by the instrument. A preferred installation includes a bypass and bleed system in which a sample is continuously drawn from and returned to the stream at a constant head by means of a separate pump.

Referring to quantitative data on the pressure/temperature/density relations of the sample under test, it may appear to be paradoxical to ask for the very data that one may in fact be required to measure. But in practice it is almost impossible to define the density of a substance without a simultaneous knowledge or control of the other variables such as temperature or pressure. The relation between temperature and density is seldom rectilinear; a study of tables issued by the Petroleum Institute will illustrate this. The problem becomes particularly acute where mixtures of solids in suspension, base liquid, and soluble substances are involved; the viscosity may change rapidly for small increases in temperature resulting in pressure variations where constant flow rates are being demanded. Such variables as those described can usually only be determined by laboratory tests to enable some secondary form of instrumentation to be added for applying corrections if required.

Various forms of density-measuring instruments have become available in recent years and the principles of operation are briefly:

'Bubbling Tube' and Differential-pressure Transducer. In this instrument density is related to the pressure required to bubble air at two points in the liquid separated by a known vertical height. Accuracy limited by pressure transducer and imperfections in temperature control of long vertical column (3–10 m). Accuracy in the order of 1%.

Float-type Densitometer. This may use displacement or force-balance principles. Satisfactory for liquids of uniform solution (not solids in suspension) and in a static state, i.e. not flowing. Ingenious weir systems have been devised to preserve the datum of the liquid surface. Accuracy of 1 part in 1000 obtainable under laboratory conditions. A typical example is manufactured by Sangamo Weston.

Nucleonic Radiation. This relies on the use of radio-isotopes and the fact that the absorption of radiation by a substance may be related to mass per unit area and hence relative density. For high accuracy, a transmission gauge is used in which the material being checked is interposed between the source and the detector. Accuracies of 1 part in 1000 have been obtained when measuring liquid densities over relatively small spans and under carefully controlled conditions, i.e. following on-the-spot calibration against a laboratory sample.

Continuous Weighing. Density is obtained by weighing a sample circulated in a U-tube of known volume. For balance restoration. Accuracy of 1 part in 1000 obtainable under controlled flow conditions. Typical examples are the Rotameter Gravitrol and the Sperry Gravitymaster.

Vibrating Beam. Density is obtained by measuring the natural frequency of the mass comprised in a container of known diameter filled with sample liquid. Accuracy of 1 part in 1000 obtainable under most conditions and 1 part in 10 000 under controlled flow conditions. Examples are the Solartron Densitometer, and the Agar vibrating-spool fluid-density meter.

Advances in the last two forms of instrument have enabled a choice between analogue or digital readout to be obtained from a basic standard instrument. The examples described below have been chosen to illustrate the broad principles involved.

Sperry Gravitymaster. The basic Gravitymaster is shown schematically in *Figure 59*, and comprises a U-tube through which the process liquid is circulated. The tube is arranged to pivot horizontally on a cross-leaf suspension, so as to constitute the weight of a conventional beam balance.

A 'suppressed-nominal' technique is used in which a calibrated mass is used to counterbalance the tube when filled with a reference liquid such

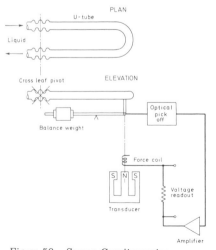

Figure 59   Sperry Gravitymaster

as water of density 10^3 kg/m^3. Any deviation in weight of liquid causes a displacement of the beam; this is detected optically by a special pick-off producing a voltage which is then amplified to drive a current through a coil suspended in a magnetic field. Movement of the coil returns the beam to horizontal, thereby reducing the output of the pick-off to zero, i.e. it is a null-seeking servo system. The gain of the amplifier is made high so that the magnitude of the restoring force is effectively equal to the error force.

One virtue of the 'suppressed-nominal' technique is that a servo 1% accuracy on a span of 10% change in density will yield an accuracy of density measurement of 0.1% assuming that the stability of the counterbalance is better than 0.1%. In practice a figure ten times as accurate is achieved. In the basic analogue version the restoring current is fed through a calibrated readout resistor, and the voltage appearing across the resistor is the analogue of the density variation from nominal, positive or negative.

In the digital version of the Gravitymaster, shown in *Figure 60*, the suppressed nominal technique is restrained, but instead of measuring deviations above or below nominal, only positive deviations are measured in order to simplify the decision-making logic. This does not limit the span of the instrument because the 'nominal' is offset to the negative end of the scale and the dynamic range is doubled by modification to the force unit circuit.

Unlike conventional analogue-to-digital converters, the need to generate precision voltages is obviated by supplying binary weighted currents to

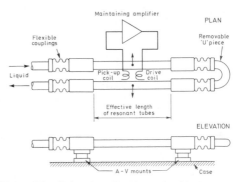

Figure 61   Solartron vibrating-tube densitometer

a second force coil suspended in the same magnetic circuit as the first. The analogue force-balance circuit is used for damping short-term disturbances only. The readout resistor is replaced by a centre stable differential relay whose threshold sensitivity is designed to correspond to half the minimum resolution required.

The foregoing technique results in an analogue-to-digital conversion process that is completed within the closed loop of a force-balance servo system. One attraction of the Sperry Conversion is the added facility of analogue readout for conventional display or control, if required, as a back-up to a digital system, at relatively low additional cost per instrument. Excellent long-term stability is achieved because in the limit the system is dependent on two magnetic circuits. These comprise the centre-stable relay used as an error detector to define the minimum digit level, and the moving-coil transducer used to convert binary related currents into units of force to restore unbalance of the beam.

Solartron Vibrating-tube Densitometer. The Solartron instrument provides an a.c. voltage output whose frequency is a function of the density of the liquid circulating in a resonant tube. The classic relation between frequency, 'stiffness' and inertia is used to derive the value of density.

The 'stiffness' term is a constant and obtained from the dimensions and Young's Modulus of a sampling tube. The inertia term is a function of the volume of the sample and the density of the liquid. Consequently, the only relevant variable in the above relation is the density of the liquid. In the Solartron instrument great care is exercised to define the length of the tube by machining and mechanical constraint. The instrument is shown schematically in *Figure 61*. The instrument can be mounted in any position and at any angle. Consequently, it may be secured vertically, thereby minimizing the precipitation of sludge and formation of bubbles that might otherwise contribute significantly to errors in measurement.

The readout of the instrument can be made to display either the actual density or density deviation. A digital binary-coded decimal output is

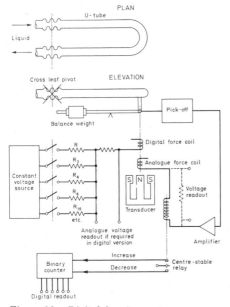

Figure 60   Digital Gravitymaster

available for transmission purposes. A simple digital-to-analogue converter is available should it be required, to connect the instrument to an existing analogue control loop.

Agar Vibrating-spool Densitometer. An interesting development of the frequency-modulated transducer approach to density measurement is the vibrating-spool fluid density meter by Joram Agar. The principle of operation is shown in *Figure 62*. Here the sensing element is a tube thickened at the two ends to form a spool. In operation it is set in oscillation in a circumferential mode, as a bell rings. Inductive pick-off and excitation coils and an amplifier maintain oscillation.

The fluid to be examined surrounds the tube and is thus similarly set in oscillation. As in the vibrating-tube densitometer, the frequency of oscillation depends on the tube stiffness and the total oscillating mass, which in the case of the spool comprises the tube walls and surrounding fluid. The output of the sustaining amplifier is monitored by a frequency meter calibrated to read units of density; an increase of density results in a reduced frequency of oscillation.

An accuracy of one part in 10^4 of full-scale deflection is claimed, with excellent immunity to plant and other noise problems by virtue of the high mechanical Q factor (gain at natural frequency) of the spool (a few hundred). By suitable selection of spool material the temperature coefficient may approach that of a tuning fork. A typical calibration curve is shown in *Figure 63*.

The wide range of change in periodic time from vacuum to trichloroethylene is defined by:

$$o = o_0 [(T \times T_0)^2 - 1]$$

where o = measured fluid density, o_0 = scale factor, T = measured time oscillation, and T_0 = periodic time at vacuum.

The advantages of measuring periodic time of frequency-modulated transducers instead of frequency are twofold. First, the relation between periodic time and density tends to be closer to rectilinear, thus requiring less correction. Secondly the output may be measured in less time. For

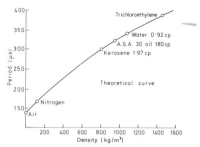

Figure 63   Calibration curve — vibrating-spool density meter

example, if the carrier frequency is a few kilohertz (a practical figure is 3 kHz) the measuring time is about one second to resolve 1 part in 3000. However, if the range of frequency is only 10% of the carrier frequency, i.e. 300 Hz, the measuring time will require to be 3 s to resolve 1 part in 1000. If periodic time is used however, this accuracy can be achieved in a fraction of this time. Accordingly, the signal is counted (averaged) over say 10 vibrations, which takes only 3.3 ms. During this period, a 1 MHz 'clock' will fill up the information storage register with 3300 pulses — 3 times the required 1000.

Linearizing Problems. When linearizing density readings from any form of density meter several important factors must be borne in mind to ensure that the meaning of linearizing is clearly defined:

The relation between density and temperature of the fluid, the means whereby this is obtained, and the accuracy with which it is defined.

The relation between actual density of the fluid and indicated measure of density provided by the instrument at some suitable reference temperature and for the range of operating temperatures.

The indicated density/temperature relations for the instrument over the working range.

Accurate linearizing will involve the continuous calculation of at least two variables, and in the limit it may be argued that it is the accuracy obtainable from the computing process that limits the ultimate accuracy of result; or the fundamental limitation of ultimate accuracy may be the predicted accuracy of the density/temperature relation of the liquid being sampled.

Systems based on digital calculation, involving frequency-modulated transducers, binary rate multipliers and totalizing counters, would appear to offer a more predictable result than analogue computing solutions.

In operation a digital calculator for processing a frequency-modulated signal would be based on a binary rate multiplier in which a 'pulse dropping' technique of computation is performed with binary rate multipliers which operate on say 1 MHz clock pulses, rather than on the incoming information, which is much slower. Thus, by measuring the periodic time of the incoming frequency-modulated

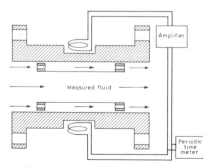

Figure 62   Principle of operation of the vibrating-spool density meter

signal it is possible to linearize and display the information in less than 5 ms.

Smoke Density Measurement

The problem of smoke density measurement has hitherto been mainly a matter of ensuring compliance with the law in one form or another, the emphasis being on keeping within the law rather than on avoiding loss of efficiency. Growing realization that minimum smoke production and maximum fuel efficiency tend to be complementary has focussed attention on the desirability of devising equipment that will control combustion as a result of smoke monitoring. Until such equipment is available and its adoption has become standard practice it is necessary for the industrial fuel user to have at his command devices or equipment that will go some way at least towards reducing the production of smoke to the desirable 'faint brown haze' now accepted as a safe indication that combustion is proceeding satisfactorily. In the UK at least, the consequences of the Clean Air Act are such that smoke-measuring devices and smoke alarms must be accepted as essential instruments.

1. The best known simple device for measuring the density of smoke issuing from a factory chimney is the well-known Ringelmann Chart described in Ch.18.3. The equipment for using it consists of shade cards with grilles or lattices of black lines of different intensities, numbered 0 to 4, in accordance with the proportion of the surface covered by the black lines, number 4 indicating 80% coverage.

The charts were originally meant to be observed from a distance of 50 feet against the background of the smoke plume, the shade of grey of the plume being matched against the card whose shade most closely approximated to it. The cards are still used by smoke inspectors and by furnace operators as a check on the emission of smoke. For use, it is convenient if two observers take part in the operation, one to hold the cards in the required position between the chimney and his colleague and in such a position as to enable the latter to make a comparison.

Apart from the difficulty of applying the method in built-up areas, it suffers from a number of obvious disadvantages, particularly the state of the weather and the unpredictable human element. Furthermore, many smokes are not grey (i.e. 'shades' of black); some may even appear almost white.

2. The density of smoke can be measured at a convenient point in the flue, stack, or uptake and the detecting element consists of two parts, a light projector on one side of the flue or stack and a receiver containing the light-sensitive cell, on the other. The source of light is an electric bulb fed from a power supply through a constant-voltage transformer. The beam of light is directed through the duct and on to the receiver by a lens; a second lens is in some cases used to concentrate the beam on to the light-sensitive element. The electrical output from the light-sensitive cell is proportional to the intensity of the light falling on its surface, which is determined by the density of the smoke in the intervening space. The electric output from the cell is thus inversely proportional to smoke density (*Figure 64*). The output from the cell is measured on an electrical measuring instrument, indicating or recording, which may, if desired, be calibrated on site in terms of the Ringelmann scale only if the smoke is neutral grey or black.

Two types of cell are used. The first, known as the barrier-layer or rectifier cell, consists of a metal plate covered with a light-sensitive material, e.g. selenium or cuprous oxide. This form of cell has the property of generating a small current flow, proportional to the intensity of light falling on it. The magnitude of the electrical current is very small, since all the energy to produce it is derived from the light source, but it is sufficient to be measured without amplification, e.g. on a direct deflection-type indicator.

For various reasons it is not convenient to amplify the current from a rectifier cell. Therefore, to obtain a powerful output, the second type of cell, known as the emission cell, is used. This is a true photoelectric cell and consists of a glass envelope under a vacuum or filled with an inert gas at very low pressure. The glass envelope contains two electrodes, the cathode and the anode. The cathode, which is light-sensitive, is usually a plate, and the anode a single rod or ring placed so as not to obstruct the light falling on the cathode.

The photoelectric cell does not generate a current flow, but if it is coupled into an electric circuit fed with a d.c. or rectified a.c. potential, with the cathode connected to the negative potential and the anode to the positive potential, a current flow is obtained in the anode circuit proportional to the intensity of light falling on the cathode. The current flow is minute and, therefore, an electronic amplifier must be connected into the circuit. This gives an electric output very much

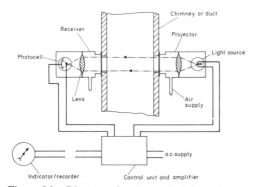

Figure 64 Diagram showing principles of operation of smoke-density meter

References p 480

greater but in direct proportion to the input, which can be measured and indicated or recorded in smoke-density units.

In some designs of equipment of this type, two photoelectric cells are used to give automatic compensation for any deterioration of the lamp which serves as the light source. The second photocell is mounted behind the lamp in the projector, the intensity of illumination falling on it being adjusted by a shutter, so that when there is no smoke the output from the two cells is equal. In this case the instrument measures the ratio of the two outputs, the logarithm of which is proportional to smoke density. Any deterioration of the lamp affects both photocells by the same amount.

To install the equipment, two holes are required on opposite sides of the flue or stack. These would be 5 cm to 10 cm in diameter. If possible, a position should be chosen at a point in the duct where there is suction, so that the surrounding air will be drawn into the duct past the glasses protecting the lenses, and so keep them clean. Alternatively, clean dry air from a suitable source can be continuously blown over the optical parts. These surfaces must be kept quite clean; facilities therefore are usually provided whereby, as an item of maintenance routine, manual cleaning can be carried out.

The ambient temperature at the projector and receiver should not generally exceed 45°C. The measuring element, which can be fitted with an alarm, or can consist of an alarm only, can be mounted at a distance from the projecting and receiving units.

The definition of the terms optical density and obscuration, and recommendations concerning the use of smoke density measuring equipment, are given in BS 2740 [12], 'Simple Smoke Alarms and Alarm Metering Devices' and BS 2811 [13], 'Smoke Density Indicators and Recorders'.

18.2 Automatic control systems

Two fundamental conditions must be taken into account when considering control systems. First, at least two separate systems are involved, the one to be controlled and the one imposing control. Secondly, all the operations and responses occur continuously in real time. To these must be added means for monitoring the control system to check that it is functioning correctly as required.

We shall review the three main prerequisites of any control system and the means whereby they are achieved: stability, i.e. ability to sustain a required operating datum without oscillation or hunting; stiffness, i.e. capability of maintaining a datum under external disturbance forces; and response, i.e. the speed with which a new datum

may be reached and maintained when it receives a new demand. The benefits of digital control techniques will be discussed.

In any control system, the control is based on the regulation of a physical quantity which may be predominantly any one of the physical variables such as temperature, pressure, flow, etc. The limitations associated with the measurement of these variables have been described; it is now relevant to review the remaining elements comprising a control system to form a summary of the functional processes involved.

Any type of controller, manual or automatic, needs three continuous inputs.
1. A signal representing the value of the quantity being controlled (Measured Variable, or MV).
2. A required level of the measured variable (Desired Value, or DV). This is usually in a 'Memory', either that of a human operator, or some index or pointer set to the DV on a scale. This is not always true; in some cases, such as metallurgical heat treatment, a time-dependent programme of temperatures must be followed ('programme control'), or the output of one controller may be the DV of another ('cascade control'), a form of passing on instructions.
3. A power supply. This is usually human in manual systems, unless the human is servo-assisted; pneumatic, electric, or hydraulic in automatic systems. The controller always performs an algebraic subtraction of MV and DV, thus determining the 'error', usually designated by θ:

$$\theta = \pm \, (\text{MV} - \text{DV})$$

This is the amount by which the measured variable has departed from the desired value, together with the direction of the departure. After operating on this mathematically in ways to be described later, the controller applies a signal to the process which it is controlling in one of two distinct ways, known as 'Open-loop' or 'Closed-loop' control.

Open-loop control is also known as 'Feed-forward' control. This technique requires a very precise knowledge, expressible mathematically, of the response of the system to be controlled to disturbances arriving via specified routes, e.g. ambient temperature changes, feedstock quality changes, load (i.e. output of process) changes, etc. Disturbances arriving at the process are measured, since they are uncontrolled variables and cannot be predicted, and after mathematical computation appropriate adjustments are made to controlled variables to compensate for the changes.

The measured and computed information moves forward to the process with the incoming material or energy, hence the name 'feed-forward'. A typical example is to be found in a method of central-heating control practised in North America. The thermostat sensing element is outside the building, detects temperature changes that are due to sun, cloud, wind change, diurnal variations, etc.

and transforms these temperature changes into signals regulating the output of heat from the boiler. Clearly the response of the whole building to external temperature changes must be known, and incorporated in the mathematical transformation of exterior temperature measurements. This example illustrates three properties of open-loop control.

1. The controller is never informed of the results of its actions.
2. Because of this, the system can never go into oscillation.
3. Downstream disturbances cannot be compensated for; for example, if a fire be lit in a room in a building with open-loop temperature control, the room will become too hot.

In a perfect open-loop system, disturbances would be compensated for before affecting the output of the process, and no error would ever appear. Owing to the very great difficulty of finding out and expressing the behaviour of any piece of industrial plant, and the considerable complication of the calculations necessary in the controller, open-loop control is not commonly used. Quite simple mathematical operations suffice in the case of closed-loop systems, and these are dealt with below.

Two-Position. This is known as 'Bang-bang', and, in a common form, as 'On-off'. Control is typified by the behaviour of a simple electric thermostat, or by a safety valve. The desired value is set into the device, and when the measured variable exceeds that value the controller output goes to one extreme (heat turned off, or safety valve opens) and when, as a result of this action (closed loop) the measured variable falls below the desired value, the controller provides the other extreme condition (heat on, or safety valve closes). The controlled variable thus follows a sawtoothed or approximately sinusoidal pattern with time. This type of control is only suited to large capacity long-time-constant systems, the capacity serving to average the oscillation, and the time-constant avoiding overwork of the controller. Compensation for load changes in this system is achieved by varying the ratio of 'on' to 'off' time.

Closed-loop control is also known as 'feedback' control. The elements comprising a typical system are represented by the block diagram *Figure 65*, and constitute a conventional closed-loop control system in which deviations from a preset datum are detected by comparing the output of the transducer with an electrical signal corresponding to the required datum, and using the resulting error signal to effect the necessary controlling action.

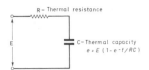

Figure 66   Simulation of thermal time-constant of temperature probe by electrical analogy

Incoming disturbances are allowed to enter the process, and measurement is made of the actual quantity to be controlled. The controller feeds back a signal to some incoming material or energy flow to correct for the disturbance, hence the name 'feedback'. An example is provided by the European type of central-heating control, in which the thermostat sensor is placed inside the building whose temperature is to be controlled. Three features of closed-loop control are apparent from this example.

1. The controller 'sees' the result of its actions.
2. Because of this, it is possible for the system to oscillate if the controller action is too great for a unit error. A finite time must elapse between controller action and the result of the action arriving at the sensor, and the controller may have overcompensated. This condition is known as 'hunting'.
3. Some departure from the desired value must occur to produce an error to initiate controller action.

In any control system relying on negative feedback, the ultimate degree of control of the system relies on sufficient gain from each of the stages and the stability of the loop depends largely on the frequency characteristics of each stage in the loop. The most commonly encountered frequency-dependent variables are often referred to as transfer lags. Transfer lags may be defined as that part of a transmission characteristic, exclusive of distance-velocity lag (propagation lag of control signal), which modifies the time—amplitude relation of a signal and thus delays the full manifestation of its influence (BS 1523 [14], Section 2).

The most useful analogy to apply in assessing the performance of a control loop is the use of electrical a.c. theory applied to resistance-capacity stages. A classic example in temperature measurement occurs in the case of a resistance-thermometer sensing element and the thermal lags introduced by the surrounding filling and picket. These may be lumped together and likened to electrical capacitance and the thermal resistance of the heat transfer film between the process and pocket to electrical resistance, as illustrated schematically in *Figure 66*.

Thus, a step change in temperature may be compared with the equivalent step change in voltage across the terminals of the RC circuit from the well-known equation:

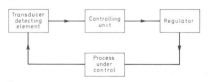

Figure 65   Elementary control loop

$$RC \frac{\mathrm{d}T}{\mathrm{d}t} + T = T_2 - T_1 \tag{5}$$

where R = thermal resistance, C = total thermal capacitance, T_2 = initial temperature of medium, T_1 = final temperature of medium, T = bulb temperature, and t = time. For this equation the solution is:

$$T = T_1 + (T_2 - T_1)(1 - e^{-t/RC}) \qquad (6)$$

If we take T_1 as the datum, we can write $T_1 = 0$ and the equation reduces to $T = T_2 (1 - e^{-t/RC})$. The product RC is the time constant of the system (*Figure 66*); the temperature displacement when $t = RC$ is shown in *Figure 67*.

Referring again to *Figure 65*, four principal forms of control may be applied: proportional, proportional plus integral, proportional plus derivative, proportional plus integral plus derivative. The principles of each and the relevant problems of introducing digital signalling to any one of them will be briefly reviewed.

In a simple control system the potential correction θ is proportional to the deviation of the output from the required value, illustrated in *Figure 68* for a proportional-speed-control system, $\theta = \mu\theta_e$ where μ is the proportionality factor and θ_e is the instantaneous deviation. Such a system achieves equilibrium with a small deviation from the required datum, known as offset, as shown in *Figure 68*. Thus, for any given load condition and assuming that the proportionality factor μ is unchanged, the offset will increase. The offset can only be decreased by increasing μ (the gain of the system), and a limiting value is reached above which the system bursts into oscillation.

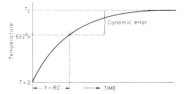

Figure 67   Temperature/time response of temperature probe

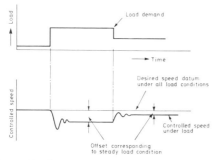

Figure 68   Proportional-speed-control system, response under load

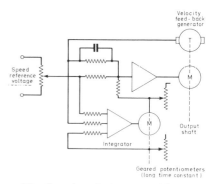

Figure 69   Speed-control system with integral control

In a digital proportional control system, the minimum bit level corresponding to the smallest practical increment of measurement and control determines the backlash in the system. Thus, the system will tend to hunt about the offset datum by an amount corresponding to ±1 bit of measured datum.

The controller may be arranged to be displaced by a fixed amount so as to reduce the offset to zero, and such an adjustment may be obtained by means of integral control. Integral control relies on correcting changes at a rate proportional to the deviation, i.e. $d\theta/dt = -y\theta$, and by integrating $\theta = y\int\theta\,dt$, where y is the integral control factor. A simple practical application of integral control is illustrated in *Figure 69* for a d.c. motor driving a load. Speed changes corresponding to a change of load are detected by comparing the output of a d.c. tachogenerator with a d.c. voltage preset to correspond to the required speed datum. The error voltage is amplified and used to rotate a motor-driven potentiometer to produce a voltage $V = y\int e\,dt$, where e is the error voltage. The voltage V may now be used to adjust the field excitation of the motor so as to increase or decrease speed until the error voltage is reduced to zero.

Integral control based on digital techniques may be compared with an analogue solution involving the build-up of a charge on a capacitive integrator or increasing voltage on a motor-driven potentiometer. Integration may be obtained by counting increments corresponding to error, either in an electronic counter or by the direct integration of pulses in an electro-mechanical stepper motor or similar inching mechanism. The technique may be based on a proportional error signal that may either originate from a d.c. voltage converted to frequency in a voltage-to-frequency converter, or as the difference in two frequencies, one representing the required datum, the other obtained by sensing the datum obtained at the output of the system. An example of the latter is the pulse generator used as a speed sensor in digital tachometry.

If integral and proportional control are combined, the offset problem is cured and, as with simple proportional-control correction, is effected more spontaneously than if integral control only is used. The control equation becomes:

$$\theta = -\mu\theta_e - y\int\theta_e dt$$

(7)

i.e. $d\theta/dt = -\mu(d\theta_e/dt) - y\theta_e$

The constants μ and y both require adjustment in amplitude and both have different phase relations with the deviation. The phase angle between the deviation θ_e and proportional correction $\mu\theta_e$ is π, and the phase angle between the integral correction $y\int\theta_e dt$ and $\mu\theta_e$ is $\pi/2$.

The use of integral control, while eliminating the offset present with simple proportional control, does not assist the short-term stability of the system, i.e. it does not contribute in any way towards minimizing the tendency for the system to hunt about the required datum or burst into divergent oscillation if the proportional control μ is increased. What the system lacks is damping in the form of control proportional to the rate at which the deviation changes, and this is particularly important in systems with large inherent transfer lags. The use of a derivative term enables higher values of proportional control to be applied because the higher the instantaneous value of correcting force the higher the derivative action and hence the damping:

$$\theta = \psi d\theta/dt \text{ where } \psi = \text{derivative control factor.}$$

For a sinusoidal waveform, i.e. $\theta = x \sin wt$ then $\theta = \psi wx \cos wt$, from which it is apparent that the amplitude of the potential correction is directly proportional to the frequency as well as the derivative control factor. Derivative control is also referred to as 'phase advance' or the 'rate' term. The combined equation for proportional plus derivative control is:

$$\theta = -\mu\theta - \psi d\theta/dt$$

(8)

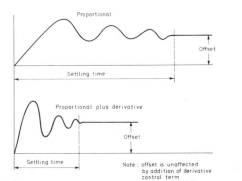

Figure 70   Effect of derivative term on settling time

References p 480

The effect of increasing the potential correction is to increase the frequency of oscillation from the classic equation:

Natural frequency $f = \frac{1}{2}\pi$ (stiffness/inertia)$^{0.5}$ where the stiffness term corresponds to the mechanical gain of the system and depends on the value of μ. The benefit of increased natural frequency is to reduce the stabilization time as shown in *Figure 70.*

A digital signal corresponding to a derivative term, if taken literally, is not practicable because a derivative term involves the differentiation of displacement with respect to time and a digital signal has a finite value which increases or decreases by fixed increments or bits to some new absolute value. However, if a digital signal is obtained as a result of some secondary process such as counting an analogue-derived pulse train, i.e. frequency, then a derivative term may be approximated by measuring the elapsed increment of time between successive pulses, while the position may be obtained by counting the number of pulses for a fixed time interval.

Applying a digital derivative term may prove to be troublesome because, unlike a d.c. analogue voltage that may be summed with a voltage corresponding to displacement, the addition of two digital signals may only be carried out by a successive additive operation unless two pulse trains of random spacing are to be added as in the application of binary rate multipliers[1]. Having regard to the importance of observing the correct time phasing of a derivative term with respect to the displacement term in equation (8) to ensure stability of the control, it is implicit that any form of digital computation to obtain a resultant control demand must be carried out at a speed many times in excess of the natural frequency of the system, so that lags introduced by the computation time are at least an order smaller than the response time of the system.

A combination of proportional, integral and derivative control provides the most comprehensive solution to control problems requiring some desired datum to be sustained in the presence of external disturbing forces, because it combines means for directly correcting an error, and means for achieving rapid restoration of the disturbance in the shortest time consistent with stability. The equation is:

$$\theta = -\mu\theta - y\int\theta dt - \psi d\theta/dt$$

(9)

Note that, although the integral and derivative terms are opposed, the amplitude of the integral potential correction is inversely proportional to frequency whereas the amplitude of the derivative potential correction is directly proportional to frequency. Thus if the integral term is increased excessively the system will tend to hunt at a lower frequency than the theoretical natural frequency of the system.

Very roughly, proportional action is adjusted to give stability to integral action to compensate for load changes, and derivative action to allow for measurement lags.

The application of digital techniques to a 'three-term control system' extends the problems already described. This is especially so in respect of those techniques used for obtaining a derivative term and the means for summing the displacement derivative and integral terms in a time interval that does not materially add to the transfer lags of the process.

The calculation of control demands by any step-by-step process involving digital measurement and subsequent arithmetical operations, if carried out repetitively, requires the addition of filtering to prevent chatter of the system, or a high sampling rate to minimize increments of change. The inclusion of such filtering assumes that the time constant introduced does not damp the system to the extent that successive corrections are impeded, and relies on digital-to-analogue conversion before signalling a demand to the system actuator.

In traditional analogue controllers, the three types of control action are computed pneumatically or electronically from signals from transducers, and from manually set desired values inserted by pressure regulators (pneumatic) or potentiometers (electronic). The various functions may be widely separated. A transducer on one part of an industrial plant may transmit a signal to a controller at another point, and the human operator may insert his desired value from yet another point (perhaps in a control room). The controller output drives a valve or actuator at some other point on the plant.

Since desired values are inserted as pneumatic pressures or electric signals, and the outputs or controllers are the same, one controller may have its output connected to the desired value input or another controller. Either controller may have one, two, or three term functions. Use of this 'cascade' system may eliminate some upstream disturbances arriving at a plant. For example, if a tank of liquid is heated by a steam coil, variations in steam pressure will cause changes in steam flow which will affect the temperature of the liquid.

The simplest control is to provide a temperature transducer in the tank connected to a controller provided with a desired value of temperature ('set point') and operating a valve in the steam main. Such a controller would probably be three-term, compensating for thermometer lag and for load changes. Variations in steam pressure would cause the temperature of the tank contents to change and eventually appear as a change in the output of the temperature controller. However, a 'slave loop' may be installed in which the rate of steam flow is metered and the rate signal is passed to a controller which operates the valve in the steam line. This controller receives its DV from the temperature controller (master controller). In this way disturbances due to steam main pressure variations never reach the tank, and the temperature con-

troller has only to correct for load and ambient changes. Further, a flow transducer in the output of hot liquid from the tank could be used, through a suitably characterized relay, to send a signal to the steam flow controller warning of a load change, and leaving the master temperature controller only to trim the tank temperature. Such examples can be multiplied and lead to the concept of control loops as a rudimentary nervous system, acting as 'instinctive' responses to stimuli. A simpler type, actually an open-loop cascade system, is the flow-ratio control system, very common in combustion applications. The problem is to keep the flow rates of at least two materials in a fixed ratio, as in fuel/air ratio control. Here one component, probably the air flow since being electrically pumped it is liable to sudden failure, is metered; the transducer output is increased or decreased by an adjustable relay and applied as desired by value to a flow control on the fuel line. If the air flow fails the fuel is automatically cut off. If the fuel flow were the master, and air flow the slave, air failure would only result in a temperature drop in the combustion space; and if this temperature were automatically controlled by operating on the fuel flow, a dangerous increase in unburned fuel to the combustion space would result. A refinement of this system uses an oxygen analyser in the flue gases to provide a desired-value signal to the scaling relay, thus adjusting air/fuel ratio.

Automatic Valves

Automatic valves for control of the flow of liquids or gases can be of the solenoid, electric motor, hydraulic or pneumatic operated type. The choice depends largely upon the type of control system used, but the valve bodies are substantially the same.

For the control of hot water flow to a space-heating installation, an on/off butterfly valve is normally used, operated by a small electric motor fixed to the valve body. For the control of high-pressure hot water, particularly for process heating and for space or process heating calorifiers, the valve must be able to withstand the more arduous conditions imposed by high-pressure hot water. The valve may be of either the on/off or modulating type. For the control of steam flow, the valve may be of either the on/off or modulating type, depending upon the installation.

Modulating valves for the control of water or steam are intended to restrict the flow through the pipe so that the load requirement is satisfied in all conditions and the heat input is continuous, being changed, increased or decreased, by a suitable controlling thermostat. The application of modulating valves, for the control of steam or water flow, demands a good deal of thought if the valve is to modulate properly throughout the whole of its operating range, and it is necessary to size the pipe correctly according to the service required.

The range of controls available is considerable, extending as it does to all types of thermostats, thermostatic controllers and automatic valves operated electrically, pneumatically or hydraulically and being able to give on/off control for the simplest installations or modulating control for the more complicated ones.

Valves

The most important of all calculations in the design of a control loop is that of valve size. Any system through which a fluid is passing will have a pressure differential across it specific to each flow rate. Insertion of the control valve implies a concentration of part of the differential pressure across the valve, leaving the remainder available to drive the fluid through the plant. When the control valve is fully opened it can perform no regulation. A generally accepted rule is that not less than 60% of total pressure drop across a plant must be available at the control valve during operation at normal design level. This means:
1. That the pumping-power, or pressure head, available must be increased when automatic control valves are used.
2. That the nominal bore of the control valve will always be less than that of the conduit in which it is installed. Sizing control valves correctly is difficult, and the designer must always use the tables, nomographs, or slide rules provided by the manufacturer of the valves being used. These will produce a 'valve coefficient' C_V, which is defined as

$$C_V = Q(P/\rho)^{0.5}$$

where Q is flow rate (gal/min), ρ is relative density of the fluid, P is the pressure drop across the valve (psi). A C_V value of 1 is, therefore, a valve that will pass 1 gal/min of water at a differential pressure of 1 psi. The C_V number obtained from the maker's data is used to identify the valve size required in the maker's catalogue.

Rangeability and Turn-down. These terms are used in control-valve technology, although the value of the concepts defined is doubtful:

Rangeability = maximum flow/minimum flow at constant pressure drop.

The minimum flow is defined as the 'minimum flow at which design characteristics are maintained'. For example, a double-seated valve with a leakage of 2% in the closed position, but otherwise conforming fully to its specification, would have a rangeability of R = 100%/2% = 50.
    Turn-down is based more on the performance of the valve when in use on a plant:

Turn-down = normal flow (% of full design flow)/minimum flow (% of full design flow)

An example would be a valve designed to work, when the plant is running at its designed capacity at 80% of its maximum, and with a leakage of 2%; its turn-down would be 80/2 = 40.

Control Performance

Since the purpose of a control system is to maintain a measured variable as closely as possible at a desired value, the measure of its success can be the area between the desired value and the measured variable on a plot of these values against time. Offset, as experienced with proportional-only control, would show a permanent error with load change, and a steadily increasing area. Two-position control, producing a saw tooth oscillation about the desired value, might be acceptable if the excursions were not too great, but the true desired value is only achieved intermittently and momentarily.
    Two-term (proportional plus integral) control would produce a better result, but might be slow in returning the variable and here addition of derivative action would be made. These considerations apply to a system which has received a disturbance, and whose controller is working to remove its effects. However, if a controller with proportional action has too great a gain (U) the controller itself may cause oscillation about the desired value. A proportional controller whose gain is too great for the system to which it is applied produces 'hunting', a condition exactly like that produced by a two-position control.
    Let us assume that a sinusoidal signal is being measured on the line leaving the process. On passing through the proportional-only controller its phase is altered by $180°$ (this is the function of a proportional controller) and its amplitude changed from a to a', this signal now being fed to the regulator. If the process is such that it produces a $180°$ phase lag in disturbances passing through it, the signal from the controller will be in phase with the incoming disturbance. If the ratio a'/a is unity the disturbance will continue unchanged. If a'/a is less than unity the disturbance will die out, but this low gain ($U = a'/a$) may produce an unacceptable offset. If U is increased in order to reduce the offset, when it reaches unity the disturbance will be perpetuated, and when it exceeds unity the oscillations will continue with increasing amplitude — a dangerous condition. If integral action is used with proportional action the $180°$ phase change will have a phase angle subtracted from it, dependent in value on the value of y. If derivative action is used a phase advance W, related to ψ, equation (9), is added to the $180°$. Various combinations of U, y and W permit matching of a controller to plants with different phase-changing behaviour, but instability has to be avoided, as well as too long and too great a departure from the desired value. Mostly the controller adjustments are made by experience as follows: proportional only is used, increasing U until the process starts to oscillate, when it is reduced and

integral action is then gradually introduced (y), U probably being altered at the same time. Finally, derivative action W is added and U and y are readjusted.

Formal methods exist for 'setting up' or matching controllers to processes. One, 'Frequency Response Analysis', requires apparatus for injecting sinusoidal perturbations into the system at varying frequencies and calculating settings from the results, plotted in a form known as a Bode Diagram.

It is possible with complicated equipment, and access to a computer, to obtain Bode Diagrams from a single step-change input to a process, so reducing the period, which may be many hours, over which oscillations have to be introduced to the plant in order to obtain the data.

Analogue Control Systems

In conventional control systems, Analogue Computing Units provide simple and reliable control solutions. Pneumatic or electric signals are easily operated on by units which add, subtract, multiply, divide, raise to powers and extract roots. Thus a chemical process requiring x,y,z, parts of three components, the ratios perhaps subject to variation with flow rates, temperatures, pressures, or analysis of products or by-products, may be controlled. The chemical equation for the reaction is established and transducers measuring properties of x, y and z provide signals for scaling, computing, and sending desired-value signals to controllers operating on appropriate variables. This approach is much cheaper, sounder, and safer than the application of a single digital computer to overall control. Failure of any component only affects a small part of the whole process, and nowadays nearly all the pneumatic analogue computers, and some electronic computers, have a facility for plugging-in replacement units without shutting the plant down, although care must be exercised with electronic units in hazardous areas.

Extremum Controllers

The maintenance of desired values for the many variables in a process is the prime object of automatic control. However, the desired values themselves will change with plant conditions, such as corrosion, ageing of catalysts, deposits in vessels etc., and performance is judged by criteria external to the plant. It may be necessary to maximize yield of a product, or to minimize costs, etc. Changing states of the plant or raw materials may require changing operating conditions. If the sequence of undesired change is known, a 'mathematical model' of plant behaviour may be used to adjust the desired values of some or all control loops. This is clearly open-loop overall plant control. The same effect may be achieved in closed-loop fashion by installing a device which can produce small changes (perturbations) in the desired value of selected controllers. If the mean effect is to improve conditions the device continues

its small steps in the same sense. If the measured effect is a decrease in the desired condition, the sense of the steps is reversed. Eventually such a device will bring the plant to an optimum operating state. This type of 'optimizing' controller is called a 'blind hill climber'. It may be modified in several ways to make its attainment of the peak condition more rapid.

Other Types of Control

The desired value setting of the controller may be altered by a function generator, usually a cam driven by a constant-speed motor and producing an output which varies with time. This is called a 'Programme Controller', and may be set to follow sequences such as slowly raising a furnace temperature, holding a fixed value for a set time, changing to another value, or cooling at a set rate.

Another type is called 'Cycle' control. This is a constant speed cam or cams arranged to trip pneumatic or electronic switches at chosen intervals, so initiating or terminating sections of a process; or stopping and later restarting the cam of the programme controller so that long flat parts, that is no change in output, may be avoided. Cycle controllers may also start and stop other cycle controllers, permitting the introduction of sub-routines in the process.

Trip Levels

Simple on—off controllers may be arranged to provide instantaneous action when some value reaches a pre-determined level. Such controllers would operate quick-acting valves and are commonly used to ensure an orderly sequence of shut-down actions on a plant when a dangerous level of a measured variable has been reached. As each valve strokes fully it operates a pneumatic or electric switch which passes a signal to the next valve. Pneumatic forms of such circuits become quite complex, often including 'memory' devices which register transient changes whose effect may be important if other conditions arise later. Such circuits must be irreversible so that the plant remains shut down after an abnormal condition is recognized by the trip-level controller, even though the condition disappears in the subsequent shut-down procedure.

Control Rooms

With modern pneumatic or electric transducers it is usual to place the instruments necessary for human operator supervision in a control room. The instruments are usually mounted on large panels, the layout being such that the more critical indications for recordings are grouped together. It is also common to have a 'mimic diagram' above the panel, on which a simplified diagram of the controlled plant is displayed with insertion of operation signals, such as semaphores for valves, lamps to show pumps running, and alarm lights for off-normal conditions. Such diagrams are of use mainly to personnel unfamiliar with the process. Very often a control desk or console is placed in front of the panel, and an operator at this desk has important measurements, manual over-rides, telephone, etc. in front of him.

Annunciators

These are a valuable addition to a control room and consist of frames with windows which, when illuminated, display warning legends, such as 'Number 4 pump stopped', 'Flame extinguished', or 'Steam temperature low', etc. The illumination is initiated by trip-level devices and is accompanied by an audible alarm, which may be manually cancelled. If this is done the alarm legend light changes from steady illumination to flashing. Supervisors then know that an alarm has occurred, what it is, and that the operator has received it by cancelling the audible alarm.

Data Loggers

A further refinement is the use of a 'data logger'. This is a complex electronic device which scans the plant measurements at a high speed, digitizes the analogue values, converts them to decimal notation, and types out, at chosen intervals, complete plant log sheets on prepared forms, taking carbon copies if required. This apparatus is particularly useful during plant disturbances or breakdowns when the operator is fully engaged and not able to write a manual log. Data logging may be carried a further stage by manually inserting high and low permissible levels for critical variables. If these are exceeded a second typewriter is immediately activated and records the condition without waiting for the main scan, which may be at hourly or longer intervals. The record so produced is known as an 'Off-normal log'. The system may be refined by having sub-routines so that scanning of chosen critical measurements takes place more frequently than others. If a digital computer is used for logging it may be programmed to perform all these functions. However, the plant/computer interface may well be larger and more expensive than the computer itself. This 'interface' is the equipment in which measurements of flow, temperature, pressure, etc. are conditioned to make them acceptable to the computer.

Digital System Applications

Digital-control systems, by definition, depend on translation of primary measurements into a digital code that enables an increment of adjustment to be made; the magnitude of the adjustment is determined by:
1. The resolution or minimum 'bit level' of the primary data, and
2. The corresponding response or step change of the output actuator.

Digital-control systems may range in complexity from the contact-operated motor described above to self-optimizing systems. In the latter, calculations are repeatedly made on the primary physical variables associated with a process, and corrections thereby computed to achieve the criteria defining a pre-programmed 'optimum' result.

Having due regard for the 'analogue' nature of the majority of physical process variables and the exponential nature of the control equations to be satisfied for any control system to stabilize, it is seldom practicable to rely on a system that is literally 100% digital, although it may be expedient for certain elements (particularly those involving calculations or requiring adjustment to some preset datum) to be instrumented digitally. This point is discussed in more detail in the context of digital control with analogue back-up facilities.

It is apposite to review digital-control techniques by first reviewing analogue-control fundamentals to indicate the problems associated with the substitution of digital control terms, and to follow with examples of digital-control systems developed from step-by-step replacement of analogue-control elements by their equivalent digital counterparts. It will be shown that within certain limitations, the datum-holding capability of a typical control system may be improved by one or more orders of accuracy by the addition of digital measuring and control techniques.

The first point to note is that by the very nature of the time-dependent variables in any control problem, digital techniques (i.e. the use of incremental or discontinuous measuring and control processes) can only be effective if the time constant of the system is long compared with the digital sampling time. For this reason the use of digital techniques tends to be confined to the calculation of proportional-control signals or offset or integral terms.

A second point is that because any system involving proportional corrections is basically an analogue system, then digital measurement or control signals have to be converted to analogue form sooner or later and invariably result in a hybrid solution. As discussed previously, the increased resolution obtainable from a digital transducer, compared with an analogue solution, coupled with the benefit of improved repeatability of digital storage for datum set points, enables the accuracy of a conventional analogue control loop to be considerably enhanced by the addition of digital reference systems.

Digital controllers or controller sub-systems are generally based on three main elements that can be identified regardless of the system application. These are:
1. A uni-directional or bi-directional (reversible) counter having as many decades as is necessary to accommodate the span of control required.
2. A time-base reference for gating the counter; for example, in one of the several modes of operation detailed below.

3. A digital storage reference (such as edge switches, diode pinboard, core store, magnetic or punched tape) for simulating a numerical datum to which the counter may be referred. Either method enables the number corresponding to the required datum to be represented in parallel-coded binary form in whatever code is preferred.

Associated with the above three main elements will be the necessary interfaces between signal input and control output, such as a digital-to-analogue (d-a) converter, and system logic such as coincidence detection, gating and data transfer.

The form of counter used will depend on the application, but consideration of a few examples of the alternative ways of using counters might now be appropriate. The simplest form of counter is an asynchronous or 'ripple through' uni-directional counter that will continue to totalize a train of pulses presented at the input until the train of pulses is discontinued or the counter input is clamped in response to some external control signal such as a timing signal. The counter stages may be preset to some required limiting value and coincidence logic added to enable this value to be detected when reached or exceeded. This arrangement is popular for batch counting and presetting a datum, in certain counting and presetting indicating-and-control systems, such as numerical coincidence, totalizing and off-limit detection in alarm supervisory systems.

As an alternative to the use of external preset data that may be adjustable, the counter may simply be required to totalize and clear to zero when a full-scale reading is obtained. One useful feature of the latter mode of operation is in representing a 'suppressed nominal' datum; the counter is programmed to count a signal frequency input against a time base and arranged so that the output of the counter is ignored until after it has filled and recycled to count for the second time; the second count corresponds to the error from the required nominal represented by full scale of the counter.

A number obtained after recycling corresponds to a positive difference signal, and if the counter fails to total, i.e. it does not recycle, the complement is taken to obtain a negative difference signal. The nominal may be represented by 3 decades, i.e. 999, and a count of 1000 units resets the counter to zero. Thus, in a finite gating time of, say, 1 second, a frequency of 1020 Hz will show as a positive error of 20 bits and a frequency of, say, 980 Hz will show as a negative error of 20 bits.

A bi-directional counter is more expensive than a uni-directional counter but may be preferred for certain applications where a continuous or semi-continuous difference signal is required. There are two principal modes of operation that enable a reversible counter to be used to obtain a difference or error signal, or a signal that may be used as the integral of an error. In the first, the counter is arranged to count up in one direction from one input source for a fixed time interval, then count down in the reverse direction from the second input source in an identical time interval, the resulting number being the difference between the two.

In the second mode of operation the counter is used as an integrator. Both inputs are presented simultaneously so that the counter is adding or subtracting, depending on the respective frequencies of the inputs. Where there is a likelihood of the two pulses arriving simultaneously, as indeed they would if the system eventually reached perfect equilibrium with no frequency difference, a special input circuit eliminates pulses that are coincident to prevent the counter from locking up. The design of such a coincident pulse-canceller circuit is critical however, because the tolerance of pulse widths determines the ultimate resolution of the difference count; whereas the former mode of operation, in which the counter is cycled alternately in one direction and then the other, produces an absolute difference count.

When digital control is added to improve the accuracy of the datum position, the simplest form of control (i.e. integral term) may be obtained by totalizing in a reversible counter; but for applications requiring a position reference to be obtained from a frequency source such as a digital tachometer, the controller will require to be related to absolute time and a crystal timebase or similar time reference will be required, depending on the accuracy required. The accuracy of a crystal may be typically a few parts in a million which may be improved by temperature control. The accuracy of an electromechanical time reference such as tuning fork may be one order smaller but adequate for many applications. The electromechanical solution generally tends to be cheaper because the natural frequency of electromechanical devices is relatively low, and less division is required to produce suitably spaced timing signals (e.g. of the order of 0.1 to 1 s for tachometry).

Metering-pump-speed Control System

Examples of hybrid control systems may be obtained by the addition of digital-control techniques to a conventional analogue speed controller. A step-by-step development of a speed-control system embodying digital proportional, plus integral, control terms may be analysed by reference to block schematic diagrams, *Figures 71* and *72*.

In *Figure 69* a conventional analogue speed-control system comprises a difference amplifier, drive motor, tachogenerator, and integrator. The speed reference voltage is obtained from a potentiometer simulating the required datum, and the tachogenerator provides the source of voltage corresponding to the measured speed, enabling a proportional error signal to be obtained. The voltage corresponding to the integral term is obtained from a potentiometer driven in response to the difference between the reference voltage and measured voltage; consequently any offset caused by prolonged loads on the shaft is eventually backed off to zero by an equal and opposite off-set on the integrator potentiometer. To prevent momentary instability of the control loop while responding to major changes or required datum, or load changes, the integral term may be temporarily switched out. Additional damping may be pro-vided by phase advance of the tachogenerator out-put as indicated by the capacitive resistance network in the diagram. In *Figure 71* the speed reference is obtained from a crystal with suitably switched binary scaling dividers to provide the required frequency, and associated with the switches are binary weighted resistors to provide a digital-to-analogue conversion for replacing the speed-reference potentiometer voltage of *Figure 69.*

The speed of the output shaft is measured with a pulse generator, and the resulting frequency, together with the reference frequency applied simultaneously to the up–down inputs of a revers-ible counter, to result in a count corresponding to the speed difference. This is converted in a d-a converter and fed to the input of the difference amplifier as in *Figure 71.*

In the example quoted, the d-a converter may comprise an input register followed by solid-state switches to control currents through binary weigh-ted resistors fed from a constant-voltage source. The output is buffered by means of a suitable amplifier so that loading effects on the resistor do not introduce errors. The input to the d-a converter is in parallel form from a preceding counter, but an alternative to the use of a solid-state counter for d-a conversion is the use of a stepper motor driving a precision potentiometer. Stepping rates of 200 per second are typical and one advantage of such an electromechanical solu-tion for some applications is that the combined motor/potentiometer assembly and attendant fric-tion ensure an inbuilt memory in the event of power failure. This feature is particularly import-ant where a time-shared digital data processor is used to correct more than one output device relying on a d-a conversion.

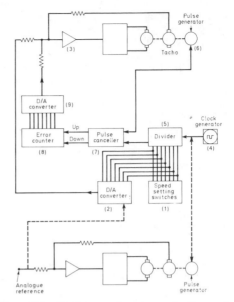

Figure 72   Digital speed and speed-ratio control

If the counters shown in *Figure 71* do not advance faster than the limiting speed at the stepper motor, direct connection between the first stage of the counter and stepper motor is possible, with consequential economy of electronic circuitry.

Figure 72 shows a typical speed control system in which the basic control loop is analogue with a superimposed digital loop to improve accuracy and repeatability. This approach is used because it is a relatively simple matter to stabilize the system and provide good transient response with the analogue loop, whereas to perform the whole operation digitally would require much more complex digital circuitry.

Consider the operation of the system. The analogue control loop performs in the normal way, the reference being derived from the speed setting switches (1) via a d-a converter (2); a separate d-a converter may in fact be dispensed with by connecting the switches via suitable summing resis-tors direct to the error amplifier (3) input. The clock generator (4) provides the stable frequency reference and this frequency is divided by the reference divider (5) as required to correspond with the frequency of pulses generated by the pulse generator (6) at the desired speed. The reference pulse train and the feedback pulse train are fed, via a coincident pulse canceller (7), to the 'add' and 'subtract' inputs of the error counter (8). This counter integrates the difference between the two pulse trains and provides the input to a d-a converter (9). Note that the range of the d-a converter need only be sufficient to correct for the maximum possible error in the analogue loop.

Suppose that the drive is running slightly slower than the set speed. Fewer pulses will be fed back

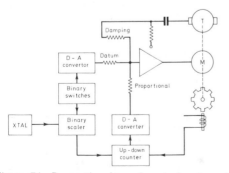

Figure 71   Proportional-speed-control system with crystal frequency reference

from the pulse generator than will be fed from the reference divider, and the error counter will count up. The output from the d-a converter will then move in such a direction as to speed up the drive. Note that the error counter is never reset, so that, in the steady state with the two pulse trains exactly equal, the output from the error d-a converter is steady and exactly offsets the error in the analogue system.

If the required system accuracy is ±0.01% of full speed than a change of one bit in the error counter must result in an analogue correction signal not greater than 0.005% of full speed. If we assume further that the error from all sources in the analogue loop is not greater than ±2% then the error counter will require to have a range of at least 800 bits. The nearest binary number would be 1024, which requires a ten-stage binary counter. Note that irrespective of any other system requirements, the counter may count in pure binary code. The analogue loop will be set up at the nominal centre of the tolerance band with the error counter at half full scale (512 bits). The digital loop can then correct up to ±512 bits or ±2.56% full speed.

A system of this type is referred to as a continuous digital-control loop, and provided that the drive is running at nominally constant speed the loop should be stable. Consider however what happens when there is a sudden increase in load and therefore a transient drop in speed. The error counter will integrate the drop in speed and the drive must run overspeed for a period of time in order to return the counter to the steady state. Thus, any transient speed change in one direction must be compensated by a transient in the other direction. We can see, therefore, that instability may occur.

Now it may well be that the accurate speed control is only required when the drive is operating at constant speed under fixed load conditions, where it is only necessary to correct for slow drifts in the analogue loop. Two devices may be employed here to avoid the problem. First, the digital loop may be switched out when switching on the drive or when making deliberate speed changes. Secondly, the digital loop may be arranged to operate intermittently, only correcting

the error every few seconds. Because the drift rates in the analogue loop are slow, a slow response digital loop will be no disadvantage. Faster digital systems can of course be designed, but the circuitry and stabilizing techniques required are outside the present scope.

In process-line speed control it may well be important to control accurately the speed of one drive relative to another, irrespective of whether or not the speed of the latter is accurately controlled. This is known as speed-ratio control and is of particular importance on drives for continuous tube mills and the like. A simple modification to the scheme described above will provide this control.

One drive is chosen as the master drive and a pulse generator linked to this drive provides the master frequency reference for the second drive. Referring to *Figure 72*, it can be seen that the reference frequency divider can now be used to set the required speed ratio. Note that the analogue reference must also be kept in step with the master reference.

Flow Control

Figure 73 shows a simple flow controller. The flow transducer (1) generates a pulse train the frequency of which is proportional to speed. A second pulse train is generated by the reference oscillator (2) and the error counter (3) stores the difference between the numbers of pulses from these two sources. An analogue of this difference is generated (4) which controls the position of the valve (5) and thus maintains the flow at the desired rate. The operation is very similar to that of the speed-control loop described earlier.

A blending system in, for example, an oil refinery may require several such control loops operating together. *Figure 74* shows in outline a control scheme for such a system. All the control loops are driven from one master oscillator (1) but each has its own reference divider (2) to produce a pulse rate corresponding to the percentage required of the constituent in the final blend which can, therefore, be, closely controlled. The frequency of the master oscillator controls the flow rate of the final product.

The error counters in the individual loops can be used to advantage in the event of some malfunctions in the system. Suppose, for example, that the flow of one constituent is restricted by a partial blockage causing the corresponding error counter to start to fill. If the counter exceeds a present level, action may be taken to reduce the frequency of the master oscillator until the error counter reverses, thus reducing the total flow demand to a level at which the restricted constituent flow is sufficient to maintain the desired percentage. If, in spite of this action, the error counter continues to fill, indicating a complete blockage, the plant can be arranged to shut down.

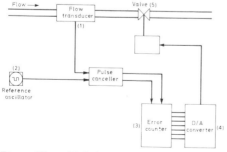

Figure 73   Digital flow control

Note that the error counter stores the total shortfall in the constituent concerned so that when the plant is restarted this shortfall is automatically made up and the accuracy of the blend is not affected.

A similar approach may be used to control the blending of solids. *Figure 75* shows a system for blending a percentage of high-quality product with a run of uncontrolled product to maintain a predetermined final quality. Such a system might be used, for example, to control the quality of coal from a mine by adding cleaned coal to run-of-mine coal in order to maintain a predetermined ash content. A pulse train proportional in frequency to tons per hour of uncontrolled product acts as the reference pulse train (after passing through a controlled divider) for the high-quality-feeder conveyor. The actual division ratio is determined by sampling the final product by an on-line quality monitor.

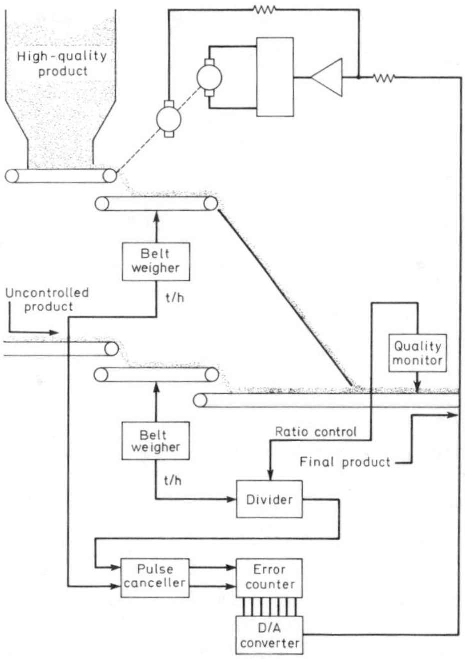

Interfacing Pneumatic Controllers with Digital Computers

Instrumentation for connecting pneumatic control systems to process control computers varies to suit the input/output concepts employed by these computers. Since computers are general-purpose in nature, a variety of input/output signals and schemes exists. There is no guarantee that the interface instrumentation in one computer control system is completely compatible with another system — even when a computer of the same manufacture is used. In recent years, however,

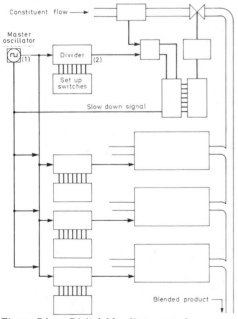

Figure 74   Digital blending control

Figure 75   Solids blending control

popular methods have emerged to allow a meaningful presentation of hardware which comprises the computer-instrument interface.

Basically, the hardware consists of pressure-to-electric transducers for analogue inputs and motorized devices which accept digital outputs. The computer accepts a voltage or current input and provides a pulse-train or pulse-duration output. Variations in computer input/output schemes dictate minor, but important, differences in the connected hardware.

The amount and location of hardware which makes up the instrument—computer link in the overall control loop is fairly well established. Standardized hardware exists to satisfy most requirements established by familiar supervisory, DDC and back-up control concepts. *Figures 76—78* represent conventional techniques. Brief mention is given to variations which, understandably, must take into account peculiar conditions — such as existing control techniques and space limitations.

Supervisory Loop (Figure 76)

A single control station provides three control modes — computer automatic, local automatic, and manual. Both of the automatic modes provide for setpoint changes to a pneumatic controller. The operator manipulates either the setpoint (local) or the valve (manual). Variations made to the basic supervisory scheme include stations without either the manual or local automatic back-up mode, and stations with either a computer or external setpoint with no local setpoint.

All supervisory loops require the computer to receive the process variable. In most instances, the computer requires the setpoint value also. The need for this latter signal is determined chiefly by the form that the control programme takes. Simply stated, setpoint feedback eases programming and increases reliability. Some programmes require the setpoint feedback to match the slope of the process input signal, even though the full-scale values are not identical.

DDC Loop (Figure 77)

The control station has three control modes — DDC, local automatic, and manual. In DDC the station makes a direct computer-to-pneumatic valve loading conversion. Synchronized local automatic and manual back-up modes allow operator setpoint or valve-loading options, respectively. Simpler versions of the same station permit operator valve-loading back-up only.

Once again the computer requires the process variable. Feedback of the valve signal depends on the control programme. Since most DDC programmes duplicate proportional-plus-reset analogue

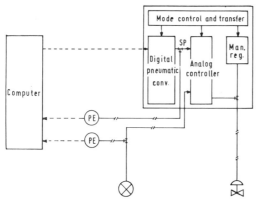

Figure 76   Supervisory Control Loop

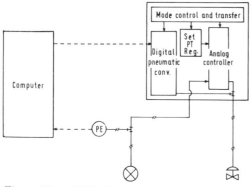

Figure 77   DDC (Direct Digital Control)

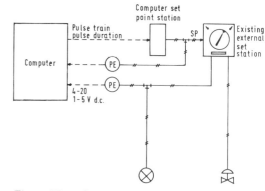

Figure 78   Supervisory Converted Loop

controllers, valve-pressure feedback is not usually required.

Supervisory Converted Loop (Figure 78)

Existing loops can be converted to supervisory computer control by the inclusion of a setpoint station and external-set capability in the existing control station. After the conversion, the loop resembles that of *Figure 76* except that the local-automatic mode may be obtained from either station.

It often occurs that panelboard space for a setpoint station does not exist. A back-of-panel 'black' box overcomes this problem. The box contains a digital-to-pneumatic converter to derive the computer-controller setpoint signal. This solution is not practicable unless a means of aligning the computer and local setpoint is available to prevent an upset during control mode transfer. Automatic tracking techniques can be used to permit fool-proof stepless switching.

In either of the above schemes the typical computer control programme requires the process and setpoint feedback to update the setpoint to a calculated target.

In conclusion, every controller in every loop is an analogue computer, usually solving a differential equation in real time. A plant fully equipped with controllers such as those described above may be said to be 'Computer Controlled'. However, the term is usually applied to the use of an electronic digital computer, and this is arranged in one of two ways.

The safer way is to use conventional control loops and let the computer scan them sequentially, adjusting desired values as necessary to achieve some chosen objective such as rate of output, preferential use of a raw material, achievement of a desired (and measured) quality of product or to optimize some part of the process etc. If the computer fails, the analogue loops continue to function and the plant is safe. The other method, called 'Direct Digital Control' (DDC) uses the computer as a multi-input, multi-output controller, sequentially interrogating measuring instruments, comparing the values with stored desired values specific to each loop, and sending control signals directly to control valves. Failure of the computer is more serious in this type of application as all controlled variables immediately become uncontrolled. However, use of a computer in one or other of these ways permits use of an optimizing programme.

18.3 Special measurements used in fuel technology

Temperature

Gas Temperature

It is possible to obtain the temperature of a gas by measuring some temperature-dependent property (e.g. the density of the gas or the velocity of sound in the gas). The best known commercially available instrument uses gas density: the density of the hot gas entering a water-cooled probe is measured by a venturi restriction and again later by a second venturi in series with the first, at a much cooler known (measured) temperature. From these three measurements the hot-gas temperature can be computed automatically. A suitable probe is shown in *Figure 79*. This instrument has been found particularly useful in dust-laden gases[15]; since no part of the probe has to reach the hot-gas temperature, the maximum that can be measured in industrial applications will depend only on the efficiency of the water cooling.

Metal Temperature

The accurate determination of temperatures in metal components of plant is often complicated by large variations in temperature of fluids local to the site of immediate interest. The temperature distribution, with respect to position and time, in heat transfer elements is of obvious relevance to investigations of thermal efficiency, corrosion or gas-side fouling; the temperature of any part of the metal will depend on the temperatures and velocities of fluids in contact with it, the radiant heat flux and its own thermal conductivity.

It is sometimes possible to correlate the values given by thermocouples in the position of interest with those for which the errors are small or are known accurately, for example in 'dead-spaces' of boilers where there is negligible heat transfer. Often the sites of interest are at abnormally high temperatures or are subject to corrosive conditions, when the life of thermocouples is likely to be short. In these circumstances such correlations should be made quickly, over as wide as possible a range of operating conditions likely to be relevant later.

In measuring metal temperatures, often by using attached or inserted thermocouples, considerable errors can be introduced by heat flow along the wires and the more bulky protective sheaths. These must be minimized.

In determining the surface temperatures, thermocouple wires and sheaths act as cooling or heating fins, especially if large excrescent pads are used to protect the thermo-junction. When small, adequate corrections for such errors can be applied; and they can be minimized[16] by making the thermocouple and its sheath integral with the tube wall near the junction, and separating the lead so far from the junction that its fin effect becomes insignificant. Such mountings will indicate the surface temperature with an error of less than $10°C$ almost independent of the heat flux, whereas metal blocks several millimetres thick around a thermocouple junction can often result in indicated temperatures high by $20-70°C$ according to the local heat flux.

Mid-wall temperatures may be calculated, or a thermocouple may be mounted in a special manner[16], particularly if in a position of high radiant-heat flux.

For high-temperature ranges, say above $500°C$, chromel—alumel thermocouples are used for reliability, with sheaths of heat-resisting alloy such as $25:20$ Cr:Ni steel.

Total Heat Flow

A number of instruments have been developed to measure the total (convection plus radiation) heat absorbed by a receiving surface. These instruments, if they are to give accurate readings[17], should have the same surface temperature, surface absorptivity and geometrical shape as the heat sink under investigation. Heat flow meters may be fixed to the heat sink or they may be portable, for insertion into the furnace for short periods of time. In the former case they are likely to be simple robust instruments capable of withstanding stringent conditions.

One heat flux meter available commercially is that developed by the Central Electricity Research Laboratories of the CEGB (UK). This meter[18] (*Figure 80*) consists of a radial disc supported in a relatively large body which is firmly attached to the heat sink. The finite thermal resistance of the disc gives rise to a temperature difference (ΔT) between its centre and periphery. Thermocouple junctions are formed by wires attached to the radial disc and to the body respectively. It

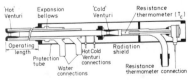

Figure 79　Section of Land venturi pneumatic pyrometer. Type VPP15

follows that the body should be made from the
same material as the connecting wires, which for
convenience are commercially available sheathed
thermocouple wires. With certain assumptions the
heat flux q is given by:

$$q = K\Delta T\theta /r^2$$

where K is a constant for the instrument, θ is the
disc thickness and r its radius. In practice it is
usual to calibrate the equipment. A similar
device[19] uses the temperature gradient in a metal
pad welded to two adjacent cooled tubes (*Figure
81*). In this case the disc radius is replaced in the
equation by the distance between the inner and
outer thermocouples. In places where it is not
possible to weld or braze a heat flow meter to the
heat sink, a strap that can be clamped to the tubes
has been developed[20]. Also for situations where a
local cooled tube is not available, conductivity
plug-type heat flow meters and circulating-water
heat flow meters have been developed[15]. The
principles and operation of the first meter are the
same as those described earlier except that the
heat flows along the conducting metal rather than
radially (*Figure 82*). Circulating-water heat flow
meters are well known simple devices in which the

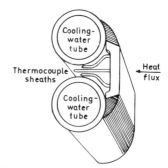

*Figure 81   Isometric section of a heat-flux pad.
Courtesy J. scient. Instrum. (D. Anson and A. M.
Godridge, 1967, Vol.44, 54). Institute of Physics
copyright*

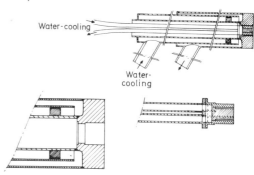

*Figure 82   Conductivity plug- type heat-flow meter.
From* Industrial Flames, *Eds. Beer and Thring, Vol.
1 Measurements in Flames by J. Chedaille and Y.
Braud (Edward Arnold, London), 1972*

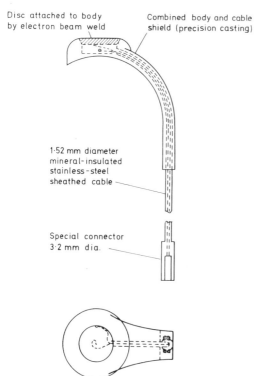

*Figure 80   CERL design of boiler-tube heat flux
meter.   Courtesy Instn of Mechanical Engrs,
London (J. A. Hitchcock and E. W. Northover
Proc. Instn mech Engrs 1966, Vol.180, Pt 3J)*

temperature rise is measured in a known flow of
water. These have the disadvantage that the receiving surface of the instrument is slightly warmer
than the cooling water but usually very much
colder than the surfaces likely to be under investigation.

Combustion Control

For both the efficient combustion of fossil fuels
and the protection of the environment, flue-gas
analysis is of major importance. In the past the
measure of efficiency has been first in terms of
carbon dioxide concentration and subsequently in
terms of oxygen. With lower limits of detection
using modern instruments it is now possible to
measure uncombusted species also, and these can
be as significant as oxygen both as an indicator of
efficiency and in environmental protection. While
percentage oxygen can be a reliable and fundamental guide to the percentage of excess air for
combustion, the gases sampled may contain varying amounts of 'tramp air' caused by casing leaks
and this can result in incorrect air/fuel ratios at

the burner if oxygen analysis of flue gas only is used.   Lower-than-design air/fuel ratios at the burner will often cause emission of undesirable species such as black smoke or particulate matter; this effect is particularly important in the more sophisticated plant designed for low-excess-air combustion. *Figure 83* shows how the various species will change with respect to oxygen; too little oxygen will result in both loss of energy through uncombusted fuel and atmospheric pollution, and too much oxygen will result in excessive sensible heat loss in the effluent gas, additional fan power, and atmospheric pollution through some increase in nitrogen oxides and significant increase in sulphur trioxide. Thus there is an optimum oxygen range (both with respect to energy and pollution) in which steam-raising plant ought to be controlled and which will need to be individually determined for each unit depending on its type and condition; the lower end of this is marked by the onset of carbon monoxide emission.

The accuracy of any system depends on the normal gas sampling principles (cf. App.D1), with special emphasis for flue gases on the following:

The gas should reach the analysis instrument unchanged in composition, which may require rapid quenching, a minimum time delay, and selection of sampling probes and lines to prevent catalysis or permeation.

The sample must be clean and free from condensible moisture.

For on-line monitoring or control, the composite sample should give a true estimate of the average composition.

Both the environmental temperature of and the electrical voltage supplied to the analysing instruments may require control within narrow limits.

For combustion control in a boiler plant the sample of flue gas should be taken preferably before the economizer or between that and the air heater. For carbon monoxide monitoring or control, the sample is sometimes taken at the induced-draught fan outlet where the gases are thoroughly mixed. For overall heat-balance calculations the sample should be taken at the point where the final gas temperature is measured.

The more usual form of sampling probe is of uncooled metal, with sometimes a filter of refractory material on the inner end to reduce blockage. A very strict filtering system is necessary between probe and instrument to protect the instrument; much apparent unreliability arises from failure to provide adequately this protection. It is common practice to cool the gases close to $0°C$, remove the condensate, and then raise them to ambient temperature (now well above the dewpoint) at which the instruments are normally operated.   Electrically operated pumps are to be preferred to water operated aspirators and where low-excess-air operation is practised essential.   Oil is attacked by flue gas and for continuous on-line operation a metal or rubber diaphragm-type pump is strongly recommended. Where particular gases require special probes this will be indicated in the relevant paragraph.

Methods can be based on a variety of techniques such as chemical reaction or adsorption, chromatography, paramagnetism or infrared absorption. Gas chromatography is the most versatile and selective method available today and it can be used both as a laboratory tool and as a process instrument. It suffers from being discontinuous, and for combustion control the analysis time is too great. For investigation of boiler atmospheres including gases not normally monitored, the gas chromatograph, with its wide selectivity, can be purchased as a laboratory instrument for a fraction of the capital cost of all the specific instruments for individual gases. Principles and details of techniques and equipment can be found in standard works[21]. Concentrations less than 100 ppm v/v can be measured on relatively cheap equipment.

Oxygen

Oxygen is generally measured by using the paramagnetic quality of this gas. Most gases are diamagnetic (i.e. repelled from the stronger parts of a magnetic field) and the only other gases likely to be found in flue gases that are paramagnetic are nitric oxide and nitrogen dioxide.

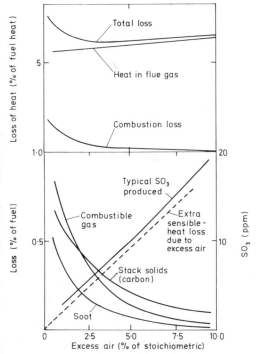

Figure 83   Typical variation of combustion losses and SO_3 concentration with excess air

References p 480

These are considerably less paramagnetic than oxygen and are present in considerably lower concentrations so that their effect can generally be discounted. The influence of the magnetic field on a gas is equal to the ratio of intensity of magnetic field to intensity of magnetism induced in the gas. This effect, called volume susceptibility, can also be shown to be proportional to $1/T^2$ where T is absolute temperature and so is of particular importance to the thermal magnetic method.

This method uses heat to destroy the induced paramagnetism and is often known as the magnetic wind principle. Other types of paramagnetic instrumentation use either direct measurement of susceptibility or are based on pressure susceptibility (Quincke effect). Electrochemical instruments are also beginning to be used to measure oxygen; for example, the zirconia cell for high-temperature measurement[22,23] is sometimes favoured because it eliminates sampling problems and reduces maintenance. Even more modern electrochemical transducers are totally self-contained faradaic devices capable of operating at both stack and ambient levels for a number of gases including oxygen, carbon monoxide, sulphur dioxide, hydrogen sulphide and nitrogen oxide[24].

Magnetic-wind Type. Sample gas is passed through a cell containing a heated element in a strongly magnetic field. The magnetic field causes oxygen in the gas to be strongly attracted into it. As the oxygen in the gas passes through this cell heat is applied to it and the paramagnetism of the oxygen is progressively destroyed, causing cool incoming

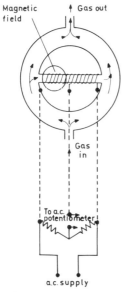

Figure 84   Measuring cell of magnetic-wind-type oxygen meter

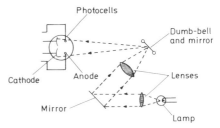

Figure 85   Principle of operation of force-type paramagnetic oxygen meter

gas which is more strongly paramagnetic to displace it. This flow of gas, whose velocity is conditioned by oxygen concentration, passes over a temperature-sensitive element, cooling it and altering its resistance. *Figure 84* shows a ring chamber layout in which the heat to the gas is provided through one arm of a Wheatstone bridge surrounded by a strong magnetic field; the other arm forms a temperature-sensitive element. When oxygen is absent there will be no flow across the ring from left to right. Other systems use thermal-conductivity cells whose filaments are connected to a Wheatstone bridge in parallel in which one arm is a cell influenced by a strong magnetic field (detecting) and one is a reference cell. In the absence of oxygen each cell is cooled to the same extent by gas diffusing into the cell. Presence of oxygen causes increased diffusion into the detecting cell and the detector element is cooled, thus unbalancing the bridge. This system is not suited to gases containing significant quantities of hydrogen. The temperature of the analyser can be controlled by a thermostat or an electric resistance thermometer incorporated in the electrical circuit.

Force Type. This type is sometimes known as the dumb-bell because the measuring cell contains the detecting element in the form of a dumb-bell with glass spheres at either end filled with the non-paramagnetic nitrogen. This dumb-bell is suspended in a homogeneous magnetic field by a vertical torsion fibre through the centre. Sample gas is passed through this magnetic field and over the dumb-bell; oxygen in the gas made paramagnetic by the field will displace the dumb-bell from the strongest part of the field. The deflection of the detecting element is measured by deflecting a beam of light on to two photoelectric cells (*Figure 85*). As the detecting element deflects, the ratio of light intensity falling on the two cells alters the output from these cells. This out-of-balance potential is amplified in terms of percentage oxygen. This method, directly proportional to susceptibility and therefore to oxygen percentage, gives a rectilinear relation and is virtually unaffected by gas composition changing in other constituents. Only at levels of 0.1% oxygen do cross-sensitivities significantly interfere (for calibration with nitrogen and air).

Pressure Type. This system, shown in *Figure 86*, requires a flow of inert reference gas such as nitrogen. The reference gas is split into two parts; most of this, together with sample gas, leaves by a common outlet. At the exit of one reference limb into the common discharge is an inhomogeneous magnetic field. Where the sample contains no paramagnetic gases, the flow in each reference limb will be the same; but the presence of such gases will create a pressure at the outlet of the right-hand limb, forcing a proportion of reference gas to flow over the sensing element. This element is part of a Wheatstone bridge, which will be unbalanced by the differential cooling and consequent change in resistance. The output from the Wheatstone bridge reflects the difference in oxygen concentration between sample and reference gases and this method, being inherently very sensitive, can therefore be used to compare accurately gases of similar composition for a small difference in oxygen content. For more conventional applications, it has the disadvantage of requiring a reference gas of reproducible composition.

Carbon Oxides and Hydrocarbons

The traditional methods have used chemical absorption and/or catalytic combustion, or thermal conductivity; while these methods are still available they have been largely superseded by infrared analysis which is highly selective and has a higher degree of sensitivity. Typical of this equipment is the 'Mono' which can be used to measure carbon dioxide, carbon monoxide and/or carbon monoxide plus hydrogen. The volumetric 'Mono' instrument is fully described in earlier publications[25].

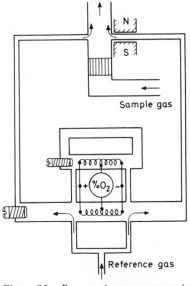

Figure 86   Pressure-type oxygen meter

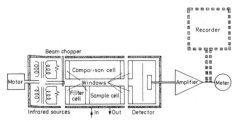

Figure 87   Infrared dual-beam non-dispersive analyser

Infrared. Heteroatomic gases which selectively absorb infrared radiation of a particular bandwidth have enabled a wide and specific range of on-line instrument technology to be developed. Use can be made of this technology to measure the percentage of uncombusted gases, principally carbon monoxide, but also methane and other hydrocarbons if desired, at their specific absorption wavelengths, or to measure carbon dioxide and water vapour, themselves the products of combustion. Energy from an infrared source is absorbed in the gas to an extent proportional to concentration and path length. The cell then is usually chosen to give a reasonable signal in the rectilinear-response section of the curve. The infrared energy absorption in the region $1-23\ \mu m$ is caused by vibrational and rotational energy changes in the heteroatomic gas molecules and gives rise to a band spectrum specific to the individual molecule absorbing. A typical double-beam non-dispersive infrared analyser is shown in *Figure 87*. An infrared source, usually an electrical heating element, emits infrared radiation which is reflected alternatively through two cells (reference and sample) into the detector. This radiation is controlled mechanically by a rotating arm (called the chopper) driven by a low-speed electric motor. The reference cell normally contains a sealed volume of nitrogen that is unresponsive to infrared radiation, providing a reference under similar conditions to those of the sample, and the sample cell is continually flushed with sample gas. All the infrared radiation emitted by the source passes through the reference cell to the detector but only the portion of the infrared that is not absorbed by the sample gas passes through the sample cell to the detector. This detector is filled with the gas being detected and has a flexible diaphragm which is part of a condenser and will move as the gas heats and expands. The capacity of the condenser will change as the gas pressure alters; when none of the particular gas being measured is present in the sample this capacity will be constant, but when it is present the capacity will vary in synchronous fashion with the movement of the chopper. An a.c. signal whose frequency is equal to that of the chopper and whose magnitude is proportional to the amount of gas in the sample will result.

Absorption by some molecules overlaps absorption by others, e.g. the carbon dioxide and carbon monoxide bands overlap. Changes such as varying carbon dioxide concentrations would have on a carbon monoxide detector are eliminated by adding to the sample beam a filter cell which is filled with the interfering gas. Thus absorption effects by carbon dioxide are eliminated before the infrared radiation reaches the sample and detector. The cells and detector require windows that allow only infrared radiation to pass: these are usually of calcium fluoride set in araldite. Minimum ranges (full-scale deflection) for up-to-date instruments are usually 0—100 ppm v/v carbon monoxide and 0—50 ppm v/v carbon dioxide, with response times varying from 1 to 10 s. Minimum range full-scale deflection for less sensitive instruments generally is 0—1000 ppm for carbon monoxide and 0—500 ppm for carbon dioxide. Higher ranges of up to 100% are equally feasible. Newer developments include solid-state electronics and optical filters.

Thermal Conductivity. This method is based on the fact that the thermal conductivity of a gas is dependent on its composition, the component to be measured having an appreciably different thermal conductivity from its carrying medium. It is therefore suitable for carbon dioxide (0.61 relative to air at $0°C$; air at $100°C$ is 0.70 on this scale), the concentration of which in combustion systems, depending upon the fuel, is likely to have values from 10% by volume upwards. Water vapour, which has a relative thermal conductivity of 0.78 at $100°C$ and is present in substantial quantities in flue gas, can therefore influence the reading. These variations must be eliminated by either drying the gas or saturating it to a constant humidity before analysis. Hydrogen too can cause an appreciable effect and where substantial (>1%) and varying quantities of this gas are present alternative means of analysis must be used. The relative thermal conductivity for hydrogen at $0°C$ is 10.15, approximately 10-fold greater than carbon dioxide. With good combustion, however, we would not expect more than 30 ppm v/v of hydrogen; while for combustion this effect has little relevance, in blast furnace exit gases, for example, its effect is significant. The method of measurement (which incidentally is probably the oldest instrumental technique and was first used in Germany in the last century) is as follows. The gas is drawn through a measuring cell containing a platinum resistance wire heated by a stable electric a.c. supply. The temperature of this wire will depend on the rate at which it loses heat to the surrounding gas, and since the resistance of the wire depends on its temperature the resistance will be directly related to the thermal conductivity of the gas flowing in the cell and therefore proportional to the percentage of the carbon dioxide it contains. This platinum resistance wire forms one arm of a Wheatstone bridge, the other arm being a reference cell filled with the reference gas, usually

air or nitrogen, that has similar thermal-conductivity properties to the remainder of the sample gas. Out-of-balance effects on the bridge caused by the presence of carbon dioxide are registered as percentage carbon dioxide by a direct-deflection type electrical meter or similar instrument.

Catalytic combustion can be used to measure carbon monoxide and hydrogen, and the products of combustion, carbon dioxide and water, can be measured by thermal conductivity as described above. Typical full-scale ranges are 0—3% CO and 0—0.5% hydrogen.

Nitrogen Oxides

A variety of methods have been used to measure these gases, a number of which were reviewed by Allen[26]; the factors which control their formation were discussed by Shaw[27]. The oxides of interest in combustion technology are nitric oxide (NO) and nitrogen dioxide (NO_2); both have been measured by continuous on-line instruments, the former using non-dispersive infrared and the latter by non-dispersive ultraviolet radiation (Halstead et al[28]). Interference by water vapour and the small concentration of NO_2 usually found in gases are disadvantages of these methods, and have given rise to a general preference for chemiluminescence. This is both specific, relatively free from interference, and proportional to the concentration of the reacting species. Nitric oxide is reacted with ozone at reduced pressure: use is made of the second and third reactions (Clyne et al[29], Clough and Thrush[30,31]).

$$NO + O_3 \rightarrow NO_2 + O_2 \tag{1}$$

$$NO + O_3 \rightarrow NO_2^* + O_2 \tag{2}$$

$$NO_2^* \rightarrow NO_2 + h\nu \tag{3}$$

$$NO_2^* + M \rightarrow NO_2 + M \tag{4}$$

M is any gas species present that can deactivate the excited NO_2 molecule, and NO_2^* represents the electronically excited molecule. Most of the reaction proceeds according to equation (1), and only a small proportion, about 7% at ambient temperature, proceeds by equation (2). Some of the excited NO_2^* will be destroyed by reaction (4), the amount varying according to the pressure. The overall concentration of NO_x^* found is governed by the relation:

$$[NO_2^*] = \frac{k_2[NO][O_3]}{(k_3 + k_4[M])}$$

where the ks are rate constants relative to the above reactions. The light emitted by reaction (3), which is in the region 0.6—3 μm, is measured by a photomultiplier and the signal is electronically converted and amplified to read concentration of

NO in ppm by volume. Nitrogen dioxide, which is destroyed by heat, can also be measured by passing the sample gas through a converter furnace whose surface is specially conditioned at temperatures up to 650°C, depending on the instrument, to convert it to the oxide. By use of solenoid valves the gas sample flow can be made either to by-pass the converter furnace (and the instrument to read NO) or to pass through the furnace and read NO + NO$_2$, commonly referred to as NO$_x$. The interference effect of other gases present (for combustion applications these are CO$_2$ and H$_2$O) owing to quenching of the reaction (reaction 4) can be considerably reduced by good design, and some instruments show less than 2% quenching at 15% CO$_2$ in the sample gas cooled to ambient temperature. Care must be taken in using some converter systems with reducing gases and in such cases some modifications may be necessary to the gas sample, such as adding a measured quantity of oxygen to the sample to prevent deactivation of the converter surface. A typical flow diagram of the reactants, and the principal components of a chemiluminescent analyser, are shown in *Figure 88*. Time of response (90%) can vary from 1 to 10 s, depending on the make and duty required, and full-scale deflection can range from 0.01 to 10 000 ppm v/v according to the instrument rating.

Sampling of gases at elevated temperatures through a metal probe containing an excess quantity of reducing species can lead to loss of NO and NO$_2$ (Halstead *et al*), and in addition metal probes at high temperatures will also tend to act as converters thus changing the relative concentrations of NO and NO$_2$. Allen *et al* suggest that silica may be used uncooled up to 750°C without affecting the sampled gas, but stainless steel may only be used up to this temperature provided that the sampled gases contain excess oxygen and the NO/NO$_2$ ratio is not being accurately determined.

Gas-borne Particles in Flue Gases

Gas-borne particles in flue gases consist of three basic materials which vary according to the nature of the fuel being burned and the kind of appliance. These are fly-ash (derived from minerals in the fuel), cokes or chars, dusts or grits (incompletely combusted fuel particles) and carbon or soot particles (formed by gas-phase reaction between air and fuel). Soot particles are generally smaller than 1 μm in diameter while fly-ash and coke particles are usually from about 5 μm upwards in diameter. Fly-ash particles often fuse or sinter into agglomerates which pass into the ash pit. The larger coke particles, having terminal velocities of some 0.3 m/s, tend to deposit in horizontal flues, or if emitted from the chimney fall in the nearby vicinity; and the smaller coke particles, having terminal velocities between 0.3 m/s and 70 μm/s, if emitted, can be deposited over a wide area around the chimney depending on the wind direction and particle size and density.

Coke Particles, Sampling and Measurement

Higher dust values such as those associated with coal-fired plant (0.2 g/m^3 and above) are generally measured by optical dust monitors and several designs are available. The standard Radio-visor consists of a lamphouse mounted on one side of the duct and the photocell mounted on the other, the reduction in light transmitted across the duct being a function of the dust burden. The CERL dust monitor uses centrifugal action to separate the coarser particles which are deposited on a glass window; a lamp and photocell are used to measure the optical obscuration on the glass. The window is mechanically cleaned at set intervals. The SEROP, somewhat similar to the Radio-visor, uses a common optical bench in the form of a stainless-steel tube with two large slots cut in its length to permit the flue gases to flow between lamp and photocell. The Safety in Mines Research Establishment (UK) has recently produced an instrument for use in mines using a light-scattering technique which may also have applications in flue gases, but field experience is limited. Of the above, only the CERL instrument is known to have been used to measure solids in flue gases from oil firing (usually significantly below 0.2 g/m^3).

All these methods are relative and require calibration. No satisfactory absolute measuring instrument is yet in general use. British standards require that for measurement of grit and dust emission (particulate matter) the total number of sampling points be at least 24 in centres of equal area (BS 3405 permits some relaxation), although suggested new BS proposals ask for a smaller number in ducts of less than 2.0 m^2 area; for ducts above 50 m^2 area, only one sampling point per m^2. Samples must be taken isokinetically at a point corresponding to an air velocity in the duct of not less than 3 m/s at 200°C, and each sample taken over a period of not less than 10 min. The dust concentration is determined by withdrawing a representative concentration of the gas, separating the particulate matter, and measuring the weight

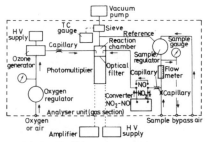

Figure 88   Typical chemiluminescent NO$_x$ instrument
- - - - Electrical connection; —— gas connection

References p 480

collected relative to the total volume of gas sampled. Two main types of probe have been used to do this: one in which the separation is carried out in or near the nozzle and the other where separation takes place at the outlet from the probe. The second type requires heating up to the point of separation, sufficient to maintain the gases above the dewpoint. Depending on the nature and quantity of particles involved, both cyclones and glass- or silica-wool filters have been used separately or in combination. For oil-fired plant where quantities are small and SO_3 in the gas will react, only silica-wool filters are recommended (Rendle[32]). Tyglass filter bags as specified in the CEGB Mk. 3 sampler have been used, however, for oil-fired particulate measurements and compared favourably with silica-wool filters both in laboratory tests and on plant measurements (Godridge and Cunningham[33]), but in general glass-wool must be avoided because of its sodaglass content. For more detailed information readers should consult the work of Hawksley, Badzioch and Blackett[34] and the relevant British Standards[35].

The most promising prototype instrument currently being tested in Europe and North America is a system particularly suitable for low levels, which draws the particle-laden gas through a moving high-efficiency filter tape on which the particles are collected. The particulate matter collected is exposed to low-energy nuclear (β) radiation, and this energy is absorbed at a rate largely dependent on the mass of matter collected per unit area. By measuring the amount of energy attenuated the result can be related to the mass of particulate matter per unit volume of gas, and yields a more absolute result than optical-type dust monitors.

Smoke

The best known device for measuring smoke is the Ringelmann Chart which consists of shade cards with grilles or lattices of black lines of different intensity numbered zero to four and which are intended to be indicative of the obscuration of light caused by smoke, No.4 for example representing 80% coverage. These charts are intended to be held in a vertical plane, where possible the chart (to the observer) being in line with the chimney top, placed so that chart and smoke from chimney have a similar sky background. The number of the shade which matches the darkness of the chimney smoke most closely is the Ringelmann smoke-number measurement. Optical measuring instruments exist to monitor this on a continuous basis (*Figure 64* in Ch.18.1), and BS publications[36] describe the requirements of these instruments and their calibration and use in relation to the Ringelmann chart. Smoke has in the past sometimes been used as a monitor of combustion control, and *Figure 83* shows how smoke and carbon monoxide follow a similar trend. Carbon monoxide is however a more absolute measurement,

and the major role of smoke-density measurement is for environmental protection of the atmosphere.

In principle, smoke meters are similar to certain of the optical dust monitors, are generally mounted at a convenient point in the flue or boiler outlet duct, and consist of two parts, the light projector mounted on one side of the flue or duct and the receiver containing the light-sensitive cell on the other. A beam of light is directed through the duct on to the receiver by a lens system. The output from the light-sensitive cell is proportional to the intensity of the light falling on its surface, which is in turn controlled by the density of smoke in the duct, and thus the cell output is inversely proportional to the smoke density. This output can then be displayed on both indicating and recording instruments in terms of the Ringelmann scale. This comparison is only valid if the smoke is neutral grey or black, as in combustion gases; if the smoke is white or coloured this will affect the output from the photo-electric cell.

Chimney Emissions

Measurement of the following parameters relating to combustion products released into the atmosphere is relevant to local pollution (Ch.19.3) and sometimes to fouling or corrosion of the plant (Ch.20):

Total solids concentration;
Coarse solids concentration;
Light absorption and reflection by plume;
Sulphuric acid concentration;
Acidic agglomerated solids ('acid smuts');
Sulphur dioxide concentration;
Nitrogen oxides concentration (see above).

Coarse Solids

A special instrument, the CERL dust monitor[37], has been developed for measuring the relative concentration of solid particles which sediment rapidly from the gas. It measures, by light absorption, the amount of dust falling onto a horizontal glass surface in a predetermined time interval. A discontinuous trace is obtained, the indication reverting to zero when the glass plate is periodically cleaned by a blast of hot air. The magnitude of the signal is dependent on the particle-size distribution of the settled dust as well as its amount.

Opacity of Gases

A simpler type of instrument in common use measures the decrease in optical transmission of a fixed light path that is caused by solids in the flue gas. This continuously indicating system, also subject to variations in the particle-size distribution, usually displays results on the Ringelmann scale[38]

The opacity of chimney plumes at temperatures above the dewpoint depends on the concentration and particle-size distribution of the suspended solid particles. At most concentrations legally permissible for more than a few minutes solids do not cause significant light absorption, but when the gases are below the dewpoint of either water vapour or sulphuric acid the small liquid droplets cause very dense and reflective plumes. By reflected light, these usually appear white, but the light transmitted through them may be deficient in components of various wavelengths (most often at the shorter end of the visible range) and appear brown, grey or black; less dense plumes often appear bluish, though the colour observed depends on the angle of incident light and on contrast with the background[39]. Because judgment of opaque plumes is usually made by humans only, perhaps the best instrumentation for plume monitoring is a colour-registering optical or television camera. Research into plume opacity has revealed its dependence on too many variables, such as those indicated above, for any simpler objective system.

Sulphuric Acid

It is not usually practicable to control the water-vapour content of combustion products, so there is little point in measuring it. However, even concentrations of a few ppm v/v of sulphuric acid greatly raise the dewpoint temperature and cause considerably opaque plumes; they also give rise to strongly acidic dust agglomerations in chimneys (disintegrating and emitted as acid smuts) and to fouling and corrosion of low-temperature parts of the plant (see Ch.20.1 and 20.2). In the emitted plume, sulphur dioxide is oxidized to sulphuric acid, a process which is catalysed by nitrogen oxides, and if this occurs more rapidly than the diminution of sulphuric acid concentration by plume dispersion, a long persistent opaque plume is formed.

Instruments for the measurement of sulphuric acid concentration in flue gases have been developed for discontinuous manual measurements and for automatic continuous indication and recording.

In the manual measurements, two alternative techniques are used to collect the sulphuric acid

or sulphur trioxide without significant oxidation of the sulphur dioxide, which is also present in 100—1000-fold concentration. In one method (*Figure 89*) isopropyl alcohol is used as an oxidation inhibitor; in the other, selective condensation of the sulphuric acid at about 60°C is achieved. Apparatus for the former is commercially available, and both of these methods are described in the latest British Standard on flue-gas analysis[40]; chemical methods for the determination of sulphur dioxide are included; four or five determinations hourly are usually possible.

The isopropyl alcohol technique is used in the automatic continuously recording instrument[41]. Flue gas is drawn through the heated sampling probe and silica-wool dust filter at a constant rate. Into it is injected a constant flow of isopropanol/water solution, and any sulphuric acid mist formed in the gas is coalesced by passage through a No.4 porosity sintered-glass membrane, which is continuously rinsed by the rest of the liquid. This solution, containing all the hexavalent sulphur in the gas as sulphate ions, is passed through a porous bed of crystals of barium chloranilate, which is quantitatively converted to barium sulphate, releasing an equivalent concentration of acid chloranilate ions. The peak optical absorption at 535 nm of this solution is measured in a continuous-flow photometer. The sensitivity of this system is quite adequate to measure to less than 1 ppm v/v of sulphuric acid: the instrument is usually calibrated using standard solutions of sulphuric acid in the isopropanol/water medium.

Precautions to be observed in measuring these low concentrations of sulphuric acid in gases are: to ensure that all parts of the sampling system are chemically inert and are above the acid dewpoint (180—200°C is normal); and to ensure that solid sulphates are removed by the silica-wool filter.

Deposition of Acid

Condensation of sulphuric acid onto plant surfaces causes local corrosion (Ch.20.2) and the emission of acid smuts. Two instruments are available for measuring the acid-dewpoint temperature and the rate of deposition of acid at lower surface temperatures. The BCURA dewpoint meter[42] measures the electrical conductivity between two platinum electrodes set in a glass surface, which is exposed to the flue gas at the end of a probe and cooled internally by a current of air (*Figure 90*). An arbitrary limit of electric current flow, normally 25 μA at an imposed 10 V, is used as the criterion of the dewpoint. At temperatures below the dewpoint, the rate of increase of conductivity with time, due to the accumulation of a film of sulphuric acid, is measured in μA/min and quoted as the 'rate of build-up'[43].

The acid deposition probe[44] permits collection of the material deposited on a test surface held at

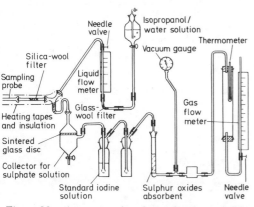

Figure 89   Apparatus for determination of SO_2 and SO_3

a known temperature, above or below the dew-point, for a known period (*Figure 91*). The deposit, often invisible, is usually washed off and, for example, sulphuric acid is titrated with standard alkali solution to the endpoint of phenol-phthalein indicator. The test surface is a known area of the exterior of the stainless-steel body of the probe, which is cooled by compressed air: a range of surface temperatures normally of about $20°C$, over the working length of the instrument, is measured by thermocouples embedded in the tube walls. Results are usually expressed as a rate of deposition; e.g. $\mu g\ SO_3/cm^2\ h$. Deposition of sulphuric acid is rectilinearly dependent on time in the range 10 s to 120 min, and on velocity in the range 5 to 20 m/s. As the probe is of all-metal construction, it survives the rigorous conditions of plant testing, and is in routine use on large boiler plant.

Fireside Fouling and Corrosion

In Ch.20.1, indications are given of the dependence of troubles experienced in boiler plant on factors that can be measured using instruments available for other purposes, such as measurement of oxygen or carbon monoxide or sulphuric acid in flue gas. The only instrument specifically designed to assess the deposit-forming properties of combustion products is the deposition probe. The form used for measuring acid deposition at temperatures around and below the dewpoint has already been described (*Figure 91*). For modelling surfaces at higher temperatures, another version of this, similar in construction, has been used for collecting deposits of fused, sintered and bonded ash and condensates of materials such as alkali-metal salts in combustion chambers and superheater passes of boilers[45]. In order to obtain sufficient sample for detailed examination, it is sometimes necessary to make exposures of more than 1 h duration, and an automatic temperature controller is then used. This instrument, like the acid deposition probe, was developed for research purposes, but is also useful for routine plant testing.

Though not strictly instruments as such, special test probes for measuring the corrosion rates of materials in particular environments should be considered here[46,47]. Air or steam are most frequently used as temperature-control media. Such equipment is designed specifically for the plant

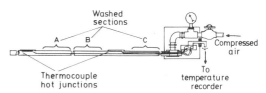

Figure 91   Acid deposition probe (part section)[44]

to be tested, but mention should be made of the important parameters, which are:

Geometry and surface finish of test pieces;
Geometry of presentation of test surface to environment;
Temperature of corroding surface;
Rate of heat transfer, where relevant;
Correct representation of period of exposure in relation to corrosion rate/time dependence;
Equal (for all specimens) and correct representation of the variable chemical and physical parameters.

Water and Steam Purity

Indicators and continuously recording instruments are available for measuring most of the more important properties: electrical conductivity, pH, dissolved oxygen, sodium, reactive silica, chloride, ammonia, and hydrazine. It is also possible to measure continuously metallic ions such as copper, iron and nickel. Currently this field is under active development, and some instruments which are not suitable for all circumstances are being replaced by others using better techniques. Much development work has been done within the CEGB on sampling apparatus and its applications; for details, the reader is referred to the appropriate CEGB publications[48].

References

1   Ref.(A)
2   BS 1828
3   BS 4937
4   Land, T. and Barber, R., 'Suction pyrometer in theory and practice', *J. Iron Steel Inst.* 1956 (Nov.), Vol.184, 289
5   Ref.(C), Vol.3, Part 1, p 329
6   BS 1904
7   Ref.(B)
8   BS 1042
9   *Fundamentals of Industrial Instrumentation*, Honeywell Controls Ltd, 1973
10   *Process Control Instrumentations*, Foxboro-Yoxall Ltd, 1973
11   Spiers, H. M. (Ed.), *Technical Data on Fuel*, 5th edn, World Energy Conf., London
12   BS 2740
13   BS 2811
14   BS 1523
15   Ref.(H)
16   CEGB (UK), Draft Standard 'Installation of thermocouples for conventional boilers'

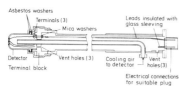

Figure 90   Probe for sulphuric acid dewpoint measurement[42]

17   Morgan, E. S. *J. Inst. Fuel* 1974, Vol.47, 113
18   Hitchcock, J. A. and Northover, E. W. *Proc. Instn mech. Engrs* 1966, Vol.180, Pt 3J
19   Anson, D. and Godridge, A. M. *J. scient. Instrum.* 1967, Vol.44, 541
20   Lucas, D. H. *J. Inst. Fuel* 1963, Vol.36, 207
21   Ref.(I)
22   Record, R. G. H., 4th Industrial Process Heating Conference, Manchester, 1969
23   Record, R. G. H. *Metals & Materials* 1971 (Jan.), Vol.5, No.1
24   Marcote, R. V. 'Evolution of portable electrochemical monitors, A.I.Ch.E. 68th National Meeting, 1971
25   Ref.(B), pp 758—760
26   Allen, J. D. *J. Inst. Fuel* 1973, Vol.46, 123
27   Shaw, J. T. *J. Inst. Fuel* 1973, Vol.46, 170
28   Halstead, C. J., Nation, G. H. and Turner, L. *Analyst* 1972, Vol.97, 55
29   Clyne, M. A. A., Thrush, B. A. and Wayne, R. P. *Trans. Faraday Soc.* 1964, Vol.60, 359
30   Clough, P. N. and Thrush, B. A. *Trans. Faraday Soc.* 1967, Vol.63, 915
31   Clough, P. N. and Thrush, B. A. *Trans. Faraday Soc.* 1969, Vol.65, 23
32   Rendle, L. K. *J. Inst. Fuel* 1964, Vol.37, 25
33   Godridge, A. M. and Cunningham, A. T. S., personal communication
34   Hawksley, P. G. W., Badzioch, S. and Blackett, J. H., *Measurement of Solids in Flue Gases*, BCURA, Leatherhead, 1961
35   BS 893:1940 and 3405:1971
36   BS 2811:1969, 2740:1969, 2742:1969 (addendum No.1, 1972)
37   Lucas, D. M. and Snowsill, W. L., Paper to National Society for Clean Air, Harrogate, 1968
38   BS 2742:1969
39   Jarman, R. T. and de Turville, C. M. *Atmospheric Environment* 1969, Vol.3, 257
40   BS 1756:Part 4:1975
41   An early version is described in: Jackson, P. J., Langdon, W. E. and Reynolds, P. J. *J. Inst. Fuel* 1970, Vol.43, 10

42   Flint, D. *J. Inst. Fuel* 1948, Vol.21, 248
43   Corbett, P. F., Flint, D. and Littlejohn, R. F. *J. Inst. Fuel* 1952, Vol.25, 246
44   Alexander, P. A., Fielder, R. S., Jackson, P. J. and Raask, E. *J. Inst. Fuel* 1960, Vol.33, 31
45   Jackson, P. J. and Raask, E. *J. Inst. Fuel* 1960, Vol.34, 275
46   Edwards, A. M., Jackson, P. J. and Howes, L. S. *J. Inst. Fuel* 1962, Vol.35, 16
47   Holland, N. H., O'Dwyer, D. F., Rosborough, D. F. and Wright, W. *J. Inst. Fuel* 1968, Vol.41, 206
48   CEGB (UK), Standards for Sampling Equipment and Instrumentation for Water in Power Stations

General Reading

(A)   Wightman, E. J., *Instrumentation in Process Control*, Butterworths, London, 1972
(B)   *The Efficient Use of Fuel*, 2nd edn, HMSO, London, 1958
(C)   *Temperature, its Measurement and Control in Science and Industry*, Reinhold, 1962
(D)   *The Units and Standards of Measurement Employed by the National Physical Laboratory*, HMSO, London
(E)   Foster and Parker *Fluidic Components and Circuits*, Wiley, New York, 1971
(F)   *The Instrument Manual*, United Trade Press Ltd (recent)
(G)   BS 4462:1969
(H)   International Flame Research Foundation, *Industrial Flames*, (Eds. Beér and Thring), Vol.1, *Measurements in Flames*, by J. Chedaille and Y. Braud; Edward Arnold, London 1972.
(I)   Jeffrey, P. G. and Kipping, P. J., *Gas Analysis by Gas Chromatography*, Pergamon, Oxford, 1964

Environmental aspects

Chapter 19

Plant safety: pollution and its prevention

19.1 Environmental control

Effects of energy industries on the environment
Internalization of environmental costs (polluter-pays principle)
Cost—benefit aspect
Difficulties in applying basic principles
Environmental criteria, standards and objectives
Responsibility for environmental control
Planning
Fuel economy

19.2 Health and safety on the plant. Legislation

Statutory legislation
Definition

19.3 Air pollution

19.4 Chimney and flue design

Chimney design

19.5 Water pollution and energy

Effect of pollution
Prevention of water pollution
Advanced treatment
Water pollution from the nuclear industry

19.6 Thermal effects on pollution

19.7 Noise

Acceptable scales of measurement
Acceptable criteria in buildings and dwelling areas
Methods of assessing nuisance
Typical noise levels

19.8 Solids

19.9 Waste and reject fuels

Wood waste
Miscellaneous solid waste fuels
Liquid wastes
Waste gases

19.10 Environmental legislation

19.1 Environmental control

Effects of Energy Industries on the Environment

Production and consumption of energy give rise to solid, liquid and gaseous waste products which require control to prevent unacceptable pollution of air and water and interference with the use of land or enjoyment of the natural landscape. Air and water pollution can cause damage to health and agriculture and to the availability of natural water supplies. Also, uncontrolled operation of many types of plant gives rise to objectionable levels of noise. The environmental control problems associated with the production and use of energy are concerned with:

Gaseous and particulate combustion products (18.3, 19.3):
   carbon dioxide and monoxide; sulphur oxides; nitrogen oxides; smoke, dust and fumes.

Liquid effluents (19.5):
   mine waters; coal washery effluents; phenolic and ammoniacal carbonization liquors; oily waste cooling water from refineries; effluents from gas treatment plants; oil spills and seepages from refineries and storage installations.

Solid wastes (19.8):
   colliery refuse; ash and clinker from combustion plants; sludges from coal cleaning and oil refining.

Waste heat (19.6, App.C7):
   power station and refinery cooling water; cooling tower plumes.

Noise (19.2, 19.7):
   operation of all types of plant.

Damage to visual amenity and encroachment on landscape:
   chimneys; cooling towers; spoil heaps; storage and handling installations; electric power transmission lines and equipment; road transport of solids, liquids and gases.

Internalization of Environmental Costs (Polluter-Pays Principle)

Many forms of energy production and use are potentially damaging to the environment or to health or amenity, and it is now generally recognized that the full costs of avoiding such damage ought to be paid by the user of heat, light or power. The principle of 'internalizing' such costs, so that they are included in the price of energy to the consumer, is generally known as the 'polluter-pays principle' and has been adopted in the environmental policy of major international organizations (e.g. OECD, EEC).

Cost—Benefit Aspect

Practicable solutions, however, require due regard to economic and political realities and priorities.

The concept of 'zero pollution' is economically impracticable and the ideal solution in socio-economic terms is one which minimizes the sum of the total cost (to the community) of environmental damage and the cost (primarily to industry, but ultimately to the user of the product) of environmental protection.

Difficulties in applying Basic Principles

The correct application of these principles in national policy and legislation is exceedingly difficult, and is currently under close study in the international sphere by technologists, economists and legal experts.

The principal difficulties are:

1. The 'polluter-pays' principle has not been observed in the past and industry has developed on the basis of energy prices which have not included environmental costs. Moreover some industries now have a backlog of environmental damage to reinstate (e.g. coal mining).
2. The impact of costs and benefits of environmental protection falls on widely differing sections of the community.
3. No satisfactory method has been found of evaluating social and amenity costs in economic terms, though this has been attempted in connection with noise, recreational use of open space, and land values.
4. Environmental damage may cross international frontiers (international waterways, long-range atmospheric drift) with consequent problems of regulation and enforcement. Moreover different countries have different approaches to standards and methods of enforcement.
5. Prevention of accidental polluting emissions and environmental control of normal industrial operation require radically different types of regulation.

Environmental Criteria, Standards and Objectives

As a matter of international agreement, the term 'criteria' is now used to refer to dose/response or exposure/damage functions. 'Standards' are the limits to pollutant concentrations as defined in regulations. 'Objectives' are the wider spectrum of desirable conditions to which environmental control is directed. Types of standards adopted in different countries include:

(*a*) Air or water quality standards. These imply that the total concentration of a specified pollutant in, e.g., the ambient air must not exceed a certain level, irrespective of the source of the pollution. At present no such standards apply in the UK.

(*b*) Emission standards, which specify the maximum amount of pollution permitted to be added to the environment from each individual source, irrespective of the existing degree of pollution of the area. Emission standards apply in the UK to certain noxious gases and particulates (e.g. HCl, SO_3, smoke, flue dust).

(*c*)   Fuel quality standards. Examples in the UK are definition of smokeless fuel under the Clean Air Act, 1956; limitation of sulphur content of liquid fuel in the City of London (Various Powers) Act 1972; and restrictions on the lead content of motor gasoline.

The regulatory approach adopted in the UK, however, largely avoids the application of rigid standards in favour of a more flexible system. For example, the Alkali Act requires the use, by the scheduled processes, of the 'best practicable means' (i) to prevent the emission of noxious or offensive gases and (ii) to render harmless and inoffensive those gases necessarily discharged. This is held to enable the regulatory authority to enforce, in cooperation rather than conflict with industrial management, high and advancing standards of control with minimum interference with industrial production.

The reader is urged to ascertain the current situation in his own country.

Responsibility for Environmental Control

A number of countries have adopted national (or federal) environmental protection legislation with responsibility delegated to local (or state) authorities for setting and enforcing regional standards.

Responsibility for environmental control in UK includes (a) direct control of specified sectors by central government exercised by, e.g., the Alkali and Clean Air and Nuclear Inspectorates, and (b) control delegated to local government or regional bodies under national legislation, e.g. the Clean Air Acts 1956 and 1968, the Rivers (Prevention of Pollution) Acts 1951—61 and the Water Act 1973.

Some product standards operating in the UK have been referred to above. More general controls include the powers of local authorities under the Public Health Act 1936 and the rights of individuals at common law against proved damage or nuisance.

Planning

Protection of the environment from the impact of industry on general amenity and from encroachment on the natural landscape is provided for in Town and Country Planning legislation in which local planning authorities are directly responsible, subject to approval of the Secretary of State for the Environment. Consents for new and extended installations by the nationalized energy industries are matters for the Secretary of State for Energy and include deemed planning approval by the Secretary of State for the Environment.

Fuel Economy

It is self-evident that the more economically energy is used, the greater is the potential for reducing the environmental impact of its production and use, in all aspects. Fuel economy may have different meanings according to the party considered. But monetary economy can be quantitatively different from energy economy (in the sense of efficiency). Consequently maximum energy or fuel efficiency does not necessarily ensure minimum harm to the environment, as in such early ideas as those which equated smoke unburnt with fuel wasted (a proposition obviously true within limits).

Further Reading

General

World Health Organization, 'Air Quality Criteria and Guides for Urban Air Pollutants', *Tech. Rep. Series* No.506, Geneva, 1972

Schurr, S. H. *Energy, Economic Growth and the Environment. Resources for the Future*, Johns Hopkins, Baltimore, 1972

Berkowitz, D. A. *et al*, *Power Generation and Environmental Change*, MIT Press, Cambridge, Mass. and London, UK, 1971

World Energy Conference, Report of the Committee on Environmental Effects: Environmental Conservation and the Energy Producing Industries, London, UK, May 1972 (Paper to U.N. Conference on Human Environment, 1972)

Scorer, R. S. *Pollution in the Air*, Routledge & Kegan Paul, London, 1973

United Kingdom

Institute of Fuel, Papers to Int. Conf. 'Fuel and the Environment', Eastbourne, Nov. 1973

Institution of Heating and Ventilating Engineers, *Conservation and Clean Air: A Study of Atmospheric Pollution in Great Britain*, Curwen Press, London, 1973

National Society for Clean Air, 'Sulphur Dioxide: an Air Pollutant', Report by the Technical Committee, 1971

Royal Commission on Environmental Pollution, First Report, HMSO, Cmnd 4585, 1971

National Survey of Air Pollution, 1961—1971, Vol.1, Warren Spring Laboratory, Department of Trade and Industry, May 1972: HMSO, 1972

'Energy and the Environment', Report of Working Party set up jointly by Committee for Environmental Conservation, Royal Society of Arts and Institute of Fuel, 1974

19.2 Health and safety on the plant. Legislation

This could be a book in itself. The UK legislation, cited as an example, is typical of an industrial country. Equivalents should be looked for in other industrialized countries and the list may be checked for these; for the less developed nations it may direct attention to the appropriate legislation, or at least to a need for it.

Legal aspects of health and safety on the plant fall into three categories:

Contract Law. Local agreements and conditions of employment negotiated between employer and employee. They take due account of the requirements of both common and statutory law.

Common Law. Employers' duty of care towards employees, visitors and third parties entails:
1. Provision of a safe place of work.
2. Provision of safe equipment.
3. Organization and maintenance of a safe system of work.
4. Employment and training of competent workers.

Must be considered against a background of tradition and of Court decisions.

Statutory Law. Acts of Parliament, Statutory Rules and Orders (S.R. & O), Statutory Instruments (S.I.) and Orders in Council (O.C.). Employers must comply with those pertaining to their particular business; these requirements should be regarded as the minimum acceptable standards of health and safety.

Statutory requirements may also be considered under three headings: plant operation and processes; services and construction; process effluents.

The third of these is the subject of Ch.19.10. There can be interaction between quality of process effluent and plant conditions, e.g. local exhaust ventilation and its influence on concentration in the exhausted air within a broader environment.

A checklist of some major items that may be appropriately included under the first two main headings follows, with brief comment as necessary.

Statutory Legislation

Plant Operation and Processes

General.
1. Factories Act 1961. The topics dealt with include: *Health.* Cleanliness, heating, lighting, ventilation, and noise. *Safety.* Guarding machinery, testing lifting apparatus and pressure vessels, dangerous substances and fume, entry into confined spaces, means of escape and current certificate from the fire authority. *Welfare.* Notification of accidents and industrial diseases. Special applications and extensions which include application for electrical stations, building and engineering construction, processes involving lead in places other than factories. The Act is mainly administered by the Factory Inspectorate and in part by local authorities.
2. Offices, Shops and Railway Premises Act 1963. Sets general standards of safety, health and welfare in offices.
3. Health and Safety at Work etc. Act 1974. Main

features: applicability to all persons at work; dependence on approved Codes of Practice; administration by Health and Safety Executive and enforcement by an integrated inspectorate; responsibility for health and safety of an employee by employer, himself and fellow employees. Under an amended form of the Fire Precautions Act 1971 general fire precautions are dealt with by fire authorities and the Home Office. 'Process risks' are the responsibility of the Safety and Health Commission and its Executive.
4. Employer's Liability (Defective Equipment) Act 1969 extends to responsibility for safety of equipment even if defect is caused by a third party.
5. Employer's Liability (Compulsory Insurance) Act 1971. Insurance against liability for damages for bodily injury to employees (other than by contract) sustained in the course of their work.
6. Diseases arising from occupations are classified as notifiable and prescribed.
(a) *Notifiable diseases.* Notifiable by the employer to the Department of Employment. Disease induced by aniline, arsenic, carbon disulphide, lead, manganese, mercury and phosphorus. Chronic benzene poisoning, chrome ulceration, compressed-air illness, toxic jaundice, toxic anaemia, epitheliomatous ulceration, anthrax. In 1967, diseases arising from beryllium and cadmium were included.
(b) *Prescribed diseases.* Must be reported to the Ministry of Health and Social Security. Include notifiable diseases and numerous occupational diseases. They must be identifiable as arising from a particular occupation and previously associated with it.
7. Employment Medical Advisory Services Act 1972. The medical branch of the Factory Inspectorate was separately constituted thereby, and remains to provide an advisory service under the Health and Safety Act 1974.
8. Noise Abatement Act 1960. Local Authorities are empowered to abate noise or vibration that constitutes a nuisance. A Code of Practice defining levels of exposure of employed persons to noise (1972) has been issued. A further Draft Code of Practice defining noise levels from new machinery aims to reduce noise at its source.
9. Fire Precautions Act 1971. Will eventually apply to safety from fire in all buildings with particular reference to means of escape. The Act is administered by fire authorities; 'process' risks remain within the oversight of the Health and Safety Inspectorate.

To these main enactments must be added numerous Statutory Instruments (S.I.) and Statutory Regulations and Orders (S.R. & O.) relating to specific trades and processes, such as:
Patent Fuel Manufacture (Health and Welfare) Special Regulations S.R. & O. 1946 No.258: Precautions to be taken to minimize the risk of occupational dermatitis, viz. control of dust and ventilation, provision of suitable washing facilities,

medical supervision and examination, provision of suitable skin and eye protection.

Clay Works (Welfare) Special Regulations S.I. 1948 No.1547: Require the provision of shelters for kiln workers, protective clothing, accommodation for clothing, washing facilities, rest rooms, ambulance rooms, canteens and mess rooms and supervision of cleaning of the facilities provided.

The final example has far-reaching effects — The Asbestos Regulations 1969, widely applicable to processes involving asbestos, particularly in relation to exhaust ventilation and protective equipment, cleanliness of premises and plant, storage and distribution of asbestos. Technical Data Note 13 sets out threshold limit values for both chrysotile and crocidollite. In addition to the legislation, advisory documents such as Technical Data Notes, and Health and Safety at Work booklets are available. These also indicate how the Inspectorate interprets the regulations. Such are Advice on Asbestos; Health Precautions in Industry (Health and Safety at Work Booklet No.44) and Standards for asbestos dust concentrations for use with the Asbestos Regulations 1969 (Technical Data Note 13).

Information for other toxic materials in the working environment are set out in Threshold Limit Values Technical Data Note 2 (issued annually), based on values issued by the American Conference of Governmental Industrial Hygienists. Internationally there is general agreement only on these; the International Labour Office issues national lists where available. Further advice is available in Dust and Fume in Factory Atmospheres: Safety, Health and Welfare Booklet No.8.

The regulations may be split into two groups. First, those that apply to Dangerous Trades and include, in addition to those already given, the Chemical Works Regulations 1922 and the Iron and Steel Regulations 1947. They provide standards for factories involved with specific materials or using specific processes. Secondly, there are Welfare Orders made under successive major items of factory legislation.

Specialized Legislation.   Examples will be given here of controls related to particular plants or processes.

1. Boiler Explosions Act 1882.

Factories Act 1961.

Examination of Steam Boiler Regulations 1964. These cover steam and hot-water boilers, air and steam receivers, steam pipes in respect of quality of both materials and construction, requirements for valves and fittings, period between tests, and notification of explosions.

2. Petroleum (Consolidation) Act 1928. Defines conditions of and quantities allowed in storage of petroleum spirit (closed flash point $<23°C$). Scope was extended to 82 other substances by the Petroleum (Inflammable liquids and other dangerous substances) Order 1947. Latterly the criterion of applicability has been changed to a closed flash point of $<32°C$ and to combustibility at $<50°C$ irrespective of origin of material. This was as set out in the Highly flammable liquid and liquefied petroleum gas regulations 1972. Other Orders made under the Act of 1928 are:

Petroleum (Carbide of Calcium) Order 1929.

Petroleum (Compressed Gases) Order 1930.

Petroleum (Liquid Methane) Order 1957.

Petroleum Spirit (Conveyance by Road) Regulations 1957, 1958, 1968.

Some conditions for transference, storage and transportation of volatile, flammable materials are stipulated in legislation and codes of practice. Plant failures leading to unconfined vapour-cloud explosions require further consideration; legislative control bearing on such events may be required, so that siting of plants and routing of loads are under closer surveillance.

3. Explosives Act 1875, amended by Explosives Act 1923. The use of explosives is subject to control in respect of handling, storage and use in building operations. Orders and regulations under this Act were made in 1919, 1937, 1947, that relate to acetylene. By these, acetylene when compressed above 9 psig (62 kPa gauge) is defined as an explosive. It may be stored at up to 300 psig (2.07 MPa gauge) under approved conditions when used for the manufacture of organic chemicals. Reference to various aspects of the control of a single material is often made in several separate places. Thus the use of carbon disulphide in the cold-cure of vulcanizing for proofing cloth is controlled under the India-rubber Regulations 1955 [Explosives Act] and its handling, movement and storage under the Petroleum Spirit (Conveyance) Regulations [Petroleum Consolidation Act]. A comprehensive search of the body of legislation is therefore essential.

4. Radioactive Substances and Nuclear Installations. There is a small body of legislation concerned with the control of these materials and plants.

Radioactive Substances Act 1960. Authority must be obtained from the Ministry of Housing and Local Government (or their successors) for the purchase, accumulation or disposal of radioactive substances.

Nuclear Installations Act 1959 and 1965. Impose a duty to licence nuclear reactors and to insure them adequately.

Ionizing Radiations (Sealed Sources) Regulations 1961. Apply to the control of radiography and thickness gauging in premises where the Factories Act is operative.

Ionizing Radiations (Unsealed Radioactive Substances) Regulations 1968. Administered by H.M. Factory Inspectorate and relate to handling and storage of unsealed sources of radioactivity.

Services and Construction

Electricity.   The Electricity Regulations 1908 and the Electricity (Factories Act) Special Regulations 1944 apply to installations in general. Main requirement is that every aspect of the installation must comply with the current edition of Regulations for the Electrical Equipment of Buildings (the I.E.E. Regulations). These may be divided into two parts:

Part I. Requirements for safety. 'Good workmanship and the use of proper materials are essential for compliance with the regulations'.

Part II. Means for complying with the Regulations. Sections (A—H) *inter alia* deal with choice of materials, installation, testing, inspection, and fault protection.

Further there are Codes of Practice relating to electrical apparatus in factories, e.g.:

CP 321 102 Installation and maintenance of electrical machines, transformers, rectifiers, capacitors and associated equipment.

CP 1003 Electrical apparatus and associated equipment for use in explosive atmospheres of gas or vapour other than mining applications. This code is divided into two parts; the first relates to flameproof and intrinsically safe equipment and also involves, for example, B.S. 229, 1259 and 4683; the second to alternative methods of meeting the hazard. The whole field is presently changing rapidly in relation to both requirements and classification, and will continue to do so as international standards evolve.

CP 1008   Maintenance of electrical switchgear.
CP 1013   Earthing.
CP 1015   Electrical equipment of industrial machines.

Water. Main requirements in plant operations are as follows:

(i)   Supply of hot and cold water adequate to the needs of operatives for washing.
Factories Act 1961.

(ii)   Disposal of liquid effluents must not adversely affect the sewer or processes of sewerage treatment.
Public Health Act 1936, 1961.
Public Health (Drainage of Trade Effluent) Act 1937.

(iii)   Disposal of liquid effluents must not damage rivers or aquifers or endanger public health.
Salmon and Fresh Water Fisheries Act 1923.
Water Act 1945
Gas Act 1948.
Rivers (Prevention of Pollution) Act 1951, 1961.
Clean Rivers (Estuarial and Tidal Waters) Act 1960.
Water Resources Act 1963.
Water Act 1972.
Control of Pollution Act 1974.

Disposal of solid waste is a matter of plant safety and may affect the water supply. Its control is included in this last enactment. Powers of control were originally given under the Deposit of Poisonous Waste Act 1972.

Gas. The Gas Act 1945 and Schedules relate to control and provision of supplies. Special control and precautions are necessary when extraneous gases or compressed air are used in conjunction with the electrical mains supply.

The Gas Safety Regulations 1972 set out requirements for the installation and use of gas-burning appliances.

Construction. Building operations taking place on a plant are subject to the provisions of the Factories Act 1961 and the Building (Safety, Health and Welfare) Regulations 1948, 1972, 1973. Within these regulations the various activities occurring during building are considered, e.g. responsibility and supervision, safe access, safe lifting apparatus, and various specific operations such as demolition and excavation. In addition there are further regulations relating to ventilation and protective clothing, set out in the Construction (General Provisions) Regulations 1961.

Definition

Threshold Limit Value (TLV): The concentration of the substance in air to which nearly all persons may be repeatedly exposed day after day without adverse effect. Units: solids in gas mg m^{-3}, mppcf (million particles/ft^3); gaseous ppm v/v (parts per million by volume); solids, liquids ppm w/w (parts per million by weight).

19.3 Air pollution

The pollutants which most need to be considered are black smoke, the oxides of sulphur, the oxides of nitrogen, and ash in the form of fine particles. Black smoke is in a special category and will not be considered further because it is both feasible and economic to reduce its emission to very low levels (see especially Chs.4—6 and 10).

The four major methods of reducing pollution of the atmosphere are:

1. To select fuels of low pollutant characteristics for regions of severe pollution.
2. To remove the pollutant from the fuel before combustion.
3. To remove the pollutant from the gases after combustion.
4. To use tall chimneys.

The first is widely used with coal and oil, particularly for SO$_2$. The second is not used on coal on a large scale but anti-pollution legislation may make it necessary to desulphurize fuel oil in some places. The third is widely used to remove ash from flue gases, and is under very active study, particularly in the United States for the removal of sulphur oxides. The fourth is also used widely and has been the subject of intensive study. We shall deal with it before considering methods of removal.

The tall chimney (Ch.9.4) is almost invariably emitting hot flue gases, and it is common to refer to the effective height as the actual chimney height plus the thermal rise. By the time the flue gases reach ground level they have become mixed with about 10 000 times their own volume of air, and this is usually sufficient to bring the ground-level concentration to a satisfactory figure. The physical processes of thermal rise and of gaseous diffusion over the full range of weather conditions are immensely complicated. They are now however sufficiently well understood to render it possible to specify a satisfactory chimney height for any given situation. This necessary height is increased appreciably by rising ground nearby, and some particularly difficult sites may be uneconomic for this reason. In general, however, chimneys can provide very adequate control for gaseous pollutants without a marked increase in the total capital cost of a plant.

In addition to tall chimneys, the control of dust emission often requires some method of removal, particularly for pulverized-coal plant. The principal types of dust collector for boiler plant are the mechanical collector and the electroprecipitator. Mechanical collectors range from simple settling chambers, suitable for particles larger than about 100 μm, via bag filters, to high-efficiency cyclones having efficiencies of 85—90% for 10 μm particles. Cyclones have relatively low capital cost but may have a fairly high running cost (ref.A).

In electroprecipitators the dust-laden gas is passed through parallel earthed tubes or between plates, known as collecting electrodes, alongside wires raised to a negative potential of 40—50 kV which emit a corona discharge. The dust particles receive charges from the corona and migrate to the collecting electrode. The accumulated dust is transferred to hoppers by periodically rapping the electrodes. Precipitators are capable of very high efficiencies, even with particles as small as 0.01 μm, and running costs are low. The capital cost is higher than that of a cyclone but small compared with that of the combustion plant.

There is little prospect of developing an economic process for removing sulphur compounds from coal, but the mineral sulphur can be reduced by physical cleaning processes. Considerable effort has gone into desulphurizing oil by treating it with hydrogen under pressure, which releases the sulphur as hydrogen sulphide gas. Full-scale plant is now being installed to a limited degree, principally in Japan and the Caribbean.

Combustion in a fluidized bed containing ground limestone has been shown on pilot scale to retain an appreciable proportion of the sulphur originally in the coal or oil.

Flue gases contain about 0.2% sulphur oxides, mainly the dioxide with about one hundredth of this as the trioxide. Nitrogen oxides vary between 0.01 and 0.06% depending on the type of combustion and the fuel.

Full-scale removal of sulphur oxides from flue gases is limited to a relatively few plants in the UK, USA and Japan. The technical problems encountered in attempts to develop processes using a dry absorbent that do not destroy the thermal lift of the gases have led to an almost complete reversion to wet-washing processes. There are about twenty operating in all including those at Battersea and Bankside power stations in the UK, satisfactory over 20 years, and about sixty more under construction or commissioning, using aqueous slurries or solutions (limestone, lime, magnesium sulphite). The washed gases are commonly reheated. Removal by this method is expensive. The capital cost for an existing station in 1973 was estimated at about 60 $/kW. As the technology advances, costs should come down but are likely always to penalize heavily the use of carbonaceous fuels.

Nitrogen oxides are not removed by these processes and the only present possibility of reducing their concentration is by controlling the temperature of combustion (steam or water injection), or by staged combustion.

19.4 Chimney and flue design

Additional notes for small flues are given in App.C7.

Chimney Design

A chimney is essentially a flue designed to exhaust waste gases and any entrained solid matter at sufficient height to allow acceptable dilution and dispersal before they return to earth. The design must also ensure acceptable pressure conditions on the plant served to permit its effective operation over the entire range of requirements. Most chimneys are also building structures in their own right. In many cases the structural member is considered separately from the gas flue, each using the materials of construction best suited to the function served. Typical materials of construction of the structural shell are brickwork, concrete, mild steel or GRP (Glass Reinforced Plastic). The flue or liner can also use these materials or refractory brick, acid-resisting brick, insulating brick, Corten steel, stainless steel, glass-lined mild steel, refractory concrete, etc. Each must be used within its limitations, e.g. tolerance to overheating or abrasion.

Where only one item of plant has to be served a simple single-bore chimney is the only type to be considered, but this can take various forms: free-standing, self-supporting; free-standing guyed; supported partly or completely by a building or some other structure; or in-built as a composite part of a building structure, as are most domestic chimneys.

Where several appliances have to be served there are the choices of an individual chimney for each appliance; two or more appliances served by each chimney, or a multi-flue chimney with one or more appliances discharging into each flue. Again any of these chimneys can take various forms as previously mentioned.

The most common chimney application is to serve a combustion appliance such as a boiler or furnace, where in addition to normal structural loads such as wind loading and weathering, the flue or chimney liner is also subjected to chemical attack, abrasion, temperature, thermal expansion and possibly thermal shock.

The overall design of a chimney, therefore, involves all branches of engineering science and the most satisfactory solution for a particular set of conditions, with the materials at present available, is generally a compromise between pollution control, flow and pressure conditions, effective life, and economics.

Choice of Materials and Form of Construction

Before detailed chimney design can be undertaken, a preliminary selection is made of the form of construction using materials suitable for the chemical properties and temperature of the gases to be handled. *Table 1* gives a rough indication of some of the factors influencing this choice for single-flue chimneys to serve small- to medium-sized industrial plant.

There is a limit to the range of gas flows that can be accommodated by a chimney bore of fixed dimensions without imposing unacceptable pressure variations on the plant served. For plant such as shell boilers, which may have to meet significant load variations, this limit may imply not more than two boilers per flue. This number could be increased at the capital and running cost of installing an induced-draught fan and pressure-control devices. A more economic solution is to install a multi-flue chimney. This can be in the form of a pre-cast concrete shell with integral insulating concrete liner divided into compartmental flues. The alternative construction mainly used for larger chimneys is an outer structural windshield of mild steel, reinforced concrete cast *in situ*, or pre-cast reinforced concrete, housing a number of flues of any material compatible with the gases handled. The mild-steel windshield is probably the most economic up to a height of about 38 m, but *in situ* concrete becomes progressively more competitive for greater heights and larger diameters. Since the windshields can have a life of 30 to 50 years and the inner flues between a half and a quarter of this, it is generally economic to make provision for future replacement of the inner flues and provide access for regular inspection.

Chimney Problems

Corrosion. Most exhaust gases contain acidic products of combustion. All contain water vapour and solid particulate matter such as unburnt carbon or ash from the fuel. If, at any point in the waste-gas system, the surfaces in contact are allowed to cool to the temperature (see later and Ch.20.2) at which the acid vapours condense — the acid dewpoint — the chimney structure will be attacked. If negligible acid vapours are present, as in the combustion products from natural gas, the problem does not arise until the surfaces cool to a lower temperature — the water dewpoint — when large quantities of water can condense.

Smuts. If condensation occurs in the chimney surface, solid particles can be trapped and build up in loose layers on the inner skin. Changes in load can then dislodge these deposits which will be discharged from the chimney as smuts. Particularly if acidic, these smuts falling out in the vicinity of the chimney can cause damage to buildings, cars, clothing and vegetation.

Downwash. Downwash occurs when low-velocity gases leaving the chimney are drawn down into the low-pressure area on the lee side of the structure caused by passage of the wind round the chimney. This reduces the height to which the gas plume can rise, limiting dispersal, and can cause staining of the chimney structure.

Inversion. Inversion occurs when cold air flows down the inside of the chimney bore on the windward side, cooling both the chimney lining and the flue gases. This can give rise to acidic condensation, corrosion, and smut formation, and again reduces the thermal rise of the gas plume. Both downwash and inversion increase in severity with increasing wind speed. The worst conditions obtain with minimum plant loading. In theory both can be avoided by maintaining the waste-gas efflux velocity sufficiently above wind speed (see Design Parameters).

Note on Chimney-Design References

Comprehensive coverage of most aspects of chimney design is given in the conference publication of the Chimney Design Symposium, Edinburgh, 1973[1]. Two simple guides to the thermal,chemical and flow design aspects of chimneys for small to medium size plants are the Shell booklet, 'Chimneys for Industrial Oil-Fired Plant'[2] and the Brightside Chimney Design Manual with Supplement[3] (which is sufficiently accurate for small chimneys). The former reduces the evaluation to a few simple calculations and presents the relevant parameters and factors required in a number of Tables and graphs. The Brightside approach is almost entirely graphical using a series of curves and nomograms.

Principal Design Parameters

Chimney Height. For waste gases with significant SO_2 content, e.g. from coal or oil firing, height can be calculated from the Clean Air Act memorandum 'Chimney Heights'[4]. For products from gaseous fuels, heights may be obtained from the British Gas paper in the Chimney Design Symposium[1]. An official guide is in course of preparation. For natural-draught systems a greater height than calculated from the above may be necessary to induce sufficient air for combustion of the fuel, adopting the procedures in ref.2 or 3. The proposed chimney height in the UK has to be approved by the Local Authority or Alkali Inspectorate.

Cross-sectional Areas. A preliminary assessment of duct and chimney areas can be based on the full-load gas quantity to be handled, assuming hot-gas

Table 1   Single-flue Chimneys — up to approximately 46 m high

Basic materials	Forms of construction	Applications or operating condition limits	Advantages and weaknesses	Life (years)	Relative cost (1974, UK)
(a) Mild steel 8 mm thick +	Single skin, uninsulated. All welded, or in flanged and bolted sections	Only applicable where no risk of gases cooling to dewpoint. Small LPG- or gas-fired plant and certain coal-fired plant. Low turndown range. Furnaces with gas temp. <500°C	Inexpensive. Rusting, mainly when not working	5 to 10	1.0
(b) Mild steel 8 mm thick + Dense lining	Single skin lined with refractory brick or concrete, backed if necessary with insulating brick	High temperature furnaces, inlet gas >500°C. Not suitable for positive gas pressures	Spalling or cracking of lining. Rusting of M.S. Brick lining only possible if diameter is large enough	5 to 10	2.3
(c) Mild steel 8 mm thick +	Single skin, unlined with annular cold-air bleed at chimney base	High-temperature furnaces. Inlet >500°C	Low-cost alternative to above. Correct design of air bleed essential		1.1
(d) Mild steel 8 mm glass lined	Mild steel, 2 mm glass lining	Chemical plant. Resistant to all chimney acids other than hydrofluoric. 350°C service limit	Expensive and subject to mechanical damage. Generally only used for small bores	5+	5.5
(e) Mild steel, aluminium clad	20 mm air gap adequately spaced and weather-sealed at joints and top. Alternatively 50–100 mm mineral wool insulation replacing air gap	Application where waste gas temperature never less than 260°C with air gap. Coal- or gas-fired boilers or oil-fired with exit temp. >150°C with mineral wool insulation	Differential expansion difficult to accommodate, rivet failure eventually. Weather-sealing difficult to maintain. Maintenance awkward.	6+	1.5 to 2.0

(f) Mild steel structural shell with inner bore	Inner skin can be M.S.; M.S. aluminium-sprayed, Corten or S.S. welded or flanged. 50–200 mm mineral wool or granular insulation between shells	All boiler applications. Can deal with all gas temps from 160 to 600° C+ depending on inner skin and degree of insulation	Structural member not in direct contact with flue gases. Maintenance of single inner bore difficult. Gas leakage at flanges of bore can cause corrosion of structural shell	10 to 20	2.0 to 3.0
(g) GRP	Single skin or double skin with air gap, with independent supporting structure or integral M.S. reinforcing	Gas- or gas-oil-fired boilers. Applications where chemical resistance is important. Chimney extensions. Upper temperature limit 280° C depending on resin. Could be affected by boiler failure or ignition of carbon build-up. Selection of type for the particular application is essential	Impervious to acid attack. Insulation properties inadequate for heavy-oil firing. Abrasion resistance inadequate for coal. Subject to hardening or softening depending on resin. Large thermal expansion. Can be economic if supported from building	Not known	2.5 to 3.5
(h) Brick structural shell	Unlined, or lined with insulating, refractory or acid resisting brick and possibly insulating brick backing, according to service conditions	Suitable for most temperature conditions. Seldom used apart from brick kiln and some furnace and coal-fired-boiler applications. Not suitable for positive gas pressures	High cost. Long construction period. Subject to cracking with overheating, spalling and weathering. Poor thermal properties without lining. High thermal storage	30 to 50	2.5 to 3.5
(i) Precast concrete outer shell	Integral insulating concrete liner. In interlocking sections with mortar suitable for conditions	Can be designed to suit most temperature requirements. Widely used for boiler plant, particularly coal- or gas-fired	Amenable to aesthetic treatment. Subject to spalling, cracking and mortar attack	Shell 45. Bore?	

velocities of around 8 m/s for natural-draught systems and up to 18 m/s for systems with either forced- or induced-draught fans. These areas may require to be modified when pressure conditions are evaluated.

Thermal Design. The main aim is to maintain the inner-skin temperature of the chimney above the acid and water dewpoints of the flue gases. The highest practicable gas outlet temperature also aids natural draught and the thermal-plume height achieved for dispersal of the products. All fuels which could be used over the life of the chimney should be considered and design based on that with the highest acid dewpoint (Ch.20.2). This will vary from $150°C$ for small to medium-sized plant burning heavy fuel-oil, $121°C$ for plant burning British coals, down to $66°C$ for gas-fired plant. This last is but slightly above the water dewpoint since the combustion products contain negligible acidic vapours. The limiting design factor is the volume and temperature of flue gases discharged into the chimney system at minimum plant loading.

In many cases this leaves little scope for loss of temperature through the system. The degree of thermal insulation required to maintain the inner-skin temperature at chimney top above the design dewpoint can then be assessed by relating the system heat losses for a specific design to the heat input in flue gases.

Pressure Design. The object is to arrive at the maximum practical exhaust-gas velocity consistent with acceptable back-pressure on the plant served. This ensures economy of design. The highest possible velocity at chimney outlet also gives the nearest practical approach to (*a*) avoiding down-wash and inversion and (*b*) the most effective plume rise for dispersal of the products.

A preferred figure for minimum efflux velocities is 13 m/s but this cannot always be achieved in practice. This would avoid inversion or downwash (in the UK) for about 97% of the year. The limiting factors are the full load gas flow and temperature discharged into the chimney systems; the available fan power on the combustion appliance to deal with the pressure variations imposed by the chimney system and the operating turndown range of the plant served.

Limitations of Basic Design Methods. The basic thermal and pressure design procedure is given in refs 2 and 3 enabling dimensions, degree of insulation, velocities and pressure conditions to be established for any particular plant served. The analysis is however based on equilibrium conditions, and the results will only be adequate in the rare cases where the plant operates continuously within its modulating range.

Special consideration has to be given where:
The load on the plant is low involving on/off operation.

Cyclic, overnight or weekend shut-downs occur.
More than one unit is served by a common chimney or a common bore of a multi-flue chimney.
Provision has to be made for the addition of future appliances to the chimney system.

Checklist

General. Ideally one item of plant per chimney flue; insulation to maintain waste gas above dew-point[2]; preferred efflux velocity at minimum load 13 m/s subject to acceptable pressure variation at plant[2].

Ductwork. Given a common chimney flue, fit individual connecting ducts to plant items; minimize changes in shape or area by gradually tapering transition pieces; avoid sudden changes in direction; branch-to-chimney flue angled upwards. If plant operation is intermittent fit gas-tight damper to conserve heat storage; if pressure control is necessary, automatic multiblade damper preferred to air-bleed draught stabilizer; provision for grit and dust measurement[5]; adequate gas-tight cleaning doors at bends; all joints gas-tight at maximum pressure possible.

Chimney. No sudden change in area at entry branch; bore blanked off immediately below duct entry; sealed provision for draining start-up condensation and rain; gas-tight door for soot removal; no heat-conduction paths between inner and outer shell, e.g. at flanges or stays; if size permits, access for internal inspection; adequate electrical earthing for lightning protection[6]; consider protection from wind-excited oscillations[7]; top section of bore in acid-resisting material, or readily replaceable; no cap, hood or ornamental device to defeat plume rise; reconsider adequacy of design if fuel, plant or method of operation is changed.

Inspection. Mild steel — uninsulated: Inspect annually; measure shell thickness; check protective coatings, rivet and bolt condition and tightness, foundation bolt security; check guy wires and anchors, air-to-earth resistance of lightning conductor.

Mild steel insulated, concrete, or brick: Inspect every three years: items above where applicable, also condition of lining and windshield; spalling; cracking; acid attack; mortar condition; leakage from or into inner shell; internal inspection of top section; check metal thickness at top section and adjacent to flanges on bore; check water collection.

Maintenance. As required by inspections; also once or twice annually remove all collections of soot; touch up damaged protective coatings; periodically check combustion efficiency of plant served and adjust air/fuel rates as necessary.

19.5 Water pollution and energy

The production of any form of waste indicates inefficient use of energy, whether the waste be low-grade heat as in cooling water discharged from the condensers of a power station or the disposal of by-products and unrecovered product from a chemical plant.   The provision of special means for removing waste products from effluent streams without any effective recovery of materials for further use is clearly a secondary waste of energy. Since no processes are perfect, some wastes must always be produced — contrary to statements sometimes made, zero waste production and complete recovery and use of waste products is an engineer's pipe-dream, although as an unachievable target to promote greater efforts to reduce the quantities of wastes and promote their re-use it may have value.

If the production of wastes is considered to constitute an inefficient use of energy, then it is necessary to deal briefly in this Section with the wide range of wastes discharged in aqueous form and the ways in which any pollution arising from their discharge can be minimized.

Effect of Pollution

Pollution can be defined as any harmful or deleterious effect on the water and any animals or plants associated with it through the addition by direct human agency of any materials changing the nature or composition of the water.   A non-polluted water, then, is one which can support a normal flora and fauna and which is not offensive to man's senses.

The effect of pollution can be direct or indirect. For example, the high temperature of a cooling-water discharge may directly affect temperature-sensitive fish, or because of the enhanced temperature even after dilution in the receiving water may reduce dissolved-oxygen concentration directly and by promoting greater microbiological activity in decomposing organic material. Similarly the discharge of a slurry of non-toxic inorganic waste, e.g. gypsum, could under extreme conditions kill fish by blocking the gill membranes and impairing respiration; or, by deposition on the river or sea bed, it could destroy the organisms that form an essential part of the food chain. Since microscopic plants (plankton) constitute the basic food of fish, impairment of light penetration leading to reduced photosynthesis can also produce an adverse effect.

Direct toxicity of the constituents of wastes is of two types: first acute toxicity caused by the presence of a substance at a concentration greater than its lethal threshold level, and secondly by absorption and accumulation from the surrounding water or from food organisms of sufficient amounts to distort normal physiological processes, resulting in chronic disorder and impaired vitality. The effect of pH is to some extent an exceptional case, since as well as normally being toxic to fish outside the range 4.0 to 9.5 it can modify the toxic effect of other materials present — for example, an increase in toxicity with falling pH with water containing sulphides and the converse for ammonia (where the free un-ionized component is toxic).

Chemicals used in the treatment of process and boiler-feed waters are frequently toxic in relatively high concentrations, although harmless on dilution. But those used to prevent organic growths (bacterial and algal slimes, mussels, etc.) in cooling waters are employed because of their toxicity; consequently further treatment of the purge may be required to prevent harm to the organisms in the receiving water. Residual chlorine is highly toxic, but is normally rapidly removed by the intervention of any organic matter present. If it is present in the form of chloro-amines (formed by reaction with ammonia) the toxic effect is more persistent.

However, the majority of hazards arise from the discharge of process wastes from manufacturing industry, particularly those producing or employing manufactured chemicals on a large scale. The oil-refining industry, and to a lesser extent the petrochemical industry, falls into a somewhat different category since in addition to the water-borne chemical wastes produced the presence of oil is a complicating factor. Even in relatively small quantities, oil not only adversely affects amenity values but many of its constituents are toxic. The water pollution problems of the coal mining industry are technically simple to solve since essentially they are concerned only with the presence of finely-divided coal which can be removed by relatively simple physical methods. The coking industry (including the preparation of town gas from coal carbonization) can present severe pollution problems since the wastes produced are not only toxic to aquatic organisms but also, like some other chemical wastes, seriously inhibit the biological treatment processes in a sewage works.

Prevention of Water Pollution

Without doubt, the best method of minimizing water pollution is to employ a production process which generates the smallest possible quantity of wastes: here, of course, economic factors are crucial.   Potentially polluting industrial aquatic wastes are, in most industrialized countries, subject to a legally constituted controlling authority; most such wastes can be treated to an acceptable level by physical, chemical and biological processes (normally in combination) either on the manufacturing site or after discharge to a public sewer in admixture with domestic sewage at a sewage treatment works. Where the nature of the waste and the geographical position of the manufacturing site are suitable, the rapid dilution and dispersion available makes discharge to estuaries or to the sea by pipe-line a convenient, economic and safe method of disposal of an untreated or partially treated waste. Other methods of treatment or disposal are less commonly employed; for instance, complete combustion or chemical oxida-tion are normally used only for extremely con-centrated wastes arising from a few specific pro-cesses, while disposal to underground strata by deep borehole is only suitable if not objectionable on geological considerations.

Physical processes of treatment normally employed consist of the use of gravity methods for removal of oil and similar materials (for example by mechanically removing the surface-layer film on standing, the use of parallel-plate and similar sepa-rators, or adsorption on specially prepared adsor-bents) and suspended solids (by the use of settle-ment tanks, filters of various types, centrifuges, etc.). Many physical processes can be improved by the use of specially developed chemicals, such as flocculants, filtration aids, sludge conditioners, etc. Chemical processes frequently involve neutra-lization by acids or alkalies (particularly lime and chalk). Most other chemical processes involve the removal of specific substances, for instance by pre-cipitation, oxidation, etc.

For any waste which contains substances that can readily be broken down microbiologically, and which are not inhibitory to microbiological processes at reasonable concentrations, biological treatment is employed. For domestic sewage treatment, gravity filters randomly packed with either natural or plastic media, or activated sludge processes, are customary. For the treatment of industrial wastes by biological methods great care has to be taken that toxic or inhibitory substances are never present in sufficient concentration to impair bacterial activity: this may involve special pretreatments and/or the use of balancing tanks to permit a more constant flow to the treatment plant. Where adequate domestic sewage is not available for dilution, it is normally necessary to supply nitrogen and phosphorus as inorganic nutrients to promote acceptable treatment efficiency.

Advanced Treatment

Chemical and physical treatments can be employed directly where it is economic to recover raw mate-rials or by-products for further use. Owing to the increasing demands for water and the natural limita-tion of supplies, water recovery and recycle are be-coming increasingly accepted as an essential part of waste treatment. For many purposes (e.g. washing, hydraulic transport, etc.) the effluent from a biological treatment plant filtered on high-rate sand filters and sterilized by chlorination can be used. Where a more highly purified water is required, for example for process purposes, addi-tional treatment is normally desirable: carbon adsorption and such membrane processes as reverse

osmosis are now playing a greater part. It is likely that the one area of extensive development in the next few years will be the introduction of improve-ments of these and other advanced methods of treatment to recover water of a high degree of purity for re-use.

Water Pollution from the Nuclear Industry

The nuclear industry (see also Ch.14) is a very special case owing to the high toxicity of such materials as plutonium and of the accumulative harmful effect of even small doses of radiation. In consequence, the harmful effects of its effluents are only approached by a very small number of extremely toxic organic chemicals, where similar, although less severe, precautions must be taken. Such wastes, when they have reached a handle-able stage, are usually disposed of to the ocean deeps in specially designed sealed containers. The most harmful wastes, such as those arising from the production of fissionable material, are at present stored in specially built chambers.

19.6 Thermal effects on pollution

The cooling water required by a thermal power station for the efficient usage of its fuel via the Rankine cycle is heated in its passage through the condensers to something like $10°C$ above ambient. This water may be obtained directly from rivers, estuaries or coastal waters. If plenty of water is available cooling towers will not be used and the problem will then be to design the intakes and out-falls so that recirculation does not take place (which would reduce station efficiency) and also to ensure that local warm areas are not formed which might affect the local eco-system. Undue biological growths resulting from these warm areas have been known to choke the circulatory system.

The ecological effects of waste heat in natural waters depend on many factors, not least of which is the actual temperature rise and the geographical area. For example, a $2°C$ temperature rise in the tropics may be lethal to animals and plants already living near their lethal limits, whereas in temperate regions temperature rises of up to $15-20°C$ will not be lethal in winter.

In these temperate regions animals and fishes are rarely killed by thermal discharges, though localized increases in water temperature may cause alterations in growth, in the life-cycles or in the distribution of a species. At present considerable use is being made of cooling-water after use, for accelerating the growth of fish in fish farms.

If there is insufficient water for direct cooling, cooling towers may be used and some improve-ment in water quality occurs as a result of increased dissolved oxygen, oxidized ammonia and scrub-bing of carbon dioxide. Because a lot of water is evaporated in the towers impurities may have their concentration increased. A proportion must be bled off and replaced by fresh water, and care must be taken with the purge water so as not to produce any degradation of local water quality.

Where generation is combined with district heating by condensing the steam at higher temperature, the cooling/heating water is chemically treated and contained in a closed system, so that thermal pollution of waterways should not arise.

19.7 Noise

Noise is now becoming increasingly regarded as another form of industrial pollution. It can affect the work efficiency of people and, at worst, excessive noise can cause permanent loss of hearing.

A noise source radiates at a given power (sound power), but the noise received by the ear (sound-pressure level) depends primarily on the distance from the source and the acoustic environmental conditions. It is not practicable to measure sound power but it can be calculated from the measurable sound-pressure level. This is important in that sound-pressure levels given by manufacturers of equipment cannot be compared unless acoustical test conditions are identical, but sound-power levels are independent of the acoustic condition and are therefore comparable. The actual sound energy is very small, but it can lead to appreciable waste of energy through the impairment of the efficiency of staff.

Acceptable Scales of Measurement

The ear becomes less sensitive as the sound pressure increases, and the ear is not equally sensitive to all frequencies. A simple decibel (dB)* reading is therefore not a true indication of the subjective loudness of a sound. To take into account the changing sensitivity of the ear, instruments are designed with weighting networks to give a frequency bias similar to that of the ear. Three approved weighting scales, with results designated dBA, dBB and dBC, were at one time used for measurement at various sound-pressure levels but it was found that the A weighting network (dBA) gave reasonable results for most applications and it is now generally used. There are limitations in the use of an overall sound-level measurement that takes no cognizance of the frequency spectrum of the noise. A series of curves has been developed based on octave band measurements which take into account the sensitivity of the ear to different frequencies and the 'annoyance factors'. The most common systems in use are the noise-criterion NC[8] and the noise-rating NR[9] curves. For practical purposes NC and NR curves are interchangeable.

* Relative sound-pressure intensity at a given point, referred to a defined threshold as standard

To compare sound-pressure levels in dBA with NC/NR curves, it is necessary that the noise spectrum of the sound is similar to that used with the NC/NR curves. Where this is so, dBA values can be said to be equivalent to (NC or NR + 6) to within ±2 units.

Acceptable Criteria in Buildings and Dwelling Areas

Background NC/NR	Situation
20	Concert halls, studios for sound reproduction.
25	Churches, large conference rooms, television studios, bedrooms in private houses.
30	Living rooms, libraries, bedrooms in hotels, cinemas.
35	Classrooms, private offices, small restaurants.
40	Drawing offices, large offices, laboratories, large restaurants, shops.
45	Computer rooms, accounting-machine rooms, canteens
45—55	Light machining shops, light-industrial areas.

Methods of Assessing Nuisance

Apart from the necessity of controlling noise within buildings, consideration has to be given to the problems of external noise. British Standard 4142:1967 provides a method of rating industrial noise affecting mixed residential and industrial areas and utilizes measurements in dBA. The method is to compare the sound level, to which are applied corrections for the impulsive, tonal or intermittent character of the noise, with a background level either measured or based on a basic 50 dBA level adjusted for the type of district, time of day etc., the difference being an indication of the likelihood of complaints.

The foregoing has been concerned with noise as a nuisance. But of more serious concern is the effect of high noise levels which could lead to loss of hearing. A code of practice has been issued which lays down limits which should be regarded as the maximum acceptable noise levels. The code recommends that hearing protection or sound reduction should be carried out if levels exceed 90 dBA or the equivalent for higher levels of shorter exposure. Noise problems can be divided roughly into those associated with mechanical services creating annoyance in offices, flats, etc., and industrial noise causing possible health problems. Both can give rise to nuisance to residents in surrounding areas by noise breakout.

With mechanical services, plant rooms and associated distribution systems are the main problems. The major noise sources are fans, boilers and associated burner equipment, compressors, cooling plant and pumps. Manufacturers will supply the sound-power levels of their equipment, but in order to ensure that it remains within the specification all equipment must be kept in good mechanical condition. Particular problems arise from worn bearings, loose panels and guards, unbalanced fan motors and badly mounted plant. With air-distribution systems, noise is due to high air velocities, turbulence, obstruction in ducts, panel drumming and vibration. In the main, fan noise is transmitted along the ductwork.

In plant rooms, breakout can be a problem and care must be taken to provide seals at all points where ducts and pipes pass through walls etc. If the plant-room noise is particularly high and likely to cause complaints in adjacent areas, then consideration may have to be given to enclosing noisy machines, treating the structure walls, replacing plant with quieter type, use of local attenuators, and anti-vibration mountings. With duct systems, attenuators may have to be fitted and ductwork lined. All pumps should be fitted with flexible connections and anti-vibration mountings.

Industrial noise is basically associated with impact or vibration — presses, gears, cutting operations, drilling etc., and gas or liquid flow — fans, pneumatics, hydraulics, steam, combustion etc.

To reduce or treat noise after eliminating mechanical faults the following should be considered:

(*a*) Replace — usually impracticable and uneconomic.
(*b*) Redesign and modify — can be expensive but should be first consideration.
(*c*) Enclose — have to consider problems of access and ventilation.
(*d*) Hearing protection — usually an easy way out but does not cure problem.
(*e*) Isolate process in one area — if one type of machine in many is the problem. Hearing protection required but confined only to operators of offending machine.

Typical Noise Levels

Sound-pressure level
dBA

0	
	Threshold of hearing
	Acoustic laboratory
10	
20	
	TV studio
	Whisper at 1 m
30	Library
40	
	Average house
	Quiet house
50	Light traffic
	Large store
60	
70	Motorway traffic
	Typing pool
	Tabulating room
80	
	Plastic-injection-moulding area
	Printing-press shop
90	Boiler house
	Corrugating paper machine
	Weaving machines
100	
	Large machine shop
	Pneumatic drill at 1 m
110	Riveting
120	
	Threshold of pain

19.8 Solids

Accumulations of solid refuse resulting from the energy industries consist chiefly of tips (including both spoil heaps and lagoons) from coal-mining operations, and ash lagoons from power stations fired by pulverized coal. Increasingly, PF ash is used for mine stowage, road making, and the manufacture of tiles, which should reduce the need for storage in lagoons. Tips are mainly of environmental concern from the safety aspect, although 'landscaping' of tips is desirable primarily for aesthetic and utilitarian reasons.

In some of the principal coal-mining countries, following major disasters such as those at Aberfan[10] and Buffalo Creek[11], much attention is being given to tip safety. In the UK extensive regulations[12-14] are now in force designed to ensure that, whether in use or not, all tips associated with active or non-abandoned mines are regularly inspected and reported upon from the safety aspect, and that any remedial measures necessary are effected. The regulations are most stringent and detailed for 'classified' tips: in respect of tips consisting mainly of solid material, a classified tip is one covering an area of over 10 000 m^2, or exceeding 15 m in height, or lying on land with an average gradient of over 1 in 12. A lagoon also falls within the definition of a classified tip, provided its height exceeds 4 m or its volume exceeds 10 000 m^3. The regulations include provisions concerning the dimensions, design and construction of new tips and extensions of existing tips. Although under no statutory obligation, the National Coal Board have laid down standards of maintenance, inspection and reporting also for the tips they own at abandoned mines.

Among the factors that merit consideration in assessing the safety of an existing tip[15] are knowledge of whether the tip has been built over springs or has surface water flowing into it; the nature of the material tipped, which may include dry spoil such as run-of-mine dirt and boiler ash, or wet spoil such as washery discard and slurry; the thickness and degree of compaction of each layer of tipped material; the presence of water-tables within the tip; possible steepening of the slope of the tip through mining below or close to it, or through surface gullying caused by uncontrolled drainage; interference with drainage within the tip and/or its foundation by mining subsidence; increase in water level within the tip by seepage of surface or lagoon water, degradation or swelling of the tip material, or freezing of the tip surface; and possible adverse effects on stability of the burning of tip material.

Regulations made under earlier legislation, and affecting many tips, may necessitate conforming with requirements of shape, height and slope[16]; employing means of preventing, or minimizing the effects of, combustion of deposited mine refuse[17]; and controlling polluting drainage and the disposal of fine tailings[18].

19.9 Waste and reject fuels

It is now more necessary than ever to consider seriously the effective utilization of the energy available in waste materials; the sources are numerous and the potential considerable.

There has always been a basic divergence of opinion regarding the wisdom of burning waste as opposed to the recycling of waste for process reuse; there is pressure on both energy resources and material resources and the latter tends to increase market potential for recycling. Thus recovery of paper waste for repulping in paper mills is periodically attractive in the UK, and may then influence the calorific value of ordinary town garbage. Similarly, recovery of waste engine-sump oils for re-use as lubricant may be sound in the long term, and bagasse is also becoming increasingly valuable as a material.

In the past those attempting to store and sell waste for re-use have often been faced with an acute disposal problem as the market value has fluctuated, and many have found it not only prudent but profitable to install plant for the burning of waste combined with heat recovery.

Some idea of the range and calorific value of waste fuels available is given in *Table 2*. (Domestic refuse is discussed in Ch.6.4).

Common problems with most solid wastes are:
1. Variability of moisture content and size or bulk.
2. Difficulty of phasing supply to coincide with factory heat demand — storage is often necessary.
3. Seasonal variations and effect on supply.
4. Necessity of supplementary fuels to boost waste-heat output when waste material is deficient.

But in addition many solid waste materials are cellulose-based and high volatility can be a problem, usually calling for generous combustion-chamber design and carefully designed secondary combustion air arrangements to ensure compliance with smoke-control ordinances.

Table 2   Typical analyses of various trades waste*

Material	Moisture (%)	Gross CV (MJ/kg)	Ash (%)
Joinery-shop refuse	12.0	18	2.0
Cedar-wood chips	28.4	14.2	0.3
Wood briquettes	13.6	16.4	0.8
Hardboard offcuts	6.4	19.0	2.7
Cardboard and waste paper	8.2	15.8	7.9
Jute waste	9.9	17.6	2.7
Fibre waste	9.7	16.6	1.5
Curled hair waste	14.4	16.2	2.0
Oil cotton waste	4.6	23.6	0.8
Cotton-impregnated rubber	2.5	18.2	34.7
Bitumenized hessian	4.3	25.4	1.45
Bitumen-impregnated chipboard	2.05	27.8	7.2
Bitumen waste	0.2	38.5	2.8
Rubber waste	—	34.0	19.4
Print-works waste	29.9	20.4	14.4
Nitrocellulose waste	0.9	8.5	0.15
Spent tan	70.1	6.1	1.7
Bagasse	40.0	11.7	2.1
Spent coffee	65.0	8.1	1.5
PVC	—	22.8	2.0
Polyethylene	—	45.6	1.2
Polystyrene	—	36.8	0.5
Polyurethane	—	26.0	4.5

* Extract from *Incineration at Source*, J. R. Teale, North West Gas Board, UK, April 1970. *Courtesy British Gas Corporation*

There is also a basic difference in mode of combustion; baled or lump materials, e.g. baled paper waste, behave differently from finely divided materials, e.g. shredded or hogged waste. The burning rate of baled materials is generally low and the rate of volatile evolution is inhibited, whereas finely divided materials have a high rate of volatile release and combustion-chamber design is more stringent. Handling of baled or bulky materials however is more costly, and shredding or hogging produces a material more suitable for mechanical handling and combustion; but power costs for pulverizing or hogging can be high.

Conditions with regard to disposal costs of waste materials vary enormously depending on quantity and degree of waste gas cleaning required, and it is necessary to treat each case on its merits.

Wood Waste

Wood waste is an important source of energy, being a principal raw material in many industries where considerable waste is generally unavoidable.

In the conversion plant or sawmill where the logs are first treated, wet bark, edgings and sawdust are produced from the green timber. The moisture content may vary up to 80% depending on whether the logs are floated; in most conversion plants moisture varies from 25 to 50%, whereas kilned wood and its derived waste moisture varies from 10 to 15%. The effect on calorific value is approximately as follows:

Moisture content (%):	0	10	20
Net CV (MJ/kg):	20.6	18.2	15.8

Moisture content (%):	40	60	80
Net CV (MJ/kg):	11.1	6.4	1.65

(8850—715 Btu/lb)

The fixed-carbon content of dry wood is about 17%; clearly most of the combustion takes place in the gaseous state.

Wet wood-waste from mills used to be disposed of by incineration in external furnaces but the tendency is toward dryers to reduce moisture to about 40% to facilitate steam raising; green-wood-waste plants are normally dual fired, with oil or coal as auxiliary fuel.

Problems peculiar to wood-waste are: handling, storage and conveying of mixed materials; relatively large combustion chamber; great risk of smoke emission; high cost of refractory maintenance (external furnaces). Modern techniques overcome these satisfactorily, and often comprise a mixture of mechanical handling, drag-link, screw or belt, and pneumatic handling. Handling and burning of off-cuts etc. may be facilitated by automatic saw, hogger, pulverizer or shredder. Wet wood-waste is best dealt with by the hogger, dry materials often by the pulverizing mill. Power consumption of hogging or pulverizing mills depends on size of waste and mill and can vary from 10 to 75 kW h /ton.

The older external Dutch oven is being superseded by improved methods and many commercial boilers, both watertube and shell, are available; these may embody inclined-grate, spreader-stoker or suspension burning and are sometimes fitted with supplementary oil or solid-fuel firing.

Dust from sanding machines, mixed with air, can be very explosive and at one time this was almost always separately collected and disposed of. But it can be quite successfully and safely burned in suspension at suitable concentration in a manner similar to pulverized coal, using precautions to prevent flaming back, and flame-failure control.

Sawdust and pulverized wood-waste can also be manufactured as briquettes; several high-pressure machines are commercially available in Europe and America. Pressures up to and exceeding 138 MPa are used with preliminary drying to produce a very hard often cylindrical briquette, CV about 17.5 MJ/kg. Briquettes can also be produced at medium pressure using a binding agent such as coal-tar pitch, molasses or petroleum residues.

Miscellaneous Solid Waste Fuels

Many other wastes can be utilized for production of heat; being of vegetable origin they are chemically allied to wood (cellulose) and the CV mostly depends on moisture content.

Industrially, paper and cellulose film is commonly burned *in situ* and the waste heat used for steam-raising or hot water. The method of combustion depends greatly on the quantity and nature of the waste, but it may be burned in a baled, in a loosely compacted or in shredded form. A common method is the external combustion chamber with sloping grate followed by vertical-type smoke-tube boiler, where difficult materials are present; e.g. with bitumen-coated papers, after-burners may be necessary to prevent smoke emission.

Bagasse or residual crushed sugar cane is used extensively for producing all heat and power requirements in raw-sugar production; reciprocating inclined grates or spreader-stokers in cellular furnaces (to facilitate maintenance) are commonly used in conjunction with watertube boilers. Spent coffee bean, paddy husk etc., can also be conveniently dealt with on the inclined reciprocating grate, using the waste heat for steam production.

Liquid Wastes

These include refinery sludges, waste sump oils, paper-mill liquors, paint wastes, such as thinners, chemical plant waste-liquors, hydrocarbons etc. Waste engine-sump oil provides an excellent heat source, usually burned easily in traditional burners. For economic reasons it is often necessary to use the same boiler burner for both fuel oil and sump oil but it is preferable to store them separately so that the plant can be changed over from fuel oil to sump oil operation; this generally takes but a few minutes. Excessive waxing can take place on mixing and should be avoided. Many installations use sump oil quite successfully; lead emission is usually not sufficiently high to be serious, although higher from petrol-engine sumps (from lead-doped petrols) than from diesels.

Refinery sludges are usually high in acid content and can pose an SO_2-pollution problem unless gas washing is employed; they have also variable viscosity and composition, are sometimes semi-solid, and in general because of corrosion, call for simple cheap replaceable pipe burners such as the scent-spray type or pipe-in-pipe burner, steam-blast atomized using generous combustion chambers because of poor atomization; heat recovery is often by traditional waste-heat boiler. Modern practice tends toward the use of homogenizing pumps with circulation to slurry tanks to produce a more uniform product; combustion takes place through a wide-bore burner jet in a rotary kiln.

Waste liquids from paint manufacturing such as white spirit are often obtained mixed with other wastes such as paint, paper etc. These are usually dumped into a tank whence the liquid is strained and pumped to a holding tank with stirrer, thence by gravity to a simple blast-pipe burner which may be located in the 2nd or 3rd cell of a cellular-type incinerator, and finally followed by a waste-heat boiler.

Waste Gases

Industrial waste gases are prolific and properly include blast-furnace gas, coke-oven gas, refinery tail-gases, carbon monoxide, hydrogen, volatile carbon sulphides, and many others. In most cases output of gas is variable and insufficient for the heat requirements of process. Generally there is little problem in the combustion of such gases and they are often burned directly in boilers in conjunction with auxiliary fuel to make good the deficiency of heat due to fluctuations.

With very low-CV gases, e.g. where diluted with inert gases and when several different waste gases are available, combustion can be accomplished in multi-port combination burners which may also embody other auxiliary-fuel burners and arranged for firing into horizontal or vertical refractory chambers with waste-heat recovery systems.

Worthy of special mention is methane (CH_4) which is produced during anaerobic treatment of sewage sludge, the most widely used single process for sludge treatment in the UK. The gas produced, usually about 70% methane and 30% CO_2, has a CV of some 23.5 MJ/m^3. The amount produced is about 0.03 m^3 per person per day and can be used to produce gas-engine power for the sewage works including the air flow for aerating the activated sludge; waste heat from the prime mover is used to heat the digestion tanks (usually 35–40°C) where the methane is produced. The Beckton sewage plant in Essex includes eight gas turbines thus driven; the Mogden main drainage station in West London has had a similar system for over 30 years.

Further information can be obtained from the Water Pollution Research Laboratory, Stevenage. Anaerobic digestion of farm manure could assume greater significance in future, e.g. pig waste produces *c.* 0.24 m^3 per pig per day which could be used for generating electricity, with W.H. recovery for heating the digester.

Methane is explosive when mixed with air between the limits 5 and 14% and adequate safety precautions must be observed.

19.10 Environmental legislation

Up to ten years ago comparatively few of even the more advanced industrialized countries possessed effective or comprehensive legislation controlling the discharge of gaseous, liquid and solid wastes to the environment. And even in those countries like the UK where antipollution legislation was of long standing (to say nothing of the extensive case history in the common law) control measures were not effectively pursued except where direct hazards to health were apparent. A major exception to this was provided by legislation arising from international agreements for the disposal of radioactive wastes, where the highly toxic nature of some of the materials plus long-term hazards arising from their radioactivity caused relatively easy acceptance of efficient control measures from an early stage.

But in recent years recognition has been more general of the need to preserve reasonable purity of natural waters and the atmosphere and to prevent destruction of amenity values by the uncontrolled tipping of solid and similar wastes to land. This, together with political pressure arising from public opinion, has led to the introduction of controlling legislation in all urbanized countries. International conventions, for example the Oslo and Paris Conventions concerning the disposal of wastes to the sea, have had comparatively little impact yet upon national legislation, other than for radioactive wastes.

National legislation controlling pollution of the environment by noise, liquid, gaseous and solid wastes is so diverse and in some cases so dependent upon collateral legislation in subordinate states and provinces as to defy any but the broadest summary, even of control procedures. For example, in spite of the establishment of the Environmental Protection Agency in the United States with the overall responsibility of ensuring by the issue of permits that waste disposal does not cause environmental harm, it is significant that over 2500 separate environmental bills were introduced into the State legislatures, during the period 1971 – 1973, which supplemented and extended Federal legislation.

In industrial countries emission limits into the open air are or should be required for many pollutants; among others, particularly SO_2 and SO_3 (Ch. 18.3, 19.3, 20.1, 20.2); NO_x (Ch.18.3, 19.3); hydrocarbons; CO (Ch.18.3); Pb.

It is essential that the reader should consult recent legislation in his own country.

There is a tendency to claim that environmental laws are luxuries to which developing countries cannot aspire. This attitude is intrinsically undesirable, but can ultimately even prove more costly over a period of years should development escalate (cf. Japan).

The UK philosophy is recommended to any concerned with fresh legislation for a developing country.

Within continental Europe (and similarly, in the USA and Canada), the philosophy and practice of environmental control differ radically from those which have been applied with considerable success in the UK for many years. Where such differences are related to geographical factors, for example the need to practice extreme water conservation in South Africa, parts of Australia, India, etc., or the massive aggregation of potentially air-polluting industry in the Ruhr in West Germany, the application of extremely strict standards or requirements for discharges is understandable. In most countries legally fixed standards relating to the quantity and composition of wastes as discharged have been applied, although the standards themselves are not necessarily related to what is scientifically required to protect a given local environment. In some cases the direct application of rigorous standards has proved impossible, resulting in extensive non-observance and a consequent deleterious effect upon respect for the law. There is little doubt that the view that pollution control standards should be appropriately related to the needs and capacities of a particular environment is gaining strength among those professionally engaged in the field, although public and therefore political pressure may well prevent less rigorous systems being introduced in the near future, particularly as this type of legislation is very recent. It is significant that in the water-pollution control legislation passed by the US Congress in 1972, the Environmental Protection Agency was required to issue control regulations to the effect that by July 1977 industry should use the best practicable control technology available to treat and restrict effluents. However, zero discharge of industrial pollutants remains in principle the aim, where practicable, by 1983.

In the UK, air-pollution legislation has been based for a century upon the employment of best practicable means. Under this philosophy the measures to be taken in a particular case are determined in the light of the needs and capacities of the local environment, the financial implications, and the current state of technical knowledge. It is no less stringent a control than that of a law which prescribes generally applicable standards, but in the hands of a skilled and experienced Inspectorate it permits a realistic and progressive improvement which is also responsive to technical changes. In practice a similar philosophy underlies the UK legislation for the disposal or emission of solid and liquid wastes. The precise current regulations for the UK cannot in any event be included since the detailed pattern is changing in early 1975, during the printing of this book.

Consents are given to the discharger by the controlling authority which state the limits for quantity and composition. Such consents are reconsidered periodically, thus enabling steady progress to be made towards the reduction of any pollution. One great advantage of such a system is that disputes over provisions of a consent can be resolved by the Minister of the Crown responsible, without recourse to a court of law with the consequent long delays in reaching a decision and heavy costs.

References

1 *Chimney Design Symposium*, Conference Publication, University of Edinburgh CICL, Edinburgh, April 1973

2 Shell, *Chimneys for Industrial Oil-fired Plant*, Shell Marketing Ltd, London, 1973

3 Brightside, *Chimney Design Manual*, 1968 Revision, Brightside Heating and Engineering Co. Ltd, R & D Department, Portsmouth, 1968

4 Ministry of Housing and Local Government, *Chimney Heights, Second Edition of the 1956 Clean Air Act Memorandum*, HMSO, London, 1967

5 *Simplified Methods for Measurement of Grit and Dust Emission:* BS 3405:1971, British Standards Institution, London, 1971

6 *The Protection of Structures against Lightning*, CP 326: 1965, British Standards Institution, London, 1965

7 *Steel Chimneys*, BS 4076: 1966, amended 1972, British Standards Institution, London, 1966, pp 23—26

8 Beranek, L. L. *Y.J.A.S.A.* 1956, Vol.28, 833

9 Kosten, C. W. and Van Os, G. J. *Proc. NPL Symp. No.12 on Control of Noise*, HMSO, London, 1962

10 *Report of the Tribunal appointed to inquire into the Disaster at Aberfan on October 21st, 1966*, HMSO, London, 1967

11 Seals, R. K. *et al*, *Failure of Dam No.3 on the Middle Fork of Buffalo Creek near Saunders, West Virginia, on February 26th, 1972*, Report prepared for the Committee on Natural Disasters, National Academy of Engineering [USA], 1972

12 *Mines and Quarries (Tips) Act*, HMSO, London, 1969

13 *Mines and Quarries — the Mines and Quarries (Tips) Regulations*, HMSO, London, 1971

14 *Mines and Quarries — the Mines and Quarries (Tipping Plans) Rules*, HMSO, London, 1971

15 National Coal Board, *Spoil Heaps and Lagoons*, 2nd draft, NCB, London, 1970

16 *Town and Country Planning Act*, HMSO, London, 1962 and 1968

17 *Clean Air Act*, HMSO, London, 1956

18 *Rivers (Prevention of Pollution) Act*, HMSO, London, 1951 and 1961

General Reading

19.3

(A) Jackson, R., *Mechanical Equipment for removing Grit and Dust from Gases*, BCURA, Leatherhead, UK, 1963

19.7

(B) *IHVE Guide 1970*, Institution of Heating and Ventilating Engineers, London

(C) British Standards Institution, BS 4142: 1967, *Methods of Rating Industrial Noise affecting Mixed Residential and Industrial Areas*

(D) *Code of Practice for reducing the exposure of employed persons to noise*, HMSO, London

19.10

(E) Jenkins, S. H., *Aspects of Environmental Protection*, IP Environmental Ltd, 1974

Chapter 20

Corrosion, erosion and mineral deposits on heating and moving surfaces from fuel impurities

20.1 Fireside erosion and corrosion by fuel impurities

Coal firing
Corrosion by oil-ash deposits
Methods of reducing corrosion
Residual life considerations

20.2 Low-temperature corrosion in boiler plant

Coal-fired plant
Oil-fired plant

20.3 Fuel and air contaminants in corrosion of gas-turbine blades

In this Chapter, certain effects of relatively minor constituents of carbonaceous fuels that can be of great importance both in relation to fuel efficiency and to the life and availability of plant are discussed.

20.1 Fireside erosion and corrosion by fuel impurities

The mineral impurities in coal and residual fuel oil (metal oxides, silicates, sulphur compounds etc.) are modified in the flame and by reaction with each other and with the principal flue-gas components, i.e. H_2O, SO_2, O_2 and CO_2. The mixed oxides, silicates, sulphates and vanadates so formed are responsible for the deposits which adhere to heat-exchange surfaces and act as a key for further accumulation of ash. Silicates and aluminosilicates predominate in slags and ashes associated with the combustion of coal. The aluminosilicate particles are mostly fused in the flame and appear as glassy spheres of 5–200 μm diameter. Silica particles, which may amount to 5–20% of the total ash, are usually unmelted sharp-edged particles which can make a notable contribution to erosion if a stream of the ash particles is blown against metal tubes. With fuel oil the deposits are mostly the sulphates and vanadates of sodium together with iron, calcium, nickel and silica, and which are present in relatively small amounts. Because of this difference, oil and coal will be considered separately. However, in both cases the sulphur oxide equilibrium:

$$SO_2 + \tfrac{1}{2}O_2 \rightleftharpoons SO_3 \qquad (1)$$

plays an important part in determining the corrosion behaviour of the deposits. In the gas phase this equilibrium is not established except during the post-flame period when the gases are at temperatures of 1200°C or above and the proportion of the sulphur existing as SO_3 is less than 1% (v/v). However, at the heat-exchange temperatures (i.e. superheater and reheater tube surfaces) the proportion of SO_3 which should exist with reaction (1) at equilibrium is much higher and it is important to note that within the deposit layer the SO_3/SO_2 ratio does approach equilibrium; catalysts are present in the form of iron and vanadium oxides to ensure this, and there is adequate time. Hence the excess air used should be the minimum to ensure adequate combustion (see Ch.20.2). SO_3 can react with the deposit and contributes to the molten or viscous semi-molten nature of tube deposits. Sodium sulphate is a major constituent of deposits. It arises from reaction between SO_2/O_2 or SO_3 with the sodium chloride present in oil or coal fuel, and is the principal sodium compound to deposit on heat exchange surfaces from the

gases as they cool, except under excess-fuel or reducing conditions when the chloride will be able to deposit[1].

In coal plant the deposited sulphate is soaked up by the ash particles, whereas with fuel oil the sulphate-rich deposit can form a thin liquid film directly on the metal surfaces.

It is the sulphate salt at the metal surface which causes high-temperature corrosion, and the mechanism of its action is discussed below.

Coal Firing

Slagging or Erosion by Coal Ash

Slagging is due to partial melting and sintering of ash particles, causing the particles to stick and melt together. The flow properties of coal ashes have been studied in great detail as described in the work of Watt[2,3] on coal slags in terms of their compositions, expressed in terms of the oxides SiO_2, Al_2O_3, Fe_2O_3, CaO and MgO. This work, though relevant to cyclone firing, has much bearing on slagging problems in other furnaces using coals with low ash-fusion temperatures or where excessive local heat release can melt ash of relatively high fusion temperature. The presence of slags on metal surfaces can give rise to corrosion of the underlying metal, perhaps by causing reducing conditions to exist at the slag/metal interface. Kiss, Lloyd and Raask[4] have looked at possible solutions to slagging problems. The wetting properties of slags are discussed by Raask[5]. Another problem affecting tube life is erosion of surfaces by flying particles of ash at high gas velocities, sometimes difficult to separate from direct corrosion. Protective oxide films will be destroyed and thus erosion exacerbates a corrosion problem. Raask[6] has studied this problem in some detail in terms of the abrasiveness of ash, the velocity of impacting particles, the angle of impaction and the metal temperature. For a given amount and type of ash in the flue gas, the rate of metal loss is proportional to the impact velocity to the power 3.5. Wear rates are insignificant when the flue-gas velocity is in the range 15–20 m/s whereas at 30–40 m/s tube failures might be expected after 10 000–50 000 h service. The contribution of soot-blowing to erosion is also considered together with possible solutions. The influence of the amount of the ash and its type — e.g. sharp particles such as unmelted silica cut the steel more readily than the rounded aluminosilicate particles — is far smaller than that of gas velocity. *Table 1* gives some erosion losses with two different coals A and B. The wear can be shown to be directly proportional to ash/carbon ratio. To take account of sharpness, I, the abrasive index, or the ratio of actual wear with a given ash to maximum wear when the ash is replaced by ground quartz of the same particle size and concentration, is introduced.

Corrosion of Steam Wall Tubes

Steam-generator tubes are usually made from mild steel and because of their relatively low operating temperatures (e.g. 400°C) they normally suffer negligible corrosion. But local corrosion can occur if there is overheating owing to bad flame distribution or if there is poor coolant flow within them. If the gas atmosphere surrounding the tubes is essentially reducing because unburnt coal particles have escaped the main flame envelope, or if there is poor fuel/air mixing, sodium chloride[2] can deposit and trigger off rapid attack of the metal. There is often an incubation period of 1000–5000 h before this happens. Coals with high sodium chloride contents are known to accentuate this type of attack. Attention to mills and classifiers to give an acceptable particle size distribution is an obvious course of action when such attack is suspected, and if this fails attention to combustion equipment is suggested, e.g. air/fuel distribution and burner angle. The existence of reducing conditions can be confirmed from deposit analyses, e.g. the presence of chlorides or sulphides in deposits[1,7].

Corrosion by Coal-ash Deposits

The chlorine content of a coal is a useful and easily measured guide to the concentration of the volatile halides of sodium and potassium which react with the oxides of sulphur in the flue gas to produce the predominantly sulphate deposits; there is usually enough sulphur to convert all the volatile alkali-metal salts to sulphates. Coals of medium and high chlorine content (0.15–0.35 wt % and >0.35 wt % NaCl) are indicative of a high corrosion risk; the effect is most pronounced when metal temperatures are above 600°C. The excess sulphur makes the deposits acidic and more actively corrosive; coals with the higher sulphur contents (>2.9 wt %) give measurably more corrosive deposits than those with >1.5 wt %. But the effect on corrosion of variations of sulphur content is less striking than that of the halides. For a given concentration of sodium and sulphur, a coal of higher ash content gives less corrosive deposits,

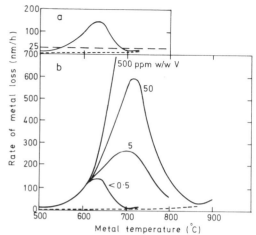

Figure 1   Corrosion of 18 Cr/12 Ni/Nb steel by:
(a) sodium sulphate deposits (2000 ppm v/v SO_2 and 1% O_2 in the flue gas);
(b) sodium sulphate/vanadium deposits (2000 ppm v/v SO_2 and 1% O_2 in the flue gas. 50 ppm w/w Na in the oil and different amounts of V).
Flue-gas temperature 1100°C.
— — 'acceptable' corrosion = 25 nm/h
- - - oxidation in air

presumably because the ash soaks up the alkali deposits.

In the presence of a sulphate deposit and in a gas containing SO_2 and O_2, all steels exhibit a sharp increase in corrosion rates at metal temperatures of about 590–620°C, and much greater than the oxidation rate of the steel in air at the same temperature. *Figure 1a* demonstrates the effect of metal temperature. Above 670°C the metal oxides become less soluble in the deposited liquid salts, and the corrosion rate decreases. Metals of higher chromium content generally have the best corrosion resistance, and the limiting acceptable rate is then reached at a higher metal temperature, e.g. 25 Cr/20 Ni% has a similar corrosion rate to type 347 stainless steel at a temperature 20–30°C higher. A duplex tube consisting of a type 347 steel inner surface with 25 Cr/20 Ni% on the outside has been shown to resist corrosion where type 347 had prematurely failed.

The principal operating factors that have a significant effect on corrosion rate are: (1) the temperature at the interface between the outer metal oxide and the innermost layer of deposit (the mid-wall tube temperature is normally measured by thermocouple and this must be allowed for); (2) the position of a tube in a bank (corrosion rate of a leading tube is often 2 to 3 times greater than that of a tube six rows in); (3) the temperature of the local flue gas and the overall heat flux.

Allowances for these factors must be taken into account in both design and plant operation. All the factors above increase in importance if the

Table 1   Typical erosion rates of mild steel with coal-ash impaction

	Metal-loss rate (μm/s)	
Flue-gas velocity (m/s)	Coal A Ash/carbon 1:2.4 (w/w) $I = 0.4$	Coal B Ash/carbon 1:4.7 (w/w) $I = 0.2$
15	6.8×10^{-6}	1.7×10^{-6}
25	4.1×10^{-5}	1.1×10^{-5}
35	1.3×10^{-4}	3.3×10^{-5}
45	3.2×10^{-4}	8.0×10^{-5}
55	6.4×10^{-4}	1.6×10^{-4}
60	8.7×10^{-4}	2.2×10^{-4}

outer metal temperature is above 600°C. Low-alloy steels, because of strength limitations, are rarely designed to operate above this temperature in modern high-pressure plant and this may explain why corrosion is usually low (<25 nm/h) unless there are serious faults in the design. However these effects are more readily observed with types 316, 321 and 347 austenitic steels. Such steels operating at 620°C with a medium-chlorine coal (0.15—0.35 wt% NaCl) at a gas temperature of 1050°C will have a corrosion rate of about 25 nm/h; if the gas temperature is increased to 1150°C, the corrosion rate will increase to 50 nm/h. An increase in metal temperature at a gas temperature of 1050°C would increase the corrosion rate from 25 to 70 nm/h; an increase in chlorine content of the coal (to above 0.35 wt % NaCl) would increase the corrosion rate to 60 nm/h. A further increase in metal temperature to 660°C at this gas temperature would increase corrosion threefold (180 nm/h).

Corrosion by Oil-ash Deposits

Residual fuel oils contain between 50 ppm w/w and 300 ppm of sodium present as the chloride and vanadium as organic compounds in amounts depending on the origin of the crude; Western Hemisphere crudes are typically about 200 ppm whereas Middle Eastern crudes average 50 ppm, but with even lower concentrations, down to 10 ppm, vanadium compounds appear in the deposits on superheaters and reheaters. Nickel and iron compounds in the oil appear on the tubes; so do clay-mineral components, but these ash-forming materials are in relatively small proportion compared with coal. Typically the deposit on the tubes will be 1—2 mm thick after a year in service. Sodium appears on the tube as sulphate, vanadium as complex vanadates. These two elements are responsible for the corrosiveness of the deposits. The ratio between them (i.e. V/Na wt ratio in the fuel) can be a useful guide to corrosiveness (see Coats[9] for the background chemistry). Other factors mentioned previously apply equally to oil and coal, except that the temperatures at which corrosion occurs are perhaps lower for oil, particularly for austenitic steels[7,10]; a 'bell-shaped' corrosion/ temperature curve has been found to apply for oil as for coal and a sharp increase in the corrosion rate occurs at 600°C with a peak between 700 and 720°C. The rate then falls until at 800—850°C the curve is almost coincident with that obtained in air oxidation without deposit. There is less firm evidence on corrosion resistance of 9% and 12% Cr ferritic steels at metal temperatures above 600°C; the reason for this is that the corrosion of ferritic steel follows a parabolic rate law and very long times are required to observe specific effects. Stainless steels follow a rectilinear corrosion/time curve.

Many comprehensive studies of corrosion in oil-fired plant are available, and the reader is referred to the extensive tests made at Marchwood power station where the gas temperature was 1100°C[10]; corresponding tests have been made at Bankside power station with a flue-gas temperature below 1000°C. On the basis of this work, the CEGB have concluded that the temperature conditions are so critical at a metal temperature above 590°C, that steam temperature will be maintained at 545°C (metal temperature c.580°C) and not increased to 565°C, as is common practice in coal-fired plant. At the present time high reliability of modern plant is usually worth more than a marginal increase in overall efficiency.

Corrosion can be combatted by the use of magnesium-based additives to react with acidic vanadium oxide[11]. Recommended rates of addition of magnesium (as MgO or $Mg(OH)_2$ usually) vary from two to five times the weight of vanadium in the fuel oil, but the average sodium content of the fuel must also be taken into account. The effectiveness of the additive will depend on its particle size and the efficiency of dispersion in the flue gas. In general the quantities of additive required will depend on the predominant factor involved, e.g. high gas temperature or high V/Na ratio. Sulphur contents of up to 4 wt % in a fuel oil do not have a significant bearing on corrosion up to 620°C. Above this temperature an increase in sulphur from 0.5 to 4.5% might double the corrosion rate.

Methods of Reducing Corrosion

Coal-fired Plant

Furnace-wall corrosion is often a localized effort owing to the proximity of a particular burner arrangement, and modification to use more expensive metals in these areas can be considered: e.g. a coextruded tube comprising an austenitic outer surface on a mild-steel base appears to be a promising solution. Similarly, local overheating of superheater and reheater tubes can be overcome (as mentioned earlier) by using a duplex tube of 25 Cr/20 Ni on, say, Eshette 1250.

Shielding of tubes suffering corrosion can be used to advantage; the outer surface temperature may thus be increased to a temperature where the bell-shaped curve is decreasing at higher temperatures. These shields can take the form of rigid shaped plates or relatively thin flexible bandages; the method will depend on the conditions of application. Suitable materials are ASSI 446, AISI 310 and Sichromal 12. 18 Cr/12 Ni steels can be used provided that the temperature is restricted to 700—800°C. Heat transfer is likely to be decreased by 10—20% on bandaged areas and this must be carefully taken into account when this course of action is taken.

Problems of hanger and support material can be overcome by the use of more exotic alloys, e.g. 25% Cr ferritic steels, 50% Cr/50% Ni (IN671 or IN657). However, design solutions should also be considered. Lugs directly subjected to flue-gas flow are more likely to suffer than those fixed to the rear of a tube where some protection is given, and they do not contribute significantly to local heat transfer.

Oil-fired Plant

Solutions via materials are more difficult here because the bell-shaped corrosion curve, *Figure 1b*, starts to decrease only at much higher temperatures than in the coal-fired case. It is best to avoid hot spots or to ensure that surfaces likely to operate at temperatures above $590°C$ are exposed to flue gas below $950°C$. Steels such as 50 Cr/50 Ni have been shown to have good resistance to attack of sodium/vanadium deposits and these can be considered for hangers and support materials. Suitable additives have been mentioned earlier.

Residual Life Considerations

In many situations where plant is subject to corrosion the simplest and cheapest solution is to let corrosion occur at a known rate and then carry out repairs as necessary at a chosen and predicted time when expensive outages will not result. Corrosion will reduce the original wall thickness, and as this occurs, the tube is less able to contain the internal pressure, this will eventually cause the tube to rupture. If corrosion is carefully monitored on an annual basis and the operating temperature, the internal stress, the original wall thickness, and the operating time are known, then the remaining life for safe operation can be predicted. It is not possible to deal with its calculation here in detail but the following can be said in general terms. A corrosion rate up to 10 nm/h can usually be tolerated without significant reduction in life below the design expectation, and corrosion rates up to 25 nm/h are often acceptable. Corrosion rates of 100 nm/h imply that there will be a significant reduction in life and a calculation of residual life should be undertaken. If the corrosion is greater than 300 nm/h remedial action must be undertaken urgently. The following 'rule of thumb' assessment can be used for a conventional boiler tube designed on the basis of 100 000-hour stress rupture: if the corrosion rate in nm/h times 200 000 is greater than the wall thickness (mm) there is a risk of premature failure.

20.2 Low-temperature corrosion in boiler plant

Coal-fired Plant

In these, condensation of concentrated sulphuric acid is usually not a serious problem. With about 15% of ash and 1–2% of sulphur in the normal UK coal, SO_3 is formed in the flame (5–20 ppm v/v) and most of it is absorbed as sulphuric acid or SO_3 in a layer on the surface of the ash particles. This film of acid is even beneficial in electroprecipitation; indeed, coals with sulphur content less than 1% require deliberate additions of SO_3 to the flue gas to make it up to 20 ppm. When this has been done, less than 0.5 ppm of SO_3 is found in the flue gas as vapour. Coals with more than 3% of sulphur do present a problem. The amount of acid formed is enough to render the ash sticky, and some condensation of liquid acid can occur. The stickiness of the ash can result in air-heater blockage which makes cleaning by soot-blowing difficult. In these situations, ammonia injection into the flue gas has been used to neutralize deposits of acid and to avoid corrosion of the electroprecipitators.

Oil-fired Plant

Most fuel oils contain 1–4% of sulphur. On burning, the sulphur is mostly converted to sulphur dioxide with 5–20 ppm v/v of SO_3. The amount of SO_3 is strongly dependent on the amount of excess oxygen present in the combustion flame at temperatures above 1670 K, and there is no evidence that SO_3 is formed in flames much below that temperature. The tendency in recent times has been to limit the excess oxygen in flames as much as possible without impairing the completeness of combustion. Many papers, e.g. refs.12–14, have been published on the chemical mechanism of SO_3 formation at high temperatures. For a given oxygen excess the degree of mixing is important, and a highly turbulent flame will give more SO_3 than a lazy diffusion flame[15].

For every plant a calibration curve should be established relating excess oxygen to SO_3 production. For diffusion flames the optimum level of excess oxygen will usually lie between 0.5 and 1.5% (v/v), whereas for well stirred reactors little more than stoichiometric combustion conditions will be needed. Suitable methods for the determination of SO_3 have been described[16–19].

As flame gases cool below $550°C$, SO_3 will increasingly react with excess water vapour to form sulphuric acid vapour. At $500°C$ about half the SO_3 will be converted. Heat-exchange surfaces throughout a boiler are coated with mixed sulphates and these come into balance with the SO_3/H_2SO_4 forming complex sulphates and bisulphates. In addition there will be some further conversion of SO_2 to SO_3 in condensed phases. With all these effects the amount of H_2SO_4/SO_3 in the gases as they cool remains fairly steady. The H_2SO_4/SO_3 which is held captive in the many tons of sulphatic ash deposits on the walls and tubes of a large boiler acts as a powerful buffer to small changes of the oxygen or sulphur in the flame.

At the cooler end of the boiler plant, deposition of the liquid acid itself can take place on plant and duct walls and on any solid particles (e.g. carbon from unburnt fuel) in the gas stream or adhering to the duct. These particles can soak up the acid (which will be of 85% w/w concentration) and cause great environmental nuisance if dislodged and carried away with the exhaust. It is usual to attempt to run plant so that the back-end temperature does not fall to the temperature at which liquid can appear, i.e. the dewpoint, but otherwise as close as possible for the sake of heat recovery. A gas temperature of 140–150°C is held in practice to be sufficiently high.

Sulphuric acid dewpoints are the subject of disagreement in the literature[20–23], with a scatter over a range of 40°C. Halstead[24] has reviewed earlier work and concludes that the figures in *Table 2* are the best available with a limit of confidence of ±10°C.

The water content of the flue gas has some effect on the dewpoint. It depends on the hydrogen content of the fuel and on the relative humidity of the combustion air. A typical low water level is 9.0 vol %. The dewpoint temperatures given in *Table 2* would be decreased by about 3°C with this amount of water vapour in the gas and increased by 3°C if the water content were at a high level of 13 vol %. It should be noted that *Table 2* refers to pure acid on clean surfaces. The presence of alkali sulphates, vanadium oxides and other impurities affects the dewpoint, but only by a few degrees. Maximum corrosion rates occur at 20–30°C below the acid dewpoint.

Operating plant above the acid dewpoint may sometimes be costly. Some designs of boiler plant use air heaters for preheating the combustion air, and the flue gases contact heat-exchange surfaces at temperatures as low as 40°C. These are certain to condense liquid acid. Cooling of the exhaust gases can also occur in the chimney stack owing to downdraughts; but only when the cross-wind speed is approximately five times the exit-gas velocity. Heat loss by radiation is another factor.

If back-end acid condensation cannot be avoided by working above the dewpoint, it will be necessary to add neutralizing material to the flue gas. Ammonia can be used[25]. It is added in rather more than the stoichiometric proportion to the H_2SO_4 content of the flue gas. If the NH_3 is thoroughly mixed with the flue gas at a gas temperature of 170°C (but below 240°C to avoid oxidation),

ammonium sulphate $(NH_4)_2SO_4$ is formed as a fine dry fume which does not deposit much and can easily be blown or washed away from surfaces. Acidic NH_4HSO_4 can be formed if SO_3 predominates. Mixtures with the sulphate begin to melt at 140°C and give rise to sticky deposits in the air heaters and elsewhere. It is difficult to avoid this happening to some extent, and this is one disadvantage in using NH_3. Another disadvantage is that the fine particles of $(NH_4)_2SO_4$ produce an intense white plume. This can be cut down to an acceptable level with properly designed electroprecipitators. MgO is another possible neutralizing agent. It can be used in the form of finely dispersed MgO in oil injected directly into the fuel. Two to four times the stoichiometric amount to neutralize the SO_3 is generally required to avoid all acid condensation at the back end. There are extensive claims by manufacturers for the effectiveness of special products, but there is a lack of published quantitative data. Cheaper additives such as finely ground dolomite ($MgCO_3$. $CaCO_3$) can be blown into the combustion chamber by air blasts; incomplete mixing of this additive with the gas stream limits its success and requires a greater amount to be used. Dolomite, or $Mg(OH)_2$ powder, may be blown into the flue gas at the air-heater inlet. However, it is claimed that dolomite is calcined if injected into the combustion chamber and that this increases its effectiveness. The selection of the correct additive and method of injection depends on the specific problem encountered[15,26–29]. The activity of a given additive is directly related to particle size and surface area and these factors both have a significant bearing on the cost. If steam is available to preheat combustion air before it enters an air heater, acid condensation is minimized. In reduced-pressure plant the leakage of air into high-temperature regions must be avoided by effective sealing, e.g. at the corners of the furnace walls or through off-load burner dampers.

The use of glass air heaters and also coatings of vitreous enamels on metals are also potential solutions to air-heater corrosion problems, but planned replacement of metal parts of the air heaters may in the end be the cheapest solution. Haneef[30] gave an account of the corrosion resistance of various metals to sulphuric acid attack which enable judgements to be made about the lives of plant items[31–34].

Table 2   Acid dewpoint curve for flue gases from an oil-fired generator containing the average concentration (11.0 vol %) of water vapour

H_2SO_4 (ppm v/v)	Dewpoint temperature (°C)
1	113
10	130
20	137
40	142
60	146
100	152

20.3 Fuel and air contaminants in corrosion of gas-turbine blades

The rate of installation of increasingly advanced gas turbines for stand-by and peaking electricity generation has grown substantially in the past decade and much experience has been gained of their operation. In this Section we deal with the importance of impurities in the air and fuel. The subject was dealt with in detail at a conference held[35] in London in 1972. We shall not consider the choice of metals for turbine blades, combustion chambers and ducts. This depends in the first instance on strength and creep resistance and it becomes the duty of the combustion engineer to tailor the combustion conditions to suit the materials. Nor do we dwell on the mechanisms of corrosion experienced in the gas turbine. There have been numerous studies of mechanisms (see, e.g., refs.36—38) but for our present purposes we need only observe that input impurities worsen corrosion experience by orders of magnitude.

What we shall describe as severe corrosion of high-temperature gas-turbine blades is marked by pitting and roughening and can occur typically in 800—2000 h even with clean, i.e. vanadium-free, fuel. When it does so, it is always associated with the presence of deposits of sodium sulphate on the stationary or moving blades and by a type of attack of the metal in which sulphides can be readily identified. High temperature in this sense means temperatures of $800°C$ or above. At blade temperatures of some 650—750°C, this type of corrosion is not encountered so severely and blades last for up to 50 000 h or more. The sulphur in the deposited sulphate comes from the fuel; even clean distillate fuel as used in aircraft gas-turbines may contain up to 0.3% sulphur by weight and the specification for land-based machines for industrial stand-by or a peaking power plant in the electricity network will usually permit up to 1%, with an average of 0.6% sulphur. Natural gas can give corrosion-free running of 50 000 h or more even at high blade temperatures.

The sodium in the deposit can originate both from the fuel and the air[39]. In the fuel the sodium may be present in the form of emulsified sea-water droplets or particles of dry salt of 1—5 μm diameter, and the amount expressed in terms of sodium lies in the range 0.05—1.0 mg/kg of oil. Fuel specifications laid down by turbine manufacturers have in the past permitted 2.0—5.0 mg/kg but the higher levels are seldom encountered. Indeed the recent tendency is to aim for a level not exceeding 0.2 mg/kg. The finer emulsified particles around 1 μm in diameter are not easy to remove reliably by filtration, though washing with clean water and centrifuging or electroprecipitation can consistently clean contaminated oil down to 0.2 mg/kg.

Because the sodium-containing particles are so small, they will be largely evaporated and the salts will dissociate in the flame zone of the combustion chamber. Thermodynamic data show that from a gas containing sodium, chlorine, excess sulphur and oxygen, sodium sulphate will form and be the first species to condense. Sodium chloride should not normally condense at all from the vapour phase. Vapour-pressure data show that with

sodium in the oil at the level of 0.1, 0.5 and 1.0 μg/kg the dewpoint of sodium sulphate is respectively 800°C, 840°C and 920°C when the sulphur in the oil is 0.6% by weight and excess oxygen in the flue gas is 15.5 vol %. This assumes total conversion of the sodium to the sulphate and suggests that condensation should occur with the lower sodium values only at the cooler roots of blades under hot running conditions. In aircraft turbines, leading-edge temperatures as high as 890°C are reached on full load[40]. This is well above the condensation temperature of an exhaust originating from 0.5 mg/kg of sodium in oil.

Air-borne salt may provide an additional source of sodium. In the UK it consists mainly of solid or perhaps dissolved sea-salt particles in the size range 0.5—20 μm in diameter. The total amount varies with the weather but usually lies in the range 2—15 μg sodium per kg of air. This is true for inland as well as coastal sites. The larger particles including rain or spray and dust particles are usually removed on woven filters. A 100 m^2 filter area would typically be required for a 50 MW gas turbine, giving 2.5 m/s air velocity. Such a filter would remove particles down to 5 μm diameter with 80% efficiency, but for smaller sizes the efficiency of removal falls off rapidly and at 1—2 μm diameter most of the particles will escape filtration. Centrifugal filters are also used with air velocities of 10—15 m/s. Particles down to 0.5 μm diameter which escape filtration are captured at high efficiency on the compressor blades, and since these account for as much as one half of the total salt in the inlet air, the build-up of salt on the blades is not without significance. There are two effects. One is a thickening layer of dry salt — sintered towards the hot end of the compressor — which can reduce the efficiency of the whole machine, and this can become noticeable after about 100 h. The other is that the accumulating deposit can come off in flakes large enough to survive evaporation and chemical reaction in the combustion chamber and collect by impaction on the concave surfaces of the turbine blades; here the chlorides react with the flue gas to sulphates and in doing so provide an extra element of 'chloride-ion attack' which can break down protective oxide films. Thus laboratory investigations have shown that metals which can withstand exposure to molten sodium sulphate for hundreds of hours without showing any penetrating attack may be triggered into a breakdown mode of corrosion immediately after brief contact with the chloride. So the build-up of chloride in the compressors must be minimized; it is not enough merely to remove the collected salt when it has thickened to the point at which it lowers the efficiency of the machine; certainly not with high-temperature machines. A typical method used to remove thickened and possibly sintered salt is to inject 5—10 kg of crushed plum stones over a 1—2 min period. That scrubs off the chloride but sends a shower of particles through the machine. But salt build-up can be avoided by water-washing at fairly short and regular intervals when the machine is cool. In this way it should be possible to eliminate flaking of the salt altogether. Unfortunately this method is not so convenient for machines running continuously, nor in fact does it markedly improve corrosion resistance with a fuel containing a high sodium concentration, 0.6 μg/kg.

The best way to ensure maximum life of a given turbine must be to keep the sodium in the oil to a low level. Analysis of an oil for 0.1 mg/kg (of sodium) is not easy. In the first place sampling is difficult because of the tendency of microdroplets of saline solution or salt in the oil to adhere to the sampling lines or bottles. Samples should be taken in a part-filled polythene bottle and the salt extracted with water with violent shaking. The sodium can then be determined with the required sensitivity by flame photometry.

The future tendency with gas turbines will be to higher blade temperatures leading to higher conversion efficiencies. The blades will probably be made of 'superalloys' protected by corrosion-resistant coatings and by internal cooling. Ceramic blades may also be used. Deposition will remain a problem, and consequent corrosion may be enhanced by the existence of a marked temperature gradient across the deposit owing to the blade cooling. Metals will no doubt continue to be treated with surface films of chromium, silica or alumina, and internal protection offered by traces of rare-earth metals in the alloy will be common. 'Dirtier' and hence cheaper fuels are often mentioned. By this is meant crude or light residual oil containing several mg/kg of vanadium and 5 mg/kg of sodium. Complex sodium vanadates readily adhere to hot blades and increase the corrosiveness of deposits. Additives based on finely divided silica and magnesia raise the fusion point of the mixed vanadates and prevent adhesion[41,42]. It is probable that the weight of the additive must exceed five times the stoichiometric equivalent of the vanadium to be entirely effective. The solid formed in the exhaust would require to be removed to avoid environmental contamination.

References

20.1

1　Halstead, W. D. and Raask, E. *J. Inst. Fuel* 1969, Vol.42, 345

2　Watt, J. D. and Fereday, F. *J. Inst. Fuel* 1969, Vol.42, 99

3　Watt, J. D. *J. Inst. Fuel* 1969, Vol.42, 131

4　Kiss, L. T., Lloyd, B. and Raask, E. *J. Inst. Fuel* 1972, Vol. 45, 213

5　Raask, E. *Fuel, Lond.* 1969, Vol.48, 366

6　Raask, E. *Wear* 1969, pp 301—315

7　Cutler, A. J. B., Halstead, W. D., Laxton, J. W. and Stevens, C. G. *Trans. ASME, J. Engng for Power* 1971, 93, Ser.A, No.3

8　Cutler, A. J. B., Hart, A. B., Laxton, J. W. and Stevens, C. G. *VGB Kraftwerkstechnik* 1974, 54, Vol.9, 611

9　Coats, A. W. *J. Inst. Fuel* 1969, Vol.42, 75

10　Parker, J. C., Rosborough, D. F. and Virr, M. J. *J. Inst. Fuel* 1972, Vol.45, 95

11　Lees, B. and Mustoe, D. H. *J. Inst. Fuel* 1972, Vol.45, 397

20.2

12　Nettleton, M. A. and Sterling, R. 'Formation and decomposition of sulphur trioxide in flames and burned gases', *12th Symp. (Int.) on Combustion,* 1969

13　Hedley, A. B., ref. (A), p 204

14　Cullis, C. F. and Mulcohy, M. F. R. *Combust. Flame* 1972, Vol. 18, 225

15　Laxton, J. W., ref. (A), p 228

16　Fielder, R. S., Jackson, P. J. and Raask, E. *J. Inst. Fuel* 1960, Vol.33, 84

17　Jackson, P. J. and Laxton, J. W. *J. Inst. Fuel* 1964, Vol.37, 12

18　Gokeyr, H. and Ross, K. *J. Inst. Fuel* 1962, Vol.35, 177

19　Jackson, P. J., Langdon, W. E. and Reynolds, P. J. *J. Inst. Fuel* 1970, Vol.42, 10

20　Gmitro, J. I. and Vermuelen, T. *A.I.Ch.E. J.* 1964, Vol.10, 740

21　Taylor, H. D. *Trans. Faraday Soc.* 1951, Vol. 47, 1114

22　Abel, E. *J. phys. Chem.* 1946, Vol.50, 260

23　Müller, P. *Chemie—Ingr—Tech.* 1959, Vol.31, 345

24　Halstead, W. D., 1974, personal communication

25　Gundry, J. T. S., Lees, B., Rendle, L. K. and Wicks, E. J. *J. Inst. Fuel* 1964, Vol.37, 178

26　Lee, G. K., Friedrich, F. D. and Mitchell, E. R. *J. Inst. Fuel* 1969, Vol.52, 67

27　Exley, L. M. *Combustion* 1970, Vol.41, 16

28　Litaudon, J. and Fiquet, J. M. 'Résolution des problèmes posés par la marche au mazout de générateur de vapeur à combustion à la centrale thermique du Havre', *Rev. gén. Therm.* 1974, 146, 131—152 (CEGB Transln No.6497)

29　Reynaud, P. *Rev. gén. Therm.* 1974, 147, 235—244

30　Haneef, M. *J. Inst. Fuel* 1960, Vol.33, 285

31　Kowaka, M. and Nagano, H. *Sumitomo Search* No.4, 1970, p.41

32　Imperial Metal Industries (Kynock) Ltd, 1968, Publn R4/MK18/23/1068

33　H. Wiggin & Co. Ltd, 1971, Publn 3564

34　Leonard, R. B. *Chem. Engng Prog.* 1969 (July), Vol.65, 84

20.3

35　Hart, A. B. and Cutler, A. J. B. (Editors), ref.(B)

36　Goebel, J. A., Pettit, F. S. and Goward, G. W., ref.(B), Paper 7, p 96

37 Erdoes, E., ref.(B), Paper 8, p 115
38 Seybolt, A. B. *Trans. AIME* 1968, Vol.242, 1955
39 Davies, I. and Polfreman, P. W. *J. Inst. Fuel*, in press
40 Tasker, F. J. D., Harris, C. W. and Musgrave, N.S., ref.(B), Paper 22, p 385
41 Elshout, A. J. *Grosskessel Milleilungen der VGB* 1969, Vol.49, 182–190
42 Lay, K. W. *ASME Paper No.*73-WA/CD-3, 1973

General Reading

(A) *The Mechanism of Corrosion by Fuel Impurities* (Eds. H. R. Johnson and D. J. Littler), Butterworths, London, 1963
(B) Hart, A. B. and Cutler, A. J. B. (Editors), *Deposition and Corrosion in Gas Turbines*, Applied Science Publishers, London, 1973

Appendixes

Appendix A
Natural fuels and their chief marketable products

Appendix B
Interconversion of fuels

Appendix C
Scientific and technological principles in the use of carbonaceous fuels

Appendix D
Sampling, analysis and testing of carbonaceous fuels

Appendix E
World energy resources and demand

Appendix F
Conspectus

Appendix A
Natural fuels and their chief marketable products

Sources, occurrence, extraction and basic transport

A1 Natural gas (methane or methane-rich gas)

A2 Petroleum; oil shales; tar sands

A3 Wood, peat, lignite, brown and black coals

Nature and varieties, classification

A4 Natural gas

A5 Petroleum; oil shales; tar sands

A6 Lignite, brown and black coals

A7 Rocket fuels and other materials

Preparation for the market, transport, storage; commercial products

A8 Natural gas

A9 Petroleum

A10 Peat, lignite, brown and black coals

SOURCES, OCCURRENCE, EXTRACTION AND BASIC TRANSPORT

A1 Natural gas (methane or methane-rich gas)

Sources

1. By-product of crude oil production; gas is dissolved in crude oil in subterranean reservoirs or associated with crude oil in overlying gas caps, serving to pressurize the oil to the surface; gas is flared or separately recovered and purified.
2. On- or off-shore subterranean gas fields not associated with oil.
3. Biological degradation of vegetable matter exuded from coal seams or their surroundings during coal cutting (drained for safety but rarely recovered).

Occurrence

Traditional areas (North America, North Italy etc, reserves now declining); recently established finds (Groningen, Lacq etc. production now static to save on reserves); new finds not yet fully exploited, UK North Sea, Australia, Bass Straits etc; also crude oil well-head gas is now being processed instead of flared (Iran etc.). Refer *Table 1*.

Extraction

E.g., from on-shore well-head gas: pressure 'choke' suppresses heavy oil; natural gasoline removed by oil scrubbing process; acidic impurities H_2S/CO_2 removed by alkali scrub or mol. sieve adsorption; higher hydrocarbons (ethane, LPG) recovered by low-temperature fractionation; water removed by absorption in glycol/water solutions or on solid desiccants (alumina, mol. sieves), or by expansion — refrigeration cycle. Dewpoint trimmed to suit ambient temperature and pipeline pressure. Refer also App.A8.

Transportation

1. Mains high-pressure pipeline (e.g. 7 MPa gauge (1000 psig), 0.915 m (36 in) diam. line) from supply to consumer area, methanol may be injected to prevent formation of solid hydrocarbon—water compounds ('hydrates') and 'slug catchers' inserted at pressure boosting stations to remove oil deposited by retrograde condensation.
2. Low-pressure pipeline (35—210 kPa gauge or 5—30 psig) within the district. Methane gas is also refrigerated to $-162°C$ at the supply point by compression and heat extraction, and the refrigerated liquid (LNG) transported at near atmospheric pressure in 12 000 t load ocean tankers to purchaser (e.g. Arzew, Algeria to Canvey Island, UK; Mersa el Brega, Libya to La Spezia, Italy). LNG regasified for use at pipeline pressure.

Table 1   Principal natural-gas sources

	10^9 normal m^3, 1970	
	Annual production	Proven and forecasted reserves
Canada, Alberta*	45	
USA		
Total	500	8000
Louisiana	120	
Texas	200	
Oklahoma	40	
Mexico	15	
South America	75	
Argentina,		
Chile, Colombia,		
Trinidad	16	500
Venezuela*	45	1000
UK North Sea,		
Leman Bank,		
Ekofisk, Frigg		
fields etc‡	35	1000
W. Europe		
Austria	1.8	30
S. France, Lacq	8	300
Italy, Po Valley*	3—4	150
N.W. Germany,		
Ems Estuary†	3	400
Netherlands,		
Groningen†	5	2000
USSR† and Eastern		
bloc	200	
Middle East‡		
(Iran etc.)	60	5000
Africa†	30	
Algeria, Hassi		
el R'Mel	5	3000
Libya,		
Zelten	12	250
Australia, Bass		
Straits etc‡	2	400

* Traditional reserves, now declining
† More recently established, steady-rate production
‡ Recent finds, growth areas

A2 Petroleum; oil shales; tar sands

Petroleum

Sources

Subterranean oil reservoirs sometimes associated with overlying natural gas.

References p 525

Occurrence

Main traditional oil fields are the Americas (Texas, Louisiana, Mexico, Venezuela etc.), Caribbean (Trinidad), Middle East (Iran, Saudi Arabia etc.), USSR, Africa (Algeria, Libya, Nigeria). Recent finds now being exploited include: UK North Sea (Forties, Brent, Ekofisk etc. fields), Australia, Bass Straits. Refer *Table 2.*

Extraction and Basic Transport

Conventional drilling techniques, crude oil stabilized at well-head by flaring or recovery of associated gas (App.A1) to low vapour pressure suitable for transportation by pipeline or ocean-going tankers (up to 500 000 t) to consumer refineries (App.A9).

Oil Shales

Natural deposits consisting of oil-producing 'kerogen' laid in laminated or fine-grained rock mixed with clay and sand, as-received CV about 7 MJ/kg.

Table 2   Principal oil-producing areas

	10^6 m^3, 1970. Rounded off	
	Annual production	Proven and forecasted reserves
North America*		
Total		7500
Alaska†		190
Texas, Louisiana etc.	570	
Mexico*	28	570
Trinidad*	8	2650
Venezuela*	210	1510
North Sea, Ekofisk†	6	130
Western Europe	4	680
USSR*, East Europe and China	760	15 100
Middle East*		57 000
Iran	230	
Saudi Arabia	210	
Other	360	
Africa*, total	3500	11 300
Algeria, Hassi Messaoud etc.	57	390
Libya	190	5000
Nigeria	60	900
Australia, Bass Straits† etc.	15	

* Traditional areas
† Recent finds

Occurrence

Widespread but only a few deposits being exploited or developed at the present time. Deposits now being worked lie in Estonia (for electric power) and Manchuria; main US deposits not yet exploited are in Green River Valley, Colorado (reserves, 174×10^9 m^3 of oil out of total estimated US shale-oil reserves of 320×10^9 m^3) but a 60 000 (short) ton/day proving plant is being developed at Parachute Creek. A shale-oil industry is planned for Brazil based on the extensive Irati deposits.

Extraction

The most promising plant designs are for rapid internal heating of the shale at 450--600°C in inclined or vertical retorts using as heat carrier either burnt gases from an earlier shale-combustion step, or hydrogen raised by partial oxidation from the heavy ends of product oil on hot ceramic pebbles (Tosco II process). Hot recirculating sand as an internal heat-transfer medium and externally heated fluid-bed retorts are other possibilities. Products will include spent shale, light and heavy oils, gas and sulphur.

Tar Sands

These are sandy deposits containing black bituminous-type oil. Main reserves not yet worked are in Athabasca, Canada (110×10^9 m^3 oil) and Eastern Venezuela ($\approx 30 \times 10^9$ m^3). Proposals for extraction include 'fire flooding' or steam purge to drive off the volatiles; condensation and secondary distillation (delayed coking or steam), or solvent extraction of the oily material.

A3 Wood, peat, lignite, brown and black coals

Wood

Main sources woods and forests; further source, waste from manufacturing industries. Occurs most countries, but only important where other fuels not readily available. Not normally transported far because of low bulk density.

Peat[1,2]

Layers in bogs, marshes, moors, fens, principally in high latitudes. Main deposits USSR, Finland, Poland, USA, Sweden, Iceland, UK, West Germany, Denmark, Japan, Ireland, Canada[3]. Extracted by hand-cutting, machine dredgers or high-pressure water jets. Owing to exceedingly high moisture content, prepared for market on site.

Lignite, Brown and Black Coals

Sources and Occurrence

Seams within stratified rocks of Devonian to Tertiary age[4]; most lignites and brown coals Tertiary or Cretaceous. Deposits widely distributed throughout world; lignites and brown coals principally USSR (66%), USA (24%), Australia (3%), West Germany, East Germany, Yugoslavia, Poland, Canada; black coals principally USSR (49%), USA (28%), China (12%), West Germany, UK, Australia, Canada, India, Poland, South Africa, Mexico, Czechoslovakia[3].

Extraction

Strip (opencast) mining where relatively shallow, otherwise usually by underground mining — principal systems room-and-pillar and longwall (advancing or retreating); auger mining of certain deposits for which normal opencast or underground methods impracticable.   Manual operations now largely superseded by mechanized mining: bulldozers, scrapers, shovels, draglines for strip mining; machines for single underground operations (drilling, cutting, loading, tunnelling, etc.); continuous miners; self-advancing and powered supports; increasing general automation. Hydraulic mining developing[5], and practised at localities in West Germany, Canada[6], USA, USSR, China[7].

Basic transport of run-of-mine coal also becoming increasingly mechanized and automated: underground by mine cars, conveyors, shuttle cars[8]; at surface by these means and road and rail transport, to coal preparation plant or installation for use of uncleaned coal.

NATURE AND VARIETIES, CLASSIFICATION

A4 Natural gas

Crude gas may be 'dry' (e.g. Groningen); associated with a light 'condensate' (e.g. Libyan, Bass Straits, North Sea); mainly (90%) methane-containing (e.g. Leman Bank, North Sea and North American gases); have a high N_2 content (e.g. Groningen, 14%) or a high H_2S content (Lacq, 15%) or a high CO_2 content (Kapuni, New Zealand, 45%) or a high C_2 etc. content (Iran, Libya, Algeria). Refer *Table 3*.

CV (purified gas) $\approx$ 34—39 MJ/m^3, requires 9.5 m^3 of air per m^3 of gas for stoichiometric combustion; limits of flammability 5—15 vol.% in air, burning velocity $\approx$ 34 cm/s (25 mm tube), ignition temp. ($^\circ$C in air) 650—750; flame temp. ($^\circ$C in air) 1920. Properties of main component, methane (CH_4): mol. wt 16.04; density (air = 1), 0.553; b.p., -161.5°C, m.p., -182.5°C; crit. temp., -82.5°C; crit. pressure (MPa), 4.8.

A5 Petroleum; oil shales; tar sands

Petroleum

Occurrence throughout the world (App.A2) — widely varying properties. Colour: straw—green—brown—black.

Table 3   Composition of natural gases

	CH$_4$	C$_2$H$_6$	C$_3$H$_8$	C$_4$H$_{10}$	C$_{5+}$	N$_2$	CO$_2$	O$_2$	H$_2$S	CV (MJ/m^3)
			(% v/v)							
France, Lacq										
Crude	69.2	3.3	1.0	0.6	0.5	0.6	9.6		15.2	
Purified	97	<--- 2 --->								36.2
Canada,										
Alberta	≈90	<--- 8 --->					0.5	0.2	1.0	
Venezuela	88.5	<--- 2.9 --->				4.6	3.8	0.2	0.3	39.1
Italy, Po										
Valley	≈98					≈1				38.0
USA	70—95	3—18	<-- 5 -->			1—14	0.1—7			≈39.1
North Sea										
(average)	90.5	3.9	0.9	0.3	0.2	3.9	0.3			37.4
Iran	73.0	21.5							5.5	
USSR	89—98	<-- 2.5 -->		1.0			7.5		1.0	39.8
Groningen	81.3	2.9	0.4	0.2		14.3	0.9			32.6
Algeria	83.8	7.1	2.1	0.9	0.4	5.5	0.2			39.1
Libya	64.5	21.0	8.4	4.2	1.9					
Australia,										
Bass Straits	93	2	1			3	1			36.8

References p 525

Sp. gr. 0.73—1.02 (typical range 0.8—0.95). Viscosity 1—1300 mm^2/s at 38°C (typical range 2—23).

Principal elements C and H, C/H ratio 6—8. Also hydrocarbon derivatives containing S, N and O and metal complexes of V, Ni and Fe. Na present usually as NaCl.

Composition

Highly complex, molecular-weight range 16—20 000. Hydrocarbons classified into paraffins, naphthenes and aromatics. Proportions determine processing needs of the refinery. Crude assays[9] provide very precise and detailed analyses of the crude oil and fractions after distillation. Preliminary analyses by U.S. Bureau of Mines Routine Method[10].

Sources of crude oil are often classified according to hydrocarbon type, e.g. heavy asphaltic, naphthenic, mixed base, etc. Plotting the specific gravity of distillate fractions against their mid-boiling points provides a useful guide. High sp. gr. for a given b.p. indicates naphthenic or aromatic structures of high C/H ratio. Examples are: paraffinic, Libya, Brazil, North Sea; naphthenic, Nigeria,

Venezuela, Canada; asphaltic, W. Venezuela, Trinidad, USSR.

Quality

Gravity — petroleum industry uses °API gravity scale. °API = $(141.5/S) - 131.5$ where S = sp. gr. 60/60°F. Distillate yield varies with gravity and hydrocarbon type (*Table 4*).

Sulphur concentrates in heavier fractions and is higher for heavier crudes from a given region (*Table 5*).

Wax content affects flow properties, particularly heavy distillates, lube fractions and residua (*Table 6*).

Nitrogen content is usually less than 0.1% in heavy gas oil fraction, up to 1.0% in residuum: e.g. Venezuelan (asphaltic/naphthenic) residua 0.5—0.6% N, Middle East, Libyan residua 0.2—0.3% N.

Metals content: V and Ni are present as stable high-molecular-wt porphyrin-type structures. They concentrate in residuum on distillation (traces sometimes appear in vacuum gas oil fraction). Ranges from 1200 ppm w/w V, 150 ppm w/w Ni in heavy naphthenic asphaltic crudes (Venezuela) to 30 ppm w/w V, 10 ppm w/w Ni (Middle East), 3 ppm w/w V, 15 ppm w/w Ni in Libyan to less than 2 ppm w/w V, Ni in North Sea crudes. Metals are catalyst poisons in many refinery processes. No economic process available for residuum demetalization. Na content variable depending on NaCl contamination from well-head and sea transportation and crude desalting processes employed. Na content of fuel oil can be removed by water washing (e.g. for gas-turbine fuel).

Oil Shales[11, 12]

Argillaceous fossil rock, often over a coal formation. Destructive distillation of kerogen content

Table 4   Distillate yields and gravity

	°API	Sp. gr.	Typical distillate yield (%)
Heavy crudes	10	1.00	20
	20	0.93	30
Medium crudes	30	0.88	50
Light crudes	40	0.83	65
	50	0.78	85

Table 5   Sulphur distribution in crude oils

	Middle East		Venezuela		Alaska	Libya	North Sea
Gravity API	39°	28°	37°	16°	28°	36°	38°
S in crude (wt %)	1.1	2.8	1.1	1.2	0.9	0.2	0.3
S in gas oil (wt %)	0.7	1.2	0.5	0.7	0.6	0.1	0.1
S in residuum (wt %)	2.4	4.9	2.5	2.1	1.9	0.4	0.5

Table 6   Effect of wax on flow properties

	Crude		Gas oil pour point (°F)	Residuum pour point (°F)
	Wax (%)	Pour point (°F)		
Venezuela	5	−30	−5	50
Middle East	7	5	5	80
North Sea	7	25	−30	80
Libya	20	75	25	105

yields shale oil and ammonia. Kerogen: 66–88% C, 7–13% H, 0.1–3.0% N, 0.1–9% S, 0.7–27% O. Sp. gr. 0.85–0.90. Minerals: silica, dolomite; possible Al recovery from dawsonite $(NaAlCO_3(OH)_2)$. Distillation of shale oil gives motor gasoline, solvent naphtha, kerosine, gas oil, paraffinic wax. (Aliphatic and alicyclic structures).

Tar Sands

Naturally occurring mixture of asphalt and loose sand terrain yielding up to 12% asphalt (heavy crude petroleum fractions).

Intensive developments in progress towards more efficient and environmentally acceptable recovery, extraction and processing for compatibility with conventional petroleum products.

A6 Lignite, brown and black coals

Petrographic Structure and Composition[13]

Humic Coals

Black coals comprise macroscopically recognizable different bands or lithotypes; lithotypes consist of microscopic associations (microlithotypes) of macerals (constituents analogous to minerals in rocks).

Lithotypes and principal component macerals:
Vitrain (very bright): vitrinite
Clarain (finely striated, bright and/or dull): vitrinite, exinite, inertinite
Durain (dull or faintly greasy): exinite, inertinite
Fusain (silky, fibrous and friable): inertinite.

Principal macerals in lignite and brown coals: huminite (precursor of vitrinite), exinite, inertinite.

Exinite is probably of spore and cuticle origin.

Sapropelic Coals (Cannels and Bogheads)

Non-banded. Consist of fragments of exinite and inertinite in groundmass of vitrinite and inertinite. Exinite probably of algal origin.

Associated Inorganic Material[14]

Comprises moisture and mineral matter. Moisture partly 'inherent' (adsorbed within fine pore structure of coal) and partly 'free' (more grossly associated, normal physical properties). Mineral matter occurs in bands, nodules, lenses, thin plates, dispersed particles; consists mainly of clay minerals (hydrated Al silicates), quartz, minerals of mica and chlorite groups, feldspars, pyrite, carbonate minerals.

Coal Rank[15–17]

Signifies degree of transformation of original organic material of a humic coal. Mean or maximum reflectance, and mineral-matter-free carbon content, of vitrinite increase with increasing rank[18]. Broad divisions of humic coals, in increasing order of rank: lignite, sub-bituminous, bituminous, carbonaceous (low-volatile steam), anthracite.

Chemical Structure and Composition[19–22]

Coal constitution is commonly referred to a structure based on a broadly aromatic cross-linked macromolecule, with much substitution by such groups as CH_3, O, OH, COOH, NH_2 and SH in aromatic rings; with increasing rank, increasing enlargement of aromatic nuclei and stacking into layers. (But alternative possibilities not yet ruled out.) Various properties of coals are related to rank (*Table 7*). Caking and swelling properties at minima in lowest-rank and highest-rank coals, at maxima in coals with 86–91% C.

N (dmmf) usually 1–2%. S (dmmf), i.e. S in organic combination, usually <c.1.0%; but total S in air-dried (a.d.) coal usually higher (normally <4.5%) because of S in mineral matter, mainly as pyrite. Cl (partly in organic association) usually <0.01–1%, P (in mineral matter) <0.001–0.1%, of a.d. coal. Trace elements include F, B, Ge, As, V, Pb, Cu, Zn, Ga.

Classification[23–25]

Most systems are based primarily on two or more parameters of coal rank (*Table 8*). Approximate correlations of main categories of coals in these

Table 7 General ranges of certain rank-related properties of humic coals

| Coal | (% dmmf*) | | | | Calorific value (MJ/kg, dmmf*) | Inherent moisture (% wet basis) |
	C	H	O	Volatile matter		
Lignite	60–77	6.4–4.9	33–15	65–38	22–30	20–15
Sub-bituminous	75–80	5.6–5.1	17–12	47–38	29–32.5	16–12
Bituminous	78–91.5	5.8–4.4	15–2.5	47–19.5	31.5–37	15–0.4
Carbonaceous	90.5–93.5	4.8–3.7	3.5–1.9	19.5–9	37–36	0.5–2.0
Anthracite	92–95	3.9–1.9	2.0–1.8	9–2	37–34.5	0.9–2.9

* dmmf — dry, mineral-matter-free basis

References p 525

Table 8 Bases of various coal classification systems

Coal classification system	Principal parameter	Subsidiary parameter(s)
American Society for Testing and Materials (ASTM)	Lower-rank coals: CV (mmmf)	Agglomerating character
	Higher-rank coals: FC (dmmf) or VM (dmmf)	Agglomerating character
International (ECE)	Soft coals: moisture (af)	Tar yield (daf)
	Lower-rank hard coals: CV (maf)	Caking and coking properties
	Higher-rank hard coals: VM (daf)	Caking and coking properties
National Coal Board (UK, NCB)	VM (dmmf)	Gray–King coke type
Seyler	C in vitrain (dmmf)	H (dmmf)
Australian	Lower-rank hard coals: CV (daf)	Crucible swelling number, Gray–King coke type, and ash (d)
	Higher-rank hard coals: VM (dmmf)	Crucible swelling number, Gray–King coke type, and ash (d)
Indian	VM (dmmf)	CV (dmmf) and moisture (mmf)

CV − calorific value; FC − fixed carbon; VM − volatile matter
d − dry basis; af − ash-free basis; maf − moist, ash-free basis; daf − dry, ash-free basis; mmf − mineral-matter-free basis; mmmf − moist, mineral-matter-free basis; dmmf − dry, mineral-matter-free basis

Table 9 Comparison of various coal classification systems

systems are shown in *Table 9*, on two calculated scales shown at side. This can be read with advantage in conjunction with *Tables 7* and *8* and App.D3.

A7 Rocket fuels and other materials

Rocket Fuels

For most conventional fuels, primary requirement is high energy release per unit mass of fuel in reaction with atmospheric air. Rockets carry both fuel and 'oxidant' (not necessarily oxygen-containing), *Figure 1* — hence primary requirement for rocket fuels is high energy release per unit mass of mixture of fuel + oxidant. This energy then converts to kinetic energy of combustion products (propellants) flowing through nozzle. Since reaction temperature is relatively high (3000°C plus), dissociation and recombination of propellants can exert overriding influence, hence overall performance is based on thrust (F) produced at nozzle exit per unit mass flow rate ($\dot{m}$) of propellants — this parameter known as specific impulse (I_{sp}) is given by $F/\dot{m}$ in units N s/kg* — equivalent to effective exit velocity of propellants, V_e, in units m/s.

Full calculation of I_{sp} is based on 'shifting equilibrium' of various components of propellants, but calculation simplified by assuming propellant components in 'frozen equilibrium' — i.e. no chemical changes — throughout nozzle gives results approximately 5% lower. I_{sp} varies with fuel/oxidant ratio, invariably peaking at a mixture fuel-rich of stoichiometric. Full expression for I_{sp} involves term $(T/M)^{1/2}$, where T = reaction temperature and M = mean value of relative molecular mass of propellants. Thus selection of fuel-oxidant pairs aims at high T and low M of combustion products. Calculated peak results for shifting equilibrium are given in *Figure 2*. As a rough guide, I_{sp} increases as relative density of reactant mixture reduces (*Figure 3*). Hydrogen-rich fuels seen to be preferable, with fluorine as

---

* 1 N s/kg = 1 m/s (=0.1020 lbf s/lb)

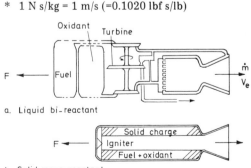

a. Liquid bi-reactant

b. Solid mono-reactant

Figure 1   Schematic diagram of chemical rockets (Derived from ref.29)

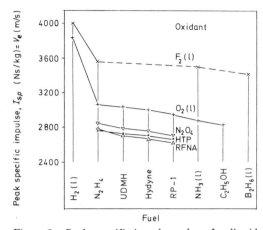

Figure 2   Peak specific-impulse values for liquid reactants
Shifting equilibrium
Chamber pressure = P_c = 68 atm
Nozzle exit pressure = P_e = 1 atm
(Derived from refs.26 and 29)

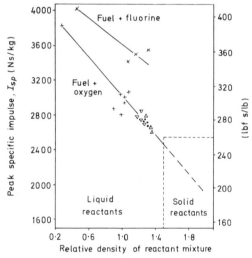

Figure 3   Relation between peak specific impulse and relative density of reactant mixture
Shifting equilibrium
P_c = 68 atm, P_e = 1 atm
(Derived from refs.26—29)

oxidant. Hydrogen (stored as a cryogenic liquid) of particular interest owing to low M, and I_{sp} improved (by about 10%) by means of tri-reactant system involving hydrogen added as propellant only, taking little part in chemical reaction. Hydrogen attractive also as propellant for future development of nuclear-fuelled rockets.

Rocket-chamber design is based on rapid and effective mixing of reactants (e.g. by impinging multi-injectors); either fuel or oxidant used as regenerative coolant for throat and nozzle prior to

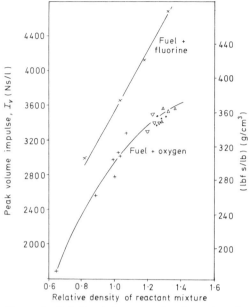

Figure 4   Relation between peak volume impulse and relative density of reactant mixture
Shifting equilibrium
$P_c = 68$ atm, $P_e = 1$ atm
(Derived from ref.29)

injection (sometimes also with film cooling). Chamber design simplified by selection of reactants that are hypergolic, i.e. ignitable spontaneously on contact (e.g. hydrazine and HTP). Energy release rates achieved up to 100 MW/l of reaction chamber volume (cf. 0.05 MW/l for average automobile engine). Reactant flow to chamber effected either by tank pressurization, or turbo-pumping energized by bleed flows of reactants.

For rocket applications within Earth's atmosphere, volume impulse I_v* (specific impulse on basis of volume of liquid reactant mixture, in units N s/l), may be more important — results in *Figure 4*. Storage volumes reduced further by use of solid reactants comprising both fuel and oxidant — must be chemically stable in storage and operation. Stored in rocket chamber as solid charge 'grain' suitably shaped to approximate to uniform rate of combustion, i.e. area exposed to flame remains constant. Grain either sheathed with inhibitor, or bonded to chamber wall (i.e. case). Mechanical strength of grain important to retain geometry during long-term storage above 66°C in varying humidity. Fuel 'system' thus simpler and inherently more reliable than with liquid reactants, but I_{sp} levels lower — see *Table 10*.

---

* 1 N s/l = 1000 kg/m^2 s = 6.3659 lbf s/ft^3

Table 10   Representative solid rocket-reactants (derived from refs.27 and 28, and other sources)

Type	Typical examples	General properties	Burning rate (cm/s)	T (K)	I_{sp} (N s/kg)
Black powder	Potassium nitrate, 70% Charcoal, 20% Sulphur, 10%	Low performance Not case-bondable	0.10	1500—3500	490—1375
Double-base colloidal	Nitrocellulose Nitroglycerine — cast plus plasticizer — extruded	Nitric esters capable of detonation, but stabilize each other. Not case-bondable	0.76 1.91	1700—2580 2360—3140	1570—2160 2010—2255
Composite	Ammonium picrate, Potassium nitrate — moulded CHO fuel binder	Abundant smoke	1.57	2030	1570—1960
	Ammonium perchlorate — cast and cured CHO fuel binder as resin or rubber	Case-bondable	0.76	1810—2760	1720—2355
Composite modified double-base	Colloidal plus ammonium perchlorate and aluminium	High-performance Case-bondable Abundant smoke	2.03	4100	2355—2550

Notes   1. Specific impulse at $P_c = 68$ atm (6890 kPa), and $P_e = 1$ atm (101.3 kPa)
2. Relative density ranges from about 1.5 to 2.1

Table 11   Reactant nomenclature

$B_2H_6(l)$	Liquid diborane
C_2H_5OH	Ethanol
$F_2(l)$	Liquid fluorine
$H_2(l)$	Liquid hydrogen
HTP	High-test peroxide (= concentrated H_2O_2)
Hydyne	60/40 mass mixture UDMH and diethylenetriamine
$NH_3(l)$	Liquid ammonia
N_2H_4	Hydrazine
N_2O_4	Nitrogen tetroxide
$O_2(l)$	Liquid oxygen
RFNA	Red fuming nitric acid
RP-1	Rocket propellant 1 (a narrow fraction kerosine, $CH_{1.97}$)
UDMH	Unsymmetrical dimethyl hydrazine

For propulsive purposes in spaceflight, rocket reactants (*Table 11*) generally selected for high I_v and relatively low cost for launching within the atmosphere (e.g. first stage of rocket vehicle) and, for manned vehicle, high I_{sp} in second and subsequent stages. For the smaller unmanned vehicle, solid reactants often suitable for final stages.

Other Materials

Other materials used as fuels in industry include the silicon in cast iron — present as an impurity — which by oxidation in the Bessemer and other processes for manufacture of steel could raise the temperature of the molten metal several hundred degrees C. Extra silicon added to raise temperature further. In the thermit process, finely divided aluminium burns in mixture with iron oxide ($3000°C$) leaving molten iron and aluminium slag — was formerly used for local heating and on-site welding.

PREPARATION FOR THE MARKET, TRANSPORT, STORAGE; COMMERCIAL PRODUCTS

A8 Natural gas (refs 30 and A, F, N)

Preparation for the Market

Purification of the crude gas includes the removal of 'condensate', water, LPG (liquefied petroleum gases), H_2S and sometimes CO_2. Refer also App. A1. Treatment of UK North Sea gas is an example of 'off-shore' gas processing. Condensate and water are removed from crude gas at the well-head in knock-out pots and separators, water discharged to sea; the condensate ('flo petrol') is returned to the shore line. At the shore terminal, 'slug catchers' remove condensate from the entry gas; a refrigerated ethylene glycol treatment plant dehydrates it to a water dewpoint of $15°F$ ($-9°C$) max.

winter ($40°F$ ($+4°C$) max. summer) and a further expansion — refrigeration cycle knocks out additional condensate to a hydrocarbon dewpoint $<5°F$ ($-15°C$) (at 7 MPa gauge mainline pressure); an ethanolamine scrub removes H_2S (→S by limited combustion in Claus kiln). N_2 (as in Groningen gas) is not usually removed (except with 0.05% He at Alfortville where Groningen gas is treated in an air-separator to make interchangeable with Lacq gas).

Storage

As gas in above-ground pressure spheres (1—1.5 MPa) or in bullets (2.5—7.0 MPa); or in batteries of welded pipe above ground or joined to gas transmission line below ground; as *gas* replacing water in underground aquifers; as LNG in cylindrical double-skin vessels near atmospheric pressure, above ground or in steel-membrane tanks with outer concrete shell; as LNG in hollows, below ground bounded by frozen soil and with insulated cover (e.g. Canvey Island and Philadelphia gas works); as gas dissolved in LPG at $-45°C$ in insulated pressure vessel, above ground.

Commercial Products

Public utility (1000 Btu/ft^3 or 37.26 MJ/m^3) gas; feedstock for 18.7 MJ/m^3 gas send-out (by catalytic steam-reforming); peak-shaving gas (LNG—air); synthesis gas ($CO + H_2$) production by Texaco partial-oxidation or Kellogg steam-reforming processes, leading to hydrogen by CO shift to CO_2 and alkali extraction of the CO_2; ammonia manufacture (Haber—Bosch, Casale etc. processes) by reaction of $H_2 + N_2$ (separated from air); methanol by synthesis-gas reaction + added CO_2; acetylene (Hüls electric-arc and Wulff flame-cracking processes); aldehydes by Oxo reaction (synthesis gas added to olefinic double bonds); carbon (channel) black by oxidation in a controlled deficiency of air. Methanol (by methane oxidation at well-head) has been proposed as an alternative industrial fuel, being cheaper to transport than LNG.

A9 Petroleum

Preparation for the Market

Refineries receive crude in large floating-roof tanks usually with associated desalters to control plant corrosion. Processes employed (App.B4) vary in complexity from basic fractionation (hydroskimming) via catalytic cracking to fully integrated refineries with coking facilities to eliminate fuel-oil yield.

Fractional distillation, often in 3 stages, to produce primary streams for processing and blending to commercial-grade products. Primary flash distillation: dry gas (methane, ethane), propane, butane and light gasoline blendstocks. Atmospheric distillation: heavy gasoline, naphtha, kerosine, light and heavy gas-oil and residuum. Vacuum distillation of residuum: lubricating oil distillates or catalytic-cracking feedstock and bitumen. General rule: straight-run fractions rarely employed exclusively for finished products.

Dry gas — used for *petrochemical feedstock* (with H_2S removal) or refinery fuel.

Propane — *petrochemical feedstock, LPG* (cold climates).

Butane — *motor gasoline* (volatility improvement), *LPG*

Motor gasoline — catalytic reforming etc. to produce components to required octane ratings. Use of TEL (tetraethyl lead) for octane no. improvement diminishing.

Aviation Turbine Fuel (AVTUR) — kerosine hydrofined for thermal stability, stringent limits on contaminants and low-temperature pumpability. *AVTAG (JP 4 in USA)* is a wide-cut gasoline turbine fuel providing greater yield but lower flash-point product.

Kerosine (Paraffin) — produced similarly to AVTUR for stoves with wicks, central heating etc.

Tractor Vaporizing Oil — kerosine fraction for spark-ignition engines, aromatic extracts for octane requirements.

High Speed Diesel Engine (DERV) fuel — tailored for antiknock (cetane rating), low smoke emission, fuel economy, low-temperature performance etc.

Heating Oil — for domestic and commercial heating, produced from gas-oil, usually hydrofined.

Marine Diesel Fuel — heavier grade of gas-oil.

Light, Medium and Heavy Fuel Oils — produced by blending gas-oil or catalytic-cracker sidestreams with atmospheric residuum to required viscosity, sulphur content etc.

Bitumen — blended to meet softening-point and penetration-test requirements, also modified by air blowing to produce oxidation grades.

See also App.B1.

Storage

For safety reasons principally governed by Petroleum Consolidation Act 1928.

LPG — pressure storage, including road transport, or refrigerated storage for large volumes (10 000 t) or combined pressure/refrigeration (App.A8). Portable cylinders or spheres (4.5, 14.5, 37.6, 680, 7112 kg capacity). Products classified according to flash point:
European model safety code — Class 0 LPG, Class I <21°C, Class II 21—55°C, Class III 55—100°C, no classification +100°C.

Storage vessels — prefabricated up to 54.5 m³ capacity.
— site-erected for larger tankage.
Underground storage customary for service stations, Class A gasoline.

Bulk storage — floating-roof tanks to exclude vapour space and vapour losses.

Fuel-oil tankage requires steam heating and heavier products including bitumen require insulated tankage.

A10 Peat, lignite, brown and black coals

Peat[31, 32]

Marketing, Transport, Products

Cut into sods or formed into pellets for drying to reduce extremely high moisture content (over 90% initially). Sometimes briquetted after drying. Tends to be used locally, as 2—2½ times bulk of coal; this factor also important in storage considerations. Dried peat (sometimes used pulverized); briquetted peat; products of carbonization (peat-coke, tar, liquor, gas); product of gasification (producer gas); products of solvent extraction (montan wax and fuel residue).

Main uses: domestic fuel, power stations, agriculture.

Lignite and Brown Coals[33]

Marketing, Transport, Products

Often sold in raw state. Preparation may consist of crushing and drying to reduce high moisture content, often followed by briquetting (with or without binder). Cleaning, when necessary, as for black coals (q.v.). Owing to low bulk density, uneconomic to transport long distances. More care needed in storage than for black coals (q.v.), owing to greater reactivity; cannot be stored for long owing to crumbling on weathering. Amount stored small whenever possible; otherwise safeguards are uniform stacking, avoiding size segregation, compacting layers, spraying with cold water, treating with oil. Underwater storage occasionally.

Commercial products: raw, dried or briquetted lignite or brown coal (sometimes pulverized); products of carbonization (coke or char, tar, liquor, gas); products of gasification (town gas, tar); products of selective solvent extraction (montan and other waxes, oils).

Black Coals

Preparation for the Market[34, 35]

Occasionally sold in raw state (run-of-mine coal).

Preparation (one or more of following operations):

1. Primary mixing (of coal from different seams and/or sites);

2. Screening, crushing and cleaning of largest material (to maximum top size);

3. Sizing (usually by screening) — smallest fraction sometimes sold without further treatment;
4. Cleaning:
   (a) Wet processes:

   For coarse and fine coal
   (i) Jigs (e.g. Baum washer)
   (ii) Dense-medium. Medium usually sand (e.g. Chance washer) or magnetite (e.g. Barvoys and Tromp washers). Including cyclones for fine coal
   (iii) Launders

   For fine coal
   (iv) Concentrator tables (e.g. Deister table)
   (v) Froth flotation

   (b) Dry processes (most suitable for fine coal):
   (i) Pneumatic jigs
   (ii) Pneumatic tables
   (iii) Pneumatic launders
   (ii and iii of little commercial importance)

5. Mechanical or thermal dewatering;
6. Recovery of fine solids from water;
7. Sizing and blending of cleaned products to market specifications;
8. Ancillary operations — dust recovery, dust-proofing (oil spraying).

Transport

Conventionally by road vehicles, rail wagons, cargo boats. Measures to meet problem of freezing in wagons include antifreeze and oil spraying; gas, infrared[36] and steam[37] heating; loosening by drills[38]. Spraying with latex/water to minimize losses in transit has been investigated[39].
Recent developments[40-44] include rapid and high-capacity bunker loading, unloading and weighing facilities; multi-size loading; unit and merry-go-round trains; often with automation. Hydraulic transport by pipeline operative at localities in USA, Poland, USSR, France[45]. Pneumatic transport under investigation.

Storage[46,47]

Problems: disintegration; loss of calorific value, caking properties and gas yield; increased friability; spontaneous ignition by heat liberated in oxidation; atmospheric pollution by wind action.
Safeguards:
(a) Surface stockpiles: layered stacking; stack height maxima for different coal ranks and sizes; prevention of undue air access; adequate ventilation (intermediate to be avoided); temperature checks.
(b) Bunkers and cargoes: avoidance of external or local internal heating; prevention of air passage through coal; good ventilation (see (a)).
Trials of plastic covers[48] and application of surface coatings[49,50] to reduce storage problems have shown promising results.

Commercial Products

Coal, either uncleaned (often used pulverized) or cleaned, and either unscreened or sized. Size grading (UK)[51] (1 in = 25.4 mm): large coal (lower limit ≥2 in normally, no upper limit); graded coals (large cobbles, screened to size range within limits 3—8 in; cobbles, within 2—5 in; trebles, within 1½—3½ in; doubles, within 1—2¼ in; singles, within ½-1½ in; peas, within ¼—¾ in; grains, within 1/8—7/16 in); smalls (upper limit normally ≤ 2 in, no lower limit). Maximum percentages of undersize (finer than lower size-limit of grade) stipulated for most graded coals.
Briquetted coal; products of low-temperature carbonization of coal or coal briquettes (semi-coke (e.g. Coalite, Rexco and Phurnacite), tar, liquor, gas) (App.B6, B8); products of high-temperature carbonization (metallurgical or domestic/industrial coke, tar, liquor, town gas) (App.B7, B8, B9, B11, B12); products of gasification (producer gas, water gas, low- or medium-CV synthesis gas) (App.B10); products of liquefaction (synthetic crude oil or liquid fractions + low-CV gas) (App.B3).

References

1   Ref.(E), pp 40—43
2   Plummer, C. E., 'Peat', *Encyclopedia of Chemical Technology*, Vol.9, The Interscience Encyclopedia, Inc., New York, 1952, 897—898
3   *World Energy Conference Survey of Energy Resources 1974*, U.S. National Committee of WEC, New York, 1974, pp 74—80
4   Ref.(D), p 219
5   Chironis, N. P. *Coal Age* 1972, Vol.77, No.3, 67
6   Grimley, A. W. T. *Can. Min. Met. Bull.* 1974, Vol.67, No.741, 44
7   Lezon, M. *Coal Age* Jan. 1974, p 31
8   Gilbert, S. *Mine & Quarry* 1972, Vol.1, No.1, 29; No.2, 21
9   Smith, N. A. C. *et al U.S. Bur. Mines Bull. No.490*, 1951
10 McKinney, C. M. and Garton, E. L. *U.S. Bur. Mines Rep. Invest. No.5376*, 1957
11 *Oil & Gas J.* 1974, Vol.72, No.8, 15
12 Cook, E. W. *Fuel, Lond.* 1974, Vol.53, 146
13 Ref.(G)
14 Ref.(H), pp 635—666
15 Ref.(D), pp 236—247
16 Ref.(H), pp 567—634
17 Ref.(J), pp 1—15
18 McCartney, J. T. and Teichmüller, M. *Fuel, Lond.* 1972, Vol.51, 64
19 Ref.(H), pp 717—753
20 Ref.(I), pp 309—480
21 Ref.(J), pp 232—295
22 Dryden, I. G. C., 'Coal', *Kirk-Othmer Encyclopedia of Chemical Technology*, 2nd edn, Vol.5, Wiley, New York, 1965, pp 627—644
23 Ref.(D), pp 246—255

24   Ref.(H), pp 361—435
25   Ref.(I), pp 10—34
26   Anon., *Theoretical Performance of Rocket Propellant Combinations*, Publn 505-X, Rocketdyne, California, 1959
27   Ref.(K)
28   Ref.(L)
29   Ref.(M)
30   Walters, W. J. (British Gas Corp.), 'Off-shore gas comes ashore', *Gas World* 1968, p 590
31   Ref.(E), pp 42—47
32   Ref.2, pp 899—901
33   Vahrman, M., 'Lignite and brown coal', *Kirk-Othmer Encyclopedia of Chemical Technology*, 2nd edn, Vol.12, Wiley, New York, 1967, pp 381—414
34   Ref.(Q), pp 5-1—14-35
35   Ref.(E), pp 137—147
36   Walsh, R. J. *Coal Age* 1970, Vol.75, No.7, 76
37   Mikhailov, N. M. *et al   Thermal Engng* 1973, Vol.18, No.1, 33
38   Shestakov, V. A. *et al   Coke & Chem. USSR* 1970, No.6, 13
39   Denton, G. H. *et al   Mining Congr. J.* 1972, Vol.58, No.9, 49
40   Broadbent, D. H. *Civil Engng & Public Wks Rev.* 1971, Vol.66, No.784, 1234
41   Ref.(Q), pp 15-12—15-14, 15-26—15-31
42   Jackson, D. *Coal Age* 1969, Vol.74, No.9, 62
43   *Modern Railways* 1970, Vol.26, No.261, 268
44   *Coal Age* 1969, Vol.74, No.12, 62
45   *Mining Magazine* 1972, Vol.126, No.4, 248
46   Ref.(E), pp 147—155
47   Ref.(Q), pp 15-1—15-11, 15-14—15-26
48   *Coal Age* 1969, Vol.74, No.10, 208
49   *Coal Age* 1971, Vol.76, No.6, 110
50   Kenyeres, J. and Takacs, P. *Publns of Hungarian Mining Res. Inst.*, No.15, 1972, pp 187—191 [in German]
51   Ref.(R)

General Reading

(A)   Tiratsoo, E. N., *Natural Gas*, 2nd edn, Scientific Press Ltd, UK, 1972

(B)   British Petroleum Co. Ltd, *Our Industry Petroleum*, 4th edn, London, 1970
(C)   McDermott, J., *Liquid Fuels from Oil Shale and Tar Sands*, Noyes Data Corp., USA, 1972 (Patent Review)
(D)   Williamson, I. A., *Coal Mining Geology*, Oxford University Press, London, 1967
(E)   Brame, J. S. S. and King, J. G., *Fuel — Solid, Liquid and Gaseous*, 6th edn, Edward Arnold, London, 1967
(F)   Medici, M., *The Natural Gas Industry*, Newnes—Butterworths, London, 1974
(G)   International Committee for Coal Petrology, *International Handbook of Coal Petrography*, 2nd edn, Centre National de la Recherche Scientifique, Paris, 1963 and (supplement to 2nd edn) 1971
(H)   Francis, W., *Coal*, 2nd edn, Edward Arnold, London, 1961
(I)   Van Krevelen, D. W., *Coal*, Elsevier, Amsterdam, 1961
(J)   Lowry, H. H. (Ed.), *Chemistry of Coal Utilization*, Suppl. Vol., Wiley, New York, 1963
(K)   Sutton, G. P., *Rocket Propulsion Elements*, 3rd edn, Wiley, New York, 1963
(L)   Sarner, S. F., *Propellant Chemistry*, Reinhold, New York, 1966
(M)   Goodger, E. M., *Principles of Spaceflight Propulsion*, Pergamon, Oxford, 1970
(N)   Lom, W. L., *Liquefied Natural Gas*, Applied Science Publishers, London, 1974
(O)   Institute of Petroleum, *Model Codes of Safe Practice in the Petroleum Industry*
(P)   Hughes, J. R., *Storage and Handling of Petroleum Liquids, Practice and Law*, Griffin, London, 1967
(Q)   Leonard, J. W. and Mitchell, D. R. (Eds.), *Coal Preparation*, 3rd edn, American Inst. of Mining, Metallurgical & Petroleum Engrs, New York, 1968
(R)   *Report of the Coal Grading Cttee on the Size Grading of British Coals*, British Colliery Owners' Res. Assocn, London, 1946

Appendix B
Interconversion of fuels

Production of rich gas from oils and coals, and coal hydrogenation

B1 Production of rich gas from oil
B2 Production of rich gas from coal
B3 Hydrogenation of coal to oil and chemicals

Refining and reforming of oils

B4 Refinery processes and operation
B5 Petrochemical operations

Coal carbonization and cokes

B6 Coal-carbonization processes
B7 Metallurgical cokes
B8 Smokeless coals, domestic cokes and
  manufactured fuels
B9 Coal gas
B10 Low-CV gases
B11 By-products and coal-tar fuels
B12 Formed coke

PRODUCTION OF RICH GAS FROM OILS AND COALS, AND COAL HYDROGENATION

B1 Production of rich gas from oil

All processes contracted for in the USA at the time of writing to supplement the dwindling supplies of natural gas are variants of low-temperature steam reforming using naphtha or natural gas condensate as feedstock, with LPG as an alternative if supplies permit. Processes include: Catalytic Rich Gas (British Gas Corporation), Methane Rich Gas (Japan Gasoline Co.) and Gasynthan (Lurgi Co.). Steps are: Vaporize feed — distil to 186°C end point (residue for furnace fuel) — Catalytic hydrodesulphurization over Ni—Co/Mo catalyst at 400°C, 273 kPa (2.7 atm) with H_2 — Purified product 0.3 ppm w/w S (max.) — Reform with steam over highly active Ni catalyst 550°C, 3.85 MPa yielding intermediate gas (CH_4 mainly, CO, H_2, CO_2) — Methanation (one or two stage) over Ni catalyst, 360°C (CO + $3H_2$ = CH_4 + H_2O) — Removal of CO_2 (amine or alkali scrub) — Removal of H_2O — Product gas 98% CH_4, 37 MJ/m^3.

Gas Recycle Hydrogenation (GRH process, British Gas Corp.) is attractive alternative, not yet proved commercially. $\approx345^{\circ}$C end-point distillate is injected with H_2 at high velocity for rapid recirculation in empty reactor: reaction conditions, 750°C, 2.1 MPa gauge; aromatic liquids are a by-product. Remaining problems include too high C_2H_4/H_2 content in product requiring secondary catalytic hydrogenation to methane, and need to produce H_2 by separate steam reforming etc. of product gas. Process has been successful producing 18.6 MJ/m^3 (500 Btu) town gas by enriching lean gas manufactured by high-temperature steam reforming of naphtha (ICI process, 750°C, 1.75 MPa gauge, Ni catalyst).

Fluid-bed Hydrogenation in pilot-plant stage only: atomized naphtha or crude-oil feedstock injected into separately fluidized bed of coke. High Btu gas produced at 750°C, 5.25 MPa gauge. Coke withdrawn continuously to maintain bed height.

Energy Refinery Concept. SNG produced from crude oil, low-sulphur fuel oil as by-product.

Crude — distil — light distillates processed by steam reforming or hydrocracking; resids, used to produce H_2 and coke fuel by partial oxidation; fuel oil hydrotreated to produce LSFO.

B2 Production of rich gas from coal

Coal gasification shown commercially in Germany and UK to produce town gas (18.6 MJ/m^3) by treating the lean product gas with LPG enriching agent and nitrogen ballast. Process now being developed in USA to produce SNG by adding a methanation step, viz:

Non-coking coal — Lurgi gasification (steam and O_2) at 2.8 MPa (ash and tarry by-products) — Fly-ash removal — Portion of yield, shift conversion (CO + H_2O = CO_2 + H_2) — Combined streams — 'Rectisol' purification and Benzole recovery (H_2S to Claus plant, S recovery) — Methanation, CO_2 and H_2O removal (see App.B1) — Product gas 36.5 MJ/m^3.

Three U.S. variants of the process, Institute of Gas Technology (Chicago), Hygas, Consolidation Coal Co., CO_2-Acceptor, and Bituminous Coal Research, Bi Gas process, are in pilot-plant stage; Bureau of Mines Synthane is still in design stage.

British Gas Corporation's process, still in research phase:

Fluidized bed of coal with H_2 at high pressure and temperature $\rightarrow$ methane and fixed carbon residue and ash; gasify residue with O_2 and steam (Lurgi Slagging gasifier) $\rightarrow$ CO + H_2 — shift reaction with further steam — CO_2 removal — H_2 recycled to fluid bed.

B3 Hydrogenation of coal to oil and chemicals

Strictly, all *industrial* systems involve hydrocracking rather than a pure hydrogenation reaction. They date from the second decade of the 20th century and have not yet been proved industrially viable (apart from emergency conditions). Much work has been done, very complex systems developed, and some developments now appear hopeful economically for suitable local conditions. A brief outline for completeness must mention the Fischer—Tropsch process (synthesis via CO + H_2) though this is at present even less economic.

For strategic reasons, Germany developed a number of small plants by 1938; research in the USA was further developed after 1945 incorporating some of the experience gained in Germany before and during War II. A test plant built by ICI in 1935 was later converted to hydrogenation of tar products, for reasons of economics and technical convenience. There was also R & D work in USSR, Australia and Japan; the first continued to operate a few German industrial plants after the war.

Fischer—Tropsch products contain much more non-aromatic material than those from coal hydrogenation or carbonization. They are of high quality and sulphur-free. The process (or a combination of F—T and hydrogenation) once viable could be an excellent basis for a chemical industry (cf. SASOL[1]).

Technical improvements based on international experience could be used to improve the prospects of reducing capital and net running costs of hydrogenation, by obtaining a high yield of high-priced primary chemicals in addition to fuel oil, economizing hydrogen needs (including recycling unwanted gaseous products via cracking) and energy, integrating secondary production of rubber, plastics and refined chemicals from the primary products, and using modern methods of plant construction, larger units, and some reduction of operating pressure. The cost gap between synthetic and natural oil could thus be reduced, so that given cheap coal some traditional processes might now be marginally viable.

Characteristics of Traditional Processes

Hydrogenation can convert a greater percentage of coal (30–40%) to aromatic hydrocarbons and phenols than carbonization (c. 3%), and is more flexible in matching product yield to markets. Since fuel oils sell relatively cheaply in large quantities and in competitive market conditions, whereas chemical products are more valuable and serve a smaller market, a balanced yield of oil and chemicals is usually more likely to be viable than an objective of mainly the one or the other alone. Carbide and Carbon Chemicals spent much money in the early stages on a pilot plant for production of chemicals, but marketing results were disappointing[2].

But coal hydrogenation requires considerable reserves of relatively cheap coal (Fischer—Tropsch

is usually not viable even then) and costly compressed H_2, of heat and electricity (or extra coal in lieu), and of available water. The product should compete in a refinery with natural crude oil, therefore is handicapped. Investment cost is considerably greater than for carbonization. Hydrogenation tends to produce initially large amounts of novel chemicals for which markets are difficult to find, and to suffer from the general market-disparity between fuel oils and chemicals.

However developments since c. 1960 have clarified process objectives.

Processes

Early (pre-War II) work:
Stage I. Crushed coal pasted with recycled vehicle oil (preferably containing H-donor such as tetralin)

$$\xrightarrow[\text{catalyst}]{H_2} \text{O-removal} \xrightarrow[\text{catalyst}]{H_2} \text{H-acquisition}$$

$$\xrightarrow{\text{filter}} \text{crude oil product}$$

Stage II. Distillate up to 325°C (middle oil)

$$\xrightarrow[\text{H_2, pelleted catalyst}]{\text{vapour phase}} \text{gasoline}$$

Conditions:

Stage I

Coal type	Temp. (°C)	Pressure (MPa)	Duration (min)	Exothermic heat (MJ/kg)
A. Brown (50% aq.)	⩾500	≈20	30–150	>bituminous
B. Bituminous (10% aq.)	⩾500	Up to 70	30–150	1.5 → 2.2

Catalysts		Oil yield (wt % of dry coal) Middle oil	(wt % of daf coal)	
			Gas yield	H_2 consumed
A. Mo compds better	) but 1–2% of	≈40	25–30	≈7
B. Sn compds better	) cheap Fe cmpds, once-through, preferred industrially	≈60	25–30	≈10

Stage II

Temp. (°C)	Pressure (MPa)	Duration (min)	Catalysts	Gasoline yield (% of middle oil)	H_2 consumed (wt % of daf coal)
≈400	≈30	2–4	WS_2 (+NiS) on activated Al_2O_3 (O, S removal), then (e.g.) WS_2 on activated Fuller's earth treated with HF	≈50 per pass	≈1

Coal treated/gasoline ratio could be about 2:1 for the two stages; but with production of the H_2, heat and power required from coal, perhaps *total* coal/gasoline as much as 4.5:1 (thermal efficiency then $\approx$25—30%). There were many problems, notably in Stage I sludge filtration, and heat transfer in coking stage (of heavy-oil let-down containing sludge).

Later simplifications have tended to concentrate on Stage I, cracking and refining the crude product to yield fuel oils and char (and some gas). They may be compared with the previously described two-stage process at up to 30 MPa yielding gasoline at $\approx$0.3 m³/t. Among the most promising[3] are the Consol and COED processes, which possess established pilot plants. Consol (now to function if continued at $\approx$29 MPa in the hydrogenation stage with $ZnCl_2$ in fluidized bed) was originally directed towards gasoline but later diverted to low-sulphur heavy oil ($\approx$ or >0.5 m³/t). COED, with staged temperature increase in fluidized beds (oil treatment at $\approx$22 MPa), produces mainly oil ($\approx$ or >0.21 m³/t) and char and some gas. Others with economic promise are the solvent-refining (Spencer, $\approx$8 MPa) and H-coal ($\approx$19 MPa in catalysis) processes, both in the design stage. The former produces ash- and sulphur-free coal (90% of the coal carbon into solution or perhaps $\approx$0.6 m³/t) for power plants, the latter, using Co molybdate catalyst in a fluidized bed, gasoline and furnace oil together with some light oil, at $\approx$ or >0.5 m³/t total liquid yield. All these treat the coal at around 400°C+. Thermal efficiencies (liquids only) are probably about 60, 25, 75 and 55% respectively, but 70% or more on *total* products. Much development research is still needed.

REFINING AND REFORMING OF OILS

B4 Refinery processes and operation

Refinery processes integrated to provide projected volumes/qualities of products to satisfy the market and also to provide flexibility in operation to respond to changes in product demand pattern. Additional flexibility by selection of crude supply.

After crude fractionation in a pipestill into the mainstream components a variety of processes may be employed to produce marketable products. Major processes are summarized below.

Catalytic Processes

Catalytic Cracking

High-boiling distillates → low-boiling gasoline components. (Fluid-bed SiO_2/Al_2O_3 catalyst 400—500°C at 200 kPa).

Hydrocracking

Refractory distillates or residua → gasoline, jet fuel or low-S fuel-oil components. Combination cracking, hydrogenation and desulphurization. (Fixed bed SiO_2/Al_2O_3 with Pt, Ni etc. 200—400°C, 1—15 MPa with H_2).

Catalytic Reforming

Low-octane naphthas → high-octane gasoline by dehydrogenation and isomerization reactions. (Fixed-bed Pt on Al_2O_3 or fluid-bed mixed metal oxides, 400—550°C and 0.5—5 MPa).

Hydrogen Treating

Removes S and N compounds with mild hydrocracking for upgrading middle distillates, lube oils, waxes, reformer feedstocks etc. and for production of lower-sulphur fuel oils. (Fixed-bed Co molybdate on Al_2O_3 carrier at 200—450°C, 0.5—10 MPa with H_2).

Isomerization

Normal C_4, C_5, C_6 paraffins → isocompounds for high-octane gasoline. (Fixed bed, $AlCl_3$ or Pt at 50—500°C, 1—7 MPa).

Alkylation

Olefins (usually from cat. cracker gases) + isobutane (from refinery gases or isomerized butane) → high-octane gasoline. (Aluminium chloride catalyst at 50—250°C, 20—70 atm).

Polymerization

Unsaturated olefin gases → liquid condensation products. (H_2SO_4, H_3PO_4 or Cu pyrophosphate 150—200°C, 1—8.5 MPa).

Thermal Processes

Viscosity Breaking

Reduction in viscosity and pour point of heavy residuum with increased distillate yields (not currently employed in Europe). (Mild cracking at 520°C, 400—800 kPa followed by oil quench and flash distillation).

Coking

Residuum → gas oil catalytic cracker feedstock + gas and coke. Fluid coking → fluidized seed coke (500—700°C at atm. pressure). Delayed coking → insulated coke drums (450°C at 200—600 kPa).

Thermal Reforming

Upgrades naphtha to high-octane gasoline (500—600°C, 3—7 MPa).

Solvent Processes

Deasphalting

Extracts lubricating oil stock from vacuum residuum. 6:1 to 10:1 liquid propane/oil ratios in continuous countercurrent process (70°C at 3–4 MPa atm). Asphalt and lube fractions stripped and propane recycled.

Selective Extraction

Removes aromatics, naphthenes, unsaturates and impurities from distillates and lube stocks. Solvents include phenol, furfural, liquid SO_2 etc.

Dewaxing

Removes wax from lubricating oils to reduce pour point. Specific solvent mixtures (e.g. methyl ethyl ketone/tolual (industrial toluene)) mixed 1:1 to 4:1 ratio, cooled (e.g. -25°C), precipitated wax filtered.

Treating Processes

Desalting (of crude oil)

Controls pipestill corrosion and minimizes product contamination. Coalescers, demulsifiers and/or electroprecipitator techniques employed.

Caustic Washing

Improves odour and colour and removes organic acids, mercaptans, phenols etc. 5–20 wt % caustic solution mixed in; settled and water-washed.

Clay Treating

Improves odour and colour and absorbs polar compounds. By percolation through coarse clay, contact with fine clay or vapour-phase contact in packed bed.

Sweetening

Converts foul-smelling alkyl mercaptans to less objectionable disulphides: $4RSH + O_2 = 2RSSR + 2H_2O$ or completely removes the mercaptans. Treatment by $NaOCl$, $CuCl_2$ and air (copper sweetening), or combination mercaptan extraction and sweetening using a Co salt catalyst (Merox process).

B5 Petrochemical operations

Primary Feedstocks (from crude oil by fractional distillation)

LPG (to 35°C, ethane, propane, butanes – (normal) butane, isobutane), naphtha ($35–200^{\circ}$C), kerosine ($150–300^{\circ}$C), gas-oil ($175–360^{\circ}$C), fuel-oil bottoms ($>360^{\circ}$C; or by further distillation, heavy gas-oil, wax, lube oil + asphalt bottoms).

First Stage Processing

Ethane → pyrolysis (825°C) → *ethylene*, similarly propane → *propylene* (U.S. practice, 800°C); butane→dehydrogenation (Houdry process) → (normal) *butenes* → *butadiene*; naphtha (European practice) → pyrolysis with steam (760°C) → *ethylene*, *propylene* etc.; naphtha (also natural gas or LPG) → catalytic steam reforming (850°C) → *synthesis gas* (CO + H_2); naphtha → hydrodealkylation ('platforming', 'powerforming', 'ultraforming' etc.) → LPG + $C_6/C_7/C_8$ *aromatics*; gas oil → fluid-bed catalytic cracking etc. → *olefinic gases*, LPG and lighter distillates; wax → steam pyrolysis (550°C) → C_6/C_{20} *olefins*; fuel oil → partial oxidation (Shell/Texaco processes, steam + O_2) → H_2 + CO → *hydrogen*.

Olefins recovered from cracked products by low-temperature fractional distillation.

Naphtha (also natural gas, App.A8) can produce *acetylene* by Huls electric arc and Wulff flame cracking processes; acetylene route to chemicals now being supplanted by *ethylene* route.

Secondary Processing

Ethylene

- Polymerize (ICI, Ziegler process) → polythene plastics
- Hydration → ethanol → dehydrogenation → acetaldehyde
- Add O → ethylene oxide → ethylene glycol → antifreeze
- Add Cl → ethylene dichloride → vinyl chloride → PVC etc.

Propylene

- Polymerize (Natta–Montecatini process) → polypropylene plastics
- Propylene glycol → polyurethane foams
- +steam/ammonia → acrolein (Shell process) → acrylonitrile fibres
- (+CO + H_2, hydroformylation, Oxo reaction) → (normal) butanol (plasticizers).

Butadiene

- Synthetic rubber (BSR styrene + nitrile type), polybutadiene resins.

Butanes/Butenes

- Liquid-phase oxidation → acetic acid + acetone (Celanese process)
- Butenes → hydration → *sec*-butanol
- Isobutylene → hydration → *tert*-butanol
- Isobutylene → polymerization → vistanex + butyl rubber
- Isobutane → oxidize → propylene oxide

Synthesis Gas

- $(CO + H_2) \rightarrow H_2$ (after CO shift + CO_2 removal) $\rightarrow$ ammonia $(N_2 + H_2)$; $(+CO_2) \rightarrow$ methanol, oxidize $\rightarrow$ formaldehyde $\rightarrow$ thermosetting plastics.

Benzene

- Hydrogenate $220°C$, 2 MPa $\rightarrow$ cyclohexane $\rightarrow$ nylon
- Add ethylene $\rightarrow$ ethyl benzene $\rightarrow$ $(-H_2)$ $\rightarrow$ styrene plastics and rubber
- Add propylene $\rightarrow$ cumene $\rightarrow$ phenol (Bakelite plastics) and acetone
- Add propylene tetramer $\rightarrow$ dodecylbenzene $\rightarrow$ domestic detergents
- Vapour phase oxidn $\rightarrow$ maleic acid $\rightarrow$ polyesters.

Toluene

- Hydrodealkylation $\rightarrow$ benzene (q.v.)
- Solvents etc.

o-Xylene

- Air oxidation $\rightarrow$ phthalic anhydride $\rightarrow$ resins and plasticizers.

p-Xylene

- Terephthalic acid $\rightarrow$ terylene fibres.

COAL CARBONIZATION AND COKES

B6 Coal-carbonization processes

The essence is:
coal $\rightarrow$ 600—1100°C $\rightarrow$ coke, semicoke or char + gas, aqueous (ammoniacal) liquor, light oil and tar.
Processes may be classified as follows.

1. *Low-temperature:* at about $600°C$[4-11]. Thermal conversion efficiency 95%; process efficiency c. 85%. Those of importance in the UK are:
(a)   Coalite: carbonization of high-volatile, weakly- or medium-caking coal in cast-iron or refractory continuous vertical retorts, externally (gas) heated to give a maximum semi-coke temperature of about $600°C$. Typical yields per t of dry coal: Coalite 0.75 t, tar 86 l, spirit 16 l, liquor 136 l, surplus gas 113 m^3 or 2950 MJ. By-products refined.
(b)   Rexco: batch and continuous carbonization of high-volatile, very weakly or weakly caking sized (lump) coal in refractory-lined cylinders (taking about 35 t of coal) by internal heating using products of combustion of gas evolved, these products being passed downwards through the charge. Typical yields per t of dry coal: Rexco 0.65 t, tar 82 l, spirit 5 l, surplus gas 690 m^3 or 3700 MJ.

Product cooled by continued recirculation of gas after air supply shut off.
Coke products, App.B8

2. *High-temperature:* at 900—1100°C[4-16]. Thermal conversion efficiency 95%; process efficiency 80—85%. High-temperature carbonization in horizontal retorts, continuous vertical retorts and intermittent vertical chambers, essentially for the production of town gas, has almost ceased in the UK (0.22 million t of gas coke was produced in 1972). The remaining processes of importance are:
(a)   Phurnacite: low-volatile Welsh steam coal briquetted with pitch; ovoids batch carbonized at $900°C$ in inclined retorts; maximum product temperature about $800°C$.
(b)   Coke ovens: the major carbonization process practised, in general and in the UK: product mainly for metallurgical uses. Coal blends (possibly containing some coke breeze) of high caking power and medium volatile matter (24—30%, dry basis) batch carbonized in units of ovens (maybe 6.5 m high × 15.0 m long × 0.45 m wide in the most modern practice) with gas-heating flues between walls. Each such oven takes 32 t of coal with maybe 100 ovens in unit; carbonizing time (blast-furnace coke) about 14 h; coke output of such a unit over 1 million t/year. Typical yields per t of dry coal: coke 0.77 t, tar 40 kg, benzole 13 l, surplus gas 215 m^3 or 4500 MJ. Coke discharged at 1000—1050°C and normally cooled by quenching with water, in which case total energy losses are about 15% of input (sensible heat of coke $\approx$ 3%). Growing interest in inert-gas dry cooling: recovery $\approx$ 80% of sensible heat, used to generate steam. Considering only the gas used for heating the ovens, about 45% of this input appears as sensible heat of coke, reduced to about 10% by dry cooling. Preheating of coal before charging to ovens now coming into application: small beneficial effect on the energy balance: main importance is improvement of oven productivity and coke quality, which can be maintained using more of the poorer coals in the blend. Coke product; App.B7. Tar product: App.B11. Gas product: App.B9.

B7 Metallurgical cokes

Process as B6, high-temperature (b)
Coal charged consistent, closely controlled quality; size 80—90% < 3 mm.
(a)   Blast-furnace coke: flue temperatures about 1250°C; coking rate (oven width/carbonizing time) 25—30 mm/h.
Requirements[17,18]:   Size-range   20—75   mm (round apertures); undersize 0—10 mm breeze for sinter-making (5% of total coke), 10—20 mm nuts for fuel (2½%); oversize may be broken to give 75 mm top size. Moisture $\not> $ 3%. Strength[19]: 1½-in shatter index $\not< $ 90; M_{40} micum index $\not< $ 75; M_{10} $\not> $ 7. Ash $\not> $ 8% (so ash of coal $\not> $ 5.5—6%).

S $\not>$ 0.6% (so S of coal $\not>$ 0.7–0.75%: not achievable throughout UK practice).

Phosphorus $\not>$ 0.01% when coke used for acid pig-iron (very little in UK practice).

(*b*) Foundry: flue temperatures about 1150°C; coking rate about 20 mm/h. Coal must be prime coking coal or equivalent blend.

Requirements: Size > 100 mm (square apertures), aiming to produce as much as possible meeting this. Moisture $\not>$ 3–5%. Strength[19]: 2-in shatter index $\not<$ 90; abrasion strength (M_{10} micum index) probably less important. Ash $\not>$ 9% (so ash of coal $\not>$ 7%). S $\not>$ 0.8–1.0% (so S of coal $\not>$ 0.95–1.2%).

B8 Smokeless coals, domestic cokes and manufactured fuels [20, 21]

Smoke (see also App.C5) is particulate tarry matter given off when coal is heated beyond the temperature of decomposition. Coals with less than about 20% of volatile matter (VM) are naturally smokeless. These include anthracites (NCB Code No.100) and low-volatile steam coal (NCB Code No.200).

Other coals, of higher VM, must be heat-treated to eliminate smoke-producing constituents, leaving a smokeless product.

Coals may be processed especially for the domestic market. A high residual VM increases the yield of saleable product and improves the reactivity — a desirable property for a free-burning domestic fuel, but the higher the VM the greater the production of smoke on combustion. For untreated coals, a broad band relates smoke production to standard VM. Using, as reference, average smoke over a test cycle in an open fire burning 2 lb (0.97 kg)/h, the average weight of smoke produced in unit time from coals of 33–42% VM dmmf has been taken as 100%. Nearly 80% reduction may be achieved with coals in the range 13–17% VM, and $\approx$90% reduction within 12–16% VM[22]. For manufactured fuels there is little correlation, but an average of 10% VM should in most cases be a safe guide to smoke reduction >90%. In practice (apart from the NCB process below) the average VM of the smokeless fuel seldom exceeds 10% because the carbonization process is uneven and some of the product with a higher VM could cause an entire batch to fail the test procedure.[23]

The average heat treatment needed in practice for a smokeless product is approximately 425°C/0.5 h. 0.5 h is only necessary because of uncertainties in the heat treatment of individual particles and the time taken for heat penetration and release of product molecules — in practice a few seconds suffice for the chemical changes for particles <10 mesh, and even milliseconds possibly for *in situ* changes within particles once they are at temperature. At this temperature very high pressures are required to convert the 'char' powder from a low-rank coal into a strong shape (NCB process). If a coking coal is available for blending, the plasticity of the mix can be improved at this temperature and normal briquetting pressures used to form smokeless briquettes suitable for the domestic market. The 'Ancit' (Dutch State Mines) process, and the 'BBF' process (Bergbauforschung–Lurgi) convert the non-coking component to a low-volatile 'breeze' before mixing with the coking coal.

Other commercial processes currently operating include Coalite and Rexco (App.B6). In these the residual VM is reduced to 8–10%. Lump coal can be coked in shaft retorts (examples are some gasworks retorts, Lurgi–Spülgas, Balfour, Otto) and briquettes in modifications of the above. Small naturally smokeless coal, or coke breeze, may be briquetted with pitch, or bitumen, and 'desmoked' either by carbonization in ovens (Phurnacite process) or retorts, or by oxidation on a moving grate at about 350°C (Anthracene process).

Surplus oven coke of small size (less than 25 mm), which is not required for the metallurgical market, may be sold to the domestic market, and is especially suitable for combustion in closed stoves.

B9 Coal gas

Process as App.B6, 900–1000°C. Approximate analysis[24]: H_2 50–52%, CH_4 20–30, CO 8–18, C_nH_m 2–4, N_2 4–6, CO_2 2–4; gross CV 16–18.5 MJ/m^3. H_2S normally restricted by legislation to approximately zero.

B10 Low-CV gases

Producer Gas

Made by blowing air through thick bed of coke[24]. Approximate analysis: H_2 11%, CH_4 0.5, CO 29, N_2 54.5, CO_2 0.5; gross CV 4.5–5 MJ/m^3.

Water Gas

Made by blowing air and steam alternately through coke bed[24]. Approximate analysis: H_2 49%, CH_4 0.8, CO 41, N_2 4.5, CO_2 4.7; gross CV 10 MJ/m^3. Carburetted by thermal cracking of oil in gas stream, raising gross CV to about 17 MJ/m^3.

Blast-furnace Gas

Is made during the reduction of iron ores. Analysis varies considerably: H_2 1.5–3.5%, CO 23–33, N_2 50–60, CO_2 8–15; CV 3–4 MJ/m^3.

B11 By-products and coal-tar fuels

Coal carbonized at 1100—1300°C or 500—700°C (App.B6) yields, in addition to gas, metallurgical coke (App.B7) or solid smokeless fuel (App.B8), crude benzole and crude tar (about 5% of coal from coke-ovens and 8.5%, low-temperature carbonization).

The crude tar is fractionally distilled in continuous stills to give the following primary fractions, some of which, as indicated, are treated to recover individual products[25]:

Coke-oven Tar

Fraction 1. Av. boiling range (5—95%), 80—160°C; average yield on crude tar, 0.5%. Desulphurized and fractionally distilled to give various grades of benzene, toluene and xylene.
Fraction 2. Av. boiling range, 160—200°C; av. yield, 3%. Acid-washed, desulphurized and fractionated; 165—180°C cut polymerized. Products are pyridine, solvent and heavy naphthas, and indene—coumarone resins.
Fraction 3. Av. boiling range, 200—240°C; av. yield, 14.5%. Treated by alkali-washing, then cooled and centrifuged (or fractionally distilled) to give phenol, cresols, naphthalene and drained naphthalene oil.
Fraction 4. Av. boiling range, 225—290°C; av. yield, 3%. Not further treated (oil A).
Fraction 5. Av. boiling range, 250—90% at 370°C; av. yield, 17.5%. Cooled and filtered to recover crude anthracene and drained anthracene oil (oil B).
Residue. Medium-soft pitch; av. yield 55% (C).

Low-temperature Tar

Fraction 1. Av. boiling range, 70—170°C; av. yield, 1%. Desulphurized and fractionated to yield benzole, toluole and xylole grades.
Fraction 2. Av. boiling range, 170—240°C; av. yield, 17.5%. Alkali-washed to recover phenol, cresols and xylenols and dephenolated carbolic oil.
Fraction 3. Av. boiling range, 240—300°C; av. yield, 20%. Alkali-washed to recover xylenols, higher-boiling tar acids and washed middle oil (A).
Fraction 4. Av. boiling range, 300—370°C; av. yield, 20%. Not further treated (B).
Residue. Medium-soft pitch; av. yield, 35%. (C).

In the USA, large quantities of crude tar are burned as a liquid fuel or, in some cases, the tar is 'topped' to 250°C and the residue burned (750 000 tons in 1973), but in the UK blends of tar oils or tar oils and pitch are preferred. These coal-tar fuels are covered by B.S. 1469:1962. They are supplied in a number of grades designated by numbers which indicate the correct atomization temperature in °F (i.e. the temperature at which the viscosity is 100 secs Redwood No.1 = 25 mm²/s).

References p 536

CTF 50 and 100 are blends of the oils designated A and B above; CTF 200, 250 and 300 are blends of these oils with pitch (C); the most widely used, CTF 250, also called 'creosote-pitch fuel', contains about 50% of pitch and 50% of oils. CTF 400 is unfluxed medium-soft pitch. A harder grade of pitch is sometimes used as a pulverized solid fuel[26].

The UK production of CTF has fallen drastically owing (a) to the reduction in the availability of crude tar — little or no tar is now produced at gas-works — and (b) to the reduction in demand for firing open-hearth steel furnaces, which are being replaced by other processes. Relevant statistics are:

Year	1960	1965	1970	1974
Crude tar produced in the UK	2799	2591	1481	924
Gas-works tar produced in the UK	1607	1451	318	2
CTF used in the UK	—	783	469	110
CTF 250 used in UK steel works	—	513	250	70
(Figures in thousands of long tons)				

Advantages of coal-tar fuels: Low ash (0.05—0.3%), low S content (<1.0%), high CV (37.5— 38.5 MJ/kg gross), high flame emissivity.

Disadvantages: Need for heated storage and transfer lines (except CTF 50), tendency to form deposits in storage tanks.

B12 Formed coke[27-30]

Formed coke may be defined as a carbonized coal product manufactured to a required shape. The major requirement for formed coke is in the metallurgical industry although there are domestic uses (App.B8). For the blast furnace, formed coke must meet a specification similar to that required for oven coke. The advantages of formcoke, by comparison with oven coke which it is intended to replace, are as follows:

(a) A uniform required size of coke is obtained.
(b) The heat treatment is consistent, thus avoiding fissures inevitable with oven coke.
(c) Throughput per unit volume is several times more than is possible with the coke oven and therefore capital and operating costs are lower; the (continuous) process permits greater automation with fewer operators.
(d) Continuous, or intermittent, operation of the plant.

(*e*)   Atmospheric pollution, which may be considerable from coke ovens when the fine coal is charged in tonnage quantities to the red-hot ovens, and also when the red-hot coke is 'pushed' into the coke car, is virtually eliminated in formcoke processes which charge and discharge continuously through sealed systems.

(*f*)   A wider range of coals can be treated, thus allowing some selection according to price, ash, and sulphur content.

The optimum size and shape may be produced either before or after carbonization. There are advantages in the former since briquetting is a well established technology. Much depends on the dominant type of feedstock coal available.

Non-coking coals, or coals with low coking properties, e.g. NCB rank Nos.100, 200, 800 and 900 (B.S. swelling no.<4), may be briquetted with bitumen, or pitch, as the binder, and carbonized in internally heated shaft retorts, or on moving grates. Coals with more of the coking constituent, when treated in this way, tend to swell on heating to a high temperature (800—1000°C), thus weakening the coke structure, and causing massive agglomeration within the carbonizing vessel. The coking properties of such coals may be modified by blending with selected additives such as iron oxide, lime, or sodium silicate. The proportion of binder may be decreased by increasing the pressure of briquette formation.

Coals with much higher swelling characteristics, e.g. NCB ranks 300, 400 and 500, may require considerable modification, e.g. by the conversion of all, or part of, the coal feedstock into coke breeze. The breeze may be made either in a fluidized bed, or in suspension in hot inert gas. In the F.M.C. (Food Machinery Corporation, USA) process the breeze is cooled, briquetted with pitch, and carbonized in a shaft furnace. In the Polish process low-rank lump coal is carbonized in shaft retorts and the product crushed to a fine breeze. This is then briquetted with pitch and cured in an air oven at about 350°C. The product is weak and easily abradable.

Reference has already been made in App.B8 to the 'Ancit' and 'B.B.F.' processes which utilize some coking coal to briquette the breeze at 450°C. The product yields about 8% VM, and carbonization to high temperature is necessary to approach the normal specification for blast-furnace coke. The Russian Sapozhnikov process is similar, but does not necessarily require an addition of coking coal. It is almost the same as the NCB process (App.B8), and hence requires high briquetting pressures to yield a briquette of sufficient strength for subsequent carbonization at high temperature.

A wide range of blast-furnace trials with formcoke has established that part or the whole of the oven coke may be replaced by formcoke. The use of pitch, or bitumen, as binder adds to the cost of a process. Some formcoke processes do not yield a high-CV gas.

References

1   Rousseau, P. E. *Chemy Ind.* 1962 (Nov.17), 1958—1966; Wld Pwr Conf. (Session 1B, Tokyo Sect. Mtg), Oct. 1966, Doc. No. 158, 13 pp
2   Ref. (F), p 59
3   Hottel, H. C. and Howard, J. B., *New Energy Technology*, MIT Press, Cambridge, Mass. and London, UK, 1973, pp 161—185
4   Ref.(L)
5   Ref.(M)
6   Ref.(E)
7   Ref.(N)
8   Ref.(O)
9   Ref.(F)
10   Ref.(P)
11   Potter, N. M. and Locke, H. B. *Trans. Instn Min. Engrs* 1966, Vol 69, 559
12   Ref.(Q)
13   Szpilewicz, A. *J. Inst. Fuel* 1969, Vol.42, 403
14   Barker, J. E. and Lee, G. W. *Min. Elect. Mech. Engr* 1967, Vol.48, 145
15   Graham, J. P. and Pater, V. J., Yearb. Coke Oven Mgrs Ass., 1972, pp 226—251
16   Graham, J. P. and Barker, J. E., Yearb. Coke Oven Mgrs Ass., 1974, pp 187—207
17   Finniston, H. M. *BCURA Gazette* 1969, No.69, pp 1—14
18   Graham, J. P., *Coke in Ironmaking*, ISI Publn 127, Iron and Steel Inst., London, 1970, pp 141—148
19   Ref.(R)
20   Potter, N. M. and Locke, H. B., 'Coal supplies and carbonization processes', *Min. Engr* 1966 (June), No.69, 559—572
21   Potter, N. M. and Martindale, J. R., 'Modern developments in smokeless fuels', *Min. Engr* 1966 (Dec.), No.75, 195—208
22   Dickinson, R. *J. Inst. Fuel* 1970, Vol.43, 75
23   British Standard 3142: Manufactured solid smokeless fuels for household use: Parts 1—3
24   Ref.(P)
25   Ref.(T)
26   Ref.(U)
27   Potter, N. M. *J. Inst. Fuel* 1970, Vol.43, 492
28   Potter, N. M. *J. Inst. Fuel* 1972, Vol.45, 313
29   Ref.(V)
30   Ref.(E), Ch.16 by D. C. Rhys Jones

General Reading

(A)   Scott Wilson, D., *Modern Gas Industry*, Edward Arnold, London, 1969
(B)   *Oil Gas J.* 1973 (June 25), 110, 123, 131
(C)   *Energy Pipelines & Systems* 1974 (March), 48
(D)   SNG Symposium 1, Inst. Gas Technology, Chicago, May 1973
(E)   Lowry, H. H. (Ed.), *Chemistry of Coal Utilization*, Suppl. Vol., Wiley, New York, 1963 (App.B3 — see Ch.22 by E. E. Donath)

(F) Dryden, I. G. C., *Carbonization and Hydrogenation of Coal*, United Nations, New York, 1973

(G) Bland, W. F. and Davidson, R. L., *Petroleum Processing Handbook*, McGraw-Hill, 1967

(H) Waddams, A. L., *Chemicals from Petroleum*, 3rd edn, John Murray, London, 1973

(I) Goldstein, R. F. and Waddams, A. L. *Petroleum Chemicals Industry*, 3rd edn, E. & F. N. Spon, London, 1967

(J) Hobson, G. D. and Pohl, W., *Modern Petroleum Technology*, 4th edn, Applied Science Publishers, London, 1973

(K) Williams, A. F. and Lom, W. L., *Liquefied Petroleum Gases: Guide to Nature, Properties and Applications*, Ellis Horwood, Chichester, U.K., 1974

(L) Roberts, J. and Jenkner, A., *International Coal Carbonization*, Pitman, London, 1934

(M) Lowry, H. H. (Ed.), *Chemistry of Coal Utilization*, Vols.1 and 2, Wiley, New York, and Chapman & Hall, London, 1945

(N) Brame, J. S. S. and King, J. G., *Fuel — Solid, Liquid and Gaseous*, 6th edn, Edward Arnold, London, 1967

(O) BP, *Gas Making and Natural Gas*, British Petroleum Co. Ltd, London, 1972

(P) Spiers, H. M. (Ed.), *Technical Data on Fuel*, 6th edn, World Power Conf., London, 1961

(Q) Mott, R. A. and Wheeler, R. V., *The Quality of Coke*, Chapman & Hall, London, 1939

(R) British Standard 1016: Part 13: 1969 (and later amendments)

(S) Smith, N., *Gas Manufacture and Utilization*, The Gas Council, London, 1945

(T) McNeil, D., *Coal Carbonization Products*, Pergamon Press, Oxford, 1966

(U) Huxtable, W. H. (Ed.), *Coal Tar Fuels*, Association of Tar Distillers, London, 1960

(V) *Proc. Int. Congr., Coke in Iron & Steel Industry*, Charleroi, 1966, and Luxembourg, 1970

Appendix C
Scientific and technological principles in the use of carbonaceous fuels

Chemistry and physics of combustion and heat release

C1 Gases

C2 Entrained liquid droplets and solid particles

C3 Suspended solid particles (fluidized combustion)

C4 Fixed-bed firing

C5 Soot formation

Specialized physical aspects of combustion

C6 Fluid flow

C7 Flow through smaller chimneys and flues, waste-gas handling problems

C8 Aerodynamic principles in combustion (including noise production)

C9 Flow measurement in practice

C10 Heat balances

Heat acceptance and transmission

C11 Heat transfer by conduction, convection and radiation

C12 Liquid-phase heating

C13 Heat exchangers

In this Appendix, abbreviated notes on the main factors of the relevant background science are given for the sake of completeness, to supplement the accounts of applications in the main chapters, and to give a lead towards more complete coverage in other publications. The style may vary from the bald checklist to a précis, as thought most appropriate.

CHEMISTRY AND PHYSICS OF COMBUSTION AND HEAT RELEASE

C1 Gases

There are two distinct modes of combustion: (1) premixed — combustible and oxidant gases are mixed before they enter the flame, and (2) diffusion — combustible and oxidant mix within the flame. A flame may be partly premixed and partly diffusion.

Premixed Flame

Exemplified by the fully-aerated Bunsen burner flame. Exhibits a blue inner cone (main combustion zone) and an outer faintly luminous zone (diffusion flame). The colour of the inner cone is due to chemiluminescence of C_2 and CH radicals; there is also ultraviolet radiation from OH radicals. When flow in the flame is turbulent the shape alters, boundaries become less sharply defined. As the amount of primary air is reduced, the outer zone of a hydrocarbon flame becomes more luminous and tip of inner cone develops a yellow luminosity due to soot formation (see App.C5). Ultimately with further reduction the whole of the outer zone becomes yellow.

Flame-stability depends on heat transfer, molecular or free radical diffusion, and the fluid flow patterns. A *burning velocity* (plane flame front advancing normal to its surface) characterizes a given fuel/oxidant mixture and temperature. Velocities are usually a maximum somewhat on the fuel-rich side of stoichiometric, are increased by increase in temperature, and reduced by dilution with an inert gas. Some *maximum burning velocities* (m/s) in air at room temperature are[1,2]: CO 0.46, H_2 2.6, CH_4 0.35, C_2H_6 0.40, C_2H_4 0.68, C_3H_8 0.39, C_4H_{10} 0.38. Increase in air temperature from 288 K to 450 K approximately doubles the burning velocity for hydrocarbons, and use of 100% oxygen instead of air results in an order of magnitude increase.

Combustion can only occur within *flammability limits*: limited ranges of concentrations of combustible gases and oxygen. For combustion of hydrocarbons and CO in air, lower limits for fuel concentration vary inversely with the heat of combustion[3], ranging from 12.5% for CO to 1.2% for octane. Expressed as a fraction of stoichiometric concentration, the lower flammability limit for hydrocarbons is about constant at 0.55, and the

upper limit varies[4] between 1.5 and 3. Buoyancy has a small effect, and ranges of flammability for upward propagation are a little wider than for horizontal, which in turn are wider tnan for downward propagation. Addition of inert gases and inhibitors reduces the width of the flammability range[5], and increase in temperature widens it.

Thickness of combustion zone depends on the chemical reaction rates. For hydrocarbons. the slowest and therefore rate-controlling step is[6] the final oxidation of CO to CO_2. The oxidation rate of CO is dependent on the concentrations of CO, O_2 and H_2O, and on the temperature. For conditions approximating to those in hydrocarbon flames, the stoichiometric combustion of CO is completed in a few milliseconds[7]. This is consistent[8] with measured combustion-zone thicknesses of about 1 mm.

Diffusion Flame

The air and fuel mix either by molecular diffusion, or by a combination of turbulent and molecular diffusion, depending upon the flow conditions (see App.C8). The combustion zone is thicker than in the premixed flame because flame is possible whenever mixing produces a composition within the flammability limits: *combustion is rate controlled by the mixing* instead of by the chemical reaction rate. The large-scale turbulence eddies cause some parts of the flame to be generally fuel-rich, and other parts fuel-lean. Combustion continues (temperature permitting) until after stoichiometric composition has been reached throughout. Fuel-rich parts in hydrocarbon flames may produce soot. *Soot* normally burns away as more oxygen is mixed in, but more slowly than a gas (see App.C5).

C2 Entrained liquid droplets and solid particles

Rapid heating of coal particle[9] gives evolution of (1) free water, (2) volatiles from thermal decomposition, of composition: CO_2 2–5%, CO 10–20%, H_2 50–65%, gaseous hydrocarbons 10–20% (as CH_4), plus vapours of oils and tars. Coke residue is left. Mass evolved may be 50–100% greater than in usual proximate analysis.

From oil, hydrocarbons evolved, decreasing in volatility as temperature rises, and finally pyrolysis leaving coke residue.

Residues from both coal and oil: carbons containing some H and O, with a fine-pore structure and/or larger cavities. May form a hollow sphere ('cenosphere'), which can be much larger than original particle[10]. Otherwise, residue before burning is similar in size to original particle or droplet.

As the evolved hydrocarbons are heated further, pyrolysis and/or oxidation occurs, depending on composition of the surrounding gas (see App.C1). If *oxygen* is *deficient*, soot is formed → high flame emissivity (see App.C5, C11 — radiation).

Residue burns more slowly than volatiles: combustion as soon as O_2 reaches surface. Reaction occurs: (a) on surface, if chemical rate $\gg$ diffusion of O_2 to surface, (b) throughout particle, if diffusion $\gg$ chemical rate. Reaction rate is proportional to oxygen concentration outside the stagnant diffusion layer surrounding the particle[11], and therefore depends on mixing processes in the flame (see App.C8). *Diffusion-controlled rate constant* is

$$K_d = 0.45 \, \phi \, T^{0.675}/x \qquad (1)$$

$$(1000 \text{ K} < T < 3000 \text{ K})$$

where K_d is kg carbon burned per m^2 external surface of particle per atm of oxygen per second, ϕ is a mechanism factor (1 when primary product is CO_2, 2 when it is CO; usually $\phi \approx 2$), T is mean temperature (K) in diffusion layer, and x is equivalent spherical particle diameter (μm). For a 50 μm particle, and $\phi = 2$, K_d increases from 1.88 to 3.89 as T increases from 1000 K to 3000 K.

Surface-reaction-controlled *(chemical) rate constant*, K_s, depends on fuel characteristics including porosity, and temperature. At 1600 K increases from 0.2 to 2.4 kg/m^2 s atm for chars from coals of increasing rank[12]. No reliable measurements relating to carbonaceous residues from oil droplets have been reported. When a combination of diffusion and chemical control operates, an overall coefficient K is given by

$$1/K = 1/K_d + 1/K_s \qquad (2)$$

and reaction rate is pK, where p is partial pressure (atm) of oxygen outside the diffusion layer.

For $<78 \mu m$ anthracite, at <2200 K, chemical control is significant[12,13], but for low-rank coals only <1600 K.

Particle temperature is governed by a heat balance[11] between combustion heat release and heat loss (convective, radiative). For oxidation to CO, the *heat release* is 9.80 MJ/kg, and to CO_2 33.05 MJ/kg.

C3 Suspended solid particles (fluidized combustion)

Gas velocity through bed high enough to support each particle individually, but velocity above bed less than terminal velocity of particle. Bed as a whole behaves as a fluid; voidage between particles typically 0.7, but increases with ratio of gas velocity to minimum fluidizing velocity. Fuel fed to bed of inert particles, and burns in fluidizing air[14]. Part of air rises as 'bubbles'.

Combustion intensity depends on: fuel concentration in bed, particle size, temperature, rate of oxygen supply. Air-flow pattern through bed

approximates to plug flow; therefore highest O_2 concentration and fastest heat release near bottom of bed. Bed temperature is uniform and controlled by heat balance: input potential combustion heat release; output net gain in sensible heat of gas plus sensible and potential heat loss in ash and coal leaving bed, plus potential heat loss in incompletely burnt gases, plus heat loss to walls and cooling surfaces in bed, plus net radiant heat loss from top of bed.

High heat transfer coefficients to surfaces in bed: combined radiative and convective coefficients 200 to 300 W/m^2 K for beds of mean particle dia. between 3 and 1 mm, rising to 600–700 W/m^2 K for 0.2 mm particles[15]. For dry ash-removal, mean bed temperature must be 200 to 300 K below the initial deformation temperature of coal ash[16], and is usually in the range 1000–1200 K. Gas residence time in bed short, therefore risk of incomplete combustion of gases (e.g. volatiles from coal or oil, natural gas) unless well distributed across airstream. *Combustion of solids* usually *chemically rate-controlled* (see App.C2).

C4 Fixed-bed firing

Much interesting research in the literature (e.g. ref.17) on the mechanism of combustion in fuel beds has little relevance to industrial conditions. In a fuel bed the oxygen is consumed in a fairly narrow band ($\succ 2$ particles) following the point where it first reaches carbon at a high enough temperature to initiate reaction. After this, the oxygen content, at the usual moderate flow rates, becomes negligible. If the gases have then to pass on through a zone of hot coke, the CO_2 first produced is reduced to CO. Since gaseous diffusion is, at the temperatures used in fuel beds, a slower process than chemical reaction at the fuel surface, combustion rate in a bed once ignited is mainly governed by diffusion; but because of its rapidity it is in practice determined by the rate of oxygen (air) entering the bed; this in turn is proportional to (pressure drop across the bed)^n where n is about 0.6 for (the usual) turbulent flow.

Approximately 11 kg of air will burn completely 1 kg bituminous coal; less air insofar as the carbon is ultimately oxidized only to CO, and the burning rate will then usually be higher. The gross effect of this can best be determined experimentally, and is small with sufficient excess air.

Other reactions to be taken into account in calculating combustion balances are:

$$H_2 \text{ (or } 2H_{fuel}) + \tfrac{1}{2}O_2 \rightarrow H_2O$$

$$S_{fuel} + O_2 \rightarrow SO_2$$

Calculation of total oxygen required must also make allowance for oxygen in the raw coal. The principal inert gaseous component is nitrogen.

The heat of formation of coal (and coke or char) from C, H, O etc. is comparatively small, and heat-release rates may be calculated using the following gross values of CV: hydrogen 144.210 MJ/kg; carbon to CO 10.12 and to CO_2 33.73 MJ/kg; sulphur 9.3 MJ/kg.

Oxygen requirements and heat-release rates can be calculated from the above data for coals of different analyses by standard procedures[18]. In a static or travelling (mechanical stoker) bed of coal, progress of devolatilization is normally faster than ignition, which tends to favour smoke production unless adequate secondary air sweeps the bed surface[19]. Increase in primary air rate ultimately leads to 'equilibrium burning' without smoke, but in practice the point of bed fluidization is reached first so that the possibility has little practical relevance.

When boiler efficiency is measured by summation of heat losses, the loss due to combustible matter in the ash is assumed to be due to unburnt carbon with a calorific value of 33.82 MJ/kg (in domestic boiler testing, an allowance is made for sulphur retention, the CV being taken as 33.50 MJ/kg). Percentage loss in efficiency is 33.82 (or 33.50) $Ab/100\ Q_1$, where A = ash (plus unburnt carbon) as % of fuel, b = percentage unburnt carbon in ash, and Q_1 = CV of fuel (gross or net CV, depending on the definition of boiler efficiency).

C5 Soot formation

Caused by pyrolysis of hydrocarbons and other compounds in vapour phase, when insufficient oxygen is present to convert all the carbon to oxides. *Quantities* usually greater than those predicted from equilibrium data[20]. Dilution of flame gases with recirculated combustion products reduces soot formation[21].

Particles spherical, about 0.05 μm dia. and composed of smaller crystallites. Chain-like agglomerates readily form. Large surface area and high particle emissivity (≈ 0.8) imparts *high emissivity* to flames, and small particle size gives high convective heat transfer rate between gas and particles (see App.C11).

Burnout: surface reaction rate coefficient (chemical coefficient) an order of magnitude lower than for anthracite chars[22]. Because of this, and small particle size, burnout is *chemically rate-controlled*.

SPECIALIZED PHYSICAL ASPECTS OF COMBUSTION

C6 Fluid flow

A very few of the essential concepts are summarized below. Detailed formulae will be found (e.g.) in

refs. (A) and (B) and other reference works. Air—steam mixtures are considered in ref. (A), flow in chimneys and flues is referred to in Chapter 19.4 and Appendix C7. Flow through granular materials in beds or fluidized is briefly treated in refs. (A) and (B); furnace aerodynamics, in App. C8.

A basic relation expressing the conservation of energy is embodied in Bernoulli's theorem. In one of its simplest forms, for an incompressible fluid, the balance of mechanical energy is written for two given points in the flow line in terms of equivalent pressures (i.e. heads, dimension length) of the fluid flowing (density ρ):

$$z_1 + p_1/\rho g + v_1^2/2g + w - F = z_2 + p_2/\rho g + v_2^2/2g \quad (3)$$

where w = work done on fluid from outside; F = frictional loss; the zs are heights of mid-stream above a datum; $p/\rho g$ are static pressure heads; $v^2/2g$ are velocity heads. p thus varies along or across the flow according to changes in v or z (and vice versa), as modified by work done (e.g. by a pump) on the fluid and by pipe friction.

Flow through Pipes

Fluids passing through pipes suffer frictional drag. At low speeds flow is laminar, i.e. parallel to the wall. At higher speeds it is turbulent: cross currents or vortices are formed which increase the drag. Obstructions, changes in direction or section also interfere with steady flow. The change from laminar to turbulent flow was shown by Reynolds to occur when the now-called Reynolds number (Re) exceeds a certain value depending on conditions and the system considered. This number is defined by the equation:

$$(Re) = Dv/\nu \quad (4)$$

where D is a characteristic dimension transverse to the flow (dia. for a circular pipe) and ν is the kinematic viscosity. The number is a dimensionless expression for the ratio of inertial to viscous forces.

It is now usual (loc. cit.) to plot values of a friction factor f against (Re) over a wide range. For pipe length l:

$$F \text{ (above)} = \frac{2flv^2}{gD} \quad (5)$$

The plot shows a sharp change of direction at the critical value (Re_c).

For a long straight commercial pipe:

$$f = 16/(Re) \quad (Re) \not> (Re_c) \quad (6)$$

$$f = 0.0035 + 0.264/(Re)^{0.42} \pm 10\%$$
$$(Re) > (Re_c) \quad (7)$$

(Re_c) for a round pipe is commonly about 2100, but the transition may not occur immediately on the upward-side. Roughness increases f, and also curvature.

Flow metering (Ch. 18, App. C9) may be achieved by inserting a resistance, e.g. a venturi throat or an orifice plate, and measuring the difference in pressure before the venturi throat and at its narrowest section, or the pressure difference before and after an orifice plate. Much less of the pressure (and hence less energy) is lost irreversibly in the former device. Flow can also be measured by traversing the cross-section with a pitot tube (comparing $v^2/2g$ with the static pressure head $p/\rho g$) and integrating the velocity data.

Nozzles

These are converging or converging-diverging tubes attached to the outlet of a pipe to convert pressure energy in the fluid more efficiently into kinetic energy. The fraction of the ideal (Bernoulli relation) jet velocity obtained in a nozzle is known as the velocity coefficient. It ranges typically from 0.85 to 0.98.

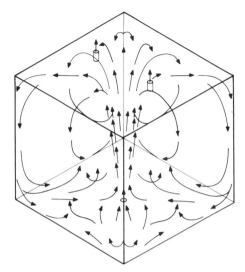

Figure 1   General flow pattern in a cube (after Bacon, Chesters and Halliday[24])

Flow from Pipes into Furnace Chambers

Most furnaces, kilns and boilers involve the use of fuel/air jets blown into chambers. Until comparatively recently it was assumed that, buoyancy apart, such jets proceeded in the direction where they were pointed, merging with the general atmosphere of the chamber and leaving by the exit flues — a condition sometimes described as plug-flow. Research, mainly from 1946 onwards, e.g. refs. 23–31, showed that plug-flow was rare: even in shapes such as cubes (*Figure 1*) and cylinders (*Figure 2*) major recirculation occurred. Visualization of flow in open-hearth furnace water-models by means of air bubbles or aluminium dust suggested that at the central cross-section one part of the flow was backwards — towards the burner — for every two parts moving forward. Results on operating furnaces showed the same effects. The incoming jet clearly entrains surrounding gases, resulting in a local drop in pressure which then leads to return flow from points further down the furnace (*Figure 3*).

a

b

Figure 2   General flow pattern in cylinder with non-swirling axial jet (after Bacon, Chesters and Halliday[24])

Nozzle dia.: 3.5 mm (9/64 in); cylinder: 98.5 mm (3⅞ in), 441 mm (17⅜ in) long; flow: 85 cm³/s. Observations were taken with liquid flowing horizontally. Tail of jet flaps violently and irregularly across horizontal plane: two patterns are shown

Gas Compression

The trend towards use of gases under considerable pressure in fuel processes increases the importance of a particular special case of the flow equation (3): w may be considerable owing to the work of compression.

(R = gas constant, T_1 initial absolute temperature, M molecular weight, p_1 initial pressure, p_2 final pressure, γ specific heat ratio).

The actual work theoretically required lies between these extremes in a practical compression

$$\text{Adiabatic: } w \left(\frac{\text{J}}{\text{kg}}\right) = \left(\frac{\gamma}{\gamma-1}\right) \left(\frac{RT_1}{M}\right) \left[\left(\frac{p_2}{p_1}\right)^{\gamma-1/\gamma} - 1\right] \quad (8)$$

$$\text{Isothermal } (T_1): w = (RT_1/M) \ln (p_2/p_1) \quad (9)$$

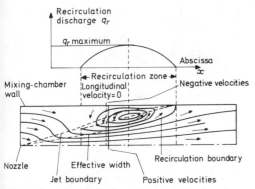

Figure 3   Recirculation for a ducted axisymmetric jet (after Barchilon and Curtet[23])

plant.   Multiple-stage compression can reduce somewhat the total.   But the theoretical figure must be divided by a pump efficiency between 50 and 75%; for safe rough estimation, calculated on the adiabatic formula.

C7 Flow through smaller chimneys and flues, waste-gas handling problems

The flow of flue gases in such systems requires consideration of most of the following:

1.   Thermal throughput
2.   Length of travel
3.   Direction changes
4.   Wall friction
5.   Intermittency of use
6.   Condensation possibilities
7.   Solids deposition
8.   Thermal properties of materials
9.   Cross-sectional area and shape
10. Height of flue or chimney
11. Air dilution at inlet
12. Wind-variation effects
13. Inlet and exit pressure conditions
14. Appearance
15. Solids fall-out
16. Local topography
17. Ground-level flue-gas concentration limits to be observed
18. Noise aspects
19. Power requirements

Theoretical and practical information is available on small flues, for example in domestic use, and also larger flues in commercial and industrial use[32-35].

Because of the complexity of detailed flues design, as illustrated above, it is necessary to

examine specific applications in order to understand their basis and operation.   The following general comments serve to illustrate the need for individual studies.

Natural Draught

Little energy from the buoyancy of hot gases is available for the operation of natural-draught flues. Small flues provide little effective draught while large flues such as large and high chimneys can provide draught up to 1 mbar (100 Pa) or greater. Because of such wide variations the design of natural-draught flues requires calculation of a precise balance between buoyancy forces generating the flow, losses, and draught requirements at the inlet of the system.   As the cross-sectional area of flues is one of the most important aspects of design, it may be useful to note that the velocities in small natural-draught flues, in for example domestic practice, range from 3–4 m/s while for larger flues 5 m/s is more widely accepted. It is emphasized that these values are given as an indication of practice and not for design purposes.

Forced Draught

Forced draught is being increasingly used for flueing and waste-gas transfer.   The range of uses now encompasses small domestic-heating appliances in addition to large industrial applications.   Fan powers range from a few watt up to several MW. Theoretical design uses accepted fluid-flow equations (App.C6) with due allowance for the fixed and variable aspects listed above.

Gas velocities range from values similar to those for natural draught up to more than 25 m/s. Design must clearly depend on specific requirements and accepting limitations on power, speeds and *noise* as relevant to individual installation.

See also Chapter 19.4 for larger chimneys.

C8 Aerodynamic principles in combustion (including noise production)

Velocity gradients are generated in a furnace because cross-section area is larger than the combined areas of the inlet ports.   The resultant shear force induces turbulence, which gives rapid mixing of fuel and air if introduced separately to the furnace (diffusion flames).   The turbulence, and friction at the walls of ducts and furnace, dissipate energy (see App.C6); also, the combustion air must be accelerated from rest.   In natural-draught furnaces, the energy is derived from buoyancy of the warm flue gases in a chimney of the required height[36]; in others, from forced and/or induced draught fans.

Mixing of fuel and air controls the rate and position of heat release by combustion in diffusion flames. For a simple round jet of fuel in still air, the change of axial concentration and velocity of jet fluid with distance from the nozzle is given by:

$$\frac{C_m}{C_0} \text{ or } \frac{V_m}{V_0} = k \left[\frac{\rho_0}{\rho_a}\right]^{1/2} \frac{d_0}{x} \text{ (for } x > kd_0) \quad (10)$$

where C_m, C_0 (V_m, V_0) are time-mean concentrations (or velocities) at points on the axis and in the nozzle fluid respectively, ρ_0, ρ_a are densities of nozzle fluid and entrained air respectively, d_0 is nozzle diameter, and x is distance from the nozzle. k is a constant, 5.0 for concentration, 6.3 for velocity.

The turbulence results in fluctuations of both concentration and velocity about their time-mean values at any point.

The transverse profiles of concentration and velocity in planes perpendicular to the jet axis are given approximately by:

$$\frac{C}{C_m} \text{ or } \frac{V}{V_m} = 0.5 \left[1 + \cos\left(\frac{\pi r}{2x \tan \beta}\right)\right] \quad (11)$$

where C, C_m (V, V_m) are time-mean concentrations (velocities) at respectively a point distance r from the axis, and on the axis; β is the jet half-angle, i.e. the angle between the axis and a line joining points where the concentrations (velocities) are half the value on the axis. For concentration, β is $6.2°$, for velocity $4.85°$.

The *rate of entrainment* by the jet is given by

$$\frac{\dot{m}}{\dot{m}_0} = 0.32 \left[\frac{\rho_a}{\rho_0}\right]^{1/2} \frac{x}{d_0} \quad (12)$$

where $\dot{m}$ is mass flowrate of fluid (including entrained air) across a plane at distance x from the nozzle, and $\dot{m}_0$ is the mass flowrate of nozzle fluid. In a furnace, two zones must be distinguished: that between the burner and the flame front, in which ρ_a is the density of the hot recirculated gases and ρ_0 is the density of the cool fuel jet; and that beyond the flame front, where ρ_0 is the density of the flame gases.

Concentric jet burners. Mixing close to the burner is faster than in simple round jets, and is particularly rapid when the velocity of the outer jet exceeds that of the inner one[37]. The width of the wall separating the jets also influences mixing rate.

Swirled jets. Centrifugal force tends to cause a wider angle of spread in swirled jets, and a radial pressure gradient tends to develop to counterbalance the centrifugal force. If pressure on the axis falls low enough, a reverse flow develops, from high pressure downstream to low pressure close to the burner. Entrainment rate in the zone between the nozzle and five nozzle diameters away is increased by swirl[38]:

$$\frac{\dot{m}}{\dot{m}_0} = 0.32 \frac{x}{d_0} + 2.2S \quad (x > 5d_0) \quad (13)$$

where S is a swirl parameter = flux of angular momentum at the nozzle divided by the flux of axial momentum and by the nozzle diameter. The jet half-angle, β, is[39]:

$$\tan \beta = 6.68S \tan \beta_0 \quad (14)$$

where β_0 is the half-angle in the absence of swirl.

Flame Noise

In turbulent flames noise results from pressure pulses due to irregularities in direction and speed of the flame front[40]. The fraction of the energy of combustion that is released as noise ranges from 1×10^{-8} for pre-mixed flames to 2×10^{-6} for a turbulent diffusion flame of a hydrocarbon in air. The sound may be represented as a collection of monopole sources of different strengths and frequencies distributed throughout the reaction zone, and its intensity depends quantitatively on the rate of change of combustion rate in the flame[41].

Periodic fluctuations of the flame front may give rise to notes of specific frequencies, as in 'singing flames'[40], and screaming combustion in tunnel burners[42]. In fuel-oil-fired boilers these fluctuations can be of low frequency and sometimes lead to structural damage, flame blowoff, and low combustion efficiency[43]. They are caused by a direct feedback relation between pressure and combustion rate, so that the amplitude increases as long as the fluctuations are in phase with the natural oscillation period (organ-pipe effect.)

C9 Flow measurement in practice

Checks needed on the following:

Fluid Conditions

Flow measurement of weight rate, based on density — correction required for density variation — wide variation may require monitoring pressure and temperature. Compensation included in some meters. Volumetric meters rated on standard conditions. Pressure is a major factor in meter selection.

Flow Rate

Meter most accurate at higher levels of capacity — some meters cannot show overload — demand usually highest at start-up. Large variation may require multiple metering for accuracy.

Accuracy

Errors incurred in meter and metering-device design — poor installation — failure to correct for non-standard conditions. Accuracy requires sensitivity — usually at higher cost. Commercial standards ±2%, with care ±1% at full flow rates.

Type of Result Required

Integrated result for cost allocation or long-period record. Indicator often sufficient for plant operation. Recorder for performance study. Units combining all three are available. Recorders require chart change — consider cost — linear or long-period charts available.

Meter/Meter-device Location

Accessibility for direct reading — need for strainer — by-pass for meter servicing. Avoid vibration, dirty or damp conditions — an advantage of panel mounting. Select pipe section — straight horizontal, unobstructed by valves, bends etc. for 10 dia. upstream 5 dia. downstream. If not possible, special precautions, e.g. straighteners, annular tappings (piezometer ring) may be necessary. Meter remote from device — 20 m about limit for hydrostatic transmission: convert to pneumatic or electrical signal for greater run. Avoid pulsating flow, e.g. reciprocating pumps.

Pressure Loss introduced in Fluid

All mechanical metering incurs some pressure loss; no loss with magnetic or ultrasonic meters, negligible loss with pitot tube. Pressure loss may increase power cost and/or reduce plant output. Normal range of loss 3—23 kPa (1—8 ft H_2O).

Differential-pressure devices (generating given head) at mid range

	Permanent pressure loss (% of diff. head)
Dall tube — long venturi	Less than 10
Dall orifice — short venturi	10 — 20
Nozzle or orifice	50 — 60

Inferential or displacement meters — loss depends on sizing to flow, usually up to 15 kPa (5 ft H_2O) — significant for liquids gravity flow — hence use weirs etc. Higher differentials — more positive result — lower first cost.

Installation

Directly mounted meters — arrange pipework to ensure flooded condition.
Pressure-differential types, pressure tapping positions:
Gases — taps on top — meter above main so that condensate runs back to main.

Steam — taps on side — meter below — air vent upwards to main.
Liquids — as for steam (since dirt accumulates at bottom of main).
If these positions are not possible, air vents, liquid drains, seal pots, as appropriate, may be required. Pressure-impulse pipes, copper or stainless steel (suit pressure), minimum slope 1:20, size 10 mm, self draining, full-bore valves — for clearing blockage, and to avoid false head. Run impulse-pipes in same temperature conditions — take precaution against freezing if exposed. Eliminate leakage of air or fluid — avoid forming pockets. Determine fluid pressure condition at upstream position. Downstream pressure on meter should be above atmospheric.

Limitations

Mercury manometers unacceptable in some applications — e.g. food factory, direct steam injection. Inferential meters inaccurate below 12% rating — but will take short-period overloading. Differential meters inaccurate below 10% rating — do not register overload. Also design and installation must be within calibration tolerances: 'm' ratio (throat or orifice area/pipe area) values — nozzle 0.2—0.55, orifice 0.05—0.7, venturi 0.05—0.5; minimum dia. — nozzle 11 mm, orifice 7 mm, venturi 20 mm. Special limitations on pipes below 50 mm dia. (2 in) and on viscous fluids.

Interpretation of Results and Sources of Error

All meters — to be corrected for conditions varying from design basis. Mechanical-meter faults due to incorrect sizing, dirt, corrosion, wear, low flow, reverse flow. Differential-meter faults due to load fluctuation which tends to give high results, wrong zero calibration, leaks on impulse pipes, choked pipes, bent pens, damaged differential device. Square root charts need care in averaging. For details of flow formulae, construction of device and correction factors refer to BS 1042:Part 1:1964.

C10 Heat balances

Alternative terminology, Heat account, or Energy audit. Useful for assessing performance of existing plant or when designing new plant. Reasonably accurate measurement and experienced assessment of data used are more important than unnecessary precision, i.e. slide-rule working is normally sufficient. Heat balance is presented either numerically, usually in percentages, or graphically, usually in heat or energy units. Graphical presentation (usually referred to as a Sankey diagram) is often the most powerful way of showing the significance of losses.

The basic heat balance is expressed simply as:

HEAT INPUT = HEAT OUTPUT

Terms on both sides are referred to a common datum temperature — usually $15°C$.

Heat Input

1. Heat energy of fuel plus sensible heat content of fuel and air.
2. Heat evolved from exothermic reactions of materials being heated.

Heat Output

1. Total heat gained by materials heated, including sensible and latent heat.
2. Heat absorbed by the materials owing to endothermic reactions.
3. Total heat of waste gases or combustion products leaving the plant or process, including latent heat of water vapour in waste gases, and loss due to presence of incompletely burnt gases, e.g. H_2 and CO.
4. Losses due to radiation and convection to the surroundings from outside surface of plant. Heat loss by conduction to the ground is also sometimes significant.
5. Miscellaneous and special losses, e.g. unburnt fuel in ash, boiler blowdown losses, radiation from apertures in furnace walls, etc.
6. Transient losses in structure, e.g. sensible heat absorbed by brickwork etc. of heat-treatment furnace on heating up from cold.

Next step: decide whether to present heat balance on gross or net calorific value-basis — both together are common. Gross balance is more correct in heat-energy terms, but net balance is often more useful in practical terms. Simplest form of heat balance is the 'inferential' balance frequently

used for a rapid assessment of the performance of modern packaged boilers.

A typical example is as follows:

Approximate heat balance (net calorific value-basis)

	%
Heat to steam (by difference)	83.5
Sensible heat lost in dry combustion products (by nomogram or Siegert-type formula*)	13.0
Radiation and unaccounted losses (statistically assessed)	3.5
Heat in fuel	100.0

* B.S. 845:1972, item 75(a)

The only measured data required for this simple heat balance are the combustion-air temperature and the temperature and CO_2 or O_2 content of the waste gases.

A much more detailed heat balance follows, for a well operated open-hearth steel furnace, presented in both numerical and graphical form (*Figure 4*).

	%	%
Heat in coke	86.2	
Heat from exothermic reactions	8.6	
Heat in steam supplied to producer from WHB	5.2	
Total heat supplied	100.0	100.0
Heat gained by steel	19.7	
Heat gained by steam at WHB	24.5	
Total of useful heat		44.2
Heat content of waste gases leaving WHB		18.6
Structure losses etc:		
At gas producer	10.7	
At furnace	13.8	
At regenerator	10.3	
At waste-heat boilers	2.4	
Total structure losses		37.2
Unaccounted for heat, errors, etc.		Nil
Total heat accounted for		100.0

The measurement and correlation of the data required for the above type of heat balance is time-consuming if great accuracy is required. The term 'Unaccounted for heat, errors, etc' should appear in all heat balances. It should be as low as possible and can be either positive or negative.

See further examples: *Efficient Use of Fuel*, 2nd edn, HMSO, London, 1958, Ch. 26; O. Lyle, *Efficient Use of Steam*, 4th impression, HMSO, London, 1954, Ch. 20.

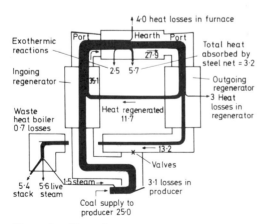

Figure 4   Heat flow diagram for open-hearth steel furnace (all figures refer to kg coal/100 kg steel)

HEAT ACCEPTANCE AND
TRANSMISSION

C11 Heat transfer by conduction, convection and radiation

Symbols

A	Area of conducting layer or radiating (absorbing) surface (m^2)
C	Concentration of suspended particles (kg/m^3)
c_p	Specific heat of fluid per unit volume at constant pressure (J/m^3 K)
D	Internal diameter of pipe (m)
d	Diameter of an equivalent spherical particle (m)
d_1	Inner diameter of a pipe covering (m)
d_2	Outer diameter of a pipe covering (m)
F	Particle absorptivity, dimensionless
f	Factor for cloud emissivity (m^{-1})
g	Gravitational acceleration
h	Heat-transfer coefficient (defined by equation 20) (W/m^2 K) (other units used in ref.47)
k	Thermal conductivity (W/m K) (other units used in ref.44)
L	Thickness of conducting layer (m)
L_h	Latent heat
l	Path length of radiation through absorbing or emitting gases (m)
n	Ratio of diameter to d_1
q	Heat flow rate (W)
q	Radiation intensity (W/m^2 sr)
S	Specific surface area of particles suspended in a gas (m^2/kg)
T	Temperature (K)
T_1	Higher temperature of a conducting or radiating body (K)
T_2	Lower temperature of a conducting or radiating body (K)
v	Fluid velocity (m/s)
a	Absorptivity of a gas towards radiation passing through it, dimensionless
δ	Absorption coefficient of a gas (m^{-1})
ϵ	Emissivity, dimensionless
λ	Wavelength (m)
v	Kinematic viscosity of a fluid (m^2/s) (different units used in references)
ρ	Density
σ	Surface tension

Subscripts

a, b, c, n	Various layers of a composite conducting covering or wall (or c = 'critical' for lagging)
B	Relating to radiating properties of a black body
g	Properties of an emitting or absorbing gas

References p 551

Conduction

Kinetic energy of vibrating molecules is transmitted to those in contact with them without change in relative mean positions.

Steady heat flow rate:

$$q = k\,(T_1 - T_2)\,A/L \tag{15}$$

Flow through a composite wall is:

$$q = \frac{(T_1 - T_2)\,A}{L_a/k_a + L_b/k_b + \ldots + L_n/k_n} \tag{16}$$

and through unit length of pipe covering:

$$q = \frac{2\pi k\,(T_1 - T_2)}{2.3 \log\,(d_2/d_1)} \tag{17}$$

Increase in insulation thickness causes two opposing effects: increase in the external area losing heat, and decrease in the heat flowing per unit area. There is a maximum value for heat loss at the 'critical diameter' d_c of the outside of the covering:

$$d_c = 2k/h_c \tag{18}$$

where h_c is the sum of convective and radiative heat loss coefficients (as a first approximation independent of d).

Further increase in lagging beyond d_c is needed to reduce the heat loss once again to that for a bare pipe. Expressing the diameters as ratios to d_1, e.g. $n_c = d_c/d_1$, this will occur at a larger lagging diameter n_3 such that $n_3/(n_3 - 1) \ln n_3 = n_c$; *only beyond this* will lagging save heat. $h_c \approx$ 15 W/m^2 K. If n_c is large, e.g. if d_1 is 1 cm and the lagging material is poor (0.25 W/m K), n_c may be ≈ 3 and n_3 10 or more. Better lagging such as powdered material (0.04 W/m K) can then reduce n_c to about 0.5 and lagging will be effective from the start. When $n_c = 2$, $n_3 \approx 5$; $n_c = 1.5$, $n_3 < 3$. To sum up, d_c may range from 0.005 m (good insulation) to 0.03 m (poor); therefore small pipes are more difficult to lag effectively because the range of n_c is higher.

Heat loss through unit length of a composite pipe-covering is:

$$q = \frac{2.73\,(T_1 - T_2)}{\log\,(d_b/d_a)/k_a + \log\,(d_c/d_b)/k_b + \ldots} \tag{19}$$

For *transient (non-equilibrium) conduction*, numerical methods[45] are convenient for solution using digital computer techniques.

Convection

Transmission of heat is effected by bulk movements of those parts of a fluid in contact with heat-receiving or donating surfaces. May be either natural or forced. Natural convection is when the movement of the fluid is caused only by buoyancy resulting from heating; forced convection, when from forces not arising from the heat transfer.

All convective heat-transfer rates can be expressed by:

$$q = hA (T_1 - T_2) = hA(\Delta T) \tag{20}$$

In natural convection, h varies with ΔT according to a fractional power (1/4 to 1/3), but in forced convection h is independent of ΔT.

Natural Convection

The size, shape and orientation of the heat-transfer surface affect h: for bodies larger than about 0.3 m the relevant power of ΔT is[46] nearer 1/3 than 1/4. Many empirical formulae have been proposed; a useful working formula is given in *Technical Data on Fuel*[47].

Forced Convection

h is proportional to the 0.8 power of fluid velocity (for fluid flowing inside a pipe) or to the 0.6 power (for fluid flowing outside a pipe). Full formulae are to be found in refs.51 and 52. The basic equation for heat transfer between a pipe and a non-viscous fluid flowing in it is[48].

$$(Nu) = 0.023 \, (Re)^{0.8}(Pr)^{0.4} \tag{21}$$

where (Nu) (Nusselt number), (Re) (Reynolds number) and (Pr) (Prandtl number) are dimensionless quantities formed as follows:

$$(Nu) = hD/k, \ (Re) = vD/v, \ (Pr) = c_p v/k$$

The basic equation for heat transfer to the outside of a single tube with its axis at right angles to the direction of flow is:

$$(Nu) = 0.26 \, (Re)^{0.6}(Pr)^{0.3} \tag{22}$$

in which (Nu) and (Re) are calculated on the *outside* diameter of the pipe and velocity used is the maximum in the tube bank. Where tubes are arranged in banks, the leading row generates turbulence which increases h for the tubes behind. The effect is greatest for staggered tubes, when the constant in equation (22) can be increased to as much as 0.35.

Turbulence and velocities can also be increased by the use of baffles[49,50]. Surface roughness, and finned tubes, increase heat-transfer rates by increasing the effective heat-transfer area[52].

Convection between Gas and a Suspended Particle

$$q = (Nu) \, k(\Delta T)/d \tag{23}$$

If no relative motion between particle and fluid and particles are far apart as in a pulverized-coal flame, then $(Nu) = 2$. Increasing particle size (and

hence, particle (Re)) increases (Nu); closer particle spacing decreases (Nu). In fluidized beds, (Nu) values as low as 0.001 for $(Re) = 10$ have been measured, owing to close particle spacing and gas by-passing the bed in bubbles (see App.C3).

Radiation

Wavelength range effective in heat transfer in most practical combustors lies between 5×10^{-5} and 5×10^{-4} cm (i.e. visible and infrared). Maximum possible radiation of energy from a body is:

$$q_B = 5.67 \times 10^{-8} AT^4 \tag{24}$$

termed *black-body radiation*. This radiation covers a continuous spectrum, the intensity of emission in a given wavelength interval being given by Planck's law[53]. At the same time the body absorbs radiation incident upon it.

Bodies frequently emit less than a black body at the same temperature; the ratio of radiant energy emitted per unit narrow range of frequency at wavelength λ to the corresponding black-body radiation is ϵ_λ, *the emissivity at wavelength λ*. The ratio of total radiation over all wavelengths to total black-body radiation is ϵ. If $\epsilon_\lambda = \epsilon$ always then it is termed *grey-body radiation*. ϵ for a number of surfaces is given in ref.54.

Black-body radiant energy emission in a direction making an angle θ with a line normal to the radiating surface, per unit area of surface and per unit solid-angle-cone of radiation is:

$$q'_B = q_B \cos \theta / \pi \tag{25}$$

When radiation falls on a surface, all or part may be absorbed, reflected or transmitted. Most reflection is non-specular. The fraction absorbed is termed the absorptivity of the surface. The absorptivity towards radiation emitted by a black body at temperature T is equal to the emissivity of the surface at T. Whatever the value of ϵ, a completely enclosing surface radiates as a black body ($\epsilon = 1$) because of multiple reflections.

The *radiant-heat transfer rate* between a grey body (A_1, T_1, ϵ_1) and a completely enclosing surface (A_2, T_2, ϵ_2) when the separating medium is non-absorbing is:

$$q = \left[\frac{A_1 A_2 \epsilon_1 \epsilon_2}{A_1 \epsilon_1 + A_2 \epsilon_2 - A_1 \epsilon_1 \epsilon_2} \right]$$
$$\times 5.67 . 10^{-8}(T_1^4 - T_2^4) \tag{26}$$

Absorbing Gases

A small volume of gas may emit and absorb radiation like a solid surface. Rate of energy emission per unit volume and per unit solid angle (i.e. radiation intensity) is given by:

$$q' = 5.67 \times 10^{-8} \delta \, T^4/\pi \tag{27}$$

δ, the absorption coefficient, is a measure of the absorbing and radiating propensity of the gas, and is defined by

$$dq'/q_i' = \delta\, dl \qquad (28)$$

where q_i' is the incident radiation intensity. Absorption may be due either to the presence of suspended particles, or absorbing gases such as CO_2 and H_2O vapour.

(a) *Particles.* Emissivity and absorptivity depend on size and refractive index of the particles. When particles are about the same size as the radiation wavelength, significant scattering occurs, and calculation is complicated. For particles of other sizes

$$\delta = \frac{1}{4}FCS \qquad (29)$$

The absorptivity (emissivity) of a path length l (m) through the cloud is

$$\alpha\,(\epsilon) = 1 - \exp\,(-\delta l) \qquad (30)$$

The total emissive power of clouds of various shapes is approximately given by replacing l in equation (30) by fl, where f is given for a number of configurations in *Table 1*.

(b) *Gases.* Emit and absorb in bands in the infrared part of the spectrum. The effective emissivities and absorptivities, ϵ_g and α_g for known combinations of path length and gas concentration and temperature can be calculated from charts based on experimental measurements[55].

(c) *Particles and absorbing gases together*

$$\alpha(\epsilon) = 1 - (1 - \epsilon_g)\exp\,(-\delta l) \qquad (31)$$

Addendum: Miscellaneous and Additional Formulae, Shortened Methods

Selected for relevance; many others omitted. See especially e.g. refs (A) and (B). Evaporation, another mechanism of cooling, treated as mass transfer (see Chem. Engng textbooks).

Radiation heat loss from freely exposed surfaces (area A), e.g. external brickwork of furnace to surroundings at T_a:

$$q\,(W) = 5.67\,A\epsilon\,[(T_1/100)^4 - (T_a/100)^4] \qquad (32)$$

(cf. equation (26)).

Approximate formula for *adiabatic flame temperature*:

$$t_f + \frac{\text{net calorific value}}{\Sigma W\rho c_p}$$

(t_f = temperature of reactants ($^\circ$C), W = weight of individual combustion products (kg). CV and ρc_p should both be in similar units, e.g. kJ/kg K and kJ/kg; CV is more commonly in MJ/kg). Dissociation must be allowed for with flame temperatures above 1650°C and hydrocarbon fuels. Trial and error solution is required (ref.(B), p 9-36). See p 123.

Formula for approximate *furnace temperature and heat transfer.* (Burner heat input − structure losses − moisture loss − hydrogen loss − unburned fuel loss − furnace heat transfer) = heat content of waste gases leaving furnace. Simplest form equates heat input less radiant heat transfer in furnace with heat in waste gases. Use $(T/100)^4$ for convenience. Emissivity used: 0.81 for coal, 0.64 for gas and 0.75 for oil. Trial and error solution required; but (ref.(B), p 9-37) gives Tables which limit number of trials necessary and factors for obtaining effective absorption area.

Table 1   Whole-cloud emissivity[56]

Values of l and f for insertion in the equation $\epsilon = 1 - \exp\,(-f\delta l)$, where δ is absorption coefficient of the medium

Cloud shape	Meaning of l	f		
		$\delta l \to 0$	$\delta l = 1$	$\delta l = 2$
General convex	Volume: surface ratio	4	3.4	3.4
Sphere	Diameter	0.67	0.64	0.61
Cylinder:				
(i) height = dia.	Diameter	0.67	0.61	0.55
(ii) height = 5 × dia.	Diameter	0.91	0.82	0.75
(iii) height ≫ dia.	Diameter	1	0.91	0.84
Rectangular parallelepiped:				
(i) 1:1:1 (cube)	Edge	0.67	0.59	0.58
(ii) 1:1:5	Shortest edge	0.91	0.77	0.68
(iii) 1:1:∞	Shortest edge	1	0.85	0.75
(iv) 1:∞:∞ (slab)	Thickness	2	1.5	1.4

Forced convection over plane walls,

$$h(W/m^2 \,^\circ C) = A + Bv^n \tag{33}$$

$v < 5.0$ m/s

	A	B	n
Smooth surfaces	5.57	3.73	1
Rolled surfaces	5.79	3.73	1
Rough surfaces	6.19	4.10	1

v = velocity of gas

$v > 5.0$ m/s

	A	B	n
Smooth surfaces	0	7.2	0.78
Rolled surfaces	0	7.2	0.78
Rough surfaces	0	7.5	0.78

v = velocity of gas

Water flowing in pipes:

h (W/m^2 $^\circ$C) = 2480 $(1 + 0.0125\, t_w)\, v^{0.8}/d^{0.2}$
(t_w = temperature of water flowing ($^\circ$C), v = velocity of water (m/s), and d = internal diameter of pipe (m). Based on density of water = 998 kg/m^3.)

Chimney gases and air in long smooth pipes:

$$h \text{ (W/m}^2 \,^\circ\text{C)} = 0.62\, v^{0.8}/d^{1/4} \tag{34}$$

Superheated steam in pipes:

h (W/m^2 $^\circ$C) =
$$(4.93 + 3.13 \times 10^{-3} t)\, v^{0.79}/(d^{0.16}\, l^{0.05}) \tag{35}$$

(t = steam temperature ($^\circ$C), v = steam velocity (STP) (m/s), l = length of pipe (m), and d = inside dia. of pipe (m)).

Condensing saturated steam
Minimum values (film condensation):

$$h \text{ (W/m}^2 \,^\circ\text{C)} = \frac{A + B\,(t_f + t_w)}{(l\,(t_s - t_w))^{1/4}} \tag{36}$$

(t_s = steam temperature ($^\circ$C), t_w = wall temperature ($^\circ$C), and l = length of vertical pipe or diameter of horizontal pipe (m)).

	A	B
Vertical pipe	8234	20.59
Horizontal pipe	35265	−74.05

Or,

$$h = 13.8 \times 10^5\, l^{-1/4} \Delta t^{-1/3} \text{ for vertical pipes} \tag{37}$$

$$= 1.07 \times 10^5\, d^{-1/4} \Delta t^{-1/3} \text{ for horizontal pipes} \tag{38}$$

in SI units (*Technical Data on Fuel*, 1961, p 76 for nomogram).
    Usually higher (partly dropwise condensation); latter gives 70–120 kW/m^2 $^\circ$C.
Boiling water outside tube:

Normal values, nucleate boiling (steam boiler):

$$\text{const.} \left[\frac{q/A}{\rho_l \nu L_h} \left(\frac{\sigma}{g(\rho_l - \rho_v)} \right)^{1/2} \right]^{1/3}$$
$$\times (Pr)^{1.7} = \frac{c_p\, \Delta t}{L_h} \tag{39}$$

Properties at saturation temperature. *But* constant can range from 0.0025 to 0.015. Rough surface favours higher values. (Symbols as at beginning of App.C11 except c_p is on mass basis; ρ_l, ν, σ and (Pr) refer to liquid phase). (Recommend J. K. Wilkinson *BCURA Mon. Bull* 1962, Vol.26, 109; origin W. M. Rohsenow *Trans. ASME* 1952, Vol.74, 969; unit change checked, I.G.C.D.)
    If $\Delta t > 25$–35°C, q becomes maximum, film boiling begins. But in boiler practice, Δt is usually 3–6°C, hence usually not relevant.

C12 Liquid-phase heating

Advantages

High temperature at atmospheric or low pressure; indirect heating; rapid response; eliminates limitations of steam and hot water.

Heat-transfer fluids (HTF) — approximate temperature ranges ($^\circ$C)

Mineral oils	−10 to 320
Diphenyl—diphenyloxide (1)	20 to 400
Chlorinated polyphenyls (2)	−20 to 315
Organosilicates (2)	−70 to 360
Metal salts	180 to 500
Molten metals	Up to 650

(1)   Boils at 260°C
(2)   Various formulations to cover this overall range

Fluid selection factors

Process temperature range. Fluid heat-transfer properties, vapour pressure, pumpability, corrosivity, effective life, thermal stability, fire resistance, toxicity, price, and availability.

System features

Expansion tank, deaerator, filter, pump, heating unit, controls, process vessel(s), insulation, draindown facilities, corrosion-resistant materials.

Packaged heaters

Shell type — HTF around fire tubes etc.
Tubular type — HTF in tubes within fired space.

Fuel

Any, but mainly oil and gas.

C13 Heat exchangers

A heat exchanger is any device that facilitates the transfer of heat from a hot medium to one of lower temperature through an intermediary wall, usually metal. Their use results in more efficient plant performance, as heat is effectively recycled within a process, minimizing the requirement for additional external heat input.

For similar (non-condensing) fluids in counter- or concurrent flow, log mean temperature difference is useful to relate heat flow to transfer coefficient:

$$\Delta t_m = (\Delta t_1 - \Delta t_2)/\ln(\Delta t_1/\Delta t_2) \qquad (40)$$

if h is nearly constant. But use with care.

Shell-and-Tube Exchangers

The most prevalent; their design has been highly developed. They can be constructed with very large heat-transfer-surface areas within relatively small volumes, can be fabricated from corrosion-resisting steels, and can be used for heating and condensing a wide variety of fluids. The simple exchanger design of one tube pass and one shell pass can be extended by the use of further tube and/or shell passes. For example a 2—4 exchanger has two shell-side and four tube-side passes, giving greater fluid velocities and consequently larger overall heat-transfer coefficients than simpler types.

An essential design feature is the use of baffles installed in the shell to increase shell-side fluid velocities, and to direct fluid flow normal to the tube bank rather than parallel with it. The resulting increases in turbulence in the vicinity of the tube bank further improves the heat-transfer coefficient.

The basic designs mentioned have the disadvantages that the tube bundle cannot readily be removed for cleaning and no provision is made for differential thermal expansion between tubes and shell. Although it is possible to use a shell with an expansion joint, a more popular method of reducing the thermal stresses is the use of a floating head. In this arrangement the one tube plate is fixed but the second is bolted to a floating head cover so that the tube bundle can move longitudinally relative to the shell. Other arrangements which allow for expansion are the use of externally packed floating heads, and of hairpin tubes.

Concentric-tube Exchangers

In this design the shell in effect encloses a single tube. Very little surface area is available for heat transfer and to obtain a useful output a number of units are generally connected in series. Particularly useful in handling corrosive fluids as glass or carbon may be used for the inner tube.

Coil-in-Bath Exchangers

These comprise a coil immersed in a tank. They are simply and cheaply constructed but very inefficient. However, this simple construction allows the use of corrosion-resisting materials.

Other Types

Special types of equipment have been developed for use under conditions where the standard forms are unsatisfactory, e.g. where the shell-side and tube-side heat-transfer coefficients are widely different owing to the relative properties of the fluids. Two examples are the finned-tube exchanger and the plate-type exchanger. Both types give enhanced heat-transfer areas for use on the side having the lower heat-transfer coefficient.

References

1 Ref. (G), pp 76—83
2 Ref. (D), pp 381—400
3 Ref. (G), p 29
4 Ref. (D), p 316
5 Ref. (D), pp 695—697
6 Hottel, H. C., Williams, G. C. and Baker, M. L. *6th Symposium (Int.) on Combustion*, The Combustion Institute, Pittsburgh, 1957, pp 398—411
7 Ref. (E), p 180
8 Ref. (H), p 257
9 Ref. (E), Ch. 4
10 Masdin, E. G. and Thring, M. W. *J. Inst. Fuel* 1962, Vol. 35, 251
11 Ref. (E), Ch. 6
12 Field, M. A. *Combustion & Flame* 1970, Vol. 14, 237
13 Smith, I. W. *CSIRO, Div. Mineral Chem., Res. Rep. 86*, Pt 1, 1970
14 Ref. (F), Ch. 5
15 Wright, S. J., Hickman, R. and Ketley, H. C. *Brit. Chem. Engng* 1970, Vol. 15, 1551
16 Ref. (F), p 46
17 Arthur, J. R. *Progress in Coal Science* (Ed. D. H. Bangham), Butterworths, London, 1950, pp 355—364
18 Ref. (A)
19 Townend, D. T. A. *Chemy Ind.* 1954, p 766
20 Street, J. C. and Thomas, A. *Fuel, Lond.* 1955, Vol. 34, 4
21 Wright, F. J. *Combustion & Flame* 1970, Vol. 15, 217
22 Ref. (E), p 183
23 Barchilon, M. and Curtet, R. *J. Basic Eng. (Trans. ASME)* 1964, 777
24 Bacon, Chesters, J. H. and Halliday, *J. Iron & Steel Inst.* 1960, Vol.195, 286
25 Chesters, J. H. *et al, J. Iron & Steel Inst.* 1949, Vol.162, 385; 1961, Vol.197, 283
26 Chesters, J. H. *et al, AIME,* 1951, *O.H. Proc.,* 34, 282

27   Boenecke, H. *J. Iron & Steel Inst.* 1949, Vol. 163, 385; 1960, Vol. 195, 286
28   Hinze, J. O. and van der Hegge Zijnen, B. G. *Appl. Sci. Res.* 1949, A1, 435
29   Fitzgerald, F. and Robertson, A. D. *J. Inst. Fuel* 1967, Vol. 40, 7
30   Gray, F. A. and Robertson, A. D. *J. Inst. Fuel* 1956, Vol. 29, 424
31   Pengelly, A. E. *J. Inst. Fuel* 1962, Vol. 35, 210
32   *IHVE Guide*, Book B, 4th edn, 1970
33   UK Clean Air Acts 1956 and 1968
34   CP 337: Flues for Gas Appliances
35   *Chimney Design Symposium*, Vols 1 and 2, University of Edinburgh, 1973
36   Ref. (A), p 53
37   Ref. (E), p 49
38   Mair, P. *J. Inst. Fuel* 1968, Vol. 41, 419
39   Ref. (E), p 68
40   Ref. (G), Ch. 7
41   Price, R. B., Hurle, I. R. and Sugden, T. M. *12th Symp. (Int.) on Combustion*, Combustion Inst., Pittsburgh, 1969, pp 1093–1102
42   Kilham, J. K., Jackson, E. G. and Smith, T. J. B. *10th Symp. (Int.) on Combustion*, 1965, pp 1231–1240
43   Mauss, F., Perthuis, E. and Sali, B. *10th Symp. (Int.) on Combustion*, 1965, 1241–1249
44   Ref. (A), pp 79, 224–235, 307–314
45   Ref. (C), pp 44–47
46   Fishenden, M. and Saunders, O. A. *The Calculation of Heat Transmission*, HMSO, London, 1932, Ch. 7
47   Ref. (A), pp 66–67
48   Ref. (C), pp 219
49   Ref. (C), pp 278–280
50   Gill, D. W. and Thurlow, G. G. *Instn Gas Engrs Publ.* No. 559, 1959, pp 30, 32
51   Ref. (A), pp 70–74
52   Ref. (B), pp 10-15, 10-16
53   Ref. (C), p 59
54   Ref. (A), p 59
55   Ref. (C), pp 83–98
56   Ref. (E), p 96

General Reading

(A)   Spiers, H. M. (Ed.) *Technical Data on Fuel*, 6th edn, World Power Conf., London, 1961
(B)   Perry, R. H. and Chilton, C. H. *Chemical Engineers' Handbook*, 5th edn, McGraw-Hill, New York, 1973
(C)   McAdams, W. H. *Heat Transmission*, 3rd edn, McGraw-Hill, London, 1954
(D)   Lewis, B. and von Elbe, G. *Combustion, Flames and Explosions of Gases*, 2nd edn, Academic Press, New York and London, 1961
(E)   Field, M. A., Gill, D. W., Morgan, B. B. and Hawksley, P. G. W. *Combustion of Pulverised Coal*, BCURA, Leatherhead, 1967
(F)   Skinner, D. G. *The Fluidised Combustion of Coal*, Mills & Boon, London, 1971
(G)   Gaydon, A. G. and Wolfhard, H. G. *Flames: Their Structure, Radiation and Temperature*, 3rd edn, Chapman & Hall, London, 1970
(H)   Fristrom, R. M. and Westenberg, A. A. *Flame Structure*, McGraw-Hill, New York, 1965
(I)   Prandtl, L. *Essentials of Fluid Dynamics*, Blackie & Son, London, 1959
(J)   Beér, J. M. and Chigier, N. S. *Combustion Aerodynamics*, Applied Science Publishers, London, 1972
(K)   Chedaille, J. and Braud, Y. *Industrial Flames. Vol. 1. Measurements in Flames*, Arnold, London, 1972
(L)   Thring, M. W. *The Science of Flames and Furnaces*, Chapman & Hall, London, 1962
(M)   Glinkov, M. S. *General Theory of Furnaces*, Metallurgizdat, Moscow, 1962
(N)   B. S. 845:1972, Acceptance Tests for Industrial Type Boilers and Steam Generators
(O)   B. S. 2885:1957, Code for Acceptance Tests for Steam Generating Units
(P)   Woods, P. C. 'Liquid-phase heating', *J. Inst. Fuel* 1970 (March), Vol.43, 108
(Q)   Ross, T. K. and Freshwater, D. C. *Chemical Engineering Data Book*, Leonard Hill, London, 1962

Appendix D
Sampling, analysis and testing of carbonaceous fuels

D1 Gases

D2 Hydrocarbon products

D3 Coal and coke

D4 Coal-tar fuels

D1 Gases

Gas Sampling

Sampling usually has three objectives: to take a small portion of gas to represent a larger bulk quantity, to place that small portion in a container suitable for transportation and storage, and to change the physical condition of the gas, e.g. temperature and pressure, into a state suitable for analysis.

It is now rarely necessary to sample gas from a static source. Analysis is usually required for a specific use of the fuel gas and it is both convenient and desirable to sample the gas from the flowing system when the gas is in use. The exception to this is liquefied petroleum gas; this is stored as a liquid and it is often necessary to obtain a composition representing the stored stock.

Definitions

A snap sample is one taken quickly at an instant in time. It only truly represents the bulk gas at that instant of time. A series of snap samples can be taken to represent the bulk gas over a period of time: the more samples taken the more accurate is the representation.

A period sample is one taken continuously throughout a period and truly represents the bulk gas during that period in as far as it reflects composition changes with time at the point of sampling. For period sampling a special system is necessary[1] unless the method of analysis is by its nature continuous.

A flow-weighted period sample would be one which represents the total gas which flows past the point of sampling. This is the ideal for measuring efficiency of utilization but in practice a period sample or a series of snap samples is satisfactory.

Nature of the Gases

Gases from different sources present different problems in sampling.

Natural Gases. These are treated at the wellhead or, in the case of submarine wells, at the coastal reception point to satisfy certain specifications[2]. They do not then present any special problems in sampling.

Liquefied Petroleum Gases (LPG). Sampling LPG is difficult because of its tendency to change into two phases of different chemical compositions. A 'full' vessel of LPG, i.e. one containing the minimum volume of gas necessary for safety 'ullage', can be sampled by removing some LPG as liquid and then vaporizing it[3]. As the vessel empties this form of sample becomes less representative but unfortunately there is no satisfactory alternative. LPG is normally vaporized prior to use and it is sometimes convenient to sample it as gas just before it enters the appliance.

Manufactured Gases. Each of these poses its own problems and these are best identified by reference to the source and treatment of the gas. For example, a series of snap samples cannot represent the gas produced by a cyclic process unless the timing of the samples is related to the process cycle. Special treatment may be necessary to remove tar or other non-gaseous materials.

Water Vapour in Gases. Water vapour is often a small but significant component in low-pressure gases and can present problems in analyses. It is best to resolve these problems at the sampling stage, either by drying the gas with a desiccant or by saturating the gas with water. The method chosen for drying or saturating the gas must not change the relative amounts of components, other than water vapour, in the sample. But high-pressure gases are by their nature virtually free of water vapour.

Mechanics of Sampling

Sample Manipulation. The method of sampling and the choice of vessel are governed by both the pressure of the source of the gas and the pressure at which the sample will be stored or transported.

High-pressure Gases. In general, the high-pressure storage of samples is recommended since, when handling for analysis, the gas is emitted from the container unaided. Higher-pressure sample vessels are normally made of stainless steel or aluminium and have either a single valve, or valves at both ends. Connectors to the sample source must be of metal or an appropriate plastic tube. It is recommended that samples are taken from probes inserted into the source; one third of a diameter inward is the traditional point for sampling from pipes. To fill the vessel the tube connection and sample container are purged and pressurized, and the sample is isolated by closing the valves in the reverse order to the flow of gas. Alternatively the vessel can be pre-evacuated, the tube connection purged, and the vessel filled to pressure. Care must be taken that the gases do not cool too much in passing through orifices or needle valves. If there is a risk of the sample containing solid or liquid particles, these should be trapped on a paper or sintered-metal filter; the filter should have a valved bypass in order that full gas pressure is not reduced across the filter.

Low-pressure Gases. These are normally taken and stored in glass gas pipettes[1]. To transfer the gas to analytical apparatus it is necessary to displace it with a liquid such as mercury, dilute acid or brine. For short-term storage, plastic beach balls may be used though these slowly lose hydrogen.

Mixing of Gases. To obtain a representative analysis it is essential that the sample be homogeneous. If the source gas is suspected of being variable in composition, positive steps must be taken to mix the sample. Unaided diffusion is effective for mixing at pressures near atmospheric but it becomes much slower at high pressures; short lengths of steel or aluminium rod placed inside the high-pressure sample vessel will assist mixing when the vessel is shaken.

A common but groundless fear is that a gas mixture will separate into its components by the influence of gravity or centrifugal forces; this it will never do.

Reactions of Sample Gases with Materials. It is essential that the materials used in sample lines, valves and vessels do not react with significant components in the sample; the use of all-stainless-steel or aluminium systems is strongly recommended. Most sample lines require a certain flow of gas to 'condition' them; but conditioning is not an acceptable way of curing losses of components from vessels.

Problems of Trace Analysis. The analysis for minor components in gases can easily be ruined by losses from reaction or by enhancement from contamination. In trace analysis, care in choice of materials for sample handling, meticulous cleanliness of surfaces which will contact the gas, and thorough purging of the sampling system are essential.

Contamination of Gas Samples. The major contaminants are likely to be air, and previous samples for which the system has been used. *All analyses should include oxygen* as a check on air contamination. The use of each sample system should be limited to a group of similar samples.

Analysis of Gases

Methods of gas analysis fall into two categories: those which produce a general description of the gas (general methods) and those which give an analysis for one component or one class of components (special methods). But some of the groups of methods discussed below can be used to a certain extent within either category. Specific measurements for special purposes (e.g. sulphur dioxide, nitrogen oxides) are mentioned in Chapter 18.3.

Classical Methods

The earlier methods of gas analysis were based on measuring the changes in volume or pressure of a sample of gas when it is subjected to chemical reactions[4]. Examples are the Bone and Wheeler and the Haldane apparatus. Such an approach is not suitable for modern fuel gas though one such method, the Orsat method, is useful in flue-gas analysis[5].

Gas Chromatography

This is the most widely used of the general methods. A sample is separated into its components by sweeping it, with the aid of a carrier gas, through a tube filled with an appropriate granular material (the column). The components are eluted from the column one by one and pass to a detector. The detector normally represents the elution of the components as peaks in a plot of detector signal against time, displayed on a strip chart recorder[6].

The time of elution of a peak identifies the component responsible for that peak; the size of the peak is a measure of the concentration of that component in the sample. Identification and estimation of components requires the use of standard gas mixtures.

Gas chromatography has wide applications outside gas analysis and appropriate equipment must be chosen[7]. A typical gas chromatograph for gas analysis consists of a gas sample-injection system, a column in a thermostatted oven and a detector (either a katharometer system, see below, or a flame-ionization detector). A supply of carrier gas and a recorder or integrator are also essential.

Choice of Columns. This is critical in the effective use of gas chromatography. The first consideration is what material will be packed into the columns; the material can be an active solid, an organic polymer (in the form of beads) or a liquid on an inert support.

Each material (stationary phase) is capable of separating only certain components, and a complex analysis will require the separate use of a number of stationary phases. Also significant in the achievement of the required separation are the dimensions and temperatures of the column and the flow rate of the carrier gas. The literature gives general guidance for the choice of columns and conditions[8], and schemes of analysis for certain gases and classes of gases are also described[9-11]. Gas chromatography can be effectively automated[12].

Mass Spectrometry

This is another general method which is widely used. Because of the higher capital cost of equipment, the greater skill required to use it and the complex procedure needed to calculate results, it is probably less suited for occasional use than gas chromatography.

In this method a sample of gas, at low pressure, is ionized by bombarding it with electrons; these ions are accelerated in a specific direction and then deflected by a magnetic or electrostatic field. Each compound in the gas forms a number of ions of characteristic mass (m) and charge (e) and the deflection sorts out the ions according to their value of m/e (mass number). The presence of a compound is shown by the presence of the appropriate mass numbers in the spectrum, and the concentration of the component can be calculated from the intensity of the peaks. Unfortunately, few compounds give unique peaks and interpretation of the spectrum is often difficult; calculation of concentration involves a matrix of simultaneous equations[13].

For gas analysis the mass spectrometer must have a gas-sample handling system. The simpler mass spectrometers are unable to distinguish between the parent peaks of carbon monoxide and nitrogen. The concentrations of these components are determined either by measuring the intensities of a number of peaks and computing the result or by measuring the carbon monoxide by non-dispersive infrared spectrometry (see below) and then calculating, by difference, nitrogen from the intensity of mass number 14.

Non-dispersive Infrared Spectroscopy

Certain gases such as carbon monoxide, carbon dioxide and hydrocarbons absorb infrared (i.r.) radiation within certain wavelengths. This can be used to measure the concentration of these components[12]. Radiation in the appropriate waveband is allowed to pass through a standard cell containing the gas. Some of the i.r. radiation is absorbed and some passes through; the amount absorbed is a measure of the concentration of the component of interest. In practice, the *transmitted* radiation is actually measured by a detector.

In the most common form of this analyser the detection is selective for a certain bandwidth of i.r. radiation and hence, to a degree, is selective for a certain component of interest. Such a detector consists of a chamber containing the component of interest; one wall of the chamber is a diaphragm which forms one plate of a capacitor. The transmitted radiation heats up the cell and causes a pressure rise, which alters the electrical capacity by moving the diaphragm.

In practice it is usual to interrupt the radiation at a fixed frequency so as to give a pressure fluctuation in the sensing cell at the same frequency, and also to construct the instrument so that the measurements from a sample-sensing pair of cells are compared with a reference-sensing pair of cells (double-beam instrument). The system is calibrated by means of standard gas mixtures which should approximate in overall composition to the samples to be analysed (especially so for trace analysis). The reference cell is normally sealed and should contain the matrix of gas to be analysed; the instrument should also be zeroed on the matrix gas.

Another form of non-dispersive i.r. analyser uses thermopiles as detectors following the sample and reference cells. This form is not selective but is robust and suitable for measuring high levels of components. Water vapour absorbs infrared radiation and in most applications it is essential that the gas sample be dried.

Other Methods

Thermal Conductivity and Pellistor Instruments. The thermal conductivity of a gas is a property of its composition and this can be used to analyse simple gas mixtures[14]. The normal instrumentation for this involves drawing a sample of the gas into a cell containing an electrically heated metal wire (katharometer). This metal wire forms part of an electrical bridge circuit in which its resistance is compared with that of a similar katharometer contained in a cell filled with a reference gas. This 'unbalance' of the bridge is a measure of the difference in thermal conductivity between the sample gas and the reference gas, hence of its composition. This method of analysis is particularly suited to measuring levels of hydrogen and helium in other gases.

If the filament in the device is intrinsically catalytic or is coated with an oxidation catalyst (Pellistor) it may be used to measure combustible gases in air. The heat produced by chemical reaction is much greater than the change in heat lost by the filament through gas thermal-conductivity changes. The detector is therefore much more sensitive than a conventional filament detector, and such devices are used for testing hazardous areas; they are then conveniently calibrated in terms of the lower explosive limit of the component of interest in air.

Gas-Detector Tubes. These consist of a glass tube packed with a solid chemical reagent. A known volume of sample is drawn through the tube, a chemical reaction occurs, and a coloured band is produced. The length of the coloured band is a measure of the amount of a certain component in the sample. There are many proprietary makes. Though designed for measuring a single component in air they can be used in other contexts, though one must beware of interferences.

There are many other methods of gas analysis for specific components; commercial instruments are available which use electrochemical principles for measuring oxygen- and sulphur-containing compounds[15]. Particularly valuable are the oxygen analysers which use the paramagnetic properties of that gas[16].

Standard Gas Mixtures

Most methods of gas analysis require standard gas mixtures. They can be prepared, with difficulty, in the laboratory[17] or are available commercially, with the option of a certified analysis. They are normally prepared at high pressure and must be thoroughly mixed. The composition of a standard mixture will sometimes change as its pressure approaches atmospheric.

Calorimetry

The calorific value of a gas mixture can be measured directly or calculated from its composition.

The use of a calorimeter is a skilled art and a calorimeter is only worth installing if it is to be used extensively. A simple and precise gas calorimeter is that designed by Boys; it is suited for spot checks[18,19]. Whether calculated or measured, the physical conditions for the calorific value (wet or dry, pressure or temperature) must be considered and reported or corrected for.

The quality of fuel gas sold by British Gas is controlled by powers scheduled in the Gas Act, 1972. The calorific value must be declared if the gas is charged for by the number of therms supplied[21], and there are other restrictions, on the uniformity of calorific value, on pressure of supply in relation to Wobbe number, hydrogen sulphide content, and smell[22].

A partial change to the SI-unit system has been made. The old therm (10^5 Btu) was equal to 105.5 MJ. A 'new therm' of 100 MJ is being used.

D2 Hydrocarbon products

The fuels to be considered in this Appendix cover the field of major liquid products produced by distilling and refining crude oil. Crude oil has too much to offer in these resulting refined products for it to be normally considered as a fuel itself. Individual crude oils are generally specific in their properties and refiners carry out the very minimum of testing, such as relative density, water and salt content, on incoming cargoes. The list of refined fuels is as follows: naphtha, petrol (gasoline), paraffin (kerosene), heating oil, diesel fuel, fuel oil.

Of the above the two most important from the aspect of this book are fuel oil and heating oil. More attention, particularly on analysis and testing, will be paid to these.

Sampling

The importance of sampling can never be over-emphasized. The results of analysis of any fuel are meaningless unless the fuel has been representatively sampled. To be sure that this is achieved care and attention must be given at all times to the procedure laid down[23].

The range of fuels listed above can require different sampling equipment depending on such things as volatility and use of fuel, but many general precautions apply to the sampling of them all.

The sampling apparatus, particularly the final receptacle, should be dry, clean and free from any contaminating substance. Proper sample containers of the correct size should be used whenever possible; if other bottles have to be used then special care must be taken to ensure cleanliness (e.g. 0.01% contamination of paraffin with a lubricating oil can significantly affect char formation).

It is usual to fill the sampling bottle and rinse the sample container and allow them to drain at least once before taking the sample. When finally filling the sample container approximately 10% ullage should be allowed to cater for any expansion of the liquid. Well fitting corks should be used in sample bottles and screw stoppers with cans. Rubber bungs should not be used with petroleum products.

Clean gloves should be worn at all times by the operator engaged in sampling. This is as much for the operator's protection, since all petroleum products can affect the skin, as for protection of the sample.

In order to obtain a representative sample from the contents of a tank or other container it is necessary to take 'top, middle and bottom (T.M.B.) samples' at points one-sixth, one-half and five-sixths of the depth of the liquid below the surface. Unless the tank is of modern design with take-off points set within the tank so that these T.M.B. samples can be taken from a point at the base of the outer wall, the samples should be drawn through dip hatches or other opening giving direct and unconfined access to the bulk of the liquid. Dip-pipes, gauge glasses or drain fittings should not be used for drawing off samples, unless specified.

The T.M.B. samples can be tested separately or mixed in proportion to give an average sample of the contents of the tank. If there is any reason to suspect the contents of the tank to be layered and non-homogeneous then samples at other levels may have to be taken and tested. It is advisable to take samples from the top first and continue downwards, so as to obtain each sample before the liquid at that level is disturbed.

Good housekeeping, i.e. regular draining of any possible sludge (water and rust) from the bottom of the container will obviate ingress of stirred-up sludge into the bottom sample. If convenient the bottom of the tank can be checked for water using a water-finding paste applied to the bottom of a dip stick. The paste contains a compound which changes colour when in contact with water — a common type being from blue to pink.

The more common method of sampling is to use a corked sampling can or bottle in a weighted cage which is lowered to the appropriate level in the liquid. The cork in the can or bottle is then removed by jerking on an attached cord. It is advisable when sampling from a large tank to anchor the cord or rope at the top of the tank, thus avoiding the loss of sampling gear if cord and gloves become slippery with product. It is important that all metal sampling gear should be made of non-spark-generating material.

The size of sample container depends on the testing to be carried out. To test fully most petroleum products against British Standard[24,25] and other specifications, a one-gal (4 l) sample is required.

Labelling is very important; it is frustrating to receive a sample with no information because the label has been lost or the writing smudged. A tie-on label protected (e.g. with sellotape) should be used and the writing preferably in block capitals using permanent ink or indelible pencil. Appropriate information is place, date, time, T.M.B. average, tank, product, grade, name of sampler etc.

Specific points of interest particular to the different types of oil are as follows.

Naphtha

It is possible that a vapour pressure (Reid) test may be required; this needs a separate sample according to instructions under 'Special Procedures' in the IP guide[23]. Otherwise a 1-l sample preferably in a polythene container should be taken, to avoid possible contamination with lead if a can was used and with sodium if a bottle was used. Because of the low flash point sparks form a special hazard.

Petrol

The same remarks apply to petrol as to naphtha on the points of vapour pressure and low flash point. Otherwise, a minimum of about 5 l is needed for full test and it is quite normal to use the standard 9 l (two UK gallon) brass-topped stock cans.

Heating Oil or Diesel Fuel

Normally an ordinary 4.5 l (gallon) can is used when a full test is needed. If, however, it is to be used as turbine fuel then a special sample can be taken in a polythene bottle to check for sodium. When this sample is taken special care is needed to ensure that no perspiration from the body enters or even comes into contact with the sample or container.

Fuel Oil

A 2-l sample is normally sufficient for most testing of fuel oils. However, individual samples from different levels of the tank may be checked for water content. With these samples, approximately 25% ullage should remain in the container, allowing for adequate mixing before taking the appropriate quantity for test.

When taking fuel-oil samples it is especially important to anchor the rope holding the sampler as this oil makes everything very slippery. The pressure exerted on the sampling can at the bottom of a large full tank of fuel oil is moreover sometimes sufficient to force the cork into the sampler so hard that it is not possible to jerk the cork out. A slightly larger cork is needed in these circumstances.

Analysis and Testing

The petroleum fuels itemized at the beginning of this Appendix are derived initially by distilling crude oil. They are listed in descending order of volatility, naphtha being the most volatile and fuel oil the least. The properties of these fuels can vary depending on the source of the crude oil as well as on the temperature range of the distillation cut and the further refining processes applied to certain fractions to give a finished product.

References p 580

These fuel products are all complex mixtures of many and varied types of individual hydrocarbons. As a result a complete analysis is generally impracticable. Having said this however, great strides have been made in gas–liquid chromatography techniques whereby certain of the lighter products can be analysed if not to actual individual components, at least to small groups of hydrocarbons similar in type.

Pending fulfilment of such advanced methods, tests of several different types have to be applied to liquid fuels in order to assess their quality. These tests can be found in standard handbooks issued annually by such bodies as the Institute of Petroleum[26] and the American Society for Testing and Materials[27].

The different types include chemical, physical, empirical and engine tests. Chemical tests include the determination of elements such as sulphur, hydrogen, carbon, sodium and vanadium. Physical tests, which also give definite values for the properties measured, include those for relative density, refractive index and viscosity. Empirical tests, many of which have grown with the petroleum industry, depend on the method and apparatus employed. Examples of these are distillation, flash point and smoke point. Finally, engine tests are best exemplified by the use of the 'ASTM–CFR Engine' (CFR stands for Cooperative Fuel Research Committee) in determination of research octane and motor octane numbers of petrol.

Some of the tests used in the analysis of liquid fuels are common to several fuels, e.g. relative density, flash point and sulphur content, but many are specific to a particular type of fuel. The more relevant tests will be listed alphabetically in order to explain their significance. Following the title of each test will appear as appropriate the IP and ASTM method numbers relating to the respective standard books[26,27]. In cases where the test is identical in both books, i.e. a joint method, the letter J will be added.

Ash (IP 4/ASTM D482J)

Briefly, this test determines the inorganic residue remaining after complete ignition. From the point of view of impurities in the fuel the quantity of inorganic ash is important, but more important are the elemental constituents of the ash. These elements derive from (1) oil-bearing strata of silica, (2) water associated with crude oil, (3) organometallic compounds which occur in solution or in combination with the fuel, e.g. vanadium, and (4) contamination from handling and storage of the fuel, e.g. iron (as rust). The significance of testing the fuels for some of these elements is briefly mentioned later.

Asphaltenes (IP 143)

This test measures the percentage of a petroleum product that is insoluble in (normal) heptane when determined under the special conditions of the test. Asphaltenes are generally considered harder to burn than the remainder of the fuel and if its content is high could lead to poor combustion with formation of stack solids. Normally, particularly with fuel oils in the UK, the asphaltene content is low — typically 2.0—2.5% — and in fact a minimum specification (usually 0.5%) is included in most UK specifications in order to comply with H.M. Customs and Excise definition of a fuel oil[28].

Aniline Point (IP 2/ASTM 611)

This test measures the temperature at which the oil just becomes completely miscible with an equal volume of aniline. The result indicates a measure of the proportion of aromatic hydrocarbons in fuels. Aniline as an aromatic substance will mix much more readily at lower temperatures with other aromatic substances than other hydrocarbon types such as paraffins and naphthenes in the same boiling range. Thus a low aniline point indicates a high aromatic content.

The aniline point is used in two mathematical formulae, one to obtain a diesel index and the other to give an aniline-gravity constant which in turn is used to calculate the calorific value of aviation turbine fuels.

Calorific Value (IP 12/ASTM D240)

This test measures the heat of combustion of the fuel and is therefore of prime importance. However, whilst the calorific value of fuels is used in calculations for heat balance, burner efficiency etc. the actual test is not often carried out for the following reasons:

1. That the calorific value is not controllable in the manufacture of fuels, except in a secondary manner by limitation of other characteristics.
2. That unlike solid fuels where wide variations in calorific value are encountered, liquid fuels over the range from paraffin to heavy bunker grades of fuel oil only vary by some 10%. If the gross values are reported on a weight basis then the paraffin figure is approximately 10% greater than that for the heavy fuel oil whereas if the values are on a volume basis the converse applies to again approximately 10%.
3. That the actual test (IP 12/ASTM D240) needs specialized expensive equipment, takes a long time to complete, can be dangerous and needs skilled laboratory personnel to carry it out with a reasonable degree of accuracy.
4. That mathematical formulae exist for calculation of calorific value:

(a)   Aniline Gravity Constant (Estimation of Net Heat of Combustion of Aviation Fuels — IP 193/ASTM D1405J). Correlations have been established between the net heat of combustion and the product of aniline point and specific gravity for aviation fuels.

(b)   Taken from British Standard 2869 [25]:

Calorific value (gross)

$$Btu/lb = (22\,320 - 3780\,S^2)(1 - (x + y + z))$$
$$+ 4050\,z$$

or

Calorific value (net)

$$Btu/lb = (19\,945 - 3780\,S^2 + 1370\,S) \times$$
$$(1 - (x + y + z)) + 4050\,z - 1053\,x$$

(multiply by 2.326×10^{-3} to convert to MJ/kg) where S = relative density at $15.6/15.6°C$ $(60/60°F)$; $*x$ = proportion by mass of water (% divided by 100); $*y$ = proportion by mass of ash (% divided by 100); and z = proportion by mass of sulphur (% divided by 100). (* For most practical purposes these two factors can be ignored).

It is accepted generally that results of calorific value calculated from the above formulae are as good as those obtained using IP 12/ASTM D240, particularly taking into account the reproducibility of the test.

Cetane Number (IP 41/ASTM D613)

The cetane number describes the ignition quality of a diesel fuel. It is a measure of the delay that occurs in a diesel engine between the commencement of fuel injection and the commencement of combustion. It is defined as the percentage by volume of cetane in a blend of cetane (rated 100) and a-methyl naphthalene (rated 0) that gives the same ignition delay under controlled engine test conditions as the fuel to be tested.

The higher the cetane number, the shorter the ignition delay and hence the easier the starting, particularly under cold conditions. A good-quality fuel for diesel-engined road vehicles shows a cetane number of 50 or above. Generally paraffinic hydrocarbons have the highest cetane numbers, aromatics the lowest and naphthenes somewhere in between; therefore aromatics have higher spontaneous ignition temperatures than paraffins with naphthenes in between.

The actual test for cetane number is carried out using a special engine. Since there are two very good indices, namely calculated cetane index and diesel index, which relate to cetane number and which can be calculated from other more readily attainable parameters, the actual cetane number test is seldom carried out.

Cetane Index (Calculated) (IP 218/ASTM D976J)

Generally, and particularly with diesel fuels currently refined in the UK, the calculated cetane index can be taken as directly equivalent to the cetane number of the fuel. The cetane index is calculated from the relative density of the fuel and the temperature at which 50% of the fuel has distilled. In practice the result can be obtained from the nomograph included in the method (IP 218/ASTM D976)

Char Value or 24-Hour Burning Test (IP 10/ASTM D187)

This test is applied to paraffins. Oil is burned for 24 h in a standard lamp under closely specified conditions. At the conclusion of the burning the oil consumption and the amount of char formed on the flat wick are measured and the char value calculated as mg/kg of oil consumed. The test is very sensitive to small volumes of heavier oils, e.g. 0.01% volume contamination of the paraffin with a heavy lubricating oil will change the char value from 10 to approximately 20.

Cloud Point (IP 219/ASTM D2500J)

This is a static test applied to diesel fuels and heating oils, that measures the temperature at which a haze of small wax crystals become visible in the fuel under test. It is a limiting test, in that at temperatures equal to or below the cloud point fine fuel-filters and small-bore fuel pipes could become clogged and therefore restrict and eventually stop the flow of fuel. (See also cold-filter plugging point).

Cold-filter Plugging Point (IP 309)

This is a relatively new test that is complementary to the cloud point, in that it is a dynamic test designed to determine the lowest temperature at which the diesel fuel or heating oil will flow. It is defined as the highest temperature at which the fuel, when cooled under prescribed conditions, either will not flow through a fine wire-mesh filter or requires more than 60 s for 20 ml to pass through. The cold-filter plugging point will invariably be lower than the cloud point in that the test permits the formation of wax crystals so long as they pass through the wire-mesh filter. The test was designed to take into account the use of chemical additives which when added to a fuel tend to prevent the agglomeration of the wax crystals and thus avoid clogging of filters and fuel pipes.

Diesel Index (IP 21)

As mentioned under cetane number the diesel index is a calculated number. It is derived from the aniline point and the relative density and normally relates closely to the cetane number, differing by a few units only, e.g. a cetane number of 52 could equate to a diesel index of 56. However, with certain oils the diesel index could differ much more and it should be treated with reserve. In general the calculated cetane index is preferred.

References p 580

Distillation (IP 123/ASTM D86J)

Distillation, as mentioned earlier, is the basic process in the separation of the different fuels from crude oil. The standard laboratory test places the distillation range of the fuel on a repeatable basis. 100 ml of oil is distilled under closely specified conditions, measuring the percentage of the oil distilled against vapour temperature. The test is applicable to all fuels except the residual fuel oils. The method can be carried out manually or automatically.

Distillation is very significant with the more volatile products, particularly petrol, for which several points in the distillation range are specified.

Flash Point (IP 34/ASTM D93J, IP 170, IP 303)

The significance of this test is to give a measure of the fire risk of an oil in bulk. The test is carried out in a standard apparatus and measures the temperature at which an oil gives off just sufficient vapour (mixed with the air in the vapour space of the apparatus) to form an explosive mixture and to flash when brought into contact with a flame.

Worldwide there are many different types of standard apparatus for measuring flash point; currently studies and work are being carried out in the UK, in Europe and in the USA to attempt to standardize or rationalize flash-point methods.

For many years two flash-point instruments used in the UK have been the Abel apparatus (IP 170 or the Statutory Method IP 33) and the Pensky—Martens (IP 34). It has generally been accepted that the Abel should be used for fuels with flash points below 49°C (120°F) and the Pensky—Martens for flash points above 120°F. However, some years ago the Abel Method (IP 170) was examined statistically over the range of 24 to 152°F and it can now be used for all fuels provided that actual flash points above 152°F are not required.

The latest instrument, gaining general acceptance worldwide, is that described in IP 303 'Rapid Tests for Flash Point'. The advantages of this method are the small quantity of fuel needed for test (2 ml) and, once the temperature of the instrument is set, the short time required for the test (2—3 min). The method is basically a pass/fail test, although definitive results can be obtained fairly quickly. Comparisons of this and other standard instruments are contained in a report issued by the I.P.[29]

The Abel apparatus is the statutory instrument detailed in the 1928 Petroleum (Consolidation) Act. This Act stipulates that petroleum oils with a flash point below 22.5°C (73°F) when tested on the Abel apparatus are considered 'Dangerous — Highly Inflammable' and are subject to strict Home Office regulations with regard to storage and transport.

The flash point determined when testing fuels is known as the closed flash point, basically because the apparatuses have lids. Open flash points are known but generally are not applied to fuels.

Pour Point (IP 15/ASTM D97J)

Pour point is the lowest temperature at which movement of oil is observed when examined under the specific conditions of the test. The oil is cooled and examined for movement at intervals of $3°C$ (or $5°F$). The pour point is still extensively used as a guide to flow properties or pumpability of the fuel. However, the previous thermal history or the wax content of a fuel can affect the pour point — pumpability relationship, and a dynamic test known as pumpability has been developed which could in time replace the pour point as a measure of pumpability.

Pumpability Test for Industrial Fuel Oils (IP 230)

This test, as explained under pour point, gives a measure of the pumpability of the oil. In fact the test is designed to measure two temperatures at which the oil has viscosities of 6 poise (0.6 Pa s) and 25 poise (2.5 Pa s), both measured at a shear rate of 9.7 s^{-1}. These represent the maximum allowable viscosities and hence minimum temperatures respectively for handling and storing fuel oils in industrial installations. Knowing the handling and storing temperatures of a fuel oil can save on heating costs.

Ramsbottom Carbon Residue (IP 14/ASTM D524J)

The carbon residue of a fuel gives an indication of its tendency to deposit a carbonaceous residue on a hot surface following vaporization. The residues can form in vaporizing pot-type burners, sleeve-type burners or even on a burner or injection nozzle.

The test consists of heating a weighed quantity of oil in a special glass bulb to a temperature of $550°C$, when the volatile matter is evaporated and a coke-like residue remains.

There have, in the past, been two methods for determining carbon residue, the Ramsbottom and the Conradson. Since the advent of 'North Sea' gas, the Conradson test has been difficult to carry out owing to lack of control of the Meker burner; it is being dropped from specifications and test manuals in favour of the Ramsbottom. The results by both methods, particularly between 0.2 and 20% by weight, are effectively the same.

Reid Vapour Pressure (IP 69/ASTM D323)

This test measures the absolute vapour pressure of the more volatile fuels such as naphtha and petrol. The test is normally carried out at $37.5°C$ ($100°F$) and at a vapour-to-liquid ratio of 4 to 1. The measured Reid vapour pressure is approximately the vapour pressure of the material in pounds force per square inch absolute under these conditions (1 psia = 6.895 kPa absolute).

As stated, the test is applied to the more volatile fuels, particularly petrol where it is needed to give a measure of the optimum amount of light products to enable a car to start easily but not develop vapour locks.

Relative Density (IP 160/ASTM D1298J)

The term relative density (rel. d) replaces the former term 'specific gravity'. The relative density t_1/t_2 of a solid or liquid substance is defined as the ratio of a mass of a given volume of the substance at a temperature t_1 to the mass of an equal volume of pure water at a temperature t_2.

Practically, the measurement of relative density is generally made by a hydrometer method (IP 69/ASTM D1298J). It is also the subject of a small booklet which is Part VII of the IP Petroleum Measurement Manual[30]. In using a hydrometer, corrections can be applied for temperature; these corrections can best be obtained by reference to the ASTM/IP Petroleum Measurement Tables[31]. However, over a limited range, temperature-correction coefficients can be used, e.g. for paraffin the coefficient could be 0.00040 per $°F$ (0.00072 per $°C$). This correction should be added for every degree above $60°F$ and subtracted for every degree below $60°F$ in order to relate the measurement of relative density of fuels to the standard $60/60°F$.

The most important use of relative density is for bulk oil measurement. It is seldom used in specifications, but is significant when properties involving weight in comparison with volume are considered (e.g. calorific value).

Research Octane Number (IP 126/ASTM D908J)

Research octane number describes a combustion characteristic of petrol. Under ideal conditions the flame initiated at the sparking plug spreads evenly across the combustion space until all the petrol has been burned. If the petrol is of poor quality the combustion could be erratic causing burning to start at many points in the hitherto unburned petrol and air mixture. This effect may be very rapid, when a detonation or 'knock' will be heard. Compression ratio, i.e. the degree to which the petrol—air mixture is compressed before it is fired, is one of the biggest factors in knocking tendency. High-compression engines need petrols of good anti-knock characteristics.

The research octane number is determined in a special test engine, when the fuel is compared for its knock tendency against standard mixtures of iso-octane (rated 100) and (normal) heptane (rated zero). The research number is determined at low speed and low temperature and is the accepted parameter in the British Standard Specification BS 4040 [24] for petrols in this country. Another octane number, namely motor octane number, is determined on the same engine at high speed and high temperature and gives a result on the same petrol some 10 numbers lower than that for research octane number.

Sediment (IP 53/ASTM D473J)

This is the residue remaining after treating a sample of fuel, in a refractory thimble, with hot toluene. It measures the quantity of carbonaceous and mineral matter that if present in relatively large quantity (greater than 0.25%) could damage burners, and block filters.

Smoke Point (IP 57)

The smoke point is defined as the maximum height in millimetres at which paraffin (or similar oil) will burn without smoking under the standard conditions of the test. In the test the flame height can be raised or lowered against a background of a graduated millimetre scale and the height at which the flame just fails to smoke is measured.

The hydrocarbon types constituting the paraffin affect the smoke point, aromatics in particular lowering the smoke point of an oil.

The significance of the test is that it indicates the limit to which a flame height can be raised and hence the limit of luminosity or heat output when burning the fuel.

Sodium Content (IP 288)

This test involves atomic absorption spectrophotometry and has only very recently been introduced into the IP Standards.

If sodium is present in fuel oil to an appreciable excess, particularly in its common form as sodium chloride, it can cause decomposition of firebrick at high temperature owing to production of silicates or aluminates.

Sodium also causes turbine-blade corrosion when present in gas-turbine fuels to a significant degree. The level of sodium in a gas-turbine fuel in the field is ≈ 0.1 ppm w/w and special precautions in sampling and special techniques in testing have to be applied.

Spontaneous Ignition Temperature

This parameter is included in order to avoid confusion between it and flash point. The spontaneous ignition temperature is the temperature at which a combustible substance will ignite, that is burst into flame, in the presence of air *without* any flame or spark being applied. Typical spontaneous ignition temperatures for fuels are: paraffin 250°C, 2-star petrol 280°C, 5-star petrol 400°C.

Sulphur Content

Sulphur is present to some extent in all petroleum fuels ranging from as low as 100 ppm w/w or less in naphtha to 4 wt% in fuel oil. The main disadvantage of sulphur in fuels is the emission of sulphur oxides into the atmosphere when burned.

For this reason low-sulphur crude oils and desulphurization plants are at a premium. Another disadvantage is corrosion by sulphuric acid following condensation of the gases from the burned fuel in cool parts of stacks or flues.

With diesel fuels an additional disadvantage is that sulphur can increase cylinder wear and creates a need, if present in excess, for detergent lubricating oils. Finally the steel industry needs special low-sulphur fuels (down to 1% or less) for use in oil-fired open-hearth melting and reheating furnaces.

Many tests are in use for determining the sulphur content of petroleum fuels. In the case of the more volatile products such as naphtha, petrol and paraffin the most common test has been the Lamp Method (IP 107/ASTM D1266J), whilst for heating oils, diesel fuels and fuel oils two tests have been used, namely the Bomb Method (IP 61/ASTM D129J) and the Quartz Tube Method (IP 63/ASTM D1551). Note: The bomb method has an advantage in that it is the same apparatus as that used for calorific value; it is not uncommon for both tests to be carried out at the same time, i.e. the residual contents from the bomb after completion of the calorific value determination can be used for the sulphur determination.

A more recent sulphur test added to the IP Standards is the Wickbold Method (IP 243). This method can be used for the full range of fuels (including liquefied petroleum gases). Other methods using X-ray fluorescence and X-ray absorption techniques find application. There is a constant effort in the industry to measure sulphur, particularly in fuel oils, as accurately as possible, because of the high premium on low-sulphur fuel oils.

Vanadium Content (IP 288)

The same test, atomic absorption spectrophotometry, as for sodium content is applied.

Vanadium concentrates in the residual fuel oil and varies with the source of crude. The products formed when oils with high vanadium content are burned cause corrosion and fouling of equipment.

Viscosity (IP 71/ASTM D445J)

This test measures the kinematic viscosity of an oil. In essence, the time in seconds taken by a fixed volume of oil in flowing through the capillary of a calibrated viscometer under a reproducible driving head and closely controlled temperature is measured. The kinematic viscosity is the product of the measured flow time in seconds and the calibration constant of the tube. The result is given in centistokes (cSt) = mm²/s.

Viscosity is important and relates to the rate of flow through a pipeline and to how it atomizes in a burner.

Some fuel oils under normal conditions of temperature and pressure behave as simple or

'Newtonian' liquids, the rate of shear being directly proportional to the shearing stress applied; others, particularly if they contain paraffin wax, may deviate unless the temperature is raised sufficiently to dissolve all the wax. A temperature of at least 49°C (120°F) is normally required for this and the B.S.I. along with the petroleum industry generally have accepted 82.2°C (180°F) as the standard temperature for determination of kinematic viscosity of fuel oils.

Following the acceptance of the more correct kinematic viscosity at 82.2°C, the previously well known Redwood viscosity which measured and quoted the time in seconds taken for a set quantity of oil to flow from a standard orifice at 100°F is being phased out.

Water Content (IP 74/ASTM D95J)

A volatile spirit immiscible with water is added to a known quantity of the oil, and the whole is distilled. Any water present in the oil accompanies the volatile spirit and is collected in a graduated trap and so measured.

Water can be present, particularly in fuel oil, either free or emulsified. Excess water in fuel oil can give rise to sporadic burning and spluttering.

Practical Application

Selective application of the above tests to individual fuels will now be discussed.

Naphtha

This product is largely used as a petrochemical feedstock or further refined to give a rich petrol blendstock. In fact most uses involve catalytic treatment, and tests for sulphur, lead content — particularly when the naphtha has been shipped along with petrols — and distillation, in particular the final boiling point, are essential tests in order to protect the catalyst.

Petrol (gasoline)

In the UK petrol is governed by a British Standard Specification BS 4040 [24] which gives four distinct grades based on research octane number (RON), namely:

Grade designation	RON value
5-star	100.0 minimum
4-star	97.0 minimum
3-star	94.0 minimum
2-star	90.0 minimum

Thus the first determination to be applied to petrol is of research octane number.

Other tests specified in BS 4040 apply equally to the four grades of petrol. Of these, distillation is important as this gives a measure of easy starting, warm up, acceleration and steady driving and, at the least volatile end, crank-case dilution.

Lead content is also checked because of current environmental concern. Legislation is moving towards lower quantities of lead in petrol. A point to note here is that any petrol containing lead has to be coloured — most petrols are coloured yellow.

Unless the product being tested has been stored for some time or is suspected of being contaminated, the RON and distillation tests are normally sufficient to check whether the petrol is satisfactory.

Paraffin

There are two grades of paraffin specified by BS 2869 [25]. This British Standard is titled 'Specification for Petroleum Fuels for Oil Engines and Burners' and covers all other grades of fuel not mentioned so far, i.e. paraffins through to heavy fuel oil. The paraffins are classed as C1 and C2, the difference being that C1 is a more refined product and is used for free-standing *flueless* domestic burners whilst C2 is used in vaporizing and atomizing burners used for domestic heating in appliances connected to flues. A comparison of the principal tests for both clases is given below:

	Class C1	Class C2
Smoke point (mm) MIN.	35	25
Sulphur content (wt %) MAX.	0.04*	0.2
Distillation, recovery at 200°C (vol %) MAX.	60	No limit
Flash point, closed, Abel (°F) MIN.	110	100
(°C)	43	37.5
Char value (mg/kg) MAX.	10	20

A check analysis for Class C2 paraffin need only include smoke point, distillation and flash point whilst for Class C1 sulphur and char value should be added.

If a paraffin is suspected of being contaminated by other products or has been held in storage for a long period all the tests in the British Standard should be applied, particularly char value.

It cannot be noted too strongly that Class C1 must be the only paraffin used in free-standing *flueless* heaters. The specification limits were chosen to take into account both the safety and amenity aspects of the fuel.

Finally, to avoid the unlawful use of paraffin in diesel road vehicles, thus evading Customs Duty, a Customs Marker is added to Class C2 paraffin imparting a slightly yellow colour whilst Class C1 paraffins are strongly coloured with distinctive dyes.

---

* This limit was formerly 0.06 but has recently been changed following medical consideration of the threshold limits of sulphur dioxide in a living room of a house

Engine Fuels (Diesel Fuels)

British Standard 2869 covers four classes of engine fuels, as follows:

Class A1:   High-quality automotive diesel fuel.
Class A2:   General purpose diesel fuel.
Class B1:   Diesel fuel for larger engines particularly those in marine practice.
Class B2:   Similar to Class B1 but less refined and can contain small amounts of residue.

A comparison of the principal tests for these four classes of fuel is given below:

	Class A1	Class A2
Sulphur content (wt %) MAX.	0.5	1.0
Cetane number, MIN.	50	45
Viscosity, kinematic at 100°F or 38°C MIN.	1.6	1.6
(mm²/s) MAX.	6.0	6.0
Cloud point, Summer (°C) MAX.	0	0
Winter (°C) MAX.	−7	−7

	Class B1	Class B2
Sulphur content (wt %) MAX.	1.5	1.8
Cetane number, MIN.	35	Not significant
Viscosity, kinematic at 100°F or 38°C MIN.	—	—
(mm²/s) MAX.	14	14
Pour point (°C) MAX.	0	0

Sulphur content and cetane number (or calculated cetane index) are the main checks on these engine fuels (except Class B2); cold temperature properties, particularly in winter, are probably needed in addition.

A property not shown in the British Standard is relative density. This test is of particular importance to engine fuels since it correlates with calorific value and is therefore reflected in the volumetric consumption of the engine in which it is burnt. Assuming satisfactory combustion, the higher the relative density the better, since fuel is normally bought and taxed on a volume basis, but appreciated by the engine on a weight basis.

Other tests can be significant, depending on the type of duty expected. Flash point is important, particularly when the fuel is used on ships when certain legal standards have to be complied with.

In this category, gas-turbine fuel should be mentioned since the product used for this purpose conforms to Class A2 of the British Standard. Of specific importance is the ash level and in particular the elemental analysis such as sodium, potassium, vanadium, calcium and lead, which if present in excess, can give rise to turbine-blade corrosion. The order of magnitude that is being put on the limits of these elements individually is 0.5--5 ppm

w/w.   To ensure this low level of impurity it is better that the customer has some equipment or technique such as a filter coalescer before the combustion chamber.

Heating Oil

This product is covered in BS 2869 as Class D and is exactly the same specification as Class A2 above except that cetane number is not mentioned. In other words this product can be used for heating purposes such as atomizing burners in domestic and industrial use, and also as a general-purpose diesel fuel. In comparison to black fuel oils, heating oil is a clean, low-sulphur product which can be easily handled at ambient temperature.

When checking a sample of heating oil the following tests should be carried out:

1. *Appearance.*   This is a test that can be very important and yet is seldom mentioned. Condition, colour and possibly smell can sometimes give all the information needed when checking fuel for contamination. Heating oil should be clear, bright and free from adventitious matter or water droplets. It is coloured red by a dye injected along with the Customs chemical markers. If the sample is hazy or wet or contains sediment further tests such as water content, sediment and possibly ash should be carried out.

2. *Flash Point.*   A low flash point indicates contamination with a more volatile substance which could possibly cause explosive situations when burned. The flash point determination is particularly effective in detecting small amounts of contamination, e.g. 0.25% petrol will lower the flash point of heating oil by approximately 10°F (5.5°C).

3. *Relative Density.*   This is generally considered important and is quoted in specifications within the petroleum industry. Generally heating oils have relative densities at 60/60°F (15.5°C) of between 0.830 and 0.840.

4. *Distillation.*   This is specified at one point, i.e. the minimum volume percentage recovered at 357°C should be 90%. This is controlling the higher boiling fractions which if present in excess are less likely to be burnt and thus more likely to cause deposits in combustion chambers and/or injectors.

5. *Sulphur Content.*   The limit is 1% by weight.

6. *Cloud Point.*   This is only carried out if ambient temperatures are low enough to give rise to wax precipitation in the fuel and thus cause fuel starvation.   A note of warning here: heating oil is manufactured to seasonal specifications and care should be taken not to purchase at the end of the summer period (September) particularly if the turnover is slow.

7. *Calorific Value.*   This may be calculated by using the formula from BS 2869 (see CV, p 560). The relative density and sulphur content are needed.

Fuel Oil

There are four classes of fuel oil specified in BS 2869, and these cover most applications in industry today. The only exceptions of any consequence are the low-sulphur fuel oils used in the steel industry and certain ships-bunker fuels of intermediate viscosity to the four classes.

Class E. Commonly known as 'Light Fuel Oil', has a maximum kinematic viscosity of 12.5 mm^2/s (cSt) at 82°C (180°F) which is better and more accurate than the old '250 second Redwood 1 at 100°F'.

Class F. Commonly known as 'Medium Fuel Oil', has a maximum kinematic viscosity of 30 mm^2/s (cSt) at 82°C (180°F) which is better and more accurate than the old '950 second Redwood 1 at 100°F'.

Class G. Commonly known as 'Heavy Fuel Oil', has a maximum kinematic viscosity of 70 mm^2/s (cSt) at 82°C (180°F) which is better and more accurate than the old '3500 second Redwood 1 at 100°F'.

Class H. Commonly known as 'Heavy Bunker Fuel', has a maximum kinematic viscosity of 115 mm^2/s (cSt) at 82°C (180°F) which is better and more accurate than the old '6000 second Redwood 1 at 100°F.'

Generally, these fuels are used in atomizing burners and normally require preheating to reduce them to the viscosity suitable for atomization and subsequent burning. Obviously more preheat temperature is needed for higher-viscosity oils, i.e. more heat is needed to atomize Class H fuel oil than Class G and so on. Also more heat is normally necessary to store and handle the heavier grades. This need not necessarily depend solely on viscosity; wax content of the fuel is important in this respect and pumpability is the best guide.

The four classes specified in BS 2869 are compared below:

	Class E	Class F
Kinematic viscosity at 82°C (180°F), (mm^2/s = cSt) MAX.	12.5	30
Flash point, closed, Pensky—Martens, MIN.	66°C or 150°F	
Water content (vol %) MAX.	0.5	0.75
Sediment (wt %) MAX.	0.15	0.25
Ash (wt %) MAX.	0.1	0.15
Sulphur content (wt %) MAX.	3.5	4.0

	Class G	Glass H
Kinematic viscosity at 82°C (180°F) (mm^2/s = cSt) MAX.	70	115
Flash point, closed, Pensky—Martens, MIN.	66°C or 150°F	
Water content (vol %) MAX.	1.0	1.0
Sediment (wt %) MAX.	0.25	0.25
Ash (wt %) MAX.	0.2	0.2
Sulphur content (wt %) MAX.	4.5	5.0

When testing fuel oils the above tests will certainly distinguish them and generally indicate whether the product is satisfactory. Much depends on whether the fuel is being checked because of inadequate or poor burning or because it is causing corrosion or other problems, in which case several other tests may be applied. Quite often the fuel is blamed and checked when the mechanism of the burner is at fault. However, when the fuel oil is to be checked for whatever reason the following tests can be applied:

1. *Viscosity.* This will check that the right grade is being used and the correct preheater temperature applied. It is worth remembering that a top, middle and bottom sample all checked for viscosity will indicate whether the fuel is homogeneous; it is not unknown for a wrong grade to be delivered, or layering of fuels of differing viscosity to occur.

2. *Flash Point.* Always worth checking with burning fuels. As mentioned earlier it only needs a small percentage of volatile material to lower the flash point and give rise to a possibly dangerous situation.

3. *Water Content.* The relative density for fuel oil is greater then 0.9, i.e. it is nearing that of water. Hence water will sometimes take time to separate from the oil, and it is therefore important to check for the presence of water.

4. *Ash.* The ash level of fuel oils is usually well below 0.1%. If a relatively high figure is obtained it is as well to check for individual elements, particularly sodium and vanadium, which can cause problems.

5. *Sulphur Content.* The British Standard specifications for sulphur for the four grades are all significantly higher than the UK petroleum industry currently supplies to customers. In fact other restraints, particularly environmental, imply limits appreciably lower than some of the British Standard specifications. As a result, the test for sulphur is assuming great importance, although it should always be remembered that the reproducibility of some of the existing methods of test are not particularly good: e.g. two single determinations carried out in separate laboratories on a 3% sulphur fuel oil could vary by up to 0.3%.

6. *Pour Point or Pumpability.* The British Standard does not specify limiting figures for either of these tests but proposes minimum storage and handling temperatures for classes E, F, and G. Obviously if the pour point or preferably the pumpability is known for the fuel a possible saving on heating costs could ensue.

Whilst discussing pour point it is as well to mention that with Class E, light fuel oil, a more serious fault can sometimes occur. This is the precipitation of wax in the form of agglomerates of crystals which can reduce and eventually stop the flow of the oil owing to filter blockages. There is evidence that if light fuel oil is stored for a lengthy period at temperatures between 21 and $32°C$ this can tend to induce precipitation of the wax crystals more than if the fuel was stored below 21 or above $32°C$. Additives are available that while not stopping the precipitation of the crystals do stop the crystals forming agglomerates and so keep the fuel flowing.

7. *Relative Density.* This is invariably needed for accounting purposes and also for use in the calculation of calorific value. The relative density of fuel oils has to be carried out at temperatures of $49°C$ or above to overcome non-Newtonian flow.

8. *Calorific Value.* For those interested in efficiency calculations, this is needed, although the calculation method using relative density and sulphur content is normally adequate.

Further tests can be applied, such as those for asphaltenes, Ramsbottom carbon, sediment, and others depending largely on the type of problem or investigation.

D3 Coal and coke

To establish the quality of a solid fuel and to determine whether it is suitable for use in a particular process, it is necessary to investigate the composition of the fuel and on occasions to survey its physical properties. For example, to determine whether a coal is suitable for coke making it is required to know to what extent it will swell during carbonization, and to determine the sulphur, moisture and ash contents, since the carbonizing conditions and the properties and suitability for industrial usage of the resultant coke are, to a considerable extent, dependent on these parameters. The size and strength of the coke, physical properties of considerable importance in metallurgical processes, must also be measured.

Large tonnages of coal and coke are consumed and manufactured in these operations. If the data relating to composition or physical properties are to be meaningful, it is essential that methods are available which can guarantee that the relatively small quantities of material subjected to testing are fully representative of the whole of the material.

Accurate systems of sampling are thus required to enable confidence to be attached to the results of tests based on the samples obtained; it is pointless to carry out accurate determinations of the major or minor constituents of solid fuels if the accuracy of the sampling is questionable. With this background in mind, systems of sampling have been devised which ensure that a representative sample of the fuel is obtained under various conditions. The essential basis of each of these sampling systems is to ensure that a sufficient weight of sample is collected by a variable number of increments, the specified weight of each increment being dependent upon the physical condition and composition of the gross sample and the accuracy to which the various determinations used in assessing the quality of the sample are to be made.

Manual methods are considered to provide the most accurate form of sampling, but in many instances automatic sampling is essential and is permitted, provided that such methods can be demonstrated to be unbiased with reference to the manual method.

It is not possible in this survey to deal adequately with the theories of sampling, and to cite illustrations or examples of particular methods employed under different conditions. This specialized information may be found in a number of references[32-34].

Sampling

General Procedures

The incremental method of obtaining a sample ensures that the whole of the solid fuel is exposed to the process and that all individual particles of the fuel have an equal chance of being included in the sample. The increments, or the quantity of fuel sampled at any one time, must be evenly spaced over the whole period during which the sample is being collected. Such a system is known as systematic sampling and is to be recommended in preference to other methods of sampling, such as random sampling or stratified random sampling; these procedures are difficult to apply as routine procedures and unsuitable for automation.

Bias, i.e. a persistent tendency towards results which are too high or too low, can easily occur during sampling, particularly if the material being sampled is unrepresentative of the whole material. It is good practice and assists in avoiding bias to ensure that the sampling equipment is of adequate dimensions to ensure that the larger particles in the mass are evenly distributed in the sample, and that it and the mass of the increment bear a relation to the maximum particle size of material being sampled.

This practice frequently leads to the collection of increments and a sample larger than required for a given precision of sampling. In such conditions it is permissible to adjust the number of increments taken subject to maintaining the required accuracy, but no other method or change in the sampling procedure is permissible.

Consideration should be given to the siting of the actual sampling point to enable the increments to be collected in safety and with the minimum physical effort. A special structure may be built, particularly where manual sampling is practised, to assist the sampler. He should be trained and conscientious in his approach and it is the duty of management to impress upon the sampler the importance of his work. Where automatic sampling is practised, provision should be made, for example by using rubber-lined chutes free from projections and steep gradients, to ensure that larger pieces are not broken in transit to the receiver.

Though the basic methods of sampling may be similar, further treatment will vary according to the testing procedure. Precautions should be taken to prevent loss of moisture from coal and coke samples, by storing them in containers (preferably plastic) covered by well-fitting lids. If the sample is to be subjected to size analysis, methods for preventing breakage of larger pieces before testing are essential.

Sampling of Coal for Ash and General Analysis

Sampling from Conveyors and Falling Streams. The incremental principle of sampling is applied under these conditions, having regard to the precautions already outlined which ensure that the sample is representative of the whole material. The whole cross-section of the stream should be sampled, and on conveyors the whole depth of the coal should be available to the sampling instrument. Where the fuel to be sampled is above 80 mm in size, it is both difficult and dangerous to attempt to obtain a sample from a moving belt or conveyor and stopped-belt sampling is recommended in these conditions. Stopped-belt sampling is always considered the reference method and should be used to check the bias of any alternative method.

The standard of accuracy required controls the number of increments of a given mass to be taken from the sample. This number varies according to the size of coal and the ash level. For a detailed statement of the number and size of increments to be taken under various conditions, reference should be made to the appropriate publications[35,36].

Sampling from Wagons, Stockpiles and Ships

Wagon Sampling. The fundamental difficulty in sampling from wagons arises from the fact that it is frequently impossible to sample the whole of the consignment owing to the inaccessibility of part of the coal, and the sample should therefore be taken during loading or discharge. Three general methods of sampling are available: (*a*) by probes from the tops of the wagons; (*b*) from wagons during discharge; (*c*) from exposed faces whilst the wagon is unloaded on a tippler. The use of probes is to be avoided wherever possible, but if their use is essential a probe which penetrates to the full depth of the coal and of a minimum dimension (diameter) not less than 30 mm should be used. The aperture of the probe should not be less than 2.5 times the upper size of the coal. The method should be checked to establish whether it is free from bias.

Where the sample is collected while the wagon is being unloaded the sampling implement should be swung across the whole width of the stream. It is frequently impossible to fulfil this requirement using manual sampling, and the position of insertion of the scoop into the stream should then be varied from wagon to wagon. Automatic methods of sampling should be used wherever possible.

If, when mounted on a tippler, the wagon is unloaded by a side-door, it should be partially emptied to expose the wagon floor. The angle of the wagon is then reduced to leave an exposed face, which is sampled by the same technique as used in wagon-top sampling, where the face of the coal is divided into equal areas and the required number of increments taken in random order from these areas. The positions from which the increments are taken should be varied from wagon to wagon.

Sampling from Stockpiles. It is virtually impossible to obtain a sample without bias from a stockpile more than about 2 m high, since the whole of the area is not uniformly accessible and the quality of the coal in the periphery may be affected by oxidation. If possible, sampling from a stockpile should be carried out during stocking or lifting, when a more representative material may be obtained from a conveyor or boom-loader.

Where sampling from a stockpile is unavoidable, the area should be carefully marked out and sampling positions selected to ensure as great a coverage of the pile as is consistent with the accuracy required. The sample is representative of that area from which the coal has been obtained.

Sampling from Ships. Where possible, the sample should be obtained from a conveyor during loading, or unloading, but if it is withdrawn from the ship's hold it is essential to employ trained personnel. Because there is a tendency for size segregation to occur in the hold of a ship, it is frequently necessary to estimate the amount of large and small coal present and sample each separately. The increments should be distributed over the face of the coal, and samples taken at various depths after the withdrawal of a portion of the coal[35].

Sampling for Total Moisture

It is sometimes necessary, and convenient, to collect a separate sample for the determination of moisture. Precautions must be taken to prevent loss of moisture by evaporation or during crushing, and after collection of the increments, spaced evenly over the consignment, the samples must be stored in closed containers with well fitting lids. Plastic dustbins are suitable.

Alternatively the moisture sample may be extracted, by increments, from the general analysis sample.

Methods for the determination of total moisture in coal are fully described in a number of National and International Standards[37-40].

Sampling for Screen Analysis

The basic principles underlying sampling for screen analysis are the same as those relating to sampling for ash and general analysis. The number of increments required may be found by reference to the appropriate Standard[41-43].

Precautions must be taken to ensure that the coal is subjected to gentle handling, thereby minimizing breakage. Mechanical screening is permitted, provided that the method minimizes breakage and is free from bias.

Sampling of Coke

Sampling of coke follows the same general principles as sampling of coal, but because of the difference in the physical properties of the two materials certain special procedures must be adopted.

The moisture content of coke is normally its greatest variable, since it changes with size. The rough, abrasive, irregular surface of coke makes it difficult to collect a sample representative of the size distribution of the consignment, and a bias in sizing is usually reflected as a bias in moisture.

The most recent information on this subject is found in an International Standard[44] where methods are described for the sampling of coke under various conditions, e.g. single consignments, regular consignments, continuous sampling and intermittent sampling. The standards of precision are based on the 95% probability limits, and instructions are given to indicate any adjustment necessary in the number of increments to attain this standard of precision in all cases.

Details are specified of the number and mass of the increments which are required when sampling for moisture, ash or physical testing; these vary considerably according to the size of coke and differ where coke is used for various processes, for example in foundries using large coke and in large central-heating boilers using much smaller coke.

Automatic sampling is permitted and is essential under many circumstances: such a system must be tested for bias, using the stopped-belt method of sampling as the reference method.

Instructions for sampling are intended to be read by the engineer or supervisor in charge, and the essential information applicable to the particular conditions obtaining in his organization should be given to the sampler responsible for the operation of the process.

The amount of coke to be crushed in a single operation should be limited to about 70 kg to prevent loss of moisture by exposure to the atmosphere. Samples of greater mass should be treated as individual sub-samples, crushed and tested separately.

Sampling for Physical Tests

The criterion upon which the sampling of coke for physical tests is based is the precision of the determination of the mean size, obtained by calculation from the size analysis of the physical test sample. Since the test samples are dried before examination, variation of the number or mass of increments to take account of the level of the moisture content of the coke is unnecessary. Generally the coke to be sampled is relatively large (>20--25 mm), and the moisture content is unlikely to be excessive. The variation, according to the upper size of the coke, of the number and weight of increments to be taken to ensure that the reported mean size of the coke is within 1/10 of the true value is shown in *Table 1*, which is derived from BS 2074:1965, *Methods for the Size Analysis of Coke*[41]. Where automatic sampling is practised, it will generally be found that the weights of the increments are larger than those given above and the total weight of the sample is greater. This will tend to promote a higher degree of accuracy.

Special Condition where a Sample is taken for the Determination of Both Moisture and Size

Under these circumstances the stipulated precision of sampling must apply to both parameters. Since a greater number of increments must be taken for the moisture sample this will enable mean size to be determined with a precision better than the 1/10 level.

Sampling under Conditions of Very High Flow Rates

Modern industrial developments involve the handling of coal and coke in much greater quantities than hitherto. For example, the delivery of coal to power stations by block trains which discharge up to 2000 t of coal in 30 min, and the projected construction of coke ovens handling over 100 000 t of coal and coke in a week demand an appropriate system of sampling.

There are no Standard specifications applicable to such conditions but Standardizing bodies are well aware of the situation and the matter is being actively pursued.

Table 1   Recommended weights and numbers of increments (derived from BS 2074:1965)

Nominal upper size of coke (mm)	180	180	120	90	60	40	30
Minimum weight of increment (kg)	14	7	4.5	2.5	2.5	1	1
Number of increments	96	96	60	48	24	24	24
Minimum weight of sample (kg)	1344	672	270	120	60	24	24

References p 580

Methods of sampling coal travelling on high-speed belts at tonnage rates of up to 3000 t/h are in use in the USA. The general principle of these methods involves the automatic incremental collection of a primary sample which is then sample-divided to give a secondary sample. This is followed by a crushing operation before a tertiary automatic sample division provides a final sample of about 10 kg of −3 mm coal.

Other systems are in operation where wagons are randomly selected from a train load and the coal from them is diverted from the main stream before it is reduced by automatic means to the required mass and particle size.

Checking the Sampling Accuracy

The principle of duplicate sampling may be employed to check the accuracy of sampling coal or coke. Essentially this procedure consists of taking two sub-samples by placing alternate increments in two bins. Suitable key determinations are subjected to mathematical analysis which indicates whether the correct precision of sampling is being attained, or whether too few or too many increments or, in the case of intermittent sampling, too few samples are being taken or examined. An excellent dissertation on the method of checking the accuracy of sampling by this method is given in the British Standard Specifications dealing with the sampling of coke[41,45].

It is worthwhile re-emphasizing that breakage must be avoided as far as possible during the handling and transporting of samples, that they should be stored in secure and sealed containers, and that labels describing the origin of the samples and the tests required on them should be clearly displayed on the inside and outside of the container. The use of plastic labels and waterproof inks for this purpose is to be strongly recommended.

The Analysis of Coal and Coke

The Origin and Nature of Coal

The diverse nature of plant remains and subsequent variation in climatic and geological effects which have contributed to the formation of carbonaceous materials have resulted in the formation of a wide variety of coals. These are briefly described in App.A6. All coals consist largely of carbon, hydrogen and oxygen. They are *not* hydrocarbons; oxygen has an important influence on their most distinctive characteristics. Relatively small amounts of (principally) sulphur and nitrogen are organically associated with the coal substance, and there may be 0–50% of mineral matter of most variable composition — some 60 elements have been recognized in it[46]. The broad variation in the composition of coals is shown in App.A6.

The term rank is frequently used to designate the degree of coalification of the original plant material, from low-rank lignite or brown coal to high-rank coking coal or anthracite. Coal-classification systems are summarized in App.A6. Those based on fundamental properties are often too complex or detailed to be entirely suitable for use in industrial conditions, and simplified systems such as those derived by the NCB, ECE, USA or the International System, which break down coals into broad groups and describe them by a digital system, are more convenient for everyday use (App.A6).

The suitability of coal for a particular purpose is generally a function of its composition. The main parameters describing the quality of coal are the moisture content, ash and calorific value. For many purposes, particularly with coking coals, the sulphur content is also of importance and a knowledge of the thermal behaviour, for example swelling characteristics, is also vital in selecting the optimum coal or blend to be used in carbonization.

To evaluate the composition of coal, tests have been devised which enable the major and minor constituents to be determined with high accuracy. Some of these, 'model' tests, are empirical in nature and for a meaningful result the method of carrying them out must be strictly controlled. Other tests demand more analytical skill and are more fundamental in character; they enable a precise indication to be obtained of the amount of individual elements present in the coal. The empirical tests are generally grouped together under the term 'proximate analysis' and it is customary for the remaining tests to be referred to as 'ultimate analysis'. Methods for the analysis of coals and cokes are fully described in many Standard Specifications. The relevant British Standard[38] is currently being revised and the individual parts of this Standard describe most comprehensively the conditions under which the various tests are carried out, the method of calculation of the results, the precision of the determination and the permitted tolerances within and between laboratories. Other Standards of which the engineer, fuel technologist or chemist should be aware are those prepared by the American Society for Testing and Materials (ASTM), Committee D-5, Coal and Coke; the DIN Standards and AFNOR Standards of Federal Germany and France respectively; the Japanese Imperial Standards (JIS); and those of the USSR (GOST).

In addition, many countries have accepted the use of International Standards (ISO) and in many cases these replace or are identical with National Standards. The reader should ascertain whether his own country has developed its own system, and if not select the most appropriate of the above for his purpose.

Coal Analysis

Proximate Analysis. The proximate analysis of coal is usually considered to include the determination of moisture, ash, volatile matter and fixed carbon.

Moisture. Moisture is an inert material and its presence reduces the fuel value of the coal. Free moisture is the term generally applied to the water which has percolated through overlying strata into the coal measures and to that remaining after the treatment of coal in a washery or after rainfall. This moisture can be removed by natural drying to leave a residual air-dried moisture, the amount of which is a function of coal rank. Air-dried moisture may be removed from the coal by further drying in oven or vacuum and is generally referred to as inherent moisture.

Total moisture may be determined by a distillation method, using a 1 kg sample of coal having a maximum particle size of 13 mm, or by drying in air at $105-110°C$. The former method is suitable for all coals; the drying method should be restricted to high-rank coals which are not susceptible to oxidation under these test conditions. The distillation method can also be applied to coals of a maximum particle size of 3 mm, a 100 g sample then being used.

An alternative method, based upon drying about 10 g of -3 mm coal in a nitrogen atmosphere at $105-110°C$, may be used for all coals.

Air-dried or inherent moisture is determined on the air-equilibrated laboratory analysis sample using 2 g of coal crushed to pass a sieve of 212 μm aperture.

Two methods are in general use, both involving the removal of moisture from the coal at a temperature of $105-110°C$ by a stream of nitrogen. The minimum-free-space oven may be used, the coal being placed in the oven in an open dish and the loss of moisture determined by weighing the dish before and after the experiment. A direct gravimetric method is also recommended; in this case the moisture removed from the coal by the nitrogen stream is absorbed in a drying agent and the increased weight of the absorption tube is a measure of the moisture content.

Although of limited significance in itself (as some indication of rank), the air-dried moisture must be determined whenever an analysis is made to enable the other results to be corrected to the dry basis or the dry mineral-matter-free basis.

Ash. Ash is present in coal in two forms, adventitious ash and inherent ash. It reduces the heating value of a coal, and a high ash level in coals used for carbonization can cause a deterioration in the resistance of the coke to abrasion. Adventitious ash can be removed from coal by cleaning, but inherent ash is frequently finely disseminated throughout the coal substance and remains after combustion.

The determination of ash is an empirical determination and care must be taken to adhere strictly to the conditions set out in the prescribed methods. In general 2 g of coal ground to pass a test sieve of aperture 212 μm is heated in an oxidizing atmosphere for a prescribed time at a temperature of $800-815°C$. It is important to ensure that the muffle furnace used for the determination is adequately ventilated. The weight of material remaining after combustion represents the weight of the ash. Coal ash is an extremely hygroscopic material, and precautions must be taken during the cooling and weighing process to prevent adsorption of moisture.

Volatile Matter. The volatile matter of coal is a complex mixture of combustible gases and tar and water derived from the chemical decomposition of the coal, and water from the thermal breakdown of shales and minerals. Individual substances present include hydrogen, methane, ethane, other hydrocarbons, and oxygen-containing compounds. This test is carried out at a temperature of $900°C$, air being excluded from the crucible. It is prudent to grind-in the crucible lid to ensure that it is a close fit within the crucible, and to keep the two units together for all determinations. The volatile matter yield of the coal is expressed as the loss of weight of 1 g of coal when heated for 7 min under these conditions. A correction for moisture is applied to this result.

Fixed Carbon. The sum of moisture, ash and volatile matter is frequently subtracted from 100 to obtain a value termed 'fixed carbon'. It should be noted that this term is hypothetical and bears no relation to the total carbon content of the coal as determined by ultimate analysis, nor does it imply the existence of uncombined carbon in the coal. There is probably no free carbon present as such in coal, and fixed carbon is no more a precise measurement of any coal constituent than the equally hypothetical volatile matter. The two are interchangeable within moderate limits, depending on the conditions of VM measurement.

Ultimate Analysis. The ultimate analysis of coal generally implies the exact measurement of the quantities of carbon, hydrogen, nitrogen and sulphur present. When the sum of the carbon, hydrogen, and nitrogen, calculated to the dry mineral-matter-free basis, is subtracted from 100, a difference remains which may be considered to represent the oxygen content of the coal. The direct determination of oxygen is a complex procedure, since the contribution made by the oxygen in the mineral matter to the total oxygen present cannot be readily assessed. A method for the direct determination of oxygen has been described[47] and can be applied directly to coals of low mineral-matter content, but for coals containing more than 5% of mineral matter it is necessary to demineralize them prior to the determination of oxygen.

Provided that the determination of carbon, hydrogen and nitrogen has been carefully carried out, the value obtained for oxygen by difference is usually within 0.5% of the determined value except for coals with high organic sulphur.

Carbon and Hydrogen. The Sheffield high-temperature method[48,49] has almost entirely replaced the Liebig method for the determination of these elements. The combustion of a 0.5 g sample is completed in 10 min at a temperature of 1350°C. Precautions are taken to prevent any gases other than carbon dioxide and water vapour from entering the absorption vessels.

Nitrogen. The nitrogen content of the majority of coals used in industrial practice will usually be found to lie between about 1.2 and 1.8% but a precise knowledge of the nitrogen content of the fuel is important in calculating heat balances in furnaces. The determination is carried out by the Kjeldahl method, in which 0.1 g of coal is digested with sulphuric acid and a catalyst. The ammonia released is absorbed in an alkaline absorbent and subsequently recovered by steam distillation followed by absorption in acid solution and determination by titration.

Modern equipment is now available in which carbon, hydrogen and nitrogen can be determined simultaneously.

Sulphur. The sulphur content of coal is particularly important in relation to the amount of oxides of sulphur produced when coal is burned and their effects upon appliance corrosion etc. and atmospheric conditions, and to the amount of sulphur remaining in coke after carbonization. In general the sulphur content of coals used for carbonization seldom exceeds 1.5%, but this figure may be increased where coals are used for electric-power generation. Two methods are in general use for the determination of sulphur: the well-tried Eschka method and the Sheffield high-temperature combustion method which uses the same basic apparatus as is used for the determination of carbon and hydrogen. In the Eschka method the coal is heated with an alkaline mixture and, after extraction with acid and removal of iron, the sulphur is precipitated as barium sulphate and determined gravimetrically. A number of publications provide evidence to show how the time required to complete a determination can be reduced to the order of 1½ h[50].

The high-temperature method takes approximately 15 min[51]. The coal is burned in a stream of oxygen at 1350°C and the acid gases are absorbed in hydrogen peroxide. Titration of this solution gives a value for total acidity that is due to the presence of both sulphuric and hydrochloric acids. A correction for the contribution of the hydrochloric acid (and therefore for the chlorine in the coal) must be made to enable the sulphur content to be determined.

Although the total sulphur is the most important parameter in relation to the industrial usage of coal, it is sometimes valuable, for example in coal-cleaning or desulphurizing processes, to establish the distribution of the various *forms of sulphur* in the coal.

There are two principal forms of sulphur in coal, inorganic and organic. The former is generally present as pyrite, and may be found in large discrete particles or finely disseminated throughout the coal. In its massive form, pyrite is amenable to

removal from coal by a variety of washing or cleaning processes.

The organic sulphur is firmly bonded to the carbon structure[52] and cannot be removed by mechanical means. It remains during carbonization and becomes firmly combined within the coke structure. No direct method for its determination is available. Methods are available by which pyritic sulphur may be determined in coal[53], these generally depending upon selective acid-extraction of the pyritic iron from the coal and calculating the pyritic sulphur on the assumption that it is present as FeS_2. The organic sulphur is generally taken as the difference between the total sulphur content and the pyritic sulphur content, though small amounts of sulphur may also be present as sulphate, especially in oxidized coals.

Other Elements. These are generally found in minor proportions, the principal substances that may also be determined in coal analyses being chlorine, phosphorus and arsenic. The former may attain a level of up to 1.2% in certain coals and is normally determined by the high-temperature combustion method described for the determination of sulphur.

Phosphorus is of importance in metallurgical processes involving the production of acid pig-iron. It is generally found in coal in combination with calcium and is normally determined by a classical extraction procedure using a colorimetric finish.

Arsenic is present in British coals in very small quantities, usually not above 10 ppm w/w. It is normally determined in fuels used in the manufacture of foodstuffs or beer, which have to conform to a strict specification for arsenic content. The method for the determination is described in the appropriate B.S. Specification[54].

Analysis of Ash. New methods of ash analysis have been developed during the last decade. These have resulted in a considerable shortening of the time required to complete the determination of the major constituents of the ash, as the oxides of iron, aluminium, silicon, calcium, magnesium, sodium, potassium, titanium, manganese, sulphur and phosphorus. Detailed methods for these determinations have been reported[55–57].

Concern with environmental problems has led to considerable research in the development of methods for measuring the concentration of trace elements present in solid fuels and their products and residues of combustion. The determination of elements such as lead, vanadium, beryllium, zinc, copper and cadmium is being actively studied by a variety of sophisticated techniques. No relevant standard methods have yet been published.

Calorific Value. The heating or calorific value is a very important characteristic of a fuel since it measures the amount of heat released by complete combustion of unit weight. The heat released varies according to the rank of the coal. For coals used in steam raising and metallurgical operations the level of CV is generally within the range

25—35 MJ/kg, expressed on the dry ash-free or dry mineral-matter-free basis.

The determination of CV has been simplified and shortened following recent investigations[58]. The principle of the test remains the same, in that unit weight of fuel is burned in a calorimeter bomb under pressure of oxygen and the subsequent heat release is measured by the rise in temperature of a surrounding water jacket. This determination, usually carried out in an adiabatic calorimeter, provides a measure of the gross CV at constant volume. Under industrial conditions the operative parameter is the CV at constant pressure, but having regard to the accuracy of the laboratory determination the two methods of expressing the CV may be regarded as identical. Full details of the determination are available[59,60].

To obtain the net CV a deduction should be made to allow for the heat of condensation of the water vapour produced in the combustion, by cooling to 15°C. This amounts for coals to about 1 MJ/kg and is deducted from the gross CV to obtain the net CV.

The gross CV can be calculated from the ultimate analysis by use of the following formula (>15%O):

$$Q \text{ dmmf (calc, MJ/kg)} = 0.336 \text{ C} + 1.42 \text{ H} -$$

$$0.145 \text{ O} + 0.094 \text{ S}$$

where C, H, and O are the percentages of carbon, hydrogen and oxygen (dmmf) present in the coal. Agreement between the calculated and determined values of the CV (the latter having been also corrected to the dmmf basis) may be considered to be satisfactory (implying that the correct standard of accuracy was obtained in all determinations) when the two values are within 0.12 MJ/kg of each other.

Calculation of Analytical Results to Other Bases. The majority of analytical determinations are carried out on coal of particle size less than 212 μm which has been air-dried until it is in equilibrium with the laboratory atmosphere. The results obtained under these conditions are reported on the air-dried (a.d.) basis, but for certain purposes it is necessary, and the results are more meaningful, if they are corrected to other bases, for example dry, dry ash-free, or dry mineral-matter-free.

The conversion of results to the dry basis (db, or mf = moisture-free) eliminates the effect of variability in the air-dried moisture content of the coal. This correction is particularly important when comparing analytical results obtained at different times or in different laboratories on the same sample.

Calculation of results to the daf basis has little scientific justification, since it assumes a hypothetical situation where coal is considered to be free of moisture and ash. The formula may be used for calorific value, carbon, hydrogen and volatile matter. When applying the correction to the determined volatile matter an allowance must be made for the amount of mineral carbon dioxide released by anthracites and coals with a high carbonate content.

The dmmf basis of reporting results also assumes a hypothetical condition of pure coal, free from all contamination. The expression of results on this basis has certain advantages over analyses corrected to the daf basis, since the former recognizes that the ash remaining after combustion differs in composition and quantity from the original mineral matter in the coal.

Ash is less in quantity than the original mineral matter by the amount of combined moisture and carbon dioxide expelled on heating clay minerals and by any loss in weight due to conversion of iron pyrite to iron oxides. Direct methods for the determination of mineral matter have been described[61,62], but it is usually more convenient to calculate the results of the analysis to the dmmf basis using one of the appropriate formulae given in the relevant literature[63]. Some of these attempt to allow for organic sulphur.

Analyses on the dmmf basis are often used in coal classification.

Caking and Swelling Tests. The caking and swelling properties of a coal are of considerable significance in the manufacture of carbonized products. To form a metallurgical coke under the conditions of carbonization in conventional slot-type ovens, a coal must first soften at a temperature of about 350—420°C, subsequently swell, and resolidify with shrinkage at a temperature of 500—550°C to form the characteristic cellular structure of coke.

Coals which exhibit this behaviour are described as coking coals and the extent of the softening, swelling and shrinkage is dependent mainly upon the rank of the coal. Prime coking coals, classified as Type 301 or 434 by the (UK) NCB or ECE systems of classification respectively, usually yield between 23 and 27% of volatile matter (dmmf), have carbon and hydrogen contents between 87.5 and 89.5% and 4.9—5.2% respectively (dmmf) and exhibit the greatest extent of softening and swelling. The coking properties of coals of lower or higher volatile matter are inferior to those of the prime coking coals and the cokes formed from them are of inferior strength and hardness compared with those formed from prime-coking coals.

The term 'caking' must not be confused with coking. The caking properties of a coal are an indication of the ability of individual coal particles to adhere to one another on heating. Because this phenomenon is markedly dependent on the rate of heating, a coal classified as a caking coal on the basis of the results of laboratory tests does not necessarily fulfil the requirements of a coking coal as understood in industrial practice.

Nevertheless, a knowledge of the caking and swelling properties of coals provides a considerable amount of guidance and assistance in the selection

References p 580

and assessment of coals for carbonization. The tests available to measure these properties can broadly be divided into two types, namely those in which swelling power is measured at slow or rapid rates of heating, with or without an examination of the residue remaining after heating to various temperatures.

Swelling Tests. The most widely used test in this category is the *Crucible Swelling Test*[64],[65] where 1 g of coal, ground to pass a 212 μm sieve, is rapidly heated under controlled conditions for a specified period. The extent of swelling of the coal is assessed by comparing the shape of the residue (the 'coke button') with that of a series of profiles depicting swelling numbers from 0 to 9. Strongly coking coals have high swelling numbers, above 7; coals which exhibit a swelling number of 4 or less are generally non-coking under industrial conditions, but may be used for the preparation of formed cokes or for the production of fuels suitable for domestic use. Exceptions to this broad differentiation are not uncommon; in particular, certain coals from Australia having swelling numbers below 6 will produce a reasonable metallurgical coke under industrial conditions of carbonization. Their behaviour in the crucible swelling test is modified by the presence of non-swelling material classified as inert by petrographic analysis, the quantity of inert material in such coals being much greater than in coals of comparable rank found in the Western hemisphere.

Dilatometer tests measure the expansion and contraction of coal over the temperature range where the material is softening and swelling. Heating rates are usually of the order of 3 K/min, which broadly corresponds to the rate of temperature increase in the coal charge when it is softening during carbonization.

The coal sample, in powdered or compacted form, is normally confined within a retort fitted with a piston which rests on the coal. As the temperature of the surrounding furnace is increased the coal contracts and expands; the piston follows this behaviour and the movement is recorded on a suitable instrument.

Examples of such tests are the Sheffield laboratory coking test[66], the Audibert—Arnu dilatometer test[67] and the Ruhr dilatometer test[68]. Data from the Audibert—Arnu test are used as one parameter for coal classification by the ECE system.

Caking Properties. The caking power of a coal may be measured using the low-temperature Gray—King assay test[64]. The appearance of the residue is compared with that of standard cokes to which the letters A—G are assigned and on this basis the coal is classified according to its coke type.

Should the coke fill the whole section of the retort tube, the test must be repeated using mixtures of known amounts of the coal and an inert material (electrode carbon or anthracite), which reduces the swelling and the volume of the resultant coke.

By trial and error, a blend is found which forms a coke of type G. If this blend contains 15 parts of coal and 5 parts of inert material the Gray—King coke type of the coal is reported as G5. This test is used as one parameter when classifying coal by the (UK) NCB system.

Casual correlations exist between the results of the various caking and swelling tests, but these are generally too imprecise to be of a meaningful nature.

The Analysis and Testing of Coke

The analysis of coke is not generally undertaken in quite as comprehensive a manner as the analysis of coal. The principal determinations include the measurement of moisture, ash, volatile matter and sulphur, and less frequently the determination of carbon, hydrogen, nitrogen and phosphorus. The calorific value of the majority of metallurgical cokes is generally within the range of 32—33 MJ/kg.

The principles underlying coal analyses are applicable to coke analysis, but because of the more unreactive nature of coke and its greater physical hardness; some modifications are made to the analytical techniques used for the corresponding determinations on coal. These are briefly summarized in the note on each determination, the detailed methods of analysis being described in the relevant Standard Specifications[69].

Proximate Analysis. Moisture. As coke is not liable to oxidation, total moisture can be determined at temperatures up to 300°C. The air-dried moisture can be determined by heating a 1 g sample for four hours in an air-oven at 200°C, or by the direct gravimetric method using a temperature of 300—340°C[70].

Ash. Coke does not burn as readily as coal, and to enable the ash determination to be completed in a reasonable time, the sample weight is reduced to 1 g, thus increasing the area of sample surface available for combustion. The same temperature of incineration is used (815°C) as for the determination of ash in coal.

Volatile Matter. To displace air from the crucible, a measured quantity of benzene is added to the coke sample before placing the crucible into the muffle. The remaining test conditions are identical with those used for coal.

Sulphur. This determination by either the Eschka or high-temperature method is carried out in an identical manner to that described for the determination of sulphur in coal. In many cases, it will be found unnecessary to apply the correction for chlorine to the result of the determination by the high-temperature method.

Ultimate Analysis.

Carbon and Hydrogen. For certain unreactive cokes, it may be necessary to increase the time of combustion in the high-temperature furnace. Full details of this procedure have been described[71].

For all other coke, the method used for coal is directly applicable to coke.

Nitrogen. The nitrogen content of coke normally varies from 1.1 to 1.5%. It should be determined when accurate heat balances are required. The semi-micro Kjeldahl method may be used for this determination, making the following modifications to the method used for coal:

(a)   The coke should be crushed to a particle size less than 66 μm.

(b)   A catalyst is used which does not include selenium, since it has been shown[72,73] that prolonged digestion in the presence of selenium results in loss of nitrogen.

Oxygen. A direct method for the determination of oxygen has been described[74] but it is not in general use. The possibility exists of modifying automatic analysers to make them suitable for the determination of oxygen in coke, and certain physical techniques may be capable of being developed for this determination.

Minor Elements. Phosphorus in coke is determined by digesting the ash from 1 g of coke in an acid mixture and completing the determination in a similar manner to that used for coal.

The chlorine content of coke is of little importance, and is normally less than 0.1%. It can be determined using the high-temperature combustion method.

Arsenic in coke is determined by a method identical with that used for coal.

Calorific Value. The CV of coke is determined by a bomb-combustion method, essentially similar to the method used for coal. Changes are made to the type of crucible used, to enable complete combustion of coke to be achieved.

Calculation of Analytical Results to Other Bases. Coke analyses may be conveniently calculated to the dry basis, or dry ash-free basis. Formulae are available[75,76] which permit the calculation of results to the dry mineral-matter-free basis, but as these involve additional analytical determinations the calculation is rarely undertaken.

Analysis of Ash. Analysis of coke ash is carried out in the same manner as for coal ash.

Assessing the Physical and Mechanical Properties of Coke. The size and strength of coke are of great importance when assessing its suitability for metallurgical purposes. Some 70% of the coke manufactured in the UK is consumed in the blast furnace where coke comprises about half the volume of the total furnace burden. Coke is the only solid material remaining in the lower part of the blast furnace; it supports the weight of the material in the blast-furnace stack, and the size and size distribution of the coke particles remaining in the lower part of the furnace are critical in relation to permeability, output and consistent furnace performance.

The principle underlying all physical tests carried out on coke is therefore to attempt to simulate the degradation which occurs during the descent of the blast furnace. This is invariably undertaken by a process which subjects the coke to breakage by impact and abrasion, using drop-type tests or drum-type tests. The extent of the two forms of breakage differs between the two types of test, standard drum-tests generally subjecting the coke to a greater degree of breakdown by abrasion than drop tests, where impact breakage is the most predominant form.

Because of the wide variety in the size, operating conditions and burden of blast furnaces, it is unlikely that a standard test can simulate coke breakage in all furnaces; in realistic terms, tests currently in use do no more than provide a comparison of the resistance of cokes to breakage under standard conditions of testing.

One major criticism that may be levelled at the majority of breakage tests is that they examine the behaviour of a specific size-fraction of coke which is seldom entirely representative of the size of the material used industrially. The extent of coke breakage in standard tests is largely dominated by its initial size; the indices of breakage are therefore size dependent. The significance of this criticism is now widely recognized and developments are currently proceeding[77] in coke testing which it is hoped will provide evidence to enable a more realistic appraisal to be made of the breakage and degradation of coke under industrial conditions of usage.

Drop-type Tests. The best known of such tests is the shatter test[78,79]. A weighed amount of coke of specified minimum size is dropped four times in succession through a height of 1.83 m and the breakage assessed by a size analysis. The shatter index is generally reported as the percentage by weight of the material greater than 60 mm in size (or 80 mm if the coke is used in foundry cupolas) which remains after the test.

Drum-type Tests. These are far more numerous than drop-type tests. The Micum test is internationally used, but many variations of drum tests are in use as National Standards; for example, the ASTM tumbler test, the Japanese standard drum test and the Sundgren test used in the USSR. The principle of all tests is similar, a weighed amount of coke of a specified lower size limit being subjected to mechanical stress by rotation in the drum and the subsequent breakage measured by size analysis.

The indices defining coke strength vary from test to test; for example the characteristic indices of the Micum test are the M_{40} and M_{10}, i.e. the individual amounts of coke greater than 40 mm and less than 10 mm remaining after the test. Typical Micum indices of a good-quality metallurgical coke may be M_{40} of 75 or above and M_{10} of 7.5 or less. The results of one test may be correlated with those of another, but in general such correlations are only statistically satisfactory when indices of similar size are compared[80].

Detailed information on drum tests will be found in the appropriate references[78,81-85].

Other Coke Properties. Tests frequently carried out include the measurement of the bulk density, the apparent relative density, and the determination of true relative density.

Bulk Density. The weight of a given volume of coke is dependent upon the apparent specific gravity, the size and shape of the coke particles, and the relation between the size of particle and the size of the container used for the determination. The sample, which should be fully representative of the size distribution of the coke, is charged, with minimum breakage, to the weighed container and the level of the coke is adjusted by eye[78] or by struck-levelling[86]. The weight of coke occupying the standard volume is determined.

The bulk density in large containers is calculated using the Lee—Mott method of calculation[87].

Apparent Relative Density. A known weight of dried coke is weighed in water and allowed to drain for a specified time which is increased for certain types of low-porosity coke. A further weighing is then made and the apparent relative density calculated from the data.

Recent work has shown that coke more representative in size of that now used industrially (40—60 mm) may be used instead of the half-oven-width pieces previously tested. The revised version of BS 1016 currently under preparation (1974) will recognize these observations.

True Relative Density. This test is carried out on the analysis sample. 2 g of coke is introduced into a specific gravity bottle and water added. An air condenser is fitted to the neck of the bottle and air expelled from the sample by boiling in a glycerine bath. The vessel is cooled and filled with water, being then maintained at a constant temperature of $25 \pm 0.1°C$ for one hour before weighing. The calculation of true relative density is carried out in the normal manner.

Porosity. Total porosity is calculated from the results of the determinations of true (R) and apparent (A) relative density using the formula:

$$P(\%) = 100 \frac{R - A}{R}$$

Coke Reactivity. Many methods have been proposed for the measurement of this controversial property of coke. They are generally based upon the determination of the rate of reaction of granular coke (1—3 mm in particle size) with oxidizing gases, particularly air or carbon dioxide. Although most tests will place cokes in order of their reactivity, they are essentially only measuring the extent of carbonization of the parent coal.

The reactions between coke and oxygen or carbon dioxide are most sensitive to the effects of catalysts and the mechanisms complex; many mineral oxides present in fuel ash are powerful catalysts. Not surprisingly, therefore, correlations between simplified 'results' of laboratory reactivity tests and the rate of consumption of coke in industrial practice can be vague and imprecise. No real attempt has been made to propose such tests as standard tests; the one exception is the CAB (Critical Air Blast) Test[78,88] which is basically an ignition test carried out at relatively low temperatures and was used to a considerable extent in assessing the suitability of hard cokes for use under domestic conditions.

D4 Coal-tar fuels

As outlined in Appendix B11, coal-tar fuels are blends of tar oils or of tar oils and residual coal-tar pitch conforming to BS 1469:1962. They are supplied in six grades designated by numbers which correspond to the temperatures in degrees Fahrenheit at which the viscosity is approximately 100 seconds Redwood No.1 (25 mm²/s). A seventh grade is a solid pulverized hard pitch. *Table 2* lists the specification requirements and the average properties of the various grades.

Uses of Coal-tar Fuels[89]

The major outlet has been and is in the steel industry for firing Siemens—Martin open-hearth furnaces; the high emissivity of the flame improves direct heat transfer to stock and permits a useful saving in working time (between starting to charge and tapping). Figures, for thermal input of large steel furnaces using CTF 250, as low as 3 million Btu per ton of steel (3100 MJ/t) have been claimed, and an improvement of 1.5 h in working time per charge is achieved[90]. The low sulphur content contributes somewhat to the latter, and in addition limits the danger of pick-up of sulphur by the molten metal, particularly in the 'acid' process. In recent years, open-hearth furnaces have been increasingly replaced by electric arc furnaces or oxygen-blowing processes and the demand for coal-tar fuels has declined very considerably from about 510 000 long tons in 1965 to 70 000 tons in 1974.

Coal-tar fuels are also used for firing forging, rolling, billet and strip heating furnaces at steelworks. CTF 200 or 250 are the usual grades; they have the advantages of producing stack gases low in sulphur oxides and of enabling a reducing atmosphere to be produced without excessive smoke.

Annealing and tempering furnaces operating at comparatively low temperatures generally use the lower grades, i.e. CTF 50 or 100, and these can also be employed satisfactorily in small boiler plant[91].

Large rotary furnaces, such as are employed in the cement, alumina and other industries, are very suitable for firing with CTF 200, 250 or 300. In these, since the transfer of heat is to a large extent dependent on flame radiation, the higher emissivity of the CTF flame is of considerable benefit, and the low ash level is an additional advantage.

Coal-tar fuels (CTF 200—400) are also used for firing pot and tank furnaces in glass making; for this use their high flame temperature, high flame emissivity, the good furnace control achieved by the use of air atomization, their low sulphur content and virtual freedom from vanadium are advantageous.

Sampling of Coal-tar Fuels

CTF 50 and 100 are specified to be completely liquid at 0°C and 32°C respectively. The other grades are more viscous but are, except for pulverized pitch, delivered in liquid form in heated or insulated road or rail tankers and stored at such a temperature that they can be pumped. Recommended storage temperatures are[92]: CTF 50, 5°C; CTF 100, 32°C; CTF 200, 27°C; CTF 250, 60°C; CTF 300, 82.5°C; CTF 400, 133°C.

When taking a sample from a delivery tanker or storage tank, the contents must first be thoroughly mixed, preferably by pump circulation, and the sample taken from the top filling manhole by an open sampling tube, a weighted sampling bottle or a closable sampling cage. The sampling device should be lowered slowly into the liquid and allowed to remain there for a few minutes before being withdrawn. When several spot samples are required to give the sample for testing, these should be taken at increasing depths so that each sample is obtained from undisturbed material.

If the tank contents are not mixed or the tank is fitted not with a top manhole but with a fill-pipe, an average sample is taken on unloading by combining a number of spot samples taken either by a pipeline-sampler installed in the delivery line or by the use of a can or ladle from the exit stream. When a pipeline-sampler is fitted, the line must be cleared by running a small amount to waste before taking the spot samples; when they are taken from the exit stream, the container must be of sufficient size to collect all of the stream over a short period.

If an acceptance sample is required and the tank is fitted with a fill-pipe, two samples are withdrawn, one from the fill-pipe by using an open sampling tube and the other from the unloading pipe; these are combined, or tested separately as a check on homogeneity.

The above methods are not used with CTF 400 since its temperature on delivery is so high that loss of volatiles is almost inevitable. It is sampled by lowering a water-cooled dip rod into the liquid pitch and leaving for a few minutes before turning on the cooling water. The rod is then moved through the pitch to increase the volume explored and withdrawn slowly so that adhering liquid pitch drains into the bulk. The sample is discharged by breaking off the solid pitch adhering to the rod, and this process is repeated until the required size of sample is obtained.

Pulverized pitch is sampled by a sampling spear, taking numerous small samples from different locations in the bulk. The size of sample taken depends on the size of the consignment; up to half a ton, the minimum is 2.5 kg, from 5 to 10 tons it is 15 kg and from 50 to 100 tons, 60 kg.

The final sample for analysis and testing (1200 ml for liquid fuels and 1 kg for solids) should be taken from the bulk sample after thorough mixing, by shaking in the case of mobile liquid samples and by warming, melting and stirring in the case of more viscous materials or liquefiable solids like CTF 400 and pulverized pitch.

Fuller details of the acceptable sampling methods and descriptions of sampling apparatus for coal-tar fuels will be found in ref.93, p 58 et seq. and in BS 616, 'Sampling of Coal Tar and its Products'.

Analysis and Testing of Coal-tar Fuels

Full descriptions of the testing procedures for coal-tar fuels are given in ref.93 and only brief details are included here. The letters and numbers in parenthesis after each test are the serial numbers of the test and the page number in ref.93.

Calorific Value

The calorific value is determined as the gross CV, expressed as 15°C calories under conditions of constant volume of the combustion products. The determination is carried out in a high-pressure bomb calorimeter in which 0.5—0.8 g of sample is electrically ignited in an oxygen atmosphere at 2.53 MPa (25 atm) pressure (CO 12-67, p 470). The net CV is approximately 1.63 MJ/kg less than the gross value for CTF 50 and 100 and 1.40 MJ/kg less for the other grades.

Viscosity

For the less viscous grades, this is determined using the Redwood No.1 viscometer (Appendix 1, p 576; CO 11-67, p 468) or the Standard Tar Viscometer (Appendix 1, p 581; RT 2-67, p 154).

In the determination of the viscosity of CTF 200 and 250 on the Redwood instrument, hysteresis effects may affect the accuracy of the results and the following procedure is recommended.

About 200 mm^3 of the sample is heated in a loosely stoppered container, filled as completely as possible, for one hour at 100°C in a suitable liquid bath. Before the fuel is poured into the oil cup of the viscometer, adjust its temperature to slightly above the temperature of the test. Make the test within one hour of the fuel reaching the desired temperature; in tests at progressively lower temperatures, a similar interval should not be exceeded. Effect any subsequent heating by a source of heat not exceeding 120°C, and in no circumstances heat the fuel over a flame prior to filling the cup or adjust the temperature of the

Table 2   Specification requirements and average properties of coal-tar fuels (see refs (OO) and (PP))

Property	CTF 50 Specification	CTF 50 Av. props	CTF 100 Specification	CTF 100 Av. props	CTF 200 Specification	CTF 200 Av. props	CTF 250 Specification	CTF 250 Av. props	CTF 300 Specification	CTF 300 Av. props	CTF 400 Specification	CTF 400 Av. props	Pulv. pitch Specification	Pulv. pitch Av. props
Gross CV (MJ/kg)	38.4 min.	38.4–40.7	38.4 min.	38.4–40.7	37.8 min.	37.8–39.5	37.2 min.	37.7–38.8	37.2 min	37.2–37.9	36.6 min.	36.6–37.9	36.6 min	36.6–37.9
Viscosity Redwood No.1 seconds	60 max. at 100°F	30–50 at 100°F	100 max. at 100°F	35–50 at 100°F	1000–1500 at 100°F 100 max. at 200°F	50–60 at 200°F	—	—	—	—	—	—	—	—
Standard tar viscometer (10 mm cup): seconds at 30°C	—	—	—	—	—	—	70–120	70–120	—	—	—	—	—	—
at 55°C	—	—	—	—	—	—	—	—	40–80	40–80	—	—	—	—
Equiviscous temp. (°C)	—	—	—	—	—	—	—	32–35	—	53–58	—	—	—	—
Softening point (°C) Ring and Ball	—	—	—	—	—	—	—	—	—	33–38	75–85 max.	78–85	110 min.	110
Water content (wt %)	1.0 max.	<1.0	1.0 max.	<1.0	1.0 max.	<1.0	0.5 max.	<0.5	0.5 max.	<0.5	0.5 max.	<0.5	1.0 max.	<1.0
Ash (wt %)	0.05 max.	Nil	0.05 max.	Nil	0.25 max.	0.1	0.3 max.	0.1	0.3 max.	0.2	0.75 max.	0.2	0.75 max.	0.3
Matter insol. in toluene (wt %)	0.5 max.	Nil	0.5 max.	Nil	15 max.	5–10	23 max.	5–15	26 max.	10–20	30 max.	10–20	—	—
Sulphur (wt %) max.	—	0.2	—	0.2	—	0.5	—	0.5	—	0.5	—	0.7	—	0.7

fuel in the cup by immersing hot bodies in it. When a series of determinations is to be made, it is permissible to carry these out all on the same sample, tests at higher temperatures being made before those at lower temperatures.

For CTF 250 and 300, which are within the usual range of road-tar consistencies, a rapid and convenient indication of viscosity is obtained by measuring the equiviscous temperature (i.e. the temperature (°C) at which the liquid has a viscosity of 50 seconds when tested on the Standard Tar Viscometer using a 10 mm cup). This determination is carried out using the E.V.T. viscometer (RT 3-67, p 156). For CTF 400 and pulverized pitch, the softening point, as determined by the Ring and Ball method (PT 3-67, p 511), is the usual viscosity parameter specified and measured.

Coal-tar fuels behave as Newtonian fluids with temperature coefficients of viscosity increasing as the viscosity level increases. The following Table, taken from the nomograph in ref.89 (Figure 8), gives average values for the kinematic viscosities of CTF 200, 250 and 300 over the temperature range 100—300°F.

Viscosity (mm²/s or cSt)	CTF 200	CTF 250	CTF 300
At 37.8°C	310	11.5 × 10³	65 × 10⁴
65.7°C	45	480	55 × 10²
93.5°C	12	70	350
121.5°C	—	20	65
149°C	—	—	20

Water Content

This is determined by the Dean and Stark method (CO 2-67, p 446; RT 4-67, p 166; PT 9-67, p 525).

Insolubles in Toluene, and Ash

These are generally determined together. A weighed quantity of the fuel is extracted with toluene at 90—100°C, the residue filtered, washed with hot toluene, dried and weighed. This gives the toluene-insoluble content plus the ash; the ash is determined by igniting the dry residue and reweighing. The amounts of sample and toluene used vary with the grade. For CTF 50 and 100, which generally contain negligible amounts of toluene-insoluble matter, 20 g of sample and 100 mm³ of solvent are used for the extraction and a further 200 mm³ of hot solvent for transferring and washing the precipitate (CO 6-67, p 458). For CTF 200, 250 and 300, 2 g of the fuel is extracted with 100 mm³ of solvent and 500 mm³ employed for transfer and washing (CT 4-67, p 122; RT 8-67, p 174). For CTF 400 and pulverized pitch these same quantities of solvent are used with 1 g of sample (PT 7-67, p 522).

If it is desired to determine the ash without determining the toluene-insoluble content, it is

Flash point (°F) (74°C)	Carbon residue (Conradson) (wt %)
150 max.	2.0 max.
>150	—
150 max.	2.0 max.
>150	—
150 max.	—
>150	—
150 max.	—
>150	—
150 max.	—
>150	—
150 max.	—
>150	—
150 max.	—
>150	—
150 max.	—
>150	—

necessary with CTF 50 and 100 to distil the sample, slowly to avoid coke formation, until only a small residue remains, and ignite a weighed aliquot of the residue in a porcelain or silica crucible (CO 7-67, p 459). This initial distillation is unnecessary with the more viscous grades from CTF 250 upwards. For these, about 2 g of sample, weighed into a porcelain or silica crucible, is heated on a hot plate or the front of a muffle furnace to remove volatile matter without ebullition, and the crucible is then heated at 650°C in the muffle to constant weight (RT 6-67, p 124; PT 8-67, p 524).

The main constituents of the ash of coal-tar fuels are ferric oxide, silica, lime, magnesia and alkali (when the tar has been doped with alkali as an anti-corrosion measure). Typical analyses of the ash of coal-tar pitch (wt% of the ash) are given in the following Table (Ref.2, p 16):

	Sample 1	Sample 2	Sample 3*
Silica	5.1	12.1	2.45
Ferric oxide	65.3	57.1	42.3
Alumina	0.6	9.4	0.3
Titanium oxide	0.5	0.15	0.2
Manganese oxide	0.6	0.35	0.6
Sulphate	12.55	12.95	22.5
Lime	9.95	2.0	2.2
Magnesia	3.8	2.8	2.1
Alkali etc.	1.6	3.1	27.2

* From alkali-doped tar

Calcium, iron, lead, sodium and zinc occur in coal-tar fuels in concentrations between 5 and 50 ppm w/w; aluminium, bismuth, copper, magnesium, manganese, potassium, silicon, sodium and tin in the 0.05 to 5.0 ppm range; whilst arsenic, boron, chromium, germanium, titanium, vanadium and molybdenum may be present as traces of less than 0.05 ppm. Antimony, barium, cadmium, cobalt, nickel, strontium, tungsten and zirconium have not been detected. Fusion temperature of CTF ash is normally between 1300 and 1500°C in a reducing atmosphere except with high-alkali samples similar to sample 3 above which would be expected to have a fusion temperature of about 1100°C.

Sulphur Content

The sulphur content of coal-tar fuels, although not specified, is usually about 0.2—0.5 wt% for CTF 50 to 250 and between 0.5 and 1.0 wt% for the others. The sulphur content is usually determined at the same time as the calorific value. After the bomb has been allowed to cool and the pressure released, the contents are carefully washed out with a measured amount of distilled water and an aliquot

of the washings titrated (CO 12-67, p 475) or the sulphur in it gravimetrically determined as barium sulphate (CO 13-67, p 479).

Where the use of the combustion bomb is inconvenient or impracticable, the sulphur content may be determined by burning 1.5—2.0 g of the sample in an excess of oxygen in the apparatus used in the Institute of Petroleum test IP 63, absorbing the combustion gases in a solution of hydrogen peroxide and determining the sulphuric acid formed gravimetrically as barium sulphate (N 8-67, p 390).

Flash Point

Flash points are determined in the Pensky—Martens closed tester following BS 2839 (CO 10—67, p 466).

Carbon Residue

The Conradson method for carbon residue involves burning 10 g of the well dried sample under standard conditions in a silica crucible which is contained in a covered iron crucible. This in turn is contained on a layer of sand in a larger sheet-iron crucible covered by a lid. This whole assembly is surrounded by an asbestos refractory ring and rests on a nichrome wire triangle which also acts as support for the refractory ring. A sheet-iron hood sits on the upper surface of the refractory surround, enclosing the crucible assembly, which is directly heated by a Meker burner until it ignites. After combustion has ceased, indicated by the absence of smoke coming from the hood, the crucible assembly is strongly heated for 7 min, the apparatus allowed to cool and the inner silica crucible removed, cooled in a desiccator and weighed (CO 8-67, p 460).

Density

The densities of CTF 50 and 100 may be determined by the use of suitable hydrometers, but for the other grades the density-bottle method is used (GP 1-67, p 11 et seq).

References

1   Ref. (A), pp 42—48
2   Cooper, L. S. *et al* 'Some Aspects of the Reception and Transmission of North Sea Gas', Gas Council Res. Commn GC150. Obtainable from Instn of Gas Engrs, London
3   Ref. (A), pp 48—49
4   Ref. (B), Part 1, *General Analysis*
5   Ref.(C), Part 2, *Analysis by the Orsat Apparatus*
6   Ref.(D)
7   Ref.(E)
8   Ref.(F)
9   Ref.(G), *Analysis of Liquefied Petroleum (LP) Gases and Propylene Concentrates by Gas Chromatography*

10 Ref.(G), *Analysis of Natural Gas By Gas Chromatography: Hydrocarbons to Pentanes and Inorganic Gases* (Proposed)

11 Ref.(B), Part 4, *Chromatographic Analysis*

12 Ref.(H), pp 113—163

13 Ref.(I), Part 18, 1974

14 Ref.(J)

15 Ref.(H), pp 187—229

16 Ref.(H), pp 49—66

17 Ref.(K)

18 Ref.(L)

19 Ref.(M)

20 Mason, D. McA. and Eakin, B. E., 'Calculation of Heating Value and Specific Gravity of Fuel Gases', *I.G.T. Res. Bull. No.32*, Inst. Gas Technology, Chicago, 1961

21 Statutory Instruments (UK), No.1878, The Gas (Declaration of Calorific Value) Regulation, 1972

22 Statutory Instruments (UK), No.1805, The Gas Quality Regulations, 1972

*23 Ref.(A), Part IV

*24 Ref.(N)

*25 Ref.(O)

*26 Ref.(G)

*27 Ref.(I), Parts 23—25, 1974

28 Ref.(P)

29 Bell, L. H. *J. Inst. Petroleum* 1971, Vol.57, 219

*30 IP, *Petroleum Measurement Manual*, 1971, Part VII. Relative Density and Density Measurement

*31 ASTM — Inst. Petroleum, *Petroleum Measurement Tables*, 1952

32 Tomlinson, R. C. *Fuel, Lond.* 1957, Vol.36, 442

33 Visman, J. *Trans. World Power Conf.* 1947, Sect.A2

34 *ASTM Special Tech. Publ. No.162*, 1955

35 Ref.(Q)

36 Ref.(R)

37 Ref.(S)

38 Ref.(T), Part 1, 1973

39 Ref.(U)

40 ASTM, D2961-71T. *Tentative Method of Test for Total Moisture in Coal reduced to No.8 Top Sieve Size (Limited Purpose Method)*

41 Ref.(V)

42 Ref.(W)

43 Ref.(X)

44 Ref.(Y)

45 Ref.(Z)

46 Ref.(AA)

47 Ref.(CC)

48 Belcher, R. and Spooner, C. E. *Fuel, Lond.* 1941, Vol.20, 130

49 Mott, R. A. and Wilkinson, H. C. *Fuel, Lond.* 1955, Vol.34, 169

50 Mott, R. A., Ruell, D. A. and Wilkinson, H. C. *Fuel, Lond.* 1955, Vol.34, 78 and 87

51 Mott, R. A. and Wilkinson, H. C. *Fuel, Lond.* 1956, Vol.35, 6

52 Blayden, H. E. and Patrick, J. W. *Fuel, Lond.* 1970, Vol.49, 257

53 Ref.(T), Part 11

54 Ref.(T), Part 10

55 Ref.(T), Part 14

56 Archer, K., Flint, D. and Jordan, J. *Fuel, Lond.* 1958, Vol.37, 421

57 Dixon, K. *Analyst* 1958, Vol.83, 362

58 Mott, R. A. *et al* 'Studies in Bomb Calorimetry': I—XI, *Fuel, Lond.* 1954, Vol.33—1959, Vol.38

59 Ref.(T), Part 5

60 Ref.(DD)

61 Radmacher, W. and Mohrhauer, P. *Brennst.-Chem.* 1955, Vol.36, 236

62 Bishop, M. and Ward, D. L. *Fuel, Lond.* 1958, Vol.37, 191

63 Ref.(T), Part 16

64 Ref.(T), Part 12

65 Ref.(EE)

66 Ref.(FF), Ch.19

67 Ref.(GG)

68 Lange, W., Radmacher, N. and Vierneisel, H. *Brennst.-Chem.* 1961, Vol.42, 312 and 385

69 Ref.(T), Parts 2, 4 and 7

70 Wilkinson, H. C. *Fuel, Lond.* 1965, Vol.44, 191

71 Wilkinson, H. C. *Fuel, Lond.* 1956, Vol.35, 39

72 Belcher, R. and Bhatty, M. K. *Fuel, Lond.* 1958, Vol.37, 159

73 Mott, R. A. and Wilkinson, H. C. *Fuel, Lond.* 1958, Vol.37, 151

74 Kirk, B. P. and Wilkinson, H. C. *Talanta* 1970, Vol.17, 475

75 Barker, J. E. and Mott, R. A. *Fuel, Lond.* 1960, Vol.39, 363

76 Leighton, L. H. and Wald, S. *Fuel, Lond.* 1960, Vol.39, 511

77 Hyslop, W. and Wilkinson, H. C. *Symposium of 1st Annual Gen. Mtg of the Metals Society*, London, April 1974

78 Ref.(T), Part 13

79 Ref.(HH)

80 Milson, A. *Yearb. of Coke Oven Managers' Assoc.* 1965, pp 360—381

81 Ref.(II)

82 Ref.(JJ)

83 Ref.(KK)

84 Ref.(LL)

85 Ref.(MM)

86 Ref.(NN)

87 Lee, G. W. and Mott, R. A. *Special Study of Domestic Heating in the UK — Past and Future*, Inst. Fuel, London, 1956, pp 43—54

88 Blayden, H. E., Noble, W. and Riley, H. L. *J. Inst. Fuel* 1934, Vol.7, 139

89 Ref.(PP), pp 92—107

90 Chesters, J. H. and Mayorcas, R. *Iron & Steel Trades Rev.* 1957 (Oct. 18), p 909

* Subject to frequent revision (several annually)

91   Brett Davies, E. and Pritchard, A. B. *Proc. Conf. on Liquid Fuels*, Inst. Fuel, London, 1959, B-50
92   Ref.(PP) p 41
93   Ref.(QQ)

General Reading

(A)   *IP Standards for Petroleum and its Products. Part IV. Methods for Sampling, Sampling Petroleum Gases*, IP 181/62, Inst. of Petroleum, London, 1965
(B)   BS 3156:1968. *Methods for the Analysis of Fuel Gases*, British Standards Instn, London
(C)   BS 1756:1963. *Methods for the Sampling and Analysis of Flue Gases*
(D)   BS 3282:1969. *Glossary of Terms relating to Gas Chromatography*
(E)   BS 4587:1970. *Recommendations for the Selection of Apparatus and Techniques for the Analysis of Gases by Gas Chromatography*
(F)   Jeffery, P. G. and Kipping, P. J. *Gas Analysis by Gas Chromatography*, 2nd edn, Pergamon Press, London, 1972
(G)   *IP Standards for Petroleum and its Products. Part I. Methods for Analysis and Testing*, IP 264/72, Inst. Petroleum, London, 1974
(H)   Verdin, A., *Gas Analysis Instrumentation*, Macmillan, London, 1973
(I)   *Annual Book of ASTM Standards.*
(J)   Daynes, M. A., *Gas Analysis by Measurement of Thermal Conductivity*, Macmillan, London, 1934
(K)   BS 4559:1970. *Methods for the Preparation of Gaseous Mixtures*
(L)   BS 3804:1964. *Determination of the Calorific Value of Fuel Gases*. Part 1. Non-recording Methods
(M)   Hyde, C. G. and Jones, M. W. *Gas Calorimetry*, Benn, London, 1960
(N)   BS 4040:1971. *Specification for Petrol (Gasoline) for Motor Vehicles*
(O)   BS 2869:1970. *Specification for Petroleum Fuels for Oil Engines and Burners*
(P)   Statutory Instrument (UK), No.1311, 1973, plus amendment No.379, 1974. Customs and Excise. The Hydrocarbon Oil Regulations 1973
(Q)   International Organization for Standardization. ISO Standard 1988-1972

(R)   BS 1017:Part 1: 1960. *The Sampling of Coal and Coke*, Part 1, Sampling of Coal
(S)   ISO Standard 589-1974
(T)   BS 1016: Parts 1—16: 1960—1973. *Methods for the Analysis and Testing of Coal and Coke*
(U)   DIN. Normen 51718, Bestimmung der Wassergehaltes
(V)   BS 1293 and 2074:1965. *Methods for the Size Analysis of Coal and Coke*
(W)   ASTM, D410-38 (Reapproved 1969). *Standard Method of Test for Sieve Analysis of Coal*
(X)   ASTM, D431-44. *Standard Method for designating the Size of Coal from its Sieve Analysis*
(Y)   ISO Standard 2309-1973
(Z)   BS 1017: Part 2: 1960. *The Sampling of Coal and Coke*, Part 2, Sampling of coke
(AA) Bethell, F. V. *BCURA Mon. Bull.* 1962, Vol.26, 401
(BB) Francis, W. *Coal*, 2nd edn, Edward Arnold, London, 1961
(CC) ISO Standard 1994-1973
(DD) ISO Recommendation R1928-1971
(EE) ISO Recommendation R501-1966
(FF) Mott, R. A. and Wheeler, R. V. *The Quality of Coke*, Chapman & Hall, London, 1939
(GG) ISO Recommendation R349-1963
(HH) ISO Recommendation R616-1967
(II) ISO Recommendation R556 (Under revision)
(JJ) DIN. Normen 51717. Bestimmung der TrommelfestigKeit und des Abreibs von Steinkohlenkoks
(KK) ASTM, *Standard Method of Tumbler Test for Coke*, D294-64 (Reapproved 1972); and D2490-70 (Small Coke)
(LL) *Determination of Drum Indices of Coke.* Japanese Imperial Standard JIS K2151-1972
(MM) *Method of Determination of the Mechanical Strength of Coke by Testing in the Drum.* All Union Standards GOST 5953-51 and 8929-58
(NN) ISO Recommendation R567
(OO) B.S. 1469:1962
(PP) Huxtable, W. (Ed.), *Coal Tar Fuels*, Assoc. of Tar Distillers, London, 1960
(QQ) Watkins, P. V. (Ed.), *Standard Methods for Testing Tar and its Products*, 6th edn, Standardization of Tar Products Test Cttee, Gomersal (UK), 1967

Appendix E
World energy resources and demand

E1 Resources

E2 Energy demand

From the time of its formation in London in 1924 the World Power Conference, now known as the World Energy Conference, recognized the importance of data on the various energy resources and their utilization. In 1929 the central office of the Conference published a book on the Power Resources of the World, Potential and Developed. Then followed a series of Statistical Year Books, which included information on resources and the available annual statistics on the production, stocks, imports, exports and consumption of the several forms of energy in the different countries. These Year Books covered the years 1933 to 1958.

In 1952 the United Nations began their 'J Series' of Statistical Papers on the annual production, trade and consumption of the various solid, liquid and gaseous forms of energy and of electricity in the different countries and geographical areas of the world. The latest available number of the Series 'J17' on World Energy Supplies during each of the years 1969–1972 was published in 1974 [1]. In the circumstances, the International Executive Council of the World Energy Conference in 1959 decided to discontinue their Statistical Year Books and to issue at intervals of six years a new series of publications entitled World Energy Conference Survey of Energy Resources. A Panel under the chairmanship of Dr A. Parker was appointed to advise on planning the details of the new series, the first of which was published in 1962 [2] and the second in 1968 [3]. The third of the series, issued in September 1974 [4], was compiled by a group of experts in the USA with the guidance of a Consultative Panel of representatives of sixteen of about seventy countries with National Committees represented on the International Executive Council. Mr C. F. Luce (USA) was chairman of the Panel.

The information in the following paragraphs is largely a brief summary of the detailed information in the World Energy Conference Survey of Energy Resources 1974, which covers 400 large pages with the text in English and French and numerous tables of data.

E1 Resources

Solid Fuels

The commercial solid fuels include coals and peat and the coals are divided into two main classes, the first including anthracite, bituminous coals and other coals of high calorific value and the second including brown coal, lignite and similar fuels of low calorific value. Each group is then divided into measured reserves considered to be obtainable under present economic conditions, total measured reserves, and additional more speculative reserves that have not been measured but are indicated or inferred from geological and other considerations.

Coals and Lignites

Estimates for anthracites, bituminous coals, brown coals and lignites are summarized in *Table 1*.

Of the total reserves of bituminous coals the percentages are 50 in the USSR and 25 in the USA and of the measured reserves the percentages are 25 in the USSR, 25 in the USA, 28 in China and 17 in Europe. Of the measured reserves of brown coals and lignites 31% is in the USSR and 38% is in Europe.

It should be recognized that the estimates of indicated or inferred reserves for most countries are somewhat of the nature of guesses of uncertain reliability. There is also some doubt about the precise meaning of the term 'classed as economically recoverable' as used for the data provided by some countries. The governments of the UK and certain countries in Western Europe, for example, have provided large funds to maintain the production of coal in their areas and it is unlikely that the amounts provided will be repaid by the coal industries. The main reason for such government financial assistance is to reduce dependence on imports and to assist overall balance of payments.

The measured amount of hard coals in the UK is 98.877 Gt, of which only 3.871 Gt is classed as economically recoverable; the quantity of indicated and inferred coal is estimated at 63.937 Gt making an overall total of 162.814 Gt.

If only one quarter of total world reserves, including indicated and inferred, could eventually be recovered for use the amount would be equal to about 950 times the consumption of coals and lignites in 1972 and could meet an increased demand for several centuries.

Peat and Non-commercial Fuels

Peat is generally a non-commercial fuel. In consequence surveys of peat resources have been systematically undertaken in only a few countries of the world. The data on the quantities of peat used annually are also incomplete as in several countries peat is taken and dried by local users and the quantities are not systematically recorded. There are even greater uncertainties in the available records of the use of other non-commercial fuels such as wood, bagasse, vegetable wastes and dung. Wood is slow-growing, of relatively low CV, and environmentally important; immense amounts are used for constructional timber, papermaking, etc.

Figures for the recorded resources of peat are generally on the basis of the quantities of the peat when dried to a water content of about 25%. Such peat has a calorific value ranging between one-third and one-half of that of bituminous coal. On these bases the total of the recorded resources of peat in the world is about 211 Gt of which nearly 60% is in the USSR and 34% in Europe. The most recent record of world use of peat in 1971 was about 195 Mt of which 68% was used for agricultural purposes and only 32% as fuel,

Table 1   Reserves of coals and lignites (Gt)

	Measured	Indicated or inferred	Total	Classed as economically recoverable
1. Anthracite	19.337	9.152	28.490	7.146
2. Bituminous	1058	7035	8093	424
3. Brown coals and lignites (1/3 to 1/2 CV of bituminous)	343	2281	2624	160
4. Energy equiv. in bit. coal (Gt) of totals of 1 + 2 + 3	1220	7994	9176	498
5. Energy equiv. in bit. coal (Gt/yr) of world consumption of coals and lignites in 1972	<————————————— 2.407 —————————————————————>			
6. Items 4 divided by item 5 in years	507	3321	3828	207

equivalent in heating value to no more than 30 Mt of bituminous coal or only 0.4% of total world energy consumption in 1972. The use of peat and all other non-commercial materials as fuel was then probably less than 4% of total world energy consumption.

It has been estimated (Ch.6.4) that the amount of garbage for disposal in the UK is in the region of 300 t per annum per 1000 inhabitants, and that the heating value of 300 t of garbage is about equal to that of 100 t of coal. The amount of garbage for the UK population of about 55 million is then equivalent in heating value to approximately 5 Mt/yr of coal. If all the garbage were incinerated, which is not the position at present, it is possible that it could supply heat equal to about 1.5% of the total coal equivalent of use (346 Mt) in the UK in 1973. In some circumstances the heat from the incineration of garbage might be used to raise steam for factories or other premises not too far away from the incinerator, provided that the steam is required throughout the year, and subject to overall costs and consideration of the effect on local amenities.

Petroleum Oil

The total of proved reserves of petroleum considered to be economically recoverable under present conditions is 91.5 Gt of which the percentages are 54 in the Middle East, 14 in Africa, 9 in the USSR, 8 in N. America, 7.5 in S. America and 1.5 in Europe, including 0.55 for the UK. World reserves estimated as economically recoverable are probably no more than about one-third of known measured reserves. With rises in the relative price of petroleum there is little doubt that a larger proportion of the measured reserves will be recovered. Various widely differing estimates based on uncertain geological information have been made of probable world oil resources. For example, in Geological Survey Professional Paper 817 [5] issued by the US Department of the Interior in 1973 the estimates range from 183.8 Gt to 1837.7 Gt.

World consumption of petroleum in 1972 [1] was about 2.50 Gt. This means that the present estimate of the quantity of petroleum that can be economically recovered is between 36 and 37 times the recent annual rate of production. Though world consumption of petroleum oil rose from 1236 Mt in 1962 to 2100 Mt in 1968 and 2500 Mt in 1972, in each of those years, with the discovery of new resources, the estimates of economic reserves have been equal to between 30 and 40 times the annual rates of consumption. There is little doubt that this will continue to be the position for many more years and that statements to the effect that petroleum oil reserves will be exhausted within about the next 40 years will be proved to be wrong.

Greater economy of use encouraged by price and conservation requirements, combined with the fact that remaining oil will be progressively more costly to win, must however tend toward a peak output before very long. Consideration of the resources discussed here suggests that this peak may occur around the year 2000 at a figure in the region of 3.5 Gt/yr, unless ultimate reserves prove to be midway between the measured and the maximum total quoted in ref.5, or unless recovery of oil from shale and tar sands proves to be less difficult and costly than expected. In such case a somewhat higher peak might occur perhaps a century later.

Oil from Natural Gas

In most resources of natural gas, in addition to methane, ethane, propane and butane, there is vaporized light oil, which is removed by condensation or other method before the gas is distributed for use. The proved reserves of this light oil are about 1.32 Gt and recent annual production is about 108 Mt, which is only about 4% by weight of the annual production of petroleum oil.

Oil from Shale and Bituminous Sands

Since the year 1900 there have been ample supplies of petroleum to meet world demand. In consequence, systematic surveys of oil shales and bituminous sands and estimates of the quantities of oil that could probably be extracted from them, if required, have been made in only a few countries. Production of oil from shale in Scotland was begun more than 100 years ago but was discontinued in recent years. There has also been production of oil from shale on a modest scale in Estonia, France, Sweden, Germany, Spain, South Africa, Australia and China and recently in Brazil. The total cumulative production over more than a century has been no more than about 70 Mt. The total of the available estimates of recoverable oil from shales and bituminous sands is about 170 Gt of which 46% is in S. America, 37 in N. America and 12 in China. There have been highly speculative estimates of other resources. The most recent annual production of such oil is about 13 Mt with 5.5 Mt in the USSR, 3.1 Mt in Canada and 1.2 Mt in Venezuela.

Natural Gas

Many areas of the world have not been surveyed for the resources of natural gas. The world total of known proved, recoverable reserves is 52 532 km³ of which the percentages are 32 in the USSR, 23 in Asia, 20 in N. America, 11 in Africa and 9 in Europe; the quantity so far estimated for the UK is 870 km³. There is no doubt that additional recoverable reserves will be found. According to the latest statistics of the UN, world consumption of natural gas in 1972 was 1203 km³, so that the known recoverable reserves are about 43 times the quantity consumed in 1972. The rate of world consumption is increasing but additional recoverable reserves will certainly be found.

Hydraulic Energy

From measurements of the gradients and rates of flow of rivers and streams in many countries, estimates have been made of the annual amounts of electricity that could be generated at the rates of flow available throughout 95% of the year (Gen 95) and at the overall average rate of flow (Genav). The Genav figures are higher than those for Gen95. The following estimates are all for Genav in TW h/yr*. The total estimate for the world from the

---

* 1 TW h/yr = 3600 TJ/yr = 114 MW$_{av}$ is equivalent to 125 kt/yr coal equivalent on UN hydroelectric basis (100% efficiency of generation); 1 TJ = 34.7 t coal equivalent. *Editor*

information available is 9802 TW h/yr, the percentages of which are about 27 in Asia, 21 in Africa, 17 in S. America, 15 in N. America, 11 in the USSR and 7 in Europe. The countries with the largest estimated resources in TW h/yr are China 1320, USSR 1095, USA 701, Zaire 660, Canada 535 and Brazil 519. The percentage now being utilized in relation to the estimated resources is about 13, at 1290 TW h/yr, which is about 23% of total world generation of electricity of 5630 TW h/yr including that from thermal, hydro, and nuclear power stations. The growth in the production of hydro-electricity will be slow because the places in which large amounts of hydropower could be made available are too remote from the areas where the hydro-electricity could be used. The amount of hydro-electricity will certainly be much less than the possible 9802 TW h/yr (1.23 Gtce/yr) for the foreseeable future.

Nuclear Energy

The production of electricity from nuclear fuels will certainly increase well above the 1972 amount of 142 TW h/yr during the next 30 years, when the most usual uranium isotope ^{235}U will be the main fuel in non-breeder reactors, though this system uses only about 1% of the potential energy. In the 1990s it is probable that breeder reactors using plutonium derived from the most common isotope of uranium ^{238}U and recovering as much as 60–70% of the potential energy will be in operation in several areas. Moreover in the early years of the next century thorium may be used to a minor extent in breeder reactors. It is also possible that fusion reactors fuelled with tritium, derived from lithium, and deuterium will at that time reach the stage of commercial use.

From the surveys that have been made, the reasonably assured resources of uranium (in terms of the element) that could be obtained at 1974 costs of not more than 26 US dollars per kg are estimated at about 984 kt with probable additional reserves of about 821 kt at costs ranging from 38 to 77 US dollars per kg. Of the reasonably assured reserves at estimated costs not exceeding 26 dollars per kg the percentages in the main areas of the world are 52 in N. America, 28 in Africa, 12 in Australia and 6 in Europe. Total uranium resources, including those at somewhat higher cost, are estimated at 4023 kt. None of these figures include the resources in the communist countries, believed to be between 100 and 370 kt. Recent annual production of uranium for the world has been estimated at about 21 kt.

It should be mentioned that considerable quantities of uranium, if necessary, could be extracted from the sea but at appreciably higher cost than from ores*.

Surveys of resources of thorium have not covered all probable areas. From the known resources the estimated quantity that could be recovered at a cost not exceeding 25 dollars per kg, in terms of the element, is 322 kt, the percentages of which are 41 in Europe, 40 in N. America and 18 in S. America. Higher-cost resources that are reasonably assured are estimated at 474 kt of the element of which the percentages are 64 in Asia, 17 in Europe and 11 in Africa. Additional resources are estimated at 543 kt including 319 in Africa.

Geothermal Energy

The average increase in temperature going down from the earth's crust in the first 100 km is about $10°C$ and it is estimated that the temperature of the central core of the earth is in the region of $4000°C$. These increases in temperature are probably the result of the slow decay of radioactive elements, mainly uranium, thorium and an isotope of potassium. There are, however, deviations in many areas from the average increase of $10°C$ in the temperature of the first 100 km below the earth's crust owing to local increases in the changes generating and/or storing heat. As a result there are areas not very far below the earth's surface with trapped water heated to various temperatures sometimes reaching the region of $300°C$. From geysers or drillings, hot water and in some areas steam can be obtained. Hot water is suitable only for local space heating but the steam can be used for the generation of electricity. Geothermal energy is used in a number of countries, for example in Iceland, Italy, Japan, Mexico, New

---

* The coal-equivalent basis used in Chapter 14 uses for comparison coal of low CV ($\approx$22.5 MJ/kg). The estimates in this Appendix also differ in detail from those given in Chapter 14, Table 4 and Figure 16, which were derived from Seaborg, *Nuclear Engng*, Jan./Feb. 1967 and the 4th Conference on Peaceful Uses of Atomic Energy, IAEA, Vienna, 1972, p 3. For a discussion of nuclear fuel reserves in relation to rate of use the reader is referred to that Chapter, although when its conclusions are cited below or in Appendix F they have been corrected to the UN basis in respect of coal-equivalent. In brief, maximum reserves (>10 Gt of U_3O_8, extracted from the sea) that might become economic for breeder reactors (at \$450/kg or £200/lb U_3O_8) would suffice for >10^{12} TJ ($\approx$100 Tt coal-equivalent) using thermal reactors, and >10^{14} TJ ($\approx10^4$ Tt coal-equivalent) using breeder reactors, probably ample for at least a century. But at prices now almost economic, and restricting to thermal reactors, supplies could suffice for some decades even if it were required to shoulder the main energy load of the world. *Editor*

Zealand and the USA, for space heating or the generation of electricity or both purposes. Surveys of economic possibilities are being made in a number of other countries. It is unlikely however that the total geothermal energy that can be recovered economically in the foreseeable future will be more than a very small fraction of world energy demand.

Tidal Power†

Many schemes have been considered in the past for the generation of electricity from tidal power, particularly in areas with comparatively high tidal rise and fall in estuaries. The highest tidal rises of about 11 metres are in the Bay of Fundy in Canada and in some areas in northwest Australia. Other examples of high tidal rises, though not so high as 11 metres, are the English Channel and the River Severn. Schemes for the estuary of the River Severn have been examined on several occasions but were not considered to be economic. The one scheme in operation is the French installation at La Rance, reported to have a capacity of 240 MW with an ultimate capacity of 350 MW$_e$, but it is questionable whether the scheme has been really economic. The economic possibilities for a capacity of 2176 MW in the upper Bay of Fundy are under investigation. It is certain, however, that within the next few decades tidal power will provide no more than a minute part of world energy demand.

Ocean Thermal Gradients

There are appreciable temperature differences between the surface and lower layers of water in some parts of the oceans. A good example is the Gulf Stream in which the vertical temperature gradient over 1000 metres is in the region of $20°C$. Water warmed by the sun in certain areas flows in the upper layers in one direction towards a colder climate and colder water is thereby displaced to flow in the lower layers in the opposite direction. The layers at the higher temperature could be used to vaporize selected fluids and the vapour expanded through a turbine and condensed by the cooler water, the system thereby generating electricity. Theoretically, the amount that could be generated in this way would be greater than present world consumption of electricity. There are, however, obvious difficulties of installation of plant and of transmission of the electricity at reasonable cost to where it is required. There is little prospect of such schemes on an appreciable scale being economic during the next 100 years. The idea is not new[4]. It was suggested by J. D. Arsonval of France in 1881 and a plant with a capacity of 22 kW was built and operated in 1929 by G. Claude of Cuba.

---

† Wave energy is itself discussed briefly in Ch.1.2

Wind Power

Before the development of the steam engine in the nineteenth century wind and water power were the main sources of energy for various purposes, including mechanical power for industry and agriculture and motive power for ships. The total wind power in the atmosphere over the land area of the world is estimated at about 650×10^9 TW h/yr ($\equiv 79\,000$ Ttce/yr on hydraulic energy basis) and the usable portion at 20 TW h/yr (2.5 Mtce/yr). The difficulties in harnessing a larger part of this power are that the wind is of very variable and low intensity generally and would require numerous installations if it is to provide a substantial amount of power. The use of wind power could be developed appreciably if there were some economic method of storing large quantities of the electricity that could be produced. Batteries and pump storage of water, and electrolysis of water to produce hydrogen on a large scale, do not seem to be economic methods of storing the energy of the wind. Though windmills are useful for meeting small local demands for power, for example for batch work, it must be concluded that for many years the wind will be a method of providing only a small fraction of world energy demand.

Solar Energy

Hydraulic energy, ocean thermal gradients and wind power are derived indirectly from solar energy. Consideration is being given to the amount of energy that could be obtained economically in the form of heat or power directly from the sun. The amount of energy reaching the earth's surface from the sun is some five thousand times the present total of world energy consumption in various forms.

Solar energy on the earth's surface is of relatively low intensity, variable in amount and not available in any region throughout the 24 hours of the day. It has long been in use in small amounts in several countries for the evaporation of water containing salt and for drying agricultural products. It is now in use to a very small extent in some countries for heating water and for space heating. Investigations are in progress on the economic possibilities of extending such uses and of generating electricity by concentration collectors including lenses, mirrors, thermocouples and photoelectric cells. There is little prospect, however, during the next few decades of large amounts of solar energy being used directly for heating water, space heating and the generation of electricity.

Energy Plantations

An American group[6] has recently studied the possibility of using energy plantations surrounding power stations in a climate such as that in the USA in which two-thirds of the land area receives an annual average of energy from the sun equivalent to 15 MJ/m^2 per day after allowing for cloud cover. The crops grown, for example corn silage, sugar cane, sycamore and conifers, would be optimized for their energy value, converting some 0.5% of the incident radiation to fuel for boilers*. From the figures they give here and elsewhere it can be calculated that the electrical power required in the USA, which was 1853 TW h in 1972, could by this method be provided from the fuel derived from plantations on an area equivalent to one acre (4046 m^2) per inhabitant (if universal, about 9% of the gross area) since the population of the USA in 1972 was about 209 million. This may be compared with the area (perhaps 300 sq. miles) required to support a 1000 ton/day pulp mill on a continuing basis; the same area could alternatively support a 400 MW_e power station, serving 0.2 M people.

Clearly such a scheme should not be permitted to diminish good land potentially suited to agriculture producing human or animal feed; the boiler ash would be used to maintain the crops, but organic wastes would not be available for recycling; on the other hand sulphur emissions would be greatly reduced.

E2 Energy demand

To enable the changes in the annual demand for the commercial forms of energy by the world and in various areas and countries to be assessed, factors are used by the UN, EEC and other organisations to convert the several forms of energy to coal or oil equivalents; but in compiling the statistics these organizations do not all use the same conversion factors. In studying the various statistics issued, therefore, it is important to take into account the particular conversion factors used. In the following paragraphs, the figures given for coal equivalents of the quantities of the several forms of energy used are derived from the statistics published by the UN in their Series J statistical papers on World Energy Supplies.

The factors used by the UN in their J series that are relevant at this point are listed in *Table 2*.

According to the UN statistics the coal equivalent of world annual energy consumption rose from 4.707 Gt in 1963 to 7.410 Gt in 1972, that is by 57.4% in 9 years. Over the same period world

---

* The solar 'constant' is about 1.5 kJ/m^2 s normal to the sun's radiation, and the earth's mean radius is about 6400 km. The projected surface illuminated at one time is one quarter of the total; up to 30% of the radiation is scattered by the atmosphere. Hence the maximum possible contribution amounts to about 10^{12} TJ/yr. This total includes all derivative processes including natural photosynthesis. *Editor.*

population rose by 19.6% from 3162 M to 3782 M. This means that the consumption of energy per head per annum increased by 31.6% from the equivalent of 1489 kg to 1959 kg. The USA had the highest consumption of energy per head, at 11 617 kg, in 1972. In the UK the corresponding consumption in 1972 was 5398 kg. From 1963 to 1972 the percentage of world energy consumption provided by solid fuels fell from 46 to 32.5 while that from oil rose from 34.9 to 43.5; the percentage provided by natural gas rose from 16.8 to 21.6 and that from hydro plus nuclear electricity only increased from 0.22 to 0.24.

If reasonable conditions of living of the peoples in the undeveloped and developing areas of the world are to be achieved their average individual demand for energy must increase. At present rates of growth world population will reach about 6600 M by the year 2000 and in the region of 48 000 M by 2100; the energy demand in the year 2000 would be an average of 4605 kg ce per individual or a world demand of about 30 Gtce*. At these rates world consumption of energy over the period 1970 to 2000 would be equivalent to about 420 Gt of coal. With increasing world demand for energy the real price per unit must rise as the reserves that can be obtained at the lower costs will be used first.

The fundamental problem is how to reduce significantly the rate of growth of the population, especially in some areas. Over the years 1958 to 1963 the compound annual rates of growth per cent were for the world 1.8, UK 0.7, Europe 0.9, N. America 1.6, Eastern Asia 1.4, Africa 2.3, Southern Asia 2.2 and Latin America 2.6. Over the years 1963 to 1972 the corresponding rates of growth were for the world 2.0, UK 0.4, Europe 0.8, N. America 1.3, Eastern Asia 1.8, Africa 2.6, Southern Asia 2.8 and Latin America 2.9 [7]. These figures show that the rate of growth of world population has increased during the period 1958 to 1972; though there have been decreases in the UK, Europe and N. America, there have been considerable increases in Eastern Asia, Africa, Southern Asia and Latin America. Unless the peoples of the world themselves significantly reduce the rate of growth of world population, hard nature will be forced to achieve it.

---

* In the year 2100 it could be much greater even without any growth of demand; and if the individual energy demand continued to follow the same rate of increase, i.e. by a factor of 21.1 between 2000 and 2100, the total energy demand could increase *ca.* 155 times to 4.7 Tt coal equivalent annually, nearly half the total likely resources of all fossil fuels today (App.E1). The point of this *exploratory calculation* is merely to illustrate how soon an impossibly absurd situation might be reached in default of a global energy policy. *Editor*

Table 2 †   Selected UN factors for coal equivalents (adapted)

Energy form	Quantity	Energy amount (GJ)	Factor (tce/ unit quantity)
Black or hard coal	1 t	28.8	1.00
Brown coals, lignite	1 tce as Table 1	–	1.00
Peat (25% moisture)	1 tce as earlier Section	–	1.00
Crude petroleum	1 t	37.44	1.30
Natural gas	10^3 m^3	38.36	1.332
Hydro, nuclear and geothermal electricity (at 100% generation efficiency)	10^4 kW h	36.0	1.25

†*Note.* Independently of the factors used by the UN in this series, S. E. Hunt (Ch.14) and M. E. Speight (*Chemy Ind.*, 1967, p 1344) have both used a large empirical unit termed Q where:

$$Q \text{ (Hunt)} = 10^{18} \text{ Btu} = 1.055 \times 10^9 \text{ TJ}$$
$$Q \text{ (Speight)} = 10^{12} \text{ kW h} = 3.6 \times 10^6 \text{ TJ}$$

Unfortunately these are related by a factor of no less than 292:1! Special units such as Q should be avoided. *Editor*

References

1 World Energy Supplies 1968–1972 Statistical Papers Series J No.17, United Nations 1974. Copies can be obtained from HMSO, London
2 Parker, A. World Power Conference Survey of Energy Resources, 1962, World Energy Conference, London
3 Parker, A. World Power Conference Survey of Energy Resources, 1968, World Energy Conference, London
4 World Energy Conference Survey of Energy Resources, 1974, The United States National Committee of the World Energy Conference, New York, N.Y. 10017. Copies can be obtained from the Central Office of the World Energy Conference, London
5 Albers, J. P. *et al*, Summary Petroleum and Selected Mineral Statistics for 120 Countries, including Offshore Areas, Geological Survey Professional Paper 817, U.S. Dept. of the Interior. Government Printing Office, Washington D.C. 1973
6 Szego, G. C. and Kemp, C. C. *Chemical Technology* 1973 (May) p 275
7 Statistical Yearbooks of the United Nations, 1960 to 1973. Copies can be obtained from HMSO, London

Appendix F
Conspectus

To sum up the long-term energy situation in very general terms, the absolute likely maxima of presently important resources seem to be:

Solid fuels	$\approx$10 Tt coal equivalent $(3 \times 10^{11}$ TJ)
Liquid fuels (as such)	$\approx$0.7 Tt coal equivalent $(2 \times 10^{10}$ TJ)
Liquid fuels (shales, tar sands)	$\approx$0.3 Tt coal equivalent $(9 \times 10^{9}$ TJ)
Natural gas	$\approx$0.1 Tt coal equivalent $(3 \times 10^{9}$ TJ)
Nuclear fission fuel (uranium from sea included)	$\approx$10^4 Tt coal equivalent $(3 \times 10^{14}$ TJ)

To these peat and other minor fuels are unlikely to add more than a few per cent of the total of fossil fuels. Contributions from hydro-electric, tidal, geothermal, and wind energy, and combustion of wastes, are annually relatively minor but have the virtue of *continuing without exhaustion*. The energy demand for 1972 was $\approx$7.5 Gtce/yr coal equivalent $(2 \times 10^8$ TJ/yr); *at the present rate of increase* this could be 30 Gtce/yr $(9 \times 10^8$ TJ/yr) in the year 2000 A.D. and over 4.5 Ttce/yr $(1.35 \times 10^{11}$ TJ/yr) in 2100, showing that an impossible situation would be reached long before this, unless nuclear fusion on the earth's surface were able to contribute additionally and substantially to the solar output.

In any event greater use must be made of the effectively inexhaustible resources now beginning to receive serious attention, such as direct use of solar radiation, power from ocean thermal gradients and, when available, nuclear fusion.

The fundamental problem, as Dr Parker has implied, is to find methods of curbing the growth of world population really effectively without wars, famine or gross epidemics. Even if this could be substantially achieved, world demand for energy would still increase for some further time because of some decades lag in response to the birth rate, and because some 70% of the world population is currently considered to be below energy subsistence level. It is most important meanwhile that urgent steps should be taken to ensure in all countries the highest practicable efficiency in the use of all forms of energy, and for some of them to *reduce* their demand.

Index

Units: Explanatory Notes

Symbols for SI units and factors are printed in ordinary roman type. Conversion factors have been rounded off for convenience. For more exact factors sometimes desirable in computer calculations, the official documents such as (1) below should be consulted. The 'factor' symbol precedes the unit symbol with no space. Unit symbols are separated in this book by a full space (a dot is also permissible).

The solidus is used in this book to distinguish numerator and denominator in compound units. Positive and negative powers are frequently used as alternative.

The following example illustrates the rule regarding the powers of (factor + unit):

$$km^2 \text{ signifies } (km)^2 = 10^6 \text{ m}^2$$

References

1. Conversion factors and Table BS 350: Part 2: 1962 (Supplement No.1, 1967). From British Standards Institution, 2 Park Street, London W1A 2BS

A fuller selection of units (though not complete), at the same level of approximation as used here, may be found in:

2. SI and related units: quick-reference conversion factors (50p each + 25% of order for surface mail) from Dr I. G. C. Dryden, 112 Sandy Lane South, Wallington, Surrey SM6 9NR, UK